AF718901

Mantle Flow and Melt Generation at Mid-Ocean Ridges

Geophysical Monograph Series

Including

IUGG Volumes

Maurice Ewing Volumes

Mineral Physics Volumes

GEOPHYSICAL MONOGRAPH SERIES

Geophysical Monograph Volumes

1 **Antarctica in the International Geophysical Year** *A. P. Crary, L. M. Gould, E. O. Hulburt, Hugh Odishaw, and Waldo E. Smith (Eds.)*
2 **Geophysics and the IGY** *Hugh Odishaw and Stanley Ruttenberg (Eds.)*
3 **Atmospheric Chemistry of Chlorine and Sulfur Compounds** *James P. Lodge, Jr. (Ed.)*
4 **Contemporary Geodesy** *Charles A. Whitten and Kenneth H. Drummond (Eds.)*
5 **Physics of Precipitation** *Helmut Weickmann (Ed.)*
6 **The Crust of the Pacific Basin** *Gordon A. Macdonald and Hisashi Kuno (Eds.)*
7 **Antarctica Research: The Matthew Fontaine Maury Memorial Symposium** *H. Wexler, M. J. Rubin, and J. E. Caskey, Jr. (Eds.)*
8 **Terrestrial Heat Flow** *William H. K. Lee (Ed.)*
9 **Gravity Anomalies: Unsurveyed Areas** *Hyman Orlin (Ed.)*
10 **The Earth Beneath the Continents: A Volume of Geophysical Studies in Honor of Merle A. Tuve** *John S. Steinhart and T. Jefferson Smith (Eds.)*
11 **Isotope Techniques in the Hydrologic Cycle** *Glenn E. Stout (Ed.)*
12 **The Crust and Upper Mantle of the Pacific Area** *Leon Knopoff, Charles L. Drake, and Pembroke J. Hart (Eds.)*
13 **The Earth's Crust and Upper Mantle** *Pembroke J. Hart (Ed.)*
14 **The Structure and Physical Properties of the Earth's Crust** *John G. Heacock (Ed.)*
15 **The Use of Artificial Satellites for Geodesy** *Soren W. Henricksen, Armando Mancini, and Bernard H. Chovitz (Eds.)*
16 **Flow and Fracture of Rocks** *H. C. Heard, I. Y. Borg, N. L. Carter, and C. B. Raleigh (Eds.)*
17 **Man-Made Lakes: Their Problems and Environmental Effects** *William C. Ackermann, Gilbert F. White, and E. B. Worthington (Eds.)*
18 **The Upper Atmosphere in Motion: A Selection of Papers With Annotation** *C. O. Hines and Colleagues*
19 **The Geophysics of the Pacific Ocean Basin and Its Margin: A Volume in Honor of George P. Woollard** *George H. Sutton, Murli H. Manghnani, and Ralph Moberly (Eds.)*
20 **The Earth's Crust: Its Nature and Physical Properties** *John C. Heacock (Ed.)*
21 **Quantitative Modeling of Magnetospheric Processes** *W. P. Olson (Ed.)*
22 **Derivation, Meaning, and Use of Geomagnetic Indices** *P. N. Mayaud*
23 **The Tectonic and Geologic Evolution of Southeast Asian Seas and Islands** *Dennis E. Hayes (Ed.)*
24 **Mechanical Behavior of Crustal Rocks: The Handin Volume** *N. L. Carter, M. Friedman, J. M. Logan, and D. W. Stearns (Eds.)*
25 **Physics of Auroral Arc Formation** *S.-I. Akasofu and J. R. Kan (Eds.)*
26 **Heterogeneous Atmospheric Chemistry** *David R. Schryer (Ed.)*
27 **The Tectonic and Geologic Evolution of Southeast Asian Seas and Islands: Part 2** *Dennis E. Hayes (Ed.)*
28 **Magnetospheric Currents** *Thomas A. Potemra (Ed.)*
29 **Climate Processes and Climate Sensitivity (Maurice Ewing Volume 5)** *James E. Hansen and Taro Takahashi (Eds.)*
30 **Magnetic Reconnection in Space and Laboratory Plasmas** *Edward W. Hones, Jr. (Ed.)*
31 **Point Defects in Minerals (Mineral Physics Volume 1)** *Robert N. Schock (Ed.)*
32 **The Carbon Cycle and Atmospheric CO_2: Natural Variations Archean to Present** *E. T. Sundquist and W. S. Broecker (Eds.)*
33 **Greenland Ice Core: Geophysics, Geochemistry, and the Environment** *C. C. Langway, Jr., H. Oeschger, and W. Dansgaard (Eds.)*
34 **Collisionless Shocks in the Heliosphere: A Tutorial Review** *Robert G. Stone and Bruce T. Tsurutani (Eds.)*
35 **Collisionless Shocks in the Heliosphere: Reviews of Current Research** *Bruce T. Tsurutani and Robert G. Stone (Eds.)*
36 **Mineral and Rock Deformation: Laboratory Studies—The Paterson Volume** *B. E. Hobbs and H. C. Heard (Eds.)*
37 **Earthquake Source Mechanics (Maurice Ewing Volume 6)** *Shamita Das, John Boatwright, and Christopher H. Scholz (Eds.)*
38 **Ion Acceleration in the Magnetosphere and Ionosphere** *Tom Chang (Ed.)*
39 **High Pressure Research in Mineral Physics (Mineral Physics Volume 2)** *Murli H. Manghnani and Yasuhiko Syono (Eds.)*
40 **Gondwana Six: Structure Tectonics, and Geophysics** *Gary D. McKenzie (Ed.)*

41 **Gondwana Six: Stratigraphy, Sedimentology, and Paleontology** *Garry D. McKenzie (Ed.)*

42 **Flow and Transport Through Unsaturated Fractured Rock** *Daniel D. Evans and Thomas J. Nicholson (Eds.)*

43 **Seamounts, Islands, and Atolls** *Barbara H. Keating, Patricia Fryer, Rodey Batiza, and George W. Boehlert (Eds.)*

44 **Modeling Magnetospheric Plasma** *T. E. Moore and J. H. Waite, Jr. (Eds.)*

45 **Perovskite: A Structure of Great Interest to Geophysics and Materials Science** *Alexandra Navrotsky and Donald J. Weidner (Eds.)*

46 **Structure and Dynamics of Earth's Deep Interior (IUGG Volume 1)** *D. E. Smylie and Raymond Hide (Eds.)*

47 **Hydrological Regimes and Their Subsurface Thermal Effects (IUGG Volume 2)** *Alan E. Beck, Grant Garven, and Lajos Stegena (Eds.)*

48 **Origin and Evolution of Sedimentary Basins and Their Energy and Mineral Resources (IUGG Volume 3)** *Raymond A. Price (Ed.)*

49 **Slow Deformation and Transmission of Stress in the Earth (IUGG Volume 4)** *Steven C. Cohen and Petr Vaníček (Eds.)*

50 **Deep Structure and Past Kinematics of Accreted Terranes (IUGG Volume 5)** *John W. Hillhouse (Ed.)*

51 **Properties and Processes of Earth's Lower Crust (IUGG Volume 6)** *Robert F. Mereu, Stephan Mueller, and David M. Fountain (Eds.)*

52 **Understanding Climate Change (IUGG Volume 7)** *Andre L. Berger, Robert E. Dickinson, and J. Kidson (Eds.)*

53 **Plasma Waves and Istabilities at Comets and in Magnetospheres** *Bruce T. Tsurutani and Hiroshi Oya (Eds.)*

54 **Solar System Plasma Physics** *J. H. Waite, Jr., J. L. Burch, and R. L. Moore (Eds.)*

55 **Aspects of Climate Variability in the Pacific and Western Americas** *David H. Peterson (Ed.)*

56 **The Brittle-Ductile Transition in Rocks** *A. C. Duba, W. B. Durham, J. W. Handin, and H. F. Wang (Eds.)*

57 **Evolution of Mid Ocean Ridges (IUGG Volume 8)** *John M. Sinton (Ed.)*

58 **Physics of Magnetic Flux Ropes** *C. T. Russell, E. R. Priest, and L. C. Lee (Eds.)*

59 **Variations in Earth Rotation (IUGG Volume 9)** *Dennis D. McCarthy and Williams E. Carter (Eds.)*

60 **Quo Vadimus *Geophysics for the Next Generation* (IUGG Volume 10)** *George D. Garland and John R. Apel (Eds.)*

61 **Cometary Plasma Processes** *Alan D. Johnstone (Ed.)*

62 **Modeling Magnetospheric Plasma Processes** *Gordon K. Wilson (Ed.)*

63 **Marine Particles Analysis and Characterization** *David C. Hurd and Derek W. Spencer (Eds.)*

64 **Magnetospheric Substorms** *Joseph R. Kan, Thomas A. Potemra, Susumu Kokubun, and Takesi Iijima (Eds.)*

65 **Explosion Source Phenomenology** *Steven R. Taylor, Howard J. Patton, and Paul G. Richards (Eds.)*

66 **Venus and Mars: Atmospheres, Ionospheres, and Solar Wind Interactions** *Janet G. Luhmann, Mariella Tatrallyay, and Robert O. Pepin (Eds.)*

67 **High-Pressure Research: Application to Earth and Planetary Sciences (Mineral Physics Volume 3)** *Yasuhiko Syono and Murli H. Manghnani (Eds.)*

68 **Microwave Remote Sensing of Sea Ice** *Frank Carsey, Roger Barry, Josefino Comiso, D. Andrew Rothrock, Robert Shuchman, W. Terry Tucker, Wilford Weeks, and Dale Winebrenner*

69 **Sea Level Changes: Determination and Effects (IUGG Volume II)** *P. L. Woodworth, D. T. Pugh, J. G. DeRonde, R. G. Warrick, and J. Hannah*

70 **Synthesis of Results from Scientific Drilling in the Indian Ocean** *Robert A. Duncan, David K. Rea, Robert B. Kidd, Ulrich von Rad, and Jeffrey K. Weisset (Eds.)*

Maurice Ewing Volumes

1 **Island Arcs, Deep Sea Trenches, and Back-Arc Basins** *Manik Talwani and Walter C. Pitman III (Eds.)*

2 **Deep Drilling Results in the Atlantic Ocean: Ocean Crust** *Manik Talwani, Christopher G. Harrison, and Dennis E. Hayes (Eds.)*

3 **Deep Drilling Results in the Atlantic Ocean: Continental Margins and Paleoenvironment** *Manik Talwani, William Hay, and William B. F. Ryan (Eds.)*

4 **Earthquake Prediction—An International Review** *David W. Simpson and Paul G. Richards (Eds.)*

5 **Climate Processes and Climate Sensitivity** *James E. Hansen and Taro Takahashi (Eds.)*

6 **Earthquake Source Mechanics** *Shamita Das, John Boatwright, and Christopher H. Scholz (Eds.)*

IUGG Volumes

1 **Structure and Dynamics of Earth's Deep Interior** *D. E. Smylie and Raymond Hide (Eds.)*

2 **Hydrological Regimes and Their Subsurface Thermal Effects** *Alan E. Beck, Grant Garven, and Lajos Stegena (Eds.)*

3 **Origin and Evolution of Sedimentary Basins and Their Energy and Mineral Resources** *Raymond A. Price (Ed.)*

4 **Slow Deformation and Transmission of Stress in the Earth** *Steven C. Cohen and Petr Vaníček (Eds.)*

5 **Deep Structure and Past Kinematics of Accreted Terrances** *John W. Hillhouse (Ed.)*

6 **Properties and Processes of Earth's Lower Crust** *Robert F. Mereu, Stephan Mueller, and David M. Fountain (Eds.)*

7 **Understanding Climate Change** *Andre L. Berger, Robert E. Dickinson, and J. Kidson (Eds.)*

8 **Evolution of Mid Ocean Ridges** *John M. Sinton (Ed.)*

9 **Variations in Earth Rotation** *Dennis D. McCarthy and William E. Carter (Eds.)*

10 **Quo Vadimus Geophysics for the Next Generation** *George D. Garland and John R. Apel (Eds.)*

11 **Sea Level Changes: Determinations and Effects** *P. L. Woodworth, D. T. Pugh, J. G. DeRonde, R. G. Warrick, and J. Hannah (Eds.)*

Mineral Physics Volumes

1 **Point Defects in Minerals** *Robert N. Schock (Ed.)*

2 **High Pressure Research in Mineral Physics** *Murli H. Manghnani and Yasuhiko Syona (Eds.)*

3 **High Pressure Research: Application to Earth and Planetary Sciences** *Yasuhiko Syono and Murli H. Manghnani (Eds.)*

Geophysical Monograph 71

Mantle Flow and Melt Generation at Mid-Ocean Ridges

Jason Phipps Morgan
Donna K. Blackman
John M. Sinton
Editors

American Geophysical Union

Published under the aegis of the AGU Books Board.

QE
522
.M27
1992

Library of Congress Cataloging-in-Publication Data

Mantle flow and melt generation at mid-ocean ridges / edited by Jason Phipps Morgan, Donna K. Blackman, John M. Sinton.
p. cm. — (Geophysical monograph : 71)
1. Volcanism—Congresses. 2. Earth—Mantle—Congresses. 3. Mid-ocean ridges—Congresses. I. Phipps Morgan, Jason. II. Blackman, Donna K. III. Sinton, John M. IV. Series.
QE522.M27 1993 93-812
551.1'16——dc20 CIP

ISSN: 0065-8448

ISBN: 0-87590-035-6

Copyright 1992 by the American Geophysical Union, 2000 Florida Avenue, NW, Washington, DC 20009

Figures, tables, and short excerpts may be reprinted in scientific books and journals if the source is properly cited.

Authorization to photocopy items for internal or personal use, or the internal or personal use of specific clients, is granted by the American Geophysical Union for libraries and other users registered with the Copyright Clearance Center (CCC) Transactional Reporting Service, provided that the base fee of $1.00 per copy plus $0.10 per page is paid directly to CCC, 21 Congress Street, Salem, MA 10970. 0065-8448/92/$01. + .10.

This consent does not extend to other kinds of copying, such as copying for creating new collective works or for resale. The reproduction of multiple copies and the use of full articles or the use of extracts, including figures and tables, for commercial purposes requires permission from AGU.

Printed in the United States of America.

CONTENTS

PREFACE

This volume provides both a comprehensive geophysical, petrological, and geochemical background on mantle flow and melt generation at mid-ocean ridges and results of current research in the field. The first part of the book contains overview papers geared towards a broad scientific audience with the dual intent of introducing new workers to the field and providing useful background and reference for those who study different aspects of oceanic spreading centers. The second set of papers presents current research on mantle upwelling and melting beneath mid-ocean ridges.

The motivation for publishing this set of papers as a monograph is to provide access to a range of currently recognized constraints and modeling approaches on the nature of mantle flow and melt generation along mid-ocean ridges. Our hope is that the multidisciplinary style of this volume will encourage communication between researchers in different fields and that this will enhance our progress towards a more complete understanding of the plate spreading and accretion process.

The overview papers include some cross referencing, but each provides a fairly self-contained portrayal of the type and resolution of the constraints that can be brought to bear on mid-ocean ridge processes from a specific field. Forsyth elucidates the strengths and inherent limitations of geophysical data that can be applied to the problem. He reviews the fundamental physical properties in the spreading/accretion system, discusses the interplay between observable quantities and the scale of parameters used in theoretical modeling, and assesses our ability to improve on current paradigms through the use of seismic and various potential field data. Hess introduces the petrological constraints provided by studies of phase equilibria in simple systems before delving into experimental results on basaltic melts and peridotites. He discusses the effect of pressure and temperature on melting systematics in the mantle and the nature of melt interaction with the residual solid. Kohlstedt presents both a theoretical framework and recent experimental results on melt distribution within olivine and basalt aggregates. He investigates the influence of melting on the rheology of the rock and interprets permeability measurements in terms of possible melt segregation characteristics. Ceuleneer and Rabinowicz use the observed structures in ophiolites to constrain models of upper mantle flow and melt migration near spreading centers. They discuss the scale and reliability of field measurements, present structural maps of parts of the Oman ophiolite, and relate these data to the results of numerical models of mantle flow. Turcotte and Phipps Morgan review the physical development of theoretical models of mantle flow, melting, and melt migration. They assess our current understanding of the physics of mantle upwelling and melt migration beneath ocean spreading centers. Langmuir, Klein, and Plank review the petrologic and geochemical systematics of partial melting of peridotite to form basaltic melt, and they present a quantitative model of melt composition based on experimentally constrained partition coefficients. They use the observed relationships between both major and trace elements to infer the depths of melt formation and the degree to which the melt interacts with its surroundings as it migrates to the surface. Global and local characteristics of mid-ocean ridge basalts are discussed, and the influence of spreading rate and large-scale mantle properties are investigated.

Specific aspects of the petrological and geophysical signatures of mid-ocean ridge processes are investigated in the second section of the volume. Grove, Kinzler, and Bryan use new experimental results and a computational model to assess the influence of fractionation on the composition of mid-ocean ridge basalts. Blackman and Forsyth determine the effect of plate thickening, due to cooling away from the spreading center, on the pattern of three-dimensional, passive mantle flow and thermal structure beneath a ridge-transform-ridge boundary. Scott presents a series of two-dimensional models that illustrate the individual and combined effects of thermal and compositional buoyancy on the relative importance of passive versus actively driven mantle flow. He discusses the possible influence of subaxial melt on viscosity structure and explores the mode of melt delivery to the upper crust. Evans and Everett present two-dimensional numerical models of the magnetotelluric signal that would be detected over a series of assumed melt distributions beneath a mid-ocean ridge.

This group of papers is an outgrowth of the RIDGE Theoretical Institute on Mantle Flow and Melt Generation at Mid-Ocean Ridges, which was held in Boulder, Colorado, in the summer of 1990. Comprising a four-day Short Course (on which the overview papers are based) and a two-week Workshop (where current research was presented), the Institute was the first in a series that constitute an integral part of the RIDGE (Ridge Inter-Disciplinary Global

Experiments) Initiative. The 1990 Institute was designed to spur our study of interlinked processes by stimulating interactions between geophysicists, petrologists, material scientists, and geochemists who work on aspects of the mantle flow/melt generation problem. Each group brings a different set of assumptions and data to bear on the question of how basaltic melt is generated beneath a mid-ocean ridge and how it is delivered to the surface to form oceanic crust. The exchange of information and ideas between the various specialties was intended to allow individuals to broaden their scope and to improve the interplay between observational and theoretical studies of this system.

Donna Blackman
Jason Phipps Morgan
John Sinton
Editors

Geophysical Constraints on Mantle Flow and Melt Generation Beneath Mid-Ocean Ridges

DONALD W. FORSYTH

Dept. of Geological Sciences, Brown University, Providence, RI 02912

INTRODUCTION

The title "Geophysical constraints on mantle flow and melt generation beneath mid-ocean ridges" can be interpreted in two ways: either as observational constraints on models, or as the constraints that physical properties and boundary conditions place on the flow itself. In this review, I concentrate primarily on the observational constraints, but to fully understand the problem of production and delivery of the magma that forms the oceanic crust we must understand the physics and chemistry of the process. All the observational tools we possess lack the resolution to define the details of the structure on a fine enough scale to expect that we can ever simply map the pattern of mantle flow or melt production. We must rely on theoretical models to provide the detail, testing the predictions of the models against observational and experimental constraints.

The primary goals of the geophysicist are to constrain the temperature and composition of the upwelling mantle, the pattern of flow of the mantle matrix and the distribution of melt. A key problem is to constrain the pattern of upwelling beneath the spreading center; because magma is produced by decompression melting of the mantle, the form of the upwelling controls the geometry and volume of melt generation. This is a difficult observational challenge because, as Sleep [1969] showed more than twenty years ago, the overall thermal structure of the ridge (Figure 1) is relatively insensitive to the width of the upwelling zone. Our primary observational constraints come from measuring variations in density, elastic properties and electrical conductivity of the mantle, which are all controlled locally by temperature, composition, fabric associated with flow, and the direct effects of the presence of melt. We can constrain variations in density through observations of sea floor depths, relying on the isostatic response of the mantle, and gravity and geoid measurements. Variations in elastic properties can be detected with measurements of seismic velocities and attenuation using both surface waves and body waves in passive and active experiments. Electrical conductivity can be constrained with magnetotelluric observations and active source experiments. Additional constraints on temperature structure are provided by heat flow observations and the flexure of the lithosphere. Perhaps the strongest constraints on melt production, melt migration and mantle flow are provided by observations of the final distribution of melt products, i.e., variations in thickness of the igneous crust and the segmentation of ridges.

The direct effects of melting may be difficult to detect *in situ*, but there is the potential for a well-constrained geophysical inverse problem. In a region in which melt is present, there are four primary variables of interest: the temperature, the melt fraction, the shape of the melt pockets, and the connectivity of the melt. There are also four primary observables: P-wave velocity, S-wave velocity, Q or attenuation, and electrical conductivity, yielding in the ideal situation four equations in four unknowns! Of course, the resolution of the experiments which would detect these physical properities must also be considered.

As a summary of the ability of geophysical measurements to constrain mantle flow and melt generation, I have constructed a table (Table 1) of the vertical and horizontal resolution of structure at a depth of 50 km, the expected center of the zone of melt generation, for each type of measurement under optimal conditions. There are fundamental physical limitations on each type of observation. For example, the isostatic response of the seafloor to deep density variations is limited by the flexure of the lithosphere and the transmission of stresses to the base of the plate through the viscous asthenosphere. Gravity and geoid measurements are limited by the natural filtering of upper continuation and inherent vertical non-uniqueness. Seismic surface waves and magnetotelluric observations are limited by the penetration depths of waves of different frequencies. The resolution of seismic observations in general are limited by the wavelength of the waves involved. No matter how many ocean-bottom seismometers are deployed and how many teleseismic events are detected,

Mantle Flow and Melt Generation at Mid-Ocean Ridges
Geophysical Monograph 71
Copyright 1992 American Geophysical Union

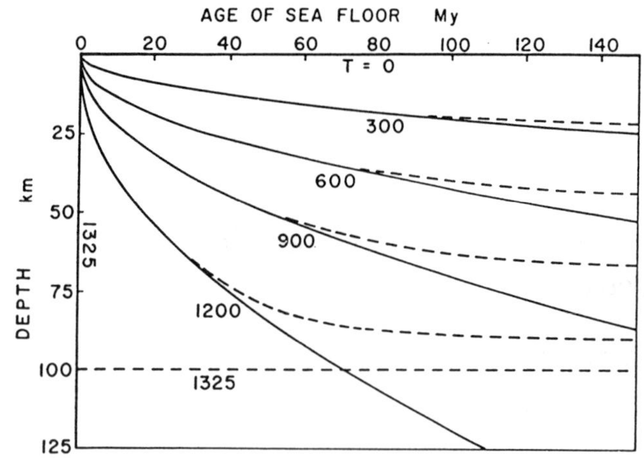

Fig. 1. Isotherms in two standard models of the temperature structure of the cooling oceanic mantle. Solid lines are isotherms for the cooling half-space model employed as a simple reference model in this review. Dashed lines are isotherms in a cooling plate model in which heat is supplied from below to maintain a constant temperature at a depth of 100 km. These are the two types of models employed in most studies of the oceanic lithosphere. Temperatures indicated are °C.

one cannot hope to resolve details of structure in a seismic tomography experiment on a scale smaller than the wavelength of the waves, or about 10 km. The scale of resolution is somewhat subjective since there is in many problems a trade-off between resolution and accuracy, but I have tried to give optimistically realistic estimates at accuracies that would provide useful constraints on flow models. The estimates of resolution given in Table 1 will be substantiated throughout this article. There may be quibbles about some of these numbers, but for those who claim that their technique will yield better resolution in the field, I will believe it when I see it. The primary point is that these observations provide constraints on theoretical models, but cannot directly illumine the melting process with enough detail to answer many of the important questions.

Before discussing some of the geophysical constraints, it is worth listing some of the specific questions we have in mind when conducting the experiments:

What are the lateral variations in composition and temperature of the upwelling mantle?

What is the relative importance of passive and dynamic upwelling?

Why are ridges segmented? Is the segmentation related to the dynamics of upwelling or to lithospheric stresses?

What is the extent of melting of the mantle? What volume is involved?

Is all the melt drained off quickly by a highly permeable network? What is the average melt concentration present beneath a spreading center?

What is the width of the upwelling region beneath mid-ocean ridges?

Why is the neovolcanic zone so narrow?

What is the mechanism for lateral migration of melt? Does the presence of partial melt affect the rheology of the matrix?

Are mantle flow and/or melt migration strongly time dependent? On what time scales?

Physical Properties

Geophysical constraints on mantle flow and melt generation depend on there being changes in the physical properties of the rocks as a function of temperature, pressure, composition, and strain. These physical changes include the effects of melting and the crystal fabric that develops during deformation. Geophysical experiments can be designed to detect or are sensitive to changes in the thermal expansion

TABLE 1. Maximum resolution of structure at a depth of 50 km beneath a mid-ocean ridge achievable with different geophysical techniques.

Method	Horizontal Resolution	Vertical Resolution
Seismic surface waves	150x150 km	30 km
Teleseismic tomography		
P waves	10x10	30
S waves	30x30	50
Magnetotellurics	50x100	40
Isostatic topography	50 - 200	50 - 150
Geoid/gravity --Average depth of compensation	50 - 200	10
Basalt chemistry	10 - 200	10
Crustal thickness	10 - 200	Melt zone
Seismic reflection?	10x10	5
Theory	< 1x1	< 1

coefficient, thermal diffusivity or conductivity, density, seismic velocities and attenuation, electrical conductivity, and the rheological response to stress. Complete interpretation of the geophysical experiments depends on an understanding of the physical properties, some of which we briefly review here.

Thermal expansion. Lateral changes in density in the mantle are primarily controlled by temperature through thermal expansion. The thermal expansion coefficient α

$$\alpha = \frac{1}{\rho}\left(\frac{\partial \rho}{\partial T}\right) \tag{1}$$

is a function of both temperature T and pressure P. The pressure dependence can be related to the density ρ through the expression [Anderson, 1967]

$$(\alpha/\alpha_0) = (\rho/\rho_0)^{-\delta_T} \tag{2}$$

where δ_T, the Anderson-Gruneisen parameter, is assumed to be independent of pressure and is given by [Anderson, 1987; Anderson and Yamamoto, 1987]

$$\delta_T = \left(\frac{\partial \ln K_T}{\partial \ln \rho}\right)_P = -\frac{1}{\alpha K_T}\left(\frac{\partial K_T}{\partial T}\right)_P = -\left(\frac{\partial \ln \alpha}{\partial \ln \rho}\right) \tag{3}$$

where K_T is the isothermal bulk modulus (incompressibility). Experimental [Copelas and Boehler, 1989; Anderson et al., 1990] and theoretical [Reynard and Price, 1990] investigations confirm that δ_T is independent of pressure, and independent of temperature above the Debye temperature (the temperature above which high frequency modes of vibration of atoms in crystals become dominant). Typical values for δ_T lie between 4 and 6 for silicate rocks. Because there is a negative coefficient for the density dependence, with increasing pressure the thermal expansion coefficient decreases. Consequently, the same temperature change generates a smaller buoyancy force and is less effective in inducing convection at great depths. Between the surface and a depth of 100 km, a decrease in α of about 15% is predicted from the pressure effect alone.

The thermal expansion coefficient is also a function of temperature. Figure 2 shows experimental measurements of α in forsterite, as summarized by Isaak et al. [1989]. The form of the temperature variation can be predicted by an approximate, statistical mechanical model, Gruneisen theory, that is a function of the Debye temperature and other thermodynamic parameters [Suzuki et al., 1979]. Typical Debye temperatures for upper mantle minerals are 400-600 °C, and the difference in thermal expansion coefficients between olivine, pyroxenes and garnet [Duffy and Anderson, 1989] are not much larger than between different experimental measures for olivine. Most simple models of the upper mantle have assumed constant values for α in the range 3.0 to 3.5 x 10^{-5} °K^{-1}, but for some purposes, it will clearly be important to consider the temperature and pressure dependence.

Composition, density and pressure. Critical factors in controlling the patterns of mantle flow and melt segregation

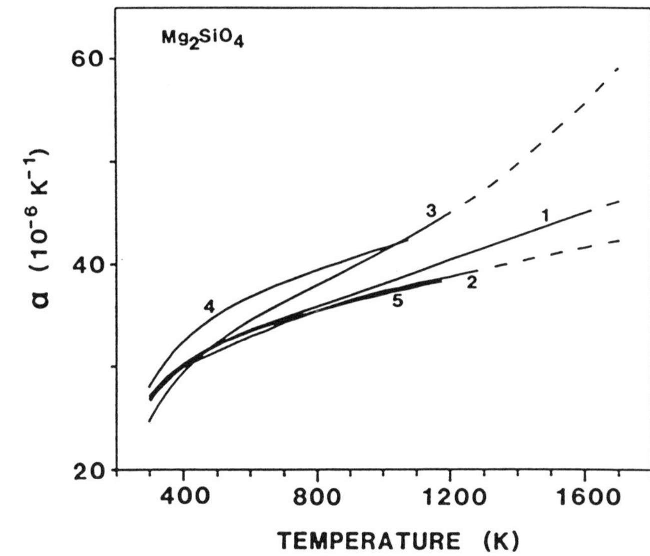

Fig. 2. Volume coefficient of thermal expansion for forsterite as a function of temperature [from Isaack et al., 1989]. Solid lines show measurements of different investigators. Dashed curves show extrapolations.

are the density of the basaltic melt and the contrast in density between unmelted mantle and the depleted residual. Niu and Batiza [1991a] have recently reviewed models and experiments for the density of melt and solids likely to be found in a mid-ocean ridge setting. The density of a silicate melt is described by

$$\rho = \frac{\Sigma X_i M_i}{\Sigma X_i V_i} \tag{4}$$

where X_i is the mole fraction of the metal oxide, M_i is the gram formula weight, and V_i is the partial molar volume which is the change in volume given an infinitesmal change in composition of some component n_i

$$V_i = \left(\frac{\partial V}{\partial n_i}\right) \tag{5}$$

V_i in general is a function of temperature, pressure and composition and has been determined experimentally by Bottinga and Weill [1970], Nelson and Carmichael [1979], and Lange and Carmichael, [1987]. Empirically, an ideal linear mixing model works very well and the only known compositional dependence for V_i is an interaction between TiO_2 and Na_2O [Lange and Carmichael, 1987]. The temperature dependence can be described by a thermal expansion coefficient which is of the same order as for solids. The critical behavior for the density of melts is the pressure dependence. Melts are much more compressible than solids, so their buoyancy in a mantle matrix is strongly dependent on depth. In fact, it has been suggested that melt would sink or olivine crystals float at depths greater than about 250 km [Stolper et al., 1981; Agee and Walker, 1988]. Beneath a mid-ocean ridge, the density contrast between melt and matrix is likely to range from about 0.25 Mg m^{-3} at a depth of 80 km to 0.50 Mg m^{-3} at the base of the crust [Niu and Batiza, 1991a].

The change in density of the residual mantle attributable to the removal of melt may have a strong effect on the dynamic (buoyant) component of upwelling beneath a mid-ocean ridge [Sotin and Parmentier, 1989; Parmentier and Phipps Morgan, 1990; Sparks and Parmentier, 1991]. As for liquids, a good first approximation to the density of a composite solid is to assume ideal, linear mixing between the mineral components. However, the fact that the bulk modulus or incompressibility of composite solids is not always best described by a simple linear average of the bulk moduli of the individual crystals [e.g., Watt et al., 1976] indicates that the linear mixing model is indeed a simplification of the true behavior. Pressure affects the density both through the compressibility of individual minerals and through pressure-induced phase changes.

The change in the aluminous phase from plagioclase to spinel to garnet with increasing pressure means that the magnitude and even the sign of the net effect of depletion on density is highly dependent on depth. There are two primary effects of the removal of a melt fraction from a peridotite: enrichment of the residual in MgO relative to FeO as MgO is highly refractory, and depletion in Al_2O_3. Although the composition of the melt is dependent on pressure as well, if we assume for example that seafloor basalts are created by approximately 15% pressure-release melting beginning at roughly 60 km depth, then the MgO enrichment of the residual decreases its density by 0.010 to 0.015 Mg m^{-3} relative to undepleted mantle under the same conditions. The depletion in Al_2O_3, CaO and Na_2O means that, at depths shallower than about 30 km, there will be less low-density plagioclase in the residual, so that the density of the depleted mantle will be greater than that of undepleted mantle at the same depth. At greater depths the aluminous phases are spinel or garnet (beyond about 60-70 km) [Ringwood, 1975], which are denser than olivine or pyroxenes, so the effect of depletion is reversed: the residual mantle is expected to be less dense than the spinel or garnet-laden undepleted mantle. The net effect of 15% depletion including MgO enrichment is an increase in density of the residuum of about 0.035 Mg m^{-3} in the plagioclase stability field, and decreases of about 0.020 and 0.040 Mg m^{-3} in the spinel and garnet stability fields, respectively. (These calculations are based on model compositions of Niu and Batiza [1991b] using experimental data of Jaques and Green[1980], Falloon et al. [1988], and Falloon and Green [1987,1988]. Similar results in the garnet stability field have been obtained by Oxburgh and Parmentier [1977] and Jordan [1979]. Up to about 25% melting, the density changes are roughly proportional to the degree of depletion.

Thermal conductivity. The effective thermal diffusivity has two components, radiative transfer and lattice thermal conductivity. Experimentally, olivines and pyroxenes have thermal conductivities in the range 3-10 W m^{-1} °K^{-1}, with lattice conductivity increasing with pressure and decreasing with increasing temperature [Beck et al., 1978]. Total thermal conductivity appears to be roughly constant under upper mantle conditions in the temperature range 500 - 2000°K [Fukao, 1969; Schatz and Simmons, 1972].

Electrical conductivity. Electrical conductivity is controlled by the abundance and mobility of charge carriers in the crystal structure. Even in olivine, the identity of the charge carriers is imperfectly known, so we clearly do not yet have a complete understanding of the possible effects of doping by impurities. However, because the electrical conductivity structure of the upper 200 km of the mantle seems to be reasonably consistent with the properties of olivine single crystals, olivine may provide a good first-order model for the conductivity variations expected under a mid-ocean ridge.

Olivine conductivity is the sum of several thermally activated processes:

$$\sigma_s = \sum_i \sigma_i e^{-A_i/kT} \qquad (6)$$

where the A_i are activation energies, σ_i are conductivities, k is Boltzmann's constant and T is absolute temperature. At least three processes have been identified by changes in slope of log conductivity versus reciprocal temperature (Figure 3), with transitions between dominant mechanisms occuring at around 720 °C and 1500 °C [Constable and Duba, 1990]. The mechanism dominant in middle temperature range is thought to be electron holes in the valence band that "hop" from one Fe site to another, transforming Fe^{+2} to Fe^{+3} [Schock et al, 1989]. Experiments on conductivity as a function of oxygen fugacity, (fO_2) suggest that there is an independent population of charge carriers that is supplemented by additional charge carriers created by the absorption of oxygen at higher (fO_2), leading to a variation of σ_2 at constant temperature of the form

$$\sigma_2 = \sigma_a + \sigma_b (fO_2)^c \qquad (7)$$

where the coefficient c has been experimentally determined to be about 0.3 [Constable and Duba, 1990]. The conductivity for this mechanism also depends on iron content with an increase of about 0.24 orders of magnitude per mole percent iron, measured over a limited range of compositions around Fo_{90}.

Variations in iron and nickel content, oxygen fugacity and the presence of impurities may thus cause roughly an order of magnitude variation in σ_s [Duba et al., 1974; Shankland and Waff, 1977; Duba and Shankland, 1982; Schock et al., 1989; Constable and Duba, 1990]. Pressure seems to have little effect on conductivity [Schock et al., 1977]. There also seems to be little difference between a polycrystalline dunite and single olivine crystals [Constable and Duba, 1990]. It is interesting to note that electrical conductivity in olivine is anisotropic, indicating different modes of conduction along different axes [Schock et al., 1989]. If the intrinsic anisotropic effects could ever be sorted out from the pseudo-

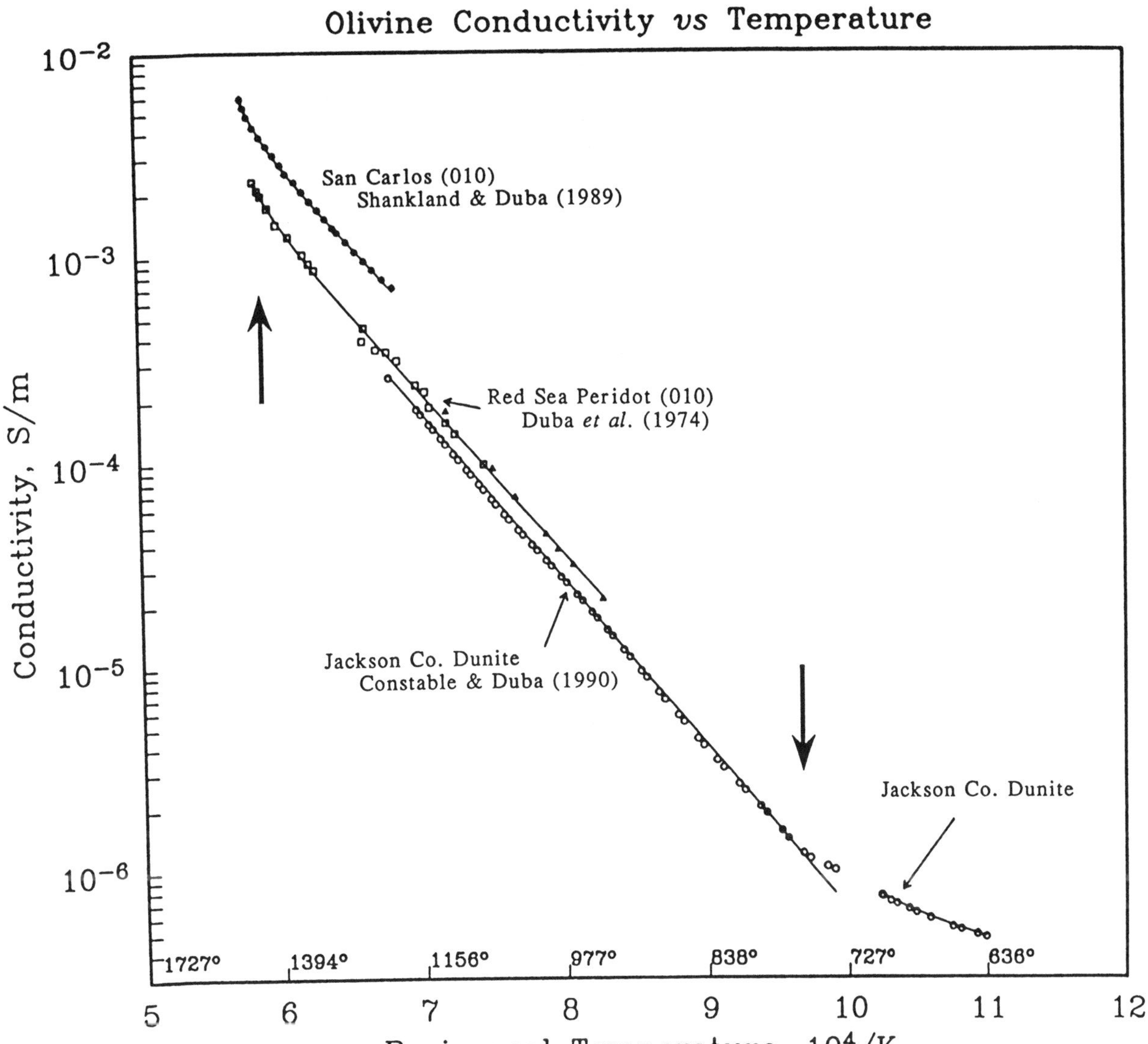

Fig. 3. Electrical conductivity of dunite versus temperature [Constable and Duba, 1990]. Vertical arrows mark change in slope of the conductivity corresponding to changes in the dominant conduction mechanism.

anisotropic effects of two- and three-dimensional structure, observations of resistivity could serve as an independent test of the alignment of olivine crystals inferred from seismic experiments.

Effect of melt on seismic properties. The influence of partial melt on seismic velocities and attenuation depends on the melt fraction and on the geometry of its distribution. Schmeling [1985a] gives an excellent review of the theory. This abbreviated review follows his development. Melt pockets assume a shape that minimizes the total free energy of the solid-solid and liquid-solid interfaces. If melt occurs within a grain, it should have a spherical geometry to minimize the surface energy. The shape of the melt pocket at a grain boundary can be described by the dihedral angle ϕ (Figure 4). If $\phi = 0$, the melt will wet all grain faces and occur in the form of thin films or ellipsoidal inclusions with small aspect ratio. For $0 < \phi < 60^\circ$ melt at grain edges and corners will form an interconnected system of tubes. When $\phi > 60^\circ$ melt will be found in pockets at grain corners and the interconnection between pockets depends on ϕ and the degree of melting [Bulau et al., 1979]. Recently, Waff and Faul [1992] have provided experimental evidence that melt at small melt fractions in olivine may be simultaneously distributed between tubes along grain edges and films wetting at least one of the flat crystal faces, in contrast to most prior experimental studies that reported only the curved interfaces at grain edges.

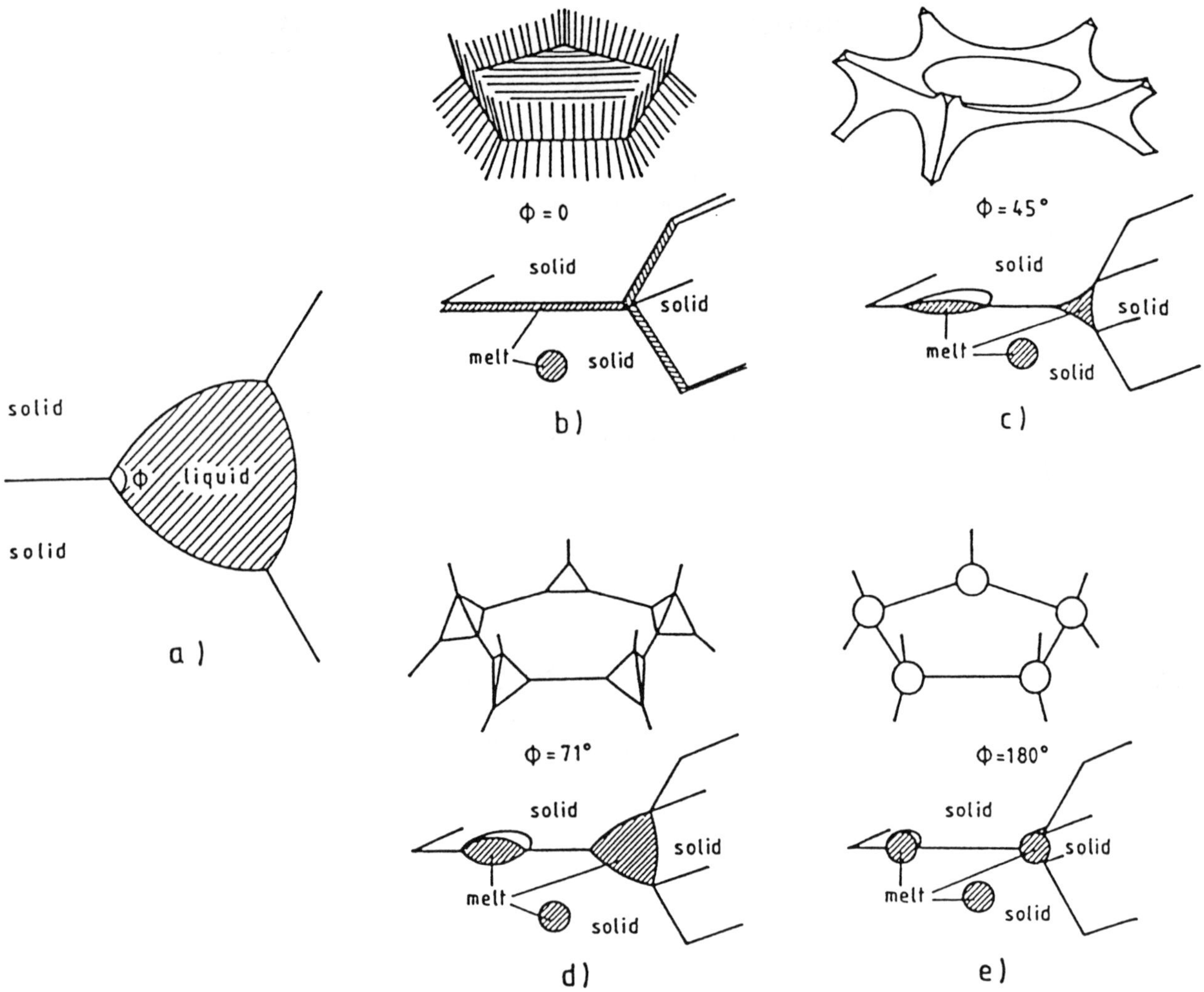

Fig. 4. Schematic diagrams of distribution of melt in partially molten, crystalline solids [Schmeling, 1985b]. (a) Geometry is controlled by the dihedral angle between grains. (b)-(e) Predicted geometries range from thin films for $\phi = 0$ to interconnected tubes to isolated melt pockets for $\phi > 60°$. Hatched areas show possible melt geometry depending on whether melt is located within grains or at grain boundaries.

Melt affects seismic velocities in two ways: a direct effect on the elastic moduli and the dispersive effect of relaxation or attenuation. In a standard linear solid representation of an attenuation mechanism (Figure 5), the transition between relaxed and unrelaxed moduli occurs at a characteristic frequency with the amplitude of the difference between relaxed and unrelaxed moduli being described by the relaxation strength. In a real viscoelastic medium, there may be several mechanisms occuring at different frequencies or relaxation times, so the result may be a gradual dispersion of seismic velocities as a function of frequency. The absorption or attenuation factor Q is bounded by the relaxation strength Δ

$$Q^{-1} \leq \Delta/2 \qquad (8)$$

for all possible distributions of relaxation processes. For small $\Delta/2$, the half relaxation strength is the integral of the Q^{-1} spectrum plotted versus $\ln\omega$ and divided by π. In a linear viscoelastic body, any of the elastic moduli can be represented by a complex modulus $M = M_1 + i\,M_2$ and with $Q = M_1/M_2$. M_1 and M_2 at any given frequency can also be related to integrals of the transition between the relaxed and unrelaxed moduli [Nowick and Berry, 1972]. For a propagating seismic wave, the amplitude A of the oscillation decays with time t from an original amplitude A_0 according to

$$A = A_0 \exp(-\pi f t/Q) \qquad (9)$$

where f is the frequency. Seismic frequencies lie in the range 0.001 to 100 Hz. Of the two primary relaxation mechanisms

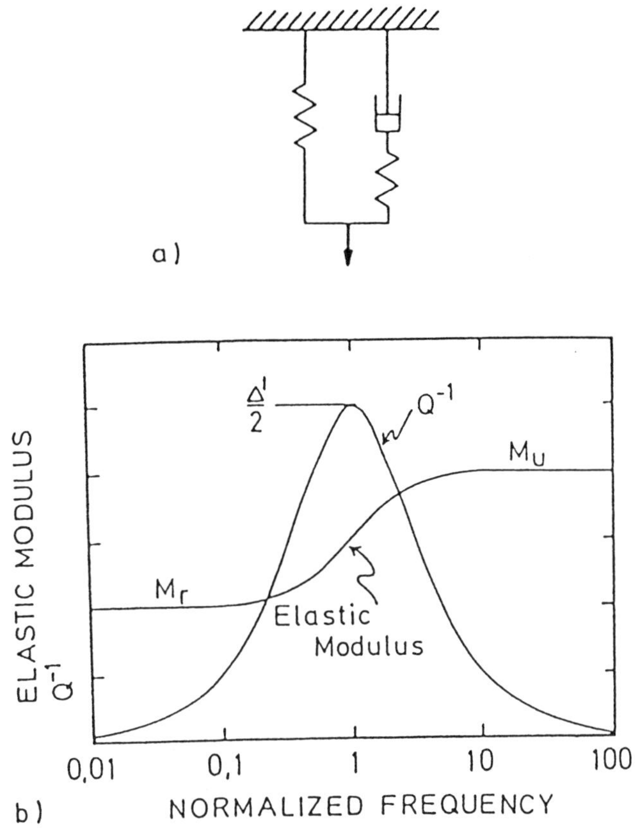

Fig. 5. Schematic elastic behavior of standard linear solid consisting of springs and dash-pots in arrangement shown in (a). A single attenuation or damping mechanism produces an attenuation peak at a characteristic frequency (b). The elastic moduli are relaxed (subscript r) at lower frequencies and unrelaxed (subscript u) at higher frequencies. Q is the quality factor, and Δ is the relaxation strength.

proposed for melt in a solid, viscous shear and melt squirt, only melt squirt is likely to be significant at these frequencies [O'Connell and Budiansky, 1977]. Melt squirt involves fluid flow driven by pressure variations, either between interconnected inclusions or within single inclusions if they deform non-uniformly. Melt squirt will only be important in the seismic frequency band if the aspect ratio α for ellipsoidal inclusions is < 10^{-2} to 10^{-3} , or if the ratio of the diameter of tubes to their length is < 10^{-2} to 10^{-3} [Schmeling, 1985]. (α has previously been used to represent thermal expansion coefficient. Rather than re-drafting figures and changing notations from that in the standard literature, I warn the reader that in this and other cases, the same symbol may be used for more than one constant.)

The elastic moduli for fluid distributed in cracks or thin films are given by O'Connell and Budiansky [1974,1977] and for tubes by Mavko [1980], with extensions or corrections given by Schmeling [1985]. The predicted bulk and shear moduli and relaxation strength are shown in Figure 6 for ellipsoidal inclusions and tubes. Tubes behave much like ellisoidal inclusions with aspect ratios 0.2 to 0.5, depending on the cross-sectional shape of the tube. Very small amounts of melt can have a substantial effect on the shear modulus and relaxation strength if it distributed in thin films, i.e., the aspect ratio is small (on the order of 0.01). This state can be recognized observationally by low shear velocities and low Q (provided the observations are near the relaxation frequency), but near normal compressional velocities. For tubes and ellipsoidal inclusions of moderate shape, relaxation strengths are relatively low, and the shear and bulk moduli behave similarly. In this case, both P and S wave velocities will decrease significantly as the melt fraction increases, but no drastic drops in Q or velocity can be expected at the onset of melting and the ratio of P wave velocity to S wave velocity may not change significantly. For example, in the limit of spheroidal inclusions, 15% melt fraction will produce roughly a 12% drop in P-wave velocity and 16% drop in S-wave velocity.

There are few experimental measurements of seismic velocities and attenuation in partially molten rocks. Murase and Kushiro [1979] and Murase and Fukuyama [1980] measured compressional and shear velocities and volume fractions of

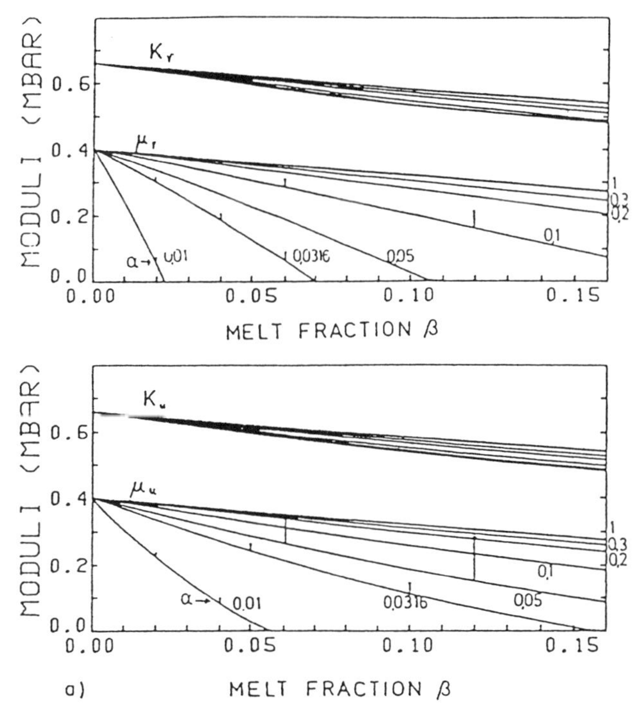

Fig. 6. (a) Relaxed and unrelaxed elastic moduli as a function of melt fraction for ellipsoidal inclusions [Schmeling, 1985]. α is the aspect ratio of the inclusion. (b) Relaxation strength as a function of melt fraction for ellipsoidal inclusions (solid lines). Dashed lines show relaxation strength for tubular inclusions with different cross-sectional shapes described by the shape parameter κ, illustrated at bottom of the figure.

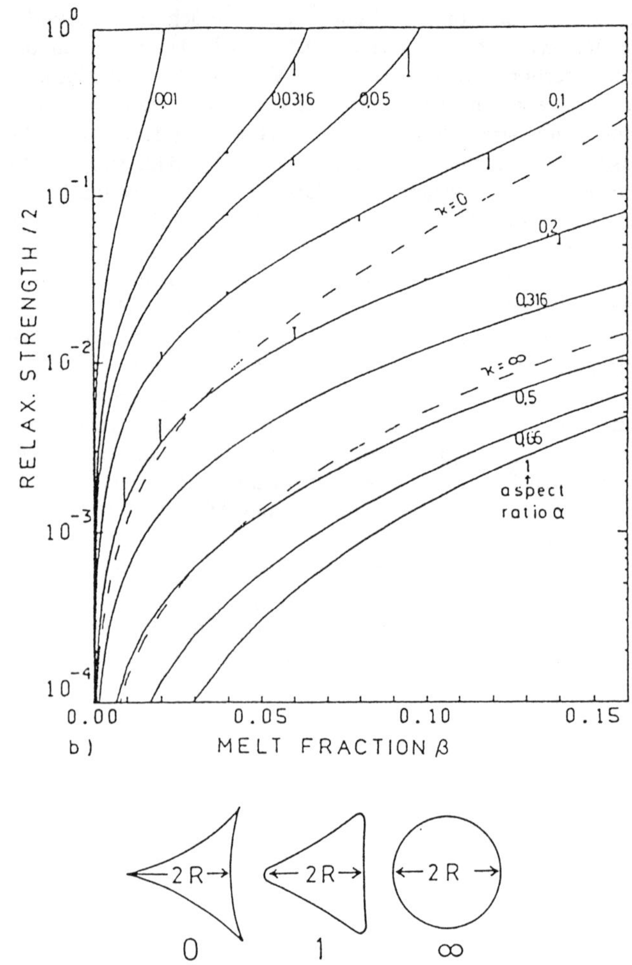

Fig. 6. (continued)

partial melt in a dry spinel lherzolite to 1 GPa and 1300 °C. Measuring the velocities and melting at two different confining pressures and hence at two different temperatures of the solidus, they found that both P and S velocity and melt fraction were a function only of the homologous temperature [Sato et al., 1988], the ratio of the temperature to the solidus temperature (°K). Their results are shown in Figure 7, including a plot of velocity as a function of melt fraction. Melting clearly causes a significant decrease in velocity, but there is no drastic drop at the onset of melting with small melt fractions. The behavior of Q as a function of temperature and pressure for the same material also seems to be describable by the homologous temperature (Figure 8), although there is no significant effect of melting on Q detected in the vicinity of the solidus [Sato et al., 1989]. To test the hypothesis of a simple dependence on melting temperature, more experiments are needed with different compositions, including trace amounts of water.

Measurements of the shear modulus and attenuation in the seismic band [Murase, Sato and others used ultrasonic frequencies) in the vicinity of the solidus were accomplished for a dunite, polycrystalline forsterite and a peridotite by Berckhemer et al. [1982] and Kampfmann and Berckhemer [1985]. The experiments were performed at 1 atm and the bulk modulus was not measured. Onset of melting was inferred from the deviation of the dunite and peridotite moduli from the forsterite modulus and the simultaneous variation of the shear modulus and electrical resistivity. The results are compatible with those of Murase, Sato and others. There is no sudden drop in velocities at the onset of melting. Q appears to be dominated by a background solid state signal that renders Q

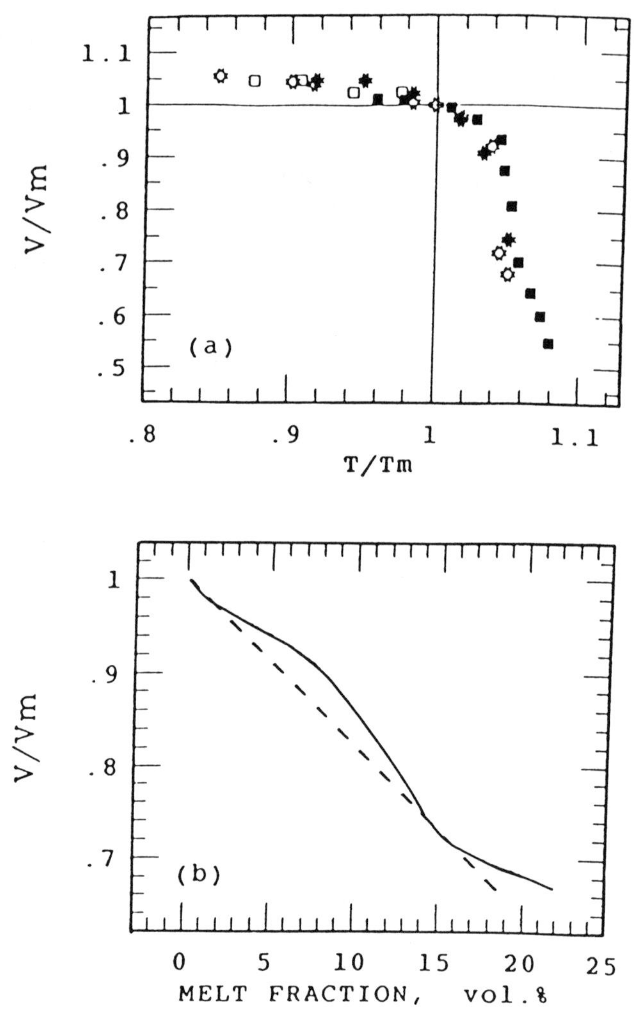

Fig. 7. Experimental measurements of seismic wave velocity as a function of temperature and melt fraction in a peridotite [Sato et al., 1988]. (a) P- and S-velocities expressed as fraction of velocity at onset of melting versus temperature as fraction of temperature at melting. Solid symbols are P-waves, open symbols are S-waves, with data taken at two different pressures with different melting temperatures. (b) Velocity as a function of melt fraction using data shown in (a) (solid line) and in a metal alloy (dashed line).

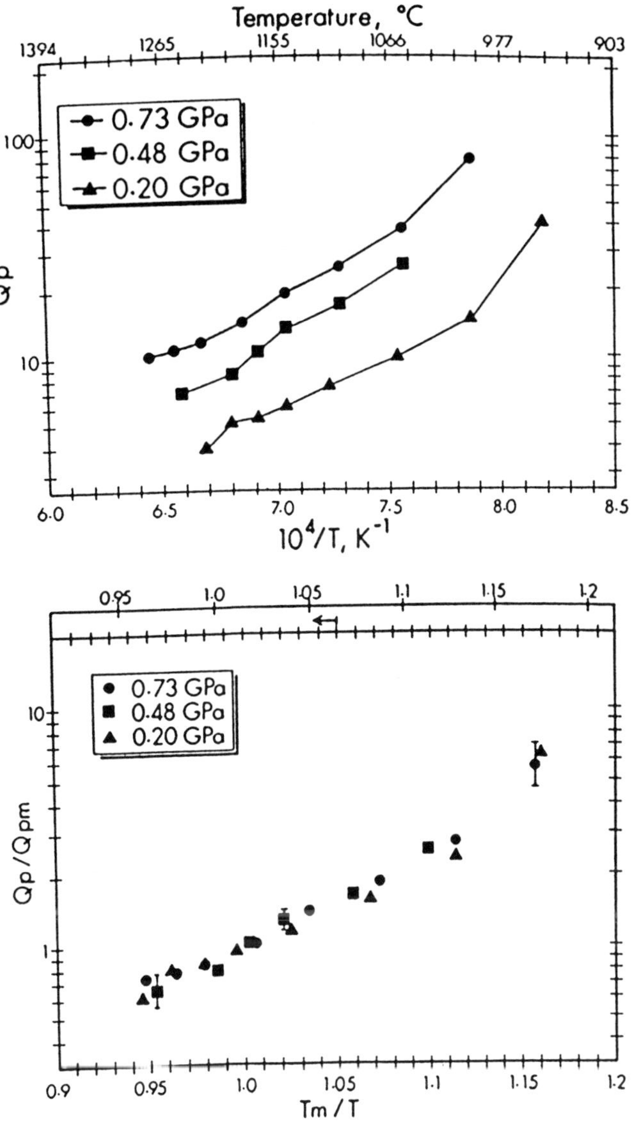

Fig. 8. (a) Compressional wave attenuation (quality factor Q) as a function of temperature in a peridotite [Sato et al., 1988] at three different pressures with different melting temperatures. (b) Same data normalized by Q at the melting point for each pressure. There is no obvious change in attenuation at the onset of melting.

temperature and frequency dependent, but there is no change in attenuation at the solidus.

The lack of change in Q above the solidus, the gradual change of velocity at the onset of melting, and the similar behavior of P and S velocity all suggest that melt is distributed in tubes or ellipsoidal pockets with only modest aspect ratios ($\alpha > .2?$) rather than as thin films wetting the grain boundaries. This is in agreement with observations of melt distribution in dunites that find dihedral angles between basalt melt and olivine between 20° and about 45° [Waff and Bulau, 1979,1982; Jurewicz and Jurewicz, 1985; Toramaru and Fujii, 1986; Daines and Richter, 1988]. (It should be noted that Waff and Faul [1991] have recently argued for the simultaneous existence of films on grain faces and tubes along grain triple junctions.) The melt is likely to be interconnected through tubes, but in a peridotite, the interconnectivity may be interrupted by orthopyroxene grains, in which the dihedral angle formed with melt is greater than 60° [Toramaru and Fujii, 1986]. The presence of water may reduce this orthopyroxene effect, improving the connectivity.

Experiments and theory suggest that seismic experiments cannot distinguish between interconnected tubes and isolated inclusions in peridotite. Whether melt is present or not cannot be determined from variations in attenuation, which is insensitive to melt fraction at small degrees of melt unless it is distributed in thin films. Because both attenuation and velocity are sensitive to temperature, but only velocities are sensitive to melt fraction, simultaneous measurement of both may help to sort out melting effects from temperature effects. A complication in interpretation of attenuation measurements is that, although the variation of Q with reciprocal temperature is quite linear in forsterite and dunite, there is a non-linear, sub-solidus variation of Q in peridotite that is frequency dependent and appears to be an absorption peak in the seismic frequency band [Kampfmann and Berckhemer, 1985]. The interconnectivity of melt that is crucial to melt migration may better be addressed with electrical conductivity measurements.

Electrical conductivity and melting. The effect of melting depends primarily on the interconnectivity of the melt, the melt fraction, and the conductivity contrast between melt and solid. Upper and lower (Hashin-Strickman) limits on the bulk conductivity can be found by minimizing and maximizing the energy dissipation in a two phase system [Waff, 1974]. If σ_m is the conductivity of the melt, σ_s is the conductivity of the solid phase, and β is the melt fraction, then the upper σ_+ and lower σ_- limits on the bulk conductivity of the rock as a whole are

$$\sigma_+ = \frac{\sigma_m + (\sigma_s - \sigma_m)(1 - 2\beta/3)}{1 + (\beta/3)[(\sigma_s/\sigma_m) - 1)]} \tag{10}$$

and

$$\sigma_- = \frac{3\sigma_s + (\sigma_m - \sigma_s)(1+2\beta)}{(2 + \beta) + (\sigma_m/\sigma_s)(1 - \beta)} \tag{11}$$

The upper limit corresponds to the interconnection of all melt in an isotropic geometry [Waff, 1974]. The lower limit is equivalent to melt being distributed in isolated spheres. Partially interconnected networks or isolated pockets of non-spherical shape fall somewhere between the limits [Hermance, 1979; Schmeling, 1986].

The effective conductivity of the rock as a whole is shown in Figure 9 for these bounds and for isolated spheroidal inclusions of various aspect ratio [Schmeling, 1985b]. The lower bound of spherical inclusions shows little effect until the fraction of partial melt exceeds 20 to 30%. Interconnected melt can increase the conductivity at very small melt fractions. The figure shows that isolated inclusions of very low aspect

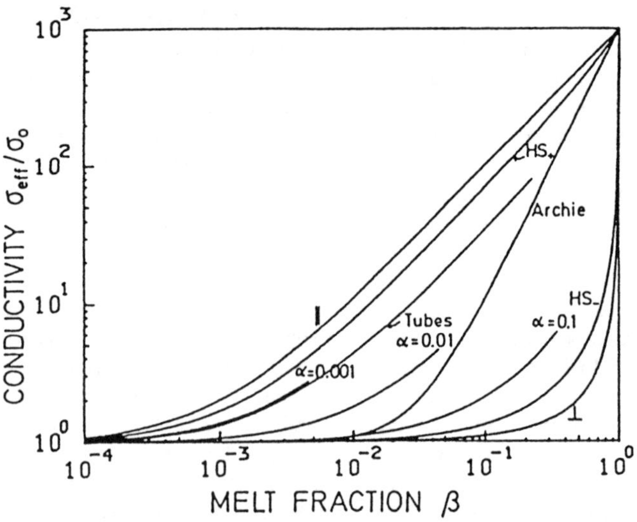

Fig. 9. Effective electrical conductivity assuming the conductivity of melt fraction is 1000 times conductivity σ of matrix for different distributions of melt [Schmeling, 1985b]: $\parallel$ and $\perp$ are thin fluid layers parallel and perpendicular to the electric current; HS+ and HS- are upper and lower Hashin-Shtrikman bounds; α = 0.001 to 0.1 unconnected ellipsoidal inclusions of varying aspect ratio; tubes are randomly oriented tubes; and Archie is the empirically-based formula called Archie's law.

ratio (α < .001) behave much like the upper limit of interconnected melt, so a significant effect at small melt fraction does not guarantee interconnection. Schmeling [1986], however, has shown that random inclusions will tend to interconnect, with the interconnections being established at small melt fractions if the aspect ratio is small. Experiments on water saturated sedimentary rocks find that σ increases as a function of β^2 [Brace et al., 1965]. For the case of partial melt, this empirical behavior is described by a modified form of Archie's law [Hermance, 1979]

$$\sigma_A = \sigma_S + (\sigma_m - \sigma_S)\beta^2 \tag{12}$$

which is consistent with increasing interconnection between melt pockets as the concentration increases [Schmeling, 1986]. Archie's law is also shown in Figure 9. A combination of resistivity and seismic observations [Shankland et al., 1981] is potentially a powerful way to address connectivity of melt because if the seismic observations require aspect ratios larger than 0.2 (or the equivalent in tubes), then melt will have little effect on resistivity at moderate melt fractions unless the melt pockets are connected.

Laboratory experiments indicate that the conductivity of molten basalt is on the order of 10 S/m [Siemens S = ohm^{-1}) [Presnell et al, 1972; Rai and Manghnani, 1978b; Tyburczy and Waff, 1983; Waff and Weill, 1975], 2 to 3 orders of magnitude larger than peridotite at the same temperature [Figure 10 from a summary by Schmeling, 1985b]. The curves for ultramafics [Rai and Manghnani, 1978a], olivine single crystals [Duba et al., 1974], dunite and lherzolite [Berckhemer et al. 1982] all show kinks that are apparently due to the onset of partial melting [Murase et al., 1977b], thus ultramafic rocks with melt removed should have conductivities about 3 orders of magnitude less than basaltic melt. The kink for olivine single crystals is not easily understood because later measurements in olivine show only a very gradual change in slope of σ vs T in measurements to 1500 °C [Schock et al., 1989]. Some of these kinks are probably caused by the instability of mineral phases in experiments with poorly controlled fO_2 or a change in conduction mechanism. Variations in iron and nickel content, oxygen fugacity and the presence of impurities plus the variation of σ_S with temperature can cause considerable uncertainty in estimating the melt fraction, although conductivities as high as 0.2 S/m in the upper mantle (Oldenburg, 1981; Eysteinsson and Hermance, 1985] would require the presence of melt.

Axial Depths and Subsidence Rates

The depth of the seafloor is a direct indicator of the vertically integrated density structure of the crust and upper mantle. The depth of the seafloor d can be given by

$$d = d_o + \delta d \tag{13}$$

where d_o is the depth for some standard reference structure and δd represents bathymetric variations or perturbations from the reference depth. Assuming local isostatic equilibrium, δd is related to the density perturbations $\delta\rho$ from the reference density structure by

$$\delta d\,(\rho_m - \rho_w) = -\int_0^{H_c} \delta\rho\, dz \tag{14}$$

where ρ_m is the mantle density and ρ_w is the sea-water density. H_c is the depth of compensation or the depth to which density variations need to be considered. Thus, variations in density structure are balanced by variations in depth to the seafloor, replacing water with denser mantle.

There are two critical assumptions in use of this formulation: the assumption of local isostatic equilibrium and the choice of compensation depth. The flexural rigidity of the lithospheric plate limits the ability of the seafloor to respond to local density variations. Local equilibrium is achieved only for perturbations at wavelengths greater than the characteristic flexural wavelength of the lithosphere, which is a function of the effective elastic thickness of the plate. The effective elastic thickness is primarily a function of the thermal state of the lithosphere. It ranges in value from less than 2 km at the crest of the East Pacific Rise (EPR) [Cochran, 1980; Madsen et al., 1984; Kuo et al., 1985] to around 6 km in the vicinity of the axis of the Mid-Atlantic Ridge (MAR) [McKenzie and Bowin, 1976; Cochran, 1980; Louden and Forsyth, 1982] to several tens of km in old seafloor [Watts, 1980; McNutt,

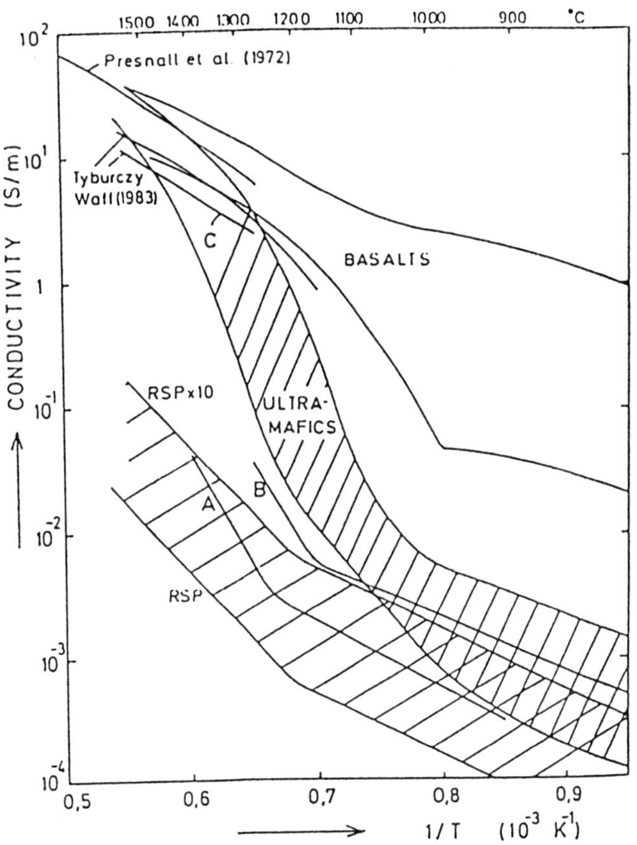

Fig. 10. Electrical conductivity of ultramafic rocks and basalts from laboratory experiments [Schmeling, 1985b]. RSP is Red Sea peridotite. Oxygen fugacity of samples not as well controlled as measurements shown in Fig. 3. Basalt and basaltic magma are much more conductive than olivine or unmelted peridotite.

1984]. The characteristic flexural wavelength thus varies from a few tens of km on the EPR to between 100 and 200 km on the MAR to as much as 500 to 1000 km in old seafloor (Figure 11).

The proper depth of compensation is less well-known. In principal, thermal expansion associated with a small temperature change over a great depth range could have the same topographic effect as a large temperature change over a small depth range. If the density anomaly lies beneath the lithosphere, however, it will induce convection in the fluid asthenosphere (Figure 12) that will deflect the surface [Runcorn, 1964; Morgan, 1965]. As the depth of the density anomaly increases, more of the compensation is likely to take place by deflection of density interfaces in the interior of the earth (Figure 13) or by viscous stresses associated with buoyant flow rather than by deflection of the seafloor [McKenzie, 1977; Parsons and Daly, 1983; Ricard, et al., 1984; Richards and Hager, 1984; Jian and Parmentier, 1985]. The surface topographic expression of density anomalies

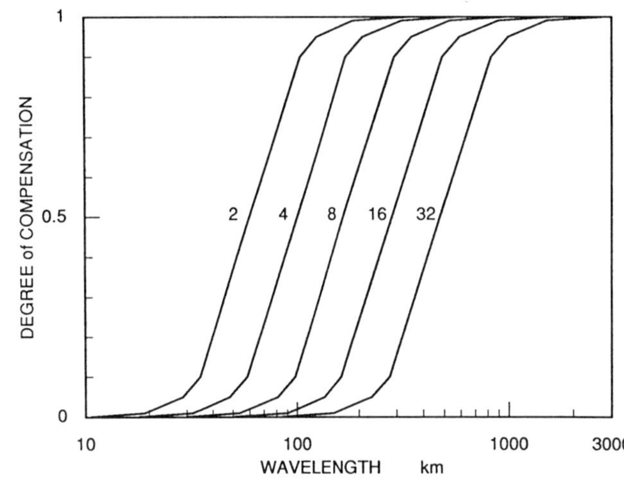

Fig. 11. Predicted transition from compensated to uncompensated topographic features. Degree of compensation = 1 indicates fully compensated, = 0 indicates uncompensated. This transition occurs at the characteristic flexural wavelength of an elastic plate and is a function of the rigidity or effective elastic thickness of the lithosphere. Numbers indicate thickness of elastic plate in km.

beneath the lithosphere can perhaps be most simply represented by

$$\delta d(k) = -\frac{1}{(\rho_m - \rho_w)} \int_0^{H_c} H(k,z)\, \delta\rho(k,z)\, dz \qquad (15)$$

where k is the wavenumber of the density variation ($k = 2\pi/\lambda$) and H is a kernel or response function that describes the fraction of the density anomaly that is compensated by deflection of the surface.

The balance between surface and interior compensation depends on the horizontal wavelength of the density anomaly, so there is actually no single depth of compensation that can be adopted for all anomalies. A low-viscosity zone, however, can be quite effective at isolating density anomalies in a convecting mantle beneath the zone from any significant topographic expression at the surface (Figure 14) and also insures that shallower anomalies will be primarily compensated by surface deflections [Richards and Hager, 1984; Robinson et al. 1987]. Beneath young sea-floor there is good evidence for a pronounced low-viscosity asthenosphere, so the appropriate depth of compensation is probably 100 to 200 km for all but the longest wavelength variations. In Table 1 listing the resolution of various geophysical techniques, isostatic topography is thus indicated as having a horizontal resolution of 50 to 200 km and a vertical resolution of about 150 km.

The dominant topographic signature in the oceans is the increase in depth away from the mid-ocean ridges. Because the increase in depth is primarily a function of the age of the

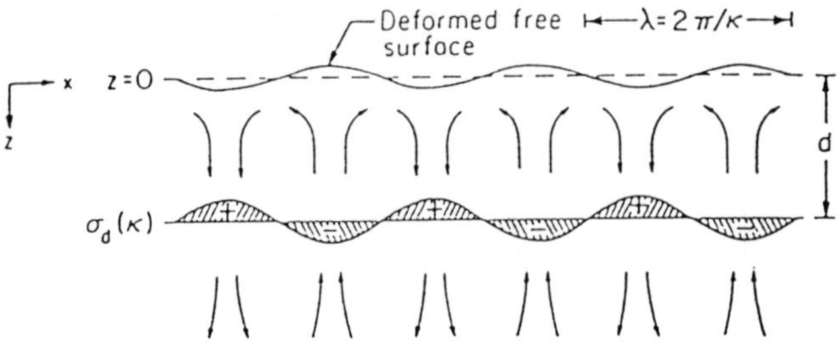

Fig. 12. Sinusoidal density anomaly at depth d within viscous medium will induce convective motion that deforms the surface [from Parsons and Daly, 1983]. Amplitude of the deformation will depend on depth and wavelength of density anomaly.

seafloor rather than distance and because heat flow also decreases with age, this decay in elevation is modelled as the isostatically compensated thermal contraction of a cooling upper mantle. To a good approximation, the topography is explained by a one-dimensional, cooling half-space (cooling thermal boundary layer) model in which heat is lost only by vertical conduction to the seafloor. The upper boundary is fixed at temperature $T = 0$ and the material is allowed to cool to infinite depth. In this model, neglecting the adiabatic temperature gradient, the solution for temperature structure as a function of depth z is

$$T = T_m \operatorname{erf}(z/2\sqrt{\kappa t}) \tag{16}$$

where κ is the thermal diffusivity, t is the age of the seafloor and T_m is the initial temperature of the half-space, equivalent to the potential temperature of the rising mantle material under the ridge axis. Isotherms for this model are shown in Figure 1. The topography due to thermal contraction of this material is by (13) and (14)

$$d = d_0 + \frac{2\alpha\rho_m T_m}{(\rho_m - \rho_w)}\sqrt{\frac{\kappa t}{\pi}} \tag{17}$$

where α is the thermal expansion coefficient and d_0 is the depth at zero age. This prediction that the seafloor will subside in proportion to $\text{age}^{1/2}$ holds very well at least to ages on the order of 80 Ma (Figure 15). The reasons that this relationship breaks down beyond 80 Ma are a subject of controversy and beyond the realm of this current review, although the cooling plate model with temperature fixed at some depth (dashed lines in Figure 1) can successfully match the subsidence data. This simple relationship ignores horizontal heat conduction, cooling of the rising material beneath the ridge axis before it begins moving horizontally, latent heat carried by transport of melt, volume changes associated with melting and the segregation of melt, solid-state phase changes induced by the cooling lithosphere, hydrothermal circulation, and small-scale convection. Nevertheless, it remains a remarkably useful and accurate first approximation.

The topography in many regions can be represented by this linear depth-vs-square root of age relationship, but with significant deviations in intercept values and/or slope from the global average values. These deviations may represent fundamental variations in the physical properties of the ridge, such as the temperature of the rising mantle material. Axial depths or intercept values range from zero on the Reykjanes ridge to a maximum of about 3700 m in the Australian-

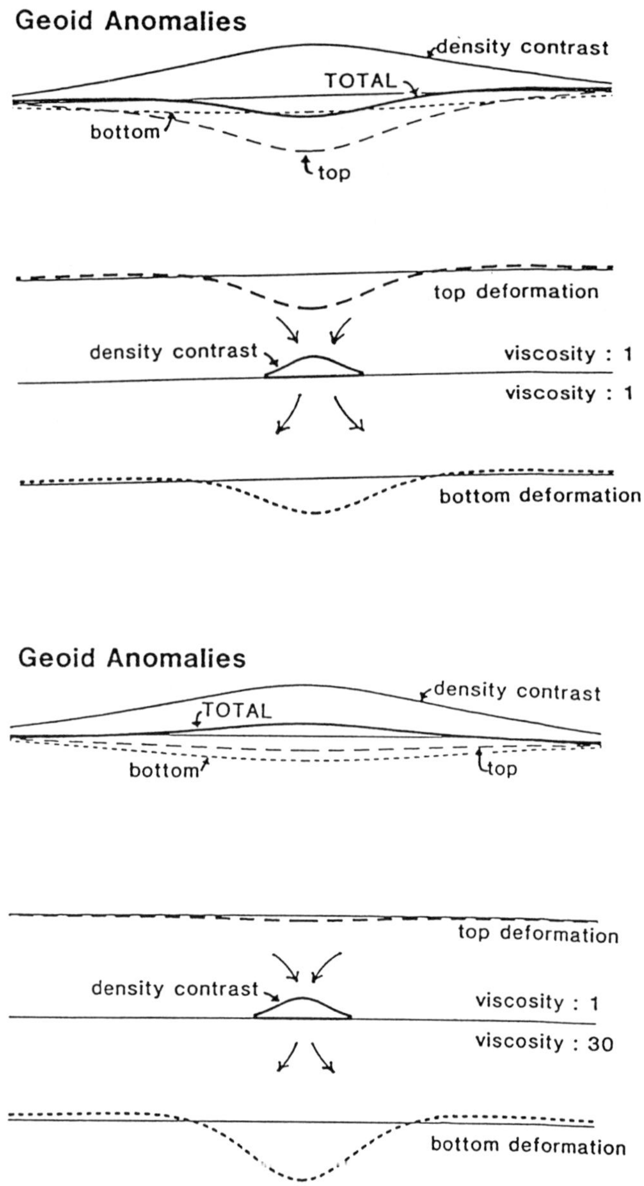

Fig. 13. Density anomalies at depth within the earth deflect both the surface and deeper density interfaces [from Richards and Hager, 1984]. If the density anomaly is beneath a low viscosity zone, the deformation of the upper surface can be quite small (bottom). The geoid anomaly is the sum of the effects of the negative mass anomalies at the surface and deeper interfaces and the positive anomaly of the anomalous body itself.

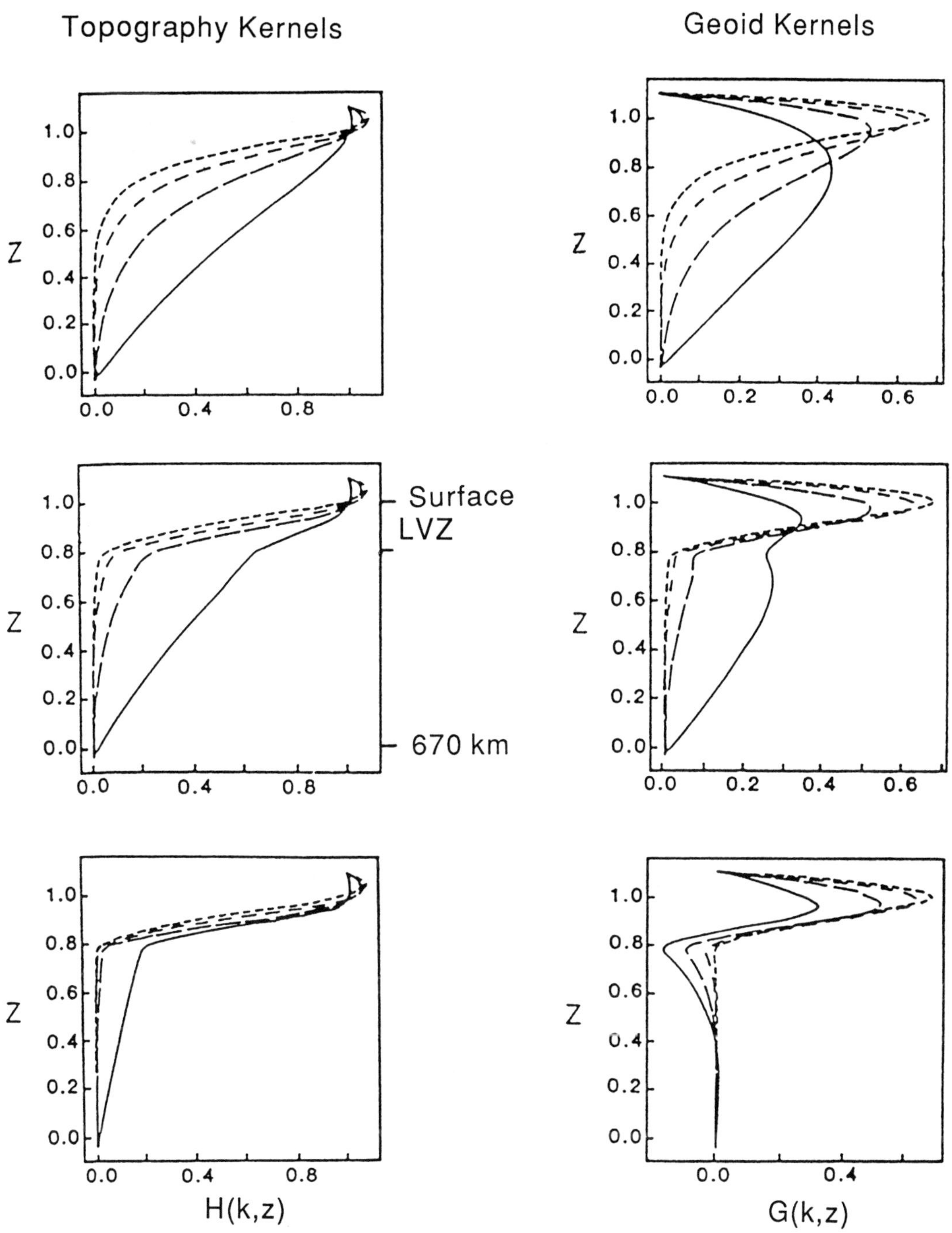

Fig. 14. Topographic and geoid signatures of sinusoidal loads within a convecting, viscous mantle [from Robinson et al., 1988]. The topography kernels show the fraction of the load that is compensated by deflection of the surface (i.e., creating compensating topography). The gravity kernels show relative amplitudes of the geoid anomaly. Z represents the depth within the viscous layer, with z = 1.0 at the top of the viscous layer at the base of a rigid, conducting lid with no-slip boundary conditions and z = 0.0 at the rigid base of the viscous layer with free-slip boundary condition. In each figure, the solid line represents a load with wavelength λ equal to eight times the thickness h of the viscous layer, the long dashed line $\lambda = 2h$, medium dashes $\lambda = h$, and short dashes, $\lambda = h/2$. Top pair of figure has no low viscosity zone, middle pair has a low viscosity zone from z = 0.8 to 1.0 with viscosity 0.1 that of underlying mantle, and bottom pair has viscosity in the low viscosity zone reduced to 0.01 of the underlying mantle.

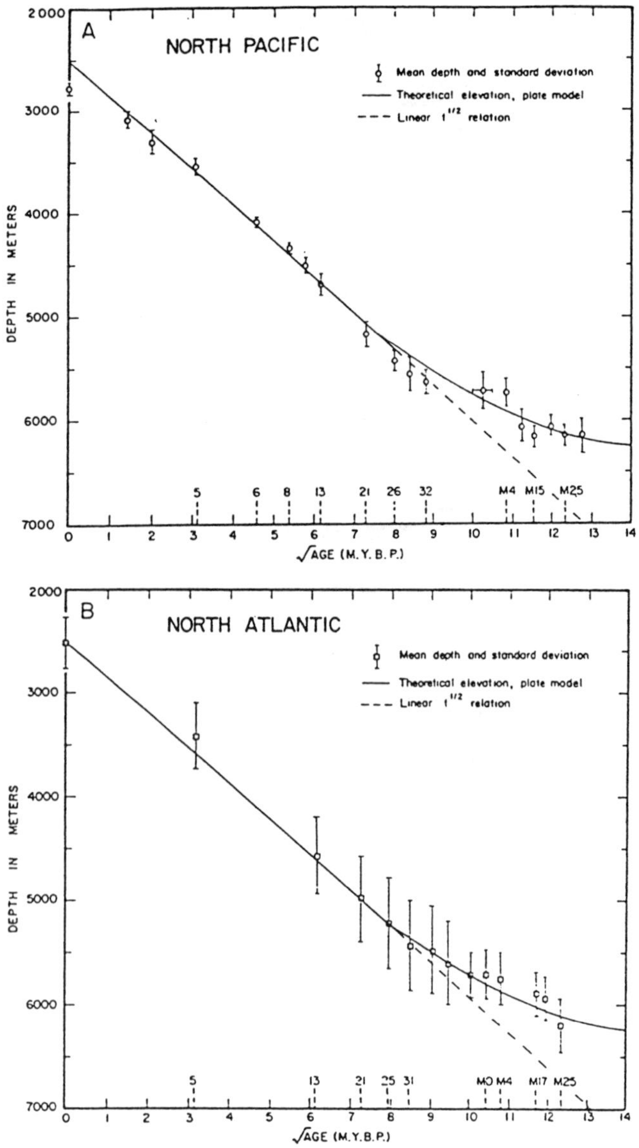

Fig. 15. Mean depths and standard deviations versus square root of the age of the seafloor for (a) North Pacific and (b) the North Atlantic [from Parsons and Sclater, 1977]. Solid line shows calculated values for a cooling plate model, dashed line for a cooling half-space model. Numbers 5, 6, M4 etc. at bottom of each Fig. refer to magnetic anomalies.

Antarctic discordance zone, with most of the world's ridge crests lying between 2500 and 3000 m (Figure 16). These values exclude the local effects of median rift valleys and fracture zones that are superimposed on the regional variations.

Reported subsidence rates vary from about 85 to 600 m/Ma$^{1/2}$ [Davis and Lister, 1974; Trehu, 1975; Parsons and Sclater, 1977; Schroeder, 1984; Cochran, 1986; Hayes, 1988; Marty and Cazenave, 1989]. However, most of the extreme values come from areas in which this assumed form of the depth-age relationship does not provide a good description of the topography, as expressed in terms of a low correlation coefficient between predicted and observed topography or large residuals, presumably due to the superposition of other processes creating formations such as seamounts, oceanic plateaus and rises. Excluding these unreliable estimates (correlation coefficients < 0.8) and estimates based on single profiles leaves a range of subsidence rates of 200 to 450 m/Ma$^{1/2}$. In some areas, particularly the EPR between 9 and 22°S and portions of the southern MAR, subsidence beginning right at the ridge axis is distinctly asymmetric. It is possible that some or all of the asymmetry arises from superposition of

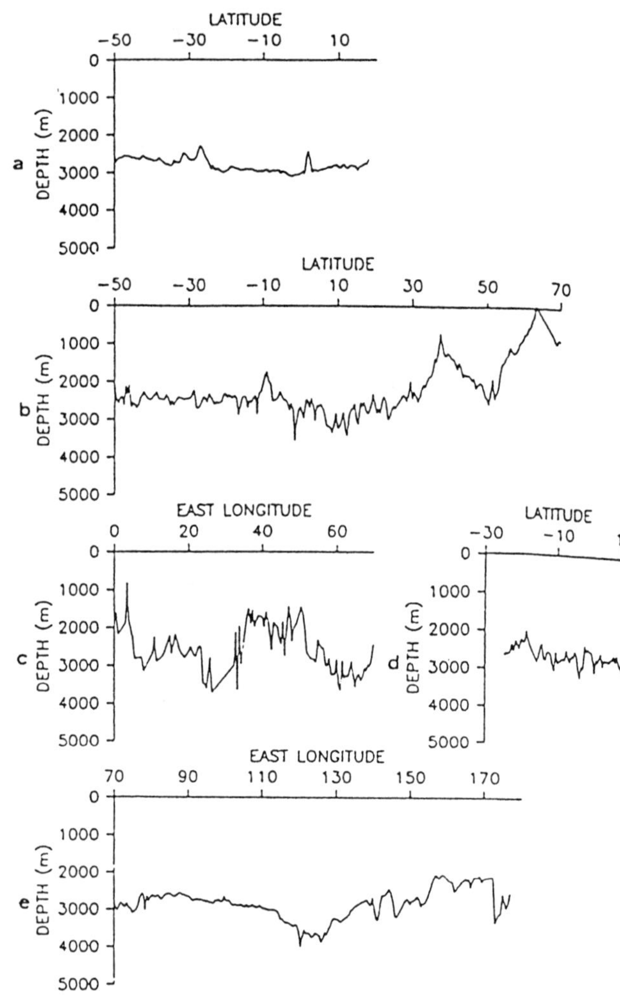

Fig. 16. Regional variations of smoothed, zero-age depth [from Marty and Cazenave, 1989]. Prominent anomalies are the Azores and Iceland hot-spot regions at about 40° and 60° N respectively, and the deep Australian-Antarctic Discordance zone at about 120° E. Profiles shown are from (a) East Pacific Rise, (b) Mid-Atlantic Ridge, (c) Southwest Indian Ridge, (d) Carlsberg Ridge and Central Indian Ridge, and (e) Southeast Indian Ridge.

relief from a (deep mantle?) process that continues across the ridge, creating a gentler slope on one side and a steeper slope on the opposite side. Perhaps only variations in the subsidence rate averaged across both sides of the ridge should be considered as representing fundamental properties of the ridge. Averaged subsidence rates vary from about 270 to 450 $m/Ma^{1/2}$. The highest subsidence rate is found in the region just east of the Australian-Antarctic discordance. Rather than representing a real subsidence rate, it may represent a time-varying ridge crest elevation, with a gradual transition from an anomalously deep spreading center in the past like that immediately to the west in the discordant zone to the current normal depth. Thus, the anomalous subsidence rate here may be caused by a transition from normal crust and lithosphere to a denser mantle and thinner crust. Excluding this observation and one other rate based on profiles all very close to the Gofar Fracture Zone [Cochran, 1986], leaves a very conservative estimate of the true variation in subsidence rates of 270 to 400 $m/Ma^{1/2}$. [Kane and Hayes [1992] report averaged profiles in the South Atlantic with subsidence rates ranging from about 200 to 450 $m/Ma^{1/2}$.)

In the context of the simple, cooling half-space model, variations in axial depth and subsidence rate can only be explained by a change in T_m or by a compositional change that affects κ, α, or ρ_m. Considering only temperature changes and a temperature of 1300°C for a typical rate of 325 $m/Ma^{1/2}$, the conservative estimate of the range in subsidence rates would require temperature variations from 1080 to 1600°C. Although in a true half-space the effect of a temperature change on elevation would be infinite, a more realistic estimate is $\alpha\rho_m H_c \delta T_m/(\rho_m-\rho_w)$. Assuming $H_c = 150$ km, $\alpha = 3.5 \times 10^{-5}$ °C^{-1} and T_m = 1300°C for a typical depth of 2600 m, the temperature range required to explain the observed range of axial depths is 1155 to 1640°C. If temperature variations are the cause, then axial depth and subsidence rate should be correlated, with the ratio of the two given by $2(\kappa/\pi)^{1/2}/H_c$ yielding, with the values adopted here, a ratio of 0.033 $m/Ma^{1/2}/m$ and $\kappa = 6.1 \times 10^{-7}$ m^2/s. Hayes[1988] and Marty and Cazenave[1989] report that the two are correlated with shallower ridge segments subsiding faster as expected. The observed ratio found by regression [Marty and Cazenave, 1989] is about triple the predicted value. The observed variance from this predicted correlation is rather large, suggesting that compositional variations must also be important or that the effective depth of compensation varies from ridge to ridge. The importance of compositional variations is also suggested by the fact that the shallowest ridge, the Reykjanes ridge, is subsiding at about 320 $m/Ma^{1/2}$, a normal rate.

The temperature range of about 500°C required to produce the observed differences in subsidence rate and axial depth may be overestimated because other temperature-dependent effects were neglected. Partial melting of the rising mantle under the ridge and extraction of the melt to form the crust replaces the fertile mantle with crust and residual mantle, both of which are less dense than the original material (Oxburgh and Parmentier, 1977; Klein and Langmuir, 1987]. The isostatic balance thus has to consider the resulting compositional stratification, not just the direct thermal expansion (Figure 17). With increased T_m, melting at greater depths and to a greater degree is expected (Figure 18) producing a thicker crust and a thicker and more depleted residual layer. The net result is an enhanced sensitivity of axial elevation to temperature change. Klein and Langmuir [1987] suggest that only one fourth of the total effect is thermal expansion, but their model overestimates the variation in crustal thickness for mid-ocean ridges excluding the Cayman trough and Iceland by a factor of two or more (see later section on crustal thicknesses).

Increasing T_m may also enhance the subsidence rate more than is predicted by the simple thermal conduction model. If small-scale convection is induced by the cooling of the plate, subsidence may still be expected to be proportional to $(age)^{1/2}$ [Buck, 1987]. Because increasing T_m increases the temperature contrast and probably reduces the viscosity, convection is more vigorous, heat flow is enhanced, and the subsidence rate is greater than in the conductive cooling

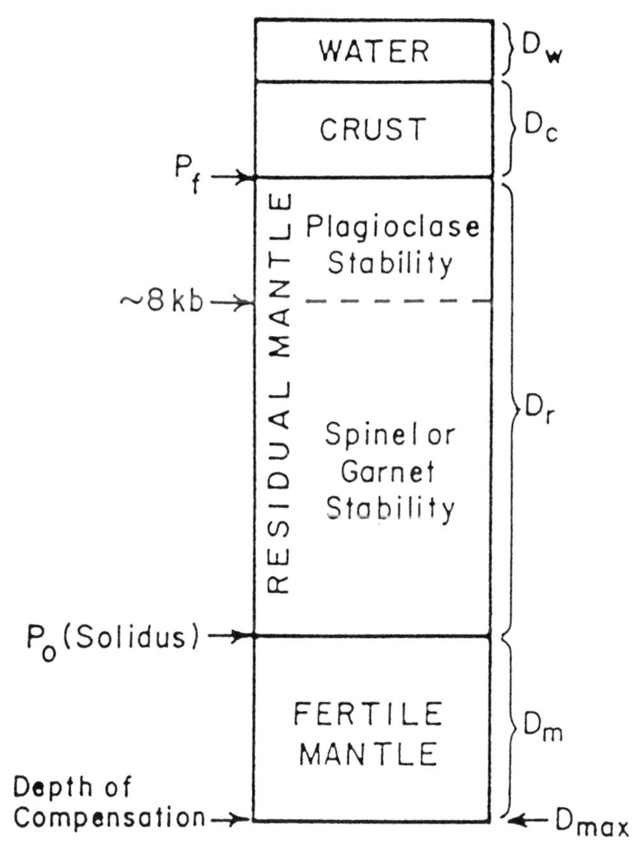

Fig. 17. Isostatic models of the spreading process must consider stratification of the crust and mantle column in addition to thermal expansion. Melt segregation yields a depleted, residual mantle that is lower in density than the fertile mantle in addition to the low density crust [from Klein and Langmuir, 1987].

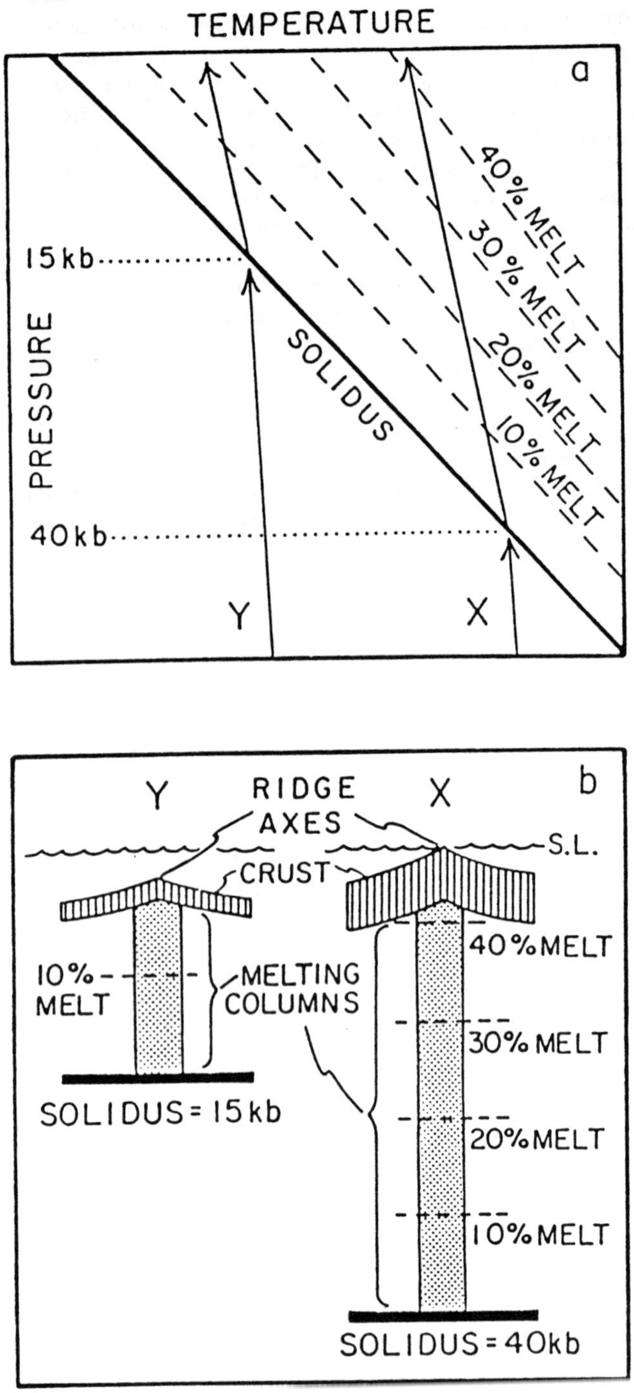

Fig. 18. Higher temperatures of upwelling (geotherm labelled X) produce shallower ridge crests through the combined effects of thermal expansion, thicker crust, and thicker, more depleted, residual mantle [from Klein and Langmuir, 1989]. Upper figure shows schematic adiabats for upwelling material beneath a ridge at two different potential temperatures. Lower figure shows resulting crustal thickness and degree of depletion in residual melting column. S.L. indicates sea level.

model. Buck[1987] estimates the enhancement to be on the order of 30 percent. Thus, if temperature variations are solely responsible for the observed variations in axial depth and subsidence rate the range in upwelling temperatures must be on the order of 300 to 350°C. Compositional variations, however, are also necessary to explain the low correlation between axial depth and subsidence rate, as well as some of the variations in basalt chemistry that correlate with axial depth [Schilling et al., 1983; Hamelin et al., 1984; Dick et al., 1984; Michael and Bonatti, 1985; Klein and Langmuir, 1987; Klein et al., 1988]. Compositional variations can affect the subsidence rate through α, κ, and the effect of viscosity on small-scale convection. Ironically, the compositional stratification caused by extraction of melt could limit the importance of small-scale convection in enhancing the flow of heat out of the mantle [Sotin and Parmentier, 1989]. Additional constraints from geoid anomalies may be the best way of assessing the importance of compositional variations and small-scale convection, as the change in effective depth of compensation may be much greater than the change in subsidence rate [Buck, 1987].

Geoid Height

The geoid provides another direct indicator of the density structure of the crust and upper mantle. In the oceans and neglecting the effects of currents, variations in sea surface height δh above a reference ellipsoid yield a direct measure of variations in the gravitational potential of the earth. Assuming local isostatic equilibrium and adopting a flat-earth approximation, δh is related to the density perturbations from the reference density structure by

$$\delta h = \frac{2\pi\gamma}{g}\int_0^{H_c} z\delta\rho \, dz \qquad (18)$$

where γ is the gravitational constant and g is the acceleration of gravity. Thus, at long wavelengths, the geoid anomaly associated with anomalous topography is directly proportional to the product of the anomalous elevation and the depth of the compensating density structure [Ockenden and Turcotte, 1977; Haxby and Turcotte, 1978, Crough, 1978]. Corrections for the sphericity of the earth have been given by Dahlen[1982] and Lister[1982]. This second integral relation involving the vertical density distribution provides a potentially powerful means for discriminating between different compensation models. For example, Buck [1987] suggests that if small-scale convection enhances the subsidence rate by 30 percent, the effect on the geoid may double because convective cooling may involve much more cooling at large depth than does conductive cooling.

The same limitations on resolution apply to isostatic geoid anomalies as for variations in depth of the seafloor. Flexural rigidity of the lithosphere limits the wavelengths that can be considered to be fully compensated. The geoid signature of density anomalies beneath the lithosphere is dependent on depth and wavelength, and can be described by a simple response function G(k,z) analogous to the kernel for surface topography

$$\delta h(k) = \frac{2\pi\gamma}{g}\int_0^{H_c} z\delta\rho(k,z)\, G(k,z)\, dz \qquad (19)$$

Because the geoid represents the sum of competing signals from the density anomaly itself and the compensating deflections of the surface and interior density interfaces (Figure 13), it is possible for the gravity kernel to change sign if the source is immediately beneath a low-viscosity zone. The form of the response function for different viscosities is illustrated in Figure 14. Again, a low-viscosity zone helps to isolate deeper anomalies, although the model shown in Figure 14 is an extreme one in the sense that all anomalies in the deeper upper mantle are considered to be compensated by a density interface at the base of the upper mantle, which is equivalent to assuming that the 670 km discontinuity is a chemical boundary across which there is no convection. If there are large lateral variations in viscosity, then the simple response function approach is not adequate [Richards and Hager, 1989]. Although the interpretation of the results in a convecting earth may thus be difficult, the resolution of an apparent depth of compensation for a ridge may be on the order of 10 km (Table 1).

For the cooling half-space, conductive model, both the anomalous topography and the depth of isotherms are proportional to the square-root of the age of the seafloor, so the geoid anomaly involving their product is directly proportional to age [Haxby and Turcotte, 1978; Parsons and Richter, 1980], providing all the density changes are compensated by deflection of the seafloor. The change in geoid height with age is given by

$$\delta h = \frac{2\pi\gamma\rho_m\alpha\kappa T_m}{g}\left[1 + \frac{2\rho_m\alpha T_m}{\pi(\rho_m - \rho_w)}\right] t \qquad (20)$$

which leads to a linear decay in geoid height with increasing distance from the ridge, given constant spreading rate. Approximately linear variations with age have been reported for slow-spreading ridges in the Atlantic and Indian Oceans [Chapman, 1977; Haxby and Turcotte, 1978; Figure 19], but in most cases it is difficult to observe because this signal is superimposed on other long-wavelength geoid anomalies with sub-lithospheric sources whose amplitude is up to 10 times that of the signal of the cooling lithosphere. To enhance the signal-to-noise ratio, Sandwell and Schubert [1980] looked at only the component that is symmetric about the ridge axis, Bowin et al. [1984] subtracted the long-wavelength components as described by a geoid model with degree and order 10 (wavelength λ > 4000 km), and Kane and Hayes [1992] solved simultaneously for a symmetric age effect and a linear or quadratic regional field. With these techniques, the linear decay is well-resolved at slow-spreading rates, but at high rates the decay is spread out over large distances and is not resolvable from the background variations. Sandwell and Schubert [1980] report geoid slopes of -0.094, -0.13, and -0.15 m/Ma for the South Atlantic, southeast Indian, and North Atlantic regions, respectively, with standard deviations on the order of 0.03 m/Ma. Kane and Hayes [1992] found large regional variations in the South Atlantic, although most estimates were between -0.05 and -0.15 m/Ma. They did not find a systematic correlation between subsidence rate and geoid rate, suggesting that a simple thermal model does not provide an adequate explanation of the variations in subsidence. The average behavior is roughly consistent with simple models. Employing the values for the physical parameters adopted in the previous section in equation (20) yields a prediction for the cooling half-space model of -0.13 m/Ma.

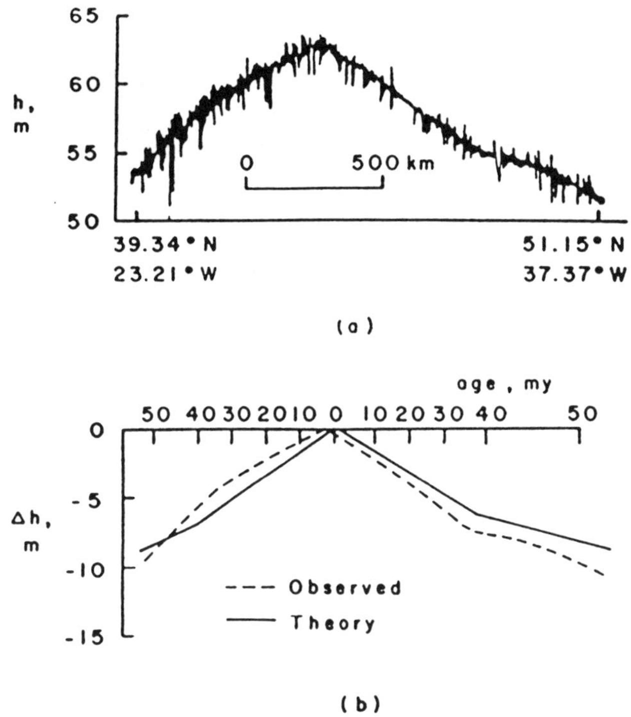

Fig. 19. (a). Observed Geos 3 geoid profile across the Mid-Atlantic Ridge at 44.5°N [from Haxby and Turcotte, 1978]. (b) Smoothed version of observed anomaly as a function of seafloor age compared to that predicted for cooling half-space model.

To avoid the problem of isolating the lithospheric signal from other long-wavelength phenomena, Crough [1979] suggested using the step offset in geoid height across fracture zones caused by the step offset in age. This geoid step is easily detected for all major fracture zones (Figure 20, Crough, 1979; Detrick, 1981; Sandwell and Schubert, 1982a; Cazenave et al., 1982; Cazenave et al., 1983; Cazenave, 1984; Vogt et al., 1984; Marsh et al., 1984; Gibert et al., 1987; Driscoll and Parsons, 1988; Marty et al., 1988; Freedman and Parsons, 1988; Marty and Cazenave, 1988; Wessel and Haxby,

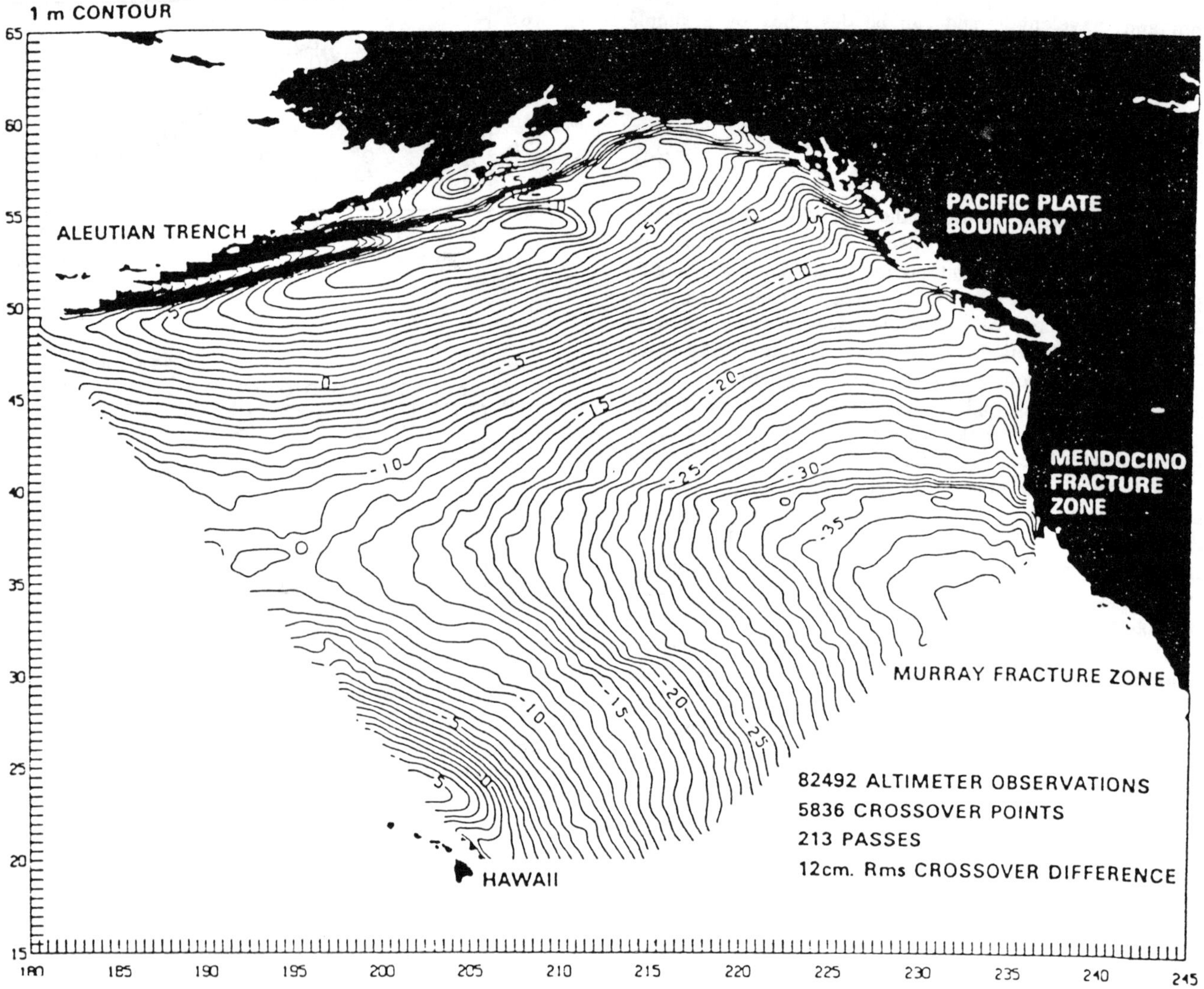

Fig. 20. Satellite geoid over the northeastern Pacific Ocean from SEASAT altimetry data [from Marsh et al., 1982]. Small steps in the seasurface elevation can clearly be seen associated with the Murray and Mendocino fracture zones. Contour interval 1 m.

1989,1990], although there still is a problem with subtracting the large regional field to obtain an accurate estimate of step size. Some typical examples of geoid profiles across fracture zones in the Pacific are shown in Figure 21. The estimate of the amount of offset will clearly depend on the way in which the regional field is removed. The method that has been most widely adopted is to first remove wavelengths longer than about 3000 km using a spherical harmonic description of the geoid, then to assume that the remaining regional trend can be represented by a low-order polynomial in distance and that the change in geoid height is a step function, solving simultaneously for the amplitude of the step function and the coefficients of the polynomial. The effects of subtracting the long-wavelengths before fitting with a polynomial are generally small [Driscoll and Parsons, 1988] as is the difference between using a second and a third order polynomial [Marty et al., 1988]. Data near the fracture zone are excluded to account for deviations in the shape of the transition from a step function. The transition is expected to be broadened by lateral heat conduction across the fracture zone [Louden and Forsyth, 1976; Detrick, 1981; Sandwell, 1984] and the shape altered by lithospheric flexure associated with differential subsidence [Sandwell and Schubert, 1982b; Sandwell, 1984] and thermal stresses [Parmentier and Haxby, 1986]. In addition, there is a local low associated with the uncompensated trough of the fracture zone. To completely avoid these effects would require eliminating data as far as 200 to 400 km from the fracture zone [Detrick, 1981; Sandwell, 1984], but it is seldom possible to find a fracture zone separated by that distance from the influence of another

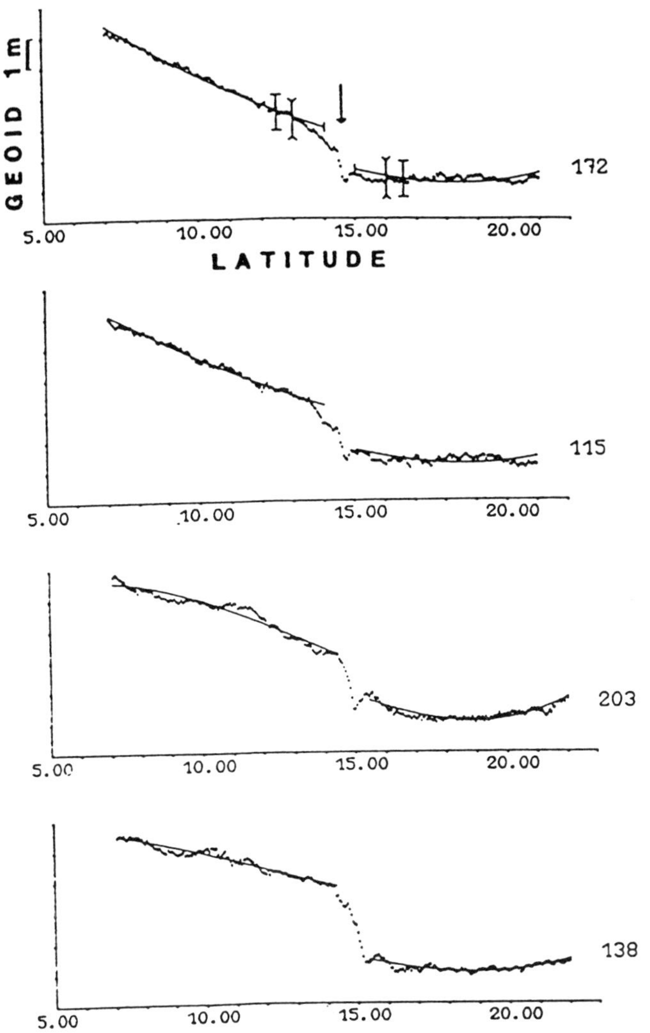

Fig. 21. Typical observed geoid profiles across a fracture zone in the Pacific, after removal of wavelengths > 3000 km [from Marty et al., 1988]. Arrow in upper lefthand figure indicates position of fracture zone. This residual signal is modelled as the sum of a step function plus a simple polynomial variation, with a one degree region around the fracture zone excluded. Model values are shown as smooth solid lines in each diagram. Numbers to right of each profile indicate orbit number.

fracture zone, and if it were possible, then the wavelengths involved are long enough that the problem of separating the signal from other sources reappears. The usual compromise is to eliminate data 100 to 200 km from the fracture zone, but it must be recognized that there will be some tendency to underestimate the size of the step, particularly in older seafloor [Detrick, 1981; Sandwell, 1984; Driscoll and Parsons, 1988; Marty et al., 1988]. A better approach may be to estimate the form of the step taking into account thermal diffusion and lithospheric flexure, then to solve for its amplitude and the shape of the regional field simultaneously [Marty and Cazenave, 1988; Wessel and Haxby, 1989,1990], although calculating the form to some extent assumes knowledge of the thermal structure that is subject of the investigation. Despite the difficulty of estimating the change in geoid with change in age across fracture zones, the results are robust enough that useful constraints on the cooling of the mantle are obtained in the Pacific where they cannot be obtained from across-ridge profiles.

For the cooling half space model with its linear relationship between geoid height and age, the geoid offset is predicted to be proportional to the age difference across the fracture zone and should be independent of the absolute ages of the sea floor (equation 20). The geoid height change δh divided by the offset in age across the fracture zone δt clearly yields an estimate of the slope of the geoid height-versus-age curve at the average age of the sea floor on either side of the fracture zone . Early studies of the Mendocino FZ in the Northeast Pacific [Crough, 1979; Detrick, 1981; Sandwell and Schubert, 1982a; Cazenave et al. 1982] using the geoid profiles of the Geos 3 and Seasat altimeter satellites found a clear decrease in $\delta h/\delta t$ with increasing age of the plate along the FZ in contrast to the prediction of the cooling half-space model of constant slope. The pattern of decrease is fit reasonably well by the cooling plate model [Parsons and Sclater, 1977] with thickness about 100 km (Figure 22). The apparent compensation depth, then, is clearly less than in the half-space model, suggesting either that there is a source of heat that retards cooling at depth or that once cooling reaches deep enough, small-scale convection within or beneath a low-

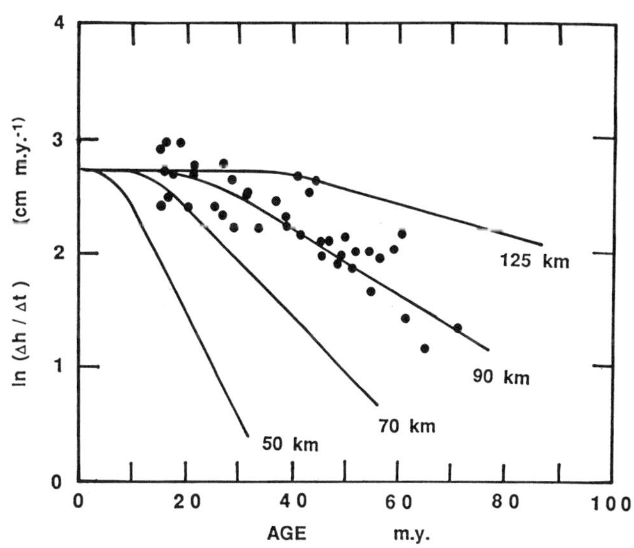

Fig. 22. Slope of the geoid vs. seafloor age curve estimated from step offsets across the Mendocino fracture zone (dots). The apparent decrease in slope with increasing age is fit by a cooling plate model with bottom boundary held fixed at a constant temperature at a depth of 90 to 100 km (solid lines). [after Cazenave et al., 1982].

viscosity zone prevents it from being detected by geoid profiles across FZs. Subsequent studies of other fracture zones, however, found much more complex patterns of δh/δt versus age that are not consistent with either the half-space model or the cooling plate model [Cazenave, et al., 1983; Cazenave, 1984; Marty and Cazenave, 1988]. These variations include oscillations with age on the Ascension FZ in the Atlantic [Gibert et al., 1987] and on the Udintsev and Eltanin FZs in the Pacific [Driscoll and Parsons, 1988], a very rapid initial decay on the Udintsev FZ [Driscoll and Parsons, 1988; Marty et al., 1988]; a reversal in sign of the offset on the Agulhas FZ in the South Atlantic [Freedman and Parsons, 1988]; high values along part of the Clarion FZ that approach twice the slope expected for a cooling half-space model [Marty et al., 1988]; and even an increase on the Mendocino FZ at great ages back to values typical of a cooling half-space [Marty et al., 1988].

A compilation of observations of geoid slope from fracture zones in the Pacific estimated using a 1 degree exclusion zone and a second order polynomial fit [Marty et al., 1988] is shown in Figure 23a. Overall, the impression is that there are large random fluctuations about a mean value which is approximately that predicted by a cooling half-space model, with the amplitude of the fluctuations much larger than the experimental error in the observations. The mean value is -0.092 ± .006 m/Ma. Wessel and Haxby [1989] report a mean value of -0.148 m/Ma for Pacific fracture zones with no trend with age (Figure 23b). The difference between the two results is attributable to the latter study explicitly including the effects of lithospheric flexure in modelling the step across the fracture zone and eliminating the Mendocino FZ from consideration because of the proximity of the Pioneer FZ. The slope of the geoid height-age curve in the Pacific based on fracture zone crossings is thus similar to that found in the Atlantic or Indian oceans from the decay away from the ridge. The fluctuations in δh/δt may be produced by small-scale convection that is induced by the temperature contrast across the fracture zone [Craig and McKenzie, 1986; Robinson et al, 1988]. The cold material adjacent to the FZ under the plate on the old side tends to become unstable and sink, inducing small-scale convection that is, at least initially, coupled to the fracture zone (Figure 24). The convection acts to erase the initial thermal step. If there is a thin, low-viscosity zone, the offset in the geoid may be preserved across the FZ, but the transition is broadened until the step would be indistiguishable from a regional trend. Without the low-viscosity layer to confine convection in the models to shallow depths, the amplitude of the geoid signal from the small-scale convection that develops would swamp the lithospheric signal and no-step would be recognizable at all [Craig and McKenzie, 1986]. Convection eventually develops beneath the low-viscosity zone, however, which can lead to large apparent oscillations in geoid step size and apparent δh/δt [Robinson et al., 1988]. It is not apparent, however, that small-scale convection in the Pacific is coupled to the FZs. Linear 'gravity rolls' [Haxby and Weissel, 1986] with amplitude in the geoid on the order of 1 m are observed in the Pacific trending obliquely to FZs

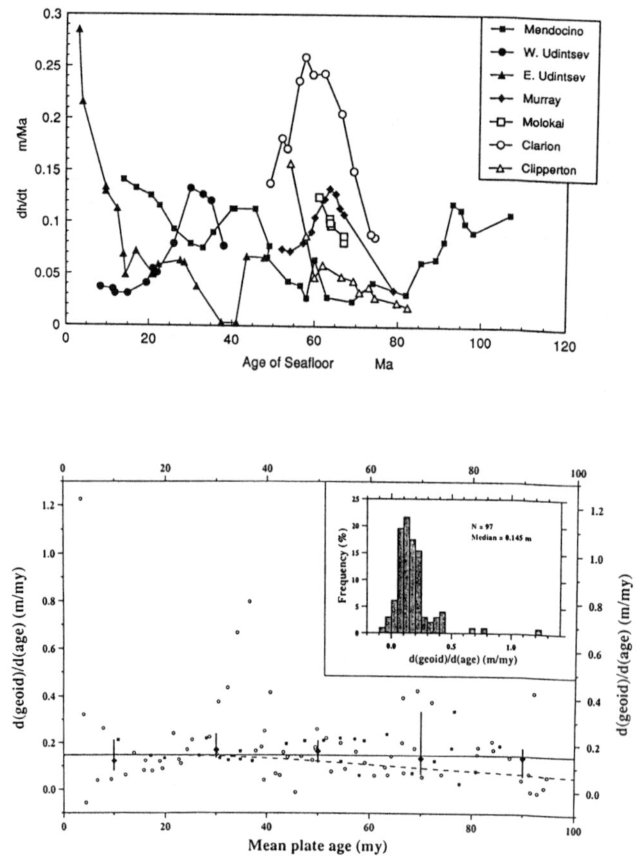

Fig. 23. Compilations of estimates of geoid-slope-versus-age from several fracture zones in the Pacific. (a) Data from Marty et al., 1988]. (b) figure. from Wessel and Haxby [1989] using a different estimation technique. Diamond symbols with error bars show averages and standard deviations over 20 m.y. age ranges. Observations are consistent with either constant average slope (solid line) or some decrease with increasing age (dashed line).

approximately in the direction of absolute plate motion in the hotspot frame. These gravity rolls may represent small-scale convection in the underlying asthenosphere, aligned with plate motion by the shear between the plate and underlying mantle. Whatever their origin, they could be responsible for some of the oscillations in δh/δt, particularly in the Central Pacific. A linear high passing obliquely across the fracture zone could first decrease, then increase the apparent step as it crossed from one side to the other of the fracture zone.

In summary, the average change in geoid height with age in seafloor less than 80 to 100 Ma old is consistent with cooling in proportion to the square root of age, as predicted by a purely conductive, cooling half-space model. Small-scale convection could be responsible for some of the cooling, however, because it may also transfer heat at a rate proportional to the conductive transfer [Buck and Parmentier; 1986; Buck, 1987] and be largely confined to a low-viscosity zone that limits the

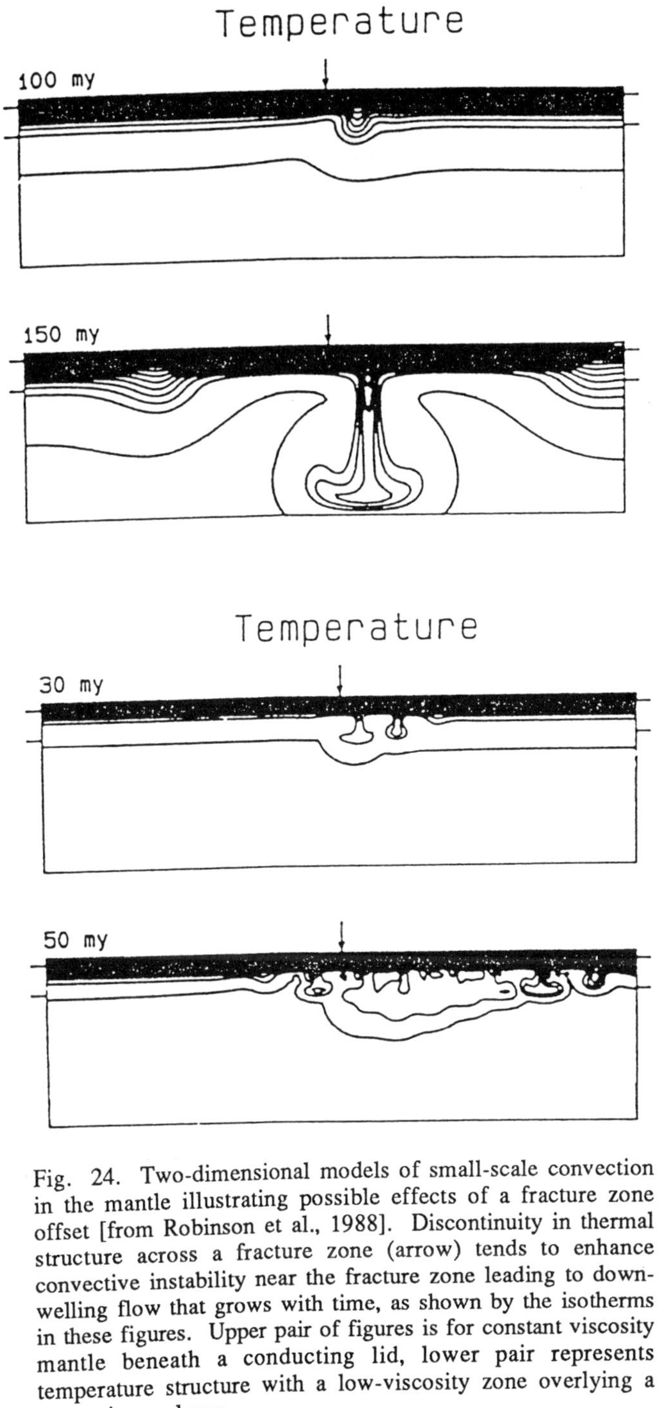

Fig. 24. Two-dimensional models of small-scale convection in the mantle illustrating possible effects of a fracture zone offset [from Robinson et al., 1988]. Discontinuity in thermal structure across a fracture zone (arrow) tends to enhance convective instability near the fracture zone leading to downwelling flow that grows with time, as shown by the isotherms in these figures. Upper pair of figures is for constant viscosity mantle beneath a conducting lid, lower pair represents temperature structure with a low-viscosity zone overlying a more viscous layer.

apparent compensation depth inferred from the geoid. Small-scale convection is suggested by the oscillations in geoid offset across fracture zones. If small-scale convection is present, it probably begins very close to the ridge because the subsidence rate and geoid slope do not change at some critical distance from the ridge that would indicate a sudden onset in convection.

Seismological Observations

There are three basic physical manifestations of flow in the mantle that can be detected seismologically. These are the temperature within the convecting medium, the distribution of melt in the upwelling zone, and the crystal fabric associated with the deformation of the mantle in shearing flow. Compressional (P) and shear (S) wave velocities are sensitive to the temperature; velocity and attenuation are sensitive particularly to small amounts of melt distributed within the medium; and the crystal fabric produces anisotropic elastic properties that can be detected in a variety of ways. The composition of the mantle also affects the elastic properties, but compositional changes in a vertically stratified medium are notoriously difficult to resolve. One of the most exciting prospects is the capability of resolving regional, anisotropic structure. Measurements of anisotropy can detect directly the change between predominantly vertical upwelling and horizontal flow. Azimuthal anisotropy can indicate the direction of shear in horizontal flow and, perhaps, the vertical extent of the deforming region.

Seismic Surface Waves

To date, surface wave studies have been our most productive seismological approach to understanding the structure of the lithosphere and asthenosphere. Seismic surface waves are sensitive to the shear velocity structure of the upper mantle to a depth of a few hundred km, depending on the period range sampled. The evolving structure of the upper mantle as a function of the age of the seafloor has been well-studied (see later discussion), particularly in the Pacific where the high spreading rates spread out the changes over great distances and there is a good distribution of seismic sources on the East Pacific Rise and at the subduction zones surrounding the Pacific Ocean. It is much more difficult to detect the effects of the cooling of the plate in the North Atlantic, where the age zones are narrow, the distribution of sources is not as good, and the ridge is sinuous. A fundamental limitation on the horizontal resolution of surface wave techniques is the wavelengths of the waves involved. Propagating seismic waves physically average the structure normal to their ray path over a region roughly the width of one wavelength. Thus, at a typical period T of 100 s and phase velocity c of 4.1 km/s, the horizontal resolution ($\lambda = cT$) can be no better than about 400 km. Of course the practical resolution depends on the path coverage and in most cases is significantly worse than the potential resolution.

Vertical resolution depends on the period range sampled, the accuracy of the data, and to some extent on the structure. There is always some trade-off between resolution and precision, in that if one is willing to accept less precise answers, more vertical layers can be resolved. In practice, however, there is relatively little leeway because in trying to improve the vertical resolution, so much precision is lost that the uncertainties swamp the vertical changes that are of interest. The resolution is best described, of course, by formal resolution kernels for each of the elastic coefficients (see following paragraphs), but crude rule-of-thumb guidelines can be useful. Fundamental mode Rayleigh waves (or spheroidal normal modes) are most sensitive to the shear velocity structure at a depth of about $\lambda/3$ or for waves in the period range we are concerned with in oceanic structure, about 4T/3 km (Figure 25), with very little sensitivity deeper than λ. The wavelength of the waves most sensitive to structure at a depth of 50 km is about 150 km, which is indicated as the limiting horizontal resolution for surface waves in Table 1. By combining data from different periods, we can obtain about $N = 1 + \log_2(T_2/T_1)$ independent pieces of information about the vertical shear velocity structure, where T_2 is the maximum period and T_1 is the minimum period observed. Thus, a typical global tomography study with periods in the range 100 to 250 s can provide only about 2+ pieces of information about the vertical velocity structure in the upper mantle at any one spot. A typical regional study with periods in the range 20 to 125 s provides 3 to 4 pieces of independent information. If the entire period range practical for mid-ocean ridges (17 to 250 s) is available, then the shear velocity can be resolved with useful precision over a vertical depth range of about 20 km at a depth of 25 km, 30 km at a depth of 50 km (Table 1), 50 km at a depth of 100 km and 300 km at a depth of 350 km. Love waves (or toroidal normal modes) in oceanic structure provide much less information because the response function is not peaked (Figure 25). Global tomographic studies of Love waves provide only one piece of information, while the entire practical period range provides only 2+ pieces of information. Addition of higher mode data can improve the vertical resolution to on the order of 300 km down to a depth of 1000 km [Cara and Leveque, 1988; Tanimoto, 1987,1988], but adds relatively little to shallow resolution and has been infrequently used because of the greater difficulty in analysis and data acquisition.

Surface waves can be used to measure anisotropy in three ways. First, Rayleigh waves that are polarized in a vertical plane containing the ray path, in an isotropic medium, sample the shear velocity structure in a fundamentally different way than Love waves that are transversely polarized. Discrepancies between the apparent isotropic velocity structures inferred from observations of the two types of surface waves, called "polarization" anisotropy or Love/Rayleigh incompatibility, is a strong indicator of anisotropic structure. In the vicinity of a mid-ocean ridge, we might expect the predominantly vertical flow to produce a strikingly different polarization anisotropy than would be found at a greater distance from the ridge where flow is probably mostly horizontal. Second, azimuthal variations in velocity of surface waves, particularly Rayleigh waves, provide an indication of the direction of horizontal shear deformation. The frequency dependence of the azimuthal anisotropy yields some control on the depth of the deforming zone. These variations have been successfully measured in regional studies in the Pacific and Indian Oceans, averaged over large areas, as well as global tomographic experiments. Third, coupling between surface wave modes in an anisotropic medium can produce polarizations of the waves that are related to the symmetry axes of the medium rather than to the direction of propagation of energy in the waves. Although particle motion studies in principle can detect this coupling, this technique has never been fully utilized for studying anisotropy in the oceans because it requires long-period, high-quality instruments located in the anisotropic medium in a low noise site on the seafloor.

In a slightly anisotropic medium, Smith and Dahlen [1973] have shown that the azimuthal dependence of the local phase velocities of Love or Rayleigh waves is of the form

$$C(\omega,\psi) - C_0(\omega) = A_0(\omega) + A_1(\omega)\cos 2\psi + A_2(\omega)\sin 2\psi + A_3(\omega)\cos 4\psi + A_4(\omega)\sin 4\psi \quad (21)$$

where ω is the angular frequency, ψ is the direction of propagation, and $C_0(\omega)$ is the velocity in a reference isotropic medium. Perturbations in travel time δt along raypaths are related to the elastic structure of the earth by

$$\begin{aligned}\delta t = \frac{\Delta}{C_p(\omega)} - \frac{\Delta}{C_0(\omega)} &= -\int_{path} A_0(\omega,\rho,A,C,F,L,N)\frac{ds}{C_0^2(\omega)} \\ &-\int_{path} A_1(\omega,B_c,G_c,H_c)\cos 2\psi\frac{ds}{C_0^2(\omega)} \\ &-\int_{path} A_2(\omega,B_s,G_s,H_s)\sin 2\psi\frac{ds}{C_0^2(\omega)} \\ &-\int_{path} A_3(\omega,E_c)\cos 4\psi\frac{ds}{C_0^2(\omega)} \\ &-\int_{path} A_4(\omega,E_s)\sin 4\psi\frac{ds}{C_0^2(\omega)}\end{aligned} \quad (22)$$

where Δ is the epicentral distance and $C_p(\omega)$ is the average velocity along path p. The parameters A, C, F, L, N, B, G, H, and E are independent linear combinations of the elastic coefficients that are functions of both depth and horizontal position. The anisotropy coefficients A_i represent vertical integrals over the structure at each point along the path with response kernels that are described in detail by Montagner and Nataf[1986] and Tanimoto[1986c]. Even with both Love and Rayleigh wave observations as a function of azimuth, there are clearly too many elastic coefficients for the structure to be uniquely constrained by surface wave observations. The A_0 term, for example, is a function of density ρ and 5 elastic parameters, yet Love and Rayleigh waves combined yield only two constraints on these 6 variables. The five coefficients each for Love and Rayleigh waves yield a total of only 10 constraints on the 21 independent elastic coefficients in the most general anisotropic medium. The A_0 term for Rayleigh waves is most sensitive to L, which is equivalent to $\rho(\beta_v)^2$,

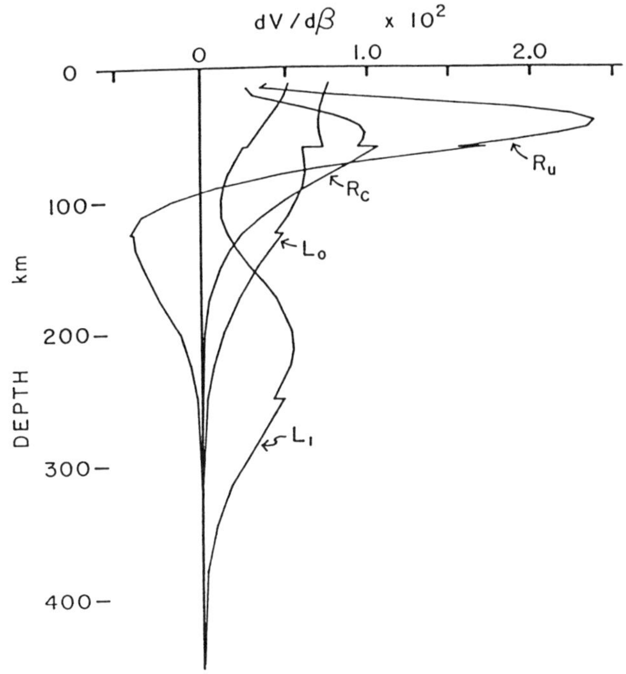

Fig. 25. Partial derivatives of surface wave velocities at 40 s period with respect to shear wave velocity β(z). Values given assume a change in velocity structure over a one km interval. Steps in the curves are due to discontinuities in the velocity model of the upper mantle and the slightly non-linear nature of the problem. R_u is the derivative of the fundamental mode group velocity for Rayleigh waves; R_c, fundamental mode, Rayleigh phase velocity; L_0, fundamental mode, Love wave phase velocity; L_1 first higher mode, Love wave phase velocity.

where β_V is the azimuthally averaged velocity of a vertically polarized shear wave, and has a depth kernel which is similar to the kernel for Rayleigh waves as a function of shear velocity in an isotropic medium (Figure 25). Similarly, the Λ_0 term for Love waves is most sensitive to the azimuthally averaged velocity β_H of a horizontally polarized shear wave. The $A_{1,2}$ terms are most sensitive to $G_{c,s}$, which describe the 2ψ variations of β_V. Although the interpretation is non-unique, tomographic mapping of the geographic variation in A_i can yield powerful constraints on the nature of crystal fabric and flow in the upper mantle.

Tomographic inversions of travel-time data for lateral variations in velocity or earth structure can be performed in a variety of ways. The earth can be divided into a small number of regions, such as geological provinces or age zones in the oceans, and the velocity coefficients in each region found that best predict the travel-times along paths that are contained in or cross these regions (the "pure-path" method). This is a very useful approach to finding the characteristic structure of different provinces, but only those geographic variations in velocity that correspond to our *a priori* knowledge of the surface of the earth can be found with the pure-path method. Alternatively, one can divide the medium into a large number of arbitrary, discrete blocks and invert for the parameters in each block [Aki et al, 1977; Zhang and Tanimoto, 1989], or represent the structure by continuous functions, solving for the coefficients in a spherical harmonic representation [Woodhouse and Dziewonski, 1984; Nakanishi and Anderson, 1983,1984; Nataf et al., 1984,1986; Tanimoto and Anderson, 1984,1985]. A combination of pure-path and spherical harmonic representations have also been employed [Nishimura and Forsyth, 1985,1988]. Some of the problems of ringing in a spherical harmonic representation can be avoided by assuming no particular functional form of the variation, but instead employing a damping or spatial covariance function that finds a continuous variation that is basically a weighted spatial average of the properties [Backus and Gilbert, 1968,1970; Tarantola and Valette, 1982; Tarantola and Nercession, 1984; Tanimoto, 1986a,b; Montagner, 1986; Montagner and Jobert, 1988; Montagner and Tanimoto, 1990]. Rather than describing the data in terms of travel-times, one can also directly invert the seismic waveforms for earth structure [Lerner-Lam and Jordan, 1983; Woodhouse and Dziewonski, 1984]. Although this approach is more difficult when there is a large number of parameters, waveform inversion is perhaps the best technique when incorporating higher modes [Tanimoto, 1987,1988]. To reduce the number of parameters, one can assume symmetry relations appropriate for olivine, for example, and solve for the orientation of the symmetry axis rather than for the elastic constants or all the anisotropic coefficients [Montagner and Nataf, 1988]. The choice of technique is relatively unimportant in regions with good spatial and azimuthal path coverage, but can lead to significant differences where velocities are less well-constrained.

Global tomographic inversions find that at moderate periods and depths (T=100 s, z = 150 km) there is a good correlation between velocity patterns and seafloor age [Woodhouse and Dziewonski, 1984; Nakanishi and Anderson, 1984; Nataf et al, 1986; Tanimoto, 1986; Dziewonski and Woodhouse, 1987; Montagner and Tanimoto, 1990]. Low-velocity regions are found along the mid-ocean ridges, particularly along the broad East Pacific Rise, and the phase velocities of both Love and Rayleigh waves increase systematically with increasing age of the sea floor. The geographic pattern for phase velocities shown in Figure 26a for 91s can be taken as representative of the geographic pattern of shear velocities integrated over a vertical response function centered and peaked at about 120 km depth and with significant contributions from the surface down to 250 to 300 km. At periods of 200 to 250 s for Rayleigh waves and depths of 300 to 400 km, the amplitude of the velocity anomalies is smaller and there is little remaining correlation with the location of mid-ocean ridges (Figure 26b). Although in some places it appears that anomalous structure associated with spreading centers extends to depths of 300 km or more, this may well be just the random superposition of deeper anomalies that are independent of the spreading process itself. Tanimoto and Zhang [1991] suggest that most of the

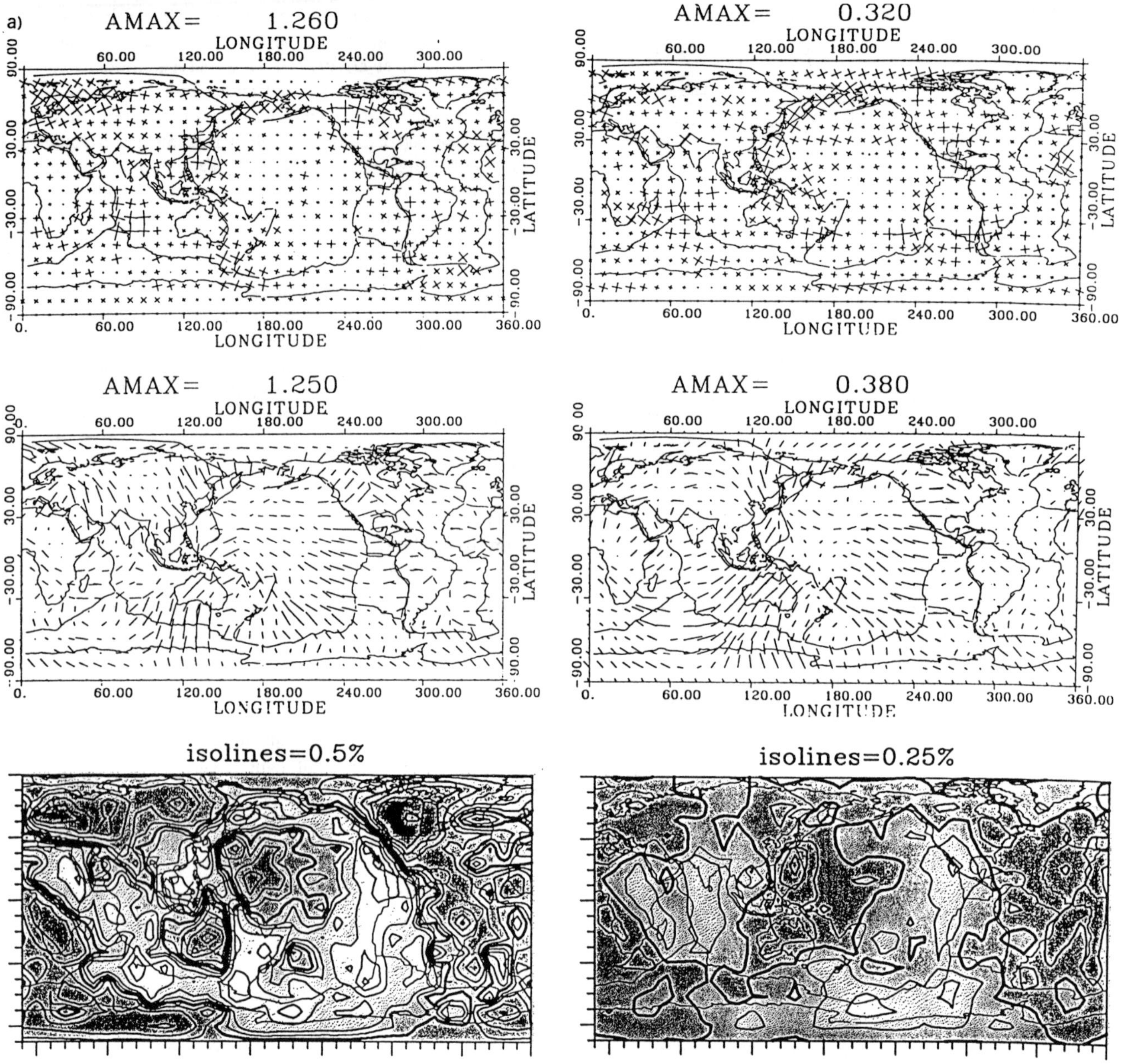

Fig. 26. Maps of the global distribution of phase velocities at different periods for Rayleigh waves [from Montagner and Tanimoto, 1990]. T is equal to 91s in (a) and 250 s in (b). In the bottom figures, the average phase velocity (A_0) is given. Note that contour intervals are twice as large in (a) as in (b). The middle figures present the maps of the 2ψ terms representing azimuthal anisotropy. The line gives the direction of maximum velocity with the length proportional to the size of the term and the maximum value indicated by AMAX at top of figures. For the 4ψ term (top figures), two lines are plotted to indicate that there are two maximum directions.

low-velocity anomalies associated with ridges occur in the depth range 0 to 100 km. Better resolution of the vertical structure at depths shallower than 200 km comes from regional studies of the Pacific and other oceans.

In the Pacific, many studies have detected a systematic increase of Love and Rayleigh wave velocities with increasing age of the oceanic plate [Leeds, Knopoff, and Kausel, 1974; Forsyth, 1975a,b; Yoshii, 1975; Schlue and Knopoff, 1977; Forsyth, 1977; Yoshida, 1978; Yu and Mitchell, 1979; Mitchell and Yu, 1980; Montagner and Jobert, 1981; Montagner, 1985; Nishimura and Forsyth, 1985,1988; Suetsugu and Nakanishi, 1987a,b; Teng and Chao, 1988].

Variations with age are also observed in the Atlantic and Indian Oceans, but with much poorer resolution and with other regional variations apparently as important as the changes with age [Weidner, 1974; Montagner, 1986; Honda and Tanimoto, 1987; Montagner and Jobert, 1988; Mocquet et al., 1989; Mocquet and Romanowicz, 1990]. The largest changes occur in young seafloor, with more gradual changes continuing into seafloor greater than 100 Ma in age (Figure 27). These phase velocities correspond to the A_0 coefficients for Rayleigh and Love waves and can be inverted in terms of an azimuthally averaged shear velocity, say β_V, and a measure of the Love/Rayleigh discrepancy, $\xi = (\beta_H/\beta_V)^2$. The inverted models are shown in Figure 28. The results for β_V qualitatively match those expected for a cooling half-space model. Initially, there are large changes at shallow depths, with the minimum velocity in the youngest zone occuring at depths of 40 to 80 km, the depths at which voluminous melting is expected to begin beneath mid-ocean ridges. At depths of 20 to 40 km, the shear velocity approaches a steady-state value of about 4.7 km/s in old seafloor. The depth at which the steepest, negative velocity gradients are found increases with increasing age, which may be the best indication of the location of the bottom of the thermal lithosphere. No horizontal change in velocity deeper than 100 to 120 km is required in the first 80 Ma, but in old seafloor, velocity increases extend to a depth of at least 150 km.

There is also an increase in ξ with increasing age, indicating that Love waves sample apparently higher velocity shear waves than do Rayleigh waves. This may represent a change to a more dominantly horizontal alignment of the olivine a-axis in older seafloor, as might be expected for a change from vertical upwelling beneath the ridge to dominantly horizontal shear beneath older seafloor (Figure 29; Nakanishi and Anderson, 1983,1984; Nataf et al. 1984,1986]. The vertical resolution for this type of anisotropy is poor because it is controlled by the vertical resolution of Love waves.

Azimuthal anisotropy has been detected for Rayleigh waves in both the Pacific and Indian Oceans in regional studies (Forsyth, 1975b, Yu and Mitchell, 1979; Okal and Talandier, 1980; Kawasaki and Kon'no, 1984; Montagner, 1985,1986; Suetsugu and Nakanishi, 1987; Montagner and Jobert, 1988] as well as in global tomographic inversions [Tanimoto and Anderson, 1984,1985; Montagner and Tanimoto, 1990]. The best resolved variations are the 2ψ variations, which are larger for Rayleigh waves than for Love waves as expected for horizontally propagating waves [Smith and Dahlen, 1973]. Only Montagner and Tanimoto [1990] have reported significant 4ψ variations, which are more difficult to resolve because of the more rapid variations with azimuth that are involved. In young seafloor in the Pacific the 2ψ variations correspond to 1 to 2% faster propagation of Rayleigh waves in the direction of seafloor spreading than perpendicular to it (Figures 26 and 30). The direction and amplitude is consistent with the preferential alignment of olivine a-axes in the direction of seafloor spreading by shearing flow [Estey and Douglas, 1986; Kawasaki, 1986; Cara and Leveque, 1987]. Inverting A_1 in terms of G_c alone indicates that the anisotropy must be present in both lithosphere and asthenosphere (Figure 31), although there is some trade-off with other anisotropic coefficients at shallower depths. Although fundamental mode studies do not have the resolution necessary to test for the presence of anisotropy deeper than 200 km, Cara and Leveque [1988, correcting earlier reports of Cara, 1979 and Cara and Leveque, 1987] employ higher mode data with resolution to 500 km to show that anisotropy probably does not extend deeper than 250 km for a path across the Pacific. In older seafloor, the amplitude of the azimuthal anisotropy is weaker and better aligned with the direction of fossil seafloor spreading than the direction of absolute plate motion [Nishimura and Forsyth, 1988]. The apparent amplitude may decrease in older parts of the Pacific due to interference

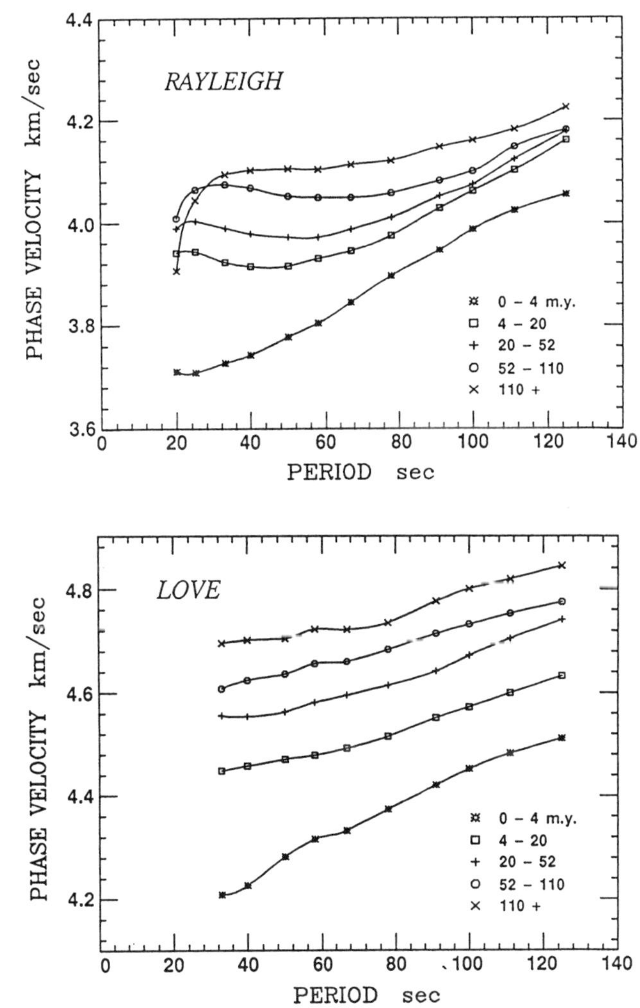

Fig. 27. Phase velocities for Love and Rayleigh waves in the Pacific as a function of period [from Nishimura and Forsyth, 1989]. Symbols indicate range of ages of seafloor for which average velocities were found.

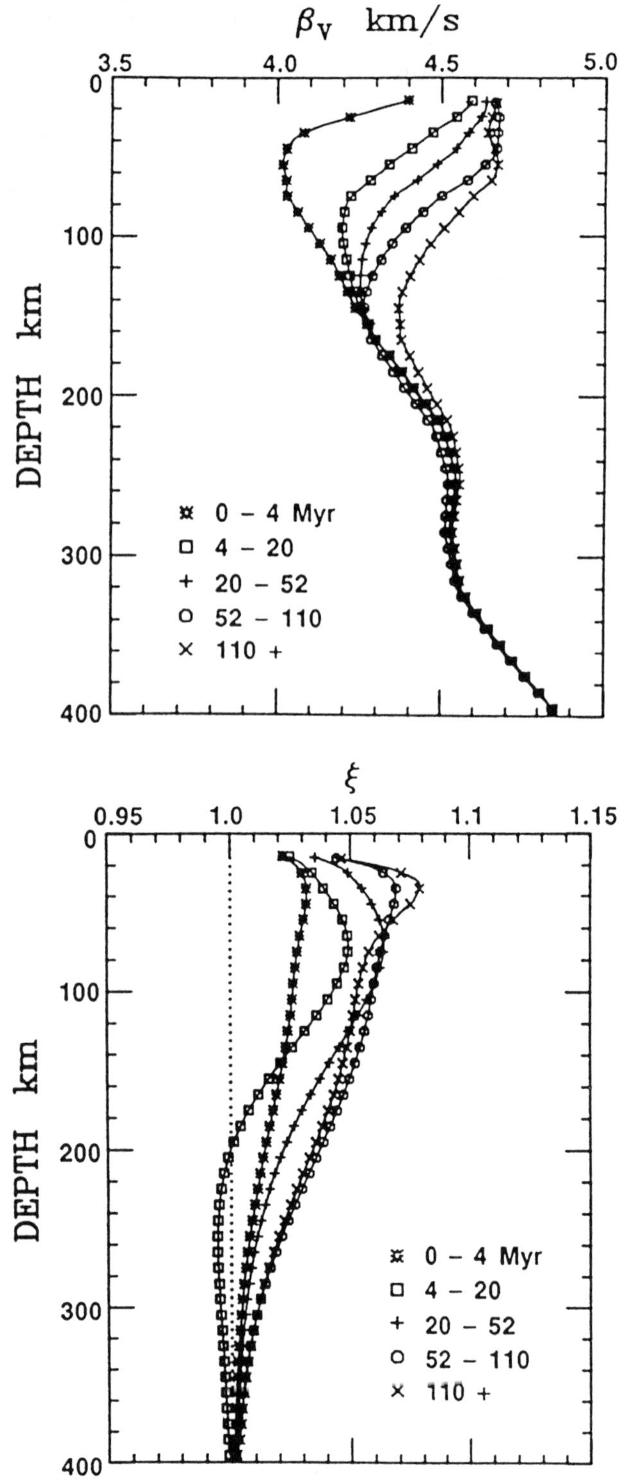

Fig. 28. Shear velocity structure of the upper mantle beneath the Pacific as a function of age of the sea floor [from Nishimura and Forsyth, 1989]. Upper panel shows increase in shear velocity with age. Lower panel shows changes in polarization (SH/SV) anisotropy with age.

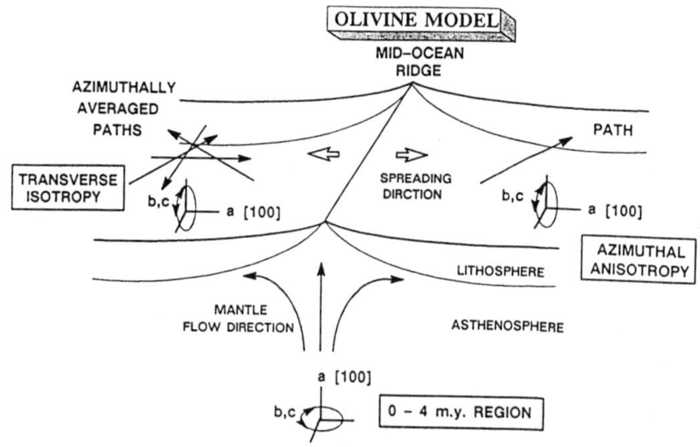

Fig. 29. Schematic diagram illustrating possible preferred orientation of olivine crystals in the vicinity of spreading centers. Where upwelling occurs beneath a ridge, vertical shear may preferentially align the olivine a-axis vertically [from Nishimura and Forsyth, 1989]. As the flow becomes dominantly horizontal, differential shear between the lithosphere and asthenosphere may align the a-axis horizontally and in the direction of plate motion relative to the underlying mantle. This may explain the origin of azimuthal anisotropy and the change in polarization anisotropy from the youngest age province to older seafloor.

between anisotropy frozen into the lower part of the lithosphere in the direction of seafloor spreading and anisotropy in the asthenosphere aligned with the absolute plate motion [Nishimura and Forsyth, 1989].

It should be noted that not all seismologists agree that azimuthal anisotropy exists as a significant factor in the earth on a gross scale. Modelling of the observed travel-times is more efficient with anisotropic parameters [Nishimura and Forsyth, 1988; Montagner and Jobert, 1988], but if sufficient lateral variations are allowed, the data can always be satisfied with heterogeneities alone. In the Pacific, however, the form of the inferred anisotropy remains stable, even when as many as 80 terms for lateral heterogeneity are introduced [Nishimura and Forsyth, 1988]. Allowing for azimuthal anisotropy does degrade the spatial resolution for the azimuthally independent terms, as there can be trade-offs between heterogeneities and anisotropy (Figure 32). To constrain azimuthal anisotropy with existing source receiver geometries requires averaging over large regions. Nishimura and Forsyth [1988] found only two statistically significant azimuthal anisotropy provinces within the Pacific and Suetsugu and Nakanishi [1987b] report optimal resolution is achieved with anisotropic provinces roughly 60° on a side. Some of the azimuthal patterns described by Montagner and Tanimoto [1990] are probably artifacts caused by trying to achieve too much horizontal resolution. They find, for example, that the 4ψ terms are as large as the 2ψ terms for Rayleigh waves, even though theory suggests they should be an order of magnitude smaller [Smith and Dahlen, 1973], and the 2ψ terms for Love waves are as

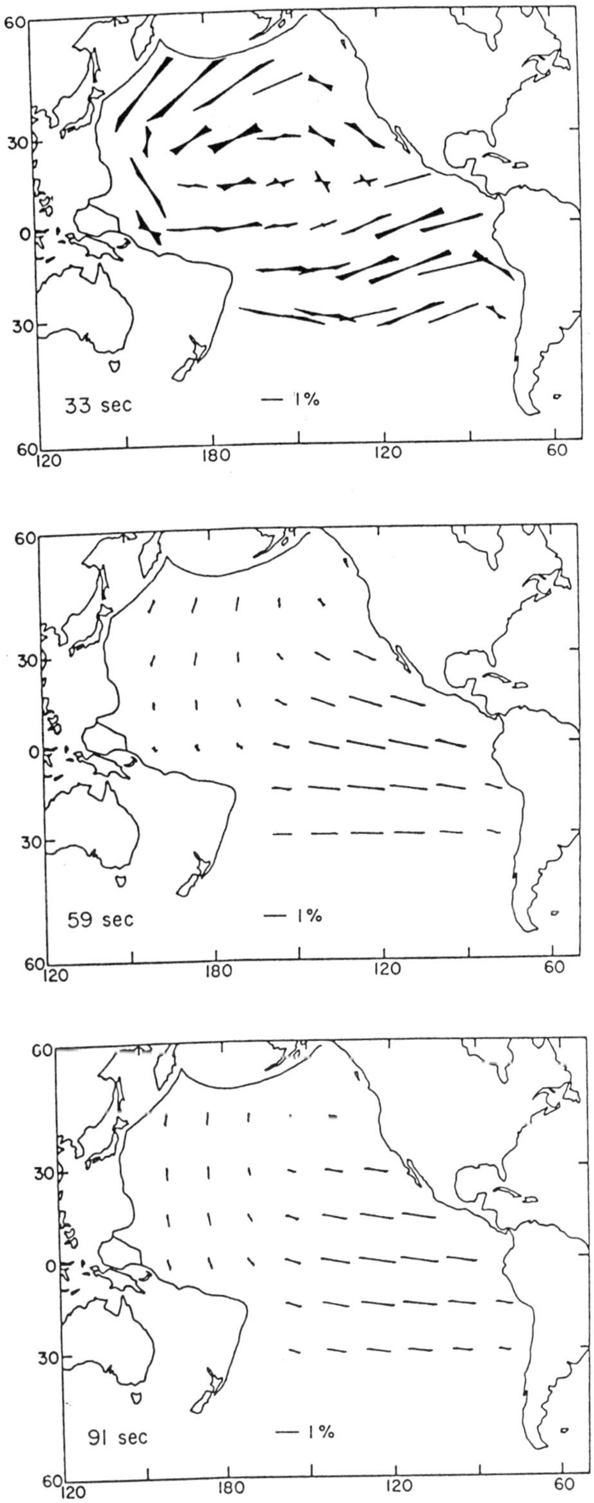

Fig. 30. Direction and degree of 2ψ azimuthal anisotropy in the Pacific for Rayleigh waves at different periods (from Nishimura and Forsyth, 1988]. The fastest direction of propagation is shown along with 1s variations.

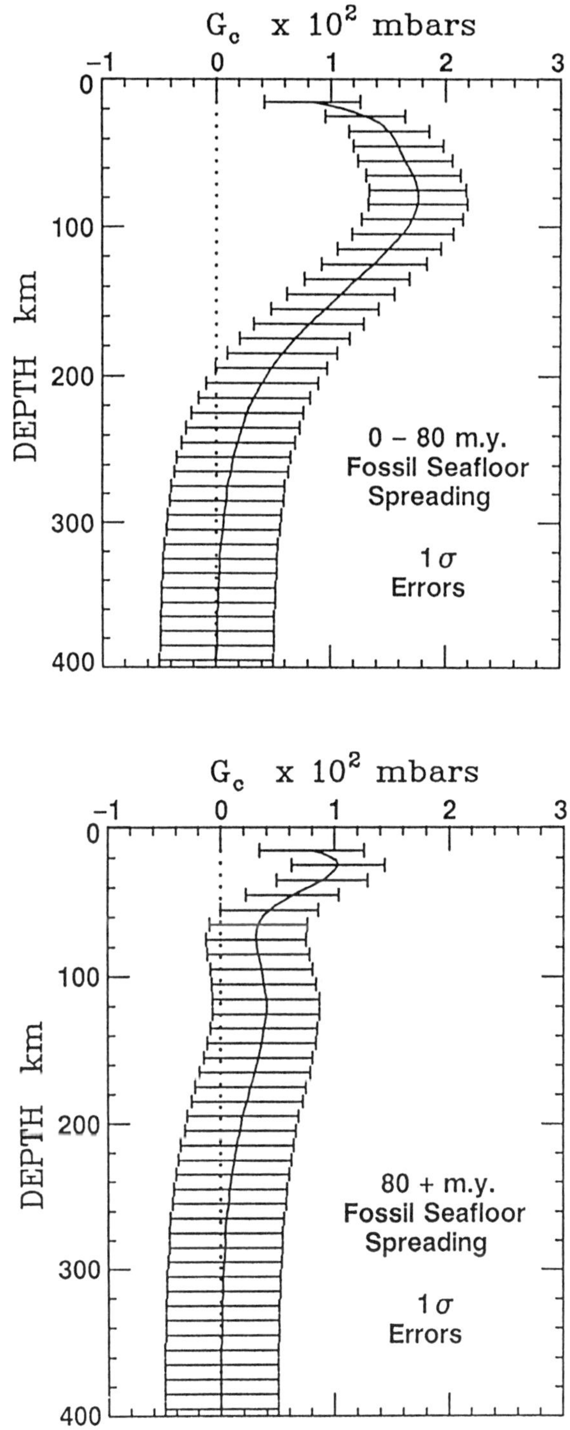

Fig. 31. Azimuthal anisotropy parameter G_C as a function of depth in two regions in the Pacific [from Nishimura and Forsyth, 1989]. For both regions, the coordinate system used to calculate the anisotropic velocity coefficients was based on the local, fossil, seafloor spreading direction rather than on the geographic or absolute plate motion coordinate systems.

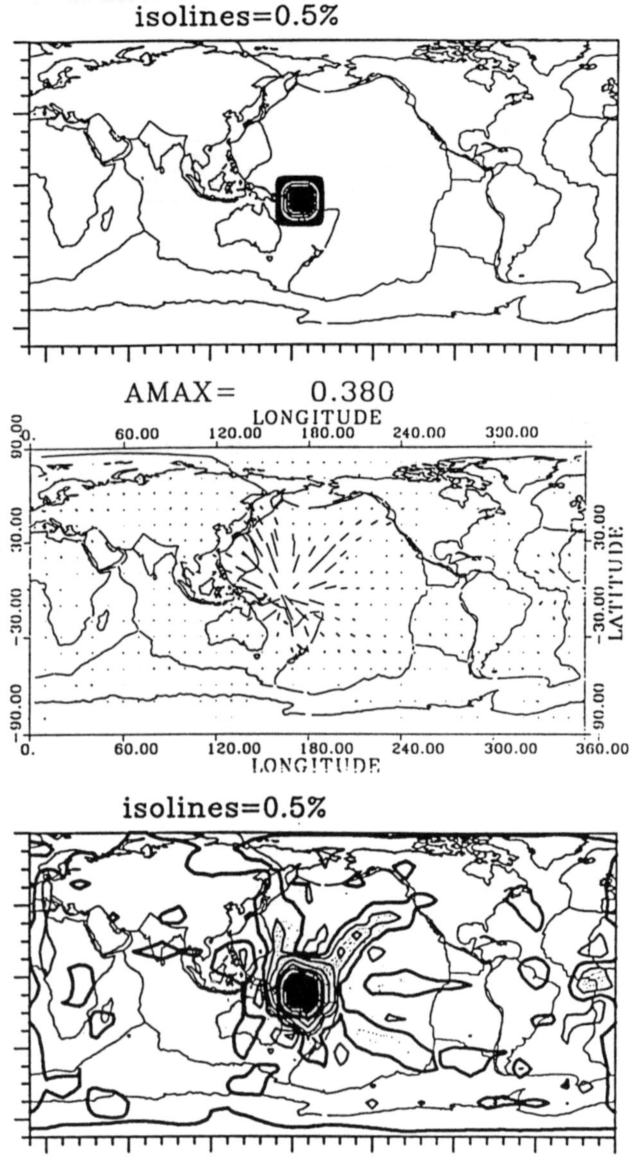

Fig. 32. Illustration of spatial resolution and trade-off between anisotropic terms and spatial variations [from Montagner and Tanimoto, 1990]. An artificial 20° by 20° velocity anomaly is introduced (top). Then, assuming the same distribution of paths of Rayleigh waves that was available for the global inversions shown in Fig. 26 and calculating the effect of this introduced anomaly on the travel times, an inversion was performed for the spatial distribution of velocities and 2ψ azimuthal anisotropy. The results show that the spatial distribution is imperfectly resolved (bottom) and that some of the spatial anomaly is misinterpreted as spatially varying azimuthal anisotropy (middle). This smearing of the spatial anomaly and trade-off with azimuthal terms that shows up primarily along a few great-circle paths radiating outward from the target area is caused primarily by a lack of earthquake sources and seismic stations in the middle of ocean basins to provide adequate azimuthal coverage.

large as the 4ψ terms, although theory again suggests they should be significantly smaller [Crampin, 1977]. Tomography studies that neglect azimuthal anisotropy, however, tend to find much more oscillatory spatial solutions [Yomogida and Aki, 1987; Zhang and Tanimoto, 1989] that may also be artifacts caused by forcing travel time variations associated with azimuthal variations into spatial variations. Ultimate resolution of this question will require seafloor observatories that can detect shear-wave splitting and provide sufficient spatial coverage to remove the trade-off between heterogeneity and anisotropy.

Lateral variations from one part of the mid-ocean ridge to another are evident in the global tomography studies (Figure 26). For example, south of Australia the low velocity region along the ridge is much less pronounced than either to the east or west. Much of the Atlantic also appears to be faster than the East Pacific Rise, but it is difficult to evaluate whether this is just a lack of resolution on this slow spreading ridge. Local or regional studies may be more effective in delineating velocity differences from region to region, but usually are limited to shorter periods than the global studies.

The extremes of the mid-ocean ridge system in terms of elevation, the Iceland hotspot area and the Australian-Antarctic discordance have been investigated in regional studies. Girardin and Jacoby [1976] report group velocities of Rayleigh waves on the Reykjanes Ridge that are indistinguishable from those on the East Pacific Rise. Evans and Sacks [1979] favor shear velocities in the low-velocity zone as low as 3.9 km/s under the Kolbeinsey ridge just north of Iceland for seafloor 0 to 5 Ma in age, slightly lower than found in the same depth range on the East Pacific Rise. It is unclear whether the velocities in the upper mantle north of Iceland really are anomalous compared to the EPR because the very short path lengths (about 600 km from source to receiver) restrict the periods to about 25 s or less and increase the uncertainty in travel time associated with uncertainty in location and origin time of the earthquake sources. The group velocities in the 20 to 25 s range are indistinguishable from group velocities for the same age range on the East Pacific Rise [Forsyth, 1975], while shorter periods are primarily sensitive to variations in water depth and crustal structure. Thus, the inference by Sato et al. [1988,1989] that there is significantly more melt present beneath the ridge north of Iceland than beneath the EPR is questionable.

In the Australian-Antarctic discordant zone (AADZ), Forsyth et al. [1987] report that in seafloor younger than 10 Ma that is more than 500 m deeper than expected from a Pacific depth-vs-age curve, Rayleigh wave phase velocities in the 20 to 100 s range are significantly faster than in seafloor of comparable age in the Pacific. The largest difference in the inferred shear velocity structure (Figure 33) is in the 20-40 km depth range. If this velocity difference is attributable solely to temperature differences, then the temperature change required depends on whether partial melt is present beneath either ridge. The fact that the velocity difference is much larger in the 20-40 km range than at other depths suggests that partial melt is present

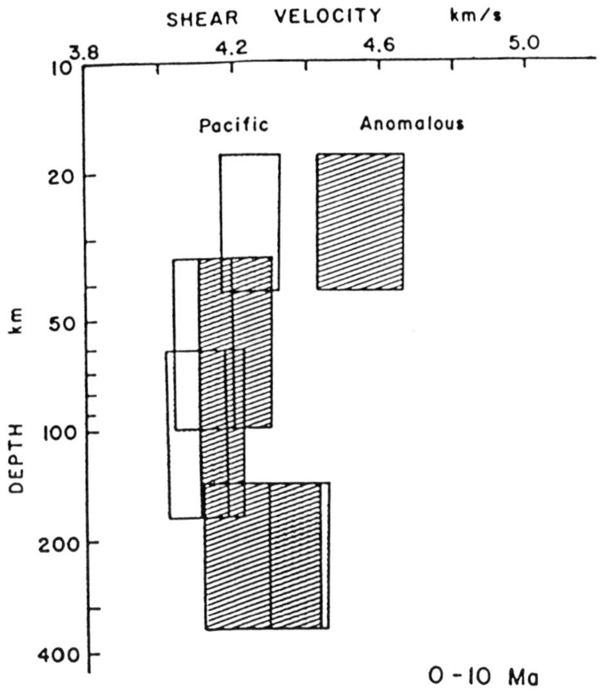

Fig. 33. Comparison of average shear velocities beneath young seafloor (<10 m.y.) in the Australian-Antarctic discordant zone (shaded, labelled anomalous) with the average velocity structure in the same depth and age in the Pacific (unshaded) [from Forsyth, et al., 1987]. Vertical extent of each rectangle indicates depth range of average and horizontal extent indicates 95% confidence limits of average.

beneath the EPR and other typical ridges, but not beneath the AADZ. If the temperature of the upwelling material were lower and no melt were involved, then the velocity difference should be nearly independent of depth. High velocities beneath the AADZ at depths of 20-40 km may thus be attributable to the disappearance of melt. The difference of about 0.35 km/s requires a temperature difference of about 900°C to explain without the involvement of melt. Using the data of Murase and Kushiro [1979] as reported by Sato et al. [1989], the velocity difference can be explained by a change in melt fraction of about 8%, requiring a temperature change of less than 100°C. If melt were distributed as thin films, the melt fraction required could be much less.

Seismic Body Waves

P and S waves propagating through the body of the earth may also be employed to sample the upper mantle beneath ridges. These waves not only provide independent information because they sample the earth in a fundamentally different way than surface waves, but they potentially have much finer spatial resolution. Travel time delays provide information about the velocity structure; amplitude and spectral measurements constrain the attenuation structure; and shear-wave splitting can detect anisotropy. Each of these approaches may be incorporated into experiments with three types of source-receiver geometry: teleseismic studies using remote earthquakes and stations on oceanic islands or the continents; passive array studies using teleseismic earthquakes but with an array of instruments deployed on the seafloor in the area of interest; and active array studies using explosions and microearthquake sources within the volume of interest. Active array studies could provide the best resolution but are limited primarily to probing the crust, because earthquakes on ridges occur within 8 to 10 km of the surface and body waves from shallow sources refract downwards away from local receivers when they encounter the low velocities in the melt production region. Crustal structure is not the primary focus of this review, but some of the results are described in a later section on crustal thickness. It may be possible to image structures in the mantle using seismic reflection if there are large enough impedance (velocity times density) contrasts. As indicated in Table 1, this difficult experiment could provide the best horizontal and vertical resolution of any technique, provided there are sharp, reflective boundaries to the top or bottom of melting zones in the mantle. Horizontal resolution for reflection is limited by the width of the Fresnel zone, which is about 10 km at a depth of 50 km for a dominant frequency of 10 Hz. Resolution of the depth to an interface is limited by the accuracy of the velocity model, which in this case would be poorly constrained by observations and is assumed to be of the order of 10% in constructing Table 1. To date, only teleseismic and passive array studies have provided any constraints on mantle structure beneath spreading centers.

Teleseismic studies. Most studies using stations on land have measured delay times of S or multiple S phases (Figure 34). With S phases, the upper mantle can be sampled only in the vicinity of events on the plate boundaries or the occasional intraplate event, and there are some errors in travel-time introduced by uncertainties in origin-time, epicentral position, and delays beneath the stations. Differential travel-times such as SS-S, or the time difference between arrival of the SS and S phases, are commonly employed because the effects of near-source and near-receiver structures and

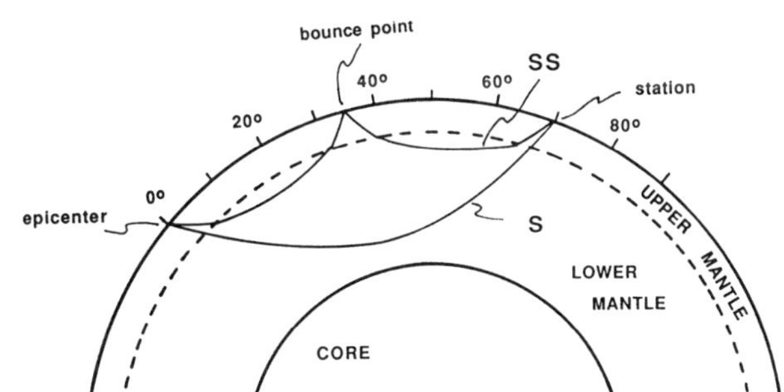

Fig. 34. Ray paths of S and SS phases for an epicentral distance of 70°. Paths are similar in vicinity of epicenter and station.

epicentral errors are nearly the same for both phases and are removed by the differencing process. The upper mantle can be sampled in the vicinity of the surface bounce point of the multiple S phase far from any station or earthquake source. In both cases, travel-time anomalies that arise along the lower mantle portion of the paths must either be ignored or corrected with a global tomographic model. True tomographic studies with the resolution required to look at the detailed structure of the melt production region are not possible because source errors and uncertainty in position of bounce points limit the horizontal resolution. Generally, averages of delay time through the entire upper mantle over regions of several hundred km or more in lateral extent are all that can be obtained with confidence. The resolution is adequate to detect clearly the cooling of the plate with age even in the slow-spreading Atlantic.

Girardin and Poupinet [1974] showed that S-waves from events on the Mid-Atlantic Ridge are delayed relative to standard earth models or to events on older seafloor in the Azores-Gibralter region. Sipkin and Jordan [1976] found a systematic variation in ScS2 - ScS differential travel times relative to the J-B tables with increasing age of the seafloor in the Pacific. Duschenes and Solomon [1977] showed that there was a roughly linear relationship between the square root of age of the seafloor and S-wave delays from oceanic, intraplate earthquakes, amounting to roughly 6s from 0 to 100 Ma, or 0.6 s/Ma$^{1/2}$. Converted to the equivalent delay for a vertical path, this coefficient, t_a, is about -0.5 s/Ma$^{1/2}$. Using a set of 65 SS-S differential times with SS bounce points in the North Atlantic, Kuo et al. [1987] reported a coefficient for the two-way travel-time of SS waves of -1.01 ± 0.21 s/Ma$^{1/2}$, or $t_a \approx$ -0.40 ± 0.10 s/Ma$^{1/2}$ for a vertical, one-way path. With a larger data set of about 350 bounce points covering more of the North Atlantic, Sheehan and Solomon [1991a] found a coefficient of -0.68 ± 0.08 s/Ma$^{1/2}$, or $t_a \approx$ -0.27 ± 0.03 s/Ma$^{1/2}$. Finally, Woodward and Masters [1991] found an SS coefficient of -0.51 ± 0.01 s/Ma$^{1/2}$ using a global data set of about 1800 oceanic SS-S differential travel times (Figure 35], or $t_a \approx$ -0.205 ±0.005 s/Ma$^{1/2}$.

If these delay coefficients are to be used as a constraint on the spreading process, the discrepancies among them must be understood. The Duschenes and Solomon estimate based on intraplate earthquakes is probably too large because it was assumed that all the events occured at the same depth. The average depth of intraplate earthquakes increases with increasing age of the seafloor [Wiens and Stein, 1983], which means that the S-waves from events in older seafloor would arrive sooner than expected. The Kuo et al. estimate is probably too high because most of the bounce points in older seafloor were located in areas that Sheehan and Solomon demonstrated were anomalously fast. The difference between the Sheehan and Solomon and the Woodward and Masters estimates may be attributable to regional differences, just as there are regional differences in subsidence rate. Sheehan and Solomon [1991a] map along-axis variations in differential travel time in the Atlantic that are nearly as large as the variations with age. This regional heterogeneity is manifested in the Woodward and Masters compilation (Figure 35) in the form of scatter of the individual age bands about the regression line. If each age band uniformly sampled the globe, then the individual points shown should fit the regression line as indicated by the standard error of estimate of the points. In the global compilation, about one quarter of the points are fit within one standard error, while in the North Atlantic (Figure 4 in Sheehan and Solomon, 1991a] about two thirds lie within one standard error of the regression line, as is expected for a normal distribution. This suggests that there are regional variations and that different parts of the spreading system are preferentially sampled within different age bands in the global compilation. The standard errors need to be multiplied by a factor of three to four to match the observed scatter and to account for the non-uniform sampling. The standard error of the regression coefficient should be multiplied by a similar factor.

Vertical, one-way S wave delays of -0.2 to -0.3 s/Ma$^{1/2}$ are consistent with the structure observed by Love and Rayleigh

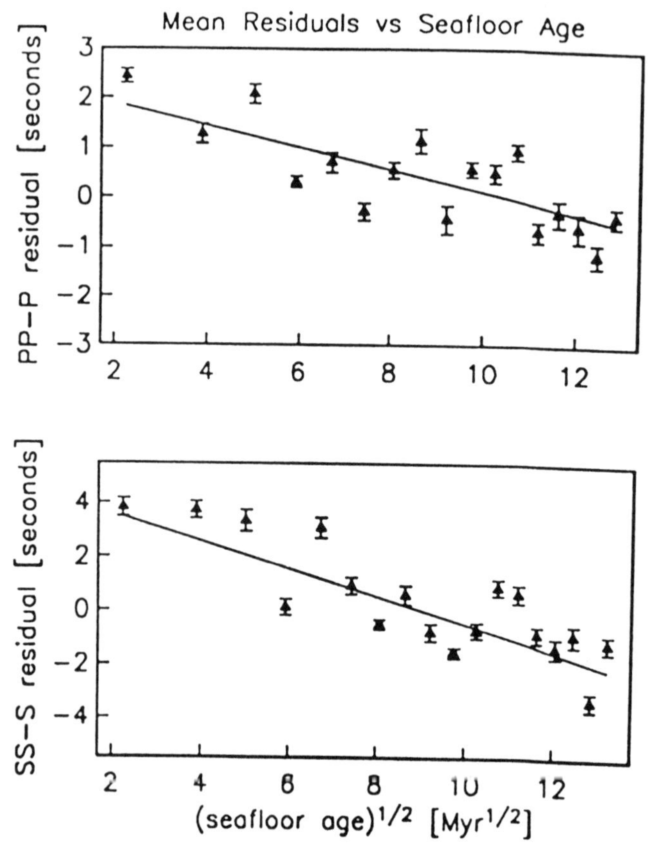

Fig. 35. PP-P and SS-S travel-time residuals as a function of the age of the seafloor at the bounce point. Each point represents the average of about 100 individual measurements from a global compilation. Straight lines represent delay as a function of square root of age of the seafloor found by linear regression. [From Woodward and Masters, 1991].

waves, except beneath seafloor younger than 10 Ma. Vertically integrating the delay times through the β_V structures shown in Figure 28a yields variations with age in the Pacific that are in excellent agreement with the global coefficient of Woodward and Masters [1991] from SS waves (Figure 36 bottom). However, only the transverse component of SS waves have been used in these differential travel-time studies, which should be more nearly comparable to the β_H structure constrained primarily by Love waves. Because the Love/Rayleigh discrepancy or polarization anisotropy

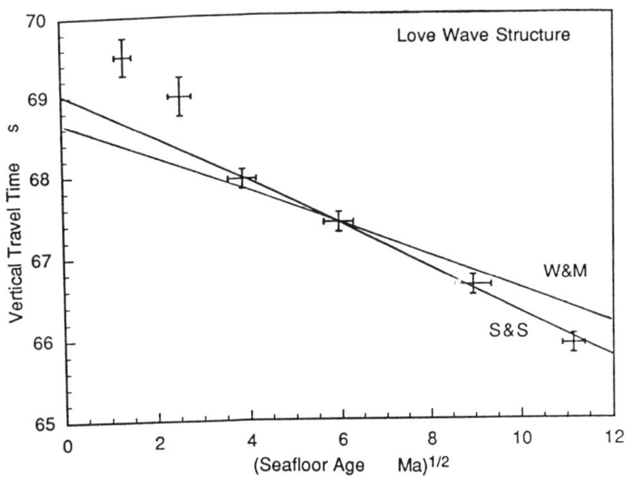

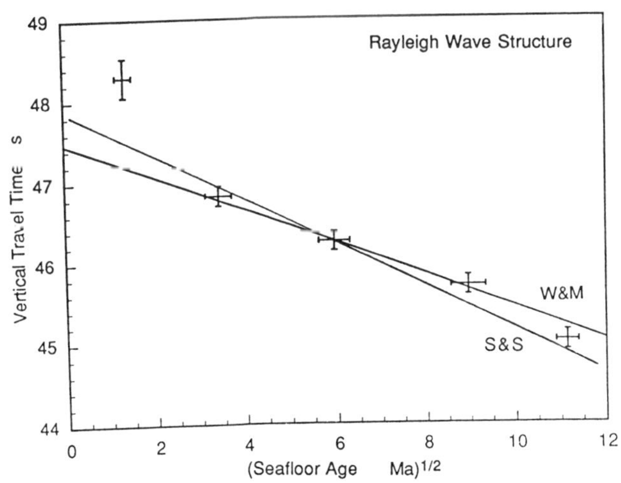

Fig. 36. Vertical travel-times for S-waves through the oceanic upper mantle calculated from the structures shown in Fig. 28. Absolute travel-times are unimportant because they depend on depth of integration. The observed, relative travel-time delays agree with the predicted delays based on SS-S studies of Sheehan and Solomon [1991], indicated by line labelled S&S, and Woodward and Masters [1991], indicated by W&M. Error bars represent $\pm 1\sigma$.

increases with age (Figure 28b), the apparent variation in β_H structure is larger (Figure 36 top), in better agreement with the coefficient of Sheehan and Solomon [1991a]. Surfaces waves detect an extra delay in the youngest provinces that is not found in the body wave studies. The SS waves may miss the anomalously slow region if it is narrow, particularly in the slow-spreading Atlantic where they could refract around it. Surface waves sample the youngest regions more uniformly and more thoroughly. There is only one age band in the Woodward and Masters compilation with an average age less than 16 Ma. There could also be some bias in the sampling: greater attenuation near ridges [Sheehan and Solomon, 1992] makes it more difficult to observe good quality SS wave forms, so those that are reported may tend to come from regions with higher than average Q and S-wave velocity. In the Pacific, variations with age in the youngest seafloor may be obscured by larger effects of lateral heterogeneity and anisotropy [Sheehan and Solomon, 1991b].

The general agreement between surface wave and body wave coefficients for variations as a function of age suggests that the S-wave delays can be attributed to cooling of a thermal boundary layer. No deep structure (> ≈200 km) is required. In a tomographic study of the S-wave structure of the entire North American plate, Grand [1987] mapped slow regions beneath the Mid-Atlantic and northern East Pacific Rise extending as deep as 320 km, but in these regions of poor depth control, the inversion technique employed would tend to spread the anomaly out as much as possible along the ray paths.

There are indications, however, that the cooling process may involve partial melting, not just the subsolidus temperature dependence of an elastic solid. First, t_a is substantially larger than predicted by laboratory experiments. The expected delay time t_d for a shear wave propagating vertically through a medium with shear velocity $\beta(z)$ is

$$t_d = \int_0^\infty \left(\frac{1}{\beta(z)} - \frac{1}{\beta_0(z)} \right) dz \qquad (23)$$

where $\beta_0(z)$ is the reference shear velocity structure. Assuming that $\beta(z)$ is a function of temperature only and that the temperature structure is a cooling half-space described by equation (16), the delay to first order is

$$t_d = \frac{-2\beta' T_m}{\beta^2} \sqrt{\frac{\kappa t}{\pi}}$$

and

$$t_a' = \frac{t_d}{\sqrt{t}} = \frac{-2\beta' T_m}{\beta^2} \sqrt{\frac{\kappa}{\pi}} \qquad (24)$$

where $\beta' = \partial\beta/\partial T$. Adopting the values found for forsterite at 1700°K by Isaak et al. [1989], $\beta = 4.48$ km s^{-1} and $\beta' = -0.44$ m s^{-1} °K^{-1}, and T_m and κ used previously, yields a predicted t_a of -0.141 s/Ma$^{1/2}$. This is nearly a factor of two smaller than that found for Love waves or by Sheehan and Solomon [1991a] and is significantly smaller than that found for Rayleigh waves or by Woodward and Masters [1991]. The elastic properties at lower temperatures predict even smaller t'_a. One possible explanation of the discrepancy is that cooling eliminates a small fraction of melt in the asthenosphere, increasing the effective β'. Another possibility is that the ultrasonic laboratory experiments are measuring unrelaxed elastic moduli, while the seismic experiments at much lower frequency are detecting relaxed or partially relaxed moduli of a thermally-activated, solid-state attenuation mechanism that have larger temperature coefficients. Supporting the interpretation that a small amount of partial melt may be involved is the ratio of S delays to P delays. Woodward and Masters [1991] found a two-way age coefficient for PP-P delays of -0.18 ± 0.01 s/Ma$^{1/2}$, which yields an SS-S to PP-P ratio of 2.83. The laboratory data of Isaak et al., [1989] yield a predicted ratio of 1.92 at 1700°K. A review by Ritzwoller et al. [1988] suggested that 2.2 might be expected for upper mantle solid phases, based on earlier laboratory experiments. In the limit of a Poisson solid that is undergoing a very small degree of partial melting that affects only the shear modulus, the expected ratio is 3.9. Thus, a value of 2.83 can be explained by a contribution from partial melting. Before concluding that partial melting is required, however, it is worth noting that high ratios are frequently reported, including in the lower mantle [e.g Jordan and Lynn, 1974; Giardini et al., 1987] where melting is not expected. The discrepancy between laboratory and seismologically-derived coefficients is still an important area of research.

Passive array studies. The spatial resolution of passive tomographic arrays is limited by the density of stations, the distribution of sources, the wavelength of the waves, and the accuracy of the velocity resolution required for useful interpretation. In a typical tomography experiment, waves travel through the region of interest at a variety of different angles, sampling different subregions or blocks (Figure 37). Each block is assumed to have some constant perturbation in velocity from the starting model [Aki et al., 1977] or perturbations in velocity in the region are described by some simply parameterized, continuous function such as linear interpolation between nodal points [Hovland et al., 1981]. The perturbations are then found which best predict the observed variations in travel-time. In the absence of sources within or beneath the volume of interest, body-wave tomography studies usually employ teleseismic sources at distances greater than 30°, because triplications in the travel-time curves and lateral variations in the upper mantle make it difficult to predict the angle of incidence of P or S waves from closer sources. Teleseismic P-waves recorded on the mid-ocean ridge are expected to have dominant frequency of about 1 Hz. Greater attenuation of S-waves means that the minimum period that could be detected is 4 to 5 s, although with high noise levels from microseisms peaked at around 6s, a more practical minimum period may be about 10 to 12 s (This discussion excludes the high frequency P_o and S_o phases that propagate in the lithosphere, because they provide little depth resolution

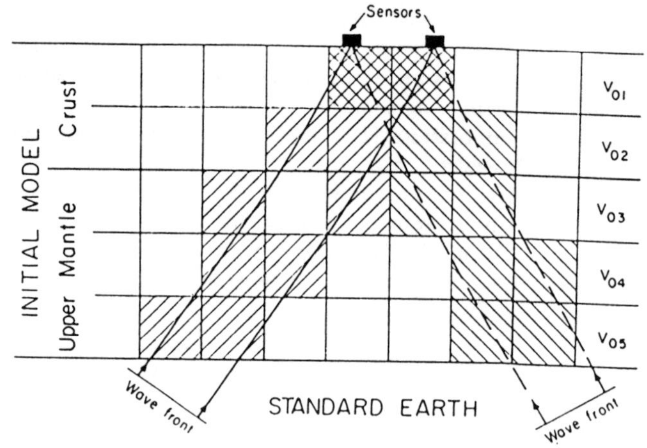

Fig. 37. Typical geometry of a teleseismic tomography experiment [from Aki et al., 1977]. Waves travelling upward at angles of 20 to 40° with respect to vertical sample different subregions beneath an array of stations. With delays from many crossing rays, it is possible to infer the velocity perturbation within each block. Outside the model region, the earth is assumed to be described by a standard, reference earth model.

and tend to be highly attenuated in the vicinity of mid-ocean ridges). These periods correspond to wavelengths of about 8 km for P phases and 20 to 50 km for S phases, limiting the maximum possible spatial resolution to similar values (Table 1).

No true passive tomography experiments have been completed successfully on a mid-ocean ridge to date. One of the primary difficulties is the length of time required to record a large number of teleseisms with good signal-to-noise ratio. A single, intermediate-focus event recorded on a small array of ocean-bottom seismometers deployed near the western intersection of the Mid-Atlantic Ridge with the Vema fracture zone (Figure 38), found a roughly 0.5 s P-wave delay within the median valley compared to stations outside the valley [Rowlett and Forsyth, 1979]. P and PKP phases from several earthquakes, ranging from shallow to deep-focus, were recorded on another array of instruments deployed near the Oceanographer fracture zone, but timing problems prevented a useful delay-time study [Forsyth, 1982]. In both cases, the phases were better recorded on hydrophones than on geophones. These observations and the delays observed in tomographic experiments in Iceland and near continental rifts suggest that the logistical effort involved in a passive tomography experiment would be worthwhile.

In Iceland, a tomographic study involving 39 stations and a total of 714 P-wave arrival times found a low-velocity zone in the 0-75 km depth range beneath the central rift of Iceland [Tryggvason et al., 1983]. The largest anomaly, 3-4% slower than average, was found in the vicinity of the active Krafla zone in northeast Iceland. Smaller anomalies, on the order of 1%, continued deeper into the 75-175 and 175-275 km depth

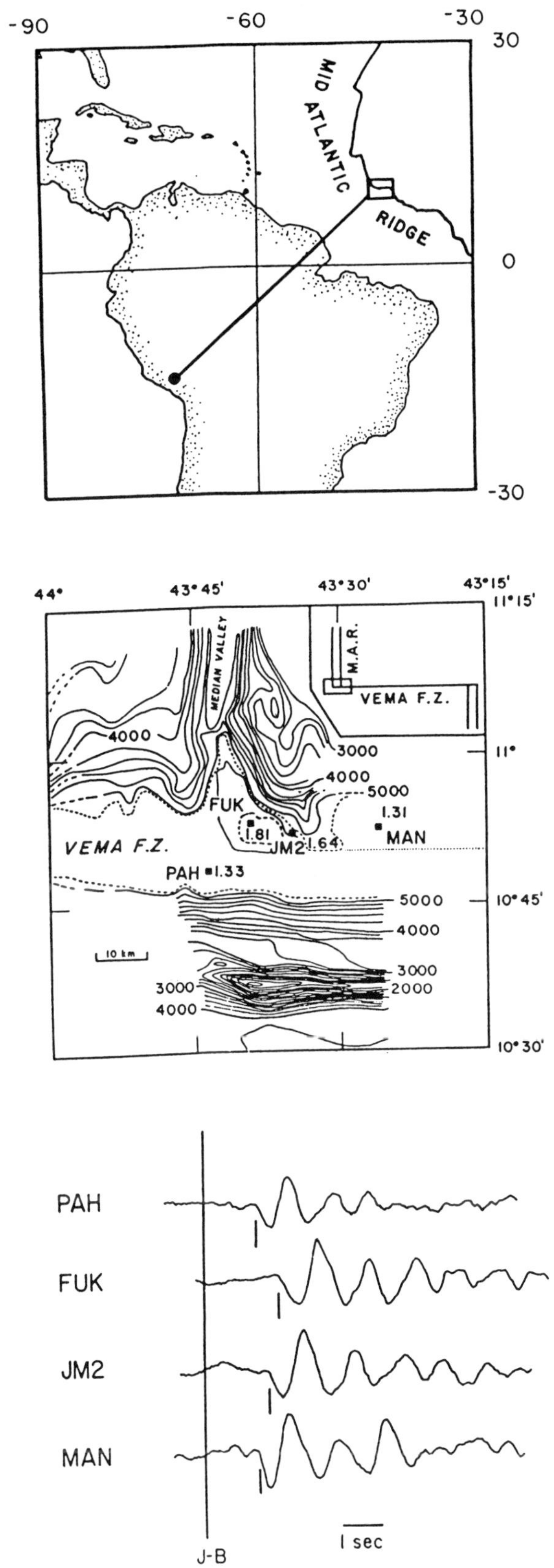

ranges in the tomographic model. Large anomalies were also found in the deepest layer included in the model, from 275 to 375 km, but these may be artifacts caused by travel-time variations elsewhere in the mantle that often can be mapped into the bottom layer with this technique. A seismic tomography experiment involving 20 well-spaced stations in a 100 x 110 km array spanning the Kenya rift just south of the intersection with the Kavirondo rift also found anomalously low velocities in the mantle beneath the rift zone [Green et al., 1991]. Travel-time variations of more than 1.0 s were reported [Green and Meyer, 1990]. In the 65-105 km depth range, there is a continuous low-velocity zone underlying the East African Rift with lateral velocity changes across the rift on the order of 12.5%. In the uppermost mantle (35-65 km), the velocity anomalies are slightly smaller and less continuous. The large velocity anomalies in the upper mantle strongly suggest that melt is present, perhaps as much as 6 to 8% [Green et al., 1991]. P-wave velocity anomalies as large as 8% were reported beneath the Rio Grande rift from data gathered on a 1000 km linear array of stations [Parker et al., 1984; Davis, 1991] crossing the rift. Halderman and Davis [1991] report variations in attenuation of P-waves that correlate with P-wave delays in both the Rio Grande and East African rifts and interpret the delays and attenuation in terms of an elevated asthenosphere with 1 to 3% partial melt. Spence and Gross [1990], on the basis of a tomographic study of data from a two-dimensional group of stations, argue that the primary low-velocity anomaly in New Mexico is actually associated with the volcanic Jemez lineament that crosses the Rio Grande rift obliquely at about the same location and perpendicular to the crossing of the linear array of stations. Although the settings are undoubtedly different than a normal mid-ocean ridge, these studies demonstrate that there are potentially large travel-time delays at considerable depth in extensional provinces that can be constrained by passive tomographic array experiments.

The Bottom Line

Is there or is there not melt present beneath young seafloor detectable by surface wave propagation? The difference between young seafloor beneath the AADZ and the EPR is one strong argument for the presence of melt (see above). A traditional argument is based on the decrease in velocity with increasing depth or on the changes with age of the seafloor. If we only examine the changes with age we avoid the problems of pressure changes or vertical compositional differences. The shear wave velocity in seafloor 0-4 Ma in age at depths of 40

Fig. 38. P-wave delay experiment on the Mid-Atlantic Ridge [from Rowlett and Forsyth, 1979]. Top, location of earthquake source (dot) and receiver array. Middle, location of ocean-bottom seismometers and delays in seconds relative to J-B travel-times, corrected for sediment thickness. Contours intervals are 200 m except dashed lines in fracture zone valley at 100 m intervals. Bottom, P-wave seismograms recorded by hydrophone channels, aligned with respect to expected J-B arrival time at each site. Late arrivals indicate slow velocities.

to 80 km reaches a minimum value that is 0.62 +/- 0.08 km/s less than the velocity in seafloor >110 Ma old (Figure 28a), a change of about 15%. Can a change this large be produced without melt being present?

The answer is an unequivocal, 'Well, probably not!?' Sato et al. [1988,1989] state that a 6% change in velocity can be achieved sub-solidus. However, they ignore the velocity changes accompanying temperature changes between 0 and 1000°C for their sample of spinel lherzolite. The shear velocity derivative coefficient $(\partial\beta/\partial T)_P$ for olivine ranges from about -0.35 to -0.44 m s^{-1}°C^{-1} as the temperature increases from 0 to 1400°C [Isaak et al., 1989; Graham et al., 1988]. At low temperatures, the derivative for orthopyroxene is about 40% larger [Frisillo and Barsch, 1972], which would enhance the temperature effect by about 15% above that of forsterite if the upper mantle is 60 to 70% olivine. If the minimum temperature at 60 km (midpoint of the 40-80 km depth range) in old seafloor is 600°C and the maximum temperature beneath the ridge at 60 km is 1500°C, then the predicted maximum possible change in shear velocity without considering the effects of melting is 0.43 km/s. At 1500°C and 60 km some partial melting would be expected in peridotite anyway, but the question is whether melt removal is so efficient that no significant amount of melt remains. The answer is yes, there must be melt present. The results of Sato et al [1989], however, suggest that $(\partial\beta/\partial T)_P$ increases rapidly in amplitude as the solidus is approached, which renders the answer equivocal. Those results could, however, be caused by the presence of very small amounts (<2%) of unrecognized melt, which is suggested by the facts that electrical conductivity and shear wave velocity vary in a similar fashion in the vicinity of the solidus [Berckhemer, 1982; Kampfmann and Berckhemer, 1985] and careful measurements of elastic properties of solids with precisely known melting temperatures show no change in temperature coefficients as the solidus is approached [e.g., Flower and Saunders, 1990]. My bottom line is that a detectable amount of melt is present beneath the East Pacific Rise in the 40 to 80 km depth range.

One of the strongest constraints on the amount of melt present is provided by vertically integrating the expected travel-time delay of S-waves through the velocity structure derived from the surface wave data (Figure 28). The delay times as a function of the square root of age form a linear trend that agrees well with the trend found for SS-S delays in old seafloor, but in the youngest zone for Rayleigh waves, there is an additional 1 s delay that cannot be explained simply by solid-state cooling of a thermal boundary layer (Figure 36). Love waves detect a similar anomaly, but the greater age resolution of this Love wave data set [Nishimura and Forsyth, 1985] provides an indication that the anomalously slow velocities, apparently indicating the presence of partial melt, extend into sea floor 4 to 10 Ma old. Depth control on the melt distribution is provided primarily by the Rayleigh wave data. Subtracting the expected change in velocity as a function of depth for a cooling half-space from the observed change from the 0-4 Ma zone to the 4-20 Ma zone yields a shear velocity

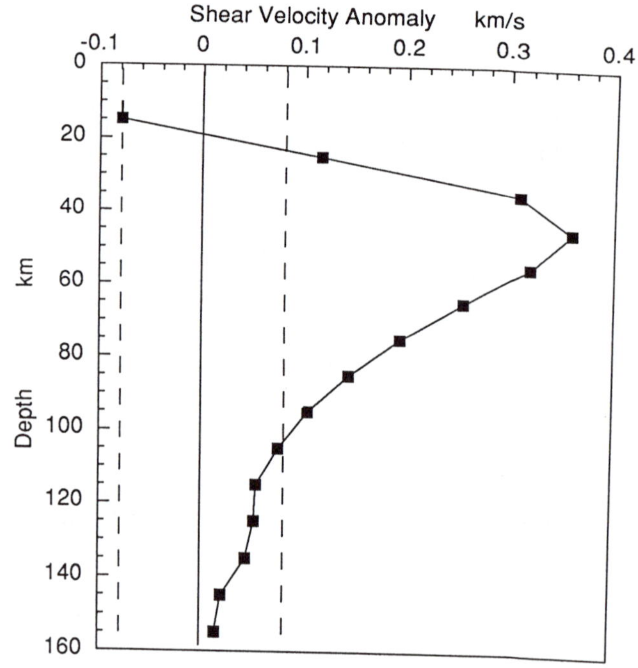

Fig. 39. Anomalous S-wave velocities beneath the East Pacific Rise. These are calculated by first finding the difference between the velocity structure for the 0-4 and 4-20 Ma age zones in Fig. 28a, then subtracting from that difference the change predicted for the cooling half-space thermal model using laboratory coefficients for $\partial\beta/\partial T$ in forsterite. Dashed lines represent $\pm 1\sigma$. These anomalously slow velocities that are not predicted by the solid-state cooling of the upper mantle are assumed to indicate the presence of melt.

anomaly strongly peaked at a depth of about 45 km (Figure 39). The velocity change from the youngest seafloor to the next age zone is significantly larger than expected from a depth of 25 km to about 100 km with a peak anomaly of about 0.35 km/s at a depth of about 45 km. Because the inversion procedure employed attempts to minimize the changes from one age zone to the next at each depth, the anomalous velocity changes in the model may be somewhat smeared out over a greater depth range than in the actual structure. The integrated delay time is well-constrained by the data, however, so that if the melt and anomalous velocities were confined to the 30 to 70 km depth range, for example, the average velocity anomaly in that range would have to be about 0.40 km/s, averaged over the 0 to 4 Ma age zone.

The percentage of melt required to be present to explain the anomalous velocities beneath the East Pacific Rise depends on the shape of the melt pockets. If the melt is distributed in tubes, as is predicted for dihedral wetting angles less than 60° in olivine, then there must be at least an average of 3 to 6% melt present at a depth of 45 km in the first 4 Ma (Figure 40). This volume of melt seems unlikely because if the distribution is governed by these dihedral angles, then the tubes should be

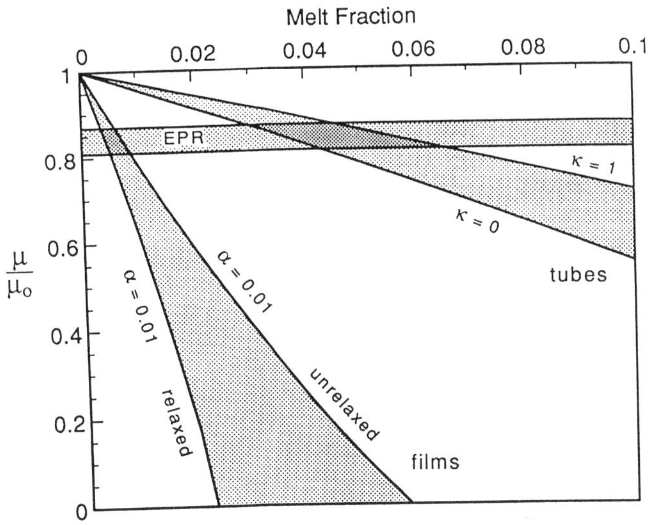

Fig. 40. Estimate of melt fraction present beneath the East Pacific Rise depends on geometry of melt distribution. Horizontal shaded region labelled 'EPR' represents ±1σ about best estimate of decrease in shear modulus at a depth of 45 km, averaged over the 0 to 4 Ma age zone. Shaded region labelled 'films' corresponds to an aspect ratio of 0.01 with bounds equivalent to completely connected or unconnected melt pockets. Bounds for tubes correspond to relaxed moduli with shape factors as illustrated in Fig. 6b.

interconnected and theory and experiment indicate that all but a very small melt fraction would be drained [see, for example, Kohlstedt, this volume]. A more plausible explanation is that some fraction of the melt is distributed in the form of thin films wetting the faces of grains. If the aspect ratio of the films were on the order of 0.01, then only about 0.5% melt is required (Figure 40). Examples of the effects of varying aspect ratios and combinations of films and tubes are given by O'Connell and Budiansky [1977], Mavko [1980] and Schmeling [1985a]. It does appear that a significant melt fraction is present in the 30 to 80 km depth range out to at least an age of 4 Ma. Simultaneous measurements of P and S-wave velocity structure could resolve much of the ambiguity concerning the form of the melt distribution and the minimum percent present.

Magnetotelluric Observations

The electromagnetic (EM) method most frequently employed to study the resistivity structure of the oceanic mantle is the magnetotelluric (MT) method, which exploits naturally occuring EM fields as sources. Wannamaker et al. [1989] give a good review of the MT method and observational techniques, which this summary follows to some extent. The ionospheric origin of the fields and the high index of refraction between air and earth make it possible to treat the incident EM fields as plane waves propagating vertically downwards. As the waves penetrate downwards into the conductor that is the earth, they decay exponentially in amplitude. In a uniform, conducting half-space, the skin depth δ, the characteristic decay distance at which the amplitude is reduced to 1/e of its surface value, is given by

$$\delta = \left(\frac{2}{\mu\omega\sigma}\right)^{1/2} \tag{25}$$

where μ is the magnetic permeability of the medium, σ is the electrical conductivity, and ω is the angular frequency, or approximately $\delta = 503\ (\rho T)^{1/2}$ in meters, where the resistivity ρ (1/σ) is given in ohm meters (Ωm) and the period T in seconds. By utilizing waves over a broad range of frequencies penetrating to different depths in the earth, the resistivity structure as a function of depth can be sampled. Because the amplitude is attenuated more rapidly with depth in a good conductor, it is difficult to probe beneath a highly conducting layer. Seawater is a good conductor (ρ = about 0.23 Ωm) so only the longest periods penetrate beneath the ocean into the underlying seafloor, limiting the period range and resolution of the MT method. Nevertheless, it is one of the few techniques available for sampling the lithosphere and asthenosphere. In the oceans, it is most successful at determining the depth to a conducting layer beneath a more resistive layer (partial melt zone or high temperature region beneath the cooler lithosphere).

The relationship between the horizontal components of the electrical (E) and magnetic (H) fields for a vertically propagating wave is described by

$$\begin{bmatrix} E_x \\ E_y \end{bmatrix} = \begin{bmatrix} Z_{xx} Z_{xy} \\ Z_{yx} Z_{yy} \end{bmatrix} \begin{bmatrix} H_x \\ H_y \end{bmatrix} \tag{26}$$

where Z_{ij} are complex impedances or transfer functions that are a function of frequency and may vary with the orientation of the x-y coordinate system. In a one-dimensional medium that varies only with depth, $Z_{xx} = Z_{yy} = 0$, and $Z_{xy} = -Z_{yx}$. The complex response of the medium is usually described by a phase function and the apparent resistivity ρ_a, which for any single measurement at a site at a particular frequency is the resistivity of the uniform half-space that would yield the observed amplitude of the transfer function. For Z_{xy} for example, the apparent resistivity is

$$\rho_a = \rho_{xy} = \frac{1}{\omega\mu_0}|Z_{xy}|^2 \tag{27}$$

where μ_0 is the magnetic permeability of free space. Over a uniform half-space, the phase is 45°. A crude estimate of the depth to a structure is obtained by substituting ρ_a into the formula for skin depth.

A two-dimensional medium sampled at any one site is anisotropic; Z_{xy} does not equal $-Z_{yx}$. Usually the coordinate axes are chosen with x parallel to structure, in which case $Z_{xx} = Z_{yy} = 0$, decoupling the total fields into two independent modes: the transverse electric (TE or E polarization) with E_x and H_y and the transverse magnetic (TM) with E_y and H_x. If the structure is actually three-dimensional, at a single site the apparent direction of strike inferred from the separation of the horizontal fields into independent modes may vary with period. In principle, useful information about 2-D structures can come from 1-D interpretation of TE data, but if there are variations along strike, then the TE data may be contaminated by TM mode behavior [Booker and Chave, 1989]. TM data are interpretable only with 2-D models, but are less contaminated by the presence of 3-D structure. 1-D inversions of TE data are often used as starting models for full 2-D inversions of MT data collected at many sites along a profile. A closely related technique is to examine the transfer function between the vertical and horizontal components of the magnetic field, an approach that is particularly useful in three-dimensional structures and one that requires only measurements of the magnetic field.

One dimensional inversions for the change in conductivity structure with age of the seafloor were executed by Oldenburg [1981] and Oldenburg et al. [1984] using data gathered by Filloux [1977, 1980] and Law and Greenhouse [1981]. The three sites were in the North Central Pacific (NCP) on seafloor about 70 Ma old, off the coast of California (CAL) on seafloor about 30 Ma old, and near the Juan de Fuca ridge (JDF) on seafloor about 1 Ma old. The data (Figure 41, plotted as $(Z_{yx})^{-1}$ instead of the traditional ρ_a) show a systematic change from NCP to CAL to JDF that corresponds to an increase in apparent conductivity or decreasing resistivity with decreasing age. Defining exactly what structure is responsible for the change is much more difficult. Figure 42 shows models derived

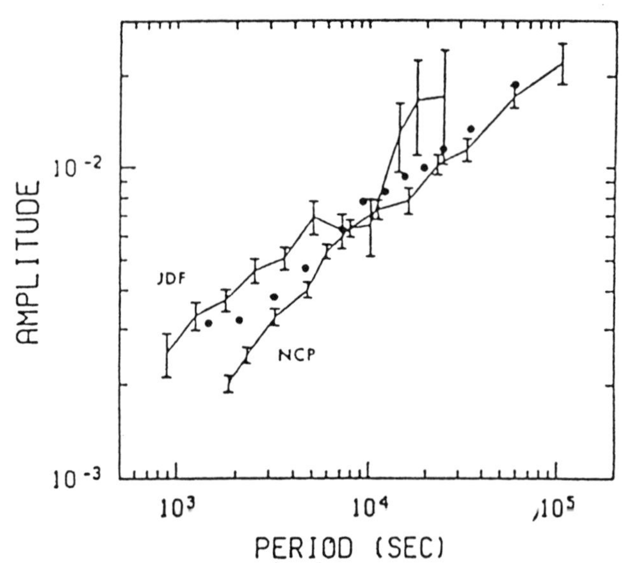

Fig. 41. Amplitude of the magnetotelluric transfer function as a function of period for three sites in the Pacific [from Filloux, 1980]. JDF represents a site near the Juan de Fuca Ridge, NCP a site on 70 m.y. old seafloor in the north central Pacific, and the dots represent data from a site off the coast of California (CAL). Vertical error bars represent one standard deviation.

using three different techniques that all fit the data acceptably. There are several reliable conclusions possible from this data set. First, the conductivity increases with increasing depth, as expected if the temperature increases with depth. Second, the depth of 35 to 50 km to a good conductor beneath JDF is shallower than beneath either CAL or NCP. Third, the average conductivity from about 50 to 100 km beneath NCP is less than beneath either CAL or JDF. The minimum conductivity in the 50 to 100 km range is high, about 0.08 S/m, which strongly suggests, but may not require, the presence of melt. Oldenburg et al. [1984] found that at least 5% melt was required. They assumed, however, a functional relationship between temperature and melt fraction based on melting of a pyrolite mantle, so that if the temperature in the model increased, the melt fraction must also. If melt is allowed to be removed from the mantle, then higher temperatures could trade-off with the melt fraction actually present.

The one-dimensional inversion illustrates the limitations in resolution of the MT method in the oceans. The period range is severely limited by the attenuation in the water column, thus limiting the possible information content, particularly at shallow depths. A crude guide is that it is possible to resolve approximately 1 piece of independent information about the conductivity structure of the earth per decade period range. The data in Figure 41 have relatively large errors. The data for JDF and CAL can be fit within the errors by a straight line, which means that a two-parameter model should be sufficient to match the data. It should not be surprising then that a two-layer model provides an adequate fit (Figure 42, right column). Inversion of MT data is in many ways similar to inversion of surface wave data, with one important difference. Linearized resolution kernels can be constructed for the non-linear problems in both cases, but these are not reliable indicators of the resolution for the MT problem. The non-linearity in both cases arises from the dependence of the response functions on the structure, but in the case of surface waves, we are confident that the variations in shear velocity are relatively small, preserving a sembleance of linearity. In the MT method, we expect order of magnitude variations in conductivity that render the linearizing assumption of small perturbations invalid.

A two-dimensional MT study of the Juan de Fuca ridge area has been carried out as part of the EMSLAB project [see Booker and Chave, 1989, for an overall description of EMSLAB]. A map of the distribution of instruments is shown in Figure 43. Most are magnetometers only, but a dense line of complete magnetotelluric instruments on land and a set of 8 sites extending out to the Juan de Fuca ridge form a profile, called the 'Lincoln line', from which a two-dimensional cross-section has been constructed [Wannamaker, 1989b]. Apparent resistivity measurements ρ_{xy} and ρ_{yx} are shown in Figure 44 for site SF7 in the middle of the Juan de Fuca plate [Wannamaker et al., 1989a]. The errors are much smaller and the period range somewhat larger than in the earlier Law and Greenhouse [1981] experiment. The effects of two-dimensionality are clear at this single site in the azimuthal

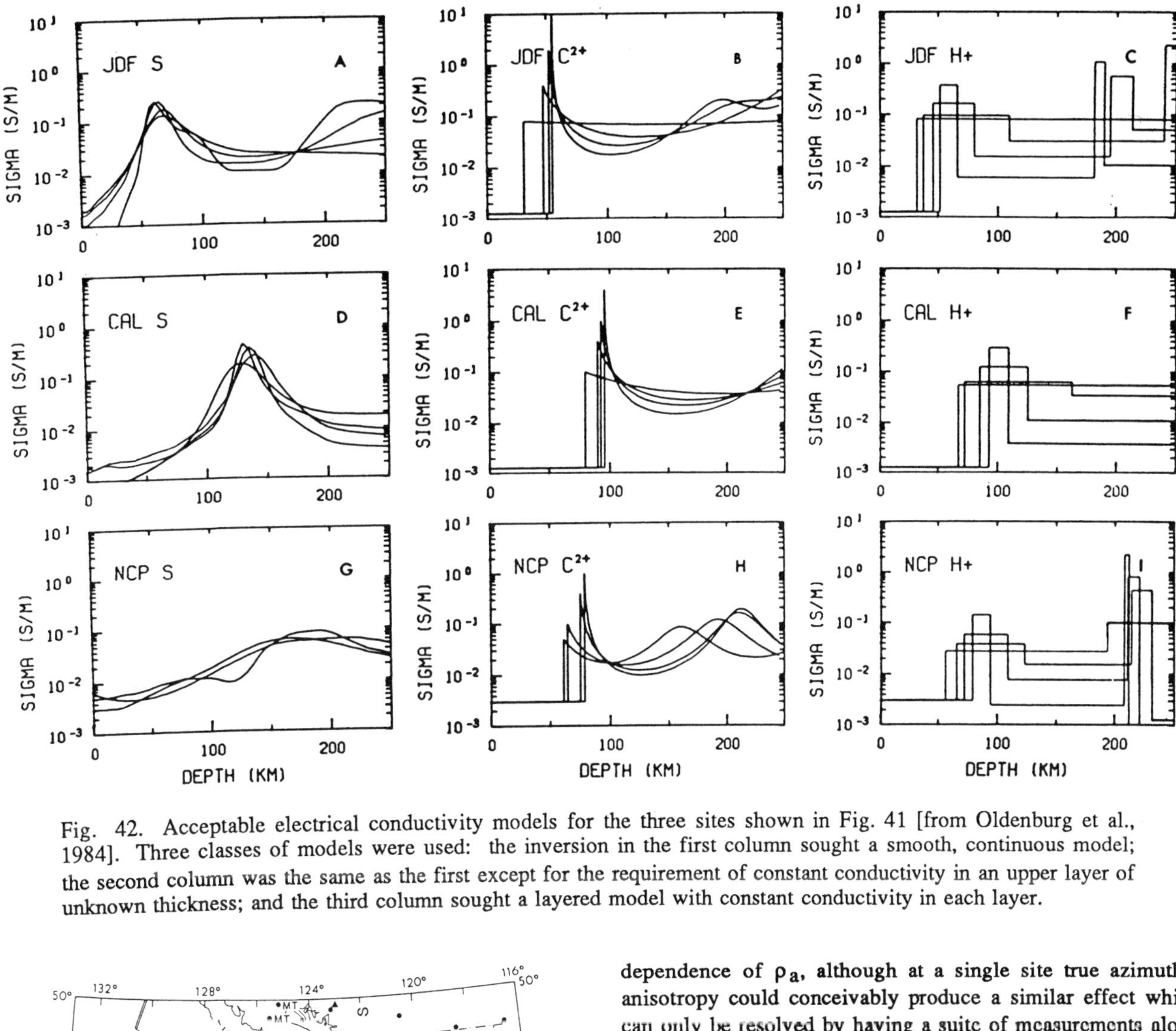

Fig. 42. Acceptable electrical conductivity models for the three sites shown in Fig. 41 [from Oldenburg et al., 1984]. Three classes of models were used: the inversion in the first column sought a smooth, continuous model; the second column was the same as the first except for the requirement of constant conductivity in an upper layer of unknown thickness; and the third column sought a layered model with constant conductivity in each layer.

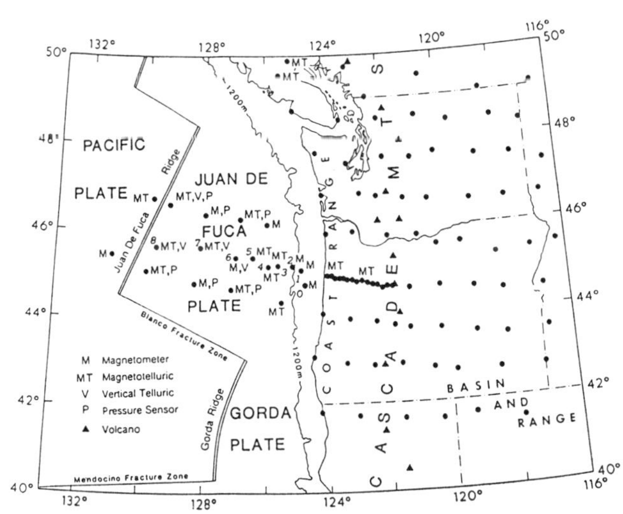

Fig. 43. Location of instruments in the EMSLAB array [from Wannamaker et al., 1989a]. Labels on dots indicate type of instrumentation according to key in lower left. The EW line of dots in western Oregon represents magnetotelluric sites in the Lincoln line. Unlabeled dots on land indicate locations of three-component magnetometers.

dependence of ρ_a, although at a single site true azimuthal anisotropy could conceivably produce a similar effect which can only be resolved by having a suite of measurements along a profile. Resistivities within the earth must be at least as low as the minimum apparent resistivity of 8 - 10 Ωm, or conductivities at least as high as 0.1 S/m. Variations in ρ_{yx} and the phase ϕ_{yx} as a function of position along the oceanic portion of the profile are shown in Figure 45. Most of these variations are attributable to the transition from oceanic structure to continental structure. A 2-D resistivity cross-section constructed from both land and seafloor data (Figure 46) shows oceanic structure similar to that found by Oldenburg et al. [1984]; a resistive lithosphere giving way to a conductive (about 0.1 S/m) asthenosphere at a depth of about 35 km [Wannamaker et al., 1989b]. This 2-D model fits the apparent anisotropy at SF 7 quite well (Figure 47). There also

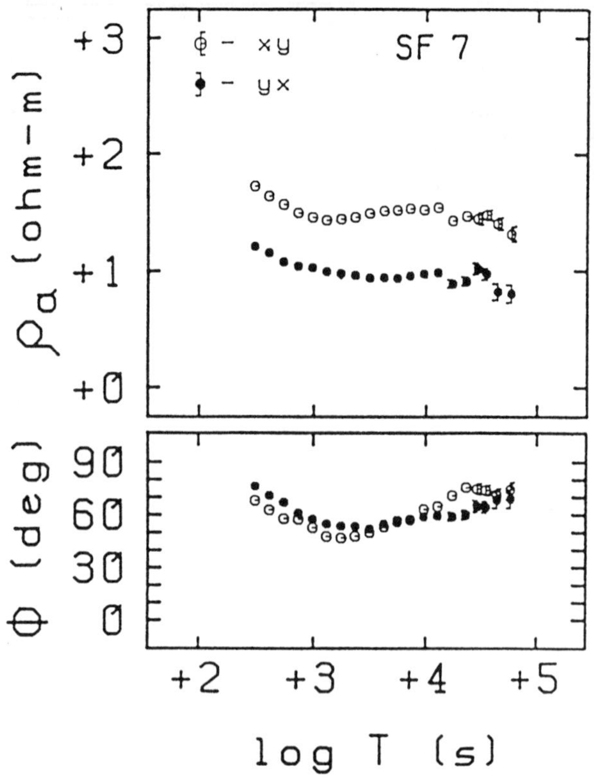

Fig. 44. Apparent resistivity and impedance phase responses at seafloor site 7 in the Juan de Fuca plate [from Wannamaker et al., 1989a]. Error bars representing one standard deviation are shown when they exceed the size of the plotting symbol. Note apparent anisotropy of apparent resistivity response for the two orthogonal components.

is a decrease in resistivity below 200 km. This decrease in resistivity at great depths is a global feature; the lower mantle is quite conductive. No one really knows why this is true, which makes the interpretation of moderately high conductivities in the 100 to 200 km depth range beneath the oceans in terms of a small fraction of partial melt more uncertain than is desirable.

There have been several MT surveys in Iceland primarily for the purposes of studying volcanic-hydrothermal systems, but extending to long enough periods to sample structure into the mantle [Beblo and Bjornsson, 1978, 1980; Haak, 1980; Hermance, 1981; Thayer et al, 1981; Beblo et al., 1983; Hersir, et al., 1984; Eysteinsson and Hermance, 1985]. Land measurements allow sampling over a broader period range and thus more structure can be seen in the raw data (Figure 48) and better resolution of shallow structure achieved. A fully 2-D inversion of data from a profile of 15 stations across the eastern neo-volcanic zone in southern Iceland yields the cross-section shown in Figure 49 [Eysteinsson and Hermance, 1985]. The conductive layer at the base of the crust is widespread across Iceland. It has been found in every survey and is generally interpreted in terms of the presence of partial basaltic melt. Note that the conductivities are particularly low beneath two active volcanic centers (Figure 49). Hermance [1981] suggests that the melt does not represent *in situ* melting of the mantle but pooling of melt formed deeper in the mantle beneath the lower density crust. Beblo and Bjornsson [1980] estimate temperatures near 1000 °C and melt fractions of about 10% to explain resistivities of about 15 Ωm. Resistivities of 5 Ωm in other surveys yield estimates of temperature of 1100-1300 °C and melt fractions of 10-20% [Hersir et al. 1984; Eysteinsson and Hermance, 1985; Schmeling, 1985b]. In this setting, reducing the estimated melt fraction by trading off with higher temperatures is not possible because higher temperatures would melt the overlying basaltic crust. The layer with melt could be thinner with higher conductivity and still yield the same fit to the data, which would imply even higher concentrations of partial melt.

CRUSTAL THICKNESS

One of the strongest constraints on mantle flow and melt generation is the resulting distribution of the melt products at

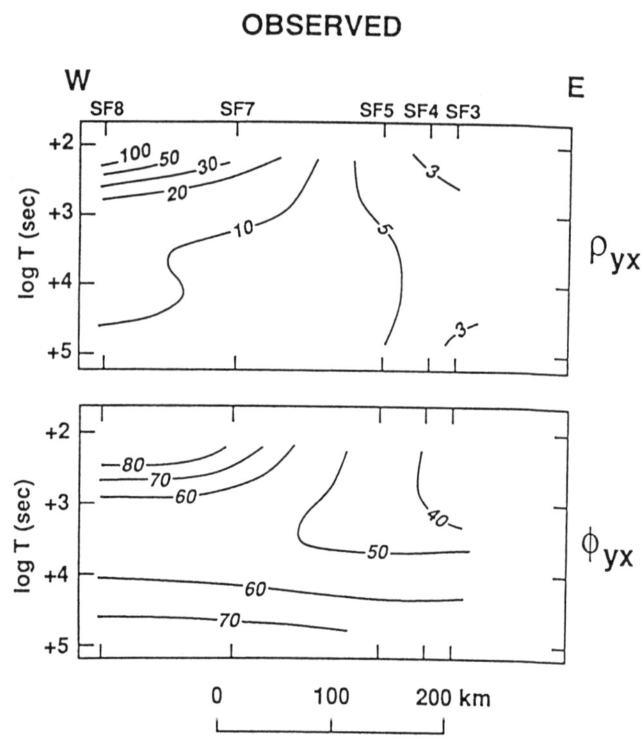

Fig. 45. Pseudosections of transverse magnetic apparent resistivity ρyx and impedance phase from the five deepwater, seafloor sites on the Lincoln line [from Wannamaker et al., 1989a]. Increasing period is a proxy for increasing depth, so that these contours of observed amplitude and phase give an indication of lateral and vertical changes in resistivity structure.

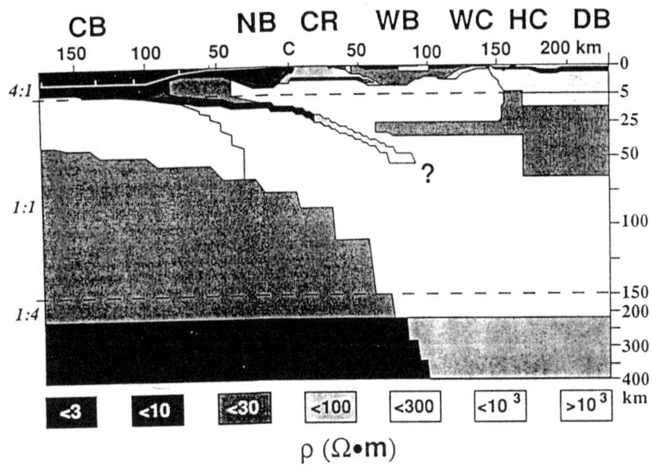

Fig. 46. East-west resistivity cross-section based on both seafloor and land stations along the Lincoln line [from Wannamaker et al., 1989b]. Data are matched by altering a series of forward models, so the uniqueness of details of structure is not established. Note changes in scale of vertical exaggeration labelled on left. Changes in structure beneath the seafloor beyond left edge of diagram are minor.

the surface. The thickness of the igneous crust provides a time-integrated measure of the volume of melt extracted from the mantle. Because melt may migrate horizontally before reaching crustal levels or flow laterally in dikes and surface flows or be trapped in pockets and pores within the mantle, there is probably not be a one-to-one mapping of crustal

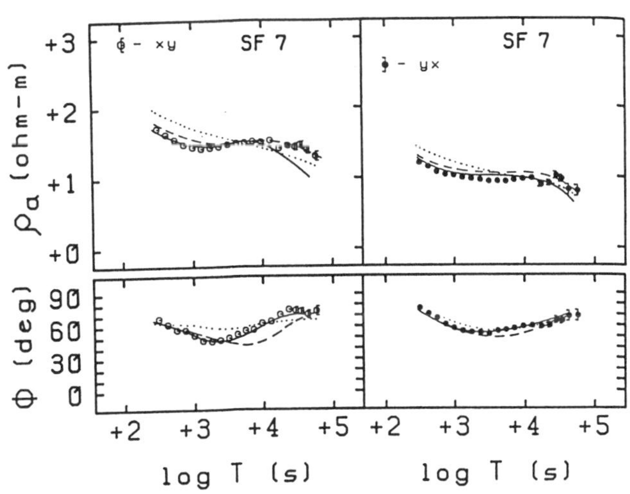

Fig. 47. Apparent resistivity and phase data at SF7 computed for the model shown in Fig. 46 (solid lines) compared to the observations at that site. Dotted and dashed lines show the effects of altering the assumed conductivity of the sediments from the trench to the coastline, indicating the importance of a large array for constraining lateral variations that influence the apparent resistivity at remote sites.

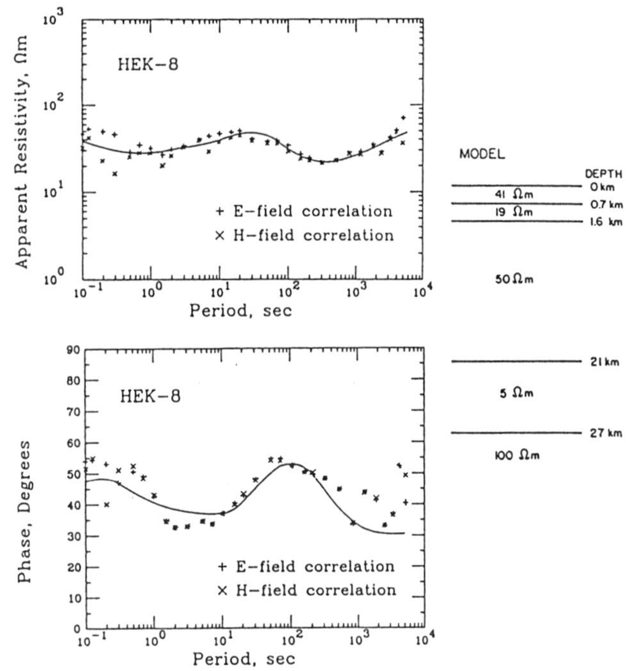

Fig. 48. Apparent resistivity and phase for transverse electric polarization at site HEK-8 located close to the eastern boundary of the neovolcanic zone in Iceland [from Eysteinsson and Hermance, 1985]. The solid lines show the response for the one-dimensional model shown at the right.

thickness to melt-production. For this reason, I have indicated in Table 1 that both basalt chemistry and crustal thickness have limited lateral resolution of processes in the melt production zone. Nevertheless, both the total volume of melt

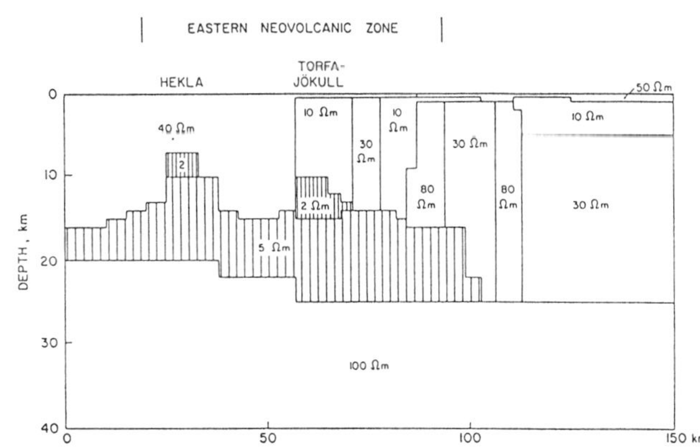

Fig. 49. Two-dimensional resistivity model across the eastern neovolcanic zone in Iceland [Eysteinsson and Hermance, 1985]. Shaded regions have resistivity so low that the presence of a connected fluid (melt) is required. This model is based on data from 14 magnetotelluric stations along the profile.

extracted and its lateral distribution provide powerful constraints on and clues to the form of mantle flow.

Crustal thickness and spreading rate. It has long been known that the oceanic crust is approximately 6 km thick [e.g., Raitt, 1963]. This in itself forms a powerful constraint on models of mantle flow and melting: for example, if melt is extracted from the upwelling mantle beginning at a depth of about 60 km, then the average degree of depletion of the residual mantle above that depth must be on the order of 10%. Except perhaps at very slow spreading rates [Reid and Jackson, 1981], the thickness is essentially independent of spreading rate (Figure 50) [Reid and Jackson, 1981; Chen, 1992]. This means that melt is produced and extracted at a rate directly proportional to the spreading rate, as might be expected if the average degree of melting is constant and the volume of mantle that undergoes decompression melting is simply proportional to the spreading rate.

The simplest theoretical models, however, predict that the crust should be thinner at slow spreading rates than at fast spreading rates. In a passive spreading model in which flow in a uniform viscosity mantle is driven by the separation of two plates with flat bottoms, the form of the flow is independent of spreading rate or viscosity and the velocity scales with the spreading rate [Reid and Jackson, 1981]. The volume of upwelling mantle then is directly proportional to the spreading rate. Some heat in the upwelling mantle is lost by conductive cooling to the seafloor. These models predict that the average degree of partial melting decreases with decreasing

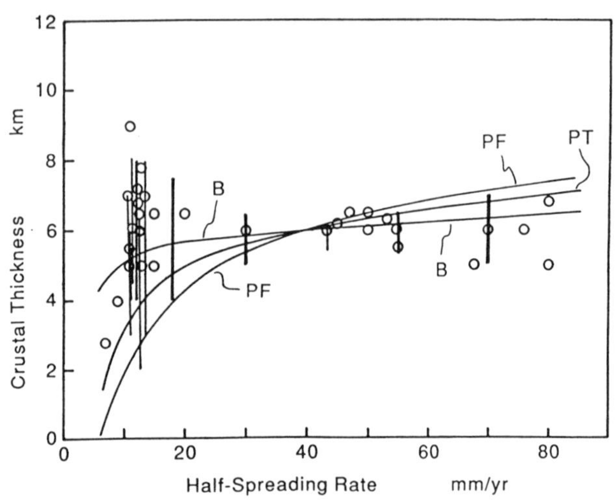

Fig. 50. Oceanic crustal thickness versus spreading rate. Observed data from Chen [1992]. Circles indicate seismic refraction measurements from individual locations, vertical bars represent range of thicknesses found within a regional survey. Solid lines represent predictions of three theoretical models: PF is a passive flow model with flat surface plates [Phipps Morgan and Forsyth, 1987]; PT is a passive flow model with plates thickening with age [Shen and Forsyth, 1992]; and B is a model including the effects of buoyancy [Sotin and Parmentier, 1989].

spreading rate (Figure 50), because with slower upwelling the conductive loss of heat to the seafloor extends to greater depth, cooling the rising mantle until the geothermal gradient is greater than the gradient of melting, terminating melt production [Reid and Jackson, 1981]. The exact form of the variation with spreading rate depends on the depth of the onset of melting, the shape of the solidus as a function of temperature, and the fraction of melt as a function of temperature above the solidus. In these models, the temperature of the upwelling material and the melting relationships are usually adjusted to match the rate of crustal production at some given spreading rate. The spreading rate dependence is reduced if melting begins at greater depth, because then a larger fraction of the melting occurs below the depth affected by conductive cooling (Figure 18). If melting is confined to depths shallower than about 80 km, the predicted spreading rate dependence clearly does not agree with the observed constancy of crustal thickness. There are several possible explanations for why the crustal thickness is nearly constant. If melting normally proceeds until all the diopside is depleted and the system involves fractional melting in which the melt is removed as fast as it is created, then the primary control could be compositional [Hess, 1991]. The upwelling matrix may already have been depleted of all the easily melted components and nearly ceased melting before it reaches depths which are affected by conductive cooling at slower spreading rates. Even if the "diopside-out" point is not reached, a non-linear relationship between melt fraction and potential temperature above the solidus in which it becomes progressively more difficult to melt as melting proceeds [e.g. McKenzie and Bickle, 1988] would decrease the dependence on spreading rate.

Other explanations appeal to upwelling velocities that are faster than the plate velocity, reducing the depth of penetration of cooling. In the flat-plate, passive model the maximum upwelling rate is just $2/\pi$ times the spreading rate. If the effective viscosity of the asthenosphere beneath the ridge is less than about 10^{20} Pa s, then buoyancy can enhance convection, increasing the upwelling rate and decreasing the dependence of crustal thickness on velocity (Figure 50 and Sotin and Parmentier, 1989]. The source of the buoyancy may be a combination of a small amount of trapped melt, chemical differentiation leaving a depleted mantle, and thermal buoyancy [Rabinowicz et al., 1984; Rabinowicz et al., 1987; Scott and Stevenson, 1989; Buck and Su, 1989; Sotin and Parmentier, 1989]. Another mechanism that can increase the upwelling rate is the thickening of the plate with age [Blackman and Forsyth, 1991b] or, equivalently, temperature-dependent viscosity [Cordery and Phipps Morgan, 1991; Shen and Forsyth, 1992]. The thickening of the plate with age narrows the region in which upwelling occurs and requires that the horizontal velocity equal the surface velocity at depth within the plate. Both of these factors require an increase in the upwelling velocity, with greater effect at slower spreading rates when the plate thickens more rapidly with distance. The net result of any increase in upwelling rate is a more constant crustal thickness (Figure 50).

Another possible way of producing faster upwelling rates and more uniform crustal thickness is episodic upwelling, possibly associated with a non-linear rheological response. At slow spreading rates, steady-state thermal models predict that there will be a lithosphere of significant thickness even at the axis of the spreading center (and no steady-state magma chamber [Sleep, 1975]. In order for this lithosphere to extend and for spreading to proceed at reasonable stress levels, there must be some mechanism of non-linear yielding, such as faulting, cracking, or plastic yielding. This non-linear response to extensional stress is likely to produce episodic extension and upwelling. Scott and Stevenson [1989] found that models including the buoyancy effect of unsegregated melt became unstable at slow spreading rates, resulting in episodic upwelling. Sotin and Parmentier [1989] also found a component of episodicity in models including compositional buoyancy caused by segregation of melt. If upwelling is episodic, then the upwelling rate during the active phase could be much faster than the average rate, producing decompression melting that would be nearly independent of spreading rate.

The density changes accompanying depletion of the mantle matrix (see earlier discussion of Physical Properties) may seem small, but a change in density of 0.01 Mg cm^{-3} is roughly equivalent to the effect of thermal expansion accompanying a 100 °C temperature change and can have a significant effect on buoyant flow. The flow pattern may be influenced in several ways by compositional buoyancy. A purely passive flow model predicts horizontal gradients in depletion in the vicinity of a spreading center (Figure 51). Below a depth of 30 km, the greater depletion beneath the axis enhances the buoyancy, which would tend to increase the upwelling rate directly beneath the axis, but decrease it off-axis, thus tending to flatten out the contours of equal depletion [Sotin and Parmentier, 1989; Parmentier and Phipps Morgan, 1990]. The reversal in sign of the density effect of depletion in the plagioclase stability field would tend to oppose this buoyant component of convection, but the across-ridge gradients in depletion are expected to be weak at depths shallower than 30 km (Figure 51), so the effect of the reversal may be small.

In a passive flow model [Phipps Morgan and Forsyth, 1988], there is expected also a strong gradient in degree of depletion from the center of a spreading segment towards a fracture zone (bottom, Figure 51). The thermal gradient along the ridge toward the fracture zone would ordinarily tend to enhance upwelling beneath the center of a segment and further decrease upwelling near fracture zones, but the strong depletion gradient at shallow depths would oppose the thermal buoyancy and may squelch the expected thermal convection. There could be a sudden threshold above which this thermal convection does become effective, however, because once established, upwelling beneath the center of the segment with flow toward the fracture zone would carry depleted mantle with it, decreasing the depletion gradient. A similar interaction between thermal and compositional buoyancy may delay the onset of small-scale convection in the cooling thermal boundary layer and produce episodic upwelling at the ridge [Sotin and Parmentier, 1989; Sparks and Parmentier, 1992].

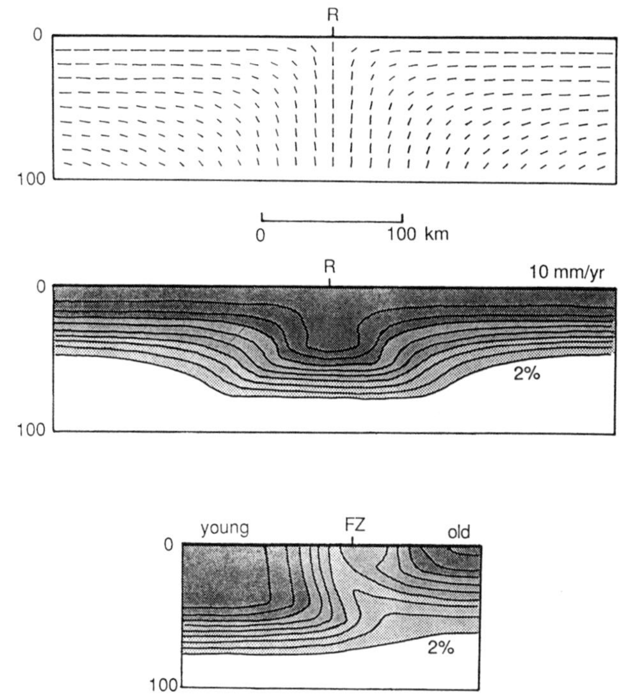

Fig. 51. Contours of depletion in the mantle beneath a spreading center predicted for pressure-release melting in a passive-spreading model [Phipps Morgan and Forsyth, 1988]. Upper diagram shows flow vectors in cross-section across ridge. Length of symbol proportional to velocity with longest symbol 10 mm/yr (assumed spreading rate). R marks location of spreading axis. Middle diagram shows contours of equal depletion in a cross-section far from any fracture zone, assuming complete removal of melt. Contour values represent maximum percentage of melting undergone by material anywhere along its flow path (2% contour interval). Bottom diagram shows depletion in a cross-section along a ridge axis (young) crossing a fracture zone (FZ) and passing into older seafloor on the other side (100 km offset assumed on FZ).

Crustal thickness, basalt chemistry and axial depth. Global correlations between the depth of the spreading axis and the Na_8, Fe_8 and CaO/Al_2O_3 content of basalts were found by Klein and Langmuir [1987]. Na_8 and Fe_8 are the Na_2O and FeO contents corrected for the effects of shallow, fractional crystallization to a reference composition of 8% MgO (by weight). Na_2O is nearly perfectly incompatible with olivine, highly incompatible with orthopyroxene, and thus resides primarily in clinopyroxene at depths greater than the plagioclase stability field [Presnall et al., 1979]. As melting proceeds, Na_2O enters the melt with an initial high concentration and then is diluted as more melt is added. Because the distribution coefficient between liquid and solid is nearly independent of pressure [Jaques and Green, 1980], Na_2O content is an excellent indicator of the extent of partial melting, with lower Na_8 corresponding to a greater degree of partial melting. The distribution coefficient for FeO is on the order of 1.0 so its concentration is not very dependent on melt fraction, but it is dependent on pressure, making Fe_8 content a useful indicator of the depth of fractionation of the melt, with higher Fe_8 corresponding to a greater depth of fractionation.

Klein and Langmuir [1987] found a positive correlation between depth of the spreading center and Na_8 and a negative correlation between depth and Fe_8. A negative correlation between Na8 and crustal thickness was also reported. These relationships are what is predicted if the potential temperature of the upwelling mantle varies, with relatively constant composition of the source (Figure 18). If the potential temperature is greater, then the solidus is encountered at greater depth during adiabatic upwelling yielding greater average depth of melting and greater average and maximum degree of partial melting. The thicker crust, thicker zone of depleted, lower density residual mantle, and thermal expansion of the hotter mantle all combine to produce a shallower seafloor.

A re-examination of the crustal thickness data with added information from more seismic refraction experiments shows, however, that the trend of crustal thickness vs. Na_8 is much weaker than predicted by Klein and Langmuir [1987] or McKenzie and Bickle [1988] for the normal range of basalts with Na_8 content between about 2.0 and 3.0% (Figure 52a). Only the points from Iceland and the Cayman trough support the predicted trends. The estimate of thickness of the crust for the latter was based on geological inference rather than seismic refraction data and Iceland is probably shifted off the general trend by an enriched mantle source. Although it is logical that there should be a relationship between degree of partial melting and crustal thickness, seismic refraction data from "normal" spreading centers exhibit relatively weak variation in average crustal thickness and do not follow the trends predicted.

The small variation in crustal thickness accompanying a variation in Na_8 content of 50% or more provides a powerful constraint on the form of mantle upwelling, strongly supporting the existence of dynamic upwelling beneath the ridge axis. Before examining the predictions of various models, it is worth considering two alternative explanations. First, the variations could be caused by compositional variations in the mantle source. Klein et al. [1991] show that upwelling with uniform temperature in a mantle with heterogeneous composition ranging from depleted to enriched cannot produce the observed negative correlation between Na8 and Fe_8. If there were a systematic negative correlation between potential temperature and enrichment, however, it might be possible to combine compositional and thermal effects to produce the observed relationships between crustal thickness and composition. But, in the asthenosphere, enriched, colder mantle would be significantly denser than depleted, hotter mantle, so they would not be expected to passively reside at the same depth and stratification would result. In the absence of any physical justification or independent observational evidence for this inverse relationship, we will concentrate on other mechanisms. Another possibility is that crustal thicknesses as measured by

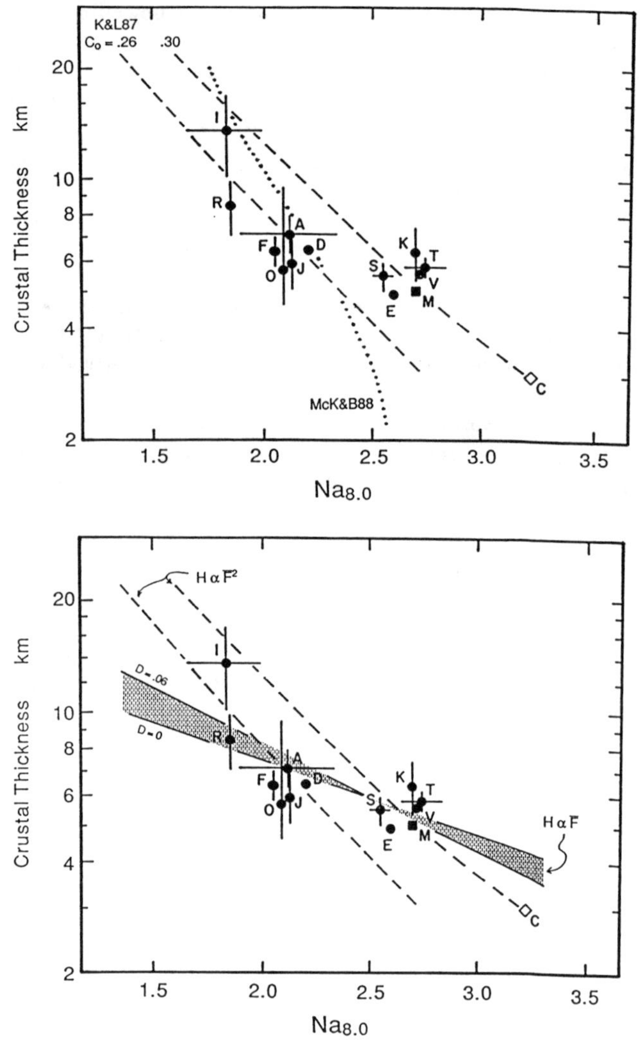

Fig. 52. Crustal thickness versus $Na_{8.0}$ at spreading centers. (a) Dashed lines indicate predictions of the column melting models of Klein and Langmuir [1987] for two different assumed initial concentrations of Na_2O in the mantle (C_O). Dotted line indicates prediction of McKenzie and Bickle [1988] model. Data are from the compilation of Klein and Langmuir except for the addition of data from spreading centers near the Oceanographer fracture zone region (indicated by letter O, from Sinha and Louden, 1983, and Ambos and Hussong, 1986]; Vema FZ [V, Louden et al, 1986]; Kane FZ [K, Purdy and Detrick, 1986]; and 33°S in the South Atlantic (T, Tolstoy et al., 1990; Michael et al., 1990]. Other points of particular importance are C (Cayman trough), I (Iceland), and R (Reykjanes Ridge). Open symbol for Cayman trough indicates that the crustal thickness is estimated from geological arguments. (b) Dashed lines are predictions of passive flow models in which variation of the potential temperature of upwelling leads to crustal thickness varying as the square of the average degree of melting. Shaded region bounded by solid lines is variation expected if crustal thickness is proportional to the average degree of melting. D represents possible solid/liquid partition coefficients for Na_2O.

seismic refraction systematically underestimate the volume of melt delivered to the crust: hotter mantle temperatures could produce magmas that crystallize greater volumes of ultramafic cumulates that are misinterpreted as residual mantle [Klein et al., 1991b]. This could be a factor, but it could not possibly account for the difference between the observations and the trends that predict a doubling of crustal thickness accompanying a change in Na_8 content from 2.7% to 2.1%.

The observed relationship between Na_8 and crustal thickness is roughly consistent with the volume of melt extracted being directly proportional to the degree of melting (Figure 52b). The Klein and Langmuir [1987] and McKenzie and Bickle [1988] trends are based on passive spreading models that predict that the thickness of crust or volume of melt is proportional to the square of the average degree of melting. In a linear melting model in which the degree of melting is controlled by the potential temperature of the upwelling mantle (Figure 18), increasing the temperature increases the average degree of melting and the depth of intersection of the adiabat with the solidus. Increasing the average degree of melting increases the volume of melt produced from each element of upwelling mantle, but in passive spreading models, the volume of mantle that undergoes melting is also directly proportional to the depth of the onset of melting (Figure 53), neglecting cooling to the surface, thus further increasing the total volume of melt and giving rise to the squared relationship between degree of melting and crustal thickness. In a dynamic upwelling model with enhanced upwelling beneath the axis and a component of downward flow off-axis, the volume of mantle undergoing melting is not simply proportional to the depth of onset of melting (Figure 53), and in the limit produces a total volume of melt directly proportional to the average degree of melting [Forsyth, 1992]. Thus, if variations in crustal thickness and composition are primarily controlled by variations in mantle potential temperature, then the small variation of crustal thickness with Na_8 content strongly favors some form of dynamic upwelling beneath the ridge.

Width of the crustal accretion zone. Almost all the oceanic crust is emplaced within a very narrow zone at the ridge axis. The neovolcanic zone at the surface is only 1 to 2 km wide, independent of spreading rate (Figure 54) [Macdonald, 1982].

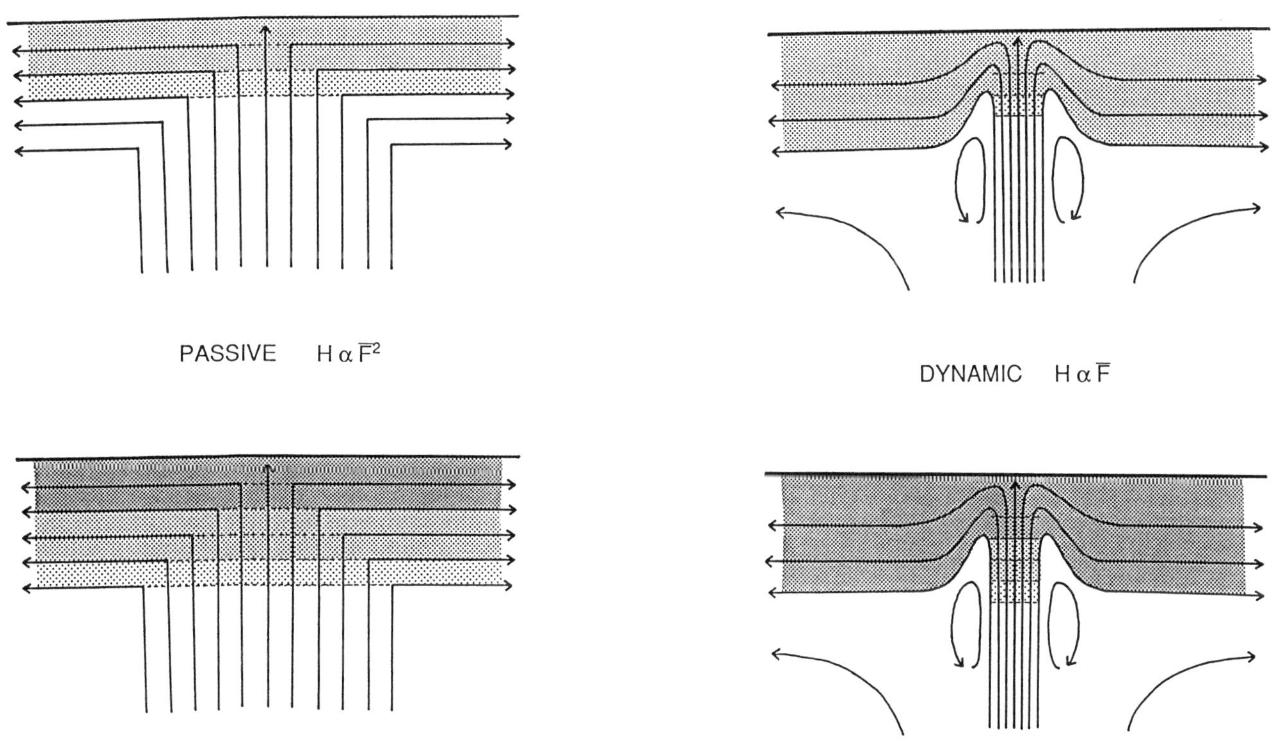

Fig. 53. Schematic depletion of the mantle in passive (left) and dynamic (right) flow models of a spreading center, assuming fractional melting. In these cross-sections of the ridge, flow lines are indicated by solid lines and maximum degree of melting experienced by each volume element within the mantle is indicated by shaded regions, with darker shading corresponding to higher percentage of melting. Ridge axis is at the center of each diagram. Top figures indicate maximum melting or depletion with low potential temperature and shallow intersection of the adiabat and the solidus (Fig. 18). Bottom figures show depletion with higher potential temperature and deep intersection of the adiabat and the solidus. Unlike the passive model, the volume of material that is partially melted in a dynamic model does not necessarily increase with increasing potential temperature.

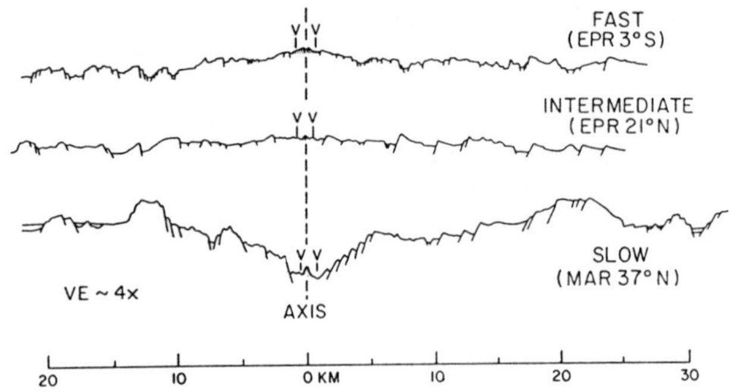

Fig. 54. High-resolution, deep-tow profiles of mid-ocean ridges at fast, intermediate and slow spreading rates (from Macdonald, 1982]. The narrow neovolcanic zone is bracketed by Vs on each profile.

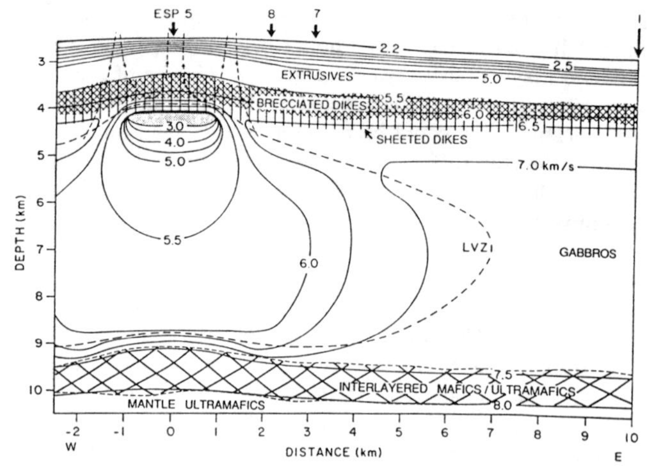

Fig. 56. Contour map of seismic velocity model of the East Pacific Rise based on seismic reflection and refraction data [from Vera et al., 1990]. Similar models have been presented by Burnett et al. [1989] and Toomey et al. [1990]. Small dashed lines and arrows indicate hydrothermal circulation cells. Shaded, low-velocity area indicates zone of magma or material with a high percentage of melt. LVZ stands for low-velocity zone, which is a zone of hot-rock with little, if any, partial melt present. Vertical arrows at top of figure indicate location of expanding spread profiles.

Within this zone, submersible observations and deep-towed cameras reveal fresh, glassy lava flows and almost no sediment cover. At both fast and slow spreading rates, there are usually axial volcanoes or ridges up to 2 km wide. The width of magnetic anomaly transitions is also consistent with a very narrow zone of accretion [Harrison, 1968; Atwater and Mudie, 1973; Klitgord et al., 1975; Macdonald, 1977; Macdonald et al., 1980; Sempere et al., 1987], although in some cases, both from magnetic anomalies and the apparent rafting off of an intact axial volcanic ridge, there are indications that the location of the neovolcanic ridge may jump as much as a few km [Macdonald, 1977; Ballard and van Andel, 1977; Sempere et al., 1987; Grindlay et al., 1991].

Seismic measurements indicate that the crust reaches nearly full thickness within 2 km of the axis. Seismic refraction measurements in median valleys on the Mid-Atlantic Ridge indicate that the crust has near-normal velocity structure and thickness essentially on axis, although sometimes the mantle velocity is lower than normal [Fowler, 1976, 1978; Louden et al., 1986; Purdy and Detrick, 1986; Toomey et al., 1988; Kong et al., 1991]. Refraction experiments on the East Pacific Rise also find normal crustal thickness beneath or within a few km of the axis, with low crustal velocities in a 5 to 10 km wide region in the mid-crust surrounding a thin, magma chamber [Orcutt et al., 1976; Rosendahl et al., 1976; Herron et al., 1980; McClain et al., 1985; Burnett et al., 1989; Vera et al., 1990; Toomey et al., 1990]. Spectacular seismic reflection profiles across the East Pacific Rise show the Moho reflector continuing nearly to the axis (Figure 55), with the only disruption being directly beneath the apparent axial magma chamber and with no apparent thinning [Detrick et al., 1987]. It is likely that the Moho would be a continuous reflector at nearly constant depth if it were not for the difficulty of imaging it through the laterally heterogeneous velocity structure of the low-velocity zone within the crust (Figure 56).

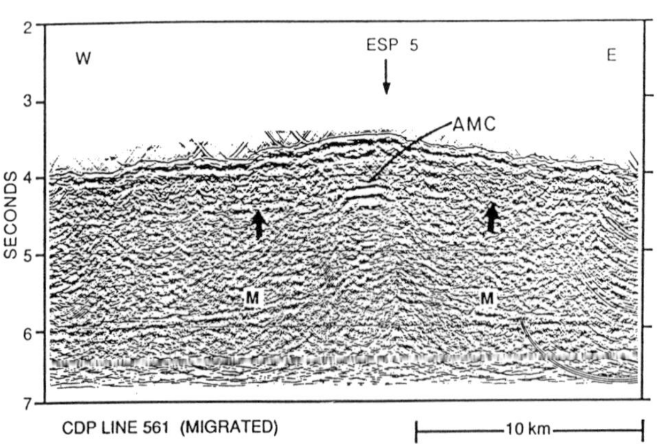

Fig. 55. Seismic reflection profile across the East Pacific Rise at 9°30'N [from Vera et al., 1990]. This common-depth-point (CDP) multi-channel data has been migrated. ESP 5 indicates the location of an expanding spread profile along the ridge axis. Axial magma chamber (AMC) and Moho (M) reflections are labelled.

The fact that the vast majority of melt delivered to the crust is emplaced within a very narrow zone at the spreading center means either that upwelling in the mantle also occurs within a very narrow zone or that there is an efficient mechanism for lateral transport of melt to the ridge axis. Buoyancy tends to cause vertical migration of the melt relative to the matrix, but passive spreading models predict that the zone of upwelling and melt production is tens to hundreds of kms wide [e.g., Reid and Jackson, 1981]. The zone of upwelling is narrowed by dynamic upwelling (Figure 57; Rabinowicz et al., 1984; Scott and Stevenson, 1989; Sotin and Parmentier, 1989] and

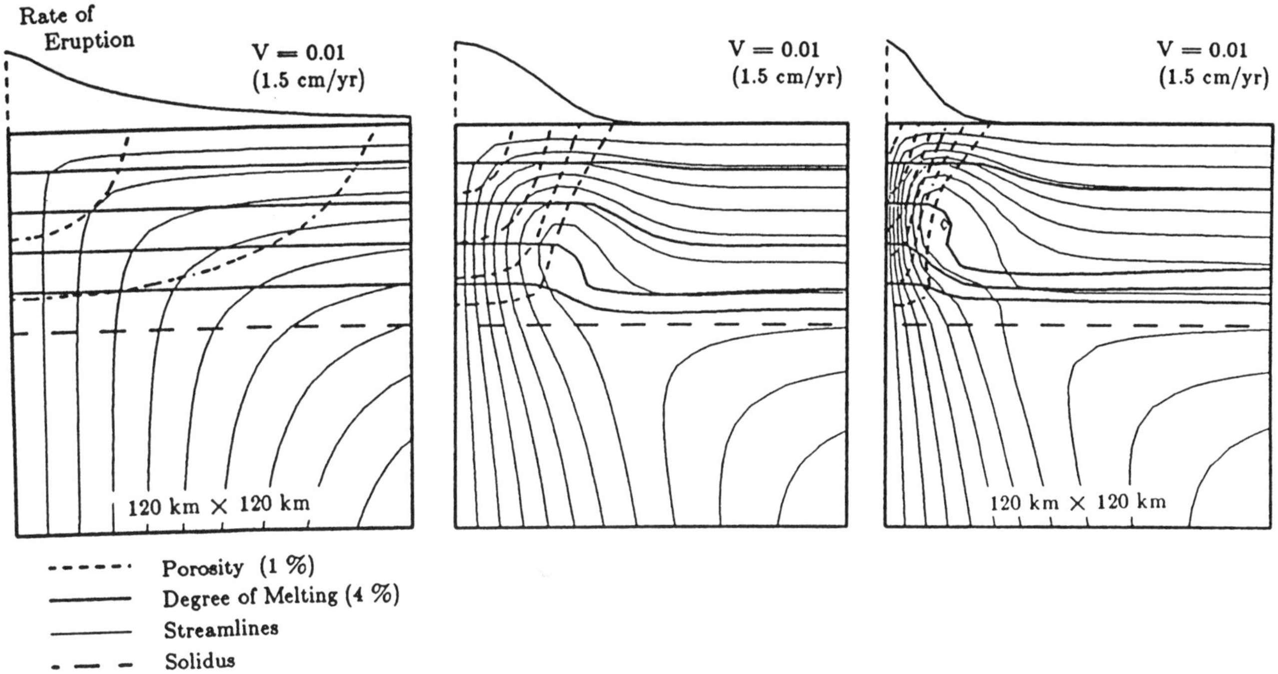

Fig. 57. Steady-state flow and distribution of liquid and solid in a model cross-section of a spreading center [from Scott and Stevenson, 1989]. Porosity is assumed to be a function of the fraction of melt present. Contours of degree of melting represent maximum degree of melting reached up to that point along the streamline, not the fraction of melt present at any instant. The curve labelled rate of eruption at the top of each diagram represents a vertical integration of the melt production beneath each point, modified slightly by lateral transport of the mantle matrix during ascent of the melt (magma driven only by vertical buoyancy). Left, passive flow; middle, buoyant flow including buoyancy of solid-liquid mixture and compositional differentiation; right, buoyant flow as in the middle figure with reduced viscosity in the partially molten region.

temperature-dependent viscosity or plate-thickening [Blackman and Forsyth, 1991; Shen and Forsyth, 1992]. Only if the melt is retained within the mantle matrix to supply added buoyancy and the viscosity is drastically reduced by the presence of melt is the predicted upwelling narrow enough to even approach the narrowness of the zone of crustal accretion [Buck and Su, 1989; Su and Buck, 1991]. Because there is little evidence that large fractions of melt are retained and because some volcanism does occur off axis, much interest has focussed on mechanisms for lateral transport of melt to the spreading center [Spiegelman and McKenzie, 1987; Phipps Morgan, 1987; Ribe, 1987; Cordery and Phipps Morgan, 1991; Sparks and Parmentier, 1991; Spiegelman, 1991]. The topic of melt migration is reviewed thoroughly by Phipps Morgan [1991] and elsewhere in this volume [Turcotte, 1991].

Not all volcanism is confined to the neovolcanic zone. Off-axis seamounts are numerous and may account for as much as a few percent of the volume of the crust in some areas although 1% or less is more typical [Fornari et al., 1987; Smith and Jordan, 1988; Batiza et al., 1988]. The volume of the seamounts is a lower bound on the volume of melt generated off-axis. Of course, much of the melt may be delivered to the spreading center, but some melt may also be trapped beneath or within the lithosphere, unable to penetrate through to the surface. In addition, there are some large sheet flows that extend as much as 18 km off-axis [Macdonald, et al., 1989]. Although most of the flows appear to originate on or near the axis [Ballard et al., 1979; Spiess et al., 1980; Macdonald et al., 1989], recent sheet flows identified by their high acoustic reflectivity have been observed as much as 70 km from the East Pacific Rise using side-scan sonar [Forsyth, et al., 1991]. The volume of these off-axis outpourings is small, with maximum thickness of the flows only a few tens of meters. Although the plumbing for seamounts and other off-axis volcanism may be largely separate from the spreading center, the presence of this volcanism is important because it indicates that there is upwelling and pressure-release melting off-axis. It is not known yet whether this off-axis upwelling is just part of the general upwelling beneath the spreading center or is independent, small-scale or hot-spot convection.

Another expression of off-axis upwelling, however, suggests that there may be a broad-zone of upwelling associated with the spreading center. At the Clipperton and other transforms on fast-spreading ridges, there are topographic highs near the ridge-transform intersections on the older plate beyond the end of the active ridges. Bottom

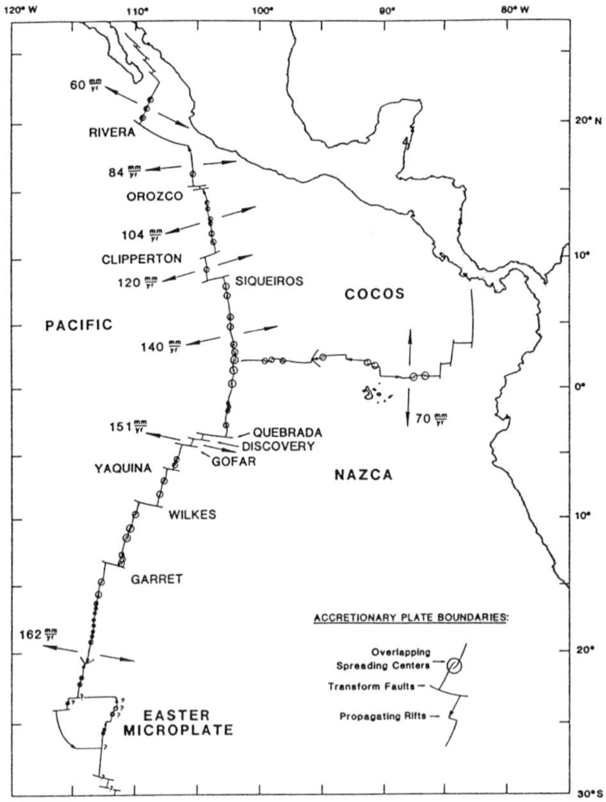

Fig. 58. Segmentation of the spreading centers in the eastern Pacific [from Macdonald et al., 1988b]. Calculated total spreading rates from zero age to anomaly 2 are given next to arrows indicating directions of relative plate motion.

photographs and side-scan sonar show recent extrusive volcanism, seismic reflection profiles reveal thickened crust, and geo-chemical and petrological analyses indicate that the chemical affinity of the recent volcanics is with the adjacent ridge, not with the ridge at which the seafloor originally formed [Kastens et al., 1986; Gallo et al., 1986; Klein et al., 1987; Barth et al., 1988; Barth et al., 1992]. All these factors demonstrate that the intersection highs represent renewed volcanism on the old plate caused by a source of magma extending beyond the end of each spreading center across the transform. The magma cannot be transported laterally in dikes from the spreading center, because the extrusive basalts on the high are much less fractionated than the spreading center basalts and they reside at a shallower depth [Barth et al., 1992]. Evidently, upwelling and pressure-release melting must extend beyond the end of the spreading center, as predicted by 3-D models of passive flow [Phipps Morgan and Forsyth, 1987]. The extension of this melt production 10 to 20 km beyond the transform suggests that the zone of melt production beneath the ridge is also much wider than the narrow zone of accretion at the spreading center.

Segmentation of mid-ocean ridges. Although to this point we have concentrated on the average or typical crustal thickness, it is clear that mantle flow, melt generation and delivery are three-dimensional processes. An obvious surface manifestation is segmentation of the spreading center on a variety of scales, ranging from large-offset transform faults to overlapping spreading centers to small bends in the axis [Langmuir et al., 1986; Macdonald et al., 1988]. First and second-order segmentation of the East Pacific Rise (EPR) and the northern Mid-Atlantic Ridge (MAR) are illustrated in Figures 58 and 59. First-order segmentation refers to offsets of the ridge by transform faults that remain in a stable location

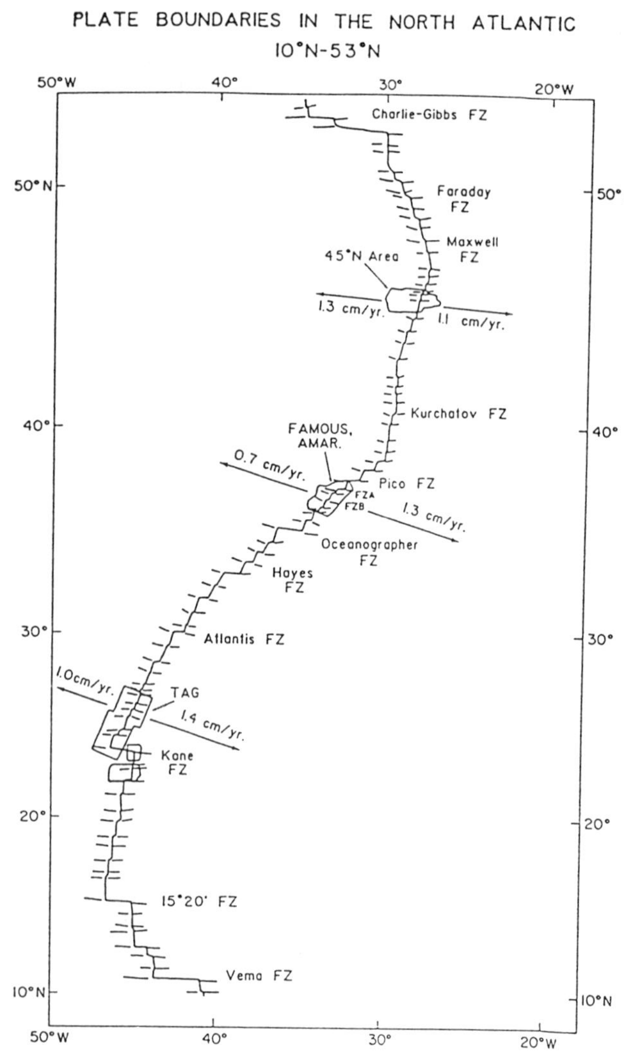

Fig. 59. Segmentation of the northern Mid-Atlantic Ridge [from Macdonald, 1986]. Tick marks next to ridge indicate approximate orientation of fracture zones and off-axis traces of non-transform offsets. Outlines around some portions of the ridge indicate areas of detailed study. Arrows indicate directions of relative motion and velocities given are half-spreading rates from magnetic anomaly identifications indicating asymmetric spreading.

for millions of years or by large-offset propagating rifts. Second-order discontinuities include large over-lapping spreading centers as well as small transform offsets that migrate with time. Accompanying this segmentation are along-axis variations in depth of the spreading axis (Figures 60 and 61).

Although the segmentation and relief are probably best described by spectra or frequency distributions since there are wide (probably continuous) ranges in discontinuity lengths and the spacings between them, the discontinuities on the slow-spreading Mid-Atlantic Ridge are typically larger in age offset and more frequent than on the fast-spreading East Pacific Rise and the accompanying topographic relief greater, producing a rougher along-axis profile on the MAR. There is general agreement on the preceding qualitative statement, but attempts to quantify the relationship between spreading rate and ridge segment length have produced disparate results primarily because of the difficulty of arbitrarily classifying a continuum. In the North Atlantic, several studies find average segment lengths ranging from 40 to 65 km [Ramberg,et al. 1977; Schouten et al., 1985; Crane, 1985; Macdonald, 1986; Sempere et al., 1990] which another author describes as having remarkable regularity [Tucholke, 1990]. Most of these studies report a characteristic length, but another emphasizes that there is a wide range of segment lengths, with the average spacing in the North Atlantic between offsets longer than 4 km estimated to be only about 30 km [Abbott, 1986]. Using a minimum offset on a discontinuity equivalent to about 2 m.y. yields an estimate of characteristic spacing of about 150 km [Sandwell, 1986]. Most studies probably under-report the number of small segment lengths because of limited bathymetric coverage and the difficulty of identifying short segments even with high resolution bathymetric data. Nevertheless, there are clearly some long-lived segments as short as 15 to 20 km [Sempere et al., 1990; Carbotte et al., 1991; Grindlay et al., 1991].

On the fast-spreading East Pacific Rise, estimates of the characteristic length of spreading segments vary from 65 km to more than 500 km [Schouten et al., 1985; Crane, 1985; Sandwell, 1986; Abbott, 1986; Sinton et al., 1991], but no matter what the definition, all agree that the length is greater than on the Mid-Atlantic Ridge. Not surprisingly, the functional form of the proposed relationship between length L and spreading rate v is a function of the author and the author's preferred theoretical explanation. Proposed forms include linear, $L = a + bv$ [Sandwell, 1986] and power law, $L = cv^{1/3}$ [Schouten et al., 1985]. Given the variations in length that occur at any given spreading rate and the difficulty in arriving at precise definitions for a continuum of behavior that has different morphological expressions at different spreading rates, we can conclude that the form of the variation with rate is not yet known. A re-examination of segmentation with the greatly increased availability of high-resolution data and with the recognition in the analysis that there may be a continuum is warranted. It should also be noted that all of these analyses ignore the Reykjanes Ridge which morphologically resembles the East Pacific Rise despite a slow spreading rate.

Several theoretical models have been proposed to explain the observed segmentation. One of the main issues is whether the segmentation arises from lithospheric sources or from fundamental segmentation of flow of the upwelling mantle and segregating melt. Citing the regularity of segmentation and the persistence of the segmentation even though the offset between spreading centers may be reduced to zero or change sign [Schouten and White, 1980; Schouten and Klitgord, 1982], Whitehead et al.[1984], Schouten et al., [1985] and Crane [1985] argued for a sublithospheric source in the form of diapiric upwelling of partially molten material (Figure 62). In the model of Schouten et al., once the wavelength of the upwelling was established, the upwelling itself would consist largely of melt. This contrasts with the diapiric upwelling models of Nicolas and Violette [1982] and Rabinowicz et al. [1984], based on field evidence in the Oman ophiolite,in which the upwelling material was envisioned to be about 5 - 10% melt. The characteristic wavelength λ in a Rayleigh-Taylor instability for diapiric upwelling of a low-viscosity (μ_1), low density fluid (partial melt region) underlying a high viscosity (μ_2) fluid is controlled by the thickness h of the layer of low-viscosity fluid and the ratio of the viscosities [Selig, 1965] and is given by

$$\lambda = \frac{2\pi h}{c}\left(\frac{\mu_2}{\mu_1}\right)^{\frac{1}{3}} \tag{28}$$

as long as the thickness of the unstable layer is small compared to the thickness of the overlying layer. c is a geometrical factor that will vary slightly depending on the shape of the low-viscosity layer [Marsh, 1979]. The key to predicting the dependence of wavelength on velocity is the model for how h grows as melt accumulates in the melt production region. Schouten et al. [1985] postulated that a cylindrical body of relatively high melt content forms beneath the axis and that the volume of the body increases uniformly with time as melt is generated by upwelling. The diameter of the body h would then increase with the square root of time. Instabilities grow exponentially with a characteristic time dependent on h. Solving for h at a critical time when finite amplitude disturbances develop leads to a prediction that $\lambda \propto v^{1/3}$, where v is spreading velocity. Once the finite amplitude disturbance develops, they envision that melt would be quickly drained and the cycle begun again. (It should be noted that Kerr and Lister [1988] dispute the Marsh claim that a cylindrical layer would have the same form of instability as a flat layer. In particular, they find that the characteristic wavelength is independent of the ratio of viscosities for a cylinder and depends only on the radius of the cylinder. Although this removes the viscosity dependence of the scale factors, it does not affect the Schouten et al. prediction of the form of the dependence of wavelength on spreading rate since they did not assume any dependence of viscosity on spreading rate.) However, this prediction is dependent on there being a mechanism for horizontal aggregation of the partial melt to a central, cylindrical region during the accumulation phase. Apart from the question of whether melt would accumulate at a

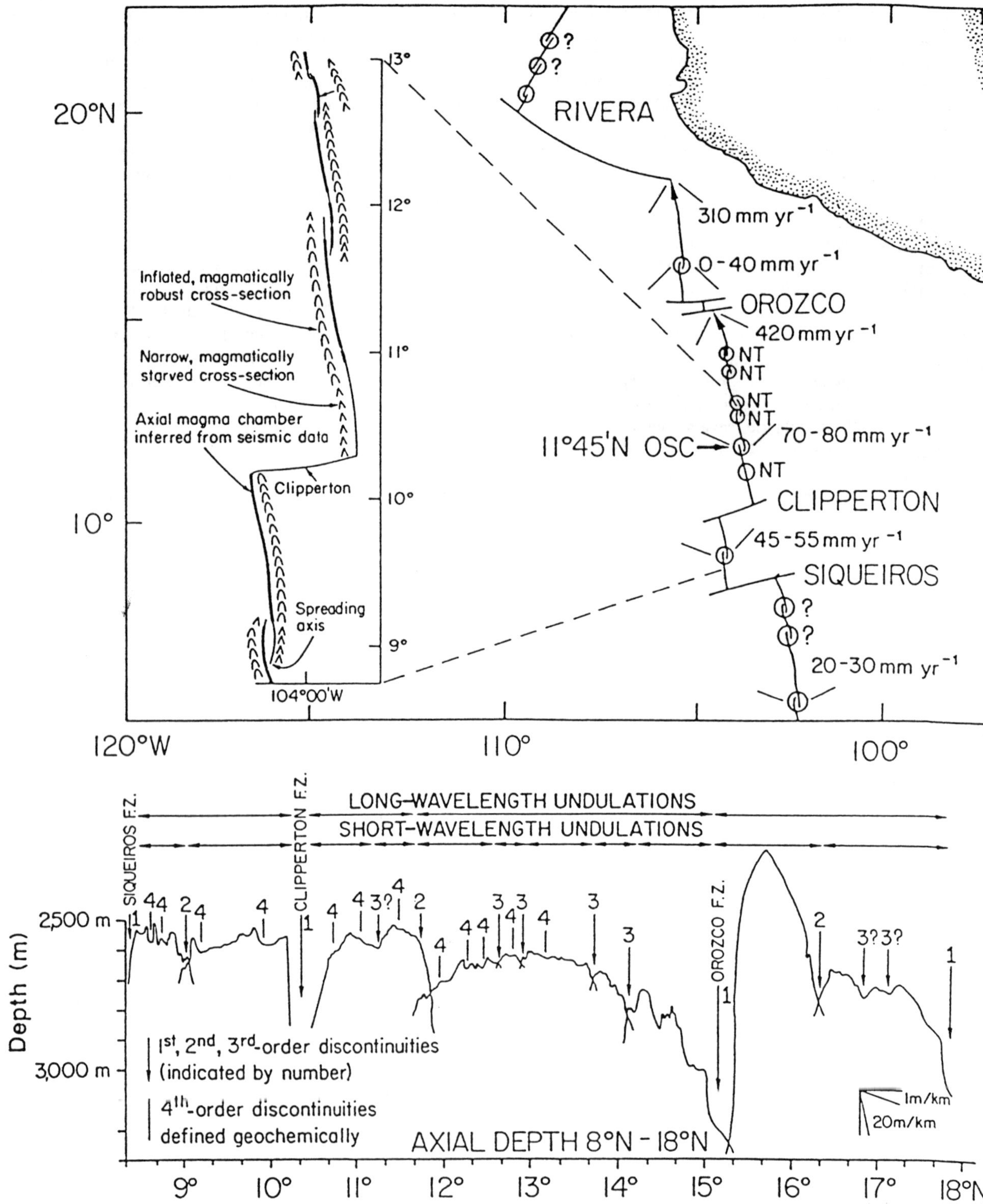

Fig. 60. Axial bathymetry of the East Pacific Rise and its relation to segmentation of the spreading center [from Macdonald, 1988a]. In upper figure, velocities given are rates of migration or propagation of overlapping spreading centers or propagating rifts, respectively. Fracture zones are characterized by large topographic deeps. Propagating rifts and large, overlapping spreading centers also often have a topographic deep associated with them.

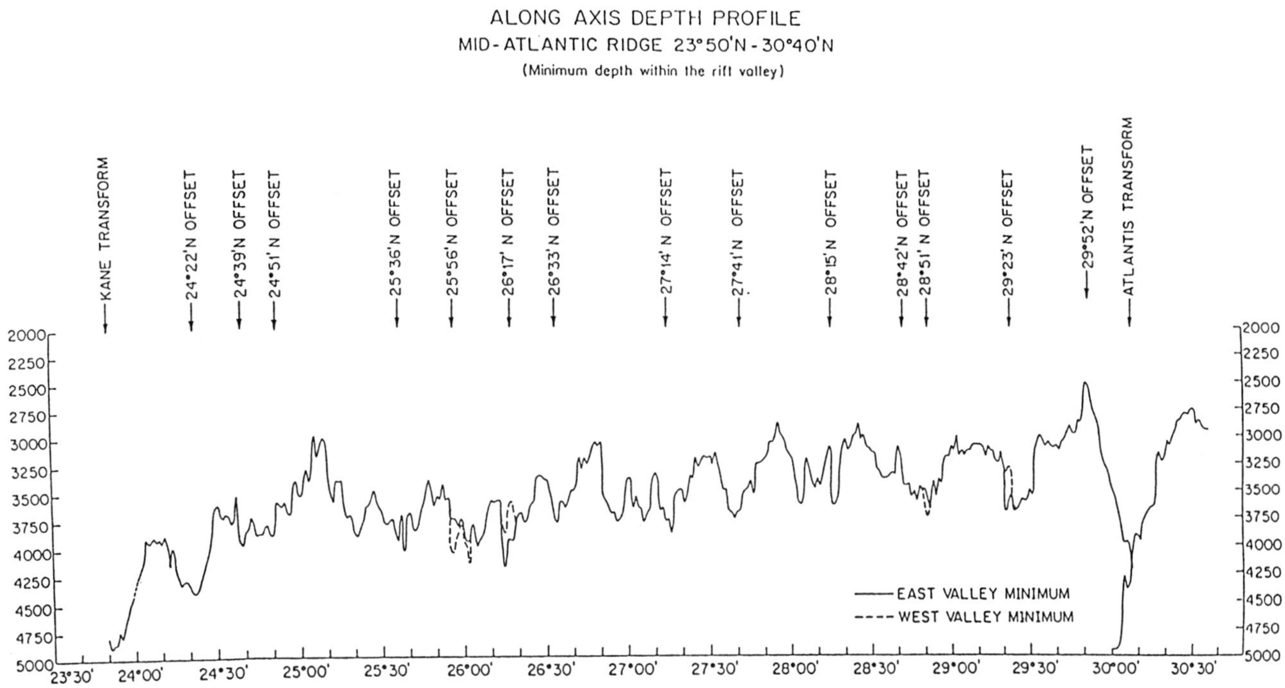

Fig. 61. Axial bathymetry of the northern Mid-Atlantic Ridge [from Sempere et al., 1990]. Note that vertical exaggeration in this figure is approximately 5 times less than in Fig. 60. Locations of fracture zones and non-transform offsets are indicated by arrows.

single depth (since melt is expected to be produced simultaneously over a thick melt zone), if the partial melt zone involved only vertical accumulation, then, assuming upwelling velocity and melt production rates are proportional to plate velocity, its thickness h would increase linearly with

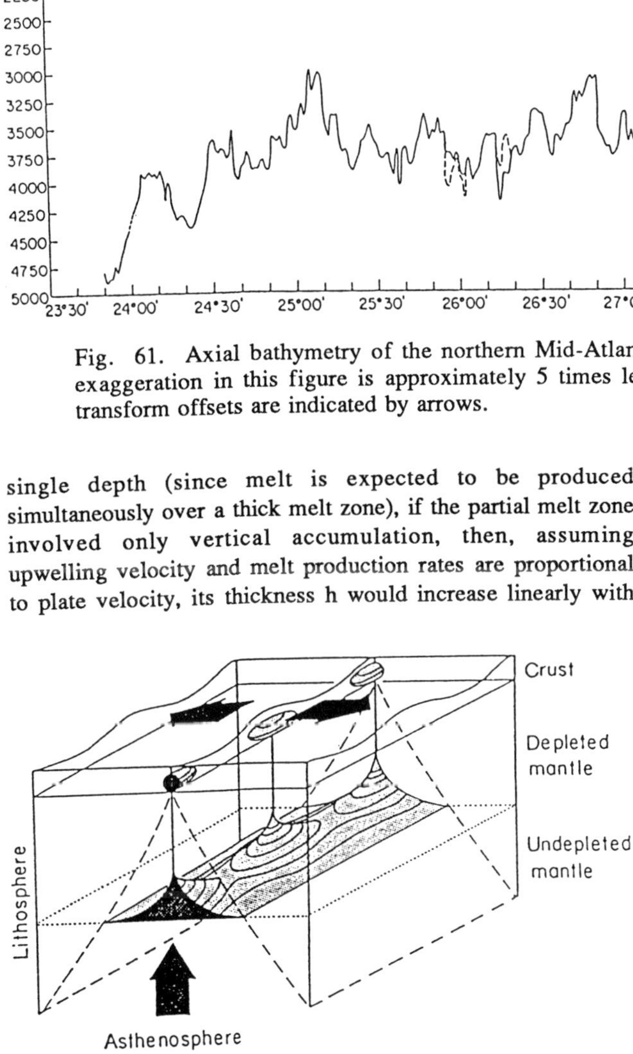

Fig. 62. Schematic mechanism for magmatic segmentation of mid-ocean ridges [from Whitehead et al., 1984]. Rising asthenosphere undergoes partial melting. The melt concentrates in a melt-rich layer, which becomes gravitationally unstable and rises rapidly to the surface in diapirs that are regularly spaced along the axis. The depleted lithosphere continues to rise slowly and becomes the lithosphere after cooling.

time and the prediction for wavelength would be $\lambda \alpha v^{1/2}$. Thus, even for a particular class of theoretical model, the predictions depend on assumptions about the geometry of the melt region that need to be verified by observation.

Segmentation may also be caused by growth of along-axis instabilities in the upwelling mantle that lead to the development of three-dimensional patterns of flow in an initially two-dimensional geometry. This could produce isolated upwelling centers beneath each spreading segment (Figure 63) in which the mantle matrix itself, not just the low-density, low-viscosity melt, rises in diapiric-like structures. Parmentier and Phipps Morgan [1990] have mapped the transition between two-dimensional, sheet-like flow and fully three-dimensional flow for the case of a linear spreading center. The passive flow driven by the separation of the plates tends to suppress the three-dimensional pattern, but the slower the spreading rate or the lower the viscosity, the easier it is for 3-D flow to develop (Figure 64). The preferred wavelength in their model was on the order of 300 km, but a large range of wavelengths was found to be stable. Pre-existing segmentation in the form of transform faults enhances the tendency for 3-D buoyant flow to develop [Sparks and Parmentier, 1991b]. The transition from sheet-flow to diapiric flow is qualitatively consistent with the change in segmentation and along-axis variations in the mantle Bouguer gravity anomalies from fast to slow-spreading ridges [Lin and Phipps Morgan, 1991]. These models predict that the Reykjanes Ridge should be segmented by 3-D buoyant flow as

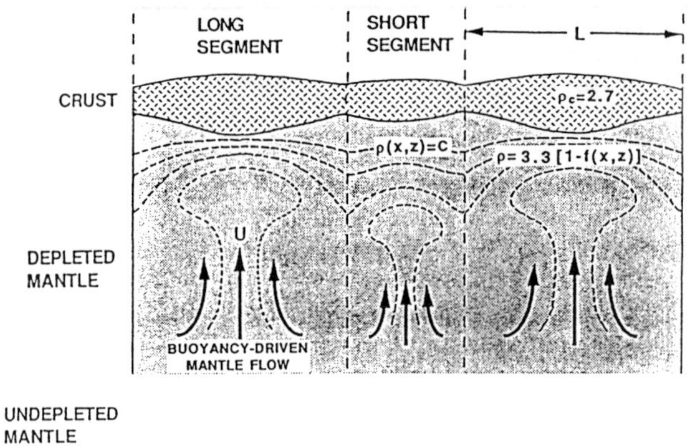

Fig. 63. Schematic mechanism for segmentation of mid-ocean ridges [from Lin et al., 1990]. Buoyant upwelling of the mantle matrix develops a three-dimensional pattern, with distinct ascending plumes beneath each ridge segment. Density variations (f(x,z)) within the mantle that produce the buoyancy arise from thermal expansion, depletion of the matrix following melt extraction, and entrapment of some melt. Mantle density variations and crustal thickness are closely related if melt migration is primarily vertical or within planes perpendicular to the ridge crest.

the spreading rate is low and the viscosity might be expected to be lower than average. The Reykjanes Ridge, however, has the morphology and mantle Bouguer signature of a fast-spreading ridge, with little indication of strong, along-axis

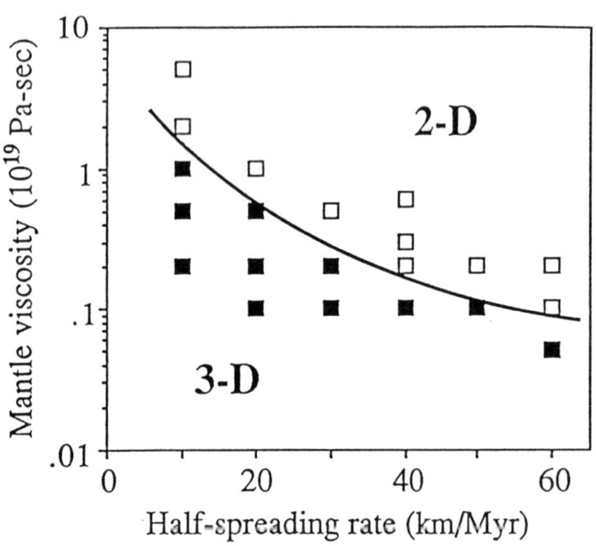

Fig. 64. Spreading-rate and viscosity dependence of buoyant segmentation of upwelling mantle [from Parmentier and Phipps Morgan, 1990]. An initially two-dimensional (2-D) sheet of upwelling mantle beneath a linear ridge develops three-dimensional (3-D) structure with along axis variability at low half spreading rates and low mantle viscosity. Under these conditions, buoyant upwelling dominates passive flow due to plate spreading.

segmentation [Bell, 1991]. It is possible that along-axis flow away from the Iceland hotspot in the asthenosphere at rates exceeding the spreading rate [Vogt, 1971; Schilling, 1973] could suppress the development of individual upwelling cells along the spreading center.

Detailed, multi-beam bathymetric surveys on the East Pacific Rise reveal small, non-transform offsets (Figure 65) that break the spreading center into 30- to 100-km-long segments [Lonsdale, 1989]. These offsets are too small to have any significant thermal effect within the mantle. Although they could conceivably be related to individual upwelling cells within the mantle (thereby destroying the argument for increased scale-length on the EPR compared to the MAR), Lonsdale [1989] argues that they are created primarily in response to changes in spreading direction and occasionally by other processes that involve formation of oblique sections of the rise crest. Some of these segments have inflated summits near their centers, deepening and narrowing towards the ends, suggesting a central magma supply, but many others are flat along profile, deepening only where adjacent segments overlap [Lonsdale, 1989]. Thus, there may be an interaction between mantle processes and this small-scale segmentation, but variations in localized magma production or delivery do not seem to be responsible for the segmentation. Other mechanisms of segmentation are also possible that involve primarily lithospheric processes rather than intrinsic upwelling or melt segregation scales. It has been suggested, for example, that stresses associated with thermal contraction of the lithosphere could produce a regular segmentation of the spreading center [Collette,1974; Turcotte, 1974; Sandwell, 1986], although Haxby and Parmentier[1988] argue that since there is no inherent, along-axis scale to the thermal stresses, it is unlikely that they alone can control the spacing. The plate thickness, however, could impose a natural scale. The overall shape of the Mid-Atlantic Ridge still largely parallels the flanking coastlines of Africa, South America, North America and Europe despite systematic asymmetric spreading on some segments, so it is clear that the inherited pattern from the time of the continental break-up persists as a controlling factor for many millions of years.

There are probably a variety of different mechanisms responsible for the segmentation of the ridges on a variety of different scales, ranging from very long-wavelength variations between major hot-spots to small fissures associated with the intrusions of individual dikes. Although the typical scales of different mechanisms may overlap, overall there may be a heirarchy of segmentation with one or two mechanisms dominating for any given length. The evidence for such a heirarchy has been recently summarized by Macdonald et al. [1988] and the dominance of different mechanisms at different scales is illustrated schematically in Figure 66.

Variations in crustal thickness accompanying segmentation. Seismic and gravity experiments have demonstrated that fracture zones in the Atlantic are often characterized by thin crust. Some crustal thinning may extend

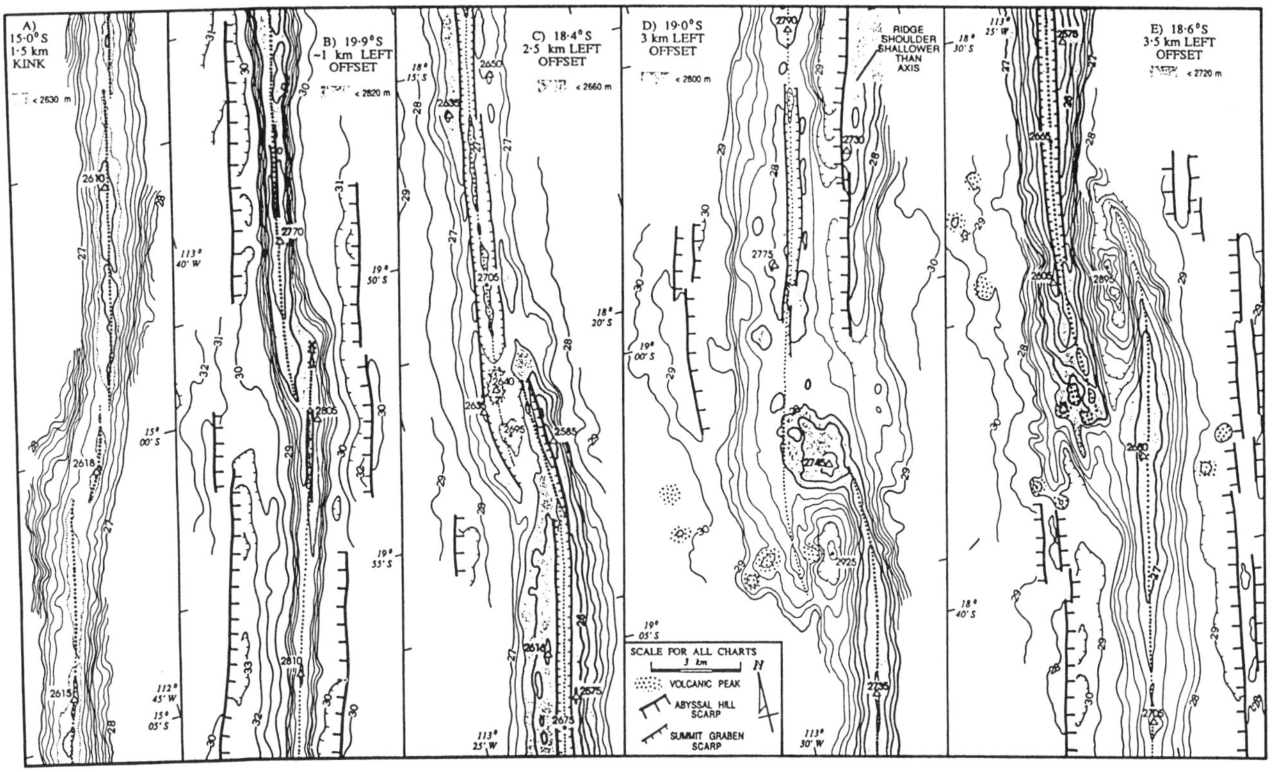

Fig. 65. Small, non-transform offsets of the East Pacific Rise south of the Garrett transform (Fig. 58) are shown in order of increasing size [from Lonsdale, 1989]. Bathymetric contours from SeaBeam data are shown at 20m intervals on the axial ridge, and 100 m elsewhere. Location of the axial rift zone inferred from bathymetry is shown as a dotted line.

30 km or more from the fracture zone, but extremely thin crust (< 3 km) is usually found only within narrow zones on the order of 10 km wide. Some typical profiles across fracture zones are illustrated in Figure 67. Results of wide-angle (conventional refraction and expanding-spread profiles) seismic experiments have been summarized by White et al., [1984], Detrick et al. [1991] and Minshull et al., [1991]. This simple picture of thin crust in fracture zones as illustrated in

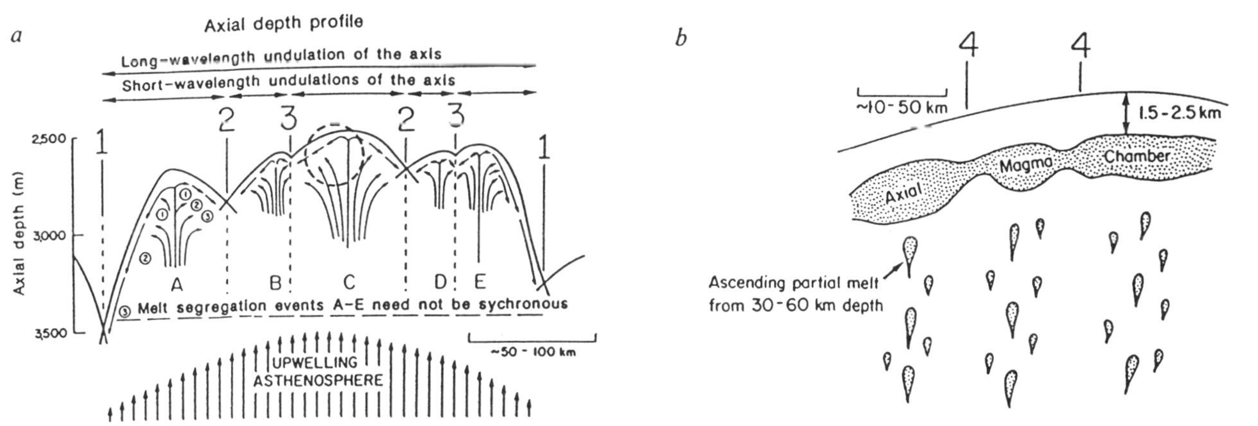

Fig. 66. Schematic model for heirarchial, magmatic segmentation of mid-ocean ridges [from Macdonald et al., 1988a]. (a) Upwelling asthenosphere at depth rises fastest near the middle of a ridge segment bounded by two fracture zones (first-order discontinuities, indicated by 1) leading to a long-wavelength undulation in bathymetry. Decompression melting during upwelling produces separate melt segregation events (A-E) at a depth of 30-60 km beneath the axis which coincide with ridge segments separated by second- or third-order non-transform discontinuities. (b) Fourth-order chemical discontinuites could arise during upward migration of melt from the melt production and segregation zone.

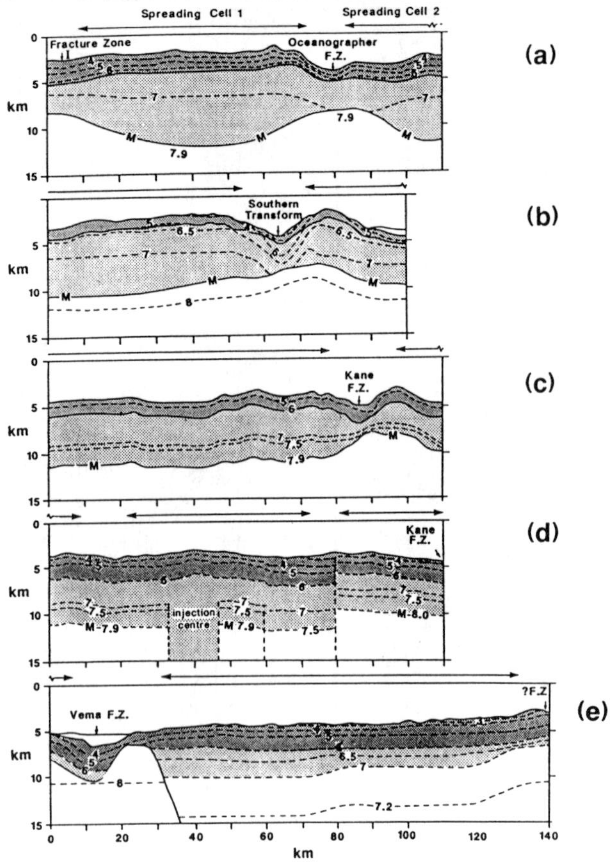

Fig. 67. Velocity-depth cross-sections of the crust along the axis of segments of the Mid-Atlantic Ridge derived from seismic refraction experiments [from White et al., 1989]. Shaded regions indicate extent of crust with M labelled where the Moho is well-developed. Extent of individual segments or spreading cells is indicated by arrows. (a) Oceanographer FZ, (b) Charlie-Gibbs FZ, (c) and (d) Kane FZ, (e) Vema FZ. Location of FZs shown in Fig. 59. Note that this profile of the Oceanographer FZ shows more extreme thickening of the crust in the middle of the segment than other profiles in the same area [Sinha and Louden, 1983; Ambos and Hussong, 1986].

Figure 67 is now widely accepted, partly because it agrees so well with preconceptions that deep topographic regions should be isostatically compensated by thin crust.

Measurements of crustal thickness as a guide to the thickness of the igneous layer and the volume of melt supplied to the surface need to be examined carefully. Mass-wasting from the steep walls of the fracture zones, cracking and hydrothermal alteration of the crust and underlying mantle may lead to systematic underestimates of the amount of crustal thinning. In major continental strike-slip fault zones, anomalously low seismic velocities and densities are often found within a few km of the active trace of the fault (Figure 68) [Healy and Peake, 1975; Feng and McEvilly, 1983; Mooney and Ginzburg, 1986; Trehu and Wheeler, 1987]. Similarly, within the Vema fracture zone (Figure 67e), the thinnest crust and highest densities are found beneath the walls of the fracture zone rather than along the active transform suggesting that the rocks along the active fault have been altered [Louden et al., 1986; Prince and Forsyth, 1988]. In oceanic fracture zones, the thin crust usually exhibits anomalously low seismic velocities and layer 3, the gabbroic layer with typical P-wave velocities of 6.5 to 7.0 km/s, is generally absent (examples are shown in Figures 67 b and e). In some regions, the thin, low-velocity crust is underlain by a layer with velocity about 7.5 km/s while in other areas there are more typical mantle velocities of about 8.0 km/s. The 7.5 km/s layer has been interpreted as partially serpentinized peridotite [Calvert and Potts, 1985; Minshull et al., 1991], but if serpeninization proceeds further, the velocities can be lowered to values even less than typical of layer 3 so that original ultramafic material would have physical properties that would ordinarily be attributed to the crust [Detrick et al., 1982; Prince and Forsyth, 1988]. The apparent thickness of the crust from seismic and gravity measurements would then be greater than the igneous thickness representing melt extraction.

Seismic refraction [Purdy and Detrick, 1986] and gravity [Morris and Detrick, 1991] surveys executed in the MARK area south of the Kane transform (Figure 69) provide an excellent example of the type of experiment carried out to study crustal thickness variations. Along the spreading axis, the water depth varies from a minimum of about 3300 m some 70 km from the ridge-transform intersection (RTI) to a maximum of over 5500 m in the nodal deep at the RTI (Figure 70a).

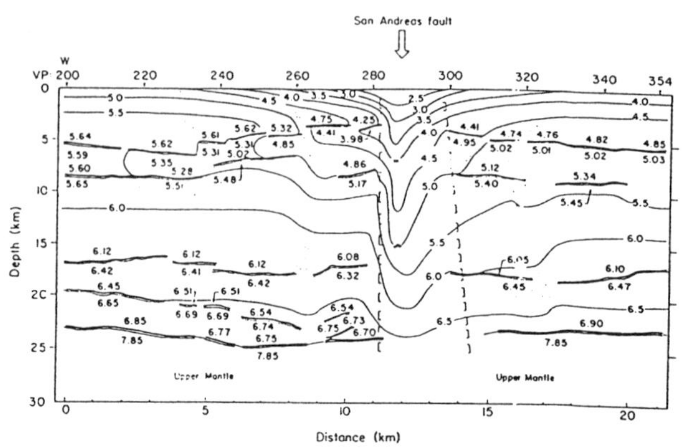

Fig. 68. Crustal velocity model of the San Andreas fault zone [from Feng and McEvilly, 1983]. Heavy lines indicate strong, continuous seismic reflections. Light lines represent seismic velocity contours based on seismic refraction and wide-angle reflection data. Near-vertical series of brackets show the approximate width of the altered fault zone, as estimated from the low seismic velocities. Numbers across top are index numbers for common depth points of the seismic reflection profile.. Small numbers immediately above and below reflections represent estimates of the P-wave velocities on either side of the reflection interface.

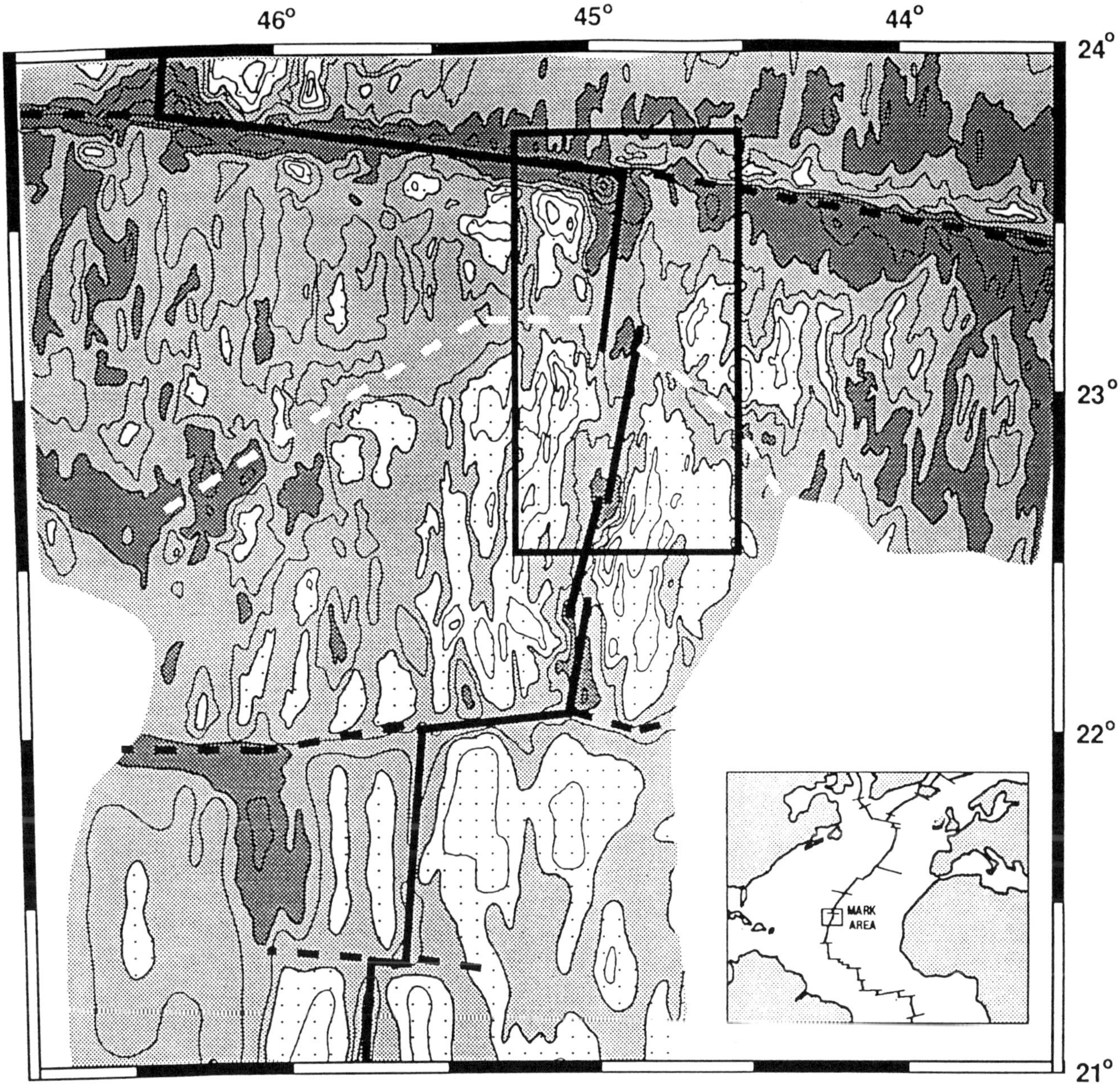

Fig. 69. Bathymetry and location of the MARK area in the central North Atlantic [from Morris and Detrick, 1991]. The plate boundary is shown as a heavy, black line. Off-axis traces of fracture zones and non-transform offsets are shown as dashed lines. Box outlines detailed study area shown in Fig. 70, including the eastern intersection of the Kane transform and the Mid-Atlantic Ridge. Dark shading, depth 4 to 6 km; medium shading, depth 3 to 4 km; and light stipple, 1 to 3 km depth.

Although the median valley is continuous for a distance of more than 150 km south of the RTI, detailed bathymetric surveys reveal several disruptions or small offsets in the neovolcanic zone, which are accompanied by significant changes in crustal structure. At 23° 15' N, for example, there is a small (<10 km) offset in the neovolcanic zone and a change in its character from a linear volcanic ridge to the north to a series of small, individual volcanoes to the south (Figure 70b). Off-axis, there may be a trace of this discordant zone in the form of a series of bathymetric deeps that form a northward pointing V suggesting migration of the offset along the axis.

A seismic refraction profile was shot within the median valley and across the Kane fracture zone (profile A in Figure 70b). A record section from one of the four ocean-bottom hydrophone (OBH) receivers is shown in Figure 71. With the aid of ray-tracing through laterally heterogeneous, two-dimensional velocity models (Figure 72), the travel-times and arrival characteristics in the record sections are interpreted in terms of simple lateral variations in structure (Figure 73). In this case, there is a significant change in crustal thickness across the discordant zone, a low-velocity, attenuating region near the base of the crust beneath the shallowest part of the spreading center, and further thinning of the crust in the immediate vicinity of the nodal deep at the RTI [Purdy and Detrick, 1986]. Other profiles parallel to the spreading center and along the axis of the Kane FZ provide some three-dimensional constraints on the structure [Cormier et al., 1984], but a more continuous indication of lateral variations in structure is provided by a gravity survey.

The free-air anomaly (FAA), shown in Figure 70c, is dominated by the effects of topography. FAA lows appear along the median and fracture zone valleys and highs are found over the inside-corner and other topographic highs. To remove the effects of topography, the gravitational attraction of the density contrast at the water-crust interface is subtracted from the FAA using the known bathymetry. In addition, because much of the short-wavelength topography in mid-

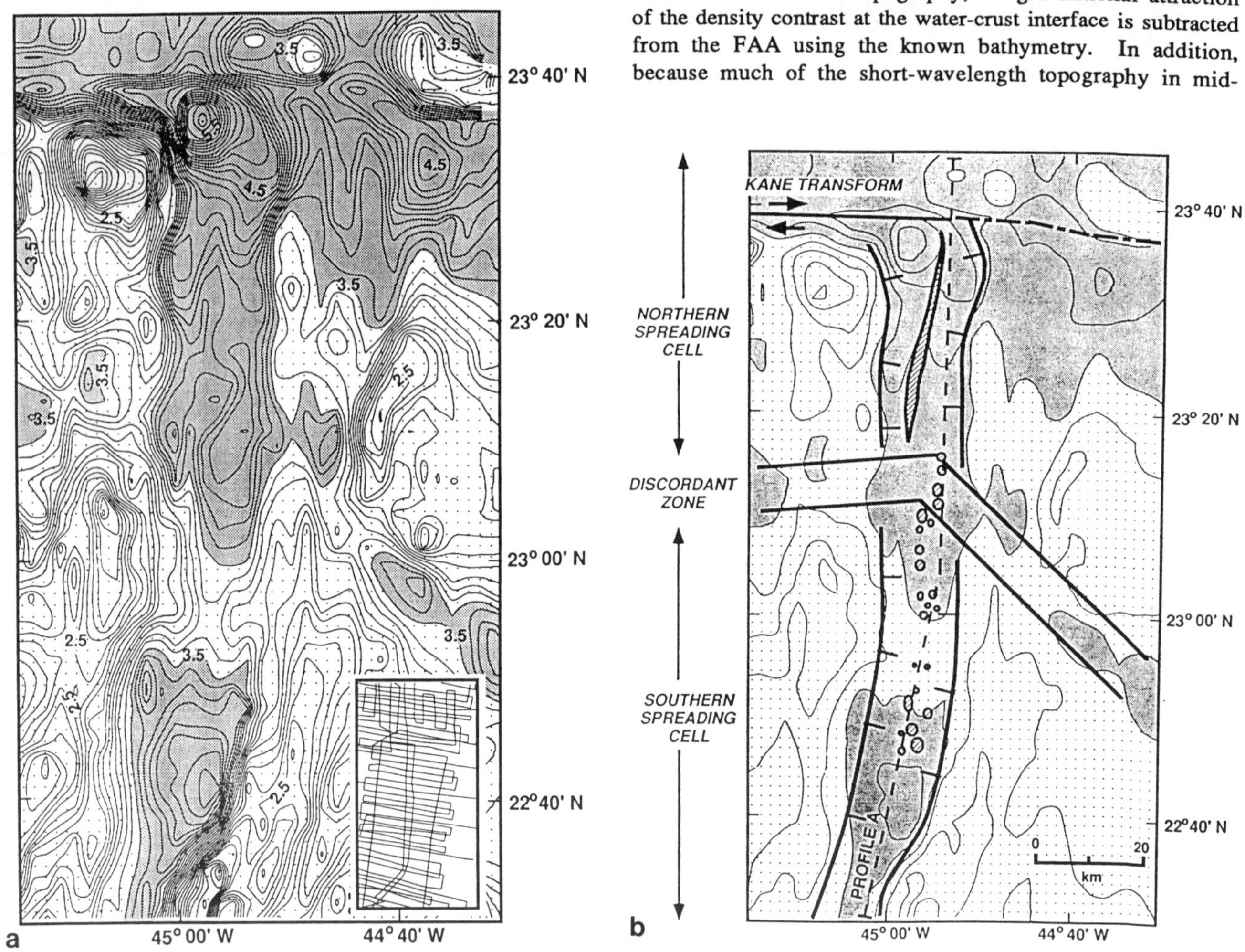

Fig. 70. Three-dimensional analysis of gravity anomalies in the MARK area of the Mid-Atlantic Ridge [from Morris and Detrick, 1991]. (a) Detailed bathymetry from Sea Beam data. Contour interval, 100m. Areas deeper than 3500 m are shaded. Inset shows track coverage for gravity. (b) Simplified tectonic map based on high-resolution bathymetry and bottom observations. Bathymetry contours shown at 500 m intervals. Diagonally shaded areas are neovolcanic ridges or volcanoes; ticked, bold lines indicate walls of the inner rift valley; other heavy lines show trace of Kane transform and the discordant zone (non-transform offset). Dashed line labeled Profile A is the location of a seismic refraction profile (Fig. 73). (c) Free air anomaly, contour interval 5 mgal. (d) Mantle Bouguer anomaly, contour interval 5 mgal. (e) Residual gravity anomaly after subtraction of predicted gravitational attraction of the mantle from mantle Bouguer anomaly. Contour interval is 5 mgal. (f) Possible variations in crustal thickness found by downward-continuing residual gravity anomaly, assuming mean crustal thickness is 6 km. Contour interval is 0.5 km.

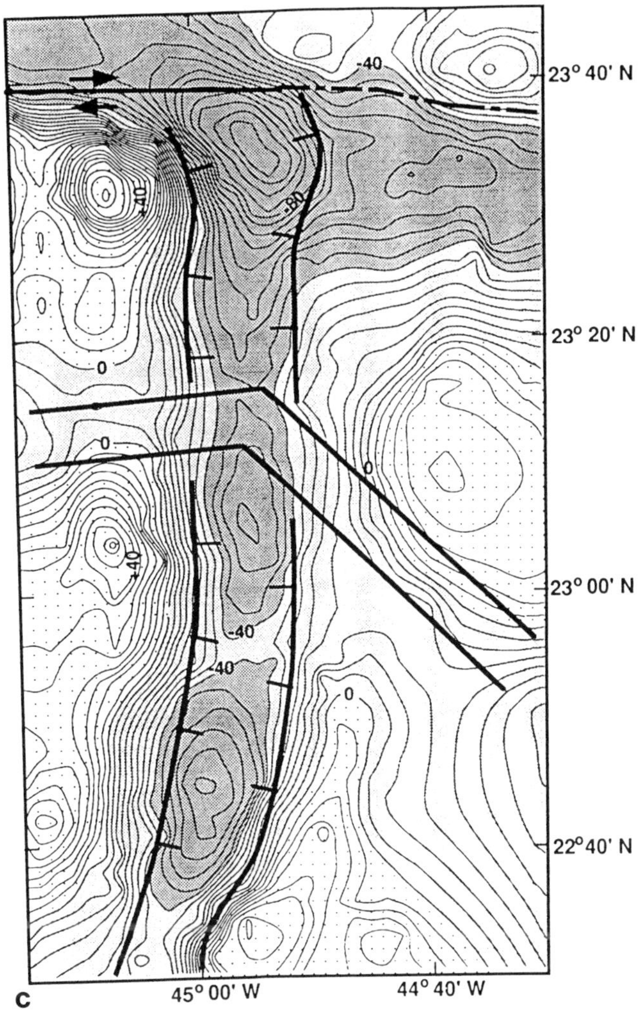

c

Fig. 70. (continued)

ocean ridge settings is uncompensated and the longer wavelength topography is dominated by subsidence associated with cooling of the lithosphere, a good first-approximation is that density interfaces within the crust and between the crust and mantle parallel the surface. Thus, to complete the correction for topography, the attraction of these deeper interfaces is calculated assuming constant crustal thickness and density and then subtracted from the FAA leaving the mantle Bouguer anomaly (MBA). The MBA (Figure 70d) then is caused by variations in crustal thickness and density and by variations in mantle density. Mantle Bouguer anomalies correlate with the age of the seafloor, with lows generally found over ridge axes gradually increasing away from the spreading centers [Madsen et al., 1984; Phipps Morgan and Parmentier, 1987; Prince and Forsyth, 1988; Kuo and Forsyth, 1988; Lin et al., 1990; Madsen et al., 1990; Blackman and Forsyth, 1991] reflecting the cooling and densification of the crust and mantle with increasing age. In this survey [Morris and Detrick,1991], the most obvious manifestation of this effect is the increase in MBA from south to north across the Kane FZ associated with passing from zero-age crust to 10 Ma crust north of the FZ (Figure 70d). The effects of cooling can be predicted with three-dimensional thermal models and subtracted from the MBA leaving residual anomalies (Figure 70e) that are caused either by variations in crustal structure or by failure of the thermal model to adequately represent the changes in mantle density. Most of the studies cited above have employed the thermal model of Phipps Morgan and Forsyth [1988] which assumes passive flow in the mantle driven by the motion of flat surface plates. The residual anomalies show a shift from negative south of the discordant zone to positive north of it (Figure 70e), in general agreement with the seismic refraction results. There is also a trail of highs parallel to the trace of the discordant zone off-axis. Translated into variations in crustal thickness, 2 km or more of crustal thinning is associated with this small discordant zone (Figure 70f). The very thin crust found by refraction in

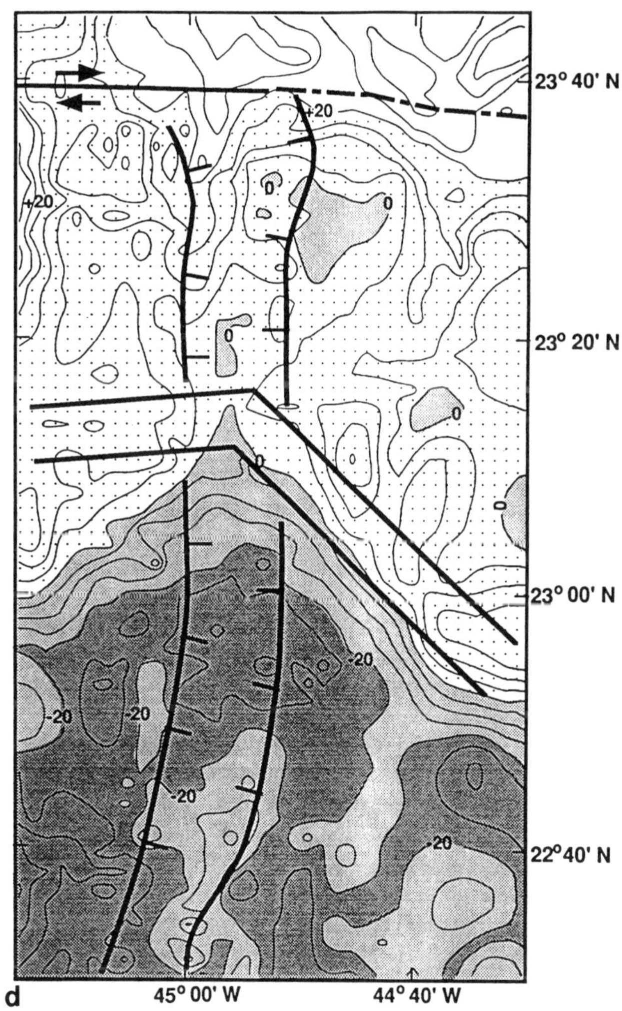

d

Fig. 70. (continued)

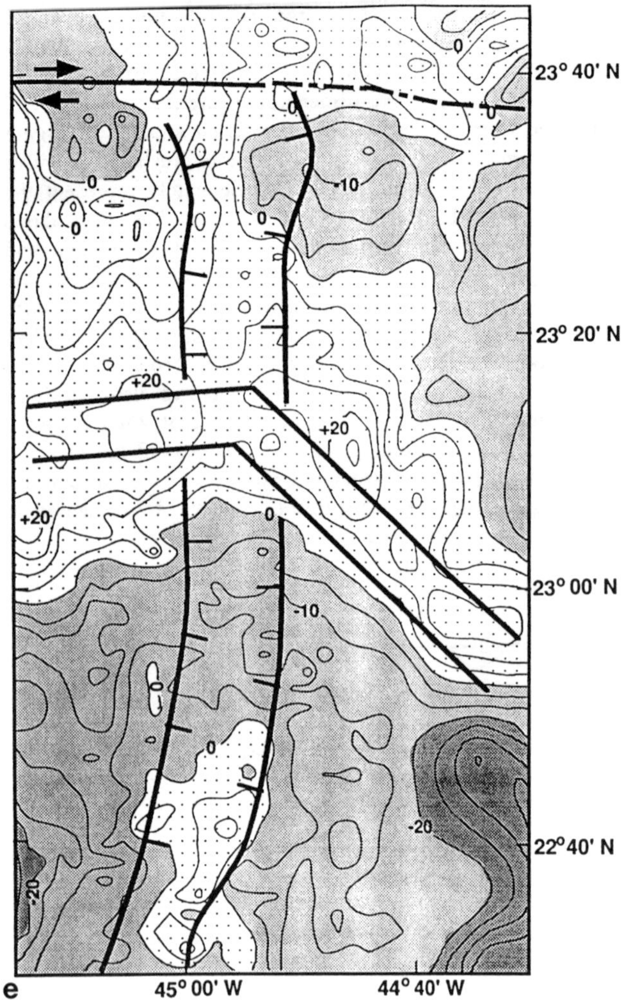

Fig. 70. (continued)

the RTI does not show up as clearly, perhaps because of its very limited lateral extent.

Crustal thinning toward fracture zones or ridge offsets in combination with the cooling of the mantle with increasing age can create circular, "bulls-eye"-shaped gravity lows in the mantle Bouguer anomalies associated with each spreading segment on slow-spreading ridges [Kuo and Forsyth, 1988; Lin et al., 1990]. These gravity lows have been suggested to be indicators of upwelling centers along the ridge axis. It is not clear how much of along-axis variation in gravity is associated with changes in mantle density instead of crustal thinning. Models involving only crustal thinning do not require impossibly thin crust, so must be considered a reasonable possibility. Since crustal thinning, however, implies a reduction in melt extraction from the mantle, it is probable that there is some variation in structure of the underlying mantle accompanying the crustal thinning. Along the axis, then, the gravity signature probably arises from coupled mantle and crustal effects (Figure 63), although the crustal variation is probably dominant.

Crustal thickness within individual fracture zones is highly variable, with changes in thickness of 1 to 3 km both along- and across-strike over horizontal distances as small as can be resolved in the various geophysical experiments (minimum of 5 to 15 km) [Cormier et al., 1984; Prince and Forsyth, 1988; Blackman and Forsyth, 1991; Minshull et al., 1991]. Thin crust is found beneath both major fracture zones and small-offset or non-transform fracture zones [White and Matthews, 1980; Mutter et al., 1984; Kuo and Forsyth, 1988; Lin et al., 1990; Morris and Detrick, 1991] with no obvious correlation in degree of thinning of the crust with length of offset of the fracture zone. At least in terms of density structure, some segments of major fracture zones such as the Vema fracture zone show no evidence of thin crust [Prince and Forsyth, 1988] and others such as the Cox transform in the South Atlantic show no thinning anywhere along their length [Kuo and Forsyth, 1988]. Some of the variability is probably

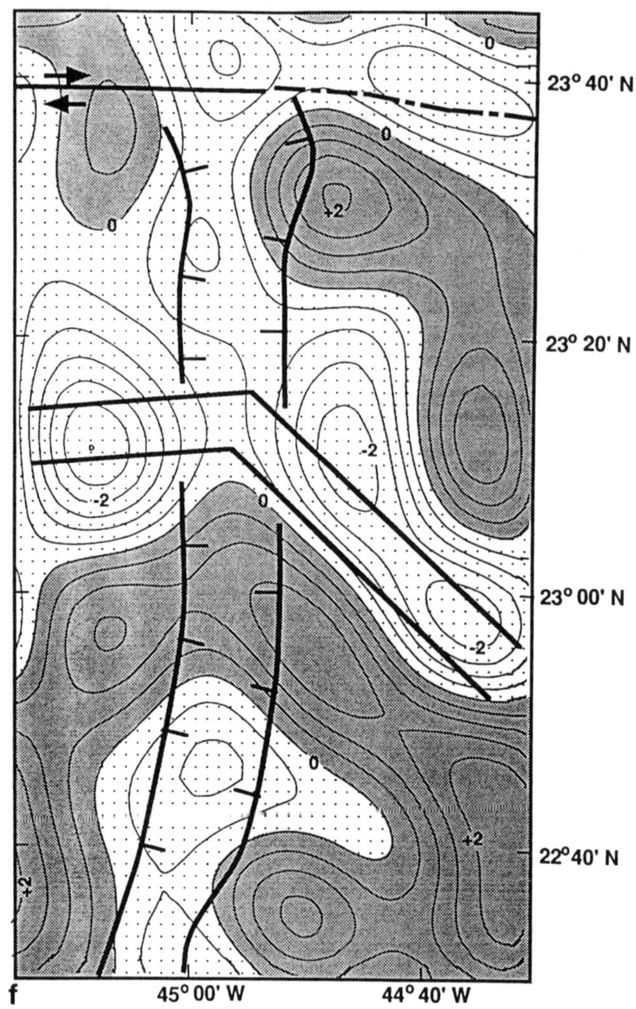

Fig. 70. (continued)

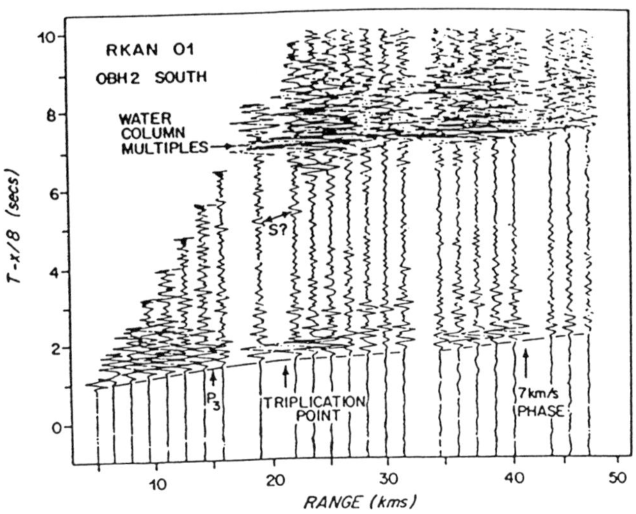

Fig. 71. Record section (corrected for topography) of the explosive charges detonated south of ocean-bottom hydrophone #2 (OBH2) on profile A south of the Kane FZ (Fig. 70b). The amplitude scale of the seismograms has been linearly increased with distance or range. [from Purdy and Detrick, 1986].

attributable to variations in degree of alteration and serpentinization and in degree of tectonic denudation, but there are probably also significant temporal and spatial variations in magma supply.

Despite the variability in structure, it is clear that the average magma supply is reduced in both large and small-offset fracture zones in the Atlantic. There have not yet been enough surveys on the fast-spreading East Pacific Rise to sort out the dependence of crustal thinning on spreading rate, but it should be noted that ultramafic rocks are seldom dredged from EPR fracture zones [Hebert et al., 1983] but are common in fracture zones on slow-spreading ridges [Bonatti, 1976]. Gravity surveys on fast [Madsen et al., 1990] and intermediate [Phipps Morgan and Parmentier, 1987] spreading rate ridges show much less along-axis variability in density accompanying the reduced variability in topographic relief suggesting a greater importance of dynamic upwelling at slow spreading rates [Lin and Phipps Morgan, 1991], but there are no studies of fracture zones with length or age offsets comparable to the Atlantic fracture zones. The lack of correlation of igneous crustal thickness with length of offset and the presence of thin crust even on small offset fracture zones that are too small for a significant thermal edge effect [Minshull et al., 1991] suggest that the primary magma supply mechanisms that control the delivery and/or production of melt lie within the

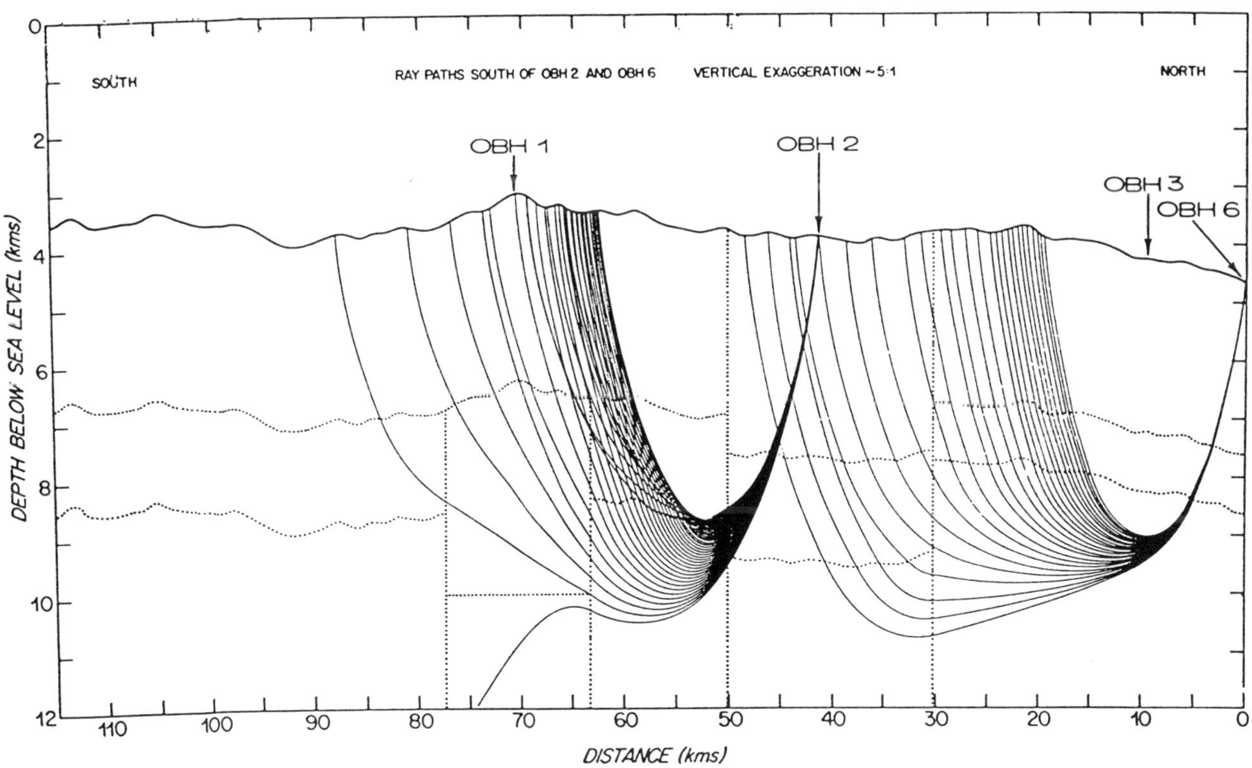

Fig. 72. Rays traced to the south of OBHs 2 and 6 through a velocity model of the crust along the ridge axis in the MARK area [from Purdy and Detrick, 1986]. Solid, wiggly line gives water depth along the profile. OBH 6 located in the ridge-transform intersection. Dotted lines separate regions with different velocity structure.

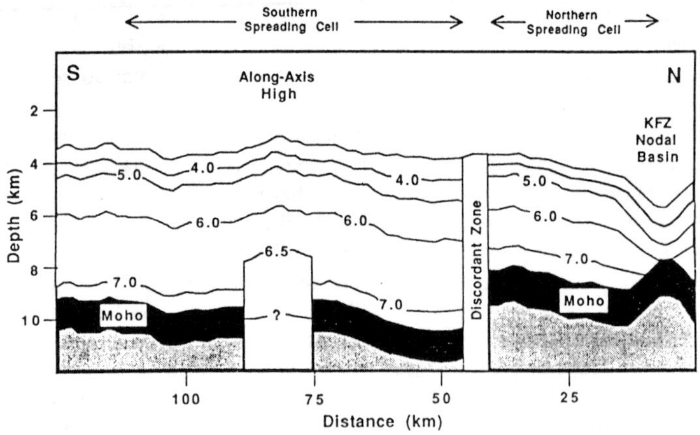

Fig. 73. Velocity model of the crust along the axis of the Mid-Atlantic Ridge in the MARK area along Profile A in Fig. 70b [from Morris and Detrick, 1991]. Moho is marked by a steep velocity gradient over a 1 km transition. Note change in crustal thickness across the discordant zone.

asthenosphere and are not totally dominated by surficial plate geometry. Again, this points to a dynamic component of upwelling.

SUMMARY

Mantle flow beneath the spreading centers must be regarded as part of the overall global convection system within the mantle. Proximity to hotspots, for example, clearly affects the morphological expression of the ridge. Flow away from hot spots may tend to suppress the segmentation of the ridge and motion of the ridge relative to the underlying mantle may influence the along-axis migration of non-transform offsets in the ridge. On a large scale, mid-ocean ridges appear to be passive features. The grossly similar morphology and chemistry of ridges no matter where they develop, the migration (and subduction) of ridges, and the lack of any deep seismological signature all point to an essentially passive role within the mantle convection system. On a more local scale, however, there is strong evidence for buoyant or dynamic flow in the upper 100 km, where thermal gradients are large and where melting and melt segregation occur.

Possible indications that buoyant upwelling and small-scale convection are important beneath spreading centers include: the variability in geoid anomalies across fracture zones; the constancy of oceanic crustal thickness independent of spreading rate; the small variation in crustal thickness with increased degree of melting as measured by Na_2O content of the basalts; the narrowness of the zone of crustal accretion; the along-axis segmentation of the ridges; and the lack of correlation of the degree of crustal thinning with the length of offset of the ridges.

Large variations in the depth of the spreading axis, the rate of subsidence of seafloor with increasing age, and shear wave velocities beneath seafloor of anomalous depth suggest substantial regional variations in the potential temperature of the upwelling mantle. The low correlation between axial depth and subsidence rate suggests that regional variations in composition are also an important influence on axial topography.

The effect of melt on seismic velocities and electrical conductivity depends on the shape of the melt pockets and their interconnectivity. To unambiguously separate the effects of melt fraction and temperature requires the simultaneous measurement of compressional and shear velocities, attenuation, and conductivity. No immediate effect on attenuation is expected for small fractions of melt. Magnetotelluric and surface wave measurements find high conductivity and low shear wave velocity beneath ridges that indicate high temperatures. The lowest velocities found at a depth of 30 to 80 km beneath young sea floor in the Pacific suggest that enough melt remains within the mantle matrix to be detected seismically. If the melt is distributed in tubes, then there must be an average of at least 3 to 6% melt present beneath seafloor less than 4 Ma old at a depth of 45 km. If the melt wets grain faces in the form of thin films, then the melt fraction could be much smaller. No evidence for large melt fractions (>10%) within the oceanic mantle has yet been found, although there is magnetotelluric evidence for the presence of pooled melt at the base of the Icelandic crust in the neovolcanic zones.

The ridges are segmented on a variety of scales, with offsets ranging from minor kinks in the spreading center to major fracture zones. Some of the offsets creating segmentation appear to arise from lithospheric mechanisms, such as changing directions of plate motion. In the Atlantic, substantial variations in crustal thickness often accompany the segmentation, but the degree of crustal thinning does not seem to correlate well with the length of the offset of the ridge. The reduction of crustal thickness by as much as 50% in some small-offset fracture zones suggests that one cause of segmentation is an intrinsic three-dimensionality associated with buoyant upwelling or magma migration.

Acknowledgment. I thank Marc Parmentier, Paul Hess, David Sparks, Jack Hermance and RIDGE Summer Institute participants for helpful discussion and comments. Two anonymous reviews were helpful in shaping the final version of the paper. This work was supported by National Science Foundation grant OCE-8911326 and by an honorarium from the RIDGE Summer Institute.

REFERENCES

Abbott, D., A statistical correlation between ridge crest offsets and spreading rate, *Geophys. Res. Lett.*, *13*, 157-160, 1986.

Agee, C.B. and D. Walker, Static compression and olivine flotation in ultrabasic silicate liquid, *J. Geophys. Res.*, *93*, 3437-3449, 1988.

Aki, K., A. Christofferson, and E.S. Husebye, Determination of the three-dimensional seismic structure of the lithosphere, *J. Geophys. Res.*, *82*, 277-296, 1977.

Ambos, E.L. and D.M. Hussong, Oceanographer transform fault structure compared to that of surrounding oceanic crust: results from seismic refraction analysis, *J. Geodynamics*, *5*, 79-102, 1986.

Anderson, D.L., A seismic equation of state II, shear and thermodynamic properties of the lower mantle, *Phys. Earth Planet. Int.*, *45*, 307-323, 1987.

Anderson, O.L., Equation for thermal expansivity in planetary interiors, *J. Geophys. Res.*, *72*, 3661-3668, 1967.

Anderson, O.L., A. Chopelas, and R. Boehler, Thermal expansivity versus pressure at constant temperature: a re-examination, *Geophys. Res. Lett.*, *17*, 685-688, 1990.

Atwater, T.M., and J.D. Mudie, Detailed near-bottom geophysical study of the Gorda Rise, *J. Geophys. Res.*, *78*, 8665-8686, 1973.

Ballard, R.D., and T.H. van Andel, Morphology and tectonics of the inner rift valley at 36°50'N on the Mid-Atlantic Ridge, *Geol. Soc. Amer. Bull.*, 507-530, 1977.

Ballard, R.D., R.T. Holcomb, and T.H. van Andel, The Galapagos Rift at 86°W, 3, sheet flows, collapse pits, and lava lakes of the rift, J. *Geophys. Res.*, *84*, 5407-5422, 1979.

Barth, G.A., J.C. Mutter, and K.A. Kastens, Increase in seismic thickness of the oceanic crust near a ridge-transform intersection of the Clipperton Fracture Zone, *EOS Trans. AGU*, *69*, 1475, 1988.

Barth, G.A., K.A. Kastens, and E.M. Klein, The origin of bathymetric highs at ridge-transform intersections: A multi-disciplinary case study at the Clipperton Fracture Zone, *Mar. Geophys. Res.*, in press, 1992.

Batiza, R., P.J. Fox, P.R. Vogt, S.C. Cande, N.R. Grindlay, W.G. Melson, and T. O'Hearn, Morphology, abundance, and chemistry of near-ridge seamounts in the vicinity of the Mid-Atlantic Ridge 26°S, *J. Geol.*, *97*, 209-220, 1989.

Beblo, M. and A. Bjornsson, Magnetotelluric investigation of the lower crust and upper mantle beneath Iceland, *J. Geophys.*, *45*, 1-16, 1978.

Beblo, M. and A. Bjornsson, A model of electrical resistivity beneath NE-Iceland, correlation with temperature, *J. Geophys.*, *47*, 184-190, 1980.

Beblo, M., A. Bjornsson, K. Arnason, B. Stein, and P. Wolfgram, Electrical conductivity beneath Iceland -- Constraints imposed by magnetotelluric results on temperature, partial melt, crust and mantle structure, *J. Geophys.*, *53*, 16-23, 1983.

Beck, A.E., D.M. Darbha, and H.H. Schloessin, Lattice conductivities of single crystal and polycrystalline materials at mantle pressures and temperatures, *Phys. Earth Planet. Int.*, *17*, 35-53, 1978.

Bell, R.E., The Reykjanes Ridge: A slow spreading ridge with the morphology and mantle Bouguer signature of a fast spreading center, *EOS, Trans. AGU, 17. Spring Meeting Supplement*, 260, 1991.

Berckhemer, H., W. Kampfmann, E. Aulbach and H. Schmeling, Shear modulus and Q of forsterite and dunite near partial melting from forced oscillation experiments, *Phys. Earth Planet. Int.*, *29*, 30-41, 1982.

Blackman, D.K., and D.W. Forsyth, Isostatic compensation of tectonic features of the Mid-Atlantic Ridge: 25-27°30'S, *J. Geophys. Res.*, *96*, 11741-11758, 1991a.

Blackman, D.K. and D.W. Forsyth, The effects of plate thickening on three-dimensional, passive flow of the mantle beneath mid-ocean ridges, *AGU Monograph*, 1991b.

Bonatti, E., Serpentinite protrusions in the oceanic crust, *Earth Planet. Sci. Lett.*, *32*, 107-113, 1976.

Booker, J.R. and A.D. Chave, Introduction to the special section on the EMSLAB-Juan de Fuca experiment, *J. Geophys. Res.*, *94*, 14093-14098, 1989.

Bottinga, Y. and D.F. Weill, Densities of liquid silicate systems calculated from partial molar volumes of oxide components, *Am. J. Sci.*, *269*, 169-182, 1970.

Bottinga, Y. and C.J. Allegre, Geophysical, petrological and geochemical models of the oceanic lithosphere, *Tectonophys.*, *32*, 9-59, 1976.

Bowin, C., G. Thompson, and J.G. Schilling, Residual geoid anomalies in Atlantic Ocean basin: Relationship to mantle plumes, *J. Geophys. Res.*, *89*, 9905-9918, 1984.

Brace, W.F., A.S. Orange, and T.R. Madden, The effect of pressure on the electrical resistivity of water-saturated crystalline rocks, *J. Geophys. Res.*, *70*, 5669-5678, 1965.

Buck, W.R., Analysis of the cooling of a variable-viscosity fluid with applications to the earth, *Geophys. J. R. Astr. Soc.*, *89*, 549-577, 1987.

Buck, W.R. and E.M. Parmentier, Convection beneath young oceanic lithosphere: Implications for thermal structure and gravity, *J. Geophys. Res.*, *91*, 1961-1974, 1986.

Buck, W.R. and W. Su, Focussed mantle upwelling below mid-ocean ridges due to feedback between viscosity and melting, *Geophys. Res. Lett.*, *16*, 641-644, 1989.

Burnett, M.S., D.W. Caress, and J.A. Orcutt, Tomographic image of the magma chamber at 12°50'N on the East Pacific Rise, *Nature*, *339*, 206-208, 1989.

Calvert, A.J., and C.G. Potts, Seismic evidence for hydrothermally altered mantle beneath old crust in the Tydeman fracture zone, *Earth Planet. Sci. Lett.*, *75*, 439-449, 1985.

Cara, M. and J.J. Leveque, Oriented olivine crystals in the upper mantle: a test from the inversion of multimode surface-wave data, *Phys. Earth Planet. Int.*, *47*, 246-252, 1987.

Cara, M. and J.J. Leveque, Anisotropy of the asthenosphere: the higher mode data of the Pacific revisited, *Geophys. Res. Lett.*, *15*, 205-208,1988

Carbotte, S., S. Welch, and K.C. Macdonald, Spreading rates, rift propagation, fracture zone offset histories during the past 5 m.y. on the Mid-Atlantic Ridge; 25°-27°30'S and 31°-34°30'S, *Marine Geophys. Res.*, *13*, 51-80, 1991.

Cazenave, A., Thermal cooling of the oceanic lithosphere: New constraints from geoid height data, *Earth Planet. Sci. Lett.*, *70*, 394-406, 1984.

Cazenave, A., B. Lago, and K. Dominh, Geoid anomalies over the northeast Pacific fracture zones from satellite altimeter data, *Geophys. J. R. Astr. Soc.*, *69*, 15-31, 1982.

Cazenave, A. B. Lago, and K. Dominh, Thermal parameters of the oceanic lithosphere estimated from geoid height data, *J. Geophys. Res.*, *88*, 1105-1118, 1983.

Chan, W.W., I.S. Sacks, and R. Morrow, Anelasticity of the Iceland Plateau from surface wave analysis, *J. Geophys. Res.*, *94*, 5675-5688, 1989.

Chapman, M.E., Geoid anomaly over mid-ocean ridges, *EOS, Trans. AGU*, *58*, 368, 1977.

Chapman, M.E., Techniques for interpretation of geoid anomalies, *J. Geophys. Res.*, *84*, 3793-3802, 1979.

Chen, Y.J., Oceanic crustal thickness versus spreading rate, Geophys. Res. Lett., in press, 1992.

Cochran, J.R., Variations in subsidence rates along intermediate and fast spreading mid-ocean ridges, *Geophys. J. R. Astr. Soc.*, *87*, 421-454, 1986.

Collette, B.J., Thermal contraction joints in a spreading seafloor as origin of fracture zones, *Nature*, *251*, 299-300, 1974.

Constable, S., and A. Duba, Electrical conductivity of olivine, a dunite and the mantle, *J. Geophys. Res.*, *95*, 6967-6978, 1990.

Copelas, A. and R. Boehler, Thermal expansion measurements at very high pressure, systematics, and a case for a chemically homogeneous mantle, *Geophys. Res. Lett.*, *16*, 1347-1350, 1989.

Cordery, M.J., and J. Phipps Morgan, Melting and mantle flow beneath a mid-ocean spreading center, *Earth Planet. Sci. Lett.*, in press, 1991.

Cormier, M.-H., R.S. Detrick, and G.M. Purdy, Anomalously thin crust in oceanic fracture zones: New seismic constraints from the Kane fracture zone, *J. Geophys. Res.*, *89*, 10,249-10266, 1984.

Craig, C.H. and D. McKenzie, the existence of a thin low-viscosity layer beneath the lithosphere, *Earth Planet. Sci. Lett.*, *78*, 420-426, 1986.

Crane, K., The spacing of rift-axis highs: dependence upon diapiric processes in the underlying asthenosphere?, *Earth Planet. Sci. Lett.*, *72*, 405-414, 1985.

Crough, S.T., Thermal origin of mid-plate hot-spot swells, *Geophys. J.R. Astr. Soc.*, *55*, 451-469, 1978.

Crough, S.T., Geoid anomalies across fracture zones and the thickness of the lithosphere, *Earth Planet. Sci. Lett.*, *44*, 224-230, 1979.

Dahlen, F.A., Isostatic geoid anomalies on a sphere, *J. Geophys. Res., 87*, 3943-3947, 1982.

Davis, E.E. and C.R.B Lister, Fundamentals of ridge crest topography, *Earth Planet. Sci. Lett., 21*, 405-413, 1974.

Detrick, R.S., An analysis of geoid anomalies across the Mendocino fracture zone: Implications for thermal models of the lithosphere, *J. Geophys. Res., 86*, 11751-11762, 1981.

Detrick, R.S., M.-H. Cormier, R.A. Prince, D.W. Forsyth, and E.L. Ambos, Seismic constraints on the crustal structure within the Vema fracture zone, *J. Geophys. Res., 87*, 10599-10612, 1982.

Detrick, R.S., P. Buhl, E. Vera, J. Mutter, J. Orcutt, J. Madsen, and T. Brocher, Multichannel seismic imaging of an axial magma chamber along the East Pacific Rise between 9°N and 13°N, *Nature, 326*, 35-41, 1987.

Detrick, R.S., R.S. White, and G.M. Purdy, Crustal structure of oceanic fracture zones, in The Tectonics of Fracture Zones, edited by P.J. Fox, Academic, San Diego, CA, in press, 1992.

Driscoll, M.L. and B. Parsons, Cooling of the oceanic lithosphere -- evidence from geoid anomalies across the Udintsev and Eltanin fracture zones, *Earth Planet. Sci. Lett., 88*, 289-307, 1988.

Duba, A., and T.J. Shankland, Free carbon and electrical conductivity in the Earth's mantle, *Geophys. Res. Lett., 9*, 1271-1274, 1982.

Duba, A., H.C, Heard, and R.N. Schock, Electrical conductivity of olivine at high pressure and under controlled oxygen fugacity, *J. Geophys. Res., 79*, 1667-1673, 1974.

Duschenes, J.D., and S.C. Solomon, Shear wave travel time residuals from oceanic earthquakes and the evolution of oceanic lithosphere, *J. Geophys. Res., 82*, 1985-2000, 1977.

Dziewonski, A.M. and J.H. Woodhouse, Global images of the Earth's interior, *Science, 236*, 37-48, 1987.

Estey, L.H., and B.J. Douglas, Upper mantle anisotropy: a preliminary model, *J. Geophys. Res., 91*, 11393-11406, 1986.

Evans, J.R. and I.S. Sacks, Lithospheric structure in the North Atlantic from observations of Love and Rayleigh waves, *J. Geophys. Res., 85*, 7175-7182, 1980.

Eysteinsson, H. and J.F. Hermance, Magnetotelluric measurements across the eastern neovolcanic zone in south Iceland, *J. Geophys. Res., 90*, 10093-10103, 1985.

Falloon, T.J. and D.H. Green, Anhydrous partial melting of MORB pyrolite and other peridotite compositions at 10 kbar: implication for the origin of MORB glasses, *Mineral. Petrol., 37*, 181-219, 1987.

Falloon, T.J. and D.H. Green, Anhydrous partial melting of peridotite from 8 to 35 kb and the petrogenesis of MORB, *J. Petrol., Special Lithosphere Issue*, 379-414, 1988.

Falloon, T.J., D.H. Green, C.J. Hatton, and K.L. Harris, Anhydrous partial melting of a fertile and depleted peridotite from 2 to 30 kb and application to basalt petrogenesis, *J. Petrol., 29*, 1257-1282, 1988.

Feng, R. and T.V. McEvilly, Interpretation of seismic reflection profiling data for the structure of the San Andreas Fault Zone, *Bull. Seism. Soc. Am., 73*, 1701-1720, 1983.

Filloux, J.H., Ocean-floor magnetotelluric sounding over north central Pacific, *Nature, 269*, 297-301, 1977.

Filloux, J.H., Magnetotelluric soundings over the northeast Pacific may reveal spatial dependence of the depth and conductance of the asthenosphere, *Earth Planet. Sci. Lett., 46*, 244-252, 1980.

Flower, S.C. and G.A. Saunders, The elastic behavior of indium under pressure and with temperature up to the melting point, *Phil. Mag. B, 62*, 311 328, 1990.

Fornari, D.J., R. Batiza, and M.A. Luckman, Seamount abundances and distribution near the East Pacific Rise 0°-24°N based on Seabeam data, in Seamounts, Islands and Atolls, *AGU Mon., 43*, Keating et al., eds., 13-21, 1987.

Forsyth, D.W., The early structural evolution and anisotropy of the oceanic upper mantle, *Geophys. J., 43*, 103-162, 1975.

Forsyth, D.W., The evolution of the upper mantle beneath mid-ocean ridges, *Tectonophys., 38*, 89-118, 1977.

Forsyth, D.W., Crustal thickness, average depth and degree of melting in fractional melting models of mid-ocean ridges, in preparation, 1991.

Forsyth, D.W. and F. Press, Geophysical tests of petrological models of the spreading lithosphere, *J. Geophys. Res., 76*, 7963-7975, 1971.

Forsyth, D.W., R.L. Ehrenbard, and S. Chapin, Anomalous upper mantle beneath the Australian-Antarctic discordance, *Earth Planet. Sci. Lett., 84*, 471-478, 1987.

Forsyth, D.W., D.S. Scheirer, K.C. Macdonald, C.M. Weiland, and M.H. Cormier, Seamount chains on young seafloor of the Pacific plate, 16 to 18°S, *EOS Trans AGU, Spring Meeting special issue, 262*, 1991.

Fowler, C.M.R., Crustal structure of the Mid-Atlantic Ridge at 37°N, *Geophys. J. R. Astr. Soc., 47*, 459-491, 1976.

Fowler, C.M.R., The Mid-Atlantic Ridge: Structure at 45°N, *Geophys. J. R. Astr. Soc., 54*, 167-183, 1978.

Frisillo, A.L. and G.R. Barsch, Measurement of single-crystal elastic constants of bronzite as a function of pressure and temperature, *J. Geophys. Res., 77*, 6360-6384, 1972.

Fujii, N., K. Osamura, and E. Takahashi, Effect of water saturation on the distribution of partial melt in the olivine-pyroxene-plagioclase system, *J. Geophys. Res., 91*, 9253-9259, 1986.

Gallo, D.G., P.J. Fox, and K.C. Macdonald, A Sea Beam investigation of the Clipperton Transform fault: The morphotectonic expression of a fast slipping transform boundary, *J. Geophys. Res., 91*, 3455-3467, 1986.

Giardini, D., X.D. Li, and J. Woodhouse, Three dimensional structure of the Earth from splitting in free oscillation data, *Nature, 325*, 405-409, 1987.

Gibert, D., C. Camerlynck, and V. Courtillot, Geoid anomalies across Ascension fracture zone and the cooling of the lithosphere, *Geophys. Res. Lett., 14*, 603-606, 1987.

Girardin, N., and G. Poupinet, Teleseismic S travel-time delay for Mid-Atlantic Ridge earthquakes, *Phys. Earth Planet. Inter., 9*, 306-313, 1974.

Girardin, N. and W.R. Jacoby, Rayleigh wave dispersion along Reykjanes Ridge, *Tectonophys., 55*, 155-171, 1979.

Graham, E.K., J.A. Schwab, S.M. Sopkin, and H. Takei, The pressure and temperature dependence of the elastic properties of single-crystal fayalite Fe_2SiO_4, *Phys. Chem. Minerals, 16*, 186-198, 1988.

Grand, S.P., Tomographic inversion for shear velocity beneath the North American plate, *J. Geophys. Res., 92*, 14,065-14,090, 1987.

Green, W.V., U. Achauer, and R.P. Meyer, A three-dimensional seismic image of the crust and upper mantle beneath the Kenya rift, *Nature, 354*, 199-203, 1991.

Green, W.V., and R.P. Meyer, Array observations of PKP phases across the Kenya rift: Implications for structure and tectonics, *Tectonophys., in press*, 1992.

Grindlay, N.R., P.J. Fox, and K.C. Macdonald, Second-order ridge axis discontinuities in the South Atlantic: Morphology, structure, and evolution, Mar. *Geophys. Res., 13*, 21-49, 1991.

Haak, V., Relations between electrical conductivity and petrological parameters of the crust and upper mantle, *Geophys. Surv., 4*, 57-69, 1980.

Halderman, T.P. and P.M. Davis, Q_p beneath the Rio Grande and East African rift zones, *J. Geophys. Res., 96*, 10,113-10,128, 1991.

Harrison, C.G.A., Formation of magnetic anomaly patterns by dyke injection, *J. Geophys. Res., 73*, 2137-2142, 1968.

Haxby, W.F. and D.L. Turcotte, On isostatic geoid anomalies, *J. Geophys. Res., 83*, 5473-5478, 1978.

Haxby, W.F. and E.M. Parmentier, Thermal contraction and the state of stress in the oceanic lithosphere, *J. Geophys. Res., 93*, 6419-6429, 1988.

Haxby, W.F. and J.K. Weissel, Evidence for small-scale mantle convection from Seasat altimeter data, *J. Geophys. Res., 91*, 3507-3520, 1986.

Hayes, D.E., Age-depth relationships and depth anomalies in the Southeast Indian Ocean, and South Atlantic Ocean, *J. Geophys. Res., 93*, 2937-2954, 1988.

Healy, J. and L. Peake, Seismic velocity structure along a section of the San Andreas fault near Bear Valley. California, *Bull. Seism. Soc. Am., 65*, 1177-1197, 1975.

Hebert, R., D. Bideau, and R. Hekinian, Ultramafic and mafic rocks from the Garret transform fault near 13°30'S on the East Pacific Rise: igneous petrology, *Earth Planet. Sci. Lett.*, *65*, 107-125, 1983.

Hermance, J.F., The electrical conductivity of materials containing partial melts: a simple model from Archie's Law, *Geophys. Res. Lett.*, *6*, 613-616, 1979.

Hermance, J.F., Crustal genesis in Iceland: Geophysical constraints on crustal thickening with age, *Geophys. Res. Lett.*, *8*, 203-206, 1981.

Herron, T.J., P.L. Stoffa, and P. Buhl, Magma chambers and mantle reflections: East Pacific Rise, *Geophys. Res. Lett.*, *7*, 989-992, 1980.

Hersir, G.P., A. Bjornsson, and L.B. Pedersen, Magnetotelluric survey across the active spreading zone in southwest Iceland, *J. Volcanol. Geotherm. Res.*, *20*, 253-265, 1984.

Hess, P.C., Phase equilibria and the origin of ocean floor basalts, *AGU Monograph*, 1991.

Honda, S. and T. Tanimoto, Regional three-D heterogeneities by the wave form inversion - Application to the Atlantic area, *Geophys. J. R. Astr. Soc.*, *91*, 737-753, 1987.

Isaak, D.G., O.L. Anderson, and T. Goto, Elasticity of single-crystal forsterite measured to 1700 °K, *J. Geophys. Res.*, *94*, 5895-5906, 1989.

Jaques, A.L. and D.H. Green, Anhydrous melting of peridotite at 0-15 kb pressure and the genesis of tholeitic basalts, Contrib. *Mineral. Petrol.*, *73*, 287-310, 1980.

Jordan, T.H., Mineralogies, densities, and seismic velocities of garnet lherzolites and their geophysical implications, *in Proceedings of the Second International Kimberlite Conference*, v. 2, eds. F.R. Boyd and H.O.A. Meyer, AGU, Washington, D.C., 1979.

Jordan, T.H., and W.S. Lynn, A velocity anomaly in the lower mantle, *J. Geophys. Res.*, *79*, 2679-2684, 1974.

Kampfmann, W. and H. Berckhemer, High temperature experiments on the elastic and anelastic behaviour of magmatic rocks, *Phys. Earth Planet. Int.*, *40*, 223-247, 1985.

Kane, K.A., and D.E. Hayes, Tectonic corridors in the South Atlantic: Evidence for long-lived MOR segmentation, *J. Geophys. Res.*, *in press*, 1992.

Kastens, K.A., W.B.F. Ryan, and P.J. Fox, Structural and volcanic expression of a fast slipping ridge-transform-ridge-plate boundary: Sea MARCI and photographic surveys at the Clipperton transform fault, *J. Geophys. Res.*, *91*, 3469-3488, 1986.

Kausel, E.G., A.R. Leeds, and L. Knopoff, Variations of Rayleigh wave phase velocities across the Pacific Ocean, *Science*, *186*, 141-143, 1974.

Kawasaki, I., Azimuthally anisotropic model of the oceanic upper mantle, Phys. *Earth Planet. Int.*, *43*, 1-21, 1986.

Kerr, R.C., and J.R. Lister, Island arc and mid-ocean ridge volcanism, modelled by diapirism from linear source regions, *Earth Planet. Sci. Lett.*, *88*, 143-152, 1988.

Klein, E.M. and C.H. Langmuir, Global correlations of ocean ridge basalt chemistry with axial depth and crustal thickness, *J. Geophys. Res.*, *92*, 8089-8115, 1987.

Klein, E.M., the Clipperton Transform Team, and the CHEPR team, Geochemistry of basalts collected during Alvin dives within and adjacent to the Clipperton transform fault (10°N East Pacific Rise), *EOS Trans. AGU*, *68*, 1540, 1987.

Klein, E.M., C.H. Langmuir, and T. Plank, Geochemical constraints on melt formation and segregation, *AGU Monograph, in press*, 1991a.

Klein, E.M., T. Plank, and C.H. Langmuir, Constraints on models for mantle melting beneath ocean ridges, *RIDGE Events*, *2*, 11-12, 1991b.

Klitgord, K.D., S.P. Huestis, R.L. Parker, and J.D. Mudie, An analysis of near-bottom magnetic anomalies: Seafloor spreading, the magnetized layer, and the geomagnetic time scale, *Geophys. J. R. Astr. Soc.*, *43*, 387-424, 1975.

Kong, L.S.L., S.C. Solomon, and G.M. Purdy, Along-axis variations in crustal structure on the Mid-Atlantic Ridge near 26°N, J. *Geophys. Res.*, *in press*, 1991.

Kumazawa, M. and O.L. Anderson, Elastic moduli, pressure derivatives, and temperature derivatives of single-crystal olivine and single-crystal forsterite, *J. Geophys. Res.*, *74*, 5961-5972, 1969.

Kuo, B.-Y. and D.W. Forsyth, Gravity anomalies of the ridge-transform system in the South Atlantic between 31 and 34.5°S: Upwelling centers and variations in crustal thickness, *Mar. Geophys. Res.*, *10*, 205-232, 1988.

Kuo, B.Y., D.W. Forsyth, and M. Wysession, Lateral heterogeneity and azimuthal anisotropy in the North Atlantic determined from SS-S differential travel times, *J. Geophys. Res.*, *92*, 6421-6436, 1987.

Lange, R.A., and I.S.E. Carmichael, Densities of Na_2O-K_2O-CaO-MgO-FeO-Fe_2O_3-Al_2O_3-TiO_2-SiO_2 liquids: new measurements and derived partial molar properties, *Geoch. Cosm. Acta*, *51*, 2931-2946, 1987.

Langmuir, C.H., J.F. Bender, and R. Batiza, Petrological and tectonic segmentation of the East Pacific Rise, 5°30' - 14°30'N, *Nature*, *322*, 422-429, 1986.

Law, L.K. and J.P. Greenhouse, Geomagnetic variation sounding of the asthensophere beneath the Juan de Fuca ridge, *J. Geophys. Res.*, *86*, 967-978, 1981.

Le Douaran, S. and J. Francheteau, Axial depth anomalies from 10 to 50° north along the Mid-Atlantic Ridge: correlation with other mantle properties, *Earth Planet. Sci*, *Lett.*, *54*, 29-47, 1981.

Lin, J., G.M. Purdy, H. Schouten, J.-C. Sempere, and C. Zervas, Evidence from gravity data for focused magmatic accretion along the Mid-Atlantic Ridge, *Nature*, *344*, 627-632, 1990.

Lin, J.. and E.M. Parmentier, Surface topography due to convection in a variable viscosity fluid: application to short wavelength gravity anomalies in the central Pacific ocean, *Geophys. Res. Lett.*, *12*, 357-360, 1985.

Lin, J., and J. Phipps Morgan, The spreading rate dependence of three-dimensional mid-ocean ridge gravity structure, *Geophys. Res. Lett.*, *in press*, 1991.

Lister, C.R.B., Geoid anomalies over cooling lithosphere: source for a third kernel of upper mantle thermal parameters and thus an inversion, *Geophys.J. R. Astr. Soc.*, *68*, 219-240, 1982.

Lonsdale, P., Segmentation of the Pacific-Nazca spreading center, 1°N-20°S, *J. Geophys. Res.*, *94*, 713-743, 1989.

Louden, K.E. and D.W. Forsyth, Thermal conduction across fracture zones and the gravitational edge effects, *J. Geophys. Res.*, *81*, 4869-4874, 1976.

Louden, K.E., R.S. White, C.G. Potts, and D.W. Forsyth, Structure and seismotectonics of the Vema Fracture Zone, Atlantic Ocean, *J. Geol. Soc. London*, *143*, 795-805, 1986.

Macdonald, K.C., Near-bottom magnetic anomalies, asymmetric spreading, oblique spreading and tectonics of the Mid-Atlantic Ridge near 37°N, *Geol. Soc. Am. Bull.*, *88*, 541-555, 1977.

Macdonald, K.C., Mid-ocean ridges: Fine scale tectonic, volcanic, and hydrothermal processes within the plate boundary zone, *Ann. Rev. Earth Planet. Sci.*, *10*, 155-190, 1982.

Macdonald, K.C., The crest of the Mid-Atlantic Ridge: Models for crustal generation processes and tectonics, in Vogt, P.R. and Tucholke, B,E, (eds.), *The Geology of North America, Volume M*, The Western North Atlantic Region, Geol. Soc. Amer., 51-68, 1986.

Macdonald, K.C., S.P. Miller, S.P. Huestis, and F.N. Spiess, Three-dimensional modelling of a magnetic reversal boundary from inversion of deep-tow measurements, *J. Geophys. Res.*, *85*, 3670-3680, 1980.

Macdonald, K.C., P.J. Fox, L.J. Perram, M.F. Eisen, R.M. Haymon, S.P. Miller, S.M. Carbotte, M.-H. Cormier, and A.N. Shor, A new view of the mid-ocean ridge from the behaviour of ridge axis discontinuities, *Nature*, *335*, 217-225, 1988a.

Macdonald, K.C., R.M. Haymon, S.P. Miller, J.-C. Sempere, and P.J. Fox, Deep-tow and sea beam studies of dueling propagating ridges on the East Pacific Rise near 20°40'S, *J. Geophys. Res.*, *93*, 2875-2898, 1988b.

Macdonald, K.C., R. Haymon, and A. Shor, A 220 km^2 recently erupted lava field on the East Pacific Rise near lat 8°S, *Geology*, *17*, 212-216, 1989.

Madsen, J.A., D.W. Forsyth, and R.S. Detrick, A new isostatic model for the East Pacific Rise Crest, *J. Geophys. Res., 89,* 9997-10015, 1984.

Madsen, J. A., R.S. Detrick, J.C. Mutter, P. Buhl, and J.A. Orcutt, A two- and three-dimensional analysis of gravity anomalies associated with the East Pacific Rise at 9°N and 13°N, *J. Geophys. Res., 95,* 4967-4987, 1990.

Marsh, B.D., Island arc development: Some observations, experiments, and speculations, *J. Geol., 87,* 687-713, 1979.

Marsh, B.D., J.G. Marsh, and R.G. Williamson, On gravity from SST, geoid from Seasat, and plate age and fracture zones in the Pacific, *J. Geophys. Res.*, 6070-6078, 1984.

Marsh, J.G., R.E. Cheney, T.V. Martin, and J.J. McCarthy, Computation of a precise mean sea surface in the eastern North Pacific using SEASAT altimetry, *EOS, Trans. AGU, 63,* 178-179, 1982.

Marty, J.C., and A. Cazenave, Thermal evolution of the lithosphere beneath fracture zones inferred from geoid anomalies, *Geophys. Res. Lett., 15,* 593-596, 1988.

Marty, J.C. and A. Cazenave, Regional variations in subsidence rate of oceanic plates: a global analysis, *Earth Planet. Sci. Lett., 94,* 301-315, 1989.

Marty, J.C., A. Cazenave, and B. Lago, Geoid anomalies across Pacific fracture zones, *Geophys. J., 93,* 1-23, 1988.

Mavko, G.M., Velocity and attenuation in partially molten rocks, *J. Geophys. Res., 85,* 5173-5189, 1980.

McClain, J.S., J.A. Orcutt, and M. Burnett, The East Pacific Rise in cross section: A seismic model, *J. Geophys. Res., 90,* 8627-8639, 1985.

McKenzie, D.P., Surface deformation, gravity anomalies and convection, *Geophys. J. R. Astr. Soc., 48,* 211-238, 1977.

McKenzie, D.P. and M.J. Bickle, The volume and composition of melt generated by adiabatic upwelling, *J. Petrol., 29,* 625-679, 1988.

McKenzie, D.P., and C. Bowin, The relationship between bathymetry and gravity in the Atlantic ocean, *J. Geophys. Res., 81,* 1903-1915, 1976.

McNutt, M.K., Lithospheric flexure and thermal anomalies, *J. Geophys. Res., 89,* 11180-11194, 1984.

Michael, P.J., B.B. Hanan, and P.J. Fox, Petrological and geochemical variations along contrasting ridge segments: 31°-34°S, MAR, *EOS, Trans. AGU, 71,* 1628, 1991.

Minshull, T.A., R.S. White, J.C. Mutter, P. Buhl, R.S. Detrick, C.A. Williams, and E. Morris, Crustal structure at the Blake Spur Fracture Zone from expanding spread profiles, *J. Geophys. Res., 96,* 9955-9984, 1991

Mocquet, A. and B. Romanowicz, Three-dimensional structure of the upper mantle beneath the Atlantic Ocean inferred from long-period Rayleigh waves, 2. Inversion, *J. Geophys. Res., 95,* 6787-6798, 1990.

Mocquet, A., B. Romanowicz, and J.P. Montagner, Three-dimensional structure of the upper mantle inferred from long-period Rayleigh waves, 1. Group and phase velocity distributions, *J. Geophys. Res., 94,* 7449-7468, 1989.

Montagner, J.P., First results on the three-dimensional structure of the Indian Ocean inferred from long-period surface waves, *Geophys. Res. Lett., 13,* 315-318, 1986.

Montagner, J.P. and N. Jobert, Investigation of upper mantle structure under young regions of the southeast Paicfic using long-period Rayleigh waves, *Phys. Earth Planet. Int., 27,* 206-222, 1981.

Montagner, J.P. and N. Jobert, Vectorial tomography - II. Application to the Indian Ocean, *Geophys. J., 94,* 309-344, 1988.

Montagner, J.P. and H.C. Nataf, On the inversion of the azimuthal anisotropy of surface waves, J. *Geophys. Res., 91,* 511-520, 1986.

Montagner, J.P. and H.C. Nataf, Vectorial tomography - I. Theory, *Geophys. J., 94,* 295-307, 1988.

Montagner, J.P. and T. Tanimoto, Global anisotropy in the upper mantle inferred from the regionalization of phase velocities, *J. Geophys. Res., 95,* 4797-4819, 1990.

Mooney, W.D., and A. Ginzburg, Seismic measurements of the internal properties of fault zones, *Pure Appl. Geophys., 124,* 141-157, 1986.

Morgan, W.J., Gravity anomalies and convection currents, 1, A sphere and a cylinder sinking beneath the surface of a viscous fluid, *J. Geophys. Res., 70,* 6175-6185, 1965.

Morris, E. and R.S. Detrick, Three-dimensional analysis of gravity anomalies in the MARK area, Mid-Atlantic Ridge 23°N, *J. Geophys. Res., 96,* 4355-4366, 1991.

Murase, T., I. Kushiro, and T. Fujii, Electrical conductivity of partially molten peridotite, *Year Book Carnegie Inst. Washington, 76,* 416-419, 1977.

Murase, T. and I. Kushiro, Compressional wave velocity in partially molten peridotite at high pressures, *Year Book Carnegie Inst. Washington, 78,* 559-562, 1979.

Murase, T. and H. Fukuyama, Shear wave velocity in partially molten peridotite at high pressure, *Year Book Carnegie Inst. Washington, 79,* 307-310, 1980.

Mutter, J.C., R.S. Detrick, and the NAT Study Group, Multi-channel seismic evidence for anomalously thin crust at Blake Spur fracture zone, *Geology, 12,* 534-537, 1984.

Nelson, S.A. and I.S.E. Carmichael, Partial molar volumes of oxide components in silicate liquids, *Contrib. Mineral. Petrol., 71,* 117-124, 1979.

Nishimura, C.E. and D.W. Forsyth, Anomalous Love-wave phase velocities in the Pacific: sequential pure-path and spherical harmonic inversion, *Geophys. J. R. Astr. Soc., 81,* 389-407, 1985.

Nishimura, C.E. and D.W. Forsyth, Rayleigh wave phase velocities in the Pacific with implications for azimuthal anisotropy and lateral heterogenetities, *Geophys. J. R. Astr. Soc., 94,* 479-501, 1988

Nishimura, C.E. and D.W. Forsyth, The anisotropic structure of the upper mantle in the Pacific, *Geophys. J., 96,* 203-229, 1989.

Niu, Y. and R. Batiza, DENSCAL: A program for calculating densities of silicate melts and mantle minerals in melting range, *Comput. Geosci., 17,* 679-687, 1991a.

Niu, Y. and R. Batiza, An empirical method for calculating melt compositions produced beneath mid-ocean ridges: application for axis and off-axis (seamounts) melting, *J. Geophys. Res., 96,* 21,753-21,777, 1991b.

Ockendon, J.R., and D.L. Turcotte, On the gravitational potential and field anomalies due to thin mass layers, *Geophys. J. Roy. Astr. Soc., 48,* 479-492, 1977.

O'Connell, R.J. and B. Budiansky, Viscoelastic properties of fluid-saturated cracked solids, *J. Geophys. Res., 82,* 5719-5735, 1977.

Oldenburg, D.W., Conductivity structure of the oceanic upper mantle beneath the Pacific plate, *Geophys. J. R. Astr. Soc., 65,* 359-394, 1981.

Oldenburg, D.W., K.P. Whittall, and R.L. Parker, Inversion of ocean bottom magnetotelluric data revisited, *J. Geophys. Res., 89,* 1829-1833, 1984.

Orcutt, J.A., B.L.N. Kennett, L.M. Dorman and W.A. Prothero, Structure of the East Pacific Rise from an ocean-bottom seismometer survey, *Geophys. J. R. Astr. Soc., 45,* 305-320, 1976.

Oxburgh, E.R. and E.M. Parmentier, Compositional and density stratification in oceanic lithosphere - causes and consequences, *J. Geol. Soc., 133,* 343-355, 1977.

Parker, E.C., P.M. Davis, J.R. Evans, H.M. Iyer, and K.H. Olsen, Upwarp of anomalous asthenosphere beneath the Rio Grande rift, *Nature, 312,* 354-356, 1984.

Parker, R.L., and D.W. Oldenburg, Thermal model of ocean ridges, *Nature, 242,* 137-139, 1973.

Parmentier, E.M. and W.F. Haxby, Thermal stresses in the oceanic lithosphere: Evidence from geoid anomalies at fracture zones, *J. Geophys. Res., 91,* 7193-7204, 1986.

Parmentier, E.M. and J. Phipps Morgan, Spreading rate dependence of three-dimensional structure in oceanic spreading centers, *Nature, 348,* 325-328, 1990.

Parsons, B., and S. Daly, The relationship between surface topography, gravity anomalies, and the temperature structure of convection, *J. Geophys. Res.*, 1129-1944, 1983.

Parsons, B., and F.M. Richter, A relation between the driving force and geoid anomaly associated with mid-ocean ridges, *Earth Planet. Sci. Lett., 51,* 445-450, 1980.

Parsons, B. and J.G. Sclater, An analysis of the variation of ocean floor bathymetry and heat flow with age, *J. Geophys. Res.*, *82*, 803-827, 1977.

Phipps Morgan, J., Melt migration beneath mid-ocean spreading centers, *Geophys. Res. Lett.*, *14*, 1238-1241, 1987.

Phipps Morgan, J., Mid-ocean ridge dynamics: Observations and theory, in *Contributions in Tectonophysics, U.S. National Report to International Union of Geodesy and Geophysics 1987-1990*, AGU, Washington, D.C., 807-822, 1991.

Phipps Morgan, J. and E.M. Parmentier, A three dimensional gravity study of the 95.5°W propagating rift in the Galapagos spreading center, *Earth Planet. Sci. Lett.*, *81*, 289-298, 1987.

Phipps Morgan, J. and D.W. Forsyth, Three-dimensional flow and temperature perturbations due to a transform offset: Effects on oceanic crustal and upper mantle structure, *J. Geophys. Res.*, *93*, 2955-2966, 1988.

Presnall, D.C., G.L. Simmons, and H. Porath, Changes in electrical conductivity of a synthetic basalt during melting, *J. Geophys. Res.*, *77*, 5665-5672, 1972.

Presnall, D.C., J.R. Dixon, T.H. O'Donnell, and S.A. Dixon, Generation of mid-ocean ridge tholeiites, *J. Petrol.*, *20*, 3-35, 1979.

Prince, R.A., and D.W. Forsyth, Horizontal extent of anomalously thin crust near the Vema fracture zone from the three-dimensional analysis of gravity anomalies, *J. Geophys. Res.*, *93*, 8051-8063, 1988.

Purdy, G.M., and R.S. Detrick, Crustal structure of the Mid-Atlantic Ridge at 23°N from seismic refraction studies, *J. Geophys. Res.*, *91*, 3739-3762, 1986.

Rabinowicz, M., A. Nicolas, and J.L. Vigneresse, A rolling mill effect in asthenosphere beneath oceanic spreading centers, *Earth Planet. Sci. Lett.*, *67*, 97-108, 1991.

Rabinowicz, M., G. Ceuleneer, and A. Nicolas, Melt segregation and flow in mantle diapirs below spreading centers: Evidence from the Oman ophiolite, *J. Geophys. Res.*, *92*, 3475-3486, 1987.

Rai, C.S. and M.H. Manghnani, Electrical conductivity of ultramafic rocks to 1820 Kelvin, *Phys. Earth Planet. Int.*, *17*. 6-13, 1978a.

Rai, C.S. and M.H. Manghnani, Electrical conductivity of basalts to 1550 °C, Proceedings of Chapman conference on Partial Melting in the Earth's Upper Mantle, edited by H.J.B. Dick, Bull. Oreg. Dep. *Geol. Miner. Ind.*, *96*, 296, 1978b.

Raitt, R.W., The crustal rocks, in *The Sea*, *vol. 3*, edited by M.N. Hill, pp. 85-102, Interscience Publisheres, New York, 1963.

Ramberg, I.B., D.F. Gray, and R.G.H. Raynolds, Tectonic evolution of the FAMOUS area of the Mid-Atlantic Ridge, lat. 35°50' to 37°20'N, *Geol. Soc. Am. Bull.*,*88* , 609-620, 1977.

Reid , I, and H.R. Jackson, Oceanic spreading rate and crustal thickness, *Mar. Geophys. Res.*, *5*, 165-172, 1981.

Reynard, B. and G.D. Price, Thermal expansion of mantle minerals at high pressures - a theoretical study, *Geophys. Res. Lett.*, *17*, 689-692, 1990.

Ribe, N.M., Theory of melt segregation - a review, *J. Volcan. Geo. Res.*, *33*, 241-253, 1987.

Ricard, Y., L. Fleitout, and C. Froidevaux, Geoid heights and lithospheric stresses for a dynamical Earth, *Ann. Geophys.*, *2*, 267-286, 1984.

Richards, M.A. and B.H. Hager, Geoid anomalies in a dynamic earth, *J. Geophys. Res.*, *89*, 5987-6002, 1984.

Richards, M.A. and B.H. Hager, Effects of lateral viscosity variations on long-wavelength geoid anomalies and topography, *J. Geophys. Res.*, *94*, 10299-10313, 1989.

Ringwood, A.E., Composition and petrology of the earth's mantle, McGraw Hill, New York, 1975.

Ritzwoller, M., G. Masters, and F. Gilbert, Constraining aspherical structure with low-degree interaction coefficients: application to uncoupled multiplets, *J. Geophys. Res.*, *93*, 6369-6396, 1988.

Robinson, E.M., B. Parsons, and S. Daly, The effect of a shallow low-viscosity zone on the apparent compensation depth of mid-plate swells, *Earth Planet. Sci. Lett.*, *82*, 335-348, 1987.

Robinson, E.M., B. Parsons, and M. Driscoll, The effect of a shallow low-viscosity zone on the mantle flow, the geoid anomalies and the geoid and depth-age relationships at fracture zones, *Geophys. J. R. Astr. Soc.*, *93*, 25-43, 1988.

Rosendahl, B.R., W.R. Raitt, L.M. Dorman, L.D. Bibee, D.M. Hussong, and G.H. Sutton, Evolution of oceanic crust, 1, A physical model of the East Pacific Rise Crest derived from seismic refraction data, J. *Geophys. Res.*, *81*, 5294-5304, 1976.

Rouland, D., S.H. Xu, and F. Schindele, Upper mantle structure in the southeast Indian Ocean: a surface wave investigation, *Tectonophys.*, *114*, 281-292, 1985.

Runcorn, S.K., Satellite gravity measurements and a laminar viscous flow model of the earth's mantle, *J. Geophys. Res.*, *69*, 4389-4394, 1964.

Sandwell, D.T., Thermomechanical evolution of oceanic fracture zones, *J. Geophys. Res.*, *89*, 11401-11413, 1984.

Sandwell, D.T., Thermal stress and the spacings of transform faults, *J. Geophys. Res.*, *91*, 6405-6417, 1986.

Sandwell, D.T. and G. Schubert, Geoid height versus age for symmetric spreading ridges, *J. Geophys. Res.*, *85*, 7235-7241, 1980.

Sandwell, D.T. and G. Schubert, Geoid height-age relation from Seasat altimeter profiles across the Mendocino fracture zone, *J. Geophys. Res.*, *87*, 3949-3958, 1982a

Sandwell, D. and G. Schubert, Lithospheric flexure at fracture zones, *J. Geophys. Res.*, *87*, 4657-4667, 1982b.

Sato, H. and I.S. Sacks, Anelasticity and thermal structure of the oceanic upper mantle: temperature calibration with heat flow data, *J. Geophys. Res.*, *94*, 5705-5715, 1989.

Sato, H., I.S. Sacks, and T. Murase, The use of laboratory velocity data for estimating temperature and partial melt fraction in the low-velocity zone: comparison with heat flow and electrical conductivity studies, *J. Geophys. Res.*, *94*, 5689-5704, 1989.

Sato, H., I.S. Sacks, T. Murase, and C.M. Scarfe, Thermal structure of the low velocity zone derived from laboratory and seismic investigations, *Geophys. Res. Lett.*, *15*, 1227-1230, 1988.

Sato, H., I.S. Sacks, T. Murase, G. Muncill, and H. Fukuyama, Qp-melting temperature relation in peridotite at high pressure and temperature: attenuation mechanism and implications for the mechanical properties of the upper mantle, *J. Geophys. Res.*, *94*, 10647-10661, 1989.

Schilling, J.G., Iceland mantle plume: Geochemical study of Reykjanes ridge, *Nature*, *242*, 565-571, 1973.

Schmeling, H., Numerical models on the influence of partial melt on elastic, anelastic and electric properties of rocks, part I., Elasticity and anelasticity, *Phys. Earth Planet. Int.*, *41*, 34-57, 1985a.

Schmeling, H., Partial melt below Iceland: a combined interpretation of seismic and conductivity data, *J. Geophys. Res.*, *90*, 10105-10116, 1985b.

Schmeling, H., Numerical models on the influence of partial melt on elastic, anelastic and electric properties of rocks, part II, Electrical conductivity, *Phys. Earth Planet. Int.*, *43*, 123-136, 1986.

Schock, R.N., A.G. Duba, H.C. Heard, and H.D. Stromberg, The electrical conductivity of polycrystalline olivine and pyroxene under pressure, in High Pressure Research: Applications in Geophysics, edited by M. Manghnani and S. Akimoto, 39051, Academic, San Diego, CA, 1977.

Schock, R.N., A.G. Duba, and T.J. Shankland, Electrical conduction in olivine, *J. Geophys. Res.*, *94*, 5829-5839, 1989.

Schouten, H. and R.S. White, Zero-offset fracture zones, *Geology*, *8*, 175-179, 1980.

Schouten, H., and K.D. Klitgord, The memory of accreting plate boundaries and the continuity of fracture zones, *Earth Planet. Sci. Lett.*, *59*, 255-266, 1982.

Schouten, H., K.D. Klitgord, and J.A. Whitehead, Segmentation of mid-ocean ridges, *Nature*, *317*, 225-229, 1985.

Schroeder, W., The empirical age-depth relation and depth anomalies in the Pacific Ocean basin, *J. Geophys. Res.*, *89*, 9873-9883, 1984.

Sclater, J.G. and J. Francheteau, The implications of terrestrial heat flow observations on current tectonic and geochemical models of the crust and upper mantle of the earth, *Geophys. J. R, Astr. Soc., 20,* 509- , 1970.

Scott, D.R., and D.J. Stevenson, A self-consistent model of melting, magma migration and buoyancy-driven circulation beneath mid-ocean ridges, *J. Geophys. Res., 94,* 2973-2988, 1989.

Selig, F., A theoretical prediction of salt-dome patterns, *Geophys., 30,* 633-643, 1965.

Sempere, J.-C., K.C. Macdonald, S.P. Miller, and L. Shure, Detailed study of the Brunhes/Matuyama reversal boundary on the East Pacific Rise at 1930'S: Implications for crustal emplacement at an ultra-fast spreading center, Mar. *Geophys. Res., 9,* 1-23, 1987.

Sempere, J.-C., G.M. Purdy, and H. Schouten, Segmentation of the Mid-Atlantic Ridge between 24°N and 34°40'N, *Nature, 344,* 427-431, 1990.

Shankland, T.J. and H.S. Waff, Partial melting and electrical conductivity anomalies in the upper mantle, *J. Geophys. Res., 82,* 5409-5417, 1977.

Shankland, T.J., R.J. O'Connell, and H.S. Waff, Geophysical constraints on partial melt in the upper mantle, *Rev. Geophys., 19,* 394-406, 1981.

Sheehan, A.F., and S.C. Solomon, Joint inversion of shear wave travel time residuals and geoid and depth anomalies for long-wavelength variations in upper mantle temperature and composition along the Mid-Atlantic Ridge, *J. Geophys. Res., 96,* 19,981-20,009, 1991a.

Sheehan, A.F., and S.C. Solomon, Lateral variations in upper mantle temperature and composition inferred from shear-wave travel time and attenuation, geoid and bathymetry, *EOS, Trans. AGU, 72,* Supplement, 514, 1991.

Sheehan, A.F., and S.C. Solomon, Differential shear wave attenuation and its lateral variation in the North Atlantic region, *J. Geophys. Res., in press, 1992.*

Shen, Y. and D.W. Forsyth, The effects of temperature and pressure-dependent viscosity on three-dimensional passive flow of the mantle beneath a ridge-transform system, *J. Geophys. Res., submitted, 1992.*

Sinha, M.C., and K.E. Louden, The Oceanographer fracure zone, I, Crustal structure from seismic refraction studies, *Geophys. J. R. Astr. Soc., 75,* 713-736, 1983.

Sinton, J.M., S.M. Smaglik, J.J. Mahoney, and K.C. Macdonald, Magmatic processes at superfast spreading mid-ocean ridges: Glass compositional variations along the East Pacific Rise 13°-23°S, *J. Geophys. Res., 96,* 6133-6156, 1991.

Sipkin, S.A., and T.H. Jordan, Lateral heterogeneity of the upper mantle determined from travel times of multiple ScS, *J. Geophys. Res., 81,* 6307-6320, 1976.

Sleep, N.H., Sensitivity of heat flow and gravity to the mechanism of sea-floor spreading, *J. Geophys. Res., 74,* 542-549, 1969.

Sleep, N.H., Formation of oceanic crust: Some thermal constraints, *J. Geophys. Res., 80,* 4037-4042, 1975.

Smith, D.K., and T.H. Jordan, Seamount statistics in the Pacific Ocean, *J. Geophys. Res., 93,* 2899-2918, 1988.

Smith, M.L. and F.A. Dahlen, The azimuthal dependence of Love and Rayleigh wave propagation in a slightly anisotropic medium, *J. Geophys. Res., 78,* 3321-3333, 1973.

Sotin, C. and E.M. Parmentier, Dynamical consequences of compositional and thermal density stratification beneath spreading centers, *Geophys. Res. Lett., 16,* 835-838, 1989.

Sparks, D.W. and E.M. Parmentier, The structure of three-dimensional convection beneath oceanic spreading centers, *Geophys. J. Int., in press,* 1992.

Sparks, D.W., and E.M. Parmentier, Melt extraction from the mantle beneath spreading centers, *Earth Planet. Sci. Lett., 105,* 368-377, 1991.

Spence, W. and R.S. Gross, A tomographic glimpse of the upper mantle source of magmas of the Jemez lineament, New Mexico, *J. Geophys. Res., 95,* 10,829-10,849, 1990.

Spiegelman, M., 2-D or not 2-D: Understanding melt migration near a sloping, freezing boundary, *EOS, Trans. AGU, Spring meeting special issue,* 265, 1991.

Spiegelman, M. and D. McKenzie, Simple 2-D models for melt extraction at ridges and island arcs, *Earth Planet. Sci. Lett., 83,* 137-152, 1987.

Spiess, F.N., K.C. Macdonald and others, East Pacific Rise: Hot springs and geophysical experiments, *Science, 207,* 1421-1433, 1980.

Stolper, E.M., D. Walker, B.H. Hager, and J.F. Hays, Melt segregation from partially molten source region: The importance of melt density and source region size, *J. Geophys. Res., 86,* 6161-6271, 1981.

Su, W. and W.R. Buck, Factors causing melt concentration at mid-ocean ridges, *EOS, Trans. AGU, Spring Meeting special issue,* 265, 1991.

Suetsugu, D. and I. Nakanishi, Three-dimensional velocity map of the upper mantle beneath the Pacific Ocean as determined from Rayleigh wave dispersion, *Phys. Earth Planet. Int.,* 205-229, 1987a.

Suetsugu, D. and I. Nakanishi, Regional and azimuthal dependence of phase velocities of mantle Rayleigh waves in the Pacific Ocean, *Phys. Earth Planet. Int.,* 230-245, 1987b.

Tanimoto, T., The Backus-Gilbert approach to the 3-D structure in the upper mantle-I. Lateral variation of surface wave phase velocity with its error and resolution, *Geophys. J. R. Astr.Soc., 82,* 105-123, 1986a.

Tanimoto, T., The Backus-Gilbert approach to the 3-D structure in the upper mantle-II. SH and SV velocity, *Geophys. J. R. Astr. Soc., 84,* 49-69, 1986b.

Tanimoto, T., Free oscillations in a slightly anisotropic *Earth, Geophys. J. R. Astr. Soc., 87,* 493-517, 1986c.

Tanimoto, T., The 3-D shear wave structure in the mantle by overtone waveform inversion - I. Radial seismogram inversion, *Geophys. J. R. Astr. Soc., 89,* 713-740, 1987.

Tanimoto, T., The 3-D shear wave structure in the mantle by overtone waveform inversion - II. Inversion of X-waves, R-waves, and G-waves, *Geophys. J., 93,* 321-334, 1988.

Tanimoto, T. and D.L. Anderson, Lateral heterogeneity and azimuthal anisotropy of the upper mantle: Love and Rayleigh waves 100-250 sec, *J Geophys. Res., 90,* 1842-1858.

Tanimoto, T. and Y.-S. Zhang, Deciphering active vs. passive upwellings from depth of low velocity anomalies, *EOS Trans. AGU, 17, Spring Meeting Supplement,* 277, 1991.

Tarantola, A. and A. Nercessian, Three-dimesional inversion without blocks, *Geophys. J. R. Astr. Soc., 76,* 299-306, 1984.

Thayer, R.E., L.A. Bjornsson, L. Alvarez, and J.F. Hermance, Magma genesis and crustal spreading in the northern neovolcanic zone in Iceland: Telluric-magnetotelluric constraints, *Geophys. J. R. Astr. Soc., 65,* 423-442, 1981.

Tolstoy, M., J.A. Orcutt, M.A.H. Hedlin, G.M. Kent, and A.E. Newell, A refraction study of the southern Mid-Atlantic Ridge using ocean bottom seismometers, *EOS, Trans. AGU, 71,* 1572, 1990.

Toomey, D.R., S.C. Solomon, and G.M. Purdy, Microearthquakes beneath the median valley of the Mid-Atlantic Ridge near 23°N: Tomography and tectonics, *J. Geophys. Res., 93,* 9093-9112, 1988.

Toomey, D.R., G.M. Purdy, S.C. Solomon, and W.S.D. Wilcock, The three-dimensional seismic velocity structure of the East Pacific Rise near latitude 9°30'N, *Nature, 347,* 639-645, 1990.

Toramaru, A., and N. Fujii, Connectivity of melt phase in a partially molten peridotite, *J. Geophys. Res.,* 9239-9252, 1986.

Trehu, A.M., Depth versus (age)$^{1/2}$: a perspective on mid-ocean ridges, *Earth Planet. Sci. Lett., 27,* 287-304, 1975.

Trehu, A.M., and W.H. Wheeler IV, Possible evidence for subducted sediments beneath central California, *Geology, 15,* 254-258, 1987.

Tucholke, B.E., Comment and reply on "North Atlantic fracture-zone distribution and patterns shown by multibeam sonar", *Geology, 18,* 911-912, 1990.

Turcotte, D.L., Are transform faults thermal contraction cracks?, *J. Geophys. Res., 79,* 2573-2577, 1974.

Tyburczy, J.A. and H.S. Waff, Electrical conductivity of molten basalt and andesite to 25 kilobars pressure: Geophysical significance and implications for charge transport and melt structure, *J. Geophys. Res.*, *88*, 2413-2430, 1983.

Vera, E.E., J.C. Mutter, P. Buhl, J.A. Orcutt, A.J. Harding, M.E. Kappus, R.S. Detrick, and T.M. Brocher, The structure of 0- to 0.2-m.y.-old crust at 9°N on the East Pacific Rise from expanded spread profiles, *J. Geophys. Res.*, *95*, 15529-15556, 1990.

Vogt, P.R., Asthenosphere motion recorded by the ocean floor south of Iceland, *Earth Planet. Sci. Lett.*, *13*, 153-160, 1971.

Vogt, P.R., B. Zondek, P.W. Fell, N.Z. Cherkis, and R.K. Perry, Seasat altimetry, the North Atlantic geoid, and evaluation by shipborne subsatellite profiles, *J. Geophys. Res.*, *89*, 9885-9903, 1984.

Von Bargen, N. and H.S. Waff, Permeabilities, interfacial areas and curvatures of partially molten systems: results of numerical computations of equilibrium microstructures, *J. Geophys. Res.*, *91*, 9261-9276, 1986.

Waff, H.S., Theoretical consideration of electrical conductivity in a partially molten mantle and implications for geothermometry, *J. Geophys. Res.*, *79*, 4003-4010, 1974.

Waff, H.S. and D.F. Weill, Electrical conductivity of magmatic liquids: effects of temperature, oxygen fugacity and composition, *Earth Planet. Sci. Lett.*, *28*, 254-260, 1975.

Waff, H.S., and U.H. Faul, Coexisting faceted and curved interfaces in partially molten olivine-basalt systems and implications for mantle partial melts, *J. Geophys. Res.*, submitted, 1992.

Wannamaker, P.E., J.R. Booker, J.H. Filloux, A.G. Jones, G.R. Jiracek, A.D. Chave, P. Tartits, H.S. Waff, G.D. Egbert, C.T. Young, J.A. Stodt, M.G. Martinez, L.K. Law, T. Yukutake, J.S. Segawa, A. White, and A.W. Green Jr., Magnetotelluric observations across the Juan de Fuca subduction system in the EMSLAB project, *J. Geophys. Res.*, *94*, 14111-14125, 1989.

Wannamaker, P.E., J.R. Booker, A.G. Jones, A.D. Chave, J.H. Filloux, H.S. Waff, and L.K. Law, Resistivity cross section through the Juan de Fuca subduction system and its tectonic implications, *J. Geophys. Res.*, *94*, 14127-14144, 1989b.

Watt, J.P., G.F. Davies, and R.J. O'Connell, The elastic properties of composite materials, *Rev. Geophys. Space Phys.*, *14*, 541-563, 1976.

Watts, A.B., J.H. Bodine, and N.M. Ribe, Observations of flexure and the geological evolution of the Pacific Ocean basins, *Nature*, *283*, 532-537, 1980.

Weidner, D.J., Rayleigh wave phase velocities in the Atlantic Ocean, *Geophys. J. R. Astr. Soc.*, *36*, 105-139, 1974.

Wessel, P., and W.F. Haxby, Geoid anomalies at fracture zones and thermal models for the oceanic lithosphere, *Geophys. Res. Lett.*, *16*, 827-830, 1989.

Wessel, P. and W.F. Haxby, Thermal stresses, differential subsidence, and flexure at oceanic fracture zones, *J. Geophys. Res.*, *95*, 375-391, 1990.

White, R.S., Asthenospheric control on magmatism in the ocean basins, in Saunders, A.D. and M.J. Norry (eds.), Magmatism in the Ocean Basins, *Geol. Soc. Amer. Sp. Publ.*, *42*, 17-27, 1989.

White, R.S., and D.H. Matthews, Variations in oceanic upper crustal structure in a small area of the north-eastern Atlantic, *Geophys. J. R. Astr. Soc.*, *61*, 401-436, 1980.

White, R.S., R.S. Detrick, M.C. Sinha, and M.-H. Cormier, Anomalous seismic structure of oceanic fracture zones, *Geophys. J. R. Astr. Soc.*, *79*, 779-798, 1984.

Whitehead, J.A, Jr., H.J.B. Dick, and H. Schouten, A mechanism for magmatic accretion under spreading centers, *Nature*, *312*, 146-148, 1984.

Wiens, D.A., and S. Stein, Age dependence of oceanic intraplate seismicity and implications for lithospheric evolution, *J. Geophys. Res.*, *88*, 6455-6468, 1983.

Woodward, R.L., and G. Masters, Global upper mantle structure from long-period differential travel times, *J. Geophys. Res.*, *96*, 6351-6378, 1991.

Woodhouse, J.H. and A.M. Dziewonski, Mapping the upper mantle: Three-dimensional modeling of Earth structure by inversion of seismic waveforms, *J. Geophys. Res.*, *89*, 5953-5986, 1984.

Yan, B., E.K. Graham, and K.P. Furlong, Lateral variations in upper mantle thermal structure inferred from three-dimensional seismic inversion models, *Geophys. Res. Lett.*, *16*, 449-452, 1989.

Yoshii, T., Regionality of group velocities of Rayleigh waves in the Pacific and the thickening of the plate, *Earth Planet. Sci. Lett.*, *25*, 305-312, 1975.

Yu, G.K. and B.J. Mitchell, Regionalized shear velocity models of the Pacific upper mantle from observed Love and Rayleigh wave dispersion, *Geophys. J. R. Astr. Soc.*, *57*, 311-341, 1979.

Zhang, Y.S. and T. Tanimoto, Three-dimensional modeling of upper mantle structure under the Pacific Ocean area, *Geophys. J. Int.*, *98*, 255-269, 1989.

Donald W. Forsyth, Department of Geological Sciences, Brown University, Providence, RI 02912.

Phase Equilibria Constraints on the Origin Of Ocean Floor Basalts

PAUL C. HESS

Department of Geological Sciences, Brown University, Providence, RI 02912

Most ocean floor basalts are not primary magmas but are descendants of more primitive magmas that were segregated from mantle at pressures exceeding 0.8 GPa. These conclusions are derived from phase equilibria constraints that show that even the most primitive ocean floor basalts have undergone near surface olivine fractionation from magmas that were generated by polybaric melting of adiabatic upwelling mantle. The absence of MgO-rich quartz tholeiite basalts requires that only small volumes of magma were produced from or even equilibrated with mantle at shallow levels.

The compositions of primary magmas are controlled by the degree of melting, the depth of segregation and the nature of the melting and segregation process. Because peridotite becomes permeable even at very small degrees of melting, melting occurs in an open system and by fractional fusion. The maximum extent of melting is controlled by the diopside-out curve where melting virtually stops once diopside is eliminated from the mantle residue. The common occurrence of diopside-free abyssal harzburgites in oceanic fracture zones demonstrates that the temperatures and pressures of melting have indeed exceeded diopside-out conditions. If it is assumed that oceanic basalts collect and segregate from mantle at 1.0 GPa and that the P-T conditions are fixed by the diopside-out curve, then it is easy to show that the maximum extent of melting approached 30% and the peridotite solidus in the suboceanic mantle is encountered at 3.0-3.5 GPa. This depth for the beginning of melting is consistent with trace element evidence that some ocean floor basalts have equilibrated with garnet peridotite.

There is an emerging paradigm that some forms of veins or dikes are required to extract melt from partially molten mantle. A model of magma segregation that first isolates magma at great depths and focuses this melt to the ridge axis is needed to reconcile the petrological and geophysical observations. A high porosity zone in a decompacting layer beneath the impermeable lid of the growing lithosphere has some of the necessary characteristics of such a model.

INTRODUCTION

Ocean floor basalts are intimately associated in space and time and share common and distinct chemical characteristics that distinguish them from basalts erupted in other tectonic environments. Such magmas are said to define a petrographic association [Turner and Verhoogan, 1960] An obvious question is to determine the factors by which the major element content of magmas acquire a common imprint within a single province, a province which, in this case, is earth-wide.

The chemical compositions= of the majority of MORB are hypersthene and olivine normative [see Hess, 1989, for a brief discussion of normative classification]; a smaller number are quartz normative. Such basalts are called olivine and quartz tholeiites, respectively.

The vast majority of the chemical analyses of ocean floor basaltic glasses contain less than 9% MgO. Such values of MgO are lower than those of basalts obtained by partial melting of mantle peridotite. Presnall and Hoover [1987] searched about 1700 analyses of glassy and fine-grained basalts and determined that only 42 (Table 1) had compositions that have the necessary (but not necessarily sufficient) chemical traits to be partial melts (primary magma) of mantle peridotite (see below for a definition for primary basalt and other petrologic terms). The compositions of the remaining samples have been influenced by processes of magmatic differentiation that have acted on these liquids following their isolation from parent mantle. The task is to identify these secondary processes, strip their effects until the "ultimate sources of the magmas are at last clearly revealed to all"

Mantle Flow and Melt Generation at Mid-Ocean Ridges
Geophysical Monograph 71
Copyright 1992 American Geophysical Union

TABLE 1

The Chemical Compositions of Primitive Ocean Floor Basalts (wt%) (Presnall and Hoover, 1987)

Samples	SiO_2	TiO_2	Al_2O_3	Fe_2O_3	FeO	MnO	MgO	CaO	Na_2O	K_2O	Cr_2O_3
519-4-1	49.07	0.74	16.44	0.89	7.97	0.16	10.15	11.65	2.13	0.07	0.00
519-4-2	48.81	0.73	16.13	0.89	8.00	0.16	10.32	11.56	2.05	0.01	0.04
522-2-1	49.65	0.87	16.05	0.87	7.85	0.15	9.82	11.76	2.25	0.09	0.06
525-5-1	49.19	0.85	16.12	0.87	7.87	0.14	10.41	11.91	2.35	0.09	0.08
525-5-2	48.93	0.84	16.14	0.88	7.88	0.15	10.49	11.84	2.41	0.09	0.09
525-5-3	48.96	0.82	15.75	0.85	7.67	0.13	9.79	11.92	1.97	0.11	0.08
526-1-1B	51.36	0.78	15.03	0.79	7.07	0.10	9.01	12.38	2.04	0.10	0.11
527-1-2	49.50	0.66	16.51	0.88	7.88	0.17	9.86	12.68	2.08	0.04	0.11
528-4-1	49.93	0.72	15.82	0.84	7.53	0.15	9.78	12.67	1.96	0.05	0.08
529-3-2	50.82	0.92	15.92	0.79	7.12	0.13	9.38	12.07	2.30	0.16	0.06
530-3-1	49.98	0.77	15.24	0.86	7.74	0.13	10.11	12.45	2.23	0.07	0.00
534-2-1	50.30	0.84	15.99	0.87	7.83	0.13	9.91	12.33	2.16	0.09	0.08
534-2-2A	50.37	0.84	15.86	0.86	7.71	0.16	9.82	12.33	2.18	0.09	0.09
ARP-73-10-3	50.60	1.05	15.00	0.90	8.07	0.18	8.70	12.50	2.43	0.11	0.00
AII-77-76-61	51.20	0.84	15.00	0.80	7.18	0.16	9.02	13.30	1.89	0.08	0.08
AII-77-76-71	51.40	0.83	15.00	0.81	7.25	0.16	9.15	13.20	1.97	0.09	0.07
AII-74-14-31	48.20	0.51	17.00	0.85	7.66	0.16	10.10	12.70	2.34	0.04	0.00
AII-74-14-33	48.20	0.50	16.90	0.85	7.65	0.15	9.91	12.80	2.30	0.04	0.01
3-14-10-1	50.00	0.79	17.40	0.82	7.40	0.13	9.81	12.90	2.28	0.02	0.00
3-18-7-1-1	50.30	0.73	16.60	0.80	7.19	0.12	10.20	13.20	2.00	0.01	0.00
3-18-7-1-4	50.40	0.79	17.10	0.80	7.17	0.00	9.88	12.90	1.62	0.02	0.00
A27-99S1	49.86	0.71	16.93	0.81	7.31	0.00	9.49	13.16	1.71	0.02	0.00
A42-96N1	50.57	0.92	15.62	0.80	7.16	0.00	9.21	13.32	1.99	0.26	0.00
72-17-44	49.68	0.86	16.75	0.77	6.90	0.13	9.19	12.23	2.31	0.05	0.00
TR-154-14-D1	51.36	1.29	16.48	0.72	6.49	0.06	8.39	12.02	2.49	0.25	0.00
ARP-74-10-16	49.47	0.82	15.23	0.82	7.34	0.14	10.66	12.21	1.94	0.16	0.09
CYP-31-35	49.57	0.85	15.26	0.83	7.45	0.14	10.63	12.11	2.11	0.19	0.09
VG-202	50.73	0.92	15.61	0.80	7.17	0.00	9.34	13.29	1.94	0.24	0.00
VG-203	50.41	0.92	15.64	0.80	7.16	0.00	9.09	13.35	2.05	0.28	0.00
VG-297	50.14	0.74	16.81	0.83	7.43	0.00	9.59	13.01	1.71	0.04	0.00
VG-416	50.73	0.73	16.88	0.82	7.40	0.00	9.84	13.33	1.68	0.04	0.00
VG-607	49.97	0.73	17.26	0.84	7.55	0.00	9.51	12.92	1.99	0.01	0.00
VG-611	49.95	0.69	17.29	0.80	7.12	0.00	9.14	13.23	1.76	0.00	0.00
VG-649	50.41	0.92	15.64	0.80	7.16	0.00	9.09	13.35	2.05	0.28	0.00
F	49.15	0.96	17.58	0.80	7.21	0.00	9.58	12.14	2.38	0.04	0.00
H	49.21	0.97	17.54	0.80	7.17	0.00	9.12	12.35	2.43	0.04	0.00
1561-1622	48.47	0.88	17.39	0.80	7.23	0.16	9.83	12.26	2.47	0.03	0.04
F2-1	49.85	0.86	17.51	0.81	7.31	0.21	9.80	12.92	2.15	0.02	0.05
F3-2	49.46	0.83	17.42	0.81	7.30	0.14	9.85	12.72	2.14	0.04	0.04
F1	47.20	0.77	16.90	0.88	7.92	0.14	10.90	12.20	2.05	0.02	0.07
Y-87-1-8	51.09	0.65	16.26	0.73	6.59	0.00	10.25	12.16	1.91	0.00	0.35
Y-87-2-8	52.63	0.58	14.50	0.74	6.65	0.00	11.68	11.14	1.66	0.00	0.41
Y-87-4-20	47.03	1.00	16.62	0.97	8.70	0.00	12.99	9.93	2.63	0.12	0.00
Y-87-5-20	46.94	0.70	15.44	0.98	8.85	0.00	14.24	10.53	2.30	0.00	0.20
Y-87-6-20	47.52	0.73	14.39	0.96	8.66	0.00	14.51	10.98	1.84	0.00	0.38
Y-87-1-10	50.90	0.81	19.16	0.68	6.14	0.00	8.52	10.48	3.31	0.00	0.00
Y-87-3-10	50.36	0.77	17.05	0.75	6.76	0.00	10.37	11.53	2.18	0.00	0.24
Y-87-6-10	50.78	0.67	15.42	0.79	7.01	0.00	11.12	12.10	1.83	0.00	0.27

[Thompson, 1987]. This paper addresses these questions from a petrologic perspective, particularly one that was born and nurtured in the work of experimental petrologists. A more complete and balanced view is gained by reading the companion papers of this series.

The paper is organized as follows. Section I contains the basic phase equilibria of simple systems needed to discuss basalt petrogenesis intelligently. Most petrologists may safely skip this section. Section II deals with the results of experiments performed on basalts and peridotite. This section is the core of the paper. It provides the data base from which models of basalt petrogenesis are born. Section III focuses on what remains of mantle materials after basalt liquids are extracted to form oceanic crust. These residues to partial melting, the abyssal peridotites, provide an additional window to the physiochemical processes that occur below oceanic spreading centers. The last section is a more speculative synthesis that tries to pull a number of disparate ideas into a coherent framework of basalt petrogenesis.

A few words should be said about what the paper does not contain. In order to avoid overlap with the companion papers, most references to ophiolites, textural equilibria, trace and isotope geochemistry and quantitative models are avoided. It is hoped, nevertheless, that the paper is informative, and readable and appropriate for an audience at the "petrological-geophysical eutectic" [Batiza, pers. com]. The reader is referred to reviews by Thompson [1987] and Elthon [1990] for additional perspectives.

SECTION I

Primary Basalt and Other Concepts

A primary MORB represents a liquid, formed by the partial melting of peridotite, whose chemical composition has not changed materially following segregation from its mantle source. Primary magmas have escaped the effects of crystal fractionation, magma mixing or other processes of magmatic differentiation. The composition of a primary MORB is controlled by the temperature and pressure at which it last equilibrated with the oceanic mantle.

Other terms commonly in use are derivative, parental and primitive magmas. A derivative magma is derived from some parental magma by the process of crystal fractionation. A parental magma may be a primary magma or any liquid derived by crystal fractionation, magma mixing or other magmatic differentiation processes from any other magma(s). A primitive magma is a nebulous term that simply implies a direction of evolution. A primitive magma produces a less primitive magma by crystal fractionation processes. The presumption is that a primitive magma may actually be a primary magma but, in fact, seldom is. A series of derivative magmas derived by crystal fractionation proceeding from a common parent magma defines a liquid line of descent. The liquid line of descent is usually expressed in the smooth and continuous variation in the chemical composition of derivative magmas. Some liquid lines of descent, however, appear to be so only in terms of the major element chemistry. Consideration of the trace and/or isotope geochemistry, often proves that a sequence of what appear to be cogenetic magmas cannot be derived from the same parental magma.

The following criteria are used to identify primary magmas. Each criterion is a necessary but not sufficient requirement of a primary basalt.

(1) A primary basalt has olivine as a sole liquidus (silicate) phase at one atmosphere because the olivine phase volume expands with decreasing pressure (see later). A primary magma in equilibrium with olivine + orthopyroxene + calcic clinopyroxene at depth will have only olivine on its liquidus at all lower pressures. Conversely, a MORB that has both olivine and plagioclase on or near the 1 atm liquidus cannot be a primary magma.

(2) A primary basalt must have olivine as one of the liquidus phases at high pressures, since olivine must coexist

with liquid at all reasonable pressures and temperatures in the basalt-forming region in the upper mantle.

(3) A primary basalt must not only coexist with olivine at depth but also with all other residual phases. A primary basalt that last equilibrated with spinel lherzolite, for example, must have spinel, olivine, orthopyroxene and calcic clinopyroxene (of appropriate compositions) on or near the liquidus at the depth of origin.

(4) The Mg* (MgO/(MgO + FeO in moles) of a primary basalt must be in Fe-Mg exchange equilibrium with mantle olivine. Olivines in lherzolite peridotite typically have Mg*=0.88 to 0.91. The Fe^{+2}-Mg exchange partition coefficient for olivine-basalt liquid is not sensitive to variations in temperature [Roeder and Emslie, 1970; Longhi et al., 1978) but increases with pressure [Ulmer, 1989]. The partition coefficient

$$K_d^{OL-LQ} = \frac{Fe^{+2}_{OL}\, Mg_{LQ}}{Mg_{OL}\, Fe^{+2}_{LQ}} \quad (1)$$

increases from about 0.30 at 1 atm to 0.37 at 3.0 GPa. The partition coefficient obtained from the chemical composition of olivine phenocrysts and the enclosing basaltic glass is somewhat lower, approximately 0.27 to 0.28 at 1 atm. This is because about 10% of the iron oxide content in MORB glasses is Fe_2O_3 rather than FeO [Christie et al, 1986]. If the analysis is corrected for the Fe_2O_3 content, a partition coefficient closer to 0.30 is obtained.

The shaded-in region of Figure 1 represents the Mg* of possible primary MORB. Primary basalts extracted from deep within the mantle have higher Mg* than those equilibrated at

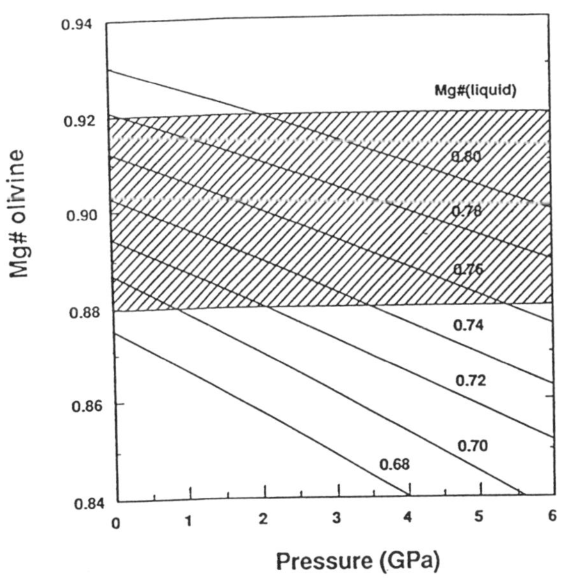

Fig. 1 Mg* values of basaltic magmas in equilibrium with olivine as a function of pressure [Ulmer, 1989]. Hatched region represents Mg* values of primary liquids in equilibrium with typical mantle lherzolites (Mg* 0.88-0.91) and harzburgite (Mg* 0.91-0.92).

shallow levels. MORB with Mg* values much below 0.68 cannot be primary basalts, unless unusually FeO-rich mantle exists. In addition, the Mg* of 1 bar liquidus olivine of a primary MORB should be 0.88 or greater. In fact, since melting in the suboceanic mantle often approaches the diopside-out curve (see later), a primary MORB should crystallize olivine with Mg* closer to 0.90 than 0.88.

(5) The NiO content of mantle olivine typically ranges from 3000 to 4000 ppm. The NiO olivine/liquid partition coefficient is a function of the temperature, pressure and MgO content of the basalt. The greater the degree of melting of peridotite at a given depth, the higher the MgO content of the liquid and the smaller the Ni partition coefficient. The following equation can be used to estimate the olivine-liquid Ni partition coefficient [Elthon, 1987]

$$D_{Ni} = \frac{wt\%\ NiO_{(OL)}}{wt\%\ NiO_{(LQ)}} = \frac{122.2}{X_{MgO}} - 2.3 \quad (2)$$

where X_{MgO} represents the weight fraction of MgO in liquid. The D_{Ni} is approximately 10 for a basalt with 10% MgO and about 5 for a picrite (a MgO-rich basalt) with 17% MgO. High pressure partitioning data suggest that the pressure effect below 3.0 GPa is not very large.

A primary basalt with 10% MgO should contain about 300-400 ppm Ni. Most basalts have lower NiO contents because even small amounts of olivine crystallization rapidly depletes the NiO contents of basalts. Fractional crystallization of approximately 10% olivine, for example, reduces the NiO content by more than half in the resulting basalt.

Phase Equilibria Constraints

Mg_2SiO_4-$CaMgSi_2O_6$-SiO_2 System. Petrologists have long recognized the importance of understanding melting and crystallization in relatively simple systems before plunging abruptly into the untidy world of real rock systems. It is true that the most important parameters of interest - the composition of liquids and solids and the extent of melting as functions of pressure and temperature - are not transferable from simple systems to real rocks. An exploration of the phase equilibria in simple systems, however, provides a guide to unravel the complex phase relations of natural magmas and rocks. A system of particular importance is described by the three components Mg_2SiO_4 - $CaMgSi_2O_6$-SiO_2, which contains about 85% of the components of typical peridotite, and also the three most important minerals of the upper mantle, forsterite (olivine), enstatite (orthopyroxene) and diopside (calcic clinopyroxene) It is shown below that this model of peridotite is more than adequate to illustrate melting of mantle and the generation of magma as functions of pressure and temperature.

Figure 2 contains the sub-solidus phase relations at 2.0 GPa (or the phase relations at temperatures just below where the first liquid appears) (Basaltic Volcanism on Terrestrial Planets, 1981). The solid line drawn between $CaMgSi_2O_6$ and $MgSiO_3$ ("the pyroxene join") describes the extent of solid solution (heavy line) between the high Ca-pyroxene (diopside) and low

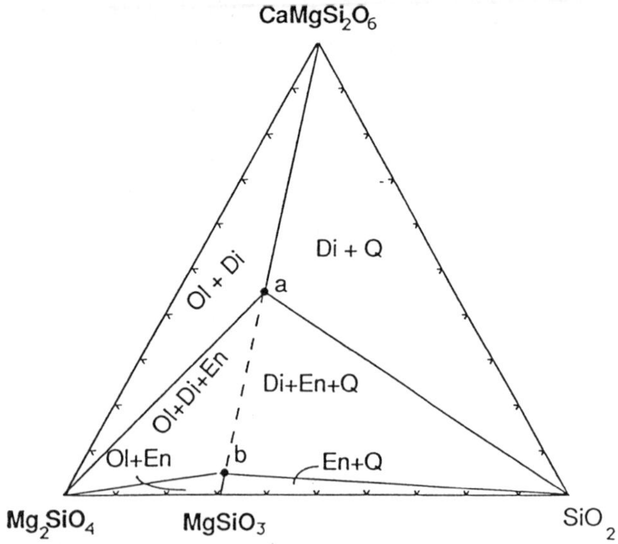

Fig. 2 Subsolidus phase relationships within the Mg_2SiO_4-$CaMgSi_2O_6$-SiO_2 system at 2.0 GPa. Dashed line (a-b) is the miscibility gap along the pyroxene join. Di= diopside En= enstatite Q= quartz, 01-forsterite.

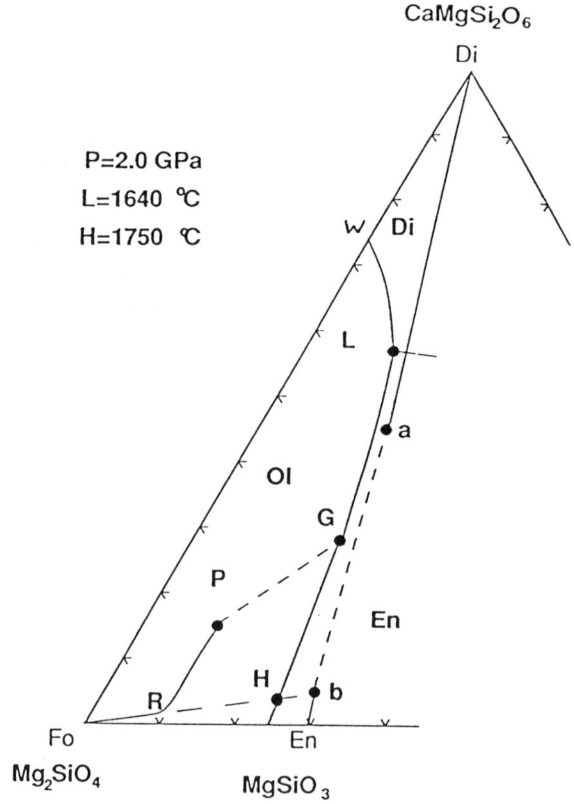

Fig. 3 Portion of the liquidus phase diagram at 2.0 GPa in the Mg_2SiO_4-CaMg Si_2O_6-SiO_2 system Fo= forsterite (see text).

Ca pyroxene (enstatite). Note that diopside can dissolve large amounts of $MgSiO_3$, whereas only small quantities of $CaMgSi_2O_6$ are soluble in the enstatite. The dashed line between a-b is a gap in which no single pyroxene exists; instead, a mixture of two pyroxenes, one of high CaO- and one of low CaO- pyroxene of compositions a and b respectively, populate this region. The solidus is thus divided by the pyroxene join into six fields; the four two-phase fields forsterite + high CaO pyroxene, forsterite + low CaO pyroxene, quartz + high CaO pyroxene, quartz + low CaO pyroxene and two three phase regions, forsterite + low CaO pyroxene + high CaO-pyroxene and quartz + low CaO pyroxene + high CaO pyroxene. The forsterite portion of the sub-solidus region is of greatest interest here since it contains the assemblages which approximate the common varieties of peridotite. It is the focus of the following discussion.

The liquidus phase diagram (Fig. 3) at 2.0 GPa contains the compositions of liquids that coexist with one or more of the minerals of peridotite. Point L (1640° C) is the only liquid at this pressure that coexists with forsterite, enstatite (or low CaO pyroxene) and diopside. Along line HL are the liquids that coexist with harzburgite (Fo + En), whereas liquids along WL coexist with wehrlite (Fo + Di). The fields denoted by Fo, En and Di are liquids which coexist with only forsterite, enstatite or diopside, respectively.

The liquidus diagram will be used to describe the melting of a lherzolite phase assemblage. Bulk compositions of lherzolite at this pressure lie near the Mg_2SiO_4 apex and within the triangle Fo-a-b. Let the bulk composition of peridotite be P; the lever rule shows that the peridotite contains 53 wt % Fo, 31 wt % En (composition b) and 16 wt % high Ca-rich pyroxene (composition a). (See Ehlers [1972] or Morse [1981] for a discussion of the lever rule and other techniques used to interpret phase diagrams; Hess [1989] has an abbreviated discussion.) Adding heat provides the energy to raise the temperature of lherzolite to the solidus and the latent heat of fusion to create the first few drops of liquid.

Melting does not begin at the lowest point of the liquidus, but rather at the isobaric invariant point L because L is the only liquid that coexists with olivine and the two pyroxenes. As long as the three minerals remain in the residue, neither the temperature of the system nor the liquid composition may change. The temperature is unchanged because the latent heat of fusion soaks up the energy added to the system, and the composition of the liquid is constrained by the phase equilibria. The melting reaction at this pressure occurs by the "peritectic" reaction [1]

[1]This invariant point represents a "peritectic" reaction because the liquid composition does not plot within the triangle formed by olivine-diopside-enstatite. If the liquid plotted within this triangle, its compositon would be expressed by a fixed amount of each of the three phases and the invariant point is an eutectic given by the reaction olivine+diopside+enstatite = melt.

(1) 0.07 forsterite + 0.93 diopside = 0.80 liquid + 0.20 enstatite

i.e. forsterite and diopside (of composition a) react to form liquid (of composition L) and enstatite (of composition b). The coefficients in (1) give the approximate weight ratios of the phases participating in the reaction (these coefficients pertain only to melting at 2.0 GPa).

Thus, it is mainly diopside that is consumed and mainly liquid that is created. Melting is therefore by "partial fusion" and represents a complex reaction between the principal phases. Melting is not "modal", i.e., the proportion of minerals entering the liquid is not the same as the proportion of minerals in the peridotite. Modal melting does not occur in peridotite, unless the modal proportions of that peridotite are equivalent to point L in Figure 3 (or an unusually pyroxene-rich rock called websterite).

Melting continues at the "isobaric invariant point" at 1640° C as long as heat is added to the system and neither olivine nor diopside is consumed by reaction (1). Producing liquid of composition L from P drives the unmelted portion of lherzolite, the residue, along the "crystal path" P-R. The residue is gradually depleted of diopside. At point R, the diopside is totally fused and the residue contains only forsterite and enstatite i.e., R lies within the olivine + enstatite field. By application of the lever rule, approximately 21% of the lherzolite has fused to liquid, L. The system is therefore composed of a diopside-free harzburgite and liquid L. Additional increments of melt are no longer constrained by reaction (1) because there is no diopside in the residue. Melting continues but at higher temperatures by the reaction

(2) 0.15 forsterite + 0.85 enstatite = liquid

and the liquid and temperature of the system progress along the LH "cotectic," until either forsterite or enstatite is consumed. It is easy to see that enstatite is consumed first, since lherzolite contains more forsterite than enstatite, and the ratio of enstatite to forsterite consumed by reaction (2) (the "cotectic ratio") is approximately 6 to 1. The "cotectic ratio" is obtained by constructing a tangent to the boundary curve and projecting the tangent to the Fo + En join. The proportion of forsterite and enstatite that enters the melt is given by the lever rule. Note that the harzburgite residue becomes progressively enriched in olivine as enstatite is preferentially melted (crystal path R-Fo) until only olivine (dunite) coexist with liquid G.

The liquid path then leaves the LH cotectic at G and trends directly to P along a path that is radial to Mg_2SiO_4 (since only forsterite is fused). At P, the system is 100% liquid and is at an extremely high temperature.

The mass of liquid generated at each stage along the melting path (L-G-P) is displayed by the dashed line in Figure 4. Several features are noteworthy. A large mass of melt of fixed composition is generated at the invariant point, approximately 21% in this example. Since the melt is enriched in $CaMg_2SiO_6$, it follows that the mass of melt produced at L is limited by the diopside content of lherzolite. While any peridotite composition within the field Mg_2SiO_4 - a - b begins to melt at the invariant point, a peridotite which is diopside

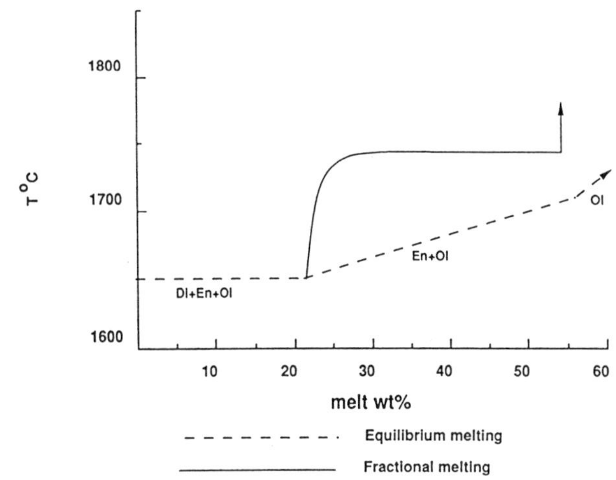

Fig. 4 Percent liquid obtained by equilibrium (dashed line) and fractional (solid line) melting of composition at 2.0 GPa.

poor and approaches an harzburgite composition (eg. any peridotite on crystal path P-R) is refractory and yields smaller amounts of melt. Once diopside is lost from the residue, further additions of melt require progressively higher temperatures and the melt composition is no longer fixed, but becomes progressively enriched in $MgSiO_3$ and diluted in $CaMg\ Si_2O_6$. There is an additional discontinuity in the melting path at G at which point enstatite is lost and olivine only is consumed.

The melt and crystal paths described above pertain only to equilibrium melting. Equilibrium melting requires that melt and the remaining crystals remain in contact, and that all phases are homogeneous (minerals are not zoned), i.e., all phases are in homogeneous and heterogeneous equilibrium. In perfect fractional melting, liquid is removed from the crystal residue as soon as it is formed. Only crystals exist in the system after each increment in temperature. The liquid path produced by perfect fractional melting is different from that obtained by equilibrium melting.

The first liquid obtained by fractional melting is at the invariant point L as long as diopside remains in the residue because it is the only liquid that coexists with all three minerals (Fig.3). The composition of the mineral residue shifts from P to R as in the equilibrium melting case. Once diopside is consumed, and the liquid is expelled from the system, the compostion of the system is at R or a diopside-free harzburgite. The liquid path now proceeds from L to H and the crystal path proceeds from R towards forsterite as more enstatite than forsterite is melted (reaction [2]). When the crystal path is at H, the residue is pure olivine and cannot be melted until the melting temperature of forsterite is exceeded. Melting, for all intents, is ended.

Note the differences between the two modes of melting. In the equilibrium mode, approximately 35% additional liquid is produced between L and G before enstatite is consumed. In the fractional melting mode, the same amount of liquid is developed but between L and H. Since the temperature of the cotectic at G is approximately 50-60°C less than at H, it follows that more melt is produced along the olivine-enstatite cotectic for an increment of temperature in the equilibrium melting mode than in the fractional melting mode. In fact, only a few percent liquid is produced by fractional melting along the cotectic from L to H because there is so little Di component in the residual hazburgite. Most of the liquid is developed near H where a high temperature, En-rich, Di-poor melt exists. Because of this, the crystal path stays near R until melt of compositon near H is obtained. The percent melt versus temperature for the fractional melting case is given schematically by the solid line in Fig. 4. It follows that the mode of the harzburgite (mode = mineral proportions) is not sensitive to the temperature of melting along the cotectic, until the temperature approaches H.

In effect, a harzburgite created by fractional melting becomes extremely refractory and undergoes only very little additional melting. Large degrees of melting require large increments of temperature. It is concluded that melting virtually ends once diopside is eliminated from the peridotite mantle provided that fractional melting is the dominant mode of melting. This effect is true even in natural peridotite (see later).

The liquidus phase diagram in Fig. 3 pertains only to 2.0 GPa. The phase boundaries, however, are dependent on the pressure. The forsterite liquidus volume expands with decreasing pressure, and contracts with increasing pressure. This has several important consequences. The composition of the liquid at the isobaric invariant point varies from relatively olivine-rich compositions at high pressures to liquids which plot on the quartz-side of the pyroxene join at one bar (Fig. 5). The dP/dT slope of the univariant equilibrium below about 8.0 GPa (formed by the collection of isobaric invariant points) is positive so that melting temperatures fall with decreasing pressure. This means that a liquid obtained at the invariant point at 2.0 GPa, segregated and isolated from the mantle and then brought adiabatically to a shallow magma chamber, will crystallize only olivine for the first decrements of temperature. Thus, primary liquids developed at depth by melting spinel or garnet peridotite should first crystallize olivine at lower pressures (Criteria [1]). A primary basalt that, by definition, has escaped near surface crystallization, can be used to determine the depth and temperature from which the melt was isolated from the mantle. Liquid L, for example, was in equilibrium with olivine-enstatite-diopside at 2.0 GPa (Fig. 3). At any other pressure, L would be in equilibrium with olivine (P<2.0 GPa) or diopside (P>2.0 GPa). Thus, if it is assumed that a primary magma was derived by relatively small degrees of melting of lherzolite, say 10%, then it is in equilibrium with olivine + enstatite + diopside at only one pressure . This can be determined by experiment, thus fixing the depth from which the primary magma was derived (Criteria [3]).

Mg_2SiO_4-$CaMgSi_2O_6$-$CaAl_2Si_2O_8$ System

The Mg_2SiO_4-$CaMgSi_2O_6$-$CaAl_2Si_2O_8$ system describes the crystallization of forsterite, anorthite and diopside, giving a relatively simple but instructive analogue for the

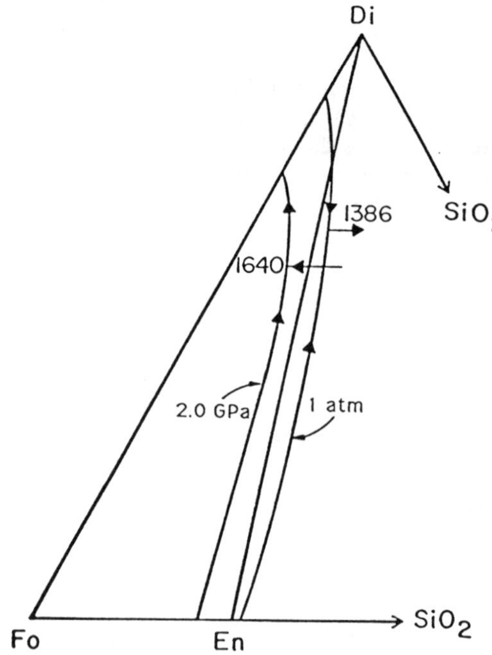

Fig. 5 Liquidus phase boundaries at 1 bar and 2.0 GPa. Note how the forsterite liquidus expands with decreasing pressure.

crystallization of ocean floor basalts (Fig. 6). The one atmosphere phase relations are discussed fully by Morse [1981] and only a brief and slightly oversimplified overview is given here. Point E is an eutectic (it is actually a "piercing point" or the trace of a boundary curve that penetrates the composition triangle in the CaO-MgO - Al_2O_3 - SiO_2 compositional tetrahedron) from which forsterite, anorthite and diopside crystallize in proportions that exactly add up to the composition at E. The three lines separating the anorthite, forsterite and diopside liquidus fields are cotectics that describe the liquids each of which coexist with two of the three minerals in question. We will not consider the crystallization of spinel at this time.

All liquids that lie on the olivine liquidus surface will crystallize olivine and follow a line (liquid line of descent) that is radial to the olivine apex [see Hess, 1989, for a brief but more detailed discussion]. Liquid A, for example, evolves to the anorthite-forsterite cotectic and then progresses towards the eutectic (see arrow) through the crystallization of anorthite and forsterite. At the eutectic, diopside, anorthite, and forsterite coexist with liquid. The liquid path is determined by the crystallization of forsterite ⟶ forsterite + anorthite⟶forsterite + anorthite + diopside, a sequence that is commonly observed in ocean floor basalts. Note that a small change in the starting bulk composition would have diopside following forsterite in the crystallization series. To my knowdege, diopside never (or at least rarely) follows forsterite in ocean floor lavas. Plagioclase always follows olivine in MORB.

Consider the phase relations of this system at higher pressures, say 1.0 GPa [Presnall et al, 1978] (Fig. 7). The

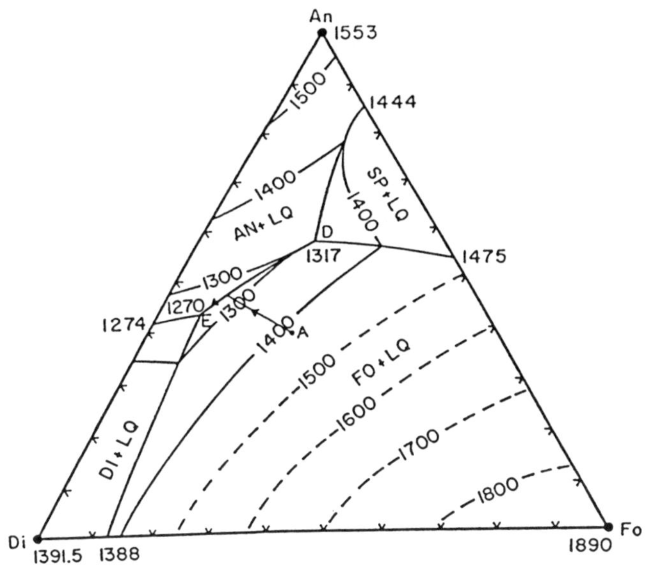

Fig. 6 Liquidus phase boundaries for the Mg_2SiO_4-$CaMgSi_2O_6$-$CaAl_2Si_2O_8$ system at 1 atm. The liquid line of descent of a liquid composition A is given by the arrows. An= anorthite; Sp= spinel.

boundary curves have been rearranged and the cotectic between anorthite and forsterite no longer exists. The liquid (liquid A) that produced the crystallization sequence forsterite⟶ forsterite + anorthite at 1 atm now crystallizaes forsterite and then forsterite + diopside. This shift in the boundary curves has interesting implications.

Suppose that the liquid reached point F on the forsterite + diopside cotectic at 1.0 GPa and then is carried adiabatically to the ocean floor. The liquid (composition F) arrives on the earth's surface in a totally crystal-free condition and on cooling crystallizes forsterite and then forsterite + anorthite. Diopside will appear only at the 1 atm eutectic. This is an interesting lesson because liquid F that has only olivine on the liquidus at 1 atm was derived from a more primitive liquid that was in equilibrium with both olivine and diopside at higher pressures. Thus, a MORB crystallizing only olivine need not be a primary basalt, and a basalt crystallizing olivine or olivine + plagioclase at low pressures may have crysallized olivine and calcic clinopyroxene at higher pressures. Indeed, there is trace element evidence that some primitive MORB may have experienced exactly this phenomenon [e.g. Schilling et al., 1983).

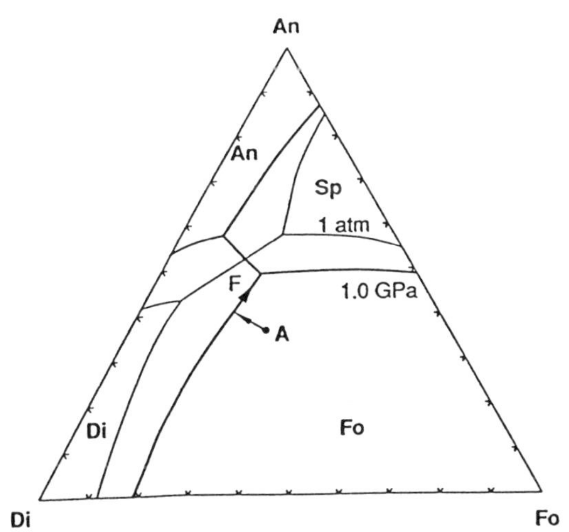

Fig. 7 Liquidus phase boundaries at 1.0 GPa (heavy line) and 1 atm (light line) for the Fo-Di- An system. Note that the crystallization sequence for A at 1.0 GPa is Fo→Di.

CaO-MgO-Al₂O₃-SiO₂ (CMAS) System

A closer approximation to peridotite is achieved by considering the system anorthite-diopside-forsterite-quartz, which includes the two systems already discussed. Presnall et al. [1979] determined the compositions of liquids in equilibrium with plagioclase lherzolite and spinel lherzolite in this system from 1 atm to 2.0 GPa. Because of the difficulty of visualization in three dimensions, they projected the liquid compositions onto various faces of the compositional tetrahedron. Before discussing the results, it will be necessary to review, albeit briefly, the mechanics and implications of projections commonly used in petrology. The phase equilibria of the ternary system, $CaAl_2Si_2O_8$-$CaMgSi_2O_6$-Mg_2SiO_4 are used to illustrate the nature of projections (Fig. 8).

A plagioclase projection of any liquid composition coexisting with anorthite is obtained by drawing a line from the $CaAl_2Si_2O_8$ apex through the liquid and to the $CaMgSi_2O_6$-Mg_2SiO_4 side. Such a line preserves the ratio of $CaMgSi_2O_6/Mg_2SiO_4$ in the liquid. The ternary system is

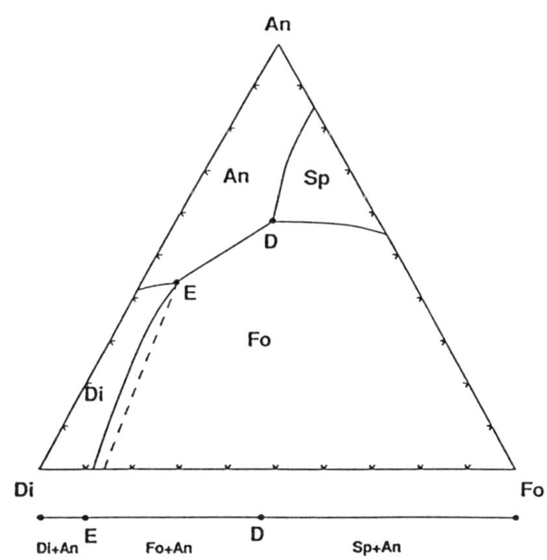

Fig. 8 The line marked-off below the compositional triangle includes the anorthite-projected liquids of all anorthite-bearing cotectics. E on this line is the projected ternary eutectic. The dashed line represents liquids which coexist with forsterite but not with anorthite. They also project on E.

replaced by a line on which is marked the diopside liquidus, the forsterite liquidus and the eutectic. All liquids coexist with anorthite. Thus, the crystallization path of any of the cotectic liquids coexisting with anorthite is easily determined and there is only one projected liquid in equilibrium with diopside and forsterite.

It is also common, however, to project phase boundaries for liquids that are not in equilibrium with the projecting phase, in this example, anorthite. The mechanics of the projections are the same; liquids are projected from $CaAl_2Si_2O_8$ onto the $CaMgSi_2O_6$-Mg_2SiO_4 join. Because the diopside-forsterite cotectic is not exactly radial to $CaAl_2Si_2O_8$, the cotectic liquids appear as a short line on the projection, reflecting the greater degrees of freedom of the phase assemblage. Moreover, the $CaAl_2Si_2O_8$ content of the cotectic liquid is no longer uniquely defined but increases as the temperature decreases towards the eutectic at E. Furthermore, a liquid composition that projects onto this cotectic is not necessarily saturated with diopside and forsterite, but may coexist with only forsterite (see dashed line in Fig. 8). Thus, the utility of such a projection is greatly diminished if the projecting phase is not one of the saturating phases. Great care must be taken in interpreting the phase boundaries of such phase-absent projections. In particular, liquids that project at the same point may have very different compositions in terms of the unplotted components.

With this caution noted, the compositions of the CMAS quaternary system are illustrated as two projections, one from anorthite, the other from diopside. The anorthite projection (Fig. 9) extends the isobaric invariant point of the anorthite-free system (Fig.3) into the $CaAl_2Si_2O_8$-bearing tetrahedron. Only liquids coexisting with anorthite lherzolite (≤0.9 Gpa) or spinel lherzolite (≥0.9 GPa) are shown, however. Liquids coexisting with anorthite lherzolite are quartz normative, (ie. the liquid compositions project to the SiO_2 side of the pyroxene join) whereas spinel lherzolite saturated liquids are olivine-normative. Increasing pressure makes the liquids richer in Mg_2SiO_4 and poorer in $MgSiO_3$ and $CaMgSi_2O_6$. The enstatite-forsterite and diopside-forsterite cotectics (not shown) also shift towards Mg_2SiO_4 with increasing pressure.

The diopside-projection (Fig. 10) describes these same liquids, but in terms of the $CaAl_2Si_2O_8$-Mg_2SiO_4-SiO_2 components. The $CaAl_2Si_2O_8$ content of the multisaturated liquids drops sharply with increasing pressure above 0.9 GPa, a consequence of the shift of all cotectics towards olivine. The effect of increasing pressure is to lower the CaO and Al_2O_3 and increase the Mg_2SiO_4 contents of the multisaturated liquids. Liquids produced at the isobaric invariant points at P > 1.0 GPa would always first crystallize olivine at shallower levels. The second silicate phase would be either diopside at high presssure or anorthite at lower pressures (≤0.7GPa)

This concludes the discussion and implications of the phase equilibria of simple systems. The next sections cover the crystal liquid equilibria of real basalts and peridotites.

SECTION II

Crystallization of MORB at Low Pressures

Phase Equilibria. The 1 atmosphere experimental crystallization sequence of primitive ocean floor basalts is [Walker et al., 1979; Grove and Bryan, 1983; Tormey et al., 1987; Juster et al., 1989]

olivine + spinel → olivine + spinel + plagioclase →
olivine + plagioclase + augite →
ıivine + plagioclase + augite + pigeonite

which substantially agrees with the sequence deduced from the phenocryst assemblages of glassy basalts. Fig. 11 describes

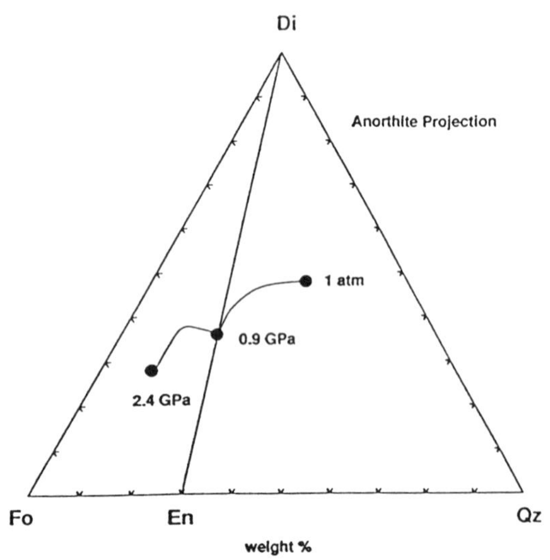

Fig. 9 Anorthite projections of liquids coexsisting with anorthite lherzolite(≤ 0.9 GPa) and spinel lherzolite (≥0.9 GPa). Note that anorthite is not stable at >0.9 GPa.

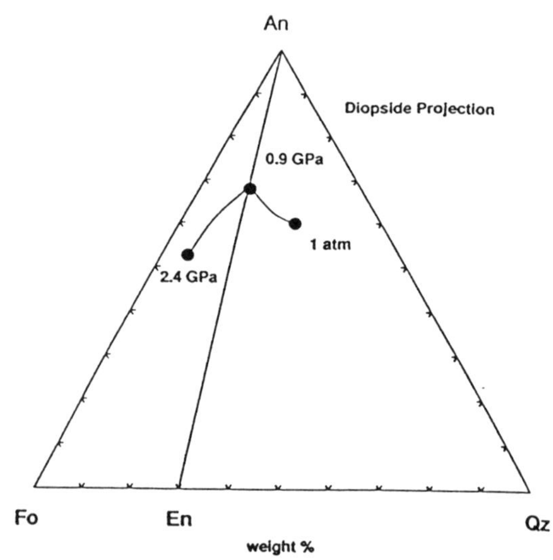

Fig. 10 Diopside projections of liquids coexsisting with anorthite lherzolite (≤ 0.9 GPa) and spinel lherzolite (≥0.9 GPa).

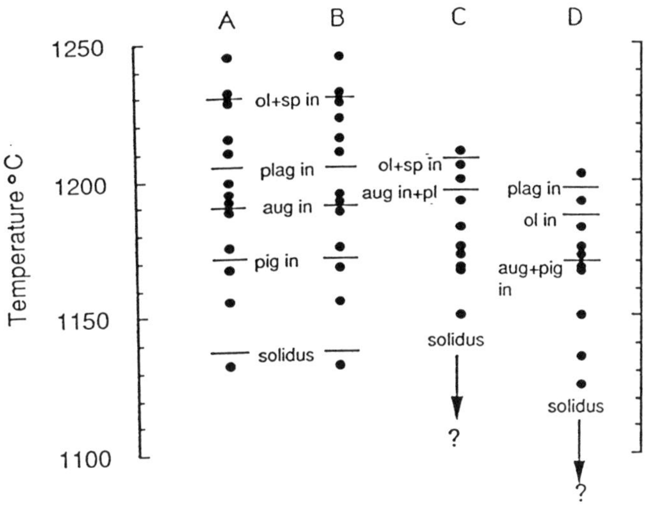

Fig. 11 Phase appearance temperatures at 1 atmosphere for various MORB compositions Compositions of the basalts (A, B, C, D) are given in Grove and Bryan [1983].

the phase appearance temperatures for each of the four sets of assemblages for a number of MORB compositions. The compositions of liquids produced by this crystallization sequence are not unique, however, because the thermodynamic variance of the multiply saturated boundaries is not low (see later).

The results of the melting experiments on ocean floor basalts are used to construct phase diagrams that display the compositions of multiply saturated liquids. Because the number of significant oxide components in oceanic basalts is seven or more, it is the practice to combine these oxides into mineral components and use these mineral components (normative components) to construct two and three dimensional phase diagrams. The chemical analyses of liquid compositions which are given as weight percent of the metal oxides are first converted to molar quantities and then are recalculated to mineral components by a prescribed recipe. The "normative components" are olivine ($[Mg,Fe]_2SiO_4$), clinopyroxene ($Ca[Mg,Fe]Si_2O_6$), silica (SiO_2), plagioclase ($NaAlSi_3O_8$-$CaAl_2Si_2O_8$), orthoclase ($KAlSi_3O_8$), magnetite (Fe_3O_4), ilmenite ($FeTiO_3$) and spinel ($MgCr_2O_4$). In most primitive basalts, the last four normative minerals listed above occur in very small amounts because the K, Fe^{+3}, Ti and Cr^{+3} contents are small. These components are ignored in most cases so that the basaltic liquids can be represented by the tetrahedron defined by the olivine-clinopyroxene-plagioclase-quartz normative components. While these approximations simplify the phase equilibria significantly, the choice of the mineral components suppress the important effects of MgO-FeO and CaO-Na_2O variations in liquids and the coexisting crystalline phases.

The phase relations of ocean floor basalts in the normative olivine-plagioclase-clinopyroxene-silica tetrahedron are complex analogues to the phase equilibria in the Mg_2SiO_4-$CaAl_2Si_2O_8$-$CaMgSi_2O_6$-SiO_2 system. In the simple system at fixed pressure, the compositions of liquids coexisting with one, two, three and four phases are represented by volumes, surfaces, lines and points respectively. The liquids lying on the the cotectic bounded by the liquidus volumes of forsterite, anorthite and diopside, for example, define a single (isobarically univariant) line at fixed pressure. Because the phase components in the ocean floor basalt tetrahedron have variable MgO/FeO or Na_2O/CaO contents, the equivalent cotectic is represented not by a line but rather by a cyclindrical volume. Likewise, planes and points in the simple system develop additional degrees of freedom. If the excluded components also are considered, it is obvious that caution should prevail when such phase diagrams are applied to a variety of basalts and their derivative liquids.

The multisaturated compositions of the experimentally derived liquids in the ocean floor basalt tetrahedron are located in a volume that generally follows the olivine-plagioclase plane and the olivine-plagioclase-augite cotectic. It is difficult, however, to examine, in detail, the liquid lines of descent in such a three dimensional display. The phase relations are condensed further by projecting liquid compositions from plagioclase or diopside onto the olivine-diopside-silica or olivine-plagioclase-silica pseudoternary systems. The projections are thermodynamically valid only for those liquids in equilibrium with the projected phase. Liquids projected onto the olivine-diopside-silica pseudoternary through plagioclase, for example, should coexist with plagioclase.

Fig. 12 is the plagioclase-projection of plagioclase-saturated liquids obtained from one atmosphere experiments [Grove and Bryan, 1983; Tormey et al., 1987]. The fields marked Ol, Au (augite), Op (orthopyroxene), and Pi (pigeonite)

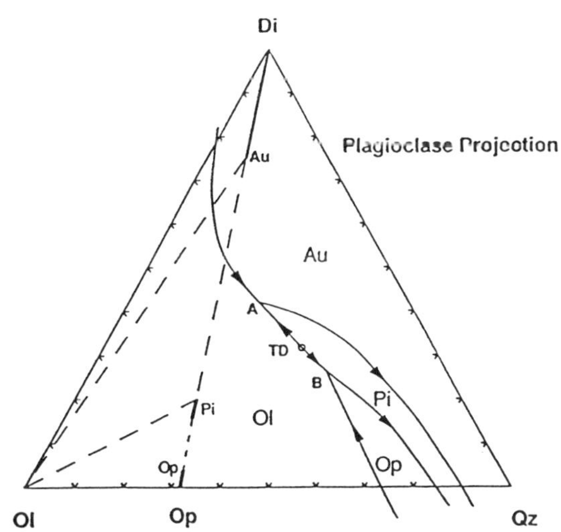

Fig. 12 Plagioclase-projection of plagioclase saturated silicate melts (given in "oxygen units") obtained from one atmosphere. Liquids within the dashed triangle Ol-Au-Pi will terminate equilibrium crystallization at point A.

are liquids which coexist with each of these phases as well as with plagioclase. The boundary lines separating two fields represent liquids in equilibrium with three minerals. The boundary curve separating the augite and olivine fields, for example, defines the liquids that coexist with olivine, augite and plagioclase. The points at the junction of three boundary curves are liquids coexisting with four crystalline phases, one from each of the bounding single phase fields plus plagioclase. Liquid A coexists with olivine, augite, pigeonite and plagioclase and liquid B coexists with olivine, orthopyroxene, pigeonite and plagioclase.

The boundary curves are not all equivalent, however. The boundary curve separating augite and olivine is an "even" boundary curve where the equilibrium is

(3) liquid = olivine + augite + plagioclase,

that is, the liquid, on cooling, crystallizes the three phases in certain "cotectic" proportions. The direction of cooling is given by the arrows on the boundary curves. The pigeonite + augite boundary curve is also "even" and is given by the equilibrium

(4) liquid = pigeonite + augite + plagioclase.

The other boundary curves are "odd" boundary curves, where there is at least one crystalline phase on each side of the equilibrium. The odd boundary curves between olivine and orthopyroxene, olivine and pigeonite, and pigeonite and orthopyroxene are fixed by the following equilibria respectivly:

(5) olivine + liquid = orthopyroxene + plagioclase

(6) olivine + liquid = pigeonite + plagioclase

(7) orthopyroxene + liquid = pigeonite + plagioclase

Points A and B are reaction points. The liquid at a reaction point, in contrast to an eutectic, has a composition which cannot be represented by the compositions of the coexisting minerals. Note that liquid A lies to the Qz side of the triangle formed by olivine-pigeonite-augite and liquid B lies to the Qz side of olivine-pigeonite-orthopyroxene. This means that a reaction must occur at these points between the liquid and one or more of the coexisting minerals. The reaction at point A is

(8) olivine + liquid = pigeonite + augite + plagioclase

and at point B

(9) olivine + liquid = orthopyroxene + pigeonite + plagioclase.

These reaction points have important implications which we will consider momentarily. One other unique point deserves mention. The point marked TD on the olivine-pigeonite (+plagioclase) odd boundary is a "thermal divide." Note that arrows on the boundary curve point away from the thermal divide. The thermal divide, therefore, directs liquids above TD to point A and liquids below TD to point B. The liquid lines of descent may thus follow very different paths, depending on where they intersect the pigeonite-olivine boundary curve.

Let's consider the liquid line of descent of a tholeiite basalt typical of the oceanic floor. Because of the nature of the projection, the liquid line of descent is recorded only after plagioclase and olivine have appeared. Olivine and plagioclase crystallization drives the liquid away from the olivine corner until it reaches the olivine-augite-plagioclase cotectic. Augite, olivine and plagioclase crystallize together and the liquid progresses down the cotectic to the reaction point A. The cotectic proportions, that is, the ratio of olivine: plagioclase: augite crystallizing from the cotectic liquids is not fixed, as indeed could be predicted from the curved boundary curve. (The cotectic proportions of olivine: augite are obtained by constructing a tangent to the boundary curve for the liquid in question and extending the tangent to the olivine-augite side of the triangle. The point of intersection and the lever rule give the required ratios. It follows, therefore, that a curved boundary curve implies that the cotectic proportions are not fixed). The cotectic ratios of olivine: plagioclase: augite are approximately 10: 55: 35 (in wt %) for liquids approaching A. At the reaction point A, the equilibrium is (in weight fractions)[Grove and Baker, 1986]:

(10) 0.15 Olivine + 0.85 Liquid = 0.49 Pigeonite + 0.20 Augite + 0.31 Plagioclase

The liquid line of descent at this stage depends on whether equilibrium or fractional crystallization prevails. In the equilibrium case, there is not enough liquid remaining to react with the olivine coexisting with liquid. The liquid is used up at the reaction point and the system is composed of olivine, plagioclase, augite and a small amount of pigeonite. It can be shown that all ocean floor basalts within the dashed region (a majority of MORB) are terminated at A under conditions of equilibrium crystallization.

Nature probably approaches the fractional crystallization mode, however. In the extreme case of perfect fractional crystallization, the system consists only of liquid at point A, i.e., it is assumed that all previously grown crystals have settled out or the liquid has been siphoned away from the crystals. In any case, reaction (10) cannot occur at A because there is no olivine to drive the reaction. The liquid, therefore, proceeds along the augite-pigeonite-plagioclase cotectic towards the quartz corner. At this stage, the liquids are so far removed from ocean floor compositions that the simplifications inherent in this projection scheme make the phase diagram less useful.

Equilibrium Crystallization

The near surface crystallization path of primitive ocean floor basalts can be divided into several segments, each characterized by a distinct suite of saturating phases. Olivine and a chrome-rich spinel are liquidus phases in the most primitive basalts extruded on the ocean floor. Because of the high Cr_2O_3 contents of spinels (~40%) and the low Cr_2O_3 contents of ocean floor basalts (~0.1%), spinel is only a minor crystallizing phase and, except for the Cr_2O_3 content, does not materially effect the major element composition of the coexisting liquids. The liquid line of descent is thus controlled largely by the crystallization of olivine. The major oxides compatible with olivine are MgO, FeO, and SiO_2,whereas the most important compatible minor oxide is NiO. The crystal-liquid

$$\left(\text{eg. } \frac{\text{MgO OL}}{\text{MgO LQ}}\right) \quad (3)$$

distribution coefficients are [Grove and Bryan , 1983] approximately 5.0 for MgO, 1.4 for FeO, somewhat less than one for SiO_2 (~0.8) and 10 for NiO [Elthon, 1987]. These distribution coefficients apply to basaltic melts with about 10% MgO, temperatures near 1200°C and if all iron is calculated as FeO. Even small amounts of crystallization of olivine severely depletes the liquid in MgO and especially NiO, but only modestly decreases the FeO content. SiO_2 increases slightly, whereas the oxides incompatible with olivine, CaO, Al_2O_3, TiO_2, Na_2O, P_2O_5 and K_2O, increase as 1/F, where F is the weight fraction of liquid remaining after (1-F) of the liquid has crystallized as olivine and spinel. In many studies, MgO is used as an index of differentiation because olivine and spinel crystallization reduces the MgO contents of the coexisting liquids. In summary, NiO, Cr_2O_3 and FeO are positively correlated with MgO, whereas the remaining oxides are negatively correlated.

Plagioclase is the third phase to appear in the second segment of the liquid line of descent. The cotectic proportions of plagioclase and olivine are not unique but are dependent on the bulk composition of the basalt. The plagioclase/olivine cotectic ratio is approximately 2/1 in the anorthite-forsterite-SiO_2 system, but more than 30/1 if albite is substituted for anorthite. However, the ratios are decreased if fayalite is substituted for forsterite. Thus, the cotectic proportions of plagioclase/olivine increases with the Ab and Fo content of plagioclase and olivine respectively. The plagioclase/olivine (in wt %) cotectic proportion for a typical N-MORB appears to be about 3/1 [Grove and Bryan, 1983; Tormey et al., 1987].

The coprecipation of plagioclase and olivine heralds a significant change in the liquid line of descent. The bulk distribution coefficients for Al_2O_3, MgO and NiO are greater than one, they are slightly less than one for Na_2O, CaO and SiO_2 but very much less than one for FeO. TiO_2 and K_2O are still highly incompatible. Al_2O_3 and NiO are positively correlated with MgO; SiO_2, CaO and Na_2O increase slightly and FeO and other incompatibles increase more noticeably with decreasing MgO. Note that Mg* numbers decrease and the CaO/Al_2O_3 ratio increases dramatically in this crystallization interval.

Augite is the fourth mineral to appear in the crystallization sequence. The cotectic defined by the coexistence of olivine, plagioclase and augite has a lower thermodynamic variance and consequently, the cotectic ratios are better constrained. The cotectic ratios are approximately OL:PL:AU = 10:55:35, and show only small variations around this value [Tormey et al., 1987; Grove and Bryan, 1983]. The bulk distribution coefficients for MgO, Al_2O_3 and CaO are significantly greater than one; they are less than one for FeO, Na_2O and other incompatible oxides and close to one for SiO_2. The liquids produced along this cotectic have nearly constant SiO_2, increasing FeO and decreasing MgO, CaO and Al_2O_3. The CaO/Al_2O_3 ratio now decreases rather than increases with advancing degrees of crystallization, a sign that augite as well as plagioclase is on the liquidus.

At the peritectic "point", olivine reacts with approximately five times its mass of liquid to produce pigeonite, plagioclase and augite (Equation 10). Although the peritectic appears as a point in projection, the reaction occurs over a finite range of temperatures and liquid compositions. The temperature interval, for example, between the onset of the peritectic reaction and the solidus is approximately 30-40° C [Grove and Bryan, 1983]. Minor phases such as ilmenite, titanomagetite and apatite eventually appear on the liquidus when the incompatible oxides TiO_2 and P_2O_5 reach appropriate saturation values.

Fractional Crystallization

The liquid lines of descent in magma chambers are likely to be controlled by fractional rather than equilibrium crystallization as forces of gravity and compositional zoning of minerals isolate crystals from the liquid. By the time the peritectic is obtained in the magma chamber, all or at least the majority of previously crystallized phases have accumulated on the floor, walls and roof of the chamber. The ratio of olivine to liquid in the largely liquid portions of the magma chamber is going to be small so that it is olivine not liquid that is consumed by the peritectic reaction. Once olivine is eliminated, the liquid then follows the plagioclase, augite and pigeonite cotectic. The initial phase proportions are PL: AU: PI = 0.55: 0.18: 0.27 [Grove and Baker, 1984]. As crystallization advances, the crystallizing assemblage contains more augite and less pigeonite and the cotectic proportions become 0.55: 0.25: 0.20. The bulk distribution coefficients are substantially greater than one for CaO, Al_2O_3 and MgO, close to unity for SiO_2, and less than unity for FeO, TiO_2 and Na_2O. There continues to be little to no SiO_2 enrichment but the incompatible oxides FeO and TiO_2 become concentrated in the liquid, eventually leading to the crystallization of ilmentite and/or titanomagnetite from iron-rich basaltic liquids (ferro- or FeTi basalts). Ilmenite crystallizes only from melts developed below the oxygen partial pressures of the QFM[1] buffer and is relatively rare in MORB, whereas titanomagnetite is favored at the NNO buffer [Juster et al, 1989]. Both phases may coexist at intermediate f_{O_2}.

The appearance of Fe-Ti rich phases has a major effect on the bulk distribution coefficients of SiO_2, FeO and TiO_2 [Juster et al., 1989].

The bulk distribution coefficient for SiO_2 is significantly less than one (initially, approximately 0.92, but then decreases to much smaller values). As a result, the SiO_2 contents of the residual liquids begin to increase dramatically

[1] QFM refers to the fugacity of oxygen that is fixed by the assemblage quartz-fayalite-magnetite at a given pressure and temperature. NNO is a similar reference curve determined for the assemblage nickel-nickel oxide. The NNO curve defines an oxygen fugacity curve about 1 order of magnitude above QFM.

and ferrobasalts quickly evolve to basaltic andesites, andesites and rhyodacites. The bulk distribution coefficients for FeO and TiO_2, in contrast, are greater than one, terminating the trend of iron enrichment and converting TiO_2 from an incompatible to a compatible oxide.

Liquid Line of Descent

Somewhat more specific chemical trends for ocean floor lavas are described in Fig. 13, where the concentrations of a select number of oxides are plotted against MgO (Wt%). The trends are those obtained by experiment on ALV-528 basalt glass (MgO 10.4% to 6%; Grove and Bryan, 1983] and a basalt from the Galapagos spreading center (MgO 6% to 1%; Juster et al., 1989]. The two basalts (Table 2) were chosen because their liquid lines of descent join. That is , the liquid line of descent for the Galapagos' basalt more or less begins where the ALV-528 liquid line of descent was terminated. The crystallization path is therefore an artificial one, since the liquid line of descent for ALV-528 should terminate at the peritectic under conditions of equilibrium crystallization. The entire sequence could be obtained, however, if the path to MgO = 7% is obtained by equilibrium processes but then the residual liquid is separated from crystals, segregated and allowed to complete its crystallization. The new solidus is that of the segregated liquid and not of the original basalt parent. The artifice doesn't create serious problems since the liquid line of descent is quite similar for fractional and equilibrium processes as long as the extent of crystallization is not extreme [e.g., Longhi, 1987]. Conversely, this trick produces increasingly poor results with advanced degrees of crystallization. The liquid line of descent, nevertheless, is used for illustrative purposes [See Longhi, 1991, for a more complete discussion of fractional vs. equilibrium crystallization].

The olivine-out reaction (reaction which consumes olivine) begins at about 1170° C and under equilibrium crystallization continues to about 1140° C producing increasing amounts of pigeonite, augite and plagioclase at the expense of olivine and the liquid. This is the peritectic "point" alluded to in the general discussion using the plagioclase projections. Approximately 50% of the liquid line of descent is contained within the peritectic reaction and major changes in the composition of the liquid are associated at this "point". This emphasizes, again, that the point is an artifact of the projection, whereas, in truth, the peritectic is a significant volume in compositional space with FeO-MgO, Na_2O-CaO and TiO_2 as compositional coordinates. This fact is more obvious in Fig. 13, where the liquid line of descent between 7.5% and about 6.0% MgO roughly corresponds to this crystallization interval.

To continue past the peritectic point, we imagine that the liquid is displaced (perfectly fractionated) to a new container free of crystals. The first crystals to appear from this liquid are pigeonite, augite and plagioclase (recall that olivine reacts with these liquids; it does not crystallize from them). Liquids obtained along this boundary curve are ferrobasalts (FeTi basalts) and may contain as much as 18% FeO and 4.5% TiO_2. The Al_2O_3 and CaO contents, moreover, are very low,

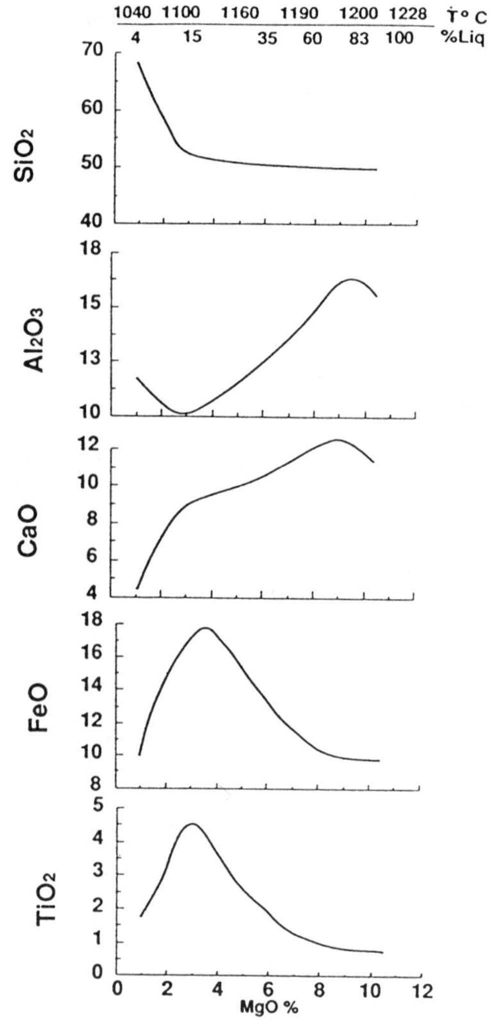

Fig. 13 Liquid line of descent obtained by the fractional crystallization of typical MORB.

approaching 10 and 9 wt %, respectively. The temperature is just below 1100° C and, using the ALV-528 basalt as the parent, the liquid is the product of more than 85% crystallization. Incompatible elements and oxides have been enhanced more than sixfold; for example, K_2O and P_2O_5 have increased from 0.11 to about 0.7 wt %.

It is easy to appreciate why such liquids are not common on the ocean floor. First, they represent only a small fraction of the parent. Two, their high FeO and TiO_2 and low Al_2O_3 contents make them dense and difficult to erupt. Finally, the magma chamber must allow such batches of liquid to segregate and be isolated from the much larger volumes of primitive basalt magmas that are replenishing the magma chamber.

Ilmenite and/or titanomagnetite saturate these liquids as the temperature drops below 1100°C. The TiO_2 and FeO contents of the liquid, thereafter, drop quickly and the SiO_2 content begins a rapid ascent. Whereas the trajectory of CaO is still downward, the Al_2O_3 trend is reversed. This implies that the cotectic ratio of augite/plagioclase is larger than obtained at higher temperatures. The liquid at 1040°C is a rhyodacite (Table 2) and represents approximately 4% of the parent. The liquid coexists with quartz, and eventually would crystallize alkali feldspar, if fractionation was allowed to extend the liquid line of descent to lower temperatures.

Experiments conducted at higher partial pressures of oxygen than above increase the Fe_2O_3/FeO ratio in the liquid and stabilize titanomagnetite at much higher temperatures [Juster et al., 1989]. The early appearance of titanomagnetite limits the enrichment of TiO_2 and terminates FeO enrichment at lower values, and at higher MgO contents.

Crystallization of MORB: High Pressure

The position of the olivine-plagioclase-calcic clinopyroxene cotectic shifts towards olivine with increasing pressure [Grove et al., 1990; Tormey et al., 1987; Fujii and Bougault, 1983; Stolper, 1980; Bender et al., 1978; Kinzler and Grove, In Press] The four phase peritectic "point" (point A in Fig. 12) not only shifts towards olivine but also towards the clinopyroxene-olivine join (Fig. 14). At pressures above 2.0 GPa, the liquid, as projected from plagioclase, lies to the SiO_2-poor side of the clinopyroxene-olivine join and hence is "nepheline" normative [Stopler, 1980; Falloon and Green, 1988].

Figure 15 compares the positions of the boundary curves for MORB produced at 1 atm and 0.8 GPa [Grove et al., 1990]. The arrows show temperature dropping towards the Qz-Di join at 1 atmosphere but towards the Ol-Di join at 0.8 GPa. This means that the peritectic equilibrium has switched from

$$\text{olivine + liquid = pigeonite + augite + plagioclase} \quad (4)$$

at 1 atm to

$$\text{low CaO pyroxene + liquid = olivine + augite + plagioclase} \quad (5)$$

at 0.8 GPa. The liquid line of descent at 1 atmosphere evolves to more SiO_2-rich liquids with decreasing temperatures, producing olivine tholeiites, quartz tholeiites

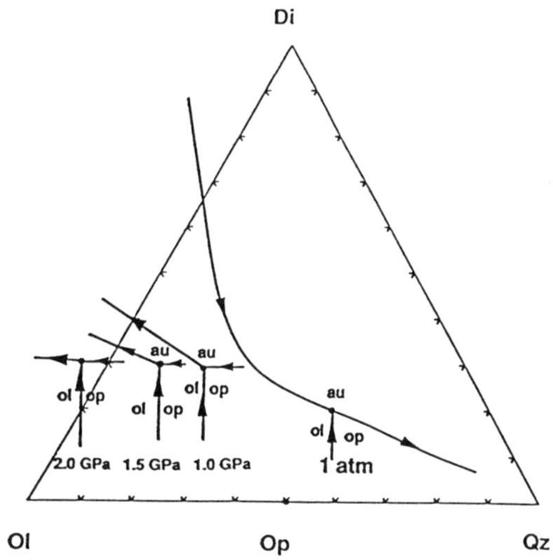

Fig. 14 Shift of the four phase peritectic (ol, op, au with either plg or sp) with increasing pressure. The cotectics are projected from plagioclase onto the Ol-Di-Qz system wt %).

and even andesite to dacite volcanic rocks, as observed in the Galapagos spreading center [see Juster et al., 1989, for extended discussion). In contrast, the liquid line of descent at 0.8 GPa produces SiO_2-poor tholeiite basalts, and by extended fractional crystallization, alkali basalts and even nepheline-normative trachytes. The cross-over between these two liquid lines of descent is not firmly established but probably lies in the 0.5-0.6 GPa range [Grove et al., 1990; Falloon et al., 1988] and certainly must depend on the bulk composition of the basalt and its volatile contents (see later).

TABLE 2

	(1)	(2)	(3)	(4)	(5)	(6)
SiO_2	49.6	49.5	50.4	51.9	51.9	62.9
TiO_2	0.67	1.1	1.5	2.1	4.6	2.2
Al_2O_3	16.0	14.3	12.6	12.8	10.2	11.1
FeO	9.6	11.8	13.6	13.5	17.7	12.6
MgO	10.4	7.5	6.6	6.1	3.1	1.5
CaO	11.9	12.7	11.2	10.4	8.7	5.6
Na_2O	1.7	2.0	1.3	2.3	2.4	2.7

Liquids derived from the 1 atmosphere equilibrium crystallization of ALV 528-1-1 (1) through (3) and GSV (4) through (6).

(1) ALV-528-1- starting composition, (2) ALV-1187°C, glass coexisting with Ol-Plg-Aug, (3) ALV-1156°C, glass with Ol+Plg+Aug , (4) GSC, 1143°C glass with Ol+Aug+Plg, (5) GSC, 1085°C glass with Plg-Aug-Ilm, (6) GSC, 1050°C glass with Plg+Aug+Plg+Ilm+TiMag+Qtz (Grove and Bryan, 1983, Juster et al, 1989)

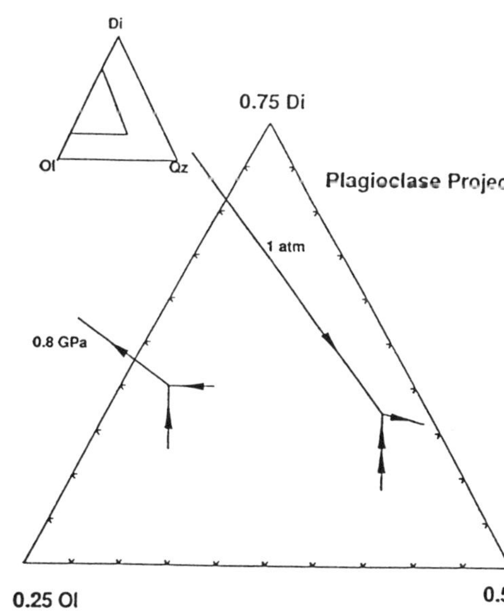

Fig. 15 Plagioclase projection of peritectic liquids at 1 atm and 0.8 GPa. Note that the arrows point in different directions implying different liquid lines of descent.

Quartz normative tholeiites cannot be produced from primary basalts evolving at pressures more than 0.5-0.6 GPa. That quartz tholeiites are a minor but nevertheless real subset of MORB [e.g., Schilling et al., 1983] indicates that some of the parent liquids have undergone crystal fractionation in shallow level magma chambers. Moreover, the majority of MORB glasses are clustered around the low pressure olivine + plagioclase + calcic clinopyroxene cotectic [Stolper, 1980]. This indicates that a large fraction of these MORB have been modified by crystal fractionation processes in shallow magma chambers. A portion of the SiO_2-poor MORB subset, however, does not superimpose on the low pressure cotectics. These MORB glasses fall between the 1.0 GPa and 1 atmosphere cotectics, and may be liquids produced by crystal fractionation at high pressures [Stolper, 1980; Grove et al.,1990; Falloon and Green, 1988], or perhaps some may even be primary magmas.

Several experimental studies have determined the liquidus phase assemblages for primitive ocean floor basalt compositions in the hope of establishing their primary character. If these compositions were equivalent to that of a primary basalt, and if the primary basalt was last in equilibrium with a mantle assemblage of olivine + orthopyroxene ± diopside ± spinel ± plagioclase, then the experimental pressure and temperature at which the liquidus assemblage consists of olivine + orthopyroxene ± diopside, spinel, plagioclase equals the conditions at which the melt was last isolated from the mantle. None of the most primitive basalt compositions studied to date [Bender et al., 1978; Fujii and Bougault, 1983; Elthon and Scarfe, 1984] have olivine and orthopyroxene simultaneously on the liquidus, indicating that these basalts are not primary magmas. These basalts have olivine only on the liquidus at low pressures and calcic clinopyroxene at higher pressures. The multisaturation point at which olivine and calcic clinopyroxene occur together has significance, however. Because the olivine liquidus field expands with decreasing pressure, a melt produced in equilibrium with olivine + orthopyroxene + calcic clinopyroxene will crystallize first olivine and then olivine with calcic clinopyroxene at lower pressures [Kinzler and Grove, In Press]. Suppose that this liquid is separated from its crystals and extruded to the surface. This derivative basalt will have olivine + calcic clinopyroxene on its liquidus at the pressure and temperature at which the liquid was separated from its crystals. The multisaturation point, therefore, has depth significance and gives the minimum depth at which the basalt was separated from mantle. The experiments to date have olivine + calcic clinopyroxene ± plagioclase on the liquidus of primitive basalts at about 1.0 GPa and 1300°C. The primary magma parental to these basalts must therefore have last equilibrated with mantle at depths greater than 30 - 35 km.

Melting Peridotite

Experimental Problems. The determination of partial melts in equilibrium with peridotite is a difficult task since a number of experimental problems must be solved. Contradictory experimental results often arise from the lack of control of the bulk composition of the charge. The problem is that it is very difficult to design a container that is totally inert for the duration of the experiment. Sealed iron capsules, for example, result in either Fe gain or Fe loss of the charge and can be used only for systems which are saturated with metallic iron, a condition inappropriate for the upper mantle. Platinum containers, which have commonly been used in the past for high P-T experiments, are notorious for alloying with iron in iron-bearing systems [e.g., Jacques and Green, 1979]. A number of studies have demonstrated that the extent of iron loss from the charge is sensitive to temperature and run duration; iron is lost preferentially from the phases in the order liquid > olivine > pyroxene. This has several unfortunate ramifications. The transfer of iron to the platinum container oxidizes the charge according to the reaction

$$FeO + Pt = (Fe, Pt) + {}^1/_2\,O_2 \tag{6}$$

and thereby stabilizing chromian spinel to very high degrees of melting [Falloon and Green, 1987]. The loss of FeO increases the Mg* value of all phases, thereby raising the liquidus temperatures and reducing the amount of melt obtained at a given P,T. This is particularly vexing, since the Mg* is one of the most important parameters used to constrain the origin of primary basalts. Increasing the Mg*, moreover, alters the phase boundaries, causing liquids to become more SiO_2-saturated and to have excessive CaO contents [Falloon and Green, 1987]. In summary, it is clear that certain aspects of the results of basalt or peridotite melting experiments performed in platinum containers generally are untrustworthy unless special precautions are taken.

The iron loss/gain problem can be mitigated by placing a graphite inner capsule within an outer Pt capsule. This assembly is effective although iron loss may still occur for long run times (>72 hrs) by diffusion through the graphite capsules [Falloon et al., 1987]. The graphite assembly is not a cure-all, however. The graphite container reacts with oxygen to form carbon dioxide. Any ferric oxide in the charge is reduced to FeO, thereby decreasing the Mg* value. If the charge is reduced, graphite reacts with FeO to produce iron metal, for example,

$$2FeO + C = 2\,Fe + CO_2 \tag{7}$$

and thereby increases the Mg* value. A large number of Takahashi and Kushiro's [1983] peridotite melting runs, for example, were done in graphite capsules and have iron metal in the residual assemblage.

Even if the container problems are resolved, the electron microprobe analyses of equilibrium partial melts of peridotite often give unreliable results. The difficulties arise from the inability to completely quench the liquid to a glass without the liquid suffering additional crystallization. The quench problems are particularly acute when the liquid is less than 20-30% of the charge. Unfortunately, this is the range of melting that generates basalts of the most useful compositions. At larger degrees of melting, the quench problems are less severe or can be overcome by using "broad beam" electron microprobe analyses which average the glass and quench crystal compositions.

Two techniques have been used to create more melt at a given T,P and therefore reduce the quench problem. Jacques and Green [1980] used peridotite compositions from which 40% olivine was subtracted. This effect `fertilizes the peridotite, since olivine remains largely unmelted for moderate degrees of melting under most relevant P-T conditions. Thus, there is more of a "basaltic component" in the peridotite, and more melt is produced at a given P and T. Ideally, the olivine subtracted must have the composition that coexists with melt at each particular P and T. Jacques and Green [1980], however, subtracted an olivine of Mg* 0.916 from the "pyrolite" peridotite starting composition. This results in a peridotite with Mg* of 0.856, a value too iron-rich for a mantle composition. An harzburgite produced by large degrees of melting of this pyrolite at 1.0 GPa, for example, has olivine of Mg* 0.87, whereas most harzburgites dredged from the ocean floor have olivines with Mg* > 0.90 [Michael and Bonatti, 1985].

A second technique is to construct a basalt-peridotite sandwich wherein the basalt layer is forced to equilibrate with a much greater mass of peridotite [Stolper, 1980; Takahashi and Kushiro, 1983; Falloon and Green, 1987, 1988]. The technique forces the basalt liquid to adjust its composition to the large reservoir of peridotite, i.e., the basalt composition is buffered by the peridotite (but see later). The glass in these experiments occurs as relatively large, homogeneous and concentrated pools. The glass suffers fewer quench problems and is chemically analyzed using a defocussed electron microprobe beam.

Experimental problems remain, however. First, the basalt component cannot be allowed to overwhelm the peridotite composition. Adding too much basalt, for example, may lower the Mg* value of the system to unrealistic values. Second, it is difficult to equilibrate the entire peridotite charge with the basalt. Minerals away from the glass layer are not compositionally uniform from core to rim [Fujii and Scarfe, 1985; Takahashi and Kushiro, 1983; Falloon and Green, 1987, 1988]. Their rim compositions are identical to the compositions of the phases within the glass layers, but the core compositions are not. This means that the composition of the equilibrated system is not known and that the effective basalt/peridotite ratio is higher than given. Finally, in runs close to the solidus or at high pressures (P > 2.0 GPa) clinopyroxene crystallization on the quench cannot be avoided and special probe techniques must be used to determine the pre-quench liquid composition [Falloon and Green, 1988]. This introduces additional uncertainties.

P-T Boundaries

There is no single mantle solidus since the initiation of melting is sensitive to bulk composition, particularly the Mg* value, the H_2O content and the concentration of incompatible oxides such as TiO_2 and P_2O_5. It is not surprising then that the experimentally determined solidi may vary by about 100° C from sample to sample. This effect is not the only source of such large variations however. Because the solidus temperature is correlated with the Mg* value of peridotite, experiments which cannot prevent exchange of iron with capsules may give aberrant melting temperatures. The 3.5 GPa melting curves of Mysen and Kushiro [1977] which were obtained in platinum capsules, for example, are about 50° C higher than those obtained in graphite capsules [Harrison, 1979] (Fig 16). The differences are explained by iron loss to the platinum capsules and an increase in the Mg* values. As was already discussed, graphite capsules are not a panacea. If the charge does not contain enough oxygen (read Fe_2O_3), the production of CO_2 will cause the reduction of FeO and the production of metal. This results in an increase in the Mg* value and, consequently, an increase in melting temperatures. Many of the melting products of Takahashi and Kushiro [1983] have iron metal, although apparently not enough to be worrisome. Melting experiments done in multianvil assemblies, however, have reported significant FeO loss by the graphite reduction mechanism [e.g., Takahashi, 1986].

It is useful to set an upper bound to the mantle solidus. This is provided by the solidus of the CMAS system in which the absence of FeO and invariant melting minimize experimental uncertainty (Fig.17). The solidus extends, albeit in curved segments, from about 1240° C at 1 atm to 1750° C at 5 GPa [Presnall, et al., 1978; Herzberg et al., 1990]. The subsolidus mineralogy is plagioclase lherzolite for 0 to 1 GPa, spinel lherzolite from 1 to about 2.1 GPa, and garnet lherzolite above 2.1 GPa. The addition of other components, particularly FeO, Na_2O and TiO_2, will depress the solidus and displace the boundaries between the different lherzolite fields.

The solidus for a very fertile spinel peridotite HK-66 (Mg*= .85) (Table 3) probably fixes the lower bound for the range of upper mantle peridotite solidi [Takahashi and Kushiro, 1983]. The solidus has cusps at about 11 and 2.6 GPa, which separate the three lherzolite stability fields. (No spinel was observed in some of the lherzolites in the intermediate field. This is

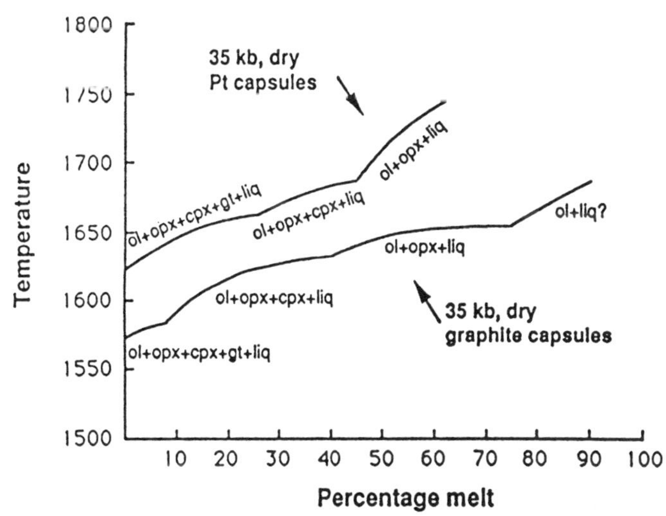

Fig. 16 Comparison of the melting of peridotite at 3.5 GPa in Pt and C capsules. Note that not only the melting temperatures but also the phase assemblages are strongly affected by the type of capsule.

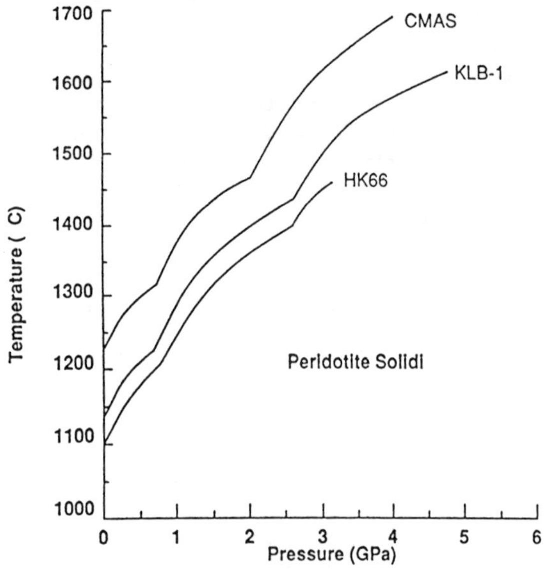

Fig. 17 Solidi of the CMAS system and other peridotites (see Table 3 for compositions). The cusps separate the anorthite lherzolite (low P) from the spinel lherzolite (intermediate P) and the garnet lherzolite (high P) fields.

TABLE 3

Peridotite-Pyrolite Compositions Used in Experiments

	(1)	(2)	(3)	(4)	(5)	(6)
SiO_2	44.3	47.2	47.9	48.0	44.5	44.7
TiO_2	0.16	0.28	1.18	0.22	0.16	0.17
Al_2O_3	4.3	7.3	5.9	4.9	3.6	4.4
FeO	9.8	7.3	8.7	9.9	8.1	7.6
MgO	36.8	30.6	28.8	32.4	39.2	38.6
CaO	3.3	5.6	5.1	3.0	3.4	3.4
Na_2O	0.39	0.66	0.95	0.66	0.3	0.4
K_2O	---	---	0.22	0.07	---	---
Mg*	0.87	0.88	0.85	0.85	.89	.90

MPY-87 (Falloon & Green, 1988), (2) MPY-90-40 (Falloon & Green, 1988), (3) HW Pyrolite-40 (Jacques & Green, 1980), (4) HK66 (Takahashi & Kushiro, 1983), (5) KLB-1 (Takahashi, 1986), (6) MPY-90 (Falloon & Green 1987)

probably due to the low oxidation states imposed by the graphite capsule, reducing Cr^{+3} to Cr^{+2}, which then dissolves in olivine and pyroxene rather than forming spinel.) Note that much higher pressures are required to stabilize garnet in real peridotites and solidus temperatures are approximately 100° C lower than those in the CMAS system.

Peridotites with higher Mg* values have higher solidi. The solidus for spinel lherzolite KLB-1 (Mg* .891) is approximately 50°C higher than HK-66 [Takahashi, 1986; Scarfe et al., 1979]. The KLB-1 solidus is probably more appropriate for the sub-ocean ridge mantle.

The melting curves (the "out curves") above the solidus mark the disappearance of the aluminum-rich accessory phases, plagioclase or garnet, and then in turn diopside, orthopyroxene and finally olivine, which marks the liquidus (Fig.18). The maximum stability of plagioclase (the"plagioclase out" curve) is not well determined but lies about 50° C above the solidus at 1 atm and intersects the solidus near 1.0 GPa, the exact position depending, of course, on the Na_2O, CaO and Al_2O_3 content of the peridotite. The stability limit of spinel is sensitive to the oxidation state, and most of the experimental spinels do not match their compositions in natural rocks. We therefore will not deal extensively with the melting relations of spinel. Garnet is the first phase to be consumed above 2.5 GPa, but the data points are few. The "garnet out" curve is within 75° C of the solidus in HK-66 and within 100° C in KLB-1 at 3.0 GPa. The "garnet out" curve shown in Fig. 18 is only a best guess.

The coordinates of the "diopside-out" curve are more important than those of the solidus, since the abundance of dredged abyssal harzburgites [Dick, 1989; Dick and Fisher, 1984; Michael and Bonatti, 1985] indicate that the conditions of melting approached or even exceeded this curve at some depth. The range of "diopside out" curves determined in several laboratories span more than 100° C at a given pressure (Fig.19). At 1.0 GPa, for example, this curve exists in the temperature interval from 1280° C in HK-66 to about 1400° C in MPY-87 [Falloon and Green, 1987; Takahashi and Kushiro, 1983]. The temperatures are correlated with the Mg* values of the liquid which range from 0.64 to 0.74. This emphasizes the point that all melting curves are elevated in depleted mantle relative to more fertile melting. This point is more than of academic interest. During fractional melting, the Mg* value of

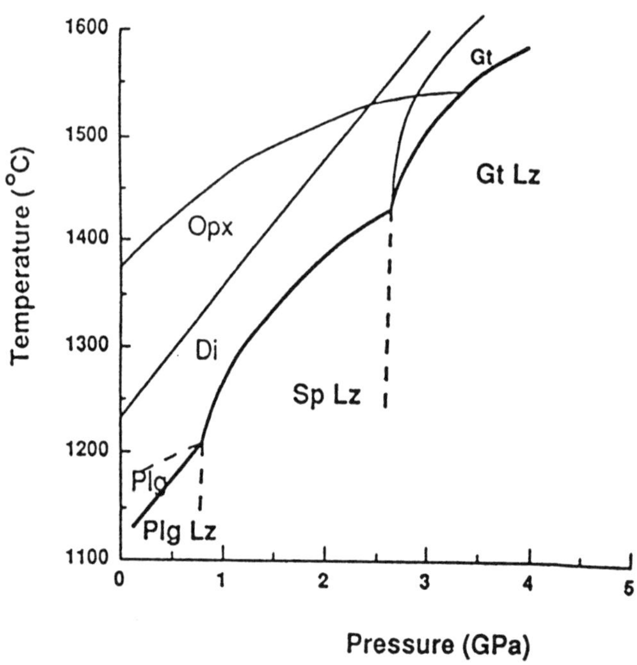

Fig. 18 The melting relations of a typical peridotite KLB1 [Takahashi, 1986]. See text.

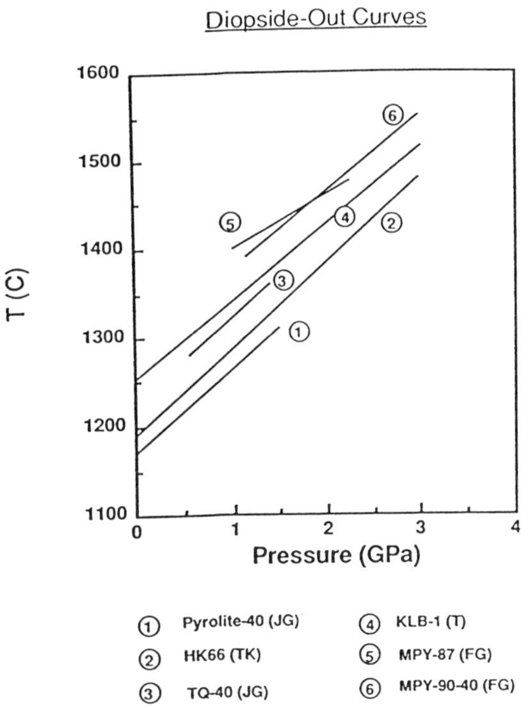

Fig. 19 Diopside -out curves for various peridotite compositions (1) and (3) Jacques and Green [1980], (2) Takahashi and Kushiro [1983]. (4) Takahashi [1986], and (5)- (6) Falloon and Green [1987, 1988].

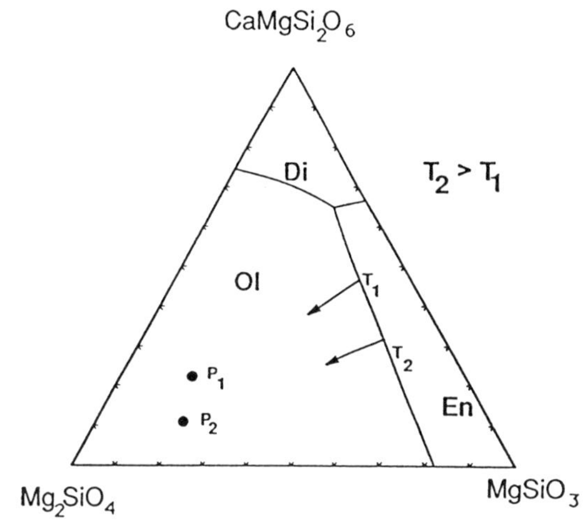

Fig. 20 Orthopyroxene out curves for peridotites P_1 and P_2, which are at temperature T_1 and T_2 respectively.

the peridotite residue becomes elevated and therefore, more refactory. Fractional melting causes the diopside-out curve temperature to rise thus making it increasingly difficult to eliminate diopside from the peridotite mantle. The "diopside-out" curve in Fig.18 is appropriate only for a relatively fertile mantle.

The "orthopyroxene-out" curve is very sensitive to the composition of the peridotite. This is illustrated with the aid of the Di-Fo-En (this is part of Di-Fo-Qz [Fig. 3]) phase diagram (Fig. 20). The orthopyroxene-olivine cotectic begins at the Mg_2SiO_4-$MgSiO_3$ binary and terminates at the invariant melting point. The temperature decreases towards the invariant point. Now consider two bulk composition with different normative diopside contents. A line from the olivine apex drawn to the peridotite bulk composition and then extended to the olivine-orthopyroxene cotectic gives the liquid compositon from which orthopyroxene disappears during equilibrium melting. Note that the "orthopyroxene out" liquid lies closer to the Mg_2SiO_4-$MgSiO_3$ binary, and therefore, at higher temperature in the diopside-poor peridotite. Since depleted mantle is relatively diopside-poor, it follows that orthopyroxene is stable to higher temperatures than in more diopside-rich fertile mantle.

An additional effect on the orthopyroxene-out curve should be noted. Because the composition of diopside expands towards enstatite at high temperatures [Lindsley, 1981], it is possible that orthopyroxene will disappear from the subsolidus mineral assemblage. This point is illustrated by the chemical compositions of coexisting olivine, orthopyroxene and clinopyroxene (Fig. 21) obtained in the experiments of Takahashi [1986]. The circle is the bulk composition of the peridotite, the heavy lines represent the compositions of pyroxene and the triangle (ABC) gives the compositon of the coexisting phases. Note how the olivine-clinopyroxene join (AB) approaches the bulk composition at high temperatures and pressures. The peridotite becomes a one pyroxene rock containing only clinopyroxene and olivine once the A-B join passes the bulk composition. Naturally, the P-T conditions at which point orthopyroxene disappears depends on the bulk composition, particularly its diopside content. A depleted mantle may retain orthopyroxene to much higher pressures and temperatures than a more fertile rock.

The "orthopyroxene-out" curve in Fig. 18 is that obtained in the experiments of Takahashi [1986] on the KLB-1 peridotite. Note that the curve intersects the solidus at pressures above 3.0 GPa. This reflects the replacement of orthopyroxene by low CaO clinopyroxene at high pressures. The "orthopyroxene curve" is displaced to higher or lower temperatures in more refractory or fertile peridotites respectively.

Melting of Peridotite Sandwiches

In the CMAS quarternary system, the coexistence of olivine, orthopyroxene, augite, spinel and liquid is an univariant equilibria and, consequently, the composition of the liquid is represented by a point at fixed P. The variance of this assemblage increases by one for each additional component so that liquid composition and temperature of this assemblage at a given pressure are not fixed uniquely. It is well known, for example, that the beginning of melting of the four phase assemblage in peridotite is typically 50 to one hundred degrees below that in the quaternary system. But to what extent are liquid compositions buffered by the lherzolite assemblage at fixed P and T? The answer is particularly relevant when

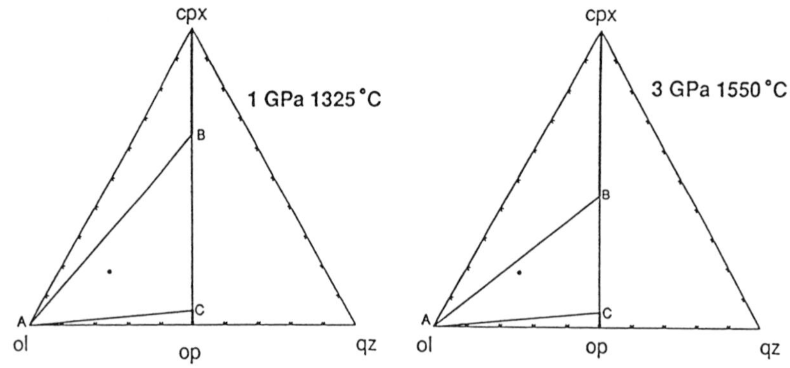

Fig. 21 Subsolidus phase relations of the F_O-Di-Q_Z system at high pressures. The line BC is the pyroxene miscibility gap. Note that the solubility of orthopyroxene (OP) in diopside (Cpx) increases with increasing temperature, causing the orthopyroxene content of peridotite (filled circle) to decrease. See text for discussion.

comparing the liquids obtained from the sandwich melting experiments to the compositions of ocean floor basalts. Specifically, how are liquid compositions affected if the bulk composition of peridotite is varied by changing the proportions of olivine, orthopyroxene and augite and/or adding a basaltic component?

It can be demonstrated through phase equilibria considerations that the major element composition of the liquid is virtually independent of the proportions of olivine, orthopyroxene and augite, provided that the composition of each phase is fixed. This proposition was tested by assembling two synthetic spinel lherzolites (Ol:30:Opx:30: Cpx:30:Sp:10 and Ol:55:Opx:20:Cpx:20:Sp:5) by combining the same minerals in different proportions and then adding 15% of the same basalt component to each charge [Fujii and Scarfe, 1985]. The liquids coexisting with these spinel lherzolite sandwiches at 1290° C are compared in Table 4. The liquids are very similar, not withstanding the very different starting compositions. The most notable differences are in the Al_2O_3 values. Whether these differences are spurious or a reflection of the lower Al_2O_3 contents of the starting composition is not clear. In any case, liquid compositions are not sensitive to the mode, provided that none of the four minerals are consumed.

TABLE 4

	(1)	(2)	(3)	(4)
SiO_2	49.1	48.9	49.3	49.0
TiO_2	0.46	0.56	0.60	0.9
Al_2O_3	16.4	16.4	16.1	15.6
FeO	6.7	6.8	7.0	7.4
MgO	12.3	12.0	11.9	12.0
CaO	12.6	12.4	12.3	12.0
Na_2O	1.8	1.9	2.1	2.1
K_2O	0.0	0.08	0.10	0.7
Mg*	.766	.759	.753	.744

Equilibrium partial melts at 10kb, 1290°C for peridotite (PMM-1) with (1) 8%, (2) 15%, and (3) 30% of added basalt of fixed composition, (4) same peridotite but with 15% more evolved basalt. (Fujii and Scarfe, 1985)

The effect of the varying basalt component was also investigated by Fujii and Scarfe [1985] by making sandwiches containing 8%, 15% and 30% of basalt but keeping the composition of basalt and peridotite constant. Liquid compositions coexisting with these spinel lherzolites at 1290° C are similar but show small variations in Mg* ratios and particularly in the contents of the incompatible oxides, TiO_2, Na_2O and K_2O (Table 4). This is hardly surprising, since the Mg* ratio is lowest in the basalt-rich sandwich, and the lherzolite phases do not have the capacity to buffer the incompatible oxides; for example, the higher the K_2O content of the sandwich, the higher the K_2O content of the liquid. Nevertheless, the liquids coexisting with spinel lherzolites are very similar and the variations that exist are easily explained. The results emphasize, however, that it is important to keep the amount of the added basalt component to a minimum and to choose the basalt component wisely. The basalt component should approach the compositions of relevant primary basalts as closely as possible.

The previous discussion showed that different mixtures of the same minerals yielded liquids of virtually indentical composition. This was true because each synthetic lherzolite had approximately the same Mg* ratio, and each phase had the same initial bulk composition. If the compositions as well as the proportions of the phases are changed, the liquid composition cannot remain constant. Table 5 contains liquid compositions obtained from three peridotite-basalt sandwiches at 1350° C and 1.0 GPa [Falloon and Green, 1987]. Each experiment had a different lherzolite component but was mixed with the same basalt in roughly equal proportions. As expected, the peridotite with the highest Mg* value produced the liquid with the highest Mg* values. The remaining components, however, vary only modestly in the three liquids.

The liquid compositions derived from the melting of various peridotite-basalt sandwiches are often recalculated into their normative components and then plotted within compositional triangles defined by the normative components diopside (Di) - olivine (Ol) - quartz (QZ) or nepheline (Ne) -

TABLE 5

	(1)	(2)	(3)	(4)
SiO_2	50.3	50.3	50.7	49.7
TiO_2	0.7	0.7	0.7	0.7
Al_2O_3	17.1	17.3	16.8	16.4
FeO	7.6	6.9	6.5	7.9
MgO	10.4	10.8	11.1	10.1
CaO	11.6	12.0	12.0	13.1
Na_2O	2.2	2.0	2.0	2.0
Mg*	0.73	0.73	0.75	.72

Equilibrium partial melts from three peridotite sandwiches with the same MORB at 10kb and 1350°C.

(1) MPY-87 + 10% MORB, (2) KLB-1 + 23% MORB, (3) MPY-90 + 24% MORB, (4) MORB composition (Falloon & Green, 1987).

olivine (Ol) - quartz (QZ). These are projections from plagioclase and diopside respectively. The general characteristics of these projections are illustrated by the 1.0 GPa sandwich melting experiments of Falloon and Green [1987, 1988] (Fig. 22). In both types of projections, liquids coexisting with olivine, orthopyroxene, augite and spinel lie along a cotectic rather than at a point as in the CMAS system. Melting begins near A. The first melts have a large normative plagioclase and a low normative quartz component. Continued melting moves the liquid compositions to B where augite and spinel are eliminated from the residue. Although the segment A-B is small, particularly in the diopside projection, there is a large variation in liquid compositions that is hidden within the normative components. The Mg* value, for example, increases from 0.69 to 0.75 from A to B, and the amount of

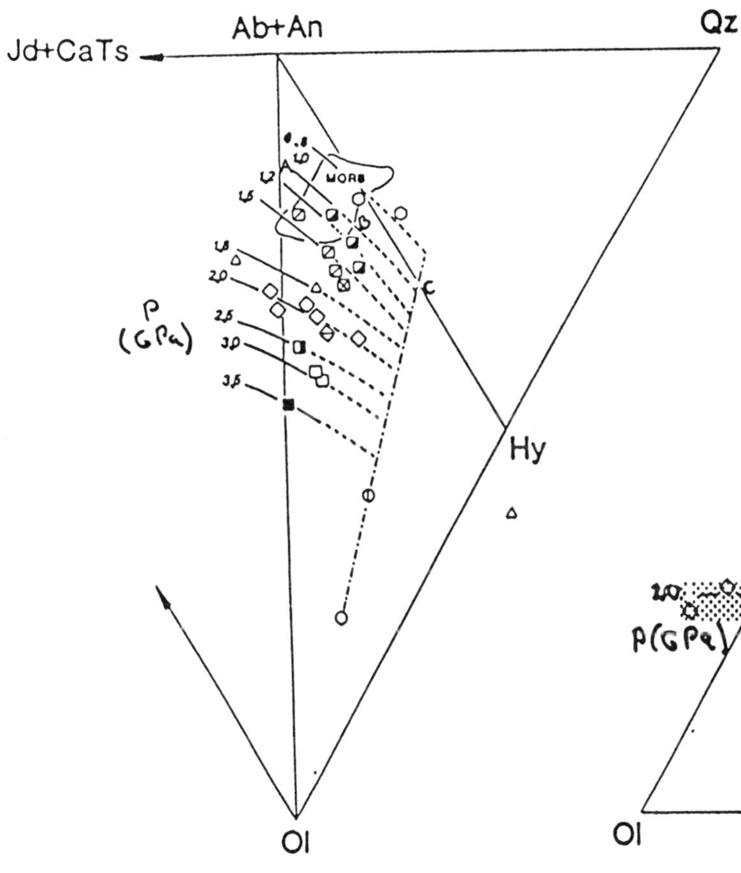

A

Fig. 22a Diopside projection of normative compositions of partial melts produced by melting peridotite at various pressures. Field of primitive MORB glasses are also shown. Open circle is peridotite composition.

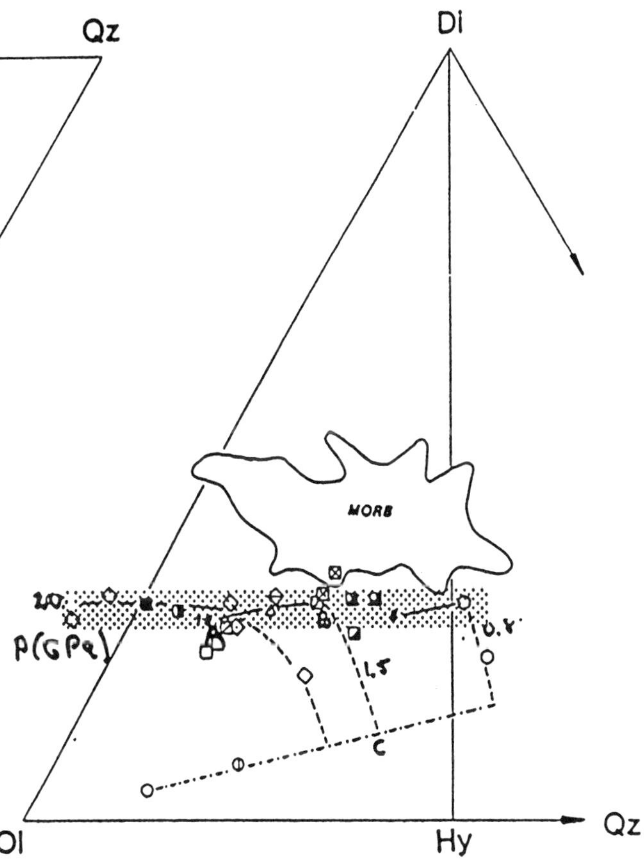

B

Fig. 22b Plagioclase projection of normative compositions of partial melts produced from melting peridotite, at 2.0, 1.5 and 0.8 GPa. Open circle is peridotite composition.

melting at B exceeds 10% (Table 6). Note, however, that the normative quartz values increase significantly during this interval (strictly speaking, it is normative hypersthene which increases).

The increase in the normative quartz values are halted by the elimination of augite. The liquids between B and C define the olivine and orthopyroxene cotectic which is roughly parallel to the Di-Hy and Plg-Hy joins. The liquids become progressively enriched in Hy and depleted in Di and Plg until orthopyroxene is totally consumed. The liquids now coexist only with olivine, and therefore lie on an olivine control line that passes through the composition of the peridotite sandwich at D. At D, the system consists only of liquid.

The results given in Table 6, while specific to the starting composition , are useful to describe the general characteristics of the melting of spinel lherzolite. Na_2O (also K_2O, P_2O_5) is the most incompatible of the major oxides being concentrated in the first increments of melt. It is then diluted as the extent of melting increases. TiO_2 is also incompatible, but less so than Na_2O. The Na_2O/TiO_2 ratio is at a maximum when melting begins and then gradually approaches the value of the peridotite at large degrees of melting. Al_2O_3 abundances are also highest in the first melts, since the crystal-liquid distribution coefficient for olivine is zero, and is much less than unity for the pyroxenes. The Al_2O_3/TiO_2 ratios, however, remain relatively constant. CaO abundances follow a different pattern. CaO first increases with the extent of melting as increasing proportions of augite are melted, but then decreases with further melting after augite is eliminated from the peridotite (harzburgite) residue. The CaO/Al_2O_3 ratio therefore increases with melting as long as augite is present as a residual phase, but then reaches a plateau when only olivine and orthopyroxene are melted (the CaO/Al_2O_3 ratio must really decrease slightly since the orthopyroxene has a CaO/Al_2O_3 ratio that is less than that of the melt; the Al_2O_3 content of orthopyroxene is relatively small, however, and this effect is minor).

TABLE 6

	1310°C	1350	1400	1420	MPY-87 + 10%
SiO_2	50.9	50.2	50.5	50.0	44.8
TiO_2	0.8	0.7	0.6	0.6	0.22
Al_2O_3	19.2	17.2	15.9	14.6	5.5
FeO^t	6.8	7.7	7.6	8.4	9.6
MgO	8.5	10.5	11.3	13.6	34.1
CaO	10.5	11.6	12.0	11.1	4.3
Na_2O	3.3	2.2	1.9	1.5	0.6
Residue	Lz	Lz	Lz	Hz	
Mg*	0.69	0.71	0.73	0.75	0.87
CaO/Al_2O_3	0.55	0.67	0.75	.76	0.78
CaO/Na_2O	3.2	5.3	6.3	7.3	7.2
Na_2O/TiO_2	4.1	3.1	3.1	2.5	2.7

Equilbrium melt compositions derived from peridotite MPY-87 plus 10% MORB glass (Falloon & Green, 1987). Note how the CaO/Al_2O_3, CaO/Na_2O and Na_2O/TiO_2 ratios approach the bulk rock composition as calcic clinopyroxene is finally eliminated from the residue. Lz - lherzolite, Hz – harzburgite.

FeO and MgO both increase with the degree of melting, although the percentage increase of MgO is greater than that of FeO. The Mg* ratio obviously must rise. In a previous study by Jacques and Green [1980], it was established that at any one pressure, SiO_2 increases with increasing extents of melting. Subsequent reversal experiments by Falloon and Green [1988] found that the compositions of the liquids estimated by Jacques and Green [1980] were in error, and that the correct compositions generally were 1-3 wt % higher in SiO_2. New results indicate that there is no SiO_2 enrichment for liquids coexisting with lherzolite (Table 6) but SiO_2 does increase along the olivine-orthopyroxene cotectic once augite is melted out [Falloon and Green, 1987].

Partial melt compositions obtained from 0.8 to 3.5 GPa from the peridotite-basalt sandwich are shown in projection in Figure 22. The olivine+orthopyroxene+ clinopyroxene (± spinel) cotectics move towards the olivine apex with increasing pressure in the Di projection and extend past the olivine-diopside join in the plagioclase projection. The olivine+orthopyroxene cotectics move towards the olivine apex in both types of projections. The movement of these cotectics results from the contraction of the olivine liquidus with pressure, as was already demonstrated in the CMAS system. Note, for example, that the olivine-orthopyroxene cotectic below 0.8 GPa lies within the quartz normative region of the basalt tetrahedron. These liquids are primitive quartz tholeiites. In contrast, the first liquids to form at pressures above 1.5 GPa lie within the nepheline-normative region of the basalt tetrahedron. These are alkali basalts. Liquids obtained by moderate degrees of melting above 0.8-1.0 GPa are olivine and hypersthene normative and are olivine tholeiites, the basalts which are dominant on the ocean floor. It is the contraction in the olivine phase volume with increasing pressure, therefore, that largely controls the evolution of primary ocean floor basalts.

Before discussing the implications of these results it is useful to generalize the results of the high pressure experiments with specific reference to the major element composition of liquids saturated with olivine+orthopyroxene+clinopyroxene. The range of liquid compositions given below are appropriate only for basalts derived from relatively fertile peridotite. [See also Falloon and Green, 1988].

SiO_2 contents: The SiO_2 contents decrease with increasing pressure from values near 53% SiO_2 at 0.5 GPa to less than 47% at 3.5 GPa.

FeO and MgO contents: For a given peridotite composition, the FeO contents will increase with pressure. At low pressures ($\lesssim$ 1.5 GPa), the FeO contents of the initial melts are less than that of the peridotite, so that FeO will increase with the extent of isobaric melting. At higher pressures, the FeO contents at low melt fractions will be about equal to somewhat higher than the coexisting residual peridotite,and the FeO contents will either remain the same or decrease with the extent of isobaric melting. The MgO contents of the liquids increase with pressure as the proportion of MgO-rich olivine in the inital melts is increased with pressure. Since the MgO contents of liquids are less than the

MgO contents of lherzolite at all pressures up to 3.0 GPa, it follows that the MgO of liquid increases with the degree of isobaric melting. Because the distribution coefficient for olivine, $K_D = (FeO/MgO)^{OL}/(FeO/MgO)^{L}$, increases from about 0.30 to 0.37 with increasing pressure up to 3.0 GPa,[Ullmer, 1988] the Mg* value of the initial liquid will also increase with pressure (see Figure 1).

Al_2O_3 and CaO: The Al_2O_3 and CaO contents will be lower at higher pressures for a given degree of melting. This is caused by the contraction of the olivine phase volume with pressure. The CaO/Al_2O_3 ratio, however, is not a strong function of pressure.

The composition and depth of origin of primary MORB is uncertain. One way to clarify the origin of these basalts is to test whether any of the basalts have compositions identical to those experimentally produced from the melting of peridotite. One method, which has been used extensively, is to compare the projected compositions of primitive MORB and experimental liquids on normative triangular diagrams. This has been done in Fig. 22 using the diopside and plagioclase projections [Falloon and Green, 1988]. In the diopside projection, the field of primitive MORB glasses largely overlies the olivine-orthopyroxene-calcic clinopyroxene 0.8-1.5 GPa cotectics. In the plagioclase projection, however, the MORB glasses plot above the cotectics, implying that the basalts are not primary melts from the melting of typical peridotite.

The equilibrium melt compositions are more "olivine"-rich, i.e., they are more picritic, than the most primitive MORB glasses. It is possible, however, to derive the MORB glasses by low pressure crystal fractionation of olivine from the picritic basalts generated at high pressures. The test is accomplished by examining whether a vector drawn from the olivine apex joins the MORB glasses and the experimentally produced picritic liquids. Many of the MORB glasses in both projections pass this test. A more rigorous test of this model is to project the MORB glasses and the olivine-orthopyroxene-calcic clinopyroxene cotectics through olivine and onto the plagioclase-diopside-quartz normative triangle (Fig. 23). The composition of a liquid that undergoes olivine-only fractionation does not change position in this representation.

Several new features are displayed in this projection. It is still true that the MORB glasses can be derived by olivine crystal fractionation of picritic liquids, but only from liquids derived by large degrees of melting of peridotite at 0.8-2.0 GPa. These liquids are generated near the diopside-out boundaries. Liquids produced at lower temperatures and smaller degrees of melting do not qualify.

Another method to test these conclusions utilizes chemical parameters that are obscured by the normative components. The compositions of primitive MORB glasses from the compilation of Presnall and Hoover [1987] are compared in Figure 24 to the equilibrium melt compositions using CaO/Al_2O_3 and Mg* as coordinates. The MORB field falls to the low Mg* side of the 1.0 GPa experimental cotectic liquids. The effect of olivine fractionation is indicated by the (Ol) arrow (CaO/Al_2O_3 remains constant as the Mg* declines). The

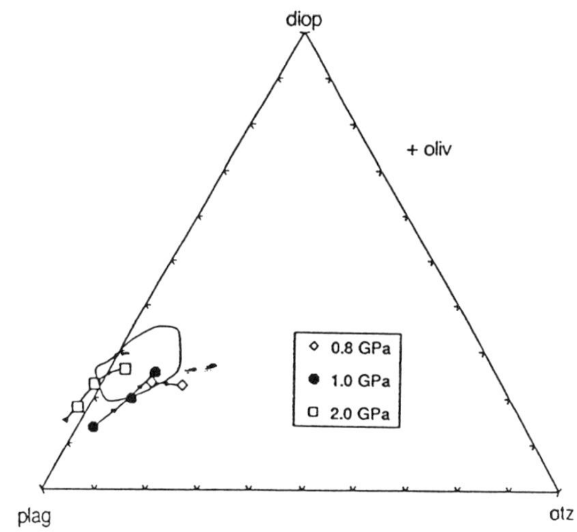

Fig. 23 Olivine projections of partial melts from peridotite coexisting with olivine +orthopyroxene+clinopyroxene. Arrows indicate increasing degrees of melt. The highest temperature-melts are at the diopside-out boundary.

MORB glass field can be reached by olivine fractionation of liquids produced at 1.0 GPa or higher. The high CaO/Al_2O_3 values, moreover, are consistent with relatively large degrees of melting, as was deduced from the normative diagrams.

Falloon and Green [1987] have taken this exercise one step further and have subjected the equilibrium partial melt compositions to simple incremental olivine fractionation using a Fe-Mg exchange partition coefficient of K_D= 0.3 and assuming that 10% of the total iron oxide is Fe_2O_3. They conclude that the majority of primitive MORB glasses can be produced by 11-25 wt % olivine-fractionation of magmas segregated from the mantle at pressures between 1.5-2.5 GPa. The parent picritic basalts were produced from between 17 to 30 wt % partial melting of a fertile peridotite composition, leaving a depleted lherzolite to harzburgite residue. Some MORB, however, have compositions similar to primary basalt magmas produced from peridotite at 0.8-1.0 GPa.

Such exercises are useful and instructive. We must recognize, however, that even primitive MORB liquids may not be derivative liquids from a single primary magma. Klein and Langmuir [1987] have argued that ocean floor basalts are mixures of primitive and/or primary magmas derived from a variety of depths. If true, none of the parent liquids to MORB would correspond to any liquid derived by the equilibrium melting of peridotite. This point is discussed later in the paper. The general conclusions remain valid, however. The most primitive MORB basalts are derived by olivine fractionation of picritic liquids. Most of the basalt components of these picritic liquids were isolated from mantle at pressures of 1.0 GPa or higher.

Melting with Volatiles

The presence of volatile components in ocean floor basalts, albeit in small amounts, indicates that the mantle solidus cannot be that of an anhydrous peridotite. Even small

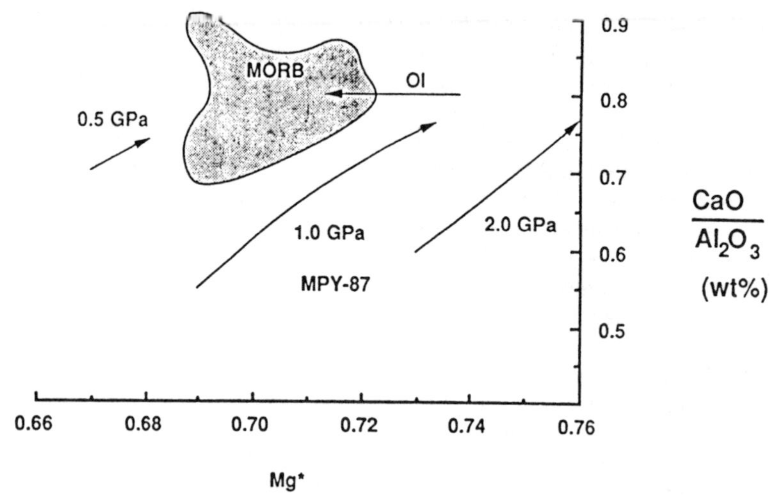

Fig. 24 Comparison of the cotectic liquids obtained by melting peridotite at 0.5, 1.0 and 2.5 GPa [Falloon and Green, 1988] to field of primitive MORB. Horizontal arrow (OL) indicates effect of olivine fractionation. Arrows on cotectic boundaries imply increasing temperatures.

amounts of CO_2 and especially H_2O have a dramatic effect on the shape and position of the solidus. We will see, however, that the extent of melting and the compositions of the basalts are not strongly influenced by the small quantities of volatiles believed to exist in the modern sub-oceanic mantle.

The beginning of melting of CO_2, H_2O-bearing peridotite is determined by 1) the composition and mass of the fluid, 3) the speciation within a fluid phase and 2) whether the volatile components are incorporated within the crystalline structures of mantle minerals, or form discreet fluid phases.

The oxidation state of the mantle exerts a fundamental control on the nature of the fluid phase. A fluid phase approximated by the C-H-O system, for example, consists of mainly CO_2 and H_2O under relatively oxidizing conditions (QFM) but mainly CH_4 and H_2O under very reducing conditions (IW*) [Green et al, 1987]. The melting relationships vary dramatically over this range of oxidation states. Unfortunately, no concensus exists about the oxidation states of various mantle reservoirs. There is good reason to argue, however, that the subocean ridge mantle is only mildly oxidizing.

Christie et al, [1986] measured the $Fe^{+3}/(Fe^{+2}+Fe^{+3})$ redox ratios of quenched glasses from the outer rims of MORB pillows. The results were consistent, yielding $Fe^{+3}/\Sigma Fe$ ratios of 0.07±0.03, which imply oxygen fugacities of 1-2 log units below the QFM buffer. Unfortunately, these may be only minimum values due to post or syn-eruptive oxidation processes, possibly due to the loss of H_2 or CO_2. Conversely, if these values are corrected for pressures appropriate for upper mantle conditions (1.0-3.0 GPa), the oxygen fugacities (fO_2) increase relative to the QFM buffer [Wood and Virgo, 1989]. A reasonable, if not infallible conclusion, is that fo_2 values in the subocenic mantle are below the QFM buffer but considerably above the IW buffer. Under these conditions, a C-H-O fluid consists predominatly of CO_2 and H_2O, and the melting relations discussed below apply to the petrogenesis of MORB.

*The I-W buffer refers to the fugacity of oxygen established by the relatively reduced assemblage iron metal and wustite (FeO) The I-W buffer defines an oxygen fugacity about 4 orders of magnitude below QFM.

The simplest case is to evaluate the melting of peridotite in the presence of a discrete water-rich (nearly pure H_2O) phase, i.e., melting with "excess water". H_2O reacts with peridotite to produce hydrated minerals, which, depending on the temperature and pressure, include serpentine, amphibole, phlogopite and possibly various hydroxylated silicates. The H_2O that is in excess of that required to form the maximum mass of the hydrous minerals appropriate for the specific mantle condition exist as an "excess fluid." How much H_2O is required in the upper mantle to generate the "excess water" state?

The principal hydrated minerals in the mantle under the conditions obtained in the upper sub-oceanic mantle are amphibole up to pressures of 3.0 GPa and phlogopite at higher pressures. A fertile peridotite can form about 20% calcic amphibole when fully hydrated. Amphibole contains about 2-3% H_2O so that amphibole- bearing peridotite can lock-up about 0.4 to 0.6% H_2O.

The "amphibole out" curve marks the maximum stability of typical calcic amphibole in the upper mantle. Its stability is bounded at low pressures (0.1-0.2 GPa), high pressures (2.5-3.0 GPa) and high temperatures (~1100° C) (Fig 25). Thus, if there is more than 0.6% H_2O in the amphibole peridotite phase region the "excess water condition" exists at temperatures below the solidus.

Phlogopite is the principal water-bearing mineral near the mantle solidus at higher pressures [Eggler, 1989]. The mass of phlogopite contained in peridotite is limited by the K_2O content (phlogopite is potassium rich mica). If fertile mantle contains 0.1% K_2O and phlogopite contains approximately 10% K_2O, a maximum of 1% phlogopite is expected. This small mass of phlogopite retains only about 0.04% H_2O. Thus, much lower H_2O contents are required to form an excess H_2O-fluid in the phlogopite-phase region than in the amphibole bearing mantle, particularly in depleted mantle where $K_2O \sim 0.01$ %.

The melting curve for peridotite with excess H_2O is depressed by three to five hundred degrees from the dry solidus in the 1.0 to 3.0 GPa range (Fig. 25). The dramatic lowering of the solidus is due to the large solubility of H_2O in melt at high pressures. The solubility of H_2O is approximately 20% by weight at 2.0 GPa [Wyllie, 1979]! This means that only the solidus melt is H_2O-saturated, and that the silicate melt must be H_2O undersaturated at temperatures above the solidus as H_2O-poor minerals are fused and added to the melt. Only small quantities of H_2O-saturated melts are generated at the solidus. If, for example, there is 0.2% by weight excess water and the solubility of H_2O in the melts is 20%, then only 1% of H_2O saturated liquid is formed at the solidus.

Primitive MORBs, however, contain only about 0.2% H_2O [Michael, 1988]. If these basalts are formed by 20% melting, and if H_2O is perfectly incompatible, then the mantle contained less than 0.04% by weight. The H_2O must be stored in amphibole, at least below 3.0 GPa, and melting does not proceed under H_2O-excess conditions. Melting begins at the temperature where amphibole melts incongruently and releases H_2O to the newly generated silicate melt. The solidus, in this case, is given by the "amphibole out" curve. The liquid, moreover, is H_2O undersaturated, and typically contains 3-4% H_2O. If the mantle contains 0.03% H_2O, then only about 1% melt containing 3-4% H_2O forms at the "amphibole out" boundary. Continued melting adds anhydrous components to the liquid, and eventually reduces the H_2O content to the observed values.

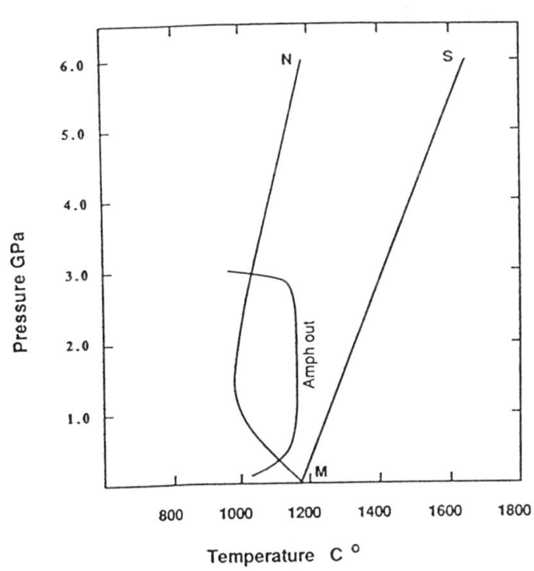

Fig. 25 H_2O-saturated solidus (MN) of peridotite compared to dry solidus (SM) [modified after Wyllie, 1979]. Amphibole-out curve only approximately located.

The melts formed by melting of fluid absent amphibole-bearing peridotite are SiO_2-poor and relatively alkali-rich [Olafsson and Eggler, 1983]. The liquid is SiO_2-undersaturated and nepheline-normative and probably resembles an alkali basalt. The melt, moreover, should contain several percent of dissolved H_2O. Only small quantities of melt (~1%) are produced near the solidus because there are only small quantities of amphibole. Continued melting by adiabatic decompression would dilute these melts with the anhydrous components of peridotite.

The role of CO_2 in the melting process is better determined than of H_2O. The mineralogy of the subsolidus mantle again determines whether melting occurs under "excess CO_2" fluid conditions. The way peridotite melts is largely determined by the dolomite forming reaction

$$2Mg_2SiO_4 + CaMgSi_2O_6 + 2CO_2 = 4MgSiO_3 + CaMg(CO_3)_2 \tag{8}$$

which adds dolomite to the mantle at greater than 2.0 GPa under subsolidus conditions [Falloon and Green, 1989]. The mass of dolomite produced by this reaction is limited by the diopside (diopside provides most of the CaO to form dolomite) and the CO_2 content of the mantle. About 5% CO_2 is needed to use up the diopside in typical peridotite [Wyllie, 1980]. If the mantle is diopside-poor or CO_2-rich, the reaction may go to completion leaving an excess CO_2 fluid. At slightly higher pressures (~0.1 GPa higher), however, a magnesite-forming reaction occurs

$$Mg_2SiO_4 + C + O_2 = MgSiO_3 + Mg\,CO_3 \tag{9}$$

and utilizes the remaining CO_2 fluid [Falloon and Green, 1989]. Thus, there is no excess CO_2-fluid in an olivine-bearing mantle for $P > 2.0$ GPa providing that the mantle is relatively oxidizing. Eggler [1983, 1989] found that the peridotite-carbonate-graphite assemblages define an fo_2 at about 1 log unit below the QFM buffer. This is consistant with the fo_2 estimated from MORB glasses. Conversely, carbonates are not stable at pressures below 2.0 GPa at mantle temperatures and there will always be an excess CO_2-rich fluid below this pressure. The phase relations for the melting of a peridotite plus 5% dolomite (or equivalent to 2.4 wt % CO_2), [Falloon and Green, 1989] are compared in Fig. 26 to those of a volatile-free peridotite. The lines refer to the carbonation reactions. The mantle below these lines is carbonate-free and contains excess CO_2-fluid, whereas magnesite and/or dolomite are stable above the lines and no CO_2-fluid exists. At pressures

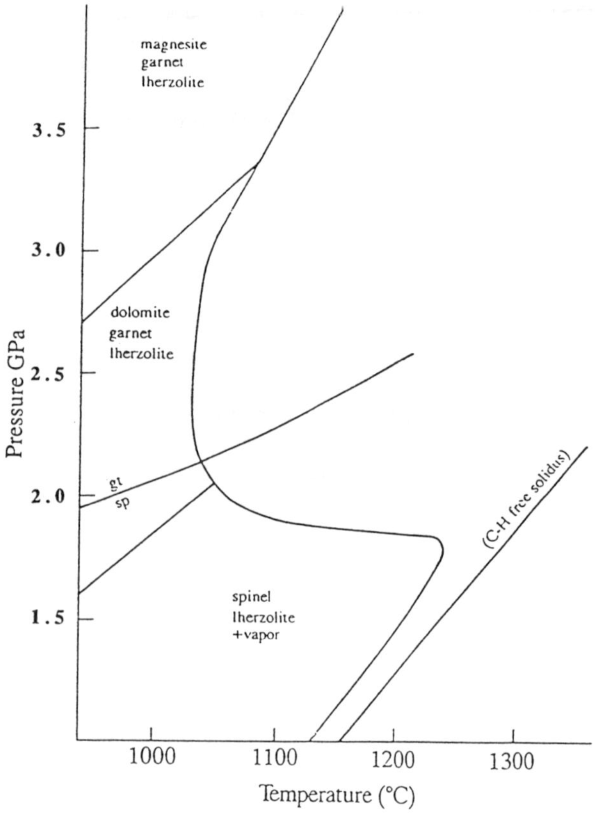

Fig. 26 Phase relations of peridotite with 2.4% CO_2. [Falloon and Green, 1989]

below 1.9 GPa, the temperature of the solidus is only slightly below (<50°C) that of the volatile-absent solidus because the solubility of CO_2 is relatively low. Pan and Holloway [1990] estimate that about 1% CO_2 is dissolved in basaltic melts at 2.0 GPa.

Above 1.9 GPa, the peridotite solidus bends sharply to lower pressures, then changes slope to become quasi-parallel to the dry solidus at higher pressures. The sharp depression of the solidus temperature is not due, as commonly assumed, to a dramatic increase in the solubility of CO_2 in silicate melts, but rather to the appearance of dolomite and the displacement of the forsterite-enstatite cotectic to magnesian carbonate-rich compositions [Wyllie, 1989]. Note that the solidus above 1.9 GPa is more than 300°C less than the volatile-free solidus.

The near solidus melts formed at pressures below 1.9 GPa are basalts similar to those produced in CO_2-free peridotite. This results from the very low solubility of CO_2 in basalts at these pressures. The near solidus melts above 1.9 GPa are very different, and are not even silicate melts. The liquids are carbonatites which are carbonate-rich magmas usually found in regions of continental rifting [Falloon and Green, 1989]. The amount of carbonatite melt that is formed depends on the carbonate content of the mantle. If, for example, the mantle contains 1% dolomite, then a little more than 1% carbonatite can form since only a limited amount of silicate minerals are fused at the solidus [Kjarsgaard and Hamilton, 1989]. Continued melting will dilute the carbonate content of the liquid with fused silicate components and depending on the CO_2 content, degree of melting and pressure, may produce kimberlites to typical basalts [Eggler, 1989; Wyllie, 1989].

MORB, of course, contain both CO_2 and H_2O [Kingsley, 1989], and since these volatiles are mutually soluble under mantle conditions, melting relations with mixed volatiles must be considered. Melting experiments with mixed volatiles are discussed by Wyllie [1980, 1989], Wallace and Green [1989], and Eggler [1989] among others, and is therefore only touched upon here. With small amounts of fluid (0.3% H_2O, 2.4% CO_2) the solidus below about 2.0 GPa is determined by the incongruent melting of amphibole in the presence of a CO_2-rich fluid (Fig. 27) [Wallace and Green, 1989]. Between 2.0 and 3.2 GPa, the mantle contains dolomite or magnesite and amphibole, but no excess fluid. At higher pressures, amphibole disappears and phlogopite or perhaps other hydrated silicates are stable (e.g., hydrated garnets). This is also a fluid absent region for the low H_2O contents estimated for the MORB mantle (see below). Near solidus melt compositions at pressures below 2.0 GPa are equivalent to those obtained in the CO_2-free mantle, whereas those above 2.0 GPa are again carbonatites.

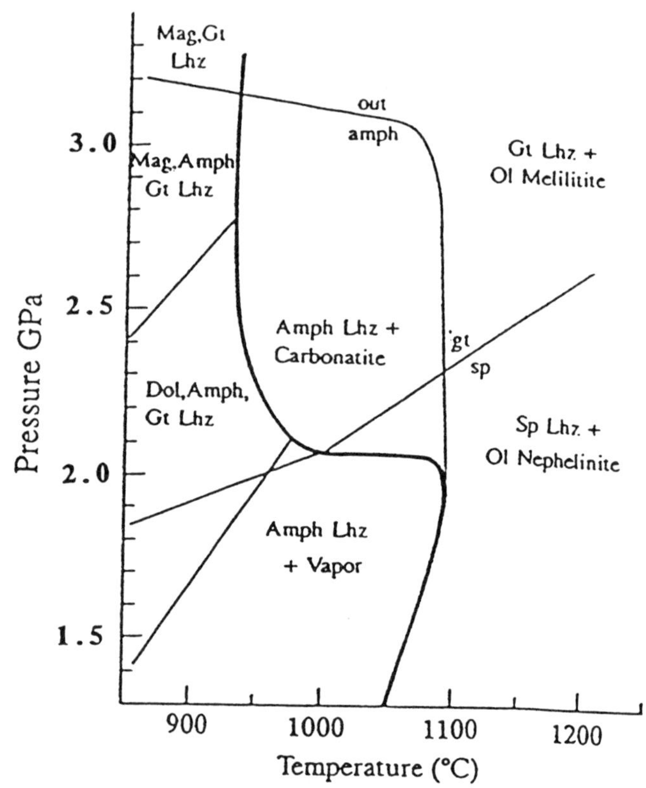

Fig. 27 Phase relations of peridotite with 0.3% H_2O and 2.4% CO_2 [Wallace and Green, 1989] Mag=magnesite, Amph=amphibole, Dol=dolomite.

TABLE 7

	(1)	(2)	(3)	(4)
SiO_2	54.1	52.6	45.8	46.9
TiO_2	0.5	0.6	1.1	0.9
Al_2O_3	11.0	14.5	11.7	11.0
FeO	8.0	7.4	9.0	7.8
MgO	13.0	11.7	20.0	19.2
CaO	10.0	11.7	10.7	12.2
Na_2O	1.7	1.7	1.4	1.2
K_2O	0.3	---	---	---
Mg*	.75	.76	.82	.81
Residue	Hz	Hz	GntLz	GntLz

(1) 5 kb, 1300°C from HW pyrolite Falloon et al, (1988) but adjusted to lower TiO_2, K_2O, (2) 8 kb, 1400°C from MPY-87 (Falloon & Green, 1988). Mg* calculated assuming 10% of iron oxide is Fe_2O_3, (3) 35 kb, 1660°C from MPY-90-40 Falloon & Green (1988), (4) 30 kb, 1550°C from KLB1 Takashashi (1986).

The H_2O contents of MORB glasses dredged from depths greater than 500 meters have an inverse relationship with the Mg* value, but a direct relationship with the incompatible elements [Michael, 1988; Kingsley, 1989]. H_2O, therefore, behaves incompatibly and the basalts are not saturated with H_2O at these depths. Since the solubility of H_2O must be greater at greater depths, it follows that the basalts could not have lost H_2O on their trip to the ocean floor. The H_2O contents of the most primitive basalts (Mg* > 0.65) range from about 800 to 3000 ppm, the low values characterizing depleted MORB, whereas the high values are from trace element rich E-MORBs. If it is assumed that the basalts are derived by about 10 to 25% partial melting of mantle, and that H_2O is perfectly incompatible, then the depleted suboceanic mantle originally contained from 80 to 200 ppm H_2O and the enriched mantle from 300 to 750 ppm H_2O. Such low concentrations of H_2O can be incorporated in mantle amphibole and suggest that melting occurred without an H_2O-rich fluid if melting begins at pressures less than 3.0 GPa.

CO_2 contents are not correlated with the indices of magmatic differentiation and basalts erupted at a given depth show variable CO_2 contents [Dixon et al, 1988, Kingsley, 1989]. The CO_2 contents, which reach values of more than 700 ppm, are greater than the experimentally determined CO_2 solubilities in basaltic liquids at comparable pressures [Stopler and Holloway, 1988]. The basalt liquids therefore are saturated to oversaturated with CO_2 so that unknown quantities of CO_2 may have been outgassed during the ascent of basaltic liquids from their source regions. The CO_2 contents of the primary magmas are unknown, but must be greater than the observed concentrations. It is safe to conclude, however, that a CO_2-rich fluid existed in the mantle at the beginning of melting at pressures less than 2.0 GPa, but the CO_2 resided in various carbonate phases above 2.0 GPa.

Michael [1988] has observed that H_2O behaves more compatibly than trace elements such as K or Rb, i.e., the H_2O content does not increase as fast as K or Rb in MORB melts with increasing degrees of fractionation. Michael [1988] concludes that this behavior is consistent with the size of the OH^- ion and the systematics of mineral/melt partitioning. An additional factor is that a CO_2 fluid is being expelled from the basalt and that H_2O must be partitioned into this fluid. Nevertheless, it is a fact that H_2O still behaves very incompatibly in MORB. This confirms that H_2O is not strongly partitioned into the CO_2 fluid and that only small volumes of CO_2 were degassed. Larger volumes of noble gasses, however, will be lost to the mantle and crust because they are more efficiently partitioned into the CO_2 fluid.

Abyssal Peridotite

Abyssal peridotites are widely believed to be the residues of pressure-release melting of adiabatically upwelling mantle beneath oceanic ridges. The peridotites are most frequently derived from fracture zones; mainly fresh to weathered basalt is recovered from rift valley floors. Although there are a few occurrences of peridotite on fast spreading ridge fracture zones, abyssal peridotites are most frequently sampled from the fracture zones at slow and very slow spreading ridges [Dick, 1989]. For the very slow spreading SW Indian ridge fracture zones, the dredge hauls are composed of 65% peridotite, 23% basalt and 12% gabbro and diabase, and for all fracture zones, the proportions are 56% peridotite, 31% basalt and only 13% gabbro and diabase in roughly equal amounts. The scarcity of gabbro is surprising given that gabbro makes up about two thirds of the classical ophiolite stratigraphy. Dick [1989] argued that the scarcity of gabbro, and also of dunite, a unit commonly flooring many ophiolite complexes, demonstrates that magma chambers rarely form beneath fracture zones. The bimodal occurrence of basalt and peridotite indicates that basalt may have been erupted directly over peridotite. A classical ophiolite model doesn't seem to apply for such environments. The Xigaze ophiolite, however, which has a thin crustal unit and lacks large masses of cumulate gabbro, may be a better paradigm of slow spreading oceanic structure [Girardeau and Mercier, 1988].

Excluding plagioclase-rich lherzolites, the average mode of abyssal peridotites is about 75% olivine, 21% enstatite, 3.5% clinopyroxene and 0.5% spinel [Dick and Fisher, 1984]. In comparison, spinel lherzolites from the island of Zabargad, which is an uplifted fragment of the sub-Red Sea lithosphere, are very clinopyroxene-rich [Bonatti et al., 1986]. Their average mode is 65% olivine, 16% enstatite, 16% clinopyroxene and 3% spinel. It is clear from this comparison that the abyssal peridotites are largely depleted of their basalt components. The averages, moreover, underestimate the extent of the depletion. More than 60% of abyssal peridotites are harzburgites and a significant fraction of these are virtually clinopyroxene-free. The abyssal peridotites from the Oceanographer Fracture Zone, for example, contain less than 0.1% clinopyroxene [Michael and Bonatti, 1984]. Moreover, some of the reported clinopyroxene may not be primary clinopyroxene. Lindsley and Anderson [1983] demonstrated that orthopyroxene can exsolve up to 9 wt % clinopyroxene under cooling conditions similar to those obtained beneath an oceanic ridge. Thus, up to about 2% clinopyroxene can originate through exsolution from orthopyroxene. This is probably an overestimate, but it may certainly account for the trace amounts observed in more than a few harzburgites.

The average spinel peridotite compositions computed from the mode and mineral compositions are highly depleted and contain little "potential basalt" component. A typical primitive ocean floor basalt, for example, contains about 2 wt % Na_2O, whereas the abyssal peridotite contains less than 0.05%. There is only sufficient Na_2O to form a maximum of 2.5% additional basalt, assuming all Na_2O is partitioned into the melt [Dick, 1989]. If the mantle initially contained 0.40 Na_2O, then about 20% melt with 2.0 % Na_2O was formed and segregated to account for the depleted peridotite.

Computed average spinel peridotite compositions [Dick, 1989; Dick and Fisher, 1984] for the American-Antarctic Ridge, SW Indian Ridge, Bullard Fracture Zone and the Islas Orcadas Fracture Zone and an average of all abyssal peridotites are plotted on a Harker-type diagram where the FeO, SiO_2, Al_2O_3 and CaO contents are plotted against the MgO contents (Fig.28). The major oxide abundance trends of the Ronda peridotite, a large alpine peridotite [Frey et al, 1985] and of spinel lherzolite xenoliths from Victoria, Australia [Nickel and Green, 1984] are included for comparison. Similar trends are obtained from a compilation of analyses of continental and oceanic spinel lherzolites [Maaloe and Aoki, 1977]. The CaO-MgO trend for oceanic lherzolites, however, lies well below the trend for continental lherzolites and the other trends in Fig.28. The discrepancy probably reflects loss of CaO during the serpentinization of oceanic mantle.

The correlations in Figure 28 are those expected of a homogeneous mantle that has undergone various degrees of melting and separation of the basalt melt. Peridotites with the lowest Al_2O_3 and CaO have suffered the greatest degree of melting. (Care should be taken to assure that the trends are real, however. Similar correlations are obtained by mixing olivine with various quantities of pyroxene, a process that can be produced by solid state differentiation processes. A useful test is to make sure that the Mg^* value of the rock, and particularly the Mg* values of the constituent minerals, increase with decreasing Al_2O_3 and CaO). Since the various trends overlap closely, it can be assumed that the parent mantle to the abyssal peridotites, the Ronda and other alpine peridotites and subcontinental spinel lherzolites are similar. This is a useful assumption, because the more complete data of the alpine-xenolith peridotite suite can be used to model the abyssal peridotites. The work done on the Ronda peridotite by Frey et al. [1985] using both major and trace elements is highly instructive in this regard.

The major element composition of the most fertile Ronda peridotite sample was used to approximate the unmelted peridotite, and the remaining more refractory samples were assumed to be residues without trapped melt that formed by

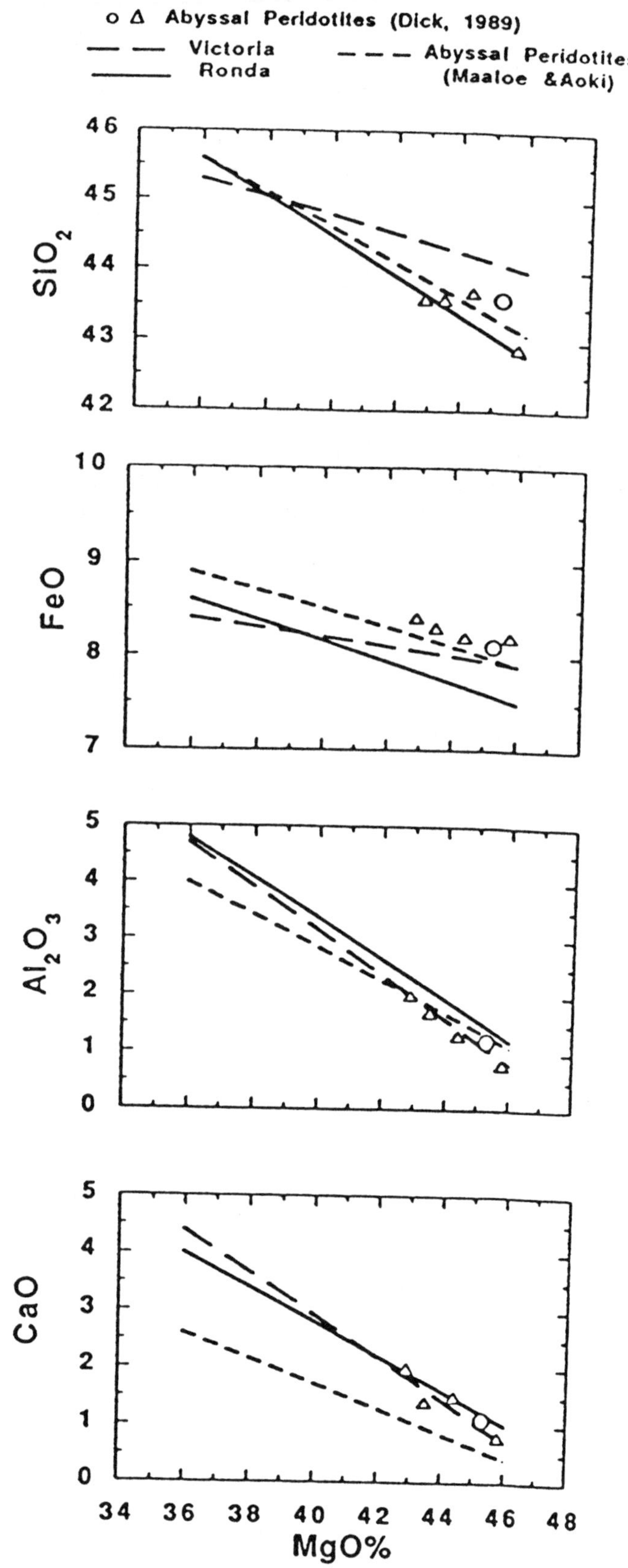

Fig. 28 Harker-type diagrams (in wt%) comparing the composition of peridotite from Ronda peridotite (solid line) [Frey et al, 1985], spinel lherzolite xenoliths from Australia (large dashed) [Nickel and Green, 1984] average oceanic spinel lherzolites [small dashes) [Maaloe and Aoki, 1977] and abyssal peridotites from various fracture zones (triangle) and their average (circle) [Dick, 1989].

melting of this peridotite. Using a mass balance procedure originated by Langmuir and Hanson [1980] and various olivine/melt FeO-MgO exchange coefficents, the entire series of peridotites can be modeled as residues from 4 to 32% melting. The inferred average melt compositions are picritic with about 16% MgO contents. Similar results are obtained using both compatible and incompatible trace elements. If the assumptions stated above are correct, then the most refractory abyssal peridotite may also be the product of 25-30% melting of the parent mantle.

The majority of abyssal plagioclase-peridotites have 2.5% plagioclase although plagioclase-rich peridotites from the Romanche Fracture Zone have up to 17% plagioclase [Dick, 1989]. The origin of plagioclase in these abyssal peridotites is debated. One model proposes that plagioclase is the product of a solid state reaction like

$$\text{Diopside} + \text{Enstatite} + \text{Spinel} = \text{Anorthite} + \text{Forsterite} \qquad (10)$$

which separates the plagioclase and spinel lherzolite facies. Others have argued persuasively that the plagioclase has been formed from the impregnation of residual peridotite by locally derived or transient basaltic melt [Dick, 1989]. If true, then the average plagioclase peridotite contains about 4% of trapped melt, and the most plagioclase-rich peridotite, up to 30% melt. More than 80% of the abyssal peridotites, however, contain very little trapped melt as evidenced by the very low contents of incompatible elements. This suggests that the melt trapped in plagioclase peridotite is transient melt which locally enriches the mantle.

The modes and bulk chemical compositions of small samples of peridotite are highly variable given the intrinsic modal herterogeneity of peridotite. If equilibrium is maintained within the system, however, the composition of each mineral will be uniform even if the distribution of mineral phases is uneven. The most reliable procedure to examine peridotites from different areas along the ridge, or to compare abyssal, alpine and ophiolite peridotites with each other is to use mineral compositions rather than bulk chemical compositions of peridotite. Even this procedure is not foolproof, since mineral compositions can undergo substantial subsolidus reequilibrium. Mass balance considerations indicate that the minerals present in lowest abundance will be most affected by such processes. Spinel compositions, for example, are very susceptible to Mg-Fe exchange with the large reservoir of olivine since the equilibrium is temperature dependent [Komor et al, 1988]. Lower temperatures will tend to reset the Mg^* values of spinel towards lower values [Engi, 1983]. Similarly, lamellae of diopside are commonly observed in orthopyroxene grains, and reflect reequilibration to lower temperatures. The original bulk composition of orthopyroxene can be estimated, however, with use of a defocused microprobe beam. All in all, even with these uncertainties, the use of mineral compositons is preferred.

Mg* values for olivine, orthopyroxene and calcic clinopryroxene generally range from 0.89-0.93. The Mg* values (Fig. 29) generally increase in the ordeı calcic clinopyroxene ≥ orthopyroxene ≥ olivine > spinel, although the differences between the three silicates are small (less than 2%). NiO contents of olivine vary from 0.22 to 0.40 [Dick et al, 1989]. Such a large range in NiO contents implies, following Frey et al. [1985], unrealistically high degrees of melting (45%). If, rather than single analyses, site averages are used, then about 30% melting is needed to explain the range of NiO contents. Site averages are more useful parameters because they smooth out the intrinsic variations in the initial NiO contents of different mantle materials.

The Al_2O_3 contents of orthopyroxene are between about 2 and 6 wt% and vary inversely with Mg* values. The Al_2O_3 contents of orthopyroxene decreases and the Mg* number increases with the degree of melting of peridotite. For example, the Al_2O_3 contents of orthopyroxene produced by the equilibrium melting of pyrolite range from 6% Al_2O_3 and Mg* = 0.88 to 3% Al_2O_3 and Mg* = 0.90 at 1.0 GPa [Falloon and Green, 1987]. This trend is produced by 20 to 25% melting of pyrolite. These numbers are appropriate only for the pressure stated. Al_2O_3 contents in orthopyroxene are higher at higher pressures for similar degrees of melting and coexisting liquids [Jacques and Green, 1980]. The Al_2O_3 content of orthopyroxene at the diopside-out curve is between 3 and 4% for 1.0-1.5 GPa, and about 1 wt% at the orthopyroxene-out curve . In comparison, alumina contents are below 3 wt% in orthopyroxene from some abyssal peridotites [e.g, Michael and Bonatti, 1985] and may, reflect melting beyond the diopside-out equilibria. Alternatively, the low values may be the product of fractional melting, a process that depletes incompatible elements more efficiently than equilibrium melting.

The CaO contents of orthopyroxene are from 1.0 to 2.9% [Smith and Elthon, 1988; Dick, 1989; Hamlyn and Bonatti, 1980]. The CaO contents of orthopyroxene coexisting with

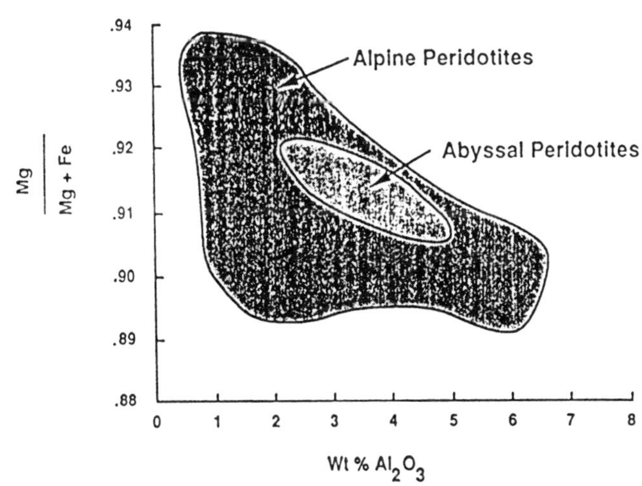

Fig. 29 Comparison of the bulk compositions of orthopyroxene from abyssal peridotites and alpine-ophiolite peridotites [Dick and Fisher, 1984]. Orthopyroxene with highest Mg* and low Al_2O_3 contents are products of the most extreme extent of melting.

diopside increases with temperature and decreases, albeit slightly, with the increases in the Mg* value and pressure. The CaO content of orthopyroxene in mantle undergoing pressure-release melting will gradually decline largely in response to decreasing temperatures. The CaO content of orthopyroxene is constrained to lie on the two pyroxene miscibility gap as long as diopside coexists with orthopyroxene. Once diopside is consumed by melting, the CaO content in orthopyroxene continues to fall as additional melting of harzburgite depletes the CaO content of the peridotite.

The CaO content of orthopyroxene coexisting with diopside is above 2 wt% for $T \geq 1200°C$ and $P \leq 1.0$ GPa [Grove et al, 1989; Falloon and Green, 1989; Lindsley, 1983]. The lower values of the abyssal peridotites may reflect additional melting beyond the diopside-out curve in harzburgites or the combined effects of exsolution and coalescence (granular exsolution) of diopside at temperatures below 1200°C. Hamlyn and Bonatti [1980], for example, measured small amounts of CaO (~ 1.5% CaO) in orthopyroxene from diopside-free harzburgites, which, however, were comparable to the CaO contents of orthopyroxenes in diopside-bearing harzburgite. It is possible that small amounts of diopside were overlooked in the harzburgite or that both orthopyroxenes have been modified by granular exsolution. Indeed, recrystallized grains of orthopyroxene (neoblasts), have lower CaO contents than uncrystallized cores of pyroxene.

Dick and Fisher [1984] observed that many orthopyroxenes have high CaO contents demonstrating saturation with diopside at near solidus temperatures even if diopside is absent from the assemblage. They conclude that melting of abyssal peridotites never proceeded far beyond the diopside-out temperatures. The average CaO contents in orthopyroxenes from abyssal peridotites from slow spreading ridges is more than 2% CaO, and confirms the analysis [Dick, 1989]. The CaO contents of orthopyroxenes in alpine peridotites and in the residual harzburgites within ophiolites are often much lower (< 1% CaO) indicating that the degree of melting is much higher than in abyssal peridotites (providing that the H_2O contents are low and similar in the two environments).

The chemical composition of calcic clinopyroxene is, in part, buffered by ubiquitous orthopyroxene and olivine, and, in part, reflects the extent of melting. The CaO content, usually expressed as a diopside or wollastonite component, decreases with increasing temperature, increases with increasing Mg* value and increases slightly with increasing pressure [Lindsley, 1983]. The temperature effect is dominant, however, so that the CaO content of clinopyroxene must increase as the mantle rises beneath the ocean ridge. The Al_2O_3 contents generally are the same or greater than that of coexisting orthopyroxene, although the opposite relations also have been observed. The Al_2O_3 contents are inversely related to the extent of melting. Na_2O and TiO_2 are incompatible elements and obviously are related to the degree of melting. The Ti contents of diopside, for example, decrease with decreasing modal clinopyroxene in abyssal peridotites [Johnson et al., 1990]. The crystal-liquid distribution coefficients for TiO_2 are about 0.4 for 1300-1400° C at 1.0 GPa [Falloon and Green, 1987] and about 0.30 at 1 atm and 1150-1190° [Grove and Bryan, 1983]. The crystal-liquid distribution coefficient for Na_2O is much lower, approximately 0.10-0.15 for the conditions given above [Falloon and Green, 1987; Grove and Bryan, 1982]. The Na_2O/TiO_2 content in pyroxene should thus decrease with the degree of melting.

Johnson et al. [1990] used these and other incompatible elements to demonstrate that diopside-bearing peridotites from the American Antarctic and Southwest Indian Ridges were residues of a near perfect fractional melting process. Peridotites containing from 2.0 to 12.0% clinopyroxene can be produced by a total of 25% to 5% melting respectively. The greatest degrees of melting found by Johnson et al. [1990] were near hotspots.

Spinel compositions are adequately described by the five component system Al_2O_3-Cr_2O_3-FeO-Fe_2O_3-MgO. The most useful chemical variation in spinel is the correlation of increasing Cr# (Cr/Cr+Al) and decreasing Mg* [Dick and Bullen, 1984]. Abyssal peridotites with spinel compositons at the low Cr end are typically lherzolites with relatively iron-rich silicates whereas harzburgites with very little diopside and relatively magnesian-rich silicates lie at the high Cr end of the spinel range. The explanation of such trends is complex and the reader is referred to Dick and Bullen [1984]. It is sufficient to know that increasing Cr# reflects increasing degrees of partial melting in the mantle. Spinel compositions of abyssal peridotites and peridotites in ophiolites have considerable overlap, but many ophiolites have chromium-rich spinels that plot well out of the abyssal peridotite field and possibly were produced by greater degrees of melting in the same or different provenance.

Pressure Release Melting

Thermodynamics of Melting. Pressure-release melting is determined by the difference in temperature, ΔT, between the temperature of the adiabatic upwelling mantle and the temperature of the solidus or

$$\Delta T = \frac{\Delta H}{Cp}(x) \tag{11}$$

where ΔH is the enthalpy of fusion, Cp the heat capacity of the mantle and X is the weight fraction of melt. The heat capacities of olivine, diopside and enstatite are between 0.30 and 0.33 cal/gm K for typical mantle temperatures and pressures [Robie et al., 1979]. The heats of fusion, however, have a greater uncertainty.

The enthalpy of fusion for the Mg* end members of relevant peridotite minerals at their melting temperature are given in Table 8. The enthalpy of fusion for diopside has been revised [Richet and Bottinga, 1986; Stebbins et al., 1984] and is almost twice as large as that listed in older references. Note that to determine the enthalpy of fusion of peridotite we must 1) correct for the melting temperature and pressure, 2) determine the proportions of minerals that enter the melt, 3) adjust the heats of fusion for the effects of solid solution, especially the FeO contents and the jadeite ($NaAlSi_2O_6$) and

TABLE 8

Heats of Fusion

	ΔH (cal/gm)	Tm(k)	Ref.	ΔH (1700)
Forsterite	195 ± 36	2163	(1)	149
Enstatite	175 ± 14	1834	(2)	162
Diopside	153 ± 2	1670	(2)	156
Pyrope	145 ± 4	1500	(2)	165
Fayalite	105 ± 1	1490	(2)	125

Enthalpies of fusion at the melting temperatures (Tm) and at 1700°K (ΔH 1700). (1) Navrostsky et al, 1989 (2) Richet and Bottinga (1986)

tschermak (MAl_2SiO_6, $M=Ca^{+2}$, Mg^{+2}) components of pyroxene, and 4) estimate the enthalpy of mixing of the melt.

The enthalpy of fusion is related to temperature by

$$\Delta H(T) = \Delta H(Tm) + \int_{Tm}^{T} \Delta Cp\, dT \qquad (12)$$

where ΔH(T) and ΔH (Tm) are the enthalpy of fusion at T°K and the melting temperature respectively and ΔCp is the heat capacity change on fusion. The ΔCp is a positive number, about 0.1 cal/gm° K [Richet and Bottinga, 1986; Navrotsky et al., 1989]. The effect of temperatures on ΔH(T) is substantial, approximately 10 cal/gm for every 100° K change. If the mean temperature in the suboceanic mantle is 1700° K, then the temperature correction makes the heats of fusion of the four phases nearly identical (Table 8). This is interesting because the melting of harzburgite or even dunite is no more difficult, from an energy balance point of view, than is the melting of lherzolite!

The effect of pressure on the heats of fusion is

$$\Delta H(P) = \Delta H(1) + \int_{1}^{P} \Delta V\,(1-\Delta\alpha T)\, dP \qquad (13)$$

where ΔH(P) and ΔH(1) are the enthalpy of fusion at pressure P and one atmosphere respectively, ΔV is the change of volume on melting, and Δα is the change in thermal expansion on melting. Since $\Delta\alpha T$ is <<1, ΔH(P) is determined by ΔV. Assuming that there is about a 10% increase in volume on melting, then $\Delta V \simeq 7 \times 10^{-4}$cal/gm bar. Thus, the enthalpy of fusion increases about 7 cal/gm for every 1.0 GPa increase in pressure.

The effect on the heats of fusion of small amounts of FeO in mantle minerals is not significant. The enthalpy of fusion for fayalite is 105 cal/gm at the melting temperature, substantially less than that of forsterite. However, if the ΔH of forsterite is recalculated to the melting temperature of fayalite, then the heat of fusion of forsterite is only about 25 cal/gm more than fayalite. Since the fayalite is only about 10% of mantle olivine, the effect of small amounts of FeO in olivine, and probably in pyroxene and garnet can be ignored. Similarly, the heats of fusion of jadeite and diopside, when compared at the same temperature, indicate that small amounts of jadeite solid solution in pyroxene will not materially affect the heat of fusion. There is no data for the tschermak components, but it is likely that the effect of modest amounts of this component also can be safely ignored. The enthalpies of mixing between diopside and anorthite, forsterite-anorthite, diopside-albite and albite-anorthite melts are respectively zero, zero, positive and negative at mantle temperatures [Navrotsky et al., 1989]. Since the enthalpies of mixing liquid pyroxenes is also likely to be nearly ideal, it appears that it is safe to set the enthalpy of mixing in liquids to small or zero values. The effect of pressure on the enthalpy of mixing also can be ignored since the volume of mixing in silicate melts is close to zero [eg. Lange and Carmichael, 1987].

In summary, the heats of fusion of peridotite at mantle temperatures and pressures can be estimated from the values in Table 8 if suitable pressure and temperature corrections are made. It is convenient that the heats of fusion are not overly sensitive to the proportions of minerals entering the liquid phase. Thus, the heats of fusion obtained along a 1700° K adiabat, for example, are going to range between 170-190 cal/gm in the pressure range of 1.0 to 3.0 GPa, and are nearly independent of the proportions of minerals fused.

Adiabatic Upwelling of Mantle. Partial melting results from the decompression of mantle as it ascends along an adiabat and crosses the mantle solidus. The percent of peridotite that undergoes melting at a given depth depends on: (1) the composition of peridotite, in particular, the CaO and Al_2O_3 contents which are expressed primarily in modal garnet and/or calcic clinopyroxene; (2) the potential temperature of the adiabat [McKenzie, 1984]; and (3) the dynamics of the melting process. The amount and the composition of basaltic melt that escapes the mantle to form oceanic crust depends on the factors already listed as well as the "shape of the melting regime" [O'Hara, 1985], how melt segregates from the mantle residue, how melt accumulates as discrete magma bodies [Klein and Langmuir, 1987], and whether mantle ascends passively or buoyantly beneath the spreading center [Scott and Stevenson, 1989].

In order to simplify this discussion, the focus is on the hottest streamline of an adiabatic mantle, either at the center of a hot plume or directly beneath the spreading axis of passively upwelling mantle. What is an appropriate adiabat? Consider the following reasonable assumptions: 1) the maximum amount of melting is 25% and 2) melting is largely terminated at 1.0 GPa, where calcic clinopyroxene is eliminated from the mantle residue. If the mean heat of fusion is 180 cal/gm and the heat capacity is 0.3 cal/gm° C, then the excess temperature, T, needed to melt 25% of the mantle is

$$\Delta T = \frac{\Delta H}{Cp}(x) = 150° \qquad (14)$$

The solidus at 1.0 GPa is approximately 1275° C, and the "diopside out curve" exists at about 1350°C. The adiabat at the beginning of melting of the upwelling mantle is obtained by adding the 150° C to 1350°C and then extending the adiabat to higher pressures (15° C/GPa, [Turcotte and Schubert [1982]). This exercise places the adiabat at the mantle solidus at about

3.2 GPa and 1530° C (Fig. 30). (The adiabat is hardly unique, however. If melting is terminated at 0.5 GPa rather than 1.0 GPa, the adiabat crosses the solidus at 2.7 GPa and 1470° C. If the diopside-out curve lies at higher temperatures as expected if the peridotite is made more refractory by fractional melting, then the adiabat would intersect the solidus at even higher pressures.) The adiabat obtained in this simple exercise crosses the solidus at depths within the garnet lherzolite facies, a depth much greater than predicted by other melting models [eg. McKenzie and Bickle, 1988]. The guiding assumptions are that MORB were extracted from mantle at pressures around 1.0 GPa, and that the maximum amount of melting, here taken at 25%, is limited by the pressures and temperatures of the diopside-out curve. The 3.0 GPa pressures for the beginning of melting apply only to the hottest adiabats that yielded the maximum amount of melting at 1.0 GPa. Colder adiabats produce less melt and do not require high pressure intersections of the solidus. The abundance of diopside-bearing harzburgites attest to the fact that many parcels of mantle never reached temperatures exceeding the diopside-out boundaries at low pressures. It is interesting, nevertheless, that MORB have $^{176}Hf/^{177}Hf$ ratios that seem to require garnet to be a residual phase in MORB melt genesis [Salters and Hart, 1989]. This observation is consistant with the existance of hot adiabats which cross the solidus at pressures exceeding 2.5 GPa and which produce maximal melting (see later).

The true mantle solidus at the beginning of melting along this adiabat is, of course, at much greater depths where traces of volatile-rich magma are produced through the partial melting of variously carbonated and hydrated mantle. The very low H_2O contents of water-undersaturated ocean floor basalts imply that

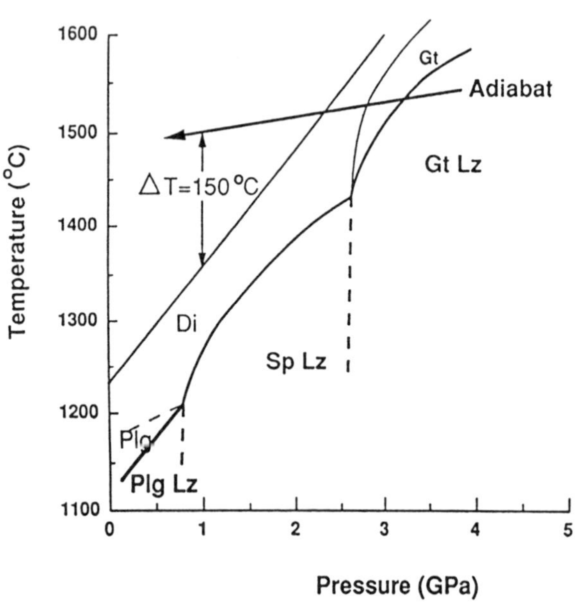

Fig. 30 Estimate of the temperature and pressure of the beginning of melting of mantle that produces approximately 25% melt at 1.0 GPa by adiabatic decompression.

the H_2O content of the mantle is 100-200 ppm, and consequently, the amount of melt produced above the dry solidus will not deviate noticeably from that in totally H_2O-free mantle. The CO_2 content of the mantle is not well-constrained; the CO_2 content must be low, however, judging by the extreme enrichment of incompatible trace elements observed in CO_2-rich igneous rocks such as kimberlites and carbonatites [Haggerty, 1989; Woolley and Kempe, 1989] which require that they are derived by very small degrees of melting even if the sources are trace element enriched. The role of this volatile- and trace element-rich melt cannot be ignored, however, as it may be an important metasomatic agent and the principal trace element reservoir in the suboceanic mantle. It is recognized by a highly fractionated trace element pattern characteristic of fluids equilibrated with garnet lherzolite.

The near-solidus melt at 3.0 GPa coexists with garnet lherzolite and is an alkali-rich picrite with about 15-18 % MgO. The rare earth elements (REE) are strongly fractionated, reflecting the effects of residual garnet, and possibly the influence of the trace element enriched melt. Melting continues as the adiabat rises above the mantle solidus selectively removing garnet from the mantle residue, according to the reaction [Kushiro and Yoder, 1974]:

0.55 Garnet + 0.37 Diopside + 0.08 Enstatite = 0.11 Olivine + 0.89 liquid (the phase proportions are estimates and must certainly vary with pressure and bulk composition). According to the melting curves of peridotite PHN 1611 at 3.5 GPa, approximately 8% melt is generated to reach the garnet-out curve, leaving a residue of olivine-enstatite-calcic clinopyroxene (Fig. 16) [Harrison, 1979].

Approximately 5-6% melt coexists with a mantle residue of olivine, calcic clinopyroxene and enstatite in this mantle at about 2.6 GPa. (The amount of melt was calculated solely from thermal considerations.) The trajectory of the adiabat to shallower levels is obtained by melting the mantle in an equilibrium mode. This model is a useful end member because it creates the maximum amount of melt along a specific adiabat. Melting within the pressure interval 2.6 GPa to 1.0 GPa selectively consumes diopside relative to olivine and enstatite, and the solidus temperature gradually increases towards the diopside-out curve with progressive melting of the ascending mantle. The diopside-out curve is crossed at 1.0 GPa and 1350° C and after 25% melting (as it must, since that is how the adiabat was chosen!).

Melts coexist only with the harzburgite at shallower levels, and enstatite is selectively melted as the adiabat approaches, but does not reach the enstatite-out curve. The melts coexisting with harzburgite below 0.5-0.6 GPa are quartz-normative but contain relatively high MgO, Mg* and low CaO, Al_2O_3 and Na_2O (Table 7). Such liquids are not found on the ocean floor, emphasizing again that equilibrium melting of mantle peridotite does not continue to shallow levels, or at the very least, is not erupted unmodified.

The composition of the liquid coexisting at each level within the upwelling mantle changes continuously with pressure and the degree of melting. The general characteristics of melt coexisting with harzburgite at low pressures have already been discussed. They reflect the movement of the enstatite-olivine boundary curve towards more enstatite-rich compositions. The liquids produced within the garnet-facies are not as well constrained, however. The initial liquids are not enriched in Al_2O_3 or CaO since both garnet and diopside phase volumes are displaced towards forsterite. The liquids derived from the melting of garnet lherzolite at 3.0 GPa, 1550°C [Takahashi, 1986; KLB peridotite] and at 3.5 GPa, 1600° C [Falloon and Green, 1988, a fertile basalt-peridotite sandwich] have the expected compositions (Table 7). Whether these compositions match the true compositions in the mantle depends on the amount and composition of volatile-rich melt that exists at these depths. For small degrees of melting (1-2 %), even the small quantities of H_2O and CO_2 that are postulated to exist in the mantle may influence melting behavior.

Continued melting of garnet lherzolite will increase the Al_2O_3 content and dilute the olivine component in the melt. At pressures less than 2.6 GPa, the mantle is within the spinel lherzolite facies and liquids coexist with diopside, enstatite, olivine ± spinel. Increased melting preferentially consumes diopside diluting Al_2O_3, lowering MgO and increasing the CaO content of the liquid (Table 7). Figure 31 illustrates this pattern, but for the case of increasing temperature at constant pressure. The CaO maximum is reached at 1.0 GPa, where the last grains of diopside are eliminated from the mantle residue. From then on, the melts are diluted by the transfer of orthopyroxene from peridotite to melt.

The question remains whether any of these melts produced by equilibrium melting of peridotite can be primary magmas. Ahern and Turcotte [1979], McKenzie [1984] and others argue that even small amounts of melt become interconnected and are extracted as soon as formed. This is certainly not inhibited by

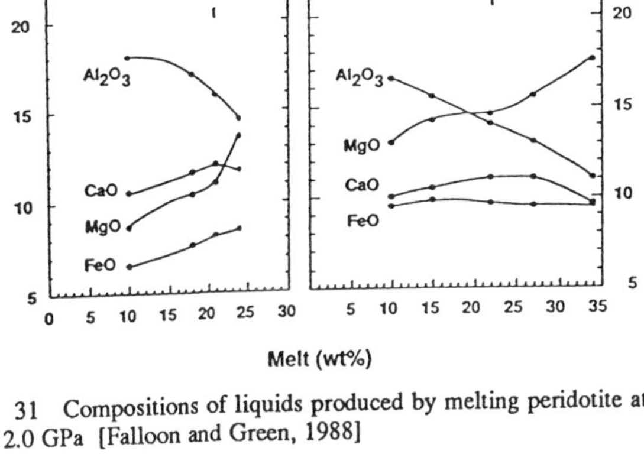

Fig. 31 Compositions of liquids produced by melting peridotite at 1.0 and 2.0 GPa [Falloon and Green, 1988]

surface energy effects, since basaltic melts readily wet and infiltrate olivine-dominated rocks [Waff and Bulau, 1978; Watson, 1982]. McKenzie [1984], furthermore, argued that melt migrates by porous flow in a compacting mantle and eventually coalesces at the top of the asthenosphere. Melt velocities of 10^{-5} to10^{-8} cm/sec are calculated [Turcotte, 1987].

The attainment of chemical equilibrium between basaltic melt and the mantle depends on the melt flow rate and the kinetics of chemical exchange between the melt and the surrounding crystals. How fast can melt flow and still maintain local equilibrium? If equilibrium times are limited by the rates of volume diffusion, then equilibrium times vary widely. Fe/Mg exchange rates are rapid in olivine, requiring equilibrium times of 1-10 years under mantle conditions, whereas up to 10,000 years may be required to equilibrate Fe/Mg ratios in garnets [Bédard, 1989]. Even longer equilibration times are required for highly charged cations such as Ti, Al and Cr in pyroxenes [Freer, 1981]. On the other hand, reequilibration rates could be much faster if dissolution-reprecipitation is involved in the melt segregation process [Bédard, 1989].

Let's focus on a droplet of melt that is born at depth and then percolates upward through lherzolite at flow rates that are much slower than the rates of chemical reequilibration. This is an example of fractional melting, that is, melt escapes its host, albeit at a slow rate, as soon as it is formed. Under these constraints, however, the melt readjusts to the boundary curves appropriate to each level in the mantle. The composition of this droplet at each depth would approach those obtained by equilibrium melting of lherzolite. Suppose that flow rates are 10^{-6} cm/s. Because Fe/Mg exchange between olivine and melt occurs on the time scale of a year or so, the melt need only pass through about 3 meters of rock to achieve exchange equilibrium with olivine. If the time of equilibration between the garnet and pyroxene is a thousand times longer, then 3 km of rock must be passed by the melt to achieve exchange equilibrium [Bédard, 1989].

This appears to be the most pessimistic case. The melts carry latent heat that can be used to melt the crystalline matrix [see also Kelemen, 1990]. The boundary curves, moreover, must migrate away from olivine with decreasing pressure, inducing olivine to crystallize from the melt droplet. The latent heat of crystallization plus the latent heat of the droplet melt is sufficient to cause partial melting of the mantle and to drive the composition of the droplet to the boundary curves appropriate to the ambient pressures.

The net effect of these series of equilibrium states depends on buffering capacity of the mantle. A droplet that remains in contact with lherzolite within a given facies, say the spinel facies, will have compositions constrained to lie on the olivine, enstatite and calcic clinopyroxene boundary. The major element composition is similar to that developed by equilibrium melting between an immobile melt and the surrounding melt. A droplet that invades dunite, however, achieves exchange equilibrium (e.g., Fe/Mg/Ni) with the olivine but is otherwise largely unbuffered. In this example, the Mg# ratio of the melt becomes decoupled from the components incompatible with olivine, CaO and Al_2O_3, for example. The extent of this decoupling depends on the rate of volume diffusion and the thickness of the dunite layer. Since the ratio of olivine to melt is large, even components such as CaO and Al_2O_3 could be affected to some degree by olivine if the entire olivine grain equilibrates with the melt. Two observations argue against this possibility. Diffusion rates of CaO and Al_2O_3 in olivine are likely to be much lower than those for FeO and MgO, requiring 100's, if not 1000's of meters of dunite to achieve equilibrium. Even if equilibration rates are faster than anticipated, there is enough CaO, Al_2O_3, and other incompatable oxides, in 1% of liquid to saturate olivine in these components. One percent of typical basaltic melt, for example, provides 0.15% Al_2O_3 to olivine if volume diffusion extends over the entire rock. In the more reasonable case, only the outermost zones of olivine are enriched in these components and quickly become saturated. It is doubtful, therefore, that dunite has the buffering capacity to significantly change the concentration of the major melt components that are highly incompatible with olivine.

Harzburgite has a greater capacity to buffer components such as Al_2O_3 and CaO since orthopyroxene is more fertile than olivine. Diopside-free harzburgites typically have orthopyroxene with 0.5 - 1.0% CaO and 1-2% Al_2O_3. Rates of volume diffusion of Ca and Al in pyroxene are very low, however, and may require 10^6 years or more to achieve exchange equilibrium with the melt. But equilibrium is also approached when melts from deeper levels infiltrate peridotite at shallower pressures. The expansion of the olivine field with decreasing pressure will cause olivine to crystallize from the melt droplet , forcing the melt composition to the local olivine-orthopyroxene boundary curve and increasing the contents of incompatible oxides. The latent heat of olivine crystallization provides energy to melt enstatite, resulting in dilution of these same incompatible oxides. Since the latent heats of fusion of olivine and enstatite are similar, the incompatible oxide content of the melt does not change materially. Thus, there will again be a decoupling, albeit not as large as within a dunite matrix, between the compatible and incompatible oxides in the melt droplet [see also Kelemen et al, 1990]. The melt composition, moreover, will not be the same as melts which are derived by equilibrium or, even fractional, melting of a homogeneous mantle.

The mantle column, moreover, is not homogeneous. Those components that are not buffered by the major mantle minerals will be controlled by the initial heterogeneities in the mantle. Consequently, a droplet many equilibrate with a number of different lithologies. The final composition, particularly for the minor elements, will represent the sum of the contributions of all the local equilibrium states that were established in the mantle column. The major element composition, however, is largely determined by the pressure and temperature at which the melt, last equilibrated with lherzolite mantle.

A different sequence of equilibrium (or quasi equilibrium) states takes place if melt segregation is achieved through channelled flow. In the ideal case, the diversion of melt into veins or dikes completely isolates the melt from the mantle, and the melt reaches the top of the asthenosphere or local magma chamber as a primary magma. If, however, there is a complex network of channels emanating from different depths in the mantle column, and like a river system, collects and mixes liquids and ultimately deposits these melts in a magma chamber, then these melts can have no simple relationship to the melts derived by equilibrium melting in the laboratory [see Klein and Langmuir, 1987]. Additional complexities arise because partial equilibration with mantle wall rock can easily be demonstrated. Wall rocks next to veins in composite mantle xenoliths [Kempton, 1987] or near dikes in alpine peridotite [Quick, 1981] show gradients in mineral and whole rock chemistry towards the vein. Dunite wall-rocks, for example, are observed in veined harzburgites [Nicolas, 1989] implying partial fusion. A long lived vein, however, should develop a depleted mantle sheath and should be a refractory and inert passageway for other mantle derived magmas.

OVERVIEW

The guiding paradigm is that most ocean floor basalts are not primary magmas, but are descendants or aggregates of primary magmas that segregated from the mantle at 0.8 GPa or more. The supporting evidence is based primarily on phase equilibria constraints. (1) Primitive ocean floor basalts have olivine and calcic clinopyroxene on their liquidi at pressures as high as 1.0 GPa [Bender et al, 1978; Fujii and Bougault, 1983]. Since crystal-liquid fractionation processes result in lowering the pressures of cosaturation, the pressures obtained by these experimental techniques yield minimum pressures for the depth of segregation. (2) Primary basalts undergo first olivine fractionation if they are cooled at pressures below those at which they last equilibrated with mantle. If the compositions of MORB are corrected for olivine fractionation, the resulting basalts have compositions that can be produced by melting of peridotite at pressures greater than 0.8 GPa. (3) The paucity of Mg-rich, quartz tholeiite basalts require that only negligible volumes of basalt magma equilibrated with mantle at shallow levels ($P<0.8$ GPa). The average pressures of segregation must therefore be greater than 0.8 GPa.

The depths at which upwelling oceanic mantle crosses the solidus are less constrained, but appear to span a significant range. The best evidence is given by the diopside contents of abyssal peridotites, which range from diopside-poor lherzolites (diopside contents 5-10%) to diopside-free harzburgites. Since about 25% -30% melting is necessary to eliminate diopside from reasonably fertile peridotite, the modal data are consistent with the formation of residual peridotites by about 5% to over 25% melting. This range of melting implies a range of potential temperatures and a range of of depths over which melting is initiated. A simple calculation illustrates this last point. Taking the slope of the mantle solidus and adiabat as 120°C/GPa and 20°C/GPa respectively and the latent heat of fusion and the heat capacity as 180 cal/gm and 0.3 cal/gm°c respectively, a melting range from 5% to 30% requires that the beginning of melting in the mantle must span more than 1.5 GPa of pressure. Stated differently, if 5% melt is obtained from an adiabat that crosses the solidus at 1.0 GPa, then 30% melt comes from mantle that crosses the solidus at 2.5 GPa.

Suboceanic mantle, therefore, has varying degrees of "hotness". This variability, however, is not expressed fully by the modal data of abyssal peridotites. It is worth recalling that abyssal peridotites are dredged not from the axial regions of the mid-oceanic ridges but mainly from the walls of oceanic fracture zones [Dick, 1989]. The oceanic fracture zones are likely to lie above cooler mantle characterized by lower amounts of melt production [Parmentier and Morgan, 1991; Morgan and Forsyth, 1988]. Larger degrees of melting occur in the upwelling regions of the subaxial mantle, particularly the mid-point portions of large spreading segments [eg. Lin et al, 1989]. It follows that subaxial peridotites should be dominated by diopside-free harzburgites and are more depleted than the typical abyssal peridotite from fracture zones. It is unfortunate that this thesis cannot be tested directly. Nevertheless, this result may explain why harzburgites at the base of some ophiolites reflect higher degrees of melting than obtained in most abyssal peridotites [Nicolas, 1989]. The highly refractory condition of the ophiolitic harzburgites simply reflects the more extreme extent of melting that occurs beneath the spreading axis of oceanic ridges.

Having argued that primary magmas are recovered from considerable depths within the mantle, and perhaps are aggregates of melts which sample many levels within the melting regime [eg. Klein and Langmuir, 1987, Salters and Hart, 1989], we are faced with the formidable task of explaining how this is done. There is an emerging scenario that density-driven porous flow mechanisms governed by Darcy's Law are not up to this task. Melts migrating solely by porous flow are not easily isolated from the mantle, but must partially reequilibrate with the mantle at all depths until the melts are siphoned into axial chambers. The direction of porous flow, moreover, is vertical yet there must be a horizontal component of migration to focus the melt to the narrow zone of crustal accretion because there is not enough melt within a column of mantle beneath the 1-2 km wide axial zone to produce 6-7 km of oceanic crust [Sparks and Parmentier, 1991]. It is necessary, therefore, that there exist some form of drainage system that efficiently collects, isolates and rapidly but episodically delivers large masses of melts to axial magma chambers. Episodicity is required by the absence of geophysical evidence of continuous magma chambers under slowly spreading ridges,

Veins and dikes may be important in extracting melt even from a partially molten mantle [Sleep, 1988]. Examples of such dikes are common in the mantle sections of ophiolite complexes. "Indigenous dikes" have irregular contacts with a wallrock composed of symmetrical screens of depleted dunite [Nicolas, 1989]. Networks of depleted dunite screens in harzburgite massifs may represent channels through which basaltic magmas have passed and reacted with the wallrock. The depleted dunites, in this view, were harzburgites from which orthopyroxene was selectively resorbed through reaction with orthopyroxene-undersaturated basaltic magmas [Sinton, 1977; Dick, 1977]. Such reactions are favored if the peridotite country rock was at or close to its solidus. These features are most common within the plagioclase lherzolite facies, ie. pressures less than 1.0 GPa [Nicolas, 1989], and do not necessarily speak to the possibility of nucleating and propagating dikes at greater depths.

As an alternative and perhaps complementary theory of melt segregation, Sparks and Parmentier [1991] find that a high porosity boundary layer might form at the interface beneath the impermeable lid of the lithosphere and a compacting column of the partially melted mantle. The porosity of this 200-300 m thick layer, is determined by the total amount of melting experienced by the mantle column minus the amount of melt which freezes onto the diverging plate. The slope of the thickening oceanic lithosphere and the along -axis component of gravity drives melt in the highly permeable layer towards the ridge axis and into the axial magmas chambers. Under certain conditions, the theory predicts periods of rapid extraction, which drain the layer, followed by periods of replenishment which recharge the boundary layer. This model accounts not only for the segregaton of melt but also the focusing of melt toward the ridge axis and the episodicity of ocean floor magmatism.

CONCLUSIONS

The physical models of magma segregation are still being developed but wherever these theories ultimately lead certain conclusions appear inescapable.

1) Partial melting in the upwelling mantle is by fractional melting with varying degrees of reequilibration of melt with mantle of varied composition and mineralogy.

2) The limit of melting is determined by the diopside out curve. If melts are segregated at depth, a diopside-free harzburgite mantle contributes only minor amounts of additional melt as it ascends to shallower levels.

3) Isotopic and chemical variability of MORB is expected on all scales if segregation is primarily by the tapping of veins, since veins individually sample some small volumes of mantle [Sleep, 1988]. Greater homogeneity is expected if magma segregation occurs within a persistent and extended low porosity zone, as, for example, described by Sparks and Parmentier [1991]. MORB formed in these environments are not obtained from a single primary magma but rather from a blend of primary magma originating at various depths from the melting regime.

Acknowledgments. Thoughtful reviews by John Longhi and Rodey Batiza are appreciated. This work was supported by NSF grant, EAR 90-17790 and EAR 87-19357.

REFERENCES

Ahern, J.L. and D.L. Turcotte, Magma migration beneath an ocean ridge, *J. Geophys. Res, 45,* 115-122, 1979

Bedard, J.H., Disequilibrium mantle melting, *Earth Planet. Sci. Lett., 91,* 359-366, 1989

Bender, J.F., F.N. Hodges and A.E. Bence, Petrogenesis of basalts from the project Famous Area: Experimental study from 0-15 kbars, *Earth Planet. Sci. Lett., 41* 277-302, 1978

Bonatti, E, C. Ottonello, and P.R. Hamlyn, Peridotites from the island of Zabargad]St. John] Red Sea: Petrology and Geochemistry, *J. Geophy. Res.*, *91*, 599-632, 1986

Bulau, J.R., H.S. Waff, and J.A. Tyburczy, Mechanical and thermodynamic constraints on fluid distribution in partial melts, *J. Geophys. Res.*, *84*, 6102-6108, 1979

Casey, J.F., and J.A. Karson, Magma chamber profiles from the Bay of Islands Ophiolite Complex: implications for crustal level magma chambers at mid-ocean ridges, *Nature*, 295-301, 1988

Casey, J.F., D.L. Elthon, F.X. Siroky, J.A. Karsons, and J. Sullivan, Geochemical and geological evidence bearing on the origin of the Bay of Islands and Coastal Complex ophiolites of Western Newfoundland, *Tectonophysics*, *116* 1-40, 1985

Christie, D.M., I.S.E. Carmichael and C.H. Langmuir, Oxidation states of mid-ocean ridge basalt glasses. *Earth Planet and Sci. Lett*, 79, 397-411, 1986

Daines, M.J., and F.M. Richter, An experimental method for directly determining the interconnectivity of melt in a partially molten system, *Geophys. Res. Lett.*, *15*, 1459-1462, 1988

Dick, H.J.B., Evidence of partial melting in the Josephine peridotite, *Oregon Dept. Geol. Mineral. Ind.*, *96*, 59-62, 1977

Dick, H.J.B., R.L Fisher, and W.B. Bryan, Mineralogic variability of the uppermost mantle along mid-ocean ridges. Earth Planet. Sci. Lett., 69, 88-106, 1984

Dick, H.J.B., and T. Bullen, Chromian spinel as a petrogenetic indicator in abyssal and alpine-type peridotites and spatially associated lavas, *Contrib. Mineral. Petrol.*, *86*, 54-76, 1984

Dick, H.J.B., Abyssal peridotites, very slow spreading ridges and ocean ridge magmatism in Magmatism in the Ocean Basins, Ed. A.D. Saunders and M.J. Norry; Geol. Soc. London Sp. Pub. 42, 71-105, 1989

Dixon, T.E., E. Stolper, and J.R. Delaney, Infared spectroscopic measurements of CO_2 and H_2O in Jaun de Fuca Ridge basaltic glasses, *Earth Planet. Sci. Lett.*, *90*, 87-104, 1988

Eggler, D. H. The effect of CO_2 upon partial melting of peridotite in the system Na_2O-CaO-Al_2O_3-MgO-SiO_2-CO_2 to 35 kb with an analysis of melting in a peridotite-H_2O-CO_2 system, *Amer. J. Sci.*, *27*, 305-368. 1978

Eggler, D. H. Carbonatites, primary melts and mantle dynamics, in *Carbonatites*, K. Bell, ed., Unwin Hyman, London, 561-579, 1989

Eggler, D. H., J. L. Holloway, and B. O Mysen,. High CO_2 solubility in mantle magmas, *Geology*, *4*, 198-199, 1976

Eggler, D.H., Upper mantle oxidation state: Evidence from olivine-orthopyroxene-ilmenite assemblages, *Geophy. Sci. Lett.*, *10*, 365-368, 1983

Ehlers, E.G., The Intrepretation of Geological Phase Diagrams, San Francisco, Freeman, 1972

Elthon, D., and C.M. Scarfe, High Pressure phase equilibria of a high-magnesian basalt and the genesis of primary oceanic basalts. Amer. Mineral., 69, 1-15, 1984.

Elthon, D., Partitioning of Ni between olivine and high -MgO basaltic liquids, *Lunar Planet. Sci. XVIII*, 258-259, 1987

Engi, M., Equilibria involving Al-Cr spinel Mg-Fe exchange with olivine, Experiments, thermodynamic analysis and consequences for geothermometry, *Am. J. Sci.*, *2834*, 29-71, 1983

Falloon, T. J. and D. H. Green, Anhydrous partial melting of MORB, pyrolite and other peridotite compositions at 10 kb: Implications for the origin of primitive MORB glasses, *Mineral. Petrol.*, *37*, 181-219, 1987

Falloon, T. J. and D. H. Green, Anhydrous partial melting of peridotite from 8-35 kb and the petrogenesis of MORB, *J. Petrol. Special Lithosphere Issue*, 379-414, 1988

Falloon, T. J. and D. H. Green, The solidus of carbonated, fertile peridotite, *Earth Planet. Sci. Lett.*, *94*, 364-370, 1989

Falloon, T. J. and D. H. Green, Solidus of carbonated fertile peridotite under fluid saturated conditions, *Geology*, *8*, 195-199, 1990

Falloon, T. J., D. H. Green, C. J. Hatton, and K. L. Harris, Anhydrous partial melting of a fertile and depleted peridotite from 2-30 kb and application to basalt petrogenesis, *J. Petrol.*, *29*, 1257-1282 1988

Fujii, T., and H. Bougault, H., Melting relations of a magnesian abyssal theoleiite and the origin of MORBs. Earth Planet. Sci. Lett. 62, 283-295, 1983.

Fujii, T., and C.M. Scarfe, Composition of liquids coexisting with spinel lherzolite at kbar and the genesis of MORBs. Contrib. Mineral Petrol., 90, 18-28, 1985.

Freer, R., Diffusion in silicate minerals: a data digest and guide to the literature, *Contrib. Mineral. Petrol.*, *76*, 440-454, 1981

Frey, F.A., C.J. Suen and H.W. Stockman, The Ronda high temperature peridotite: Geochemistry and petrogenesis, *Geochim. Cosmochim. Acta*, *49*, 2469-2492, 1985

Green, D.H, T.J. Falloon and W.R. Taylor, Mantle-derived magmas-roles of variable source peridotite and variable C-H-O fluid compositions. in Magmatic Processes: Physical Principles; Mysen, B.O. Ed, 139-154, 1987

Green, D. H. and M. E. Wallace, Mantle metasomatism by ephemeral carbonatite melts, *Nature*, *336*, 459-462, 1988

Grove, T.L. and W.B. Bryan, Fractionation of pyroxene-phyric MORB at low pressure: An experimental study, *Contrib. Mineral Petrol.*, *84*, 293-309, 1983

Grove, T.L. and M.B. Baker, Phase equilibrium controls on the tholeiitic versus calc-alkaline differential trends, *J. Geophy. Res.*, *89*, 3253-3274, 1984

Grove, T.L., R.J. Kinzler, and W.B. Bryan, Natural and experimental phase relations of lavas from Serocki volcano, Proceed. Ocean Drilling Program, Scientific Results, 106/109, 9-17, 1990

Haggerty, S.E., Mantle metasomes and the kinship between carbonatites and kimberlites, in Carbonatites, K.Bell ed, Unwin Hyman, London UK, 546-560, 1989

Hamlyn, P.R., E. Bonatti, Petrology of mantle derived ultramafics from the Owen Fracture Zone, NW Indian Ocean: Implications for the nature of the oceanic upper mantle, *Earth Planet. Sci. Lett.*, *48* 65-79, 1980

Harrison, W.J., Partitioning of REE between garnet peridotite minerals and coexisting melts during partial melting, Ann. Rep. Director Geophy. Lab, 78, 562-568, 1979

Herzberg, C., T. Gasparik and H. Sawamota, Origin of mantile peridotite: Constraints from melting experiments to 16.5 Gpa, *J. Geophy. Res 95*, 15,779-15,805, 1980

Hess, P. C. *Origins of igneous rocks*, Harvard Press, Cambridge, MA., 336 pp. 1989

Jacques, A.L, and D.H. Green, Determination of liquid compositions in experimental high pressure melting of peridotite, *Am. Mineral.* *64*, 1312-1321, 1979

Jacques, A.L., and D.H. Green, Anhydrous melting of peridotite at 0-15 kb pressure and the genesis of tholeiitic basalts, *Contrib. mineral Petrol.*, *73*, 287-310, 1980

Johnson, K.T.M., H.J.B. Dick and N. Shimizu, Melting in the oceanic upper mantle: An ion microprobe study of diopsides in abyssal peridotites, *J. Geophy. Res.*, *95* 2661-2678, 1990

Jurewicz, S.R. and E.B. Watson, The distribution of partial melt in granitic system: The application of liquid phase sintering theory, *Geochim. Cosmochim. Acta*, *49*, 1109-1122, 1985

Juster, T.C., T.L. Grove, M.R. Perfit, Experimental constraints on the generation of FeTi basalts, andesites and rhydocacites at the Galapagos spreading center, 85°W and 95°W, *J. Geoph. Res*, *94* 9251-9274, 1989

Kelemen, P.B., K.T.M. Johnson, R.J. Kinzler and A.J. Irving, High-field-srength element depletions in arc basalts due to mantle-magma interaction, *Nature*, *345*, 521-523,1990

Kempton, P.D., Mineralogic and geochemical evidence for differing styles of metasomation in spinel lherzolite xenoliths: Enriched mantle source regions of basalts, In Mantle Metasomatism, M.A. Menzies and C.J. Hawkesworth eds., p 45-89 Academic Press, London, 1987

Kingsley, R., Carbon dioxide and water in mid-Atlantic ridge basalt glasses, PhD Dissertation, Graduate School of Oceanography, University of Rhode Island, 1989

Kjarsgaard, B.A., and D.L. Hamilton, The genesis of carbonatites by immiscibility, in *Carbonatites*, K. Bell ed., London, Unwin Hyman, 388-404, 1989

Klein, E.M., and C.H. Langmuir, Global correlations of ocean ridge basalt chemistry with axial depth and crustal thickness, *J. Geophy. Res., 92,* 8089-8115, 1987

Komor, S.C., D. Elthon, and J. F. Casey, Mineralogical variations in a layered ultramafic cumulate sequence at the North Arm Mountain Massif, Bay of Islands, Ophiolite, Newfoundland, *J. Geophy. res., 90,* 7705-7736, 1985

Kushiro, I., and H.S. Yoder, Jr., Formation of eclogite from garnet lherzolite liquidus relations in a portion of the system $MgSiO_3$-Al_2O_3 at high pressures, Carnegie Insh Wash. Y6, 73, 266-269, 1974

Lange, R.A., and I.S.E. Carmichael, Densities of Na_2O-K_2O-CaO-MgO-FeO-Fe_2O_3-Al_2O_3-TiO_2-SiO_2 liquids: New measurements and derived partial molar volumes, *Geochim. Cosmochim. Acta, 51,* 2931-2946, 1987

Langmuir, C.H., and G.N. Hanson, An evaluation of the major element heterogeneity in the mantle sourses of basalt, Trans R. Soc., London, Ser. A 27, 383-407, 1980

Lin, J., G.M. Purdy, H. Schouten, J-C Sempere and C. Zervas, Evidence from gravity data for focused magmatic accretion along the Mid-Atlantic Ridge, *Nature, 344,* 627-632, 1989

Lindsley, D.H., D.J. Anderson, A two pyroxene thermometer, Proc. Lunar Planet. Sci. Conf., 13th, *J. Geophy. Res., 88,* A887-A906, 1983

Lindsley, D. H., Pyroxene thermometry, *Amer. Mineral, 68,* 477-493, 1983

Longhi, J., D. Walker and J.F. Hays, The distribution of Fe and Mg between olivine and lunar basaltic liquids, *Geochim, Cosmochim. Acta, 42* , 1545-1550, 1978

Longhi, J., On the connection between mare basalts and picritic volcanic glasses, *J. Geophy. Res., 92,* E349-E360, 1987

Maaloe, S., and K. Aoki, The major element composition of the upper mantle estimated from the composition of lherzolites, *Contrib. Mineral Petrol., 63,* 161-173, 1977

MacKenzie, D., The generation and compaction of partially molten rock, *J. Petrol., 25,* 713-765, 1984

MacKenzie, D., and M.J. Bickle, The volume and composition of melt generated by extension of lithosphere, *J. Petrol., 29,* 625-649, 1988

Michael, P.J., and E. Bonatti, Peridotite composition from the North Atlantic: Regional and tectonic variations and implications for partial melting, *Earth Planet. Sci. Lett., 73,* 91-104, 1985

Michael, P.J., The concentration, behavior and storage of H_2O in the suboceanic upper mantle: Implications for mantle metasomation, *Geochim. Cosmochim. Acta., 52,* 555-566, 1988

Morse, S.A., Basalts and Phase Diagrams, New York Springer Verlag, 1980

Mysen, B.O., and I. Kushiro, Compositional variations of coexisting phases with degree of melting of peridotite in the upper mantle, *Am. Mineral.*, 62, 843-965, 1977

Nakano, T and N. Fuji, The multiphase grain control percolation: Its implication for a partially molten rock, *J. Geophys. Res., 94,* 15653-15661, 1989

Navon, O., and E. Stolper, Geochemical consequences of melt percolation: The upper mantle as a chromatographic column, *G. Geols., 95,* 285-307, 1987

Navrotsky, A., D. Ziegler, R. Oestrike, and P. Maniar, Calorimetry of silicate melts at 1773K°: Measurements of enthalpies of fusion and mixing in the systems diopside-anorthite-albite and anorthite-forsterite, *Contrib. Mineral. Petrol, 101,* 122-130 1989

Nickel, K. G. and D. H. Green, The nature of the upper–most mantle beneath Victoria Australia, as deduced from ultramafic xenoliths, in *Kimberlites* ed. J. Kornprobst, 161–178, 1984

Nicolas, A., Structure of ophiolites and dynamics of oceanic lithosphere, Kluwer Academic Publishers, Boston, 1989

O'Hara, M.J., Importance of the "shape" of the melting regime during partial melting of the mantle, *Nature*, 314, 58-62, 1985

Olafsson, M. and D.H. Eggler, Phase relations of amphibole, amphibole-carbonate, and phlogopite-carbonate peridotite: Petrologic constraints on the asthenosphere, *Earth Planet. Sci. Lett, 64,* 305-313, 1983

Pallister, J.S., and C.A. Hopson, Samail ophiolite plutonic suite: Field variation, cryptic variation and layering and a model of a spreading ridge chamber, *J. Geophy. Res., 86,* 2593-2644, 1981

Pan, V. and J. R. Holloway, CO_2 solubility and its implications for degassing ascending magmas, *Abs. METV Workshop on Evolution of Magma Bodies on Mars,* Lunar Planet. Inst., Houston, TX, 33-34, 1990

Parmentier, E. M., and J. Phipps-Morgan, Spreading rate dependance of three-dimensional structure in oceanic spreading centres, *Nature, 348,* 325-328, 1990

Phipps-Morgan, J., and D.W. Forsyth, Three-dimensional flow and temperature perturbations due to a transform offset: Effects on oceanic crust and upper mantle structure, *J. Geophy. Res. 93,* 2955-2966, 1988

Presnall, D.C., S.A. Dixon, J.R. Dixon, T.H. O'Donnell, N.L. Brenner, R.L. Schrock and D.W. Dycus, Liquidus phase relations in the join diopside-forsterite-anorthite from 1 atm to 20 kbar: Their bearing on the generation and crystallization of basaltic magma. *Contrib. Mineral. Petrol, 66,* 203-220, 1978

Presnall, D. C., J.R. Dixon, T.H. O'Donnell, and S.A. Dixon, Generation of mid-ocean ridge tholeiites, *J. Petrol., 20,* 3-35, 1979

Presnall, D.C. and J. D. Hoover, High pressure phase equilibrium constraints on the origin of mid-ocean ridge basalts. In: Magmatic processes: Physicochemical Principles, ed., B.O. Mysen, 75-89, 1987

Quick, J.E., Petrology and petrogenesis of the Trinity peridotite, *J. Geophy. Res., 86,* 11837-11863, 1981

Ribe, N.M., Theory of melt segregation-a review, *J. Vol and Geochem. Res., 33,* 241-253, 1987

Richet P. and Y. Bottinga, Thermochemical properties of silicate glasses and melts, *Rev. Geophys., 24,* 1-25, 1984

Roeder, P.L. and R.F. Emslie, Olivine-liquid equilibrium, *Contrib. Mineral. Petrol., 29,* 275-289, 1970

Salters, V. J. M., and S. R. Hart, The Hf-paradox and the role of garnet in the source of mid-ocean ridge basalts, *Nature, 342,* 420-422, 1989

Scarfe, C.M., B.O. Mysen, and C.S. Rai, Invariant melting behavior of mantle material: Partial melting of two lherzolite nodules, *Carneg. Inst. Geophy. Lab., Yb 78,* 498-501 1979

Schilling, J. G., M. Zajac, R. Evans, T. Johnston, W. White, J.D. Devine, and R. Kingsley, Petrologic and geochemical variations along the mid-Atlantic ridge from 29°N-73°N. *Amer. J. Sci., 283,* 510-586, 1983

Scott, D.R. and D.J. Stevenson, A self-consistent model of melting, magma migration and buoyancy driven circulation beneath mid-ocean ridges, *J. Geophys. Res., 94,* 2973-2988, 1989

Sinton, J., Equilibration history of the basal alpine-type peridotite Red Mountain, New Zealand, *J. Petrol, 18,* 216-266, 1977

Siroky, F.X., D. Elthon, J.F. Casey and J.C. Butler, Major element variations in basalts and diabases from the North Arm Mountain Massif, Bay of Islands Ophiolite: Inplications for magma chamber processes at mid-ocean ridges, *Tectonophysics, 116,* 41-61, 1985

Sleep, N.H., Tapping of melt by veins and dikes, *J. Geophy. Res., 93,* 10255-10272, 1988

Smith, C.S., Some elementary principles of polycrystalline microstructure, *Met. Rev., 9,* 1-47, 1964

Smith, S. E. and D. Elthon, Mineral composition of plutonic rocks from the Lewis Hills Massif, Bay of Islands Ophiolite, *J. Geophy. Res.*,93, 3450-3468, 1988

Sparks, D.W. and E.M. Parmentier, Melt extraction from the mantle beneath spreading centers *Earth Planet.Sci.Lett..105,* 368-377, 1991

Stebbins, J.F., I.S.E. Carmichael, and L.K. Moret, Heat capacities and entropies of silicate liquids and glasses, *Contrib. Mineral. Petrol.*, *86*, 131-148, 1984

Stolper, E. A phase diagram for mid-ocean ridge basalts: Preliminary results and implications for petrogenesis. *Contrib. Mineral Petrol.* *74* 13-27, 1980

Stolper, E. and J.R. Holloway, Experimental determination of the solubility of carbon dioxide in molten basalt at low pressure, *Earth Planet. Sci. Lett.*, *87*, 397-408, 1988

Takahashi, E. and I. Kushiro, Melting of a dry peridotite at high pressures and basalt magma genesis. *Am. Mineral.*, *68*, 859-879, 1983

Takahashi, E. Melting of a dry peridotite KLB-1 up to 14 GPa: Implications on the origin of peridotitic upper mantle, *J. Geophys. Res.*, *91*, 9367-9382, 1986

Thompson R.N., Phase equilibria constraints on the genesis and magmatic evolution of ocean basalts, *Earth, Sci. Rev.*, *24*, 161-210, 1987

Toramaru, A. and N. Fuji, Connectivity of melt phase in partially molten peridotite, *J. Geophys. Res.*, *91*, 9239-9252, 1986

Tormey, D. R., T. L. Grove and W.B. Bryan, Experimental petrology of normal MORB near the Kane Fracture Zone: 22°-25°N, mid-Atlantic ridge. *Contrib. Mineral. Petrol.*, *96*, 121-139, 1987

Turcotte, D.L., Physics of magma segregation processes, in Magmatic Processes: Physiochemical Principles, ed. Mysen, B.O., Geochemical Society Spec. Pub., 1, 69-74, 1987

Turner, F.J. and J. Verhoogan, Igneous Metamorphic Petrology, New York, McGraw Hill, 1960

Ulmer, P., The dependence of the Fe^{+2}-Mg cation-partitioning between olivine and basaltic liquid on pressure, temperature and composition, An experimental study to 30 kbars. *Contrib. Mineral. Petrol.* *101*, 261-273, 1989

von Bargen, N. and H.S. Waff, Wetting of enstatite by basaltic melt at 1350°C and 1.0 to 2.5 GPA pressure, *J. Geophys. Res.*, *93*, 1153-1158, 1988

Waff, H.S. and J.R. Bulau, Equilibrium fluid distribution in an ultramafic partial melt under hydrostatic stress conditions, *J. Geophys. Res.*, *84*, 6109-6114, 1979

Walker, D., T. Shibata, and S.E. Delong, Abyssal tholeiites from the Oceanographer Fracture Zone II Phase equilibria and mixing, *Contrib. Mineral. Petrol.*, *70*, 111-125, 1979

Wallace, M. E. and D. H. Green, An experimental determination of primary carbonatite magma, *Nature*, *335*, 343-345, 1988

Watson, E.B., Melt infiltration and magma evolution, *Geology*, *10*, 236-240, 1982

Wood, B.J. and D. Virgo, Upper mantle oxidation state: Ferric iron contents of lherzolite spinels by ^{57}Fe Mossbauer spectroscopy and resultant oxygen fugacities, *Geochim Cosmochim Acta*, *55* 1277-1291, 1989

Woolley, A.R. and D.R.C. Kempe, Carbonatites: nomenclature, average chemical compositions, and element distribution in *Carbonatites*, K.Bell ed, Unwin Hyman, London UK, 1-14, 1989

Wyllie, P. J. The origin of kimberlite, *J. Geophys. Res.*, *85*, 6902-6910, 1980

Wyllie, P. J. Origin of carbonatites: Evidence from phase equilibrium studies, in *Carbonatites*, K. Bell., ed., Unwin Hyman, London, UK, 500-545, 1989

P. Hess, Brown University Department of Geological Sciences, Providence, RI 02912

Structure, Rheology and Permeability of Partially Molten Rocks at Low Melt Fractions

DAVID L. KOHLSTEDT

Department of Geology and Geophysics, Pillsbury Hall, University of Minnesota, Minneapolis, MN 55455

Experimental observations of melt distribution, creep behavior and melt migration in aggregates of olivine plus basalt place constraints on the strength and permeability of upwelling mantle at depth beneath mid-oceanic ridges. Several investigations have demonstrated that melt forms an interconnected network along three- and through four-grain junctions, but does not wet grain boundaries. For this grain-scale distribution of melt, the presence of a small amount of melt will cause a modest, but not dramatic, decrease in creep strength for texturally and chemically equilibrated samples. The decrease in strength may be associated with a change in dominant creep mechanism (e.g., dislocation creep becomes diffusional creep); however, it can also occur without a change in rate-controlling mechanism. Experiments designed to study quantitatively melt migration via porous flow indicate that permeability increases linearly with increasing melt fraction. As a result, the permeability will be relatively high at low melt fractions, permitting melt to escape readily during pressure-release melting.

INTRODUCTION

This paper presents a review of experimental investigations concerned with the rheology and permeability of partially molten aggregates, particularly as they bear on the dynamic behavior of the Earth's upper mantle beneath mid-oceanic ridges. It draws on results of laboratory studies reported in both the geological and the materials science literatures. It also summarizes theoretical models developed to describe the high-temperature creep response (rheology) and the kinetics of melt migration (permeability) of partially molten materials. These models are needed to extrapolate data obtained under laboratory conditions to processes occurring in geological environments.

An important theme in this article concerns the relationship between structure and properties, because the rheology and permeability of partially molten regions of the mantle depend critically on the grain-scale distribution of the melt† phase. Consequently, experimental measurements and theoretical constraints on the grain-scale distribution of melt (structure) are examined first.

Likewise, the rheology of a partially molten rock is directly coupled to its permeability. On the one hand, if the permeability is very low, the melt content of a mantle rock ascending under a ridge axis will continually increase as pressure-release melting progresses. Once the melt fraction exceeds the value at which the grains no longer form a complete skeletal network, the creep resistance will decrease dramatically [e.g., *Arzi*, 1978; *van der Molen and Paterson*, 1979]. On the other hand, if the permeability is high, most of the melt will be rapidly removed from the solid under the influence of buoyancy forces. As a result, the rock will behave mechanically almost like a melt-free aggregate, provided that the melt does not form a continuous pathway through the interfaces separating neighboring mineral grains.

In fact, recent chemical analyses of isotopes and trace elements in mid-oceanic ridge basalts (MORBs) [*Salters and Hart*, 1989] and in diopside in abyssal peridotites [*Johnson et al.*, 1990] indicate that melt can be effectively segregated from the residual crystalline phases at melt contents as small as 0.1%. This result is supported by laboratory measurements of the kinetics of melt migration in olivine-basalt aggregates, which yield permeabilities high enough that ≤0.1% basalt would be expected to be present at any instant in dunitic rocks ascending under a ridge axis [*Riley et al.*, 1990]. Hence, this manuscript concentrates on structure, rheology and permeability at relatively small melt fractions, for which the partially molten rock can be treated as a saturated porous medium.

†Given the context of this paper, the term melt will be used to denote the non-crystalline phase. However, fluid, liquid or amorphous phase could equally well have been applied.

Mantle Flow and Melt Generation at Mid-Ocean Ridges
Geophysical Monograph 71
Copyright 1992 American Geophysical Union

MELT DISTRIBUTION

Theoretical Considerations

Equilibrium Melt Distribution. As will be emphasized below, particularly in the section on high-temperature creep behavior, the grain-scale distribution of the melt phase (i.e., the structure of a partially molten rock) -- not merely the presence of melt -- is one of the most important factors in determining the extent to which melt influences physical properties. If the energy of the interface between two neighboring crystalline (solid) grains, γ_{ss}, and the energy of the interface between the melt and a crystalline grain, γ_{sm}, are independent of orientation, then the relative value of these two interfacial energies determines (at least under hydrostatic stress conditions) the distribution of the melt phase as expressed through the dihedral angle, θ, which is defined in Figure 1b and Equation 1:

$$\cos\frac{\theta}{2} = \frac{\gamma_{ss}}{2\gamma_{sm}} \quad (1)$$

[e.g., *Smith*, 1948; 1964; *Beere*, 1975; *Wray*, 1976; *Bulau et al.*, 1979; *von Bargen and Waff*, 1986; *Watson et al.*, 1990].

Based on Equation 1, three distinct melt distributions can be identified. First, in the limit that γ_{ss} is large relative to γ_{sm} (specifically, if $\gamma_{ss} \geq 2\gamma_{sm}$) such that melt-free interfaces are energetically unfavorable, then $\theta = 0°$. In this case, as illustrated in Figure 1a, the melt phase not only fills the three- and four-grain junctions but also forms a thin film that coats the grain boundaries, such that the crystalline grains are completely surrounded by melt. Second, for $\sqrt{3}\gamma_{sm} < \gamma_{ss} \leq 2\gamma_{sm}$ such that $0 < \theta \leq 60°$, the melt forms an interconnected network along the three-grain and through the four-grain junctions. A cross-sectional view of a three-grain junction in such a partially molten rock would have an appearance similar to the one sketched in Figure 1b. Third, in the limit that γ_{ss} is small compared to γ_{sm} (i.e., if $\gamma_{ss} < \sqrt{3}\gamma_{sm}$), melt-free interfaces are energetically favorable and $\theta > 60°$. In this case, the melt will be restricted to isolated pockets at four-grain junctions, at least at low melt fractions. No melt would be present in a cut through a typical portion of a triple junction, Figure 1c. If the melt percentage increases to a large enough value (e.g., ~2 vol% melt for $\theta = 65°$ and ~4 vol% melt for $\theta = 75°$), however, the melt pockets will become connected to one another [*von Bargen and Waff*, 1986]. If melt is able to escape from the rock, the melt pockets will become isolated again once the melt fraction falls below this critical melt fraction, which is a function of the dihedral angle.

At small but non-zero dihedral angles, *Raj* [1982] has postulated that the interface separating two neighboring grains is composed of islands of grain-to-grain contact, with a melt layer with a thickness of molecular dimensions filling the remainder of the space. The reader is referred to Figure 2 in *Raj* [1982] and Figure 2 in *Raj and Chyung* [1981] for illustrations of this island structure. *Raj* [1982] suggests that such an interface structure is thermodynamically stable [see also *Raj et al.*, 1984], much as point defects are present in crystals at thermodynamic equilibrium because the increase in Gibbs free energy due to the formation of vacancies is offset by the decrease in free energy due to their configurational entropy [e.g., *Poirier*, 1985, pp. 40-41]. No value has been given for the dihedral angle at which such an island structure might form. As will be discussed in more detail in the section on rheology, this microstructure is central to the solution-precipitation creep model of *Raj and Chyung* [1981], because it permits rapid diffusional transport through an interface and yet allows the interface to support a normal stress and a gradient in normal stress over a long period of time. To date, no direct (e.g., high-resolution transmission electron microscopy) observations of such islands have been reported.

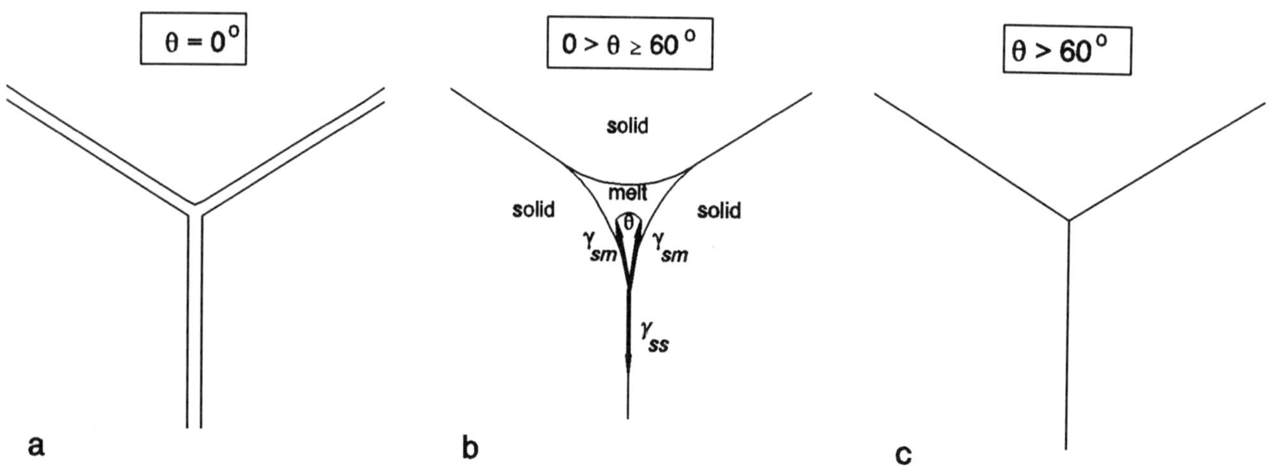

Fig. 1. Melt distribution as a function of dihedral angle for small melt fractions. (a) If $\theta = 0°$, melt wets all of the grain interfaces. (b) If $0 < \theta \leq 60°$, melt forms an interconnected network through triple and four-grain junctions. (c) If $\theta > 60°$, melt will not be present in triple junctions, except possibly in isolated pockets.

In general, neither γ_{ss} nor γ_{sm} is single-valued. In a monomineralic rock, the value of γ_{ss} for a particular grain boundary will be a function of the relative orientation of the lattices of the two contacting grains and of the orientation of the boundary itself. Likewise, the value of γ_{sm} will depend upon the orientation of the solid-melt interface relative to that of the lattice of the crystal [e.g., *Laporte and Watson*, 1992]. As a result, Equation 1 suggests that there will be a range of values distributed about a mean for the dihedral angle [*Jurewicz and Jurewicz*, 1986]. In a polymineralic rock, both grain boundaries and phase boundaries will be present. Consequently, there will be several distinct mean values for the dihedral angle. For example, if two isotropic crystalline phases are present, then three different dihedral angles can be identified, two associated with the contact between melt and the grain boundaries formed between two grains of the same phase and a third with the contact between melt and the phase boundary separating grains of different composition [e.g., *Toramaru and Fujii*, 1986; *Fujii et al.*, 1986].

Equilibrium Melt Fraction. For $0 < \theta \leq 60°$, the existence of a saturation or equilibrium melt fraction, ϕ_{eq}, has been predicted by analyses which consider the minimum interfacial free energy per unit volume of a two-phase system composed of melt plus a crystalline phase in two-dimensional [*Jurewicz and Watson*, 1985; *Riley and Kohlstedt*, 1991] and three-dimensional aggregates [*Park and Yoon*, 1985; *Laporte and Watson*, 1992]. The total

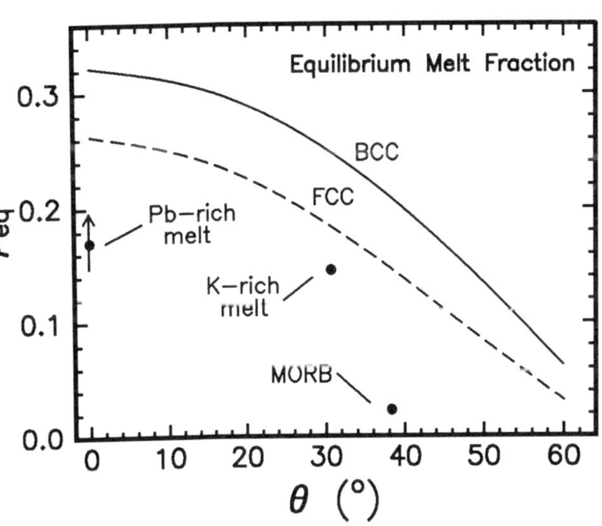

Fig. 2. Equilibrium melt fraction, ϕ_{eq}, as a function of dihedral angle, θ, for aggregates with face-centered cubic (FCC) packing (a rhombic dodecahedron) and body-centered cubic (BCC) packing (a tetrakaidecahedron) of three-dimensional grains. After *Laporte and Watson* [1992]. The solid circles are experimentally determined values for olivine plus a lead-rich melt ($\theta \approx 0°$), a potassium-rich melt ($\theta \approx 30°$) and a mid-oceanic ridge basalt ($\theta \approx 40°$). The vertical arrow on the data point for the lead-rich melt indicates that only a lower limit could be obtained for ϕ_{eq} because lead oxide dissolved in the melt was reduced to metallic lead before an equilibrium melt fraction could be attained, due to the use of iron jackets. After *Riley and Kohlstedt* [1991].

interfacial energy of such a two-phase system is made up of two terms, the solid-solid or grain boundary energy and the solid-melt energy. As illustrated in Figure 2, the volumetric equilibrium melt fraction decreases from 0.25-0.30 for θ near 0° to ~0 as θ approaches 60° [*Laporte and Watson*, 1992]. (The data points for three olivine-melt samples are discussed in the following section.)

Physically, the existence of an equilibrium melt fraction means that, if a melt-free aggregate is placed in contract with a reservoir of melt, the rock will absorb melt until the equilibrium melt fraction is reached [*Riley and Kohlstedt*, 1990]. For a closed system in which partial melting occurs, surface tension will cause the melt to be distributed uniformly along the three- and four-grain junctions for $\phi < \phi_{eq}$; pools of melt will form for $\phi > \phi_{eq}$ [*Laporte and Watson*, 1992]. For an open system, buoyancy and capillary forces will cause extraction of melt from the source region; hence, ϕ_{eq} might never be attained [*Riley et al.*, 1990].

Experimental Observations

Equilibrium Melt Distribution. The first laboratory investigations of the equilibrium melt distribution in rocks of mantle compositions were carried out on samples fabricated from mechanical mixtures of powders of dunite plus 1-2 wt% basalt [*Waff and Bulau*, 1979; 1982]. For a starting particle size of ~50 μm, the melt-solid microstructure approached textural equilibrium after 100-200 h at temperatures <1350°C and pressures <2 GPa. At 1230-1240°C and 1 GPa, dihedral angles of 47, 30 and 30° were observed for three different melt compositions in contact with olivine grains. Recently, a dihedral angle of 38° was measured for an aggregate of olivine plus a mid-oceanic ridge basalt (MORB), which was annealed at 1300°C and 300 MPa [*Riley and Kohlstedt*, 1991]. Although insufficient data are available to make a direct correlation of dihedral angle with melt composition, the dihedral angle decreased for one composition from 30° to 20° as the temperature was increased from 1230°C to 1316°C [*Waff and Bulau*, 1982]. In another study, a systematic increase in dihedral angle with increasing confining pressure was observed [*Waff and von Bargen*, 1988]. In these latter two sets of experiments, the changes in dihedral angle are probably largely due to changes in the composition of the melt that are associated with changes in temperature and pressure. Scanning electron microscopy (SEM) observations, such as those presented in Figure 3, further demonstrated that the melt is confined to three-, four- and (occasionally) five-grain junctions, while the grain boundaries are free of glass (quenched melt) to a resolution of 20 nm. This conclusion is supported by a conventional transmission electron microscopy (TEM) examination, Figure 4, which reveals a number of melt-filled triple junctions as well as one four- or five-grain junction. In both the SEM and the TEM micrographs, many of the olivine-basalt interfaces are planar, rather than smoothly curved, indicating that the solid-melt interfacial energy is not isotropic [*Cooper and Kohlstedt*, 1982; *Waff and Faul*, 1992].

The sample prepared by *Waff and Bulau* [1979] for which θ = 47° was examined at higher resolution using high-resolution transmission electron microcopy (HRTEM) and high-resolution analytical electron microscopy (HRAEM) techniques [*Cooper and Kohlstedt*, 1982; *Vaughan et al.*, 1982; *Kohlstedt*, 1990]. These high-resolution observations placed limits of <2 nm and <1 nm, respectively, for the maximum thickness of basaltic glass that could be present along the grain boundaries. A similar conclusion was reached using x-ray photoelectron spectroscopy (XPS) results obtained on grain boundaries of unaltered dunite and lherzolite xenoliths from a variety of lava flows [*Waff and Holdren*, 1981].

Experiments have also been carried out to determine the equilibrium melt distribution in aggregates of olivine plus a few percent of a carbonate melt [*Hunter and McKenzie*, 1989; *Watson et al.*, 1990]. For pressures between 0.5 and 3 GPa, temperatures of 1200-1400°C, and compositions including molten Na, Ca, K

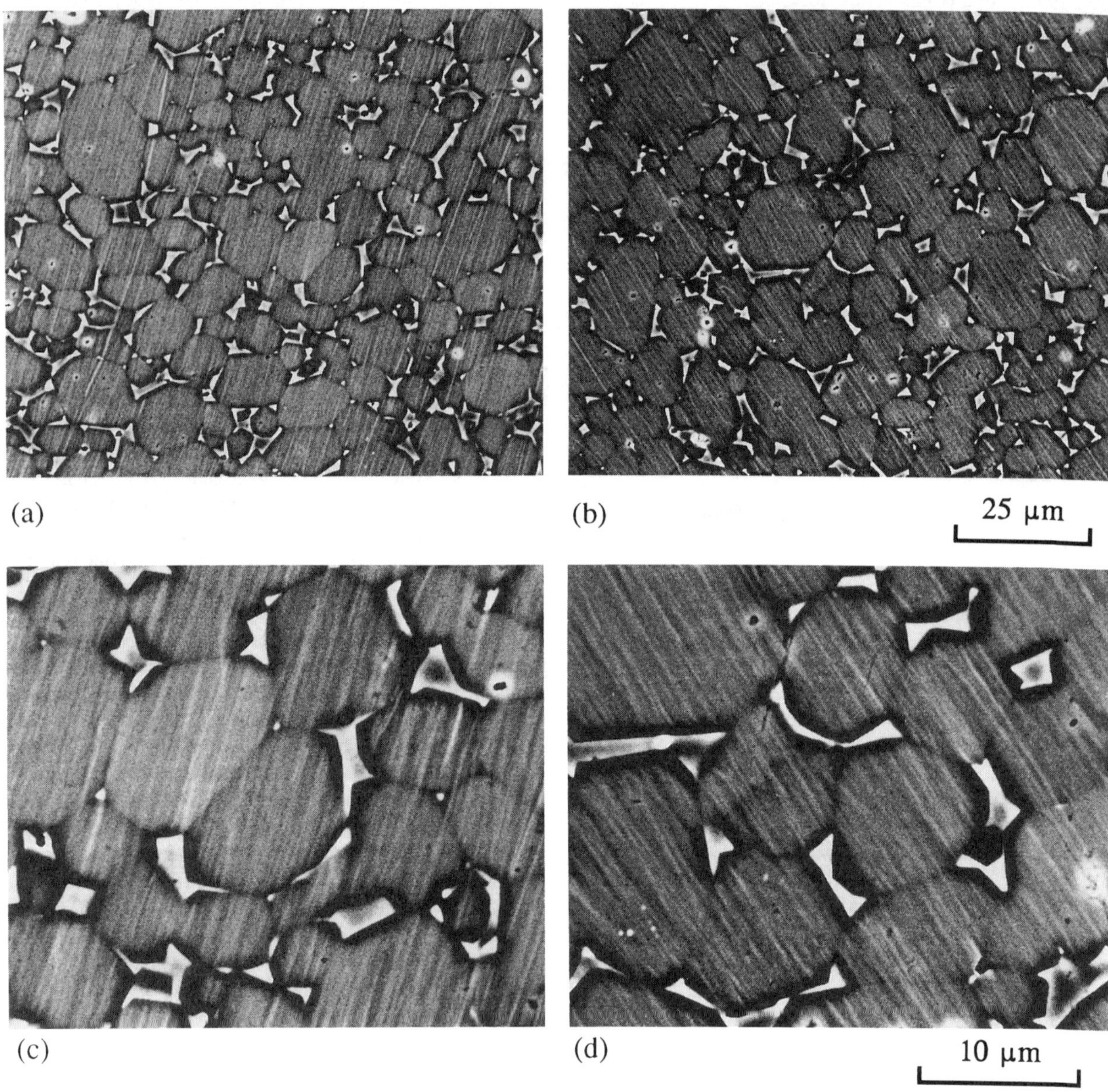

Fig. 3. Reverse-video, back-scattered electron micrographs of an aggregate of olivine plus ~7 wt% MORB, hydrostatically hot-pressed at 1300°C and 300 MPa. In the reverse-video mode, the bright regions are basalt and the gray regions are olivine; the small, black, nearly circular regions are pores. The black zone around a basalt pocket and the halo around a pore are artifacts associated with the back-scattering SEM imaging. Selected areas of (a) and (b), shown at higher magnification in (c) and (d), respectively, reveal that melt is present primarily in three- and four-grain junctions. The presence of melt in every triple junction demonstrates that melt indeed forms an interconnected network.

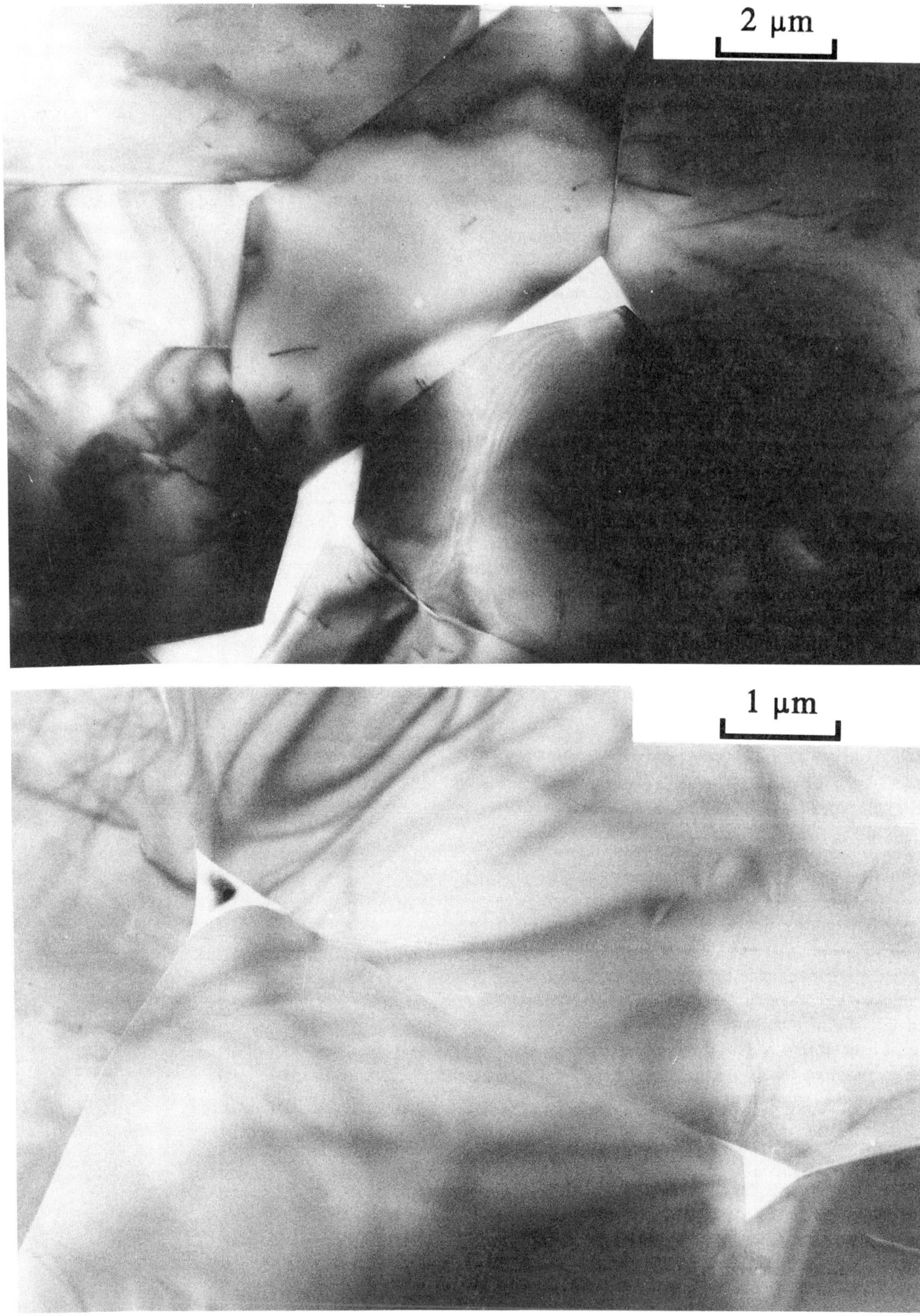

Fig. 4. Three transmission electron micrographs showing a higher resolution view of the distribution of basalt in the aggregate of olivine plus ~7 wt% MORB described in Figure 3. Images were formed primarily using electrons diffusely scattered from the glass; thus, the pockets of quenched melt appear as bright (roughly triangular-shaped) regions. The black regions within some of the triple junctions are holes introduced when the TEM samples were prepared by ion thinning.

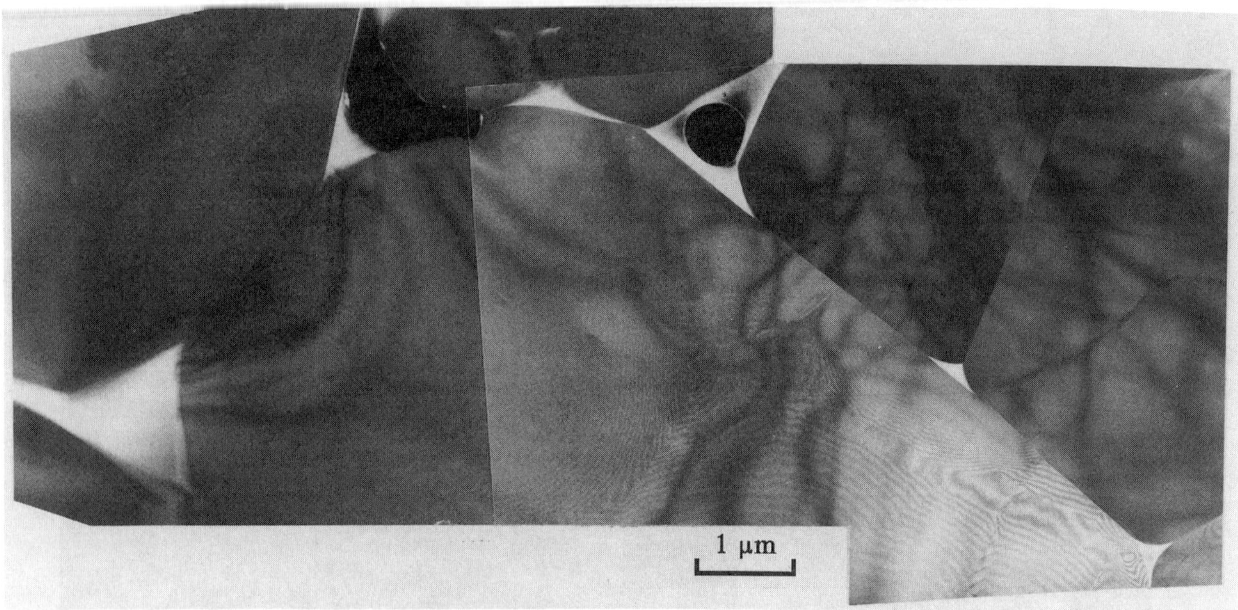

Fig. 4. Con't.

and Ca-Mg carbonates (with and without added H_2O or CO_2), the dihedral angle lay in a rather narrow range, $23 < \theta < 36°$. No systematic dependence of dihedral angle on temperature or pressure was observed.

Thus, for the temperatures, pressures and olivine-basalt or olivine-carbonate compositions investigated so far, the melt phase forms an interconnected network involving three-grain and four-grain junctions, but it does not form a thin film along grain boundaries or exist only in isolated pockets. This microstructure is schematically illustrated in Figure 5. Further verification that melt forms an interconnected network in olivine-basalt aggregates is provided by experiments in which Sm^{151} (a low-energy β emitter) that was allowed to diffuse into partially molten samples was detected using a radiographic technique in essentially all of the triple junctions [*Daines and Richter*, 1988]. This study further demonstrated that the melt network becomes interconnected at melt contents at least as low as 1 wt%.

It should be emphasized that, for any given crystalline phase, γ_{sm} depends upon the composition of the melt, as well as pressure and temperature. A dependence of dihedral angle on melt composition was noted above in a comparison of the experiments of *Waff and Bulau* [1979] with those of *Waff and Bulau* [1982]. Also, a recent investigation of the wetting of olivine by various silicate melts indicates that γ_{sm} decreases with decreasing silica content for Ca-bearing melts [*Wanamaker and Kohlstedt*, 1991], requiring a decrease in dihedral angle with decreasing silica content. In an extreme case, a Pb-rich silicate melt totally wetted the grain boundaries of olivine aggregates, such that $\theta = 0°$ [*Riley and Kohlstedt*, 1991]. However, based on presently available data for olivine plus melt of basaltic or carbonate compositions, this last situation is not expected to occur under geological conditions.

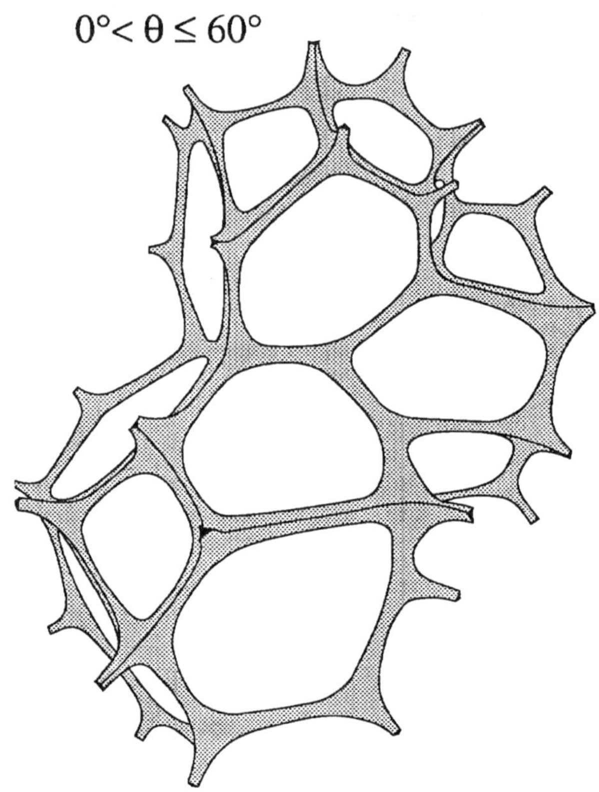

Fig. 5. Schematic drawing of melt distribution believed to characterize olivine-basalt and olivine-carbonate systems. Not shown here are small pockets of melt which may be trapped along grain boundaries. From *Riley and Kohlstedt* [1990].

In a partially molten polymineralic aggregate containing olivine and orthopyroxene with or without clinopyroxene, the dihedral angle for melt in contact with two olivine grains was significantly smaller than the dihedral angles formed for melt in contact with other solid phases [*Toramaru and Fujii*, 1986]. These authors concluded that the melt wetted only the olivine-olivine-olivine triple junctions and did not fully penetrate triple junctions involving one or more pyroxene grains. In experiments performed under dry conditions in an olivine-orthopyroxene-plagioclase system, melt was again excluded from triple junctions defined by orthopyroxene grains [*Fujii et al.*, 1986]. In contrast, if water was added, melt was present along all of the triple junctions in olivine-orthopryoxene-plagioclase and olivine-basalt systems [*Fujii et al.*, 1986; *von Bargen and Waff*, 1988]. Clearly the degree of interconnectivity of the melt phase depends critically on the composition of the melt (including the presence or absence of water) and the identity of the crystalline phases, as well as on pressure, temperature and other thermodynamic variables such as oxygen fugacity. (Note that the structure and composition of the melt depends on all of these variables.) Consequently, the permeability and rheology of partially molten mantle rocks are expected to be sensitive to these parameters, as well.

Finally, dihedral angles in the dunite plus H_2O-CO_2 system are consistently larger than 60° [*Watson and Brenan*, 1987; *Hunter and McKenzie*, 1989], indicating that these fluids will not form an interconnected network along triple junctions, at least not for small volume fractions of fluid. In this system, the dihedral angle increases monotonically from ~65° for pure CO_2 to ~90° for pure H_2O. (For complementary information on the distribution of aqueous and carbon dioxide fluids in crustal rocks, the reader is referred to papers by *Jurewicz and Watson* [1984], *Watson and Brenan* [1987], *Hay and Evans* [1988], and *Lee et al.* [1991].)

Equilibrium Melt Fraction. Values for the equilibrium melt fraction have been reported for partially molten synthetic rocks composed of olivine plus either a MORB ($\theta = 38°$, $\phi_{eq} = 0.023$) [*Riley and Kohlstedt*, 1992a], a K-rich melt ($\theta = 31°$, $\phi_{eq} = 0.145$), or a Pb-rich melt ($\theta = 0°$, $\phi_{eq} \geq 0.17$) [*Riley and Kohlstedt*, 1991]. In the experiments with the MORB and the K-rich melt, thin discs of fine-grained olivine were kept between discs of melt at elevated temperature and pressure until the melt fraction throughout the sample was uniform. In each case, the melt was saturated with olivine prior to the melt-fraction equilibration experiment. In the experiments with the Pb-rich melt, a disc of olivine was in contact with melt along only one of its flat surfaces; for this experimental arrangement, ϕ_{eq} was taken to be the melt fraction in the olivine disc at the contact between the melt and olivine. The results of this study are in qualitative, although not quantitative, agreement with the predictions of the three-dimensional model for melt fraction as a function of dihedral angle. As illustrated in Figure 2, the measured values for ϕ_{eq} are consistently smaller than the predicted values. The cause of the discrepancy between the experimental results and the theoretical predictions is not known. Part of the discrepancy may be due to the marked anisotropy in the crystal-melt interfacial energy that is manifested by the presence of not only curved but also flat crystal-melt interfaces, as observed in electron micrographs such as those in Figures 3 and 4 [see also, *Waff and Faul*, 1992]. However, examination of Figures 3 and 4 suggests that the melt fraction is larger in triple junctions containing planar interfaces than in those composed of only smoothly curved surfaces. It is also possible that the planar interfaces bias the measured dihedral angles toward higher values, leading to an artificially high value for the apparent dihedral angle. Also, as the samples are cooled to room temperature at the end of an experiment, some olivine may crystallize out of the melt, resulting in an under estimation of the melt fraction that was present at high temperature.

HIGH-TEMPERATURE CREEP BEHAVIOR

Theoretical Considerations

The high-temperature creep behavior (i.e., the rheology) of upper mantle rocks has been analyzed in terms of a power law creep equation for the strain rate, $\dot{\varepsilon}$, of the form

$$\dot{\varepsilon} = A \frac{\sigma^n}{d^r} f_{O_2}^m a_{opx}^q \exp\left(-\frac{(Q+PV)}{RT}\right) \qquad (2)$$

where A is a materials parameter, σ the applied differential stress, d the grain size, f_{O_2} the oxygen fugacity, a_{opx} the activity of orthopyroxene, Q the activation energy for creep, P the pressure, V the activation volume for creep, R the gas constant and T the absolute temperature [e.g., *Ricoult and Kohlstedt*, 1985; *Green and Borch*, 1987; *Bai et al.*, 1991]. The values of the exponents n, m, q, and r are determined by the particular diffusion or dislocation mechanism that rate limits the creep process. Typically, in dislocation creep regimes, $3 \leq n \leq 5$ and $r = 0$. Because there are presently very few experimental constraints on the values of m and q for creep of upper mantle rocks, only the dependencies of strain rate on differential stress, grain size and temperature will be considered in this paper. In addition, for the range of experimental conditions accessible in the laboratory, these latter parameters have much larger effects on strain rate than do f_{O_2} and a_{opx}.

In diffusional creep regimes, $n \approx 1$ and $r \neq 0$. If the creep rate is rate-limited by diffusion through the grains, then $r = 2$. If the creep rate is controlled by diffusion through the interface region between neighboring grains, then $r = 3$. If grain boundary diffusion provides the dominant path for mass transport but the rate of deformation is limited by the rate of an interface reaction, then $r = 1$. These dependencies of creep rate on differential stress and grain size are summarized in Table 1 and discussed in more detail below.

$\theta = 0°$: If the melt phase completely wets the grain boundaries of a partially molten rock (i.e., $\theta = 0°$), the presence of even a small amount of melt can result in a significant reduction in the high-temperature creep strength. Such a reduction in creep strength occurs primarily because diffusion through the thin film of melt which separates neighboring grains can be orders of magnitude faster than diffusion through a melt-free grain boundary.

TABLE 1. Model values for n and r in Equation 2.

Creep Mechanism	n	r	Reference
dislocation creep	3 - 5	0	*Poirier*, 1985, p.103
lattice diffusional creep	1	2	*Nabbaro*, 1949 *Herring*, 1951
grain-boundary diffusional creep	1	3	*Coble*, 1963 *Raj & Chyung*, 1981
interface-reaction-limited creep	1	1	*Raj & Chyung*, 1981

Two models have been developed to describe the rheological behavior of aggregates in which an amorphous phase forms a thin film that completely or, at least, almost completely separates neighboring grains [*Raj and Chyung*, 1981; *Pharr and Ashby*, 1983]. In addition to this microstructural constraint on the grain-scale distribution of the melt, these models require that the solid phases have at least a limited solubility in the amorphous (melt) phase, so that transport of the ions comprising the crystalline phases can indeed occur through the melt. Two constitutive equations result, one describing the case in which the rate of deformation is limited by diffusion through the melt and the other delineating the case in which the rate of creep is limited by a dissolution-precipitation (interface-reaction) step.

Diffusion-controlled creep: If ionic diffusion through the melt is slower than the rate at which ions from the grains dissolve into or precipitate out of the melt, then the rate of mass transport from regions under a relatively high compressive stress to regions under a lower compressive or tensile stress will control the creep rate. The deformation process, then, is similar to that described by *Coble* [1963] for diffusional creep rate-limited by transport of ions through grain boundaries (frequently referred to as 'Coble creep'). In fact, simple modifications of the Coble creep equation have been proposed by several authors as zeroth-order analyses of the pressure solution or fluid-phase-enhanced creep process [*Elliot*, 1972; *Stocker and Ashby*, 1973; *Rutter*, 1976; *Durney*, 1976]. However, both *Raj and Chyung* [1981] and *Pharr and Ashby* [1983] pointed out that at least some grain-to-grain contact is needed in order for a material not to disaggregate in the presence of a wetting ($\theta = 0^o$) fluid. *Raj and Chyung* [1981] dealt with this problem by invoking the existence of an island structure (see the section on equilibrium melt distribution), while *Pharr and Ashby* [1983] handled it by requiring the dihedral angle be small but non-zero. Emphasis below will be given to the former model, and the reader is referred to the latter publication for a detailed discussion of the differences between the two approaches; the resulting flow laws are similar.

The solution-precipitation creep model presented by *Raj and Chyung* [1981] largely follows the grain-boundary diffusional creep model developed by *Coble* [1963], starting with the following Coble creep equation:

$$\dot{\varepsilon} = 70 \frac{\sigma \Omega}{kT} \frac{\delta D_b}{d^3} \tag{3}$$

where Ω is the molecular volume of the crystalline phase, δ the width of the grain boundary, D_b the grain boundary diffusion coefficient, and k Boltzmann's constant. To obtain a creep equation for solution-precipitation or fluid-phase-enhanced creep, imagine that the grain boundary is replaced by a thin film of melt of thickness w (i.e., let $\delta \rightarrow w$). To avoid disaggregation of the crystalline grains into the melt and to enable the grain boundaries to sustain a gradient in normal stress, the island structure described earlier is invoked such that a fraction $(1\text{-}x)$ of the grain boundary is occupied by crystalline material and a fraction x by melt. This modification has two consequences: First, the effective thickness of the grain boundary is reduced from w to xw (i.e., $\delta \rightarrow w \rightarrow xw$); second, because all of the applied load is assumed to be carried by the crystalline islands and none by the melt, the local stress on the islands is $\sigma/(1\text{-}x)$ (i.e., $\sigma \rightarrow \sigma/(1\text{-}x)$) such that there is a stress-enhancement factor relative to the melt-free case due to the presence of melt in the grain boundaries). Finally, the grain boundary diffusivity is replaced by the effective diffusivity of the solid through the melt phase, $C_m D_m$, where C_m is the solubility or molar (fractional) concentration of the crystalline phase in the melt and D_m is the diffusivity of the solid through the melt (i.e., $D_b \rightarrow C_m D_m$). Combined, these modifications yield the following flow law for solution-precipitation creep for the case in which the rate of deformation is limited by diffusion of the ions comprising the crystalline phase through the melt which is present in the grain boundaries [*Raj and Chyung*, 1981]:

$$\dot{\varepsilon} = 70 \frac{\sigma \Omega}{(1-x)\, kT} \frac{x w C_m D_m}{d^3} \quad . \tag{4}$$

Because diffusion through the melt is typically much faster than diffusion through a melt-free grain boundary, that is, $C_m D_m >> D_b$, the rate of creep predicted by the solution-precipitation model (Equation 4) can be orders of magnitude faster than that expected for Coble creep (Equation 3).

Interface-reaction-controlled creep: If the rate of diffusion through the melt film is very rapid and/or the diffusion path is very short, then the rate of deformation can become limited by the rate at which the crystalline phase dissolves into the melt in regions of high compressive stress or the rate at which molecules precipitate out of the melt in areas of low compressive or tensile stress; that is, the rate of deformation becomes controlled by an interface reaction. *Raj and Chyung* [1981] have treated this problem in terms of the growth rate of a crystal interface in a liquid (melt), k_m, which corresponds to the velocity at which the interface would advance under a driving force equal to one kT. Their analysis yields an equation for interface-reaction-controlled creep of the form

$$\dot{\varepsilon} = \frac{\sigma \Omega}{2\,(1-x)\, kT} \frac{k_m}{d} \quad . \tag{5}$$

Again, in this equation, $\sigma \rightarrow \sigma/(1\text{-}x)$ because all of the applied load is carried by the crystalline islands and none by the melt.

The gradient in chemical potential which drives the deformation process is given by $\sigma\Omega/d$, such that the normalized thermodynamic driving force becomes $\sigma\Omega/(2kT)$. The factor of 1/2 arises because the drop in chemical potential due to dissolution and that due to precipitation are assumed to be equal. While the equations for diffusion-controlled creep and interface reaction-controlled creep yield the same dependence of creep rate on differential stress, they predict very different dependencies on grain size (i.e., d^{-3} vs d^{-1}). Thus, in analyses of experimental results, the dependence of strain rate on grain size will provide the primary method for distinguishing between the two creep mechanisms.

$0 < \theta \le 60°$: If the melt forms an interconnected network along triple and four-grain junctions but does not wet the grain boundaries (i.e., $0 < \theta \le 60°$), the enhancement in creep rate due to the presence of a small amount of melt is expected to be much smaller than if $\theta = 0°$. Two cases can be distinguished, one in which diffusional processes dominate and the other in which dislocation mechanisms dominate.

Diffusion-dominated creep: The presence of melt in all of the triple junctions enhances the rate of Coble (grain boundary diffusion) creep for two reasons [*Cooper et al.*, 1989]. First, as in the model proposed by *Raj and Chyung* [1981], the local grain-scale stress is larger than the applied stress, because at a grain boundary the load is carried only by the area of grain-to-grain contact. The introduction of melt reduces this area by an amount that depends upon both the melt fraction and the dihedral angle. For an aggregate constructed from Kelvin's three-dimensional, space-filling tetrakaidecahedrons, the stress enhancement factor causes

$$\sigma \rightarrow \sigma\left[1 - \alpha\,\phi^{1/2}\,\frac{\sin(\pi/6 - \theta/2)}{f(\theta)}\right]^{-2} \tag{6}$$

where α is a geometrical constant ≈ 1 and

$$f(\theta) = \left[\frac{(1+\cos\theta)}{\sqrt{3}} - \sin\theta - (\pi/3 - \theta)\right]^{1/2} . \tag{7}$$

Second, the melt-filled triple junctions provide a short-circuit diffusion path for ions moving from grain boundaries under the greatest compressive stress to grain boundaries under the least compressive stress. As derived by *Cooper et al.* [1989], the low resistance of the triple junction to ionic diffusion introduces a strain rate enhancement factor identical to that given by Equation 6 for the stress enhancement factor. Hence, the combined effect on the creep rate of the stress enhancement and the short-circuit diffusion is

$$\frac{\dot{\varepsilon}_2}{\dot{\varepsilon}_1} = \left[1 - \alpha\,\phi^{1/2}\,\frac{\sin(\pi/6 - \theta/2)}{f(\theta)}\right]^{-4} . \tag{8}$$

where $\dot{\varepsilon}_2$ is the strain rate of the partially molten (two-phase) aggregate and $\dot{\varepsilon}_1$ is the strain rate of the single-phase material.

In a partially molten with a the dihedral angle of 25°, the strain rate is increased, relative to the strain rate of a melt-free aggregate, by factors of approximately 2, 3, 4 and 5 with the addition of 1, 3, 5 and 7 vol% melt, as illustrated in Figure 6. It should be noted that this model assumes that the rate of deformation is limited by diffusion through melt-free grain boundaries and, thus, for example, the activation energy for creep should be identical to that obtained for Coble creep of a melt-free rock, provided either that the melt fraction does not change with changing temperature or that the effect of increasing (decreasing) melt fraction with increasing (decreasing) temperature is accounted for properly.

Dislocation-controlled creep: Within the dislocation (power-law) creep regime, the addition of a small amount of melt to an initially melt-free rock can also result in an increase in the rate of deformation at a fixed applied stress. This effect arises because the melt phase is so much weaker than the solid phase that the latter must support essentially the entire shear stress. Both an analytical technique [*Chen and Argon*, 1979] and a finite element method [*Tullis et al.*, 1991] have been used to treat the stress and strain rate distributions in the deforming phases in order to calculate the flow law for a two-phase material from those of the component phases. The finite element model demonstrates that the most important factor in determining the flow stress is the volume fraction of each phase and that the shape of the inclusions is less important for weak inclusions (such as melt, in the present analysis) than for strong inclusions. In a low-stress limit, the analytical model yields the following expression for the normalized strain rate of an aggregate containing a volume fraction ϕ of spherically shaped domains of a weak phase:

$$\frac{\dot{\varepsilon}_2}{\dot{\varepsilon}_1} = \left(\frac{3}{3 - 5\phi}\right)\left(\frac{1}{1 - \phi}\right)^{n-1} . \tag{9}$$

A similar solution was obtained for cylindrical domains.

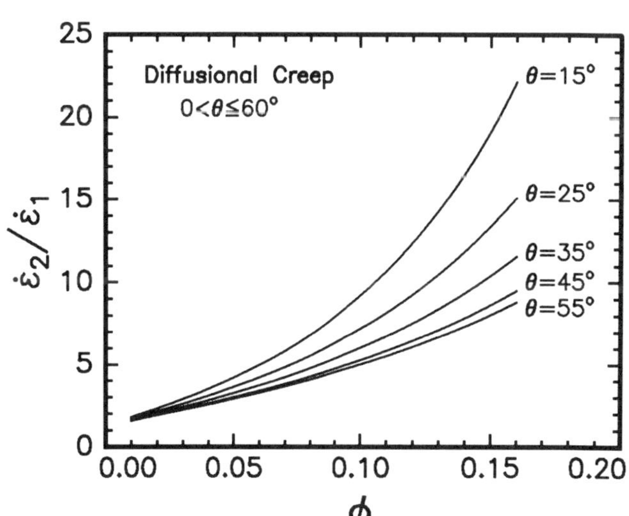

Fig. 6. Plot of strain rate of a partially molten aggregate relative to that of a single-phase aggregate, $\dot{\varepsilon}_2/\dot{\varepsilon}_1$, versus melt fraction, ϕ, for several values of dihedral angle in the range $0 < \theta \le 60°$ [*Cooper et al.*, 1989].

Equation 9 strictly applies only when both phases are undergoing power-law creep with stress exponents greater than one. If this model is applied to the case of a partially molten rock for which n = 3 for the crystalline phase, it predicts that the strain rate of a rock containing 10 vol% melt should be only a factor of ~1.5 larger than the strain rate of a melt-free aggregate.

Experimental Observations

Published experimental studies on the effect of melt or liquid on the creep behavior of materials are rather few in number and, in general, somewhat incomplete. In the summary presented below, the implications as well as the limitations of each suite of experiments are discussed.

$\theta = 0^o$. *Diffusion-dominated creep:* Possibly the first experimental evidence for a large reduction in creep strength (i.e., enhancement in creep rate) with the addition of a small amount of a melt phase came from studies on Si_3N_4 and MgO-fluxed Si_3N_4, ceramics which are used in high-temperature structural applications. High-resolution transmission electron microscopy observations reveal the presence of a thin layer of glass between the Si_3N_4 grains when MgO is added to Si_3N_4 [*Clarke*, 1987]. Creep experiments demonstrate that the two-phase aggregates deform a factor of $>10^3$ faster than the single-phase material [*Osborne*, 1975; see also Figure 6 of *Raj et al.*, 1984]. Experiments on similar two-phase samples yielded stress exponents of $n \sim 1$ for samples containing ~5 vol% glass and $n \sim 2$ for samples with 11-17 vol% glass [*Lange et al.*, 1980]. The former result was explained in terms of a liquid-phase-enhanced diffusional creep model and the latter in terms of cavitational creep mechanisms. However, the dependence of strain rate on grain size and temperature were not determined, making it impossible to distinguish between diffusion-limited and interface-reaction-limited creep. A more complete series of experiments was carried out on commercially available lithium aluminosilicate β-spodumene glass-ceramics. Two compositions were examined in some detail, one containing ~70 wt% SiO_2 (referred to here as 'high-silica') and the other containing ~65 wt% SiO_2 ('low-silica'). At 1223 K, the high-silica material displayed a linear relationship between strain rate and stress ($n \approx 1$) as well as between strain rate and inverse grain size ($r \approx 1$) [*Raj and Chyung*, 1981]. At 1410 K, the stress exponent remained at $n \approx 1$, but the grain-size exponent became $r \approx 3$ [*Wang and Raj*, 1984]. This change in the dependence of strain rate on grain size was attributed to a change from interface-reaction-controlled creep to diffusion-controlled creep with increasing temperature. In the higher temperature regime, for a given temperature and grain size, the low-silica samples, which also exhibited $n \approx 1$, deformed about a factor of 10^2 faster than the high-silica samples; no value was reported for r. The activation energies for creep were similar for the two materials. The authors imply that the rate of creep of the low-silica samples was also limited by enhanced diffusion through a liquid phase; and they suggest that the difference in creep strength between the two materials is due to a difference in viscosity, η, of the melt phase since from the Stokes-Einstein equation $\eta \propto D_m^{-1}$ in Equation 4. It should be noted that they present no direct observations of the microstructure (e.g., HRTEM) to verify that the melt indeed wetted the grain boundaries (or produces an island structure). However, the dramatic effect of melt composition on creep behavior strongly indicates that it does, at least in the low-silica samples. In fact, it has been argued that the morphology of the glass in the triple junctions of the silica-rich β-spodumene is consistent with a melt that forms an interconnected network along the triple junctions but does not wet the grain boundaries (i.e., $0 < \theta \leq 60^o$) (see in particular Figure 2 of *Cooper et al.* [1989]). No data are available on single-phase samples for comparison with the results on the partially melted materials.

An investigation of the creep behavior of KCl and sugar under both wet and dry conditions again demonstrates the marked effect that a liquid phase can have on creep strength [*Pharr and Ashby*, 1983]. For both materials, the addition of a solid-saturated liquid phase resulted in an increase in creep rate of $>10^3$ and in a transition from dislocation creep ($n \geq 3$) to diffusional creep ($n \approx 1$). Creep rate also increased linearly with increasing solubility of the solid phase in the liquid (water plus alcohol), consistent with the predictions of Equation 4. The temperature dependence of the creep rate and of the ratio C_m/η (i.e., $C_m D_m$ in Equation 4) were shown to be very similar. Although no direct observation of the grain-scale distribution of liquid was possible in these samples and the dependence of strain rate on grain size was not investigated, the mechanical results strongly suggest that the liquid wetted the grain boundaries resulting in a solution-precipitation creep process similar to that described by *Rutter* [1976], *Raj and Chyung* [1981], and other authors referenced in the previous section.

One of the most complete studies of liquid-phase-enhanced creep for a material in which the melt forms a thin film along the grain boundaries was carried out on fine-grained ($0.1 < d < 1$ μm) Y_2O_3-stabilized tetragonal ZrO_2 with and without a small amount (0.3 to 1 mol%) of CuO [*Hwang and Chen*, 1990]. The presence of a thin (1- to 2-nm thick), Cu-rich grain-boundary phase was verified by HRTEM and ESCA (electron spectroscopy for chemical analysis). The eutectic temperature, T_e, for this system is ~1130°C; below T_e the grain-boundary phase is an amorphous solid, while above T_e it is a melt. The dependencies of strain rate on stress, grain size and temperature are summarized in Table 2 for single-phase and two-phase samples containing 1 mol% CuO. The addition of even a small amount of CuO resulted in an increase in strain rate by a factor of $>10^2$, as well as by marked

TABLE 2: Values of n, r and Q for ZrO_2 + 0.3% CuO.

Sample Type	n	r	Q (kJ/mol)
melt-free samples	1.5 ± 0.1	2.7 ± 0.2	630
two-phase samples $T < T_e$	1.3 ± 0.1	1.6 ± 0.1	445
two-phase samples $T > T_e$	2.0 ± 0.1	1.2 ± 0.2	416

decreases in the grain-size dependence of the creep rate and in the activation energy. A discernable, but less pronounced, change was also observed in the stress exponent. No precipitous decrease in creep strength occurred in crossing the eutectic temperature. Based on these observations, the authors argued that the dominant deformation mechanism switched from diffusional creep controlled by grain-boundary transport (i.e., Coble creep) for the single-phase samples to liquid-phase-enhanced creep for the two-phase aggregates. Further, for the two-phase materials, they suggested that the creep rate was diffusion-controlled for $T < T_e$ but became interface-controlled for $T > T_e$ because diffusion through the melt is so fast. This point is supported by the observation that for $T < T_e$ the strain rate is more sensitive to the amount of Cu-rich phase present in the sample than for $T > T_e$. More convincing evidence for a change in mechanism associated with crossing the eutectic temperature is found in the significant increase in activation energy (233 kJ/mol → 405 kJ/mol) as the temperature is increased for samples containing 1 mol% CuO. The authors do not present any explanation for the differences between the experimentally measured values of n and r (Table 2) and those predicted by the models (Table 1). These discrepancies may indicate that more than one creep mechanism is contributing significantly to the deformation process [*Dimos et al.*, 1988].

Microcracking and failure: In compressive creep experiments on a Cu-Bi alloy (ϕ = 9.2 vol%, θ = 0°) for which over 95% of the grain boundary area was wetted by a thick (>1 μm) layer of an intergranular phase, the creep rate was a factor of 10^3 larger for temperatures slightly above the eutectic temperature than for temperatures just below the eutectic temperature [*Pharr et al.*, 1989]. In deformation experiments carried out at $T > T_e$, samples disintegrated into individual copper grains after a few percent strain. The authors associated the enhanced strain rates with rapid rates of grain boundary sliding. The grain boundary sliding led to the development of large internal tensile stresses, which resulted in intergranular cracking, disaggregation, and subsequent failure. Similar observations and conclusions were reported for deformation of KCl plus water at high compressive stresses [*Sheikh and Pharr*, 1985]. Likewise, the accelerating creep rates observed near the end of earlier experiments on KCl [*Pharr and Ashby*, 1983] may also indicate that microcracking was contributing significantly to the deformation process.

$0 < \theta \leq 60°$. *Diffusion-dominated creep:* Experiments on aggregates of olivine plus ≤10 wt% basalt, for which $30 < \theta < 45°$, have been carried out in the diffusional creep regime [*Chopra and Kohlstedt*, 1983; *Cooper and Kohlstedt*, 1984; 1986]. As summarized in Table 3, for both the single-phase and the two-phase samples, the stress exponents were $n \approx 1$, the grain size exponents were $r \approx 3$, and the activation energies were $Q \approx 380$ kJ/mol [*Cooper and Kohlstedt*, 1984]. These observations suggest that both types of samples deform by the same mechanism, namely via diffusional creep controlled by transport along grain boundaries; HRTEM and HRAEM demonstrate that the grain boundaries are indeed free of melt [*Vaughan et al.*, 1982; *Kohlstedt*, 1990]. However, the melt-bearing samples deformed a factor of 2 to 5 faster than the melt-free samples, as illustrated in Figure 7. This relatively small enhancement in creep rate was quantitatively modeled, Figure 6, by noting that the melt in the triple junctions (i) provides a short-circuit segment to the diffusion path for ions moving from grain boundaries under the maximum compression stress to those under the minimum compressive stress and (ii) results in a local enhancement in stress [*Cooper et al.*, 1989].

TABLE 3: Values for n, r and Q for olivine-rich aggregates.

Sample Type	n	r	Q (kJ/mol)
melt-free samples	1.1 ± 0.3	3.0 ± 0.2	380 ± 105
two-phase samples	0.9 ± 0.2	2.9 ± 0.4	385 ± 125

Dislocation-controlled creep: Although no measurements of n or r were reported to verify the mechanism of deformation, due to the relatively large grain sizes (100 - 200 μm) involved, a series of experiments on two-phase metallic systems [*Pharr et al.*, 1989] were probably carried out in the dislocation creep regime. For Al-Sn materials (θ = 58°) containing 3.3 vol% of an intergranular phase, data from samples deformed both above and below the eutectic temperature defined a single, continuous line on a plot of log creep rate versus inverse temperature. Creep data for pure Al samples fall on the same line. Similar behavior was observed in a comparison of a two-phase Al-Bi-Ti alloy (ϕ = 0.8 vol%, θ = 104°) with a single-phase Al-Ti alloy. Based on the model of *Chen and Argon* [1979] described in the previous section, these small amounts of melt would be expected to produce less than a factor of 1.15 enhancement in creep rate, a value well below the resolution of the experiments.

A much larger (factor of 10^4) decrease in viscosity has been reported for deformation of a partially molten (ϕ < 10%) spinel lherzolite for which n = 3 [*Borch and Green*, 1990]. The authors observed that melt was present on virtually all of the grain

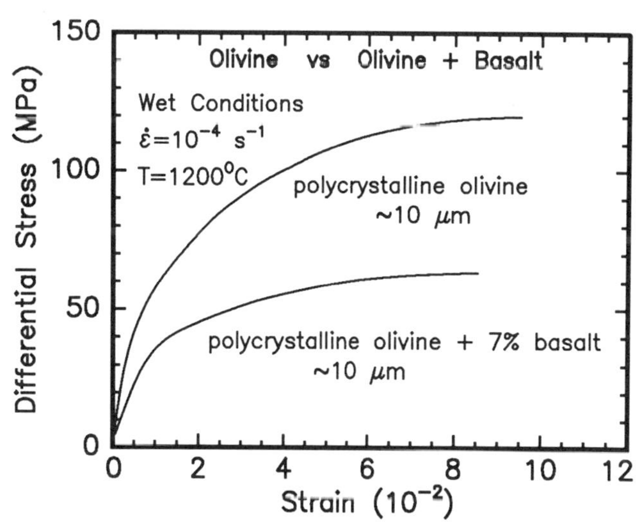

Fig. 7. Comparison of strain versus time (creep) curves for olivine and olivine plus ~7 wt% MORB. After *Chopra and Kohlstedt* [1983].

boundaries and that dynamic recrystallization was rapid. A similar microstructure and dynamic recrystallization process, but much smaller decrease in creep strength (factor of ~4), was found in samples of bischofite deformed in the presence of water [*Urai*, 1983].

MELT MIGRATION

Theoretical Considerations

Migration of melt (permeability) through a partially molten rock in response to buoyancy forces has been analyzed theoretically by an number of authors by treating the two-phase (solid-melt) system as a saturated porous medium. The first treatments of this problem used Darcy's law to describe the velocity of the fluid relative to the matrix in terms of the effective pressure gradient [e.g., *Frank*, 1968; *Ahern and Turcotte*, 1979; *Stolper et al.*, 1981; and references therein]. For one-dimensional flow in the z direction, Darcy's law can be written as

$$\frac{\partial p}{\partial z} = -\Delta\rho\, g = -\frac{\mu\phi}{K}(w - W) \tag{10}$$

where p is the pressure in the melt, $\Delta\rho$ the difference in density between the solid and the melt, g the acceleration due to gravity, μ the viscosity of the melt, K the permeability of the saturated porous medium, w the velocity of the melt, and W the velocity of the solid. This relationship applies provided that there is no net increase or decrease in melt fraction in the volume of rock under consideration.

To account for the situation in which the melt fraction is not constant with time, subsequent analyses have incorporated equations to describe deformation of the solid (matrix), a process that is necessary for melt to segregate out of or infiltrate into a rock [e.g., *Sleep*, 1974; *McKenzie*, 1984; *Scott and Stevenson*, 1986; *Ribe*, 1987; *Fowler*, 1990; and references therein]. If matrix deformation is included, one-dimensional melt migration can be described by four coupled differential equations [e.g., *Richter and McKenzie*, 1984]. The first two equations describe conservation of mass for the melt and the matrix, respectively, and can written in a form analogous to Fick's second law applied to solid-state diffusion:

$$\frac{\partial\phi}{\partial t} + \frac{\partial(\phi w)}{\partial z} = \frac{\partial\phi}{\partial t} + \frac{\partial J_m}{\partial z} = 0 \tag{11}$$

$$-\frac{\partial\phi}{\partial t} + \frac{\partial[(1-\phi)W]}{\partial z} = \frac{\partial(1-\phi)}{\partial t} + \frac{\partial J_s}{\partial z} = 0 \tag{12}$$

where J_m is the flux of melt and J_s the flux of solid measured in units of volume of melt per unit area per unit time. The quantity $(1 - \phi)$ in Equation 12 is the volume fraction of solid material in the partially molten aggregate. Equations 11 and 12 provide the necessary coupling between the flux of melt and the flux of solid. The second two equations characterize conservation of momentum for the melt and the solid phases, respectively:

$$\frac{\partial p}{\partial z} + \frac{\mu}{K}\phi(w - W) = 0 \tag{13}$$

$$\frac{\partial p}{\partial z} + (1-\phi)\Delta\rho\, g - \left(\xi + \frac{4}{3}\eta\right)\frac{\partial^2 W}{\partial z^2} = 0 \tag{14}$$

where ξ and η are the bulk and shear viscosities of the matrix, respectively. Equation 13 is Darcy's law relating the gradient in pressure in the melt to the velocity of the melt relative to that of the solid. Equation 14 gives the pressure gradient in the melt associated with the buoyancy force and the deformation of the solid (matrix). Note that if the solid phase is easily deformed (i.e., ξ and η are small), then Equations 13 and 14 can be combined to yield (for small melt fractions) the familiar form of Darcy's law given by Equation 10. To complete this system of equations, the permeability is generally written as a power-law function of the melt fraction:

$$K = \frac{d^2\phi^s}{b} \tag{15}$$

where b is a geometrical constant determined by the details of the distribution of the interconnected melt phase [e.g., *Turcotte and Schubert*, 1982, pp. 370-384].

For melts in chemical and textural equilibrium with the matrix, melt migration can be induced not only by a gravitational (buoyancy) force but also by a gradient in melt fraction (a capillary force) and a gradient in mean normal stress. Although buoyancy is clearly very important under geological conditions, it produces only an extremely small thermodynamic driving force in the relatively small samples used in laboratory experiments [e.g., *Cooper*, 1990]. Consequently, experiments designed to investigate the kinetics of melt migration via porous flow have concentrated on these last two thermodynamic driving forces.

An analysis of the role of surface tension (capillarity) as a thermodynamic driving force for melt migration concluded that the gradient in pressure in a melt due to a gradient in melt fraction could be derived by considering the surface energy (strictly, interfacial energy) of the two-phase system and arrived at the relationship [*Stevenson*, 1986]

$$\frac{\partial p}{\partial z} = \frac{\partial}{\partial z}\left(-\frac{\partial G_v}{\partial\phi}\right) . \tag{16}$$

The interfacial free energy per unit volume, G_v, incorporates contributions from both the solid-solid and the solid-liquid interfaces. Note that the term $\partial G_v/\partial\phi$ is the interfacial energy contribution to the chemical potential of the melt, such that Equation 16 expresses the thermodynamic driving force in terms of a gradient in chemical potential.

For columnar grains with hexagonal cross sections, the capillary driving force term is given by [*Riley et al.*, 1990; *Riley and Kohlstedt*, 1992b]

$$\frac{\partial}{\partial z}\left(-\frac{\partial G_v}{\partial\phi}\right) = \frac{H(\theta)\gamma_{sm}}{\sqrt{3}\,d}\phi^{-3/2}\frac{\partial\phi}{\partial z} . \tag{17}$$

The geometrical term $H(\theta)$ characterizes the relative concentration of melt-solid and solid-solid interfaces as a function of dihedral angle and is given by [*Riley et al.*, 1990]

$$H(\theta) = \frac{[1/2(\pi - 3\theta) - 2\sqrt{3}\sin(\pi/6 - \theta/2)\cos(\theta/2)]}{[\cos^2(\theta/2) - \sqrt{3}\pi/6 + \sqrt{3}/2\,(\theta - \sin\theta)]^{1/2}} \quad . \tag{18}$$

An analysis of the role of a gradient in mean normal stress as a thermodynamic driving force for melt migration was recently carried out for the case of four-point bending (flexural creep) of partially molten samples [*Cooper*, 1990]. For fine-grained aggregates in which matrix deformation occurs via Newtonian flow ($\dot{\varepsilon} \propto \sigma^1$), this experimental geometry yields a particularly simple, well-defined state of stress with $\sigma_2 = \sigma_3 = 0$ and σ_1 a linear function of z. The pressure gradient in the melt under four-point loading, flexural creep conditions is

$$\frac{\partial p}{\partial z} = -\frac{\partial\,[(\sigma_1 + \sigma_2 + \sigma_3)/3]}{\partial z} = -\frac{2L(l-a)}{h_1^3\,h_2} = \frac{2\sigma_{\max}}{3h_1} \tag{19}$$

where the load L is applied along the z-axis, l is the length of the specimen supported on the outer load points, a is the length between the inner load points, and h_1 and h_2 are the thickness and width of the specimen, respectively (see Figure 2 of *Cooper* [1990]); σ_{max} is the magnitude of the maximum principal stress.

Equation 14 can now be modified to include a gradient in melt fraction and a gradient in mean normal stress as thermodynamic driving forces for melt migration. Under laboratory conditions for which the buoyancy term is small, Equation 14 becomes

$$\frac{\partial p}{\partial z} + \frac{H(\theta)\gamma_{sm}}{\sqrt{3}\,d}\phi^{-3/2}\frac{\partial \phi}{\partial z} + \frac{2L(l-a)}{h_1^3\,h_2} - \left(\xi + \frac{4}{3}\eta\right)\frac{\partial^2 W}{\partial z^2} = 0 \tag{20}$$

First, for melt migration driven by a gradient in melt fraction (L=0 in Equation 19), comparison of Equations 13 and 20 indicates that, during infiltration of melt, viscous flow of the melt is associated with deformation of the solid; the latter is required to provide space for the increased amount of melt. Second, for melt migration driven by a gradient in mean normal stress induced by four-point bending, segregation of melt from regions of compressive stress to regions of tensile stress is resisted by deformation of the matrix which is necessary to compact the aggregate in regions which are losing melt and dilate it in areas that are gaining melt. If the load applied during flexural creep is held constant, then melt segregation will proceed until the thermodynamic driving force resulting from the gradient in melt fraction which develops across the sample is balanced by the driving force caused by the applied bending stress. If the applied load is then removed, the gradient in melt fraction that developed during deformation will cause melt to migrate back across the sample until the melt fraction is uniform throughout. Thus, the combination of the gradient in mean normal stress and the gradient in melt fraction results in anelastic (time dependent, but reversible) deformation.

Experimental Observations

Melt Migration Driven by a Gradient in Melt Fraction. Several investigations of melt migration driven by a gradient in melt fraction have been reported. In an early study on magnesia and forsterite at ambient pressure, silicate melts were observed to permeate only into aggregates with significant amounts of porosity ($\phi > 0.10$ for magnesia and $\phi > 0.15$ for forsterite), even though the dihedral angle was significantly less than 60° in both cases [*Raju et al.*, 1973]. This result should be considered with some skepticism, however, because the crystalline aggregates were not in chemical equilibrium with the silicate melts at the start of the permeation experiments. Consequently, solid phases that precipitated out of the melt near the melt-solid interface complicated the melt-infiltration process. Also, since large amounts of melt migrated into the high-porosity samples, a small amount of melt infiltration in the low-porosity samples may well have escaped detection by optical microscopy.

A subsequent study at ambient pressure of infiltration of a silicate melt into alumina resulted in the development of two regions with distinctly different melt distributions [*Flaitz and Pask*, 1987]. In the two-phase regions farthest from the initial melt-solid interface, melt was observed only in triple junctions. However, near the melt-solid interface, samples had started to disaggregate such that individual grains of alumina were surrounded by melt. These grains tended to have well-developed, planar surfaces, suggesting that they had grown by a solution-precipitation mechanism. Similar results were obtained with melts that were saturated with alumina before the infiltration experiment and with systems that were not initially in chemical equilibrium. The authors suggest that the loose grains became detached from the alumina aggregates as stresses developed during the grain-growth process. This microstructure, however, was not observed in the melt migration experiments described below which were carried out at high confining pressures.

At least three studies have been carried out at pressures ≥0.3 GPa to investigate melt infiltration in response to a gradient in melt fraction (i.e., a capillary force). Melt migration couples -- similar to diffusion couples used in interdiffusion experiments -- were formed using a disc of $Fo_{23}Ab_{39}An_{38}$ melt in contact with a disc of crystalline Fo_{100} [*Watson*, 1982] and a disc of $Ab_{50}Qtz_{50}$ against a disc of Qtz_{100} [*Jurewicz and Watson*, 1984]. In the experiments with forsterite for which $\theta < 60°$, the initial melt-solid boundary remained distinct and the maximum distance of melt infiltration increased with increasing time. Melt migration resulted in a diffusion-like penetration profile, rather than a well-defined infiltration front. In experiments with quartz ($\theta \approx 60°$), no melt migration could be detected. This observation is consistent with the fact that the capillary driving force goes to zero as the dihedral angle increases to 60° (i.e., $H(\theta) \rightarrow 0$ as $\theta \rightarrow 60°$ in Equation 20).

In contrast, in experiments on quartzite plus CO_2 and dunite plus CO_2, both systems with $\theta > 60°$, fluid was observed to penetrate the aggregates along intergranular cracks [*Brenan and Watson*, 1988]. Possibly the fluid (pore) pressure exceeded the confining pressure in these experiments.

To investigate the kinetics of melt migration in aggregates with olivine as the crystalline phase, melt migration couples were formed between either a disc of melt or a disc of olivine plus melt and a disc of olivine [*Riley and Kohlstedt*, 1990; *Riley et al.*, 1990; *Riley and Kohlstedt*, 1992b]. After annealing at high pressure and temperature, the resulting melt-migration profile (melt concentration as a function of position) was determined with an electron microprobe by mapping x-ray emissions from one of the elements (e.g., potassium) that partitions primarily into the melt. A melt-migration profile for a couple composed of a disc of olivine plus ~15 vol% of a K-Al-rich melt and a disc of olivine is presented in Figure 8. A melt-migration profile for a couple prepared from a disc of K-rich melt and a disc of olivine is displayed in Figure 9 [see also, *Riley and Kohlstedt*, 1992a]. In both cases, melt has been drawn by capillary forces out of the melt-rich side (the source) into the initially melt-free side (the sink) of the couple, yielding a melt distribution which decays approximately exponentially with increasing distance into the sink region.

To extract the permeability, melt-migration profiles determined from the experimental data obtained using partially molten aggregates as the source for melt (e.g., Figure 8) were compared to simulated profiles generated from numerical solutions of non-dimensional forms of the Equations 11-13 and Equation 20 (with L=0) [*Riley et al.*, 1990; *Riley and Kohlstedt*, 1992b].† Simulated profiles are included in Figures 8. Under the assumption that permeability and melt fraction are related via a power law, $K \propto \phi^s$ (Equation 15), the best agreement between the experimental results and the simulated profiles was obtained with a melt-fraction exponent of $s \approx 1$, a value smaller than that expected for porous flow. In addition, the exponential-like decay of the melt fraction with increasing distance into the sink cannot be reproduced by the simulations with $s = 2$ or $s = 3$. The permeability obtained from analysis of the of the data in Figure 8 is given by

$$K = 5.2 \times 10^{-15} \phi \text{ m}^2 .$$

The data in Figure 9 also yield a value of $s \approx 1$, but have not yet been analyzed in enough detail to determine the magnitude of the permeability [*Riley and Xu*, 1991, private communication].

One implication of this result, which yields relatively high permeabilities at low melt fractions, is that the melt fraction expected at any instant in lherzolite upwelling under a mid-oceanic ridge will be very small, ≤ 0.001 [*Riley et al.*, 1990]. This value is consistent with the melt fractions inferred from analyses of the rare earth and trace element characteristics of diopside grains in abyssal peridotites collected along the mid-Atlantic ridge [*Johnson et al.*, 1990] and of isotope and trace element signatures of MORBs [*Salters and Hart*, 1989].

The source of the discrepancy between the value for the melt fraction exponent determined from analysis of the melt migration profiles, $s \sim 1$, and those expected from models of Pouiselle flow, $s \geq 2$, has not yet been identified. At the onset of a melt migration experiment, the capillary force will be quite high in the sink because the magnitude of the effective pressure, $\gamma_{sm}/\Re$, is large because the melt fraction and, consequently, the radius of curvature of the triple junctions, $\Re$, are small. This high local stress might result in non-Newtonian deformation of the matrix, violating the assumption of Newtonian viscosity used in the development of the compaction theory. Also, it is possible that

† Analysis of the melt-migration profiles for runs involving discs of melt as the source are presently being carried out [*Riley and Kohlstedt*, private communication, 1992].

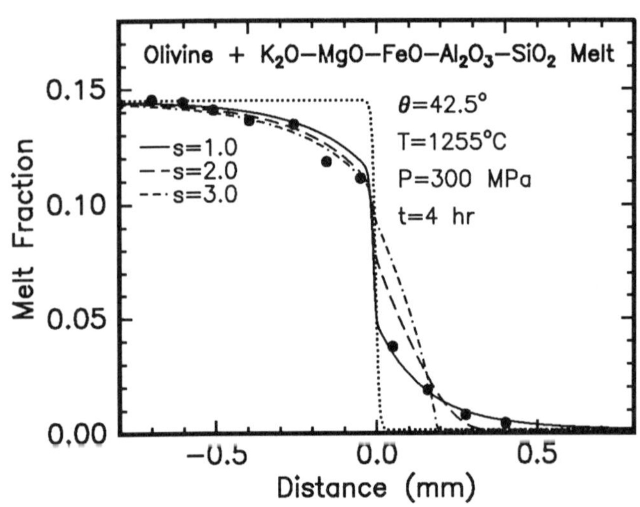

Fig. 8. Melt migration profile (solid circles) obtained using a source composed of olivine plus a K-Al-rich melt pressed against a sink of fine-grained olivine. Simulated profiles are shown for s = 1, 2 and 3. The dotted line represents the step-function profile used as the initial condition in the simulation calculations.

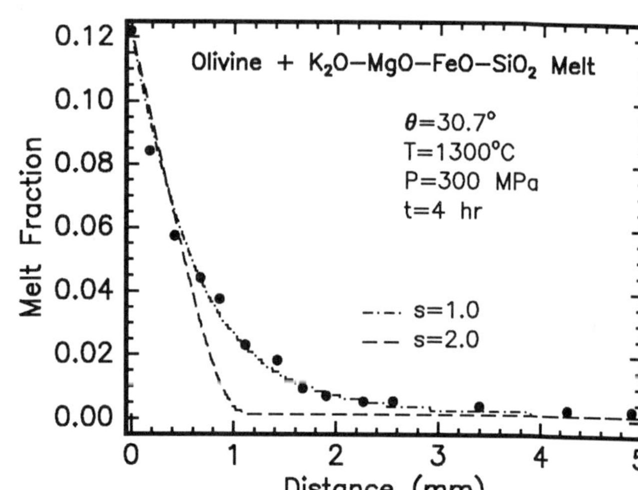

Fig. 9. Melt migration profile (solid circles) obtained using a source composed of a K-rich melt pressed against a sink of fine-grained olivine. Simulated profiles are shown for s = 1 and 2.

the rate of melt migration is limited by an interface reaction (of the type discussed in the section on theoretical considerations for high-temperature creep), rather than by diffusional creep. Again, the compaction equations would have to be modified to incorporate this behavior.

Melt Migration Driven by a Gradient in Mean Stress. To study melt migration in response to a gradient in mean normal stress, four-point bending experiments have been carried out on fine-grained samples of three different compositions -- β-spodumene glass-ceramic [*Cooper et al.*, 1989], enstatite plus a sodium-aluminosilicate glass [*Cooper*, 1990], and cobalt-magnesium olivine plus basalt [*Green et al.*, 1990]. For all of these materials, the glass forms an interconnected network along triple junctions and through the four-grain junctions (i.e., $0 < \theta < 60°$). In each case, the nonelastic deformation recorded in the strain-time (ε-t) creep curves could be modeled as the sum of two terms, the first describing steady-state plastic (time-dependent and irreversible) flow of the aggregate and the second describing anelastic (time-dependent but reversible) deformation resulting from melt migration. The steady-state creep component exhibits a Newtonian rheology ($\dot{\varepsilon} \propto \sigma$) of the type discussed in the section on diffusion-dominated creep [*Cooper et al.*, 1989] with an effective viscosity η_p for plastic flow. The anelastic component was characterized by a viscosity η_a. Expressed in terms of the mechanical analog illustrated in Figure 10, this analysis yields the following expression for the total (elastic plus inelastic) strain, ε_{tot}, in terms of the elastic (time-independent and reversible) strain, ε_e, the anelastic (time-dependent and reversible) strain, ε_a, and the plastic (time-dependent and irreversible) strain, ε_p:

$$\varepsilon_{tot} = \varepsilon_e + \varepsilon_a + \varepsilon_p = \varepsilon_e + \varepsilon_o[1 - \exp(-Bt)] + \dot{\varepsilon}_{ss}t \qquad (21)$$

where ε_o is the maximum anelastic strain produced by a given applied stress and $\dot{\varepsilon}_{ss}$ is the steady-state strain rate. The characteristic rate of anelastic deformation, B, is inversely proportional to the viscosity of the anelastic element, ($B \propto \eta_a^{-1}$). On the one hand, if compaction and dilatation of the matrix limit the rate at which melt can migrate from the compression side to the tensile side of specimen, then the anelastic viscosity is given by the steady-state creep viscosity. On the other hand, if viscous flow of the melt controls the rate at which melt can migrate in response to the gradient in mean normal stress, then the viscosity of the anelastic element will be equal to the viscosity of the melt.

As anticipated from Equation 21 and illustrated in Figure 11, the nonelastic portion of the a flexural creep curve for a β-spodumene glass-ceramic studied by *Cooper et al.* [1989] can be decomposed into two curves. One curve delineates the steady-state deformation of the matrix that occurs in response to the applied stress. The other curve describes the anelastic deformation‡ that arises due to the combined effects of the gradient in mean normal stress and the gradient in melt fraction. For the β-spodumene samples, the activation energy for steady-state creep is ~195 kJ/mol, while the activation energy for the anelastic flow is over three times larger, ~645 kJ/mol. This latter value, which is in the range measured for viscous flow of fused

‡Cooper [private communication, 1990] verified that this second component of the inelastic deformation was indeed anelastic (i.e., time dependent and reversible) rather than transient (i.e., time dependent and irreversible) deformation by showing that this strain component could be fully recovered by statically annealing a sample after it had been deformed by four-point bending.

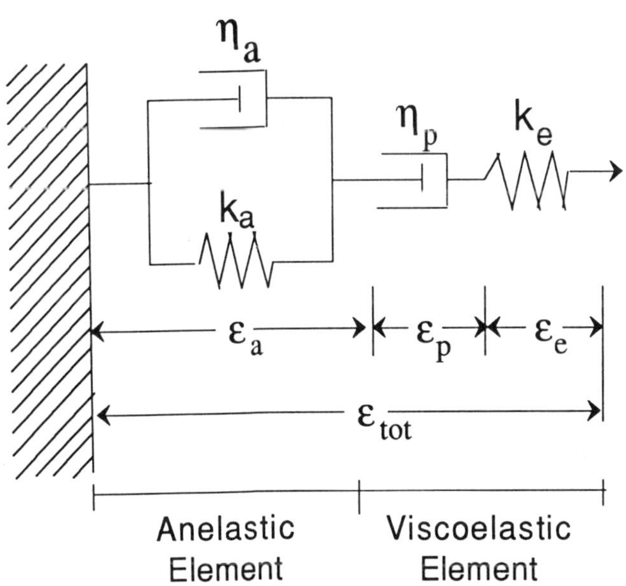

Fig. 10. Mechanical analog including an anelastic element, a plastic element, and an elastic element.

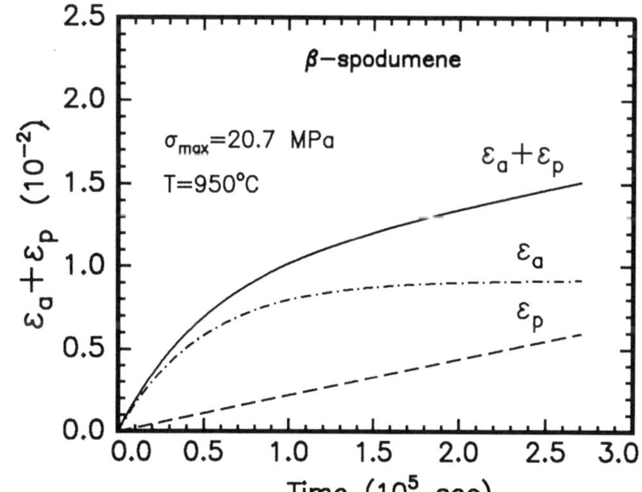

Fig. 11. Anelastic plus plastic strain versus time (creep) curve for a β-spodumene glass ceramic deformed in four-point flexure. The total creep strain has been decomposed into an anelastic contribution and a plastic contribution; the elastic contribution is too small to be resolved in this figure.

silica, suggests that the rate of anelastic deformation (melt migration) was controlled by the viscosity of the silica-rich amorphous phase in β-spodumene (i.e., $\eta_a = \mu$).

Similar strain-time flexural creep behavior was observed for an enstatite glass-ceramic [*Cooper*, 1990]. In this case, however, the activation energies for steady-state creep and anelastic deformation are approximately equal (976 vs 934 kJ/mol, respectively). Thus, deformation of the matrix appears to limit the rate of melt migration (i.e., $\eta_a = \eta_p$). Transmission electron microscopy observations verified that melt had indeed migrated from compression side to the tension side of the samples. The results from the four-point bending experiments have not yet been analyzed to determine the permeability.

Finally, the flexural creep response of a Co-Mg olivine-plus-basalt aggregate was similar to that of the β-spodumene glass-ceramic, in that the activation energies for steady-state creep and anelastic flow are markedly different [*Green et al.*, 1990]. This observation was again taken to imply that flow of the basaltic melt controls the rate of anelastic deformation. The flexural creep results were analyzed in terms of the mechanical analog in Figure 10 to obtain the attenuation due to the migration of melt. This analysis yielded a peak attenuation associated with melt migration of $Q^{-1} \approx 10$ at a frequency of $\sim 10^{-3}$ Hz. To obtain higher resolution attenuation data from flexure experiments on partially molten rocks, reciprocating four-point bending tests are now in progress [*Cooper et al.*, 1990].

GEOLOGICAL IMPLICATIONS

Introduction

In the discussion below, I have attempted to summarize the current state of understanding of the structure, rheology and permeability of partially molten upper mantle rocks. Most of the conclusions are based on the results of laboratory experiments, as supported by theoretical models. It should be emphasized, however, that there is a great deal left to learn from careful laboratory experiments on partially molten rocks. The effects of melt fraction, melt viscosity and melt distribution (i.e., melt composition) have yet to be fully explored. In such experiments, careful rheological measurements to determine the dependence of strain rate on stress, grain size and temperature must be coupled with thorough microstructural observations in order to characterize the dominant mechanism(s) of high-temperature flow as well as with careful analyses of permeability to ascertain the geologically appropriate melt fraction.

The results discussed here primarily pertain to rocks containing relatively low melt fractions in which the melt is near textural and chemical equilibrium. Since the rate of upwelling and hence melting of rocks is slow beneath a mid-oceanic ridge [e.g., *Riley et al.*, 1990], the melt distribution at the grain-scale level is expected to be determined largely by a balance between solid-solid and solid-melt interfacial energies (i.e., surface tension). Partial melting of a multiphase rock will initially produce a non-equilibrium melt distribution, because melt is generated at four-grain junctions. The rheology and permeability of such a rock far from textural equilibrium is not likely to be representative of the behavior of rocks at depth below an oceanic ridge. To illustrate the significance of this point, the amount of strain that occurs during the time required to equilibrate the melt-solid microstructure in a laboratory experiment is compared to that needed in upwelling mantle. In a partially molten rock composed of 50-μm grains, the melt distribution approaches textural equilibrium after about 10^2 h [*Waff and Bulau*, 1982]. At a laboratory strain rate of $\sim 10^{-5}$ s^{-1}, such a sample would undergo a strain of more than 3 in this time. If the grain size in the mantle is closer to 3 mm, textural equilibration would require about 10^6 h [*Cooper and Kohlstedt*, 1984]. At a geological strain rate of $\sim 10^{-14}$ s^{-1}, an upwelling rock would deform by a strain of only 10^{-4}. In laboratory experiments, therefore, it is critical to allow an adequate amount of time for the microstructure to equilibrate before the deformation experiment is initiated.

Finally, it should be noted that one major topic that has not been addressed in this article is the transition from porous flow to channelized flow in veins and dikes. While attempts have been made to analyze this problem theoretically [*Sleep*, 1988; *Stevenson*, 1989], no attempt has been made to explore it experimentally. Major changes both in rheology and permeability will undoubtedly occur once melt segregation becomes localized.

Melt Distribution

The major features of the melt distribution in dunitic rocks appear to be quite well characterized. As illustrated in Figure 5, basaltic and carbonate melts form interconnected networks along the triple junctions and through the four-grain junctions of olivine aggregates [*Waff and Bulau*, 1979]. The dihedral angle for most melts lies in the range $25 < \theta < 50°$. For extreme compositions (e.g., Pb-rich silicate melts of *Riley and Kohlstedt* [1991]), the dihedral angle is $\theta = 0°$, such that a thin film of melt separates neighboring grains. For H_2O-CO_2 fluids, the dihedral angle is $\theta > 60°$; hence, these fluids are confined to isolated pockets unless the fluid fraction becomes large or fracturing occurs [*Watson and Brenan*, 1987].

The addition of pyroxenes to a dry olivine-rich aggregate results in the formation of melt-free triple junctions at contacts involving pyroxene, thus reducing the connectivity of the melt [*Toramaru and Fujii*, 1986]. The introduction of water into such samples, however, reverses this trend [*Fujii et al.*, 1868; *von Bargen and Waff*, 1988].

Often, the surfaces of olivine grains in contact with basaltic melts are not smoothly curved, Figures 3 and 4, due to anisotropy of the solid-melt interfacial energy. Although low-index faces are generally planar, those interfaces which are smoothly curved do not appear to be faceted even when viewed by high-resolution electron microscopy techniques.

Finally, both theoretical and experimental results demonstrate that an equilibrium or saturation melt fraction exists. Defined for the range $0 < \theta \leq 60°$, the equilibrium melt fraction increases with decreasing dihedral angle [*Laporte and Watson*, 1992]. Experimental values for ϕ_{eq}, measured for three different melt compositions, are smaller than the values predicted by theory

[*Riley and Kohlstedt*, 1991]. This discrepancy may indicate that a significant amount of olivine has precipitate out of the melt upon quenching the sample at the termination of the experiment.

High-Temperature Creep Behavior

Several points deserve emphasis. First, if the melt does not wet the grain boundaries but does form an interconnected network through the triple and four-grain junctions, it will have a relatively small influence on creep behavior. In the grain boundary diffusion creep regime, the addition of 5 vol% of melt should result in less than a factor of five increase in strain rate for $\theta > 25°$, Figure 6 [*Cooper et al.*, 1989]. In the dislocation creep regime, the effect would be about a factor of 1.1. In the context of convective flow beneath a mid-oceanic ridge, laboratory measurements of permeability and geochemical analysis of MORBs and abyssal peridotites indicate that the melt fraction in any parcel of rock may be as small as 0.001 [*Riley et al.*, 1990; *Johnson et al.*, 1990; *Salters and Hart*, 1989]. Under this condition, the effect of melt on the rheology is expected to be minimal for partially molten rocks of upper mantle composition.

Second, the examples discussed above do demonstrate that the addition of a few percent of melt that totally wets the grain boundaries of an initially single-phase material can cause a large increase (greater than a factor of 10^3) in creep rate (i.e., decrease in creep strength). In the diffusional creep regime, this effect results because diffusion through the grain boundary phase can be orders of magnitude faster than through melt-free grain boundaries. In the dislocation creep regime, it is due to enhanced grain boundary sliding and associated intergranular fracture; if fracturing were suppressed, the influence of the melt phase would probably be much smaller. However, this condition of $\theta = 0°$ does not appear to apply to partially molten mantle rocks; thus, marked decreases in creep strength are not anticipated.

Finally, the addition of a liquid phase can -- but does not necessarily have to -- result in a change in creep mechanism, typically from dislocation creep to grain boundary diffusional creep. The presence of melt in mantle rocks does not necessitate that these rocks deform by diffusional creep processes.

Permeability

Both laboratory experiments on melt migration in olivine-rich aggregates [e.g., *Riley et al.*, 1990] and trace element analyses of diopside in abyssal peridotites and of MORBs [*Johnson et al.*, 1990; *Salters and Hart*, 1989] lead to the conclusion that the melt percentage in rocks upwelling beneath a mid-oceanic ridge is likely to be much less than 1%, that is, the permeability is quite high. One of the unexpected results of the laboratory experiments is that the permeability is approximately linearly proportional to the melt fraction; models typically suggest that permeability should be proportional to melt fraction squared or possibly cubed. The observed linear dependence of permeability on melt fraction leads to a relatively high permeability at low melt fractions. Consequently, melt is removed quickly after it is produced by decompression melting, minimizing the time for chemical interaction with the residual crystalline phases. Additional experiments are essential to confirm or disprove this initial observation; if it proves correct, then theoretical analyses are needed to explain it in terms of solution-transport-precipitation processes.

Also, flexural creep experiments have demonstrated that melt migration in response to a gradient in mean normal stress can be induced under laboratory conditions [*Cooper*, 1990]. The anelastic deformation associated with this stress-induced flow of melt can give rise to a substantial damping of seismic waves [*Green et al.*, 1990].

Acknowledgments. Support from the National Science Foundation through grant EAR-8916438 is gratefully acknowledged. The author is particularly indebted to Dr. Reid Cooper and Dr. Gilbert (Bart) Riley for their enthusiastic and intellectually stimulating involvement in major portions of the research described here. Help from Dr. Frank Richter with the analysis of the melt migration profiles and assistance from Ms. Yaqin Xu with computer simulations of some of the melt migration profiles are most gratefully acknowledged. Discussions with Dr. Martha Daines, Dr. Greg Hirth, Dr. Steve Mackwell, Dr. Mervyn Paterson and Mr. Mark Zimmerman were particularly valuable. Thoughtful reviews by Dr. Brian Evans and Dr. Robert Borch sharpened this paper. Finally, the endless assistance of Ms. Rupa Patel was critical in bringing this manuscript to fruition.

REFERENCES

Ahern, J.L. and D.L. Turcotte, Magma migration beneath an ocean ridge, *Earth Planet. Sci. Lett.*, *45*, 115-122, 1979.

Arzi, A.A., Critical phenomena in the rheology of partially melted rocks, *Tectonophys.*, *44*, 173-184, 1978.

Bai, Q., S.J. Mackwell and D.L. Kohlstedt, High-temperature creep of olivine single crystals: 1. Mechanical results for buffered samples, *J. Geophys. Res.*, *96*, 2441-2463, 1991.

Beere, W., A unifying theory of the stability of penetrating liquid phases and sintering pores, *Acta Metall.*, *23*, 131-138, 1975.

Borch, R.S. and H.W. Green II, Experimental investigation of the rheology and structure of partially molten lherzolite deformed under upper mantle pressures and temperatures, *EOS*, *Trans. Am. Geophys. Union*, *71*, 629, 1990.

Brenan, J.M. and E.B. Watson, Fluids in the lithosphere, 2. Experimental constraints on CO_2 transport in dunite and quartzite at elevated *P-T* conditions with implications for mantle and crustal decarbonation processes, *Earth Planet. Sci. Lett.*, *91*, 141-158, 1988.

Bulau, J.R., H.S. Waff and J.A. Tyburczy, Mechanical and thermodynamic constraints on fluid distribution in partial melts, *J. Geophys. Res.*, *84*, 6102-6108, 1979.

Chen, I.W. and A.S. Argon, Steady state power-law creep in heterogeneous alloys with microstructures, *Acta Metall.*, *27*, 785-791, 1979.

Chopra, P.N. and D.L. Kohlstedt, The influence of wet basaltic melt on the flow properties of fine-grained polycrystalline olivine, *EOS Trans. AGU*, *64*, 323, 1983.

Clarke, D.R., On the equilibrium thickness of intergranular glass phases in ceramic materials, *J. Am. Ceram. Soc.*, *70*, 15-22, 1987.

Coble, R.L., A model for boundary diffusion controlled creep in polycrystalline materials, *J. Appl. Phys.*, *34*, 1679-1682, 1963.

Cooper, R.F., Differential stress-induced melt migration: an experimental approach, *J. Geophys. Res.*, *95*, 6979-6992, 1990.

Cooper, R.F. and D.L. Kohlstedt, Interfacial energies in the olivine-basalt system, in *Advances in Earth and Planetary Sciences, 12, High Pressure Research in Geophysics*, edited by S. Akimoto and M.H. Manghnani, Center for Academic Publications, Tokyo, pp. 217-228, 1982.

Cooper, R.F. and D.L. Kohlstedt, Solution-Precipitation enhanced diffusional creep of partially molten olivine basalt aggregates during hot-pressing, *Tectonophys.*, *107*, 207-233, 1984.

Cooper, R.F. and D.L. Kohlstedt, Rheology and structure of olivine-basalt partial melts, *J. Geophys. Res.*, *91*, 9315-9323, 1986.

Cooper, R.F., D.L. Kohlstedt and K. Chyung, Solution-precipitation enhanced creep in solid-liquid aggregates which displays a non-zero dihedral angle, *Acta Metall.*, *37*, 1759-1771, 1989.

Cooper, R.F., D.H. Green and D.K. Bidner, Reciprocating four-point flexure testing at high temperature with application to attenuation in partial melts, in *The Brittle-Ductile Transition in Rocks: The Heard Volume*, edited by A.G. Duba, W.B. Durham, J.W. Handin, and W.F. Wang, American Geophysical Union, Washington, pp. 201-206, 1990.

Daines, M.J. and F.M. Richter, An experimental method for directly determining the interconnectivity of melt in a partially molten system, *Geophys. Res. Lett.*, *15*, 1459-1462, 1988.

Dimos, D. D.L. Kohlstedt, and H. Schmalzried, High-temperature creep and kinetic demixing in (Co,Mg)O, *J. Am. Ceram. Soc.*, *71*, 1988.

Durney, D.W., Pressure solution and crystallization deformation, *Phil. Trans. R. Soc. Lond. A.*, *283*, 229-240, 1976.

Elliot, D., Diffusion flow laws in metamorphic rocks, *Geol. Soc. Am. Bull.*, *84*, 2645-2664, 1973.

Flaitz, P.L. and J.A. Pask, Penetration of polycrystalline alumina by glass at high temperatures, *J. Am. Ceram. Soc.*, *70*, 449-455, 1987.

Fowler, A.C., A compaction model for melt transport in the Earth's asthenosphere. Part I: The basic model, in *Magma Transport and Storage*, edited by M.P. Ryan, John Wiley & Sons, pp. 3-14, 1990.

Frank, F.C., Two-component flow model for convection in the earth's upper mantle, *Nature*, *220*, 350-352, 1968.

Fujii, N., K. Osamura and E. Takahashi, Effect of water saturation on the distribution of partial melt in the olivine-pyroxene-plagioclase system, *J. Geophys. Res.*, *91*, 9253-9259, 1986.

Green, D.H., R.F. Cooper and S. Zhang, Attenuation spectra of olivine/basalt partial melts: transformation of newtonian creep response, *Geophys. Res. Lett.*, *17*, 2097-2100, 1990.

Green, H.W., II, and R.S. Borch, The pressure dependence of creep, *Acta Metall.*, *35*, 1301-1305, 1987.

Hay, R.S. and Evans, B., Intergranular distribution of pore fluid and the nature of high-angle grain boundaries in limestone and marble, *J. Geophys. Res.*, *93*, 1988.

Hunter, R.H. and D. McKenzie, The equilibrium geometry of carbonate melts in rocks of mantle composition, *Earth Planet. Sci. Lett.*, *92*, 347-356, 1989.

Hwang, C. and I.W. Chen, Effect of a liquid phase on superplasticity of 2-mol%Y_2-O_3-stabilized tetragonal zirconia polycrystals, *J. Am. Ceram. Soc.*, *73*, 1626-1632, 1990.

Johnson, K.T.M., H.J.B. Dick and N. Shimizu, Melting in the oceanic upper mantle: An ion microprobe study of diopsides in abyssal peridotites, *J. Geophys. Res.*, *95*, 2661-2678, 1990.

Ju :wlcz, S.R. and A.J.G. Jurewicz, Distribution of apparent angles on .ndom sections with emphasis on dihedral angle measurements, *J. ;eophys. Res.*, *91*, 9277-9282, 1986.

.irewicz, S.R. and E.B. Watson, Distribution of partial melt in a felsic system: the importance of surface energy, *Contrib. Mineral. Petrol.*, *85*, 25-29, 1984.

Jurewicz, S.R. and E.B. Watson, The distribution of partial melt in a granitic system: the application of liquid phase sintering theory, *Geochim. Cosmochim. Acta*, *49*, 1109-1121, 1985.

Kohlstedt, D.L., Chemical analysis of grain boundaries in an olivine-basalt aggregate using high-resolution, analytical electron microscopy, in *The Brittle-Ductile Transition in Rocks: The Heard Volume*, edited by A.G. Duba, W.B. Durham, J.W. Handin, and W.F. Wang, American Geophysical Union, Washington, pp. 211-218, 1990.

Lange, F.F., B.I. Davis, and D.R. Clarke, Compressive creep of Si_3N_4/MgO alloys. Part 1: Effect of composition, *J. Mater. Sci.*, *15*, 601-610, 1980.

Laporte, D. and E.B. Watson, Experimental and theoretical constraints on melt distribution in crustal sources: implications for the segregation of granitic magmas, *Phys. Earth Planet. Int.*, in press, 1992.

Lee, V.W., S.J. Mackwell, and S.L. Brantley, The effect of fluid chemistry on wetting textures in novaculite, *J. Geophys. Res.*, *96*, 10023-10037, 1991.

McKenzie, D., The generation and compaction of partially molten rock, *J. Petrol.*, *25*, 713-765, 1984.

Osborne, N.J., Mechanical properties of ceramics, *Proc. Brit. Ceram. Soc.*, *25*, 263, 1975.

Park, H.H. and D.N. Yoon, Effect of dihedral angle on the morphology of grains in a matrix phase, *Metall. Trans. A.*, *16*, 923-928, 1985.

Pharr, G.M. and M.F. Ashby, On creep enhanced by a liquid phase, *Acta Metall.*, *31*, 129-138, 1983.

Pharr, G.M., P.S. Godavart and B.L. Vaandrag, Effects of wetting on the compression creep behavior of metals containing low-melting intergranular phases, *J. Mater. Sci.*, *24*, 784-792, 1989.

Poirier, J.-P., *Creep of Crystals: High- Temperature Deformation Processes in Metals, Ceramics and Minerals*, Cambridge University Press, 1985.

Raj, R., Creep in polycrystalline aggregates by matter transport through a liquid phase, *J. Geophys. Res.*, *87*, 4731-4739, 1982.

Raj, R. and C.K. Chyung, Solution-precipitation creep in glass ceramics, *Acta Metall.*, *29*, 159-166, 1981.

Raj, R., R.L. Tsai, J.G. Wang, and C.K. Chyung, Superplastic flow in ceramics enhanced by a liquid phase, in *Deformation of Ceramic Materials II, Materials Science Research Series*, vol. 18, edited by R.E. Tressler and R.C. Bradt, Plenum Press, New York, pp. 353-378, 1984.

Raju, A.P., I.A. Aksay, and J.A. Pask, Permeation of silicates in magnesia and forsterite compacts, *Am. Ceram. Soc. Bull.*, *52*, 166-169, 1973.

Ribe, N.M., Theory of melt segregation - A review, *J. Volcanol. Geotherm. Res.*, *33*, 241-253, 1987.

Richter, F.M. and D. McKenzie, Dynamical models for melt segregation from a deformable matrix, *J. Geol.*, *92*, 729-740, 1984.

Ricoult, D.L. and D.L. Kohlstedt, Experimental evidence for the effect of chemical environment upon the creep rate of olivine", in *Point Defects in Minerals*, edited by R.N. Schock, American Geophysical Union, Washington, pp. 171-184, 1985.

Riley, G.N., Jr. and D.L. Kohlstedt, An experimental study of melt migration in an olivine-melt system, in *Magma Transport and Storage*, edited by M.P. Ryan, John Wiley & Sons, New York, pp. 77-86, 1990.

Riley, G.N., Jr. and D.L. Kohlstedt, Kinetics of melt migration in upper mantle-type rocks, *Earth Planet. Sci. Lett.*, *105*, 500-521, 1991.

Riley, G.N., Jr. and D.L. Kohlstedt, The influence of H_2O and CO_2 on melt migration in two silicate liquid-olivine systems, in *Fault Mechanics and Transport Properties in Rocks: A Festschrift in Honor of W.F. Brace*, edited by B. Evans and T.-F. Wong, Academic Press, New York, in press, 1992a.

Riley, G.N., Jr. and D.L. Kohlstedt, Thermodynamics and kinetics of melt infiltration into polycrystalline olivine, *J. Am. Ceram. Soc.*, in press, 1992b.

Riley, G.N., Jr., D.L. Kohlstedt and F.M. Richter, Melt migration in a silicate liquid-olivine system: an experimental test of compaction theory, *Geophys. Res. Lett.*, *17*, 2101-2104, 1990.

Rutter, E.H., The kinetics of rock deformation by pressure solution, *Phil. Trans. R. Soc. Lond. A.*, *283*, 203-219, 1976.

Salters, V.J.M. and S.R. Hart, The hafnium paradox and the role of garnet in the source of mid-ocean-ridge basalts, *Nature*, *342*, 420-422, 1989.

Scott, D.R. and D.J. Stevenson, Magma ascent by porous flow, *J. Geophys. Res.*, *91*, 9283-9296, 1986.

Sheikh, G.R. and G.M. Pharr, Further observation on creep enhanced by a liquid phase in porous potassium chloride, *Acta Metall.*, *33*, 231-238, 1985.

Sleep, N.H., Segregation of magma from a mostly crystalline mush, *Geol. Soc. Am. Bull.*, *85*, 1225-1232, 1974.

Sleep, N.H., Tapping of melt by veins and dikes, *J. Geophys. Res.*, *93*, 10,255-10,272, 1988.

Smith, C.S., Grains, phases and interfaces: an interpretation of microstructure, *Trans. A.I.M.E.*, *197*, 15-51, 1948.

Smith, C.S., Some elementary principles of polycrystalline microstructure, *Metall. Rev.*, *9*, 1-47, 1964.

Stevenson, D.J., On the role of surface tension in the migration of melts and fluids, *Geophys. Res. Lett.*, *13*, 1149-1152, 1986.

Stevenson, D.J., Spontaneous small-scale melt segregation in partial melts undergoing deformation, *Geophys. Res. Lett.*, *16*, 1067-1070, 1989.

Stocker, R.L. and M.F. Ashby, On the rheology of the upper mantle, *Rev. Geophys. Space Phys.*, *11*, 391-426, 1973.

Stolper, E., D. Walker, B.H. Hager and J.F. Hays, Melt segregation from partially molten source regions: the importance of melt density and source region size, *J. Geophys. Res.*, *86*, 6261-6271, 1981.

Toramaru, A. and N. Fujii, Connectivity of melt phase in a partially molten peridotite, *J. Geophys. Res.*, *91*, 9239-9252, 1986.

Tullis, T., F. Horowitz, and J. Tullis, Flow laws of polyphase aggregates from end member flow laws, *J. Geophys. Res.*, *96*, 8081-8096, 1991.

Turcotte, D.L. and G. Schubert, *Geodynamics: Applications of Continuum Physics to Geological Problems*, John Wiley & Sons, 1982.

Urai, J.L., Water assisted dynamic recrystallization and weakening in polycrystalline bischofite, *Tectonophys.*, *96*, 127-157, 1983.

van der Molen, I. and M.S. Paterson, Experimental deformation of partially melted granite, *Contrib. Mineral. Petrol.*, *70*, 291-318, 1979.

Vaughan, P.J., D.L. Kohlstedt, and H.S. Waff, Distribution of the glass phase in hot-pressed, olivine-basalt aggregates: an electron microscopy study, *Contrib. Mineral. Petrol.*, *81*, 253-261, 1982.

von Bargen, N. and H.S. Waff, Permeabilities, interfacial areas and curvatures of partially molten systems: results of numerical computations of equilibrium microstructures, *J. Geophys. Res.*, *91*, 9261-9276, 1986.

von Bargen, N. and H.S. Waff, Wetting of enstatite by basaltic melt at 1350°C and 1.0- to 2.5-GPa pressure, *J. Geophys. Res.*, *93*, 1153-1158, 1988.

Waff, H.S. and J.R. Bulau, Experimental determination of near-equilibrium textures in partially molten silicates at high pressures, in *Advances in Earth and Planetary Sciences, 12, High Pressure Research in Geophysics*, edited by S. Akimoto and M.H. Manghnani, Center for Academic Publications, Tokyo, pp. 229-236, 1982.

Waff, H.S. and J.R. Bulau, Equilibrium fluid distribution in an ultramafic partial melt under hydrostatic stress conditions, *J. Geophys. Res.*, *84*, 6109-6114, 1979.

Waff, H.S. and U.H. Faul, Coexisting faceted and curved interfaces in partially molten olivine-basalt systems and implications for mantle partial melts, *J. Geophys. Res.*, in press, 1992.

Waff, H.S. and G.R. Holdren, Jr., The nature of grain boundaries in dunite and lherzolite xenoliths: implications for magma transport in refractory upper mantle material, *J. Geophys. Res.*, *86*, 3677-3683, 1981.

Wanamaker, B.J. and D.L. Kohlstedt, The effect of melt composition on the wetting angle between silicate melts and olivine, *Contrib. Mineral. Petrol.*, *18*, 26-36, 1991.

Wang, J.G. and R. Raj, Mechanism of superplastic flow in a fine-grained ceramic containing some liquid phase, *J. Am. Ceram. Soc.*, *67*, 399-409, 1984.

Watson, E.B., Melt infiltration and magma evolution, *J. Geol.*, *10*, 236-240, 1982.

Watson, E.B. and J.M. Brenan, Fluids in the lithosphere, 1. Experimentally determined wetting characteristics of CO_2-H_2O fluids and their implications for fluid transport, host-rock physical properties, and fluid inclusion formation, *Earth Planet. Sci. Lett.*, *85*, 497-515, 1987.

Watson, E.B., J.M. Brenan, and D.R. Baker, Distribution of fluids in the continental mantle, in *Continental Mantle*, edited by M.A. Menzies, Claredon Press, Oxford, pp.111-125, 1990.

Wray, P.J., The geometry of two-phase aggregates in which the shape is determined by its dihedral angle, *Acta Metall.*, *24*, 125-135, 1976.

David L. Kohlstedt, Department of Geology and Geophysics, Pillsbury Hall, University of Minnesota, Minneapolis, MN 55455.

Mantle Flow and Melt Migration beneath Oceanic Ridges: Models derived from Observations in Ophiolites

GEORGES CEULENEER AND MICHEL RABINOWICZ

Centre National de la Recherche Scientifique, UPR 234 - GRGS, Toulouse, France

1. INTRODUCTION

"Our present views of the uppermost mantle will seem primitive by 2100, or well before that."

Jack Oliver

The tectonic and volcanic features of the Earth are largely conditioned by processes taking place in the upper mantle. However our present knowledge of the mantle is essentially derived from indirect evidence. As pointed out by Oliver [1991], the promising technologies of very deep drilling and of high resolution geophysical imagery which should improve our views of the Earth's interior are still to be invented. In the meantime, to progress in the study of geological stuffs, widespread and of easy access at the Earth surface, and in their interpretation in terms of mantle processes, is not a vain task. For example, the Earth's magmatic budget is largely dominated by oceanic ridge and hotspot volcanism [e.g. Crisp, 1984]. The basaltic melts erupted there are likely produced by pressure-release induced partial melting of a peridotitic source. Consequently, their composition and distribution are expected to be sensitive to the thermal and flow structure of the mantle. Potentially, basalts constitute a widespread and abundant pool of information about mantle convection, and one of the goals pursued by petrological studies is to decipher this message [e.g. Klein and Langmuir, 1987; McKenzie and Bickle, 1988; Watson and McKenzie, 1991; Albarède, 1992, Langmuir et al., this volume; Grove et al., this volume].

To approach a solution to this "petrological inverse problem" implies an accurate knowledge of all the processes modifying a melt composition, from the initiation of partial melting, to its final eruption at the Earth surface. The evolution of a melt composition in response to fractional crystallization in shallow crustal reservoirs is relatively well understood thanks to a century of petrological research almost entirely devoted to that topic. At the opposite extremity of the train of processes, major advances in our understanding of the phase diagram of the mantle in the melting region were made possible recently thanks to high-pressure melting experiments [e.g. Presnall et al., 1979; Takahashi and Kushiro, 1983; Takahashi, 1986, 1990; Ohtani and Sawamoto, 1987; Falloon et al., 1988; Wyllie, 1988; Wei et al., 1990; Herzberg et al., 1990; Hess, this volume]. So, we have now a reasonable knowledge of the thermodynamics involved in basaltic melt generation and evolution, compared to some other unknowns. Among them, the melts composition depends in a complex way on the fate of each incremental melt fraction once produced in a mantle upwelling. This is due to the dependence of melt composition on the melting mechanism but also to the peritectic nature of mantle melting for a wide range of the pressure and temperature conditions realized in the mantle [e.g. O'Hara, 1968, Herzberg et al., 1990]. Melt migration in a polycrystalline and polymineralic plastic medium like the mantle involves complex mechanisms down to the grain scale (a few millimeters). Up to now, these mechanisms have not been clearly elucidated in spite of considerable theoretical and experimental efforts [e.g. Waff, 1980; McKenzie, 1984; Scott and Stevenson, 1984, 1986; Cooper and Kohlstedt, 1986; Toramaru and Fujii, 1986; von Bargen and Waff, 1988; Sleep, 1988; Scott, 1988; Stevenson, 1989; Kohlstedt, this volume]. Consequently, large scale melt migration patterns are hardly predictable and self-consistent treatments of mantle convection and partial melting, assuming extremely simplified melt migration processes, give possibly even not an approximate picture of the "real world".

The difficulty originates precisely in the lack of evidence of what actually happens in the melting region of the mantle. Theoretical models and laboratory experiments have shown that some geophysical signals should be very sensitive to the presence and to the connectivity of a melt phase [e.g. Schmeling, 1985]. However, the present data set and inversion methods do not resolve unambiguously the small scale structures involved in melt migration. For example, the attenuation of seismic waves can be attributed more or less with confidence to the presence of a partially molten region but does not indicate clearly if the melt is distributed as an intergranular phase or if it forms a network of close-spaced

Mantle Flow and Melt Generation at Mid-Ocean Ridges
Geophysical Monograph 71
Copyright 1992 American Geophysical Union

veins. This particular point has important consequences, not only for the composition of partial melts, but also for the creep properties of the mantle, and then for the flow pattern and for the thermal field, which in turn will govern the evolution of partial melting higher up in the upwelling. In other words, the existence of complex *feed-back* processes between melt migration and solid-state flow are suspected in the upper mantle which could have drastic geophysical and petrological consequences.

It will never be possible, of course, to observe directly such high-temperature processes developing in the present-day mantle. Fortunately, some providential geological events have brought frozen pieces of upper mantle to the Earth surface. In spite of the probability that the information they carry are jammed to some extent by their cooling and emplacement, and by later tectonic events, they constitute the only opportunity we have to "finger" asthenospheric processes. Considering only the data of easy access through field and petrographic observations, information about two categories of processes can be gathered in mantle outcrops: one dealing with the plastic deformation (i.e. the solid-state flow), the other concerning partial melting and melt segregation. Petrographers realized very early that most of the peridotites cropping out at the Earth surface display textures and fabrics which can be ascribed to sub-solidus plastic deformation [e.g. Ernst, 1935]. A method making use of this property to reconstruct frozen flow patterns has been developed by Nicolas and coworkers [e.g. Nicolas, 1989]. It will be presented briefly in section 2. The peridotites are residue of partial melting. Their modal composition gives an idea, at least qualitative, about the extent of partial melting they suffered, or about magma/rock interaction processes [e.g. Dick, 1977; Dick et al., 1984; Berger and Vannier, 1984; Kelemen, 1990; Berger and Weber, 1991]. More exceptionally, the peridotites contain crystallization products ("cumulates") derived from the melts which circulated through them, the nature and distribution of which bring valuable information about the melt segregation and migration processes. Occurrences of a glass phase possibly ascribable to chilled partial melts have been reported but are very uncommon.

Descriptions of the various occurrences of mantle outcrops for the structural characters mentioned above have given rise to an abundant literature. The present paper makes no pretention of bringing an exhaustive overview of these data. We will rather select a few basic problems, in which we have personal practice, concerning mantle processes beneath spreading ridges and show to what extent a solution can - or cannot! - be found in the observations conducted so far in mantle outcrops. A special effort will be devoted to make a clear distinction between first order answers, directly given by the structural and petrographic data themselves, from the more indirect conclusions, inferred from theoretical reasonings inspired by the same data. Our aim is to suscitate the critical reading of the papers referenced, where data and models are published *in extenso*.

2. STRUCTURAL STUDIES IN MANTLE PERIDOTITES

2.1. Modes of occurrence

Four modes of occurrence of mantle rocks at the Earth's surface are currently distinguished:

Nodules are upper mantle fragments expelled to the surface by intraplate volcanism. In spite of their small size (< 1 m) which precludes their use for any reconstruction of flow geometries, they bring valuable information about the flow mechanisms and the stress level in the mantle at depths ranging from the Moho to about 200 kilometers [Boullier and Nicolas, 1973; Green and Gueguen, 1974; Boullier and Gueguen, 1974; Mercier and Nicolas, 1975; Francis, 1987; Berger and Weber, 1991]. Exceptionally, nodules contain glass blobs that could have developed by incongruent melting of enstatite [J.-C. Mercier, pers. com.), by anatexis in a shallow magma chamber [Maaloe and Printzlau, 1979] or by contamination with the host magma; whatever the origin of the melt, the textural relationships between the glass and the residual crystals are very informative about the mobility of an intergranular melt phase in the mantle. More frequent is the occurrence of concentrations of mafic minerals, likely precipitated from melts circulating in veins or cracks; they provide important constraints on melt composition and segregation processes at great depth in the mantle [Wilshire and Kirby, 1989; Haggerty and Sautter, 1990]. In some cases, clear spatial and/or temporal evolutions of the composition and/or structure have been put in evidence for the nodules expelled by a single, or by a group of volcanoes. This allows one to infer a corresponding zonation in the underlying mantle and to have access to large scale processes [Sen and Presnall, 1986; Nicolas et al., 1986].

Abyssal peridotites are dredged, or exceptionally drilled or sampled from submersibles, in areas of reduced crustal thickness and/or active faulting of the oceans. They are the only windows open on the present-day mantle; their properties can thus be related without any ambiguity to active spreading processes. Although most of the samples are highly serpentinized, some of their structural and petrological character can still be deciphered; they have been correlated with some success to other petrological and geophysical data collected along ocean ridges [e.g. Boudier, 1978; Nicolas et al., 1980; Auzende et al., 1984, 1989; Dick et al., 1984; Michael and Bonatti, 1985; Dick, 1989; Cannat et al., 1990a,b; Johnson et al., 1990; Mevel et al., 1991]. In spite of the difficulties inherent to undersea structural studies, they constitute a useful test of the models of oceanic lithophere derived from the detailed observations conducted in ophiolite massifs [cf. Auzende et al., 1989; Lagabrielle and Cannat, 1990].

Orogenic peridotites are large - up to several tens of kilometers in extent - thrust sheets generally associated with granulitic terrains and scattered along the suture zone between two major geological provinces. These massifs can be pervasively affected by regional metamorphism [e.g. Cordellier et al., 1981; Drury et al., 1990; Scambelluri et al., 1991] which obliterated most of their asthenospheric paragenesis and structures. Originally, they were interpreted as pieces of the continental lithospheric mantle [Jackson and Thayer, 1972]. Although they are not unambiguously associated with basic extrusive rocks and silicic sediments, their relation to some kind of oceanic spreading environment, most likely marginal basins related to active margins, is not excluded. Most of them have a relatively undepleted average composition, reflected in their lherzolitic modal composition and in their chemistry, pointing to moderate extent of partial melting [e.g. Bodinier et al., 1988; Seyler and Mattson,

1989]. When not overprinted by regional deformation and metamorphism, abundant melt migration structures are currently observed in these lherzolites. Some of them have been ascribed to incipient partial melting in a rising diapir [Nicolas, 1986a]. In that case, they would offer the unique opportunity to constrain melting models and to put in evidence the role of plastic flow in collecting the first melt fraction. More frequently, however, they bear witness to subsequent reactions between the solid mantle and a melt which has not necessarily a direct genetic relationship with the residual peridotite it crosses [e.g. Quick, 1981a,b; Bodinier et al., 1989; Kelemen, 1990]. Orogenic peridotites are also frequently layered; this layering brings information about melt segregation processes which took place at great depth, before the end of the plastic deformation, and possibly very early in the Earth history [Polvé and Allègre, 1980; Allègre and Turcotte, 1986]. Orogenic peridotites might constitue a clue to understanding the origin of mantle heterogeneities.

In **ophiolite massifs**, ultramafic rocks are clearly overlain by a largely mafic section whose structure mimics the stratified present-day oceanic crust. One of the main advantages of ophiolitic peridotites is precisely this possibility to interpret their flow structures in an oceanic spreading reference frame (section 2.2). The largest and best preserved outcrops of mantle rocks are exposed in ophiolites (almost 20,000 square kilometers of sub-continuous outcrops of harzburgites in Oman). Hence, the scale of the structures mapped in ophiolites may be comparable to those resolved along present-day oceanic ridges. Accordingly, in the present paper, which is devoted to ridge processes, we will mostly discuss ophiolitic structures. However, some limitations in the use of ophiolites as witnesses of oceanic ridges have to be noted before going further. Only the uppermost part of the mantle crops out in the ophiolite massifs; the maximum paleo-depths accessible to the observation are estimated to reach about 10 km beneath the paleo-Moho in the best cases. No direct information on processes occurring at depths where partial melting initiates in a steady-state spreading situation (greater than 50 km) can be gathered in ophiolites. Another problem frequently advanced concerns the equation between ophiolites and "normal" oceanic lithosphere. As a matter of fact, there are significant differences in chemical composition between the average present-day ocean crust and the crustal section of most ophiolites. Some authors, following Miyashiro [1973], have proposed to equate ophiolites with fossil island arcs. This hypothesis is actually consistent with some (but not all) of the geochemical characters of the ophiolites [e.g. Pearce et al., 1981] and is not unlikely in the case of the peri-Pacific belt where the geological evolution is controlled by subduction since paleozoic times [e.g. Irwin, 1979; Coleman, 1984]. However, in the case of the Tethysian ophiolitic belt, it is in complete contradiction with regional geology [e.g. Coleman, 1977, 1984]. The contemporaneity between the formation and early thrusting events has been demonstrated for some of the main tethysian ophiolites. In other words, it is the ridge itself, or at least very young oceanic lithosphere, which has been sampled during the obduction process. Accordingly, it is clear that the details (and may be more!) of the accretion geometry and melt composition is influenced to some degree by early emplacement tectonics. It is probably there that an explanation of the geochemical peculiarities of the tethysian ophiolites has to be looked for. Nevertheless, it remains that *in ophiolites we analyse a fossil environment where spreading did occur and where basaltic melt was extracted from the mantle to feed crustal magma chambers; although a naïve analogy with present-day ridges is an unacceptable approach, it is possible to deduce from the ophiolitic record some information relevant to major physical and petrological processes occurring in the shallow mantle.* The present paper illustrates this last approach. It will be mostly inspired by observations conducted in the Oman ophiolite where accurate reconstitutions of the paleo-spreading geometry are made possible thanks to its exceptional size and to the perfect outcrop conditions. References to other ophiolites will be made to test the representativity of the Oman case. A more exhaustive description of some representative ophiolites and orogenic peridotite massifs can be found in Nicolas [1989].

2.2. *Methodology*

In this section, we present a short summary of the structural methods developed for the study of mantle rocks. For a more complete overview of this topic, the reader is referred to Nicolas and Poirier [1976], Green [1976], Gueguen and Nicolas [1980], Mercier [1985], Poirier [1985] and Nicolas [1986b].

2.2.1. Textures of mantle peridotites. Plastic deformation of a polycrystalline aggregate like the mantle occurs through slip and climb of dislocations within individual crystals accompanied by dynamic recrystallization. These mechanisms confer to mantle rocks a characteristic texture which reflects to some extent the physical conditions prevailing during deformation. To look at the texture is thus a fundamental step for distinguishing between structures acquired during accretion (high-T°, i.e. close to solidus, or "*asthenospheric*" deformation) from those recorded after the lithosphere formation (low-T° or "*lithospheric*" deformation).

Ophiolitic peridotites are derived from a very shallow level of the oceanic mantle where temperature varies far more than any other physical parameter. Hence, the deformation history of ophiolitic peridotites is controlled by their thermal evolution. The deviatoric stress prevailing during deformation is itself linked to the temperature through the flow law. Beneath the ridge axis, the mantle temperature is estimated to be between 1250°C and 1300°C. At such a temperature, the flow rate of the peridotites under stresses of the order of one MPa is compatible with the rate of shearing induced by plate movements (about 10^{-14} to 10^{-15} s^{-1}). On spreading away from the axial zone, the mantle cools through heat conduction to the surface, thus transforming into lithosphere. Stress conditions prevailing during deformation of the mantle after its transformation into lithosphere are characteristically higher than those related to asthenospheric shear flow. A drop of temperature from 1200°C to 1000°C, for example, corresponds to an increase in stress of about 10 MPa, all other parameters remaining constant [e.g. Goetze and Evans, 1979; Darot and Gueguen, 1981; Kirby, 1983]. It is impossible to determine the temperature which prevailed during plastic deformation of ophiolitic peridotites from a simple petrographic investigation within the P-T field considered, which do not correspond to any phase change. However, it can be estimated indirectly. There are several ways of estimating the magnitude of the deviatoric stress which prevailed during rock deformation. Of these paleo-piezometers, the size of recrystallized grains is the most reliable being the most robust regarding eventual thermal and/or tectonic events postdating the main episode of deformation. The recrystallized grain size is related to the major deformation episode, i.e. the one responsible for the macroscopic structures measured in the field. In a compilation of these piezometers, Karato [1984] showed that one order of magnitude increase in the size of the recrystallized grains corresponds to about a one order of magnitude decrease in the deviatoric stress.

According to this outline, two main textural categories of mantle peridotites can be distinguished (fig. 1). The first includes the coarse-grained equigranular texture (fig. 1a) featuring a unimodal grain size distribution around an average of 3 - 4 mm (corresponding to stresses around 0.5 MPa). The grain boundaries are sharp, often slightly curved, and make up 120° triple junctions. The grains are equant to slightly flattened and elongated. Most of them lack optically distinguishable substructures. Only the more or less developed lattice fabric (see § 2.2.2.) shows that these rocks have undergone plastic deformation, thus distinguishing them from cumulates. Some grains are polygonized into subgrains, reflecting a very efficient migration of the dislocations toward dislocation walls. Due to the strong shear strain, most of these sub-boundaries have evolved into grain boundaries (disorientation exceeding about 15°) through progressive rotation. This mechanism is accountable for the nearly equant texture of the rocks. It requires a temperature close to the solidus as far as low stress conditions are concerned. Solidus to hypersolidus temperatures are also indirectly recorded by the presence of variously deformed interstitial clinopyroxene and plagioclase bearing witness to the presence of a basaltic melt impregnating the residual peridotite.

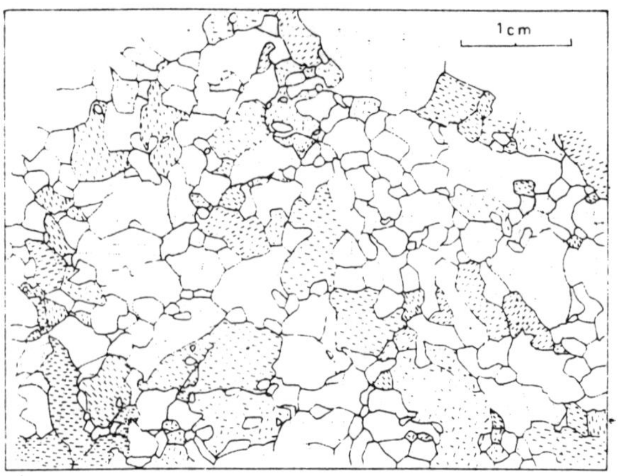

a.

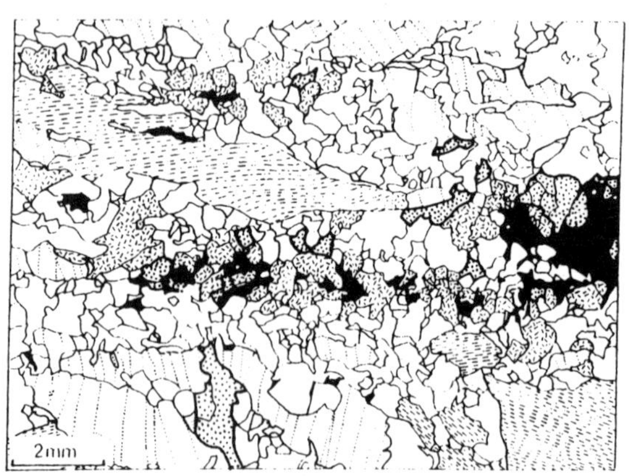

b.

Fig. 1. Microstructures of mantle peridotites. a) coarse-grained equigranular (from Boullier, 1977); b) porphyroclastic (from Mercier and Nicolas, 1975); Open areas: olivine (dottet lines are traces of sub-grain boundaries); dashed areas: orthopyroxene; dotted areas: clinopyroxene; black areas: spinel.

The second textural category is the porphyroclastic texture (fig. 1b) featuring a bimodal grain size distribution of millimeter-scale porphyroclasts being worn down into a fine-grained matrix. The grain size distribution in the matrix is often widely scattered between 100 and 200 microns to a few tens of microns, corresponding to stresses of a few tens of MPa to several hundred MPa. Where the lithospheric deformation is very intense, the porphyroclastic texture evolves into mylonitic texture. In the porphyroclastic and mylonitic textures, matrix recrystallization occurred mainly through nucleation and growth at the edge of the porphyroclasts. Most of these are very elongated and the substructure is quite dense, indicating a poor migration of the dislocations. This poor recovery means that the porphyroclastic texture developed, or at least ended, at temperatures a few hundred degrees below the solidus of the rock.

As a rule, the porphyroclastic and mylonitic textures become increasingly developed close to the basal thrust plane of the ophiolites. This evidence, together with the concordance between the deformation recorded by the peridotites there and by the amphibolites of the metamorphic sole indicate that the lithospheric deformation currently relates to the thrusting event. However, high-stress textures can also develop during the accretion stage, as witnessed by the porphyroclastic and mylonitic samples found among the abyssal peridotites; in that case, they reflect transform or normal fault tectonism (e.g. Cannat, 1990; Mevel et al., 1991].

2.2.2. Kinematic analysis of mantle peridotites Plastic deformation confers to mantle rocks a shape and a lattice fabric. Shape fabric features, foliation and lineation, correspond respectively to the average orientation of the flattening plane and to the stretching direction of the minerals that make up the rock. Although the precise microscopic mechanisms leading to the development of the lattice fabric are not definitely elucidated, it results likely from the progressive rotation of the atomic lattice of the individual crystals during deformation by intracrystalline slip: as the deformation progresses, the dominant slip planes of the dislocations tend to parallel the shear plane (in the case of simple shear) or the two conjugate shear planes (in the case of coaxial strain). The efficiency of this rotation depends on the orientation of each crystal relative to the applied stress. Recrystallization at grain boundaries appears as a minor but geometrically necessary mechanism. This has been reproduced through simple numerical models like the one conducted by Etchecopar [1974, 1977] (fig. 2a). Kinematic analysis, permitting a precise definition of the flow direction, and of the shear sense in non-coaxial strain, relies on relating the shape to the lattice fabric.

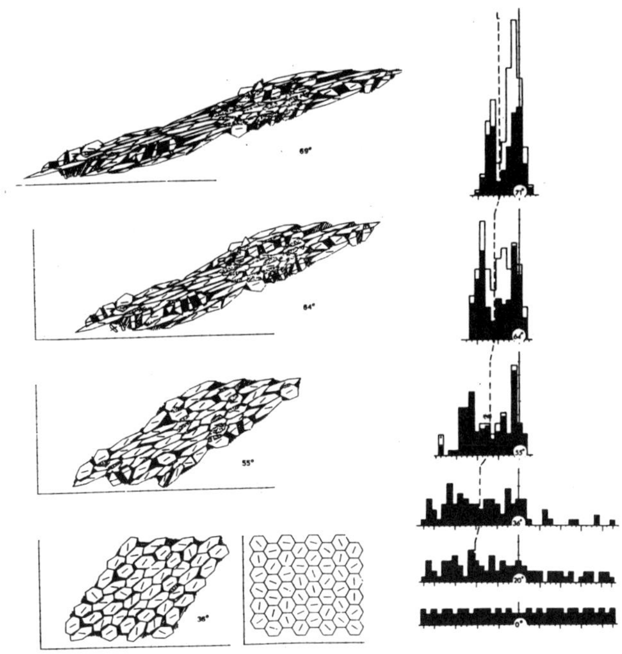

a.

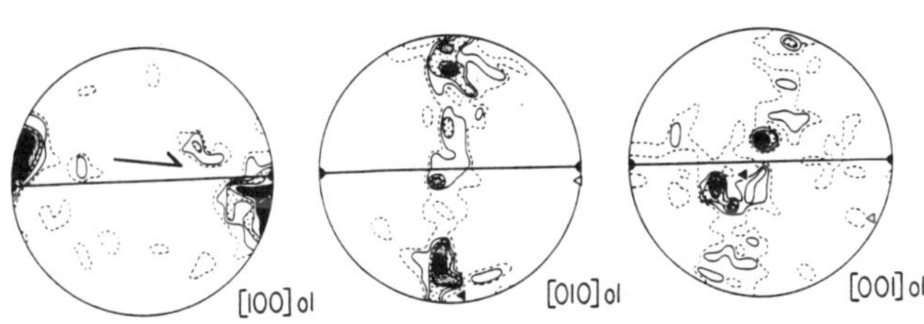

b.

Fig. 2. a) Textural changes with increasing simple shear of a planar network of hexagonals cells. The line inside the cells is the slip line. Histograms: slip lines preferred alignment with increasing simple shear corresponding to the sequence presented on the left. The graduated zero line is the shear direction. The dashed line is the mean elongation direction in the deformed aggregate. From Etchecopar (1974); b) Typical preferred orientation of olivine crystallographic axes in a coarse-grained equigranular harzburgite of Oman. Arrow: shear sense; dashed line: shear plane; horizontal line: foliation trace; dot: lineation. Equal-area projection, lower hemisphere. 100 measurements. Contours: 1, 2, 3, 4, and 5%. From Ceuleneer (1986).

For example, in the olivine fabric of figure 2b, the [100] crystallographic axis of the olivine crystals is oriented in a direction close to that of the lineation while the [010] and the [001] axes spread out in a girdle perpendicular to the lineation. The [100] axis and the (010) and {0kl} planes correspond respectively to the preferential slip direction and slip planes of the dislocations in olivine crystals in high-temperature conditions, i.e. more than about 900°C [e.g. Mercier, 1985]. As, in our example, the lattice fabric is strong and slightly oblique to the shape fabric, it may be deduced that this rock has recorded high plastic strain under simple shear (non-coaxial) regime. The average orientation of the (010) plane of the olivine crystals is equated to the shear plane, and the average [100] direction to the flow direction. The shear sense (dextral in our example) is that of the rotation the foliation would require in order to coincide with the shear plane. It is currently determined under the polarizing microscope in thin sections cut in a plane perpendicular to the foliation and parallel to the lineation, the trace of the shear plane being deduced from the orientation of the preferred extinction of the olivine crystals. The same method can, of course, be applied to the pyroxenes. Shear sense determinations based on olivine or enstatite fabrics are generally consistent [e.g. Boudier and Coleman, 1981]. Petrofabric analysis helps also to evaluate the magnitude of the asthenospheric strain. The angle α between the foliation and the shear plane is ideally related to the shear strain γ according to $\gamma = 2 \cot 2\alpha$. However, our experience shows that the use of α to define accurately a local value of γ is dubious. Only the average value of α in zones where the flow is homogeneous gives reliable results. Other criteria such as the strength of the lattice fabric, and occasionally the deformation of chromite nodules can also be used to estimate the strain magnitude.

In the field, the plastic deformation structures of the peridotites are revealed by the flattening and lengthening of pyroxene and chromian spinel crystals. The latter, scattered in the rock and constituting usually less than 1% of its volume, can be considered as passive strain markers. An oriented sample is taken on a regular, usually kilometric, sampling mesh in order to determine the shear sense and the microstructure. When not seen in the field, the foliation and lineation are measured in the laboratory on oriented specimens bleached with HCl, the black metallic colour of the spinels contrasting with the pale yellowish colour of the bleached silicates. In ophiolite massifs, the *regional* attitude of the contact between mantle peridotites and the basal cumulates of the crustal section (paleo-Moho) is taken as the paleo-horizontal. The trend of the sheeted dyke complex is assumed to be parallel to the axis of the paleo-ridge [Cann, 1974]. This hypothesis is confirmed by field studies of the emerging parts of the mid-Atlantic ridge [Helgason and Zentilli, 1985], by undersea observations [Auzende et al., 1989] and by the seismic anisotropy of the present-day oceanic crust [Shearer and Orcutt, 1985]. The flank of origin of the ophiolite can sometimes be deduced from the preferred chilled-margin facing direction in the sheeted dyke complex [Cann, 1964]. However, in the Oman ophiolite, it has been shown that most of the dykes have two chilled margins and, for those which have been cut in their middle, the designation of chilled margins to one or the other preferred facing direction is quite weak [Pallister, 1981 and personnal observations].

To sum up, the information accessible through the structural study of mantle rocks concerns the *plastic flow geometry* at the time of cessation of the flow, i.e. at the time of cooling of the asthenosphere. A map of foliation and lineations trajectories gives an image of the deformation field recorded by the peridotite massif. Provided the strain magnitude is sufficiently large, the *deformation field* is a good approximate picture of the plastic *flow field* [e.g. McKenzie, 1979]. The obliquity between the shape and the lattice fabrics is currently less than 15°, a value which makes possible such an approximation. Consequently, the asthenospheric flow direction is generally equated with the orientation of the stretching lineation. However, there is no way to determine the absolute magnitude of the flow velocity. Only relative values, i.e. the local *velocity gradients*, which are actually responsible for the plastic deformation, are preserved in mantle rocks. The direction of the local velocity gradient is given by the shear sense and its magnitude is proportional to the finite strain. In ophiolites, where mantle rocks are overlaid by the crustal section, these geometrical data can be analyzed in a reference frame related to the paleo-horizontal and to the paleo-spreading direction.

3. PLASTIC FLOW PATTERNS IN OPHIOLITES

Thanks to the persistent efforts of several geologists during the last twenty years, most of the major peridotite massifs cropping out all around the world are now described for their plastic flow structures. Among the few exceptions, the Papouasian ophiolites are probably the main ones. In spite of the hugeness of the data set (several tens of thousand field measurements), the important results are relatively easy to summarize. The reason lies in the restricted variety of contrasted flow patterns put in evidence [Nicolas and Violette, 1982; Boudier and Nicolas, 1985; Ceuleneer et al., 1988] which reflects probably the homogeneity of the spreading process itself. We will now present some of the main progress in our understanding of these processes which were triggered by these data. The following points will be discussed successively: the spreading dynamics, the upwelling geometry, and evidence for along axis variablity.

3.1. "Standard" flow pattern: an off-axis record

3.1.1. Overview of the data. The mantle flow pattern which is by far the most frequently observed in ophiolites (fig. 3) features flow planes sub-parallel to the paleo-Moho and flow lines at a large angle (on average perpendicular) to the azimuth of the sheeted dyke complex (presumed strike of the paleo-ridge axis). At first sight, this result seems somewhat disappointing. Due to the usual lack of recognized continuity with a paleo-upwelling, its interpretation in terms of axial processes is problematic. However, the richness of this pattern lies in its omnipresence in the ophiolitic record, at least in those ophiolites having a harzburgitic mantle section. The few structural features related to it, and summarized below, have been repeatedly observed in several ophiolites: New Caledonia [Cassard, 1980; Cassard et al., 1981; Prinzhofer et al., 1980; Nicolas and Prinzhofer, 1983], Newfoundland [Girardeau, 1979, Girardeau and Nicolas, 1981], some massifs in the Philippines [Violette, 1980], some Turkish ophiolites [Juteau et al., 1977; Reuber, 1985], and several Oman massifs [Ceuleneer, 1986; Ceuleneer et al., 1988; Nicolas et al., 1988]. Such a constancy confers to these observations the strength of well established facts that any model of oceanic spreading should be able to account for. Structures will be presented later which are more spectacular and informative about axial processes, but for which a few cases only - when not just a single one - are documented. They could lead to the hazardous generalization of some local and/or anecdotal features.

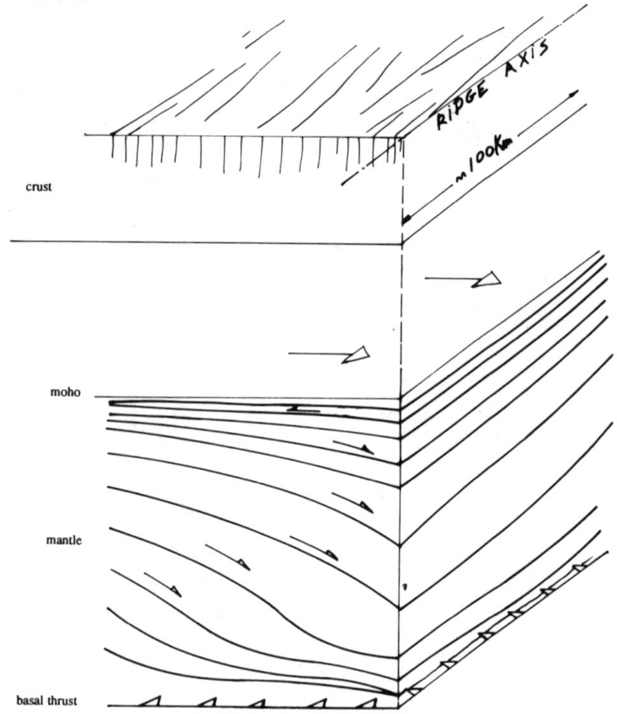

Fig. 3. Sketch summarizing the main characters of the "standard flow pattern" discussed in the text.

The first characteristic of this flow pattern consists in its great *spatial homogeneity* (fig. 3). Its maximum recognized extension is of about 40 km along the strike of the paleo-ridge axis (Hilti massif of the Oman ophiolite) but reaches 200 km when considering adjacent massifs which are now separated by post-obduction faults, from the Semail Gap to the Wadi Fizh area, in Oman [Ceuleneer, 1986; Nicolas et al., 1988]. It has been observed on a distance of 80 km in the direction normal to the paleo-ridge and down to a paleo-depth of 10 km. In smaller ophiolites, it is currently observed at the scale of an entire massif (a few tens of kilometers). To illustrate what we mean by "great homogeneity", the orientation statistics for the Hilti and Southern Fizh massifs of Oman (a 100 km-long segment of the Oman paleo-ridge) are presented on the stereonets of figure 4. The dispersion in the plastic flow orientations is remarkably weak considering (i) the size of this area (about 3,000 km^2), (ii) the occurrence of large serpentinized faults cross-cutting these massifs and which likely introduce some "noise" in the data (how much would necessitate a specific study), and (iii) the tenuous character of the plastic deformation structures, especially of the lineations, which means that a significant part of the dispersion can reflect imprecisions (or errors!) in the field measurements.

The second characteristic of this flow pattern is the *homogeneity of the shear sense* on almost all the exposed thickness of the mantle section (up to 10 km) (fig. 3). In areas where a high density of shear sense measurements were performed from thin sections carefully cut perpendicular to the

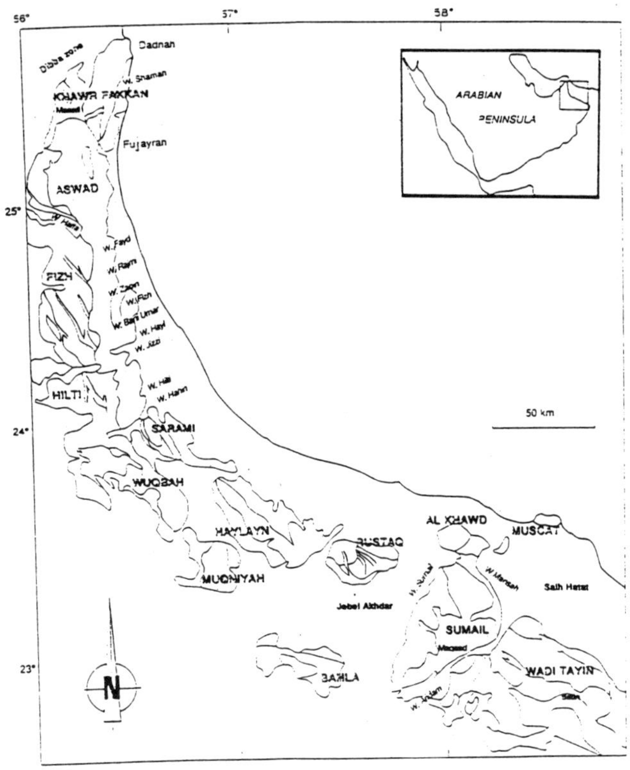

foliation and parallel to the lineation, about 80% of consistency is currently found for one or the other shear direction [Girardeau, 1979; Misseri, 1982; Ceuleneer, 1986, 1991]. This means that the simple shear deformation regime deduced from petrofabrics at the scale of a thin section can be largely extrapolated to the scale of several tens of kilometers. A *shear sense inversion* is currently observed at shallow, but variable, depth beneath the paleo-Moho (fig. 3). It occurs at about 2 km depth in the Table Mountain massif of Newfoundland [Girardeau, 1979], 1 km in the Massif du Sud of New Caledonia [Cassard, 1980], 500 m in the Hilti massif of Oman [Ceuleneer, 1986] and about 100 m in both the Coto area of the Zambales massif in the Philippines [Violette, 1980] and the Maqsad area of Oman [Ceuleneer, 1991].

The third characteristic of the standard flow pattern lies in the *vertical evolution of the shear strain* (fig. 3): as a rule, it is increasing upward from moderate values ($1<\gamma<3$) in the main part of the mantle section to very high values ($\gamma >10$) at the top of the mantle section. The most intense shear strain are realized just beneath the cumulates of the crustal section; however, the latter show virtually no sign of *plastic* strain and a sharp decrease of the shear strain within the top few meters of the mantle section has been locally observed in well exposed areas [e.g. Ceuleneer, 1986]. The zone of marked increase of the shear strain coincides roughly with the zone of shear sense inversion.

The last characteristic related to the standard flow pattern is a slight but significant increase of the flow plane paleo-dip with increasing depth (fig. 3). This has been clearly observed in Oman [Ceuleneer, 1986] and in Bay of Island [Girardeau, 1979]. At the top of the mantle section, the flow plane is definitely parallel to the paleo-Moho. In the deeper parts of the mantle section, it increases gradually downsection up to values of about 30°. This increase occurs systematically in the same direction relative to the shear sense. As illustrated in figure 3, the dip direction coincides with the direction of relative displacement of the upper section of the mantle indicated by the shear sense below the zone of inversion.

3.1.2. Discussion: implications for spreading dynamics. Our objective being to learn something about the geometry and dynamics of *active* asthenospheric flow, it is clear that we have to make some hypothesis about the way the plastic flow structures were frozen. Assuming that a drop in temperature of

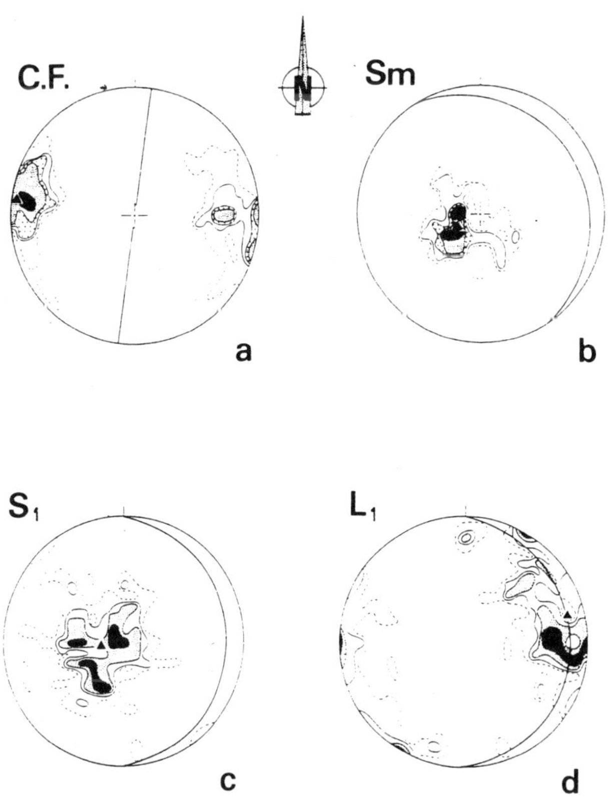

Fig. 4. Preferred orientation of the structural elements in the Hilti and southern part of the Fizh massif (Oman ophiolite). This illustrates the homogeneity of mantle structures along a 100 km long paleo-ridge segment. The significant dispersion in lineation azimuths reflects essentially a progressive rotation of the flow lines into parallelism with the paleo-ridge axis toward the northern part of this segment, and points to 3-D complexity of the upwelling upstream. Black triangle: best computed axis. a) Dykes from the sheeted dyke complex; 231 measurements. Contours: 1.3, 2.6, 4.3, 10.4, and 18.2%. b) Magmatic lineations in the basal cumulates; 46 measuremnts. Contours: 2, 4, 6, 8 and 10%. c) foliations and d) lineations in mantle peridotites displaying an asthenospheric texture. 119 measurements. Contours: 1, 2, 4, and 7%. From Ceuleneer, 1986.

about 200°C-300°C is sufficient to preclude the peridotites from flowing under the weak deviatoric stress prevailing in the asthenosphere (a few MPa), it is deduced that the mantle section of ophiolites, whose thickness never exceeds 10 km, was incorporated into the lithosphere at distances from the ridge axis which are of the order of several tens of kilometers (ages ≤ 2 Ma). This is a maximum estimate based on steady-state thermal profiles [Morton and Sleep, 1985]. Higher cooling rates can be expected in the case of a sudden shift from expansion to compression related to early emplacement tectonics. The occurrence of such a rapid inversion is clearly documented in the case of the Oman ophiolite [Ceuleneer, 1986; Montigny et al., 1988]. Accordingly, there are two schemes in which to interprete the flow structures recorded in ophiolites. Either, we consider that they were progressively frozen in response to conductive cooling of the lithosphere related to oceanic spreading; in that case, they bear witness to the flow character in the thermal boundary layer developing at the base of the lithosphere, and the deepest parts of the mantle section of the ophiolites are also those which were accreted farthest from the ridge axis. The other alternative is to consider that the cooling was virtually instantaneous, in response to the thrusting event; in that case, a three-dimensional snapshot of the axial flow pattern has been recorded and can be exposed in ophiolite massifs, provided the ridge itself is sampled by the thrusting process. Rapid cooling seems to have occurred at least in the few massifs where upwelling flow has been frozen [Rabinowicz et al., 1987]. However, the actual cooling process is likely a combination of the two extreme mechanisms envisaged above. As a matter of fact, even in the case of instantaneous cooling, it is clear that, considering the extension of most massifs *along* the spreading direction (several tens of kilometers), a significant proportion of the mantle section was already incorporated into the lithosphere by conductive cooling associated with steady-state spreading when the shift from expansion to compression occurred. The lower the spreading rate, the larger this proportion should be. The exact value remains unknown. The accurate reconstruction of a vertical profile for some mantle character is thus strongly model-dependent: it will never be possible to separate, in the sections presently exposed in ophiolites, what is actually attributable to a paleo-vertical gradient from what reflects a possible horizontal evolution along the paleo-spreading direction.

According to this outline, let us discuss the few conclusions about mantle dynamics that can be deduced from the observations summarized above. The orthogonality between the flow lines and the strike of the sheeted dyke complex shows that the asthenospheric flow geometry is conditioned by the spreading geometry at a short distance from the ridge axis. The homogeneity of this pattern over distances reaching 200 km along the ridge axis shows that any along strike flow component induced by possible small-scale 3-D flow structure beneath the ridge axis is rapidly superseded by the orthogonal flow component. On a strictly geometrical ground, this pattern confirms the information brought by the azimuthal anisotropy of seismic wave propagation in the present-day oceans: the direction of maximum seismic wave velocity, which matches the [100] axis of olivine crystals, is usually parallel to the spreading direction [Francis, 1969, Raitt et al., 1971]. It suggests that the mantle circulation associated with oceanic spreading is largely driven by the drift of the overlying plate but it does not preclude that the flow might be three-dimentional. It just indicates that asthenospheric flow rates and plate velocities have comparable magnitudes.

However, although this driving mechanism is consistent with the general mantle flow geometry, it does not predict the vertical deformation gradients recorded in ophiolites, and particularly these two unexpected characters: the shear sense inversion and the dramatic increase of the finite strain in the uppermost part of their mantle section. This last point evokes a sharp velocity gradient, and thus the existence of a dynamical boundary layer at the mantle crust interface. What is surprising is the extreme thinness of this boundary layer (several hundred meters, at most). Constant viscosity mantle flow models predict dynamical and thermal boundary layers not thinner than one tenth of the lateral flow extent [McKenzie et al., 1974; Oxburgh and Turcotte, 1978]. This leads us to consider viscosity drops of a few orders of magnitude at the top of mantle upwellings. Further evidence for this hypothesis will be provided in the next section. The shear sense inversion points to an inversion of the flow rate gradient close to this boundary layer, as illustrated in the inferred flow velocity profile of figure 5. This means that, at shallow depth and close to the ridge, the asthenospheric flow was driven by its own buoyancy forces at a higher rate than the drift velocity of the overlying plates. Beneath the zone of inversion, the shear sense could reflect either the flow rate gradient associated with the convective flow, or the one induced by the plate drift (forced flow). However, the increase with depth of the dip of the flow plane in a direction pointing away from the ridge suggests that the isothermal surfaces were slightly convex upward (fig. 5), and not downward as predicted by the conductive cooling law [e.g. Parker and Oldenburg, 1973; Forsyth, 1977]. This shape of the thermal field is observed when a hot convective plume is cooled along a cold upper boundary [e.g. McKenzie et al., 1974]. This strongly suggests that internal buoyancy forces are driving the mantle flow along the base of the lithosphere during periods of time which are commensurate with the cooling time of a 10 km-thick oceanic lithosphere (of the order of 1 Ma).

These data have important implications for mantle rheology beneath spreading centers. They suggest that, in a near-ridge environment, mantle flow forced by the plate drift coexists with a flow driven by internal buoyancy forces. The origin of this last driving force can be thermal (convection) or related to partial melting. These two mechanisms induce density contrasts of the order of one percent. Structural data in ophiolites do not allow us to decipher which one of these two forces is preponderant. However, they bring very strong contraints on the viscosity of the mantle in the area dominated by the active flow. Structural data show that velocities associated with the two types of flow are commensurate. This requires an effective viscosity of the order of 10^{18} Pa.s, which can be deduced simply from the Stokes law [Rabinowicz et al., 1984]. This value is relatively well bracketed: a lower viscosity would lead to upwellings narrower than those mapped in ophiolites, a higher viscosity would not allow the development of instabilites at a rate compatible with plate drift velocities. The existence of such a low viscosity level in the mantle has been invoked in some geophysical studies [e.g. Sleep and Rosendahl, 1979; Yuen and Fleitout, 1985; Buck and Parmentier, 1986]. Now, the problem concerns the mechanisms capable of decreasing the mantle viscosity by at

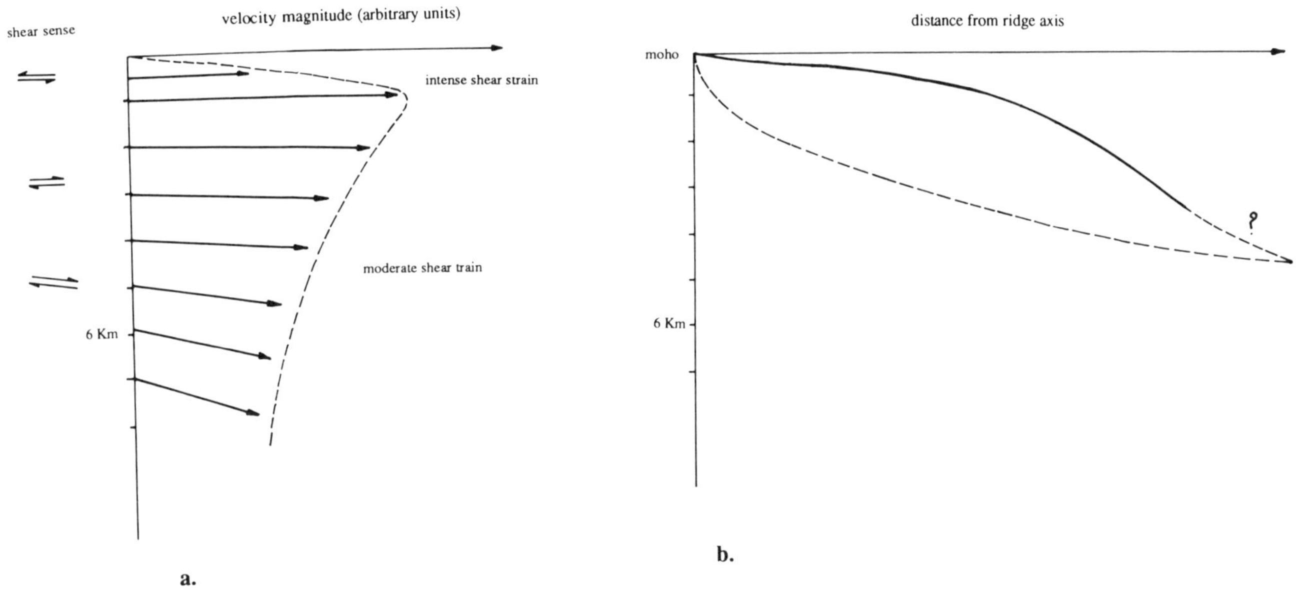

Fig. 5. a) Interpretation of the "standard flow pattern" in terms of a vertical flow velocity profile; the thickness of the upper dynamical boundary layer ranges from a few hundred meters to about 2 km from one massif to the next; b) Orientation of the accretion surface in a near ridge environment, in the case of progressive accretion of the mantle section away from the ridge axis; dashed: orientation of an isothermal surface deduced from a conductive cooling models.

least two orders of magnitude. One could propose that there are enough poorly constrained parameters in the mantle to get such low viscosity levels in the frame of a stress- and temperature-dependent rheology. However, a stress level exceeding a few MPa is not realistic in a rising diapir. It would be also in contradiction with the paleo-stresses deduced from the grain size of mantle peridotites [Karato, 1984]. In the same way, to invoke strong temperature variations in the upwelling seems inconsistent with the first order petrological homogeneity of the composition of the residue of melting and of the oceanic basalts, pointing to a relatively homogeneous temperature in the mantle beneath the ridge axis. The maximum range of *large-scale* temperature variations is estimated at 250°C, at most [e.g. Klein and Langmuir, 1987]. Local temperature variations, i.e. at a scale of several tens of kilometers, relevant to mantle diapirism beneath the ridge axis, are even lesser and cannot account for the viscosity contrasts of a few order of magnitude required by the ophiolite data [e.g. Kirby, 1983]. Accordingly, the only reasonable weakening mechanism seems to be the dependance of mantle viscosity on the melt fraction. And, at least, we are sure that melting does occur at oceanic ridges! The reduction of mantle viscosity in response to partial melting is widely admitted in the case of high/melt rock ratios. Evidence for this process will be presented in section 4.

3.2. Paleo-spreading structures and upwelling geometry

3.2.1. Overview of the data. The occurrence of vertical asthenospheric flow lines in ophiolites were first reported by Violette [1980] in the Acoje area of the Zambales ophiolite (Philippines). They were interpreted as an evidence for the existence of small-scale diapirism beneath oceanic ridges [Nicolas and Violette, 1982]. However, due to the poor quality and small extent of the outcrops, the picture which was drawn from this patchy data set remained very schematic. The term "diapir" refers to a possible mechanism accountable for the vertical flow, i.e. an upwelling due to an unstable density stratification. However, in mantle sections of ophiolites, where intrusive relationships between contrasted lithologies are not put in evidence, one should rather refer to "paleo-upwelling zones", or to "zones of paleo-vertical flow" if *upward* movement has not been demonstrated by a careful kinematic analysis (shear sense determination). The Canyon Mountain ophiolite (Oregon) can, in this respect, be considered as an exception. Misseri [1981] described there vertical flow lines clearly related to a diapiric intrusion of the asthenosphere into the crust. In spite of the richness of the structures observed in Canyon Mountain, they are not directly relevant to ocean ridge processes. As a matter of fact, this ophiolite is one of the few ones for which convincing evidence for an island arc origin can be found in both the petrology and the geological setting [Misseri, 1982; Misseri and Boudier, 1985].

The best exposed mantle upwelling described so far that is clearly related to a spreading environment crops out in the Maqsad area of the Oman ophiolite. During the course of a systematic mapping program of this huge remnant of the tethysian oceanic lithosphere, flow lines perpendicular to the paleo-Moho were shown to be correlated with particularly abundant melt discharge, now witnessed by a high density of chromite ore bodies and by some other petrological features [Ceuleneer and Nicolas, 1985]. Further work in the Maqsad area, including small-scale structural mapping and petrological studies, have confirmed this interpretation and given an accurate picture of flow kinematics and melt migration beneath a spreading axis [Ceuleneer, 1986, 1991; Ceuleneer et al., 1988; Lorand, 1988; Lorand and Ceuleneer, 1989; Leblanc and Ceuleneer, 1991; Quatrevaux et al., 1991]. Four other occurrences of possible paleo-upwelling zones have been reported in Oman: in the Batin, Wuqbah, Al Hawqain and Shamah areas [Nicolas et al., 1988; Ceuleneer et al., 1988; Reuber et al., 1991]. However, they have still not been substantiated by detailed studies, among which careful and statistically significant shear sense determinations are particularly critical.

3.2.2. An example of mantle upwelling: the Maqsad area in the Oman ophiolite. The main structural characteristics of the Maqsad area are summarized in a map (fig. 6), on a schematic cross section (fig. 7) and on stereonets of the structural data (fig. 8). The paleo-ridge axis has a N125°E orientation (fig. 8a) (the sheeted dyke complex outcrops lie out from the map area); it strikes parallel to a well-developed normal faults system (fig. 6) which was contemporaneous with high-temperature hydrothermal circulations, and associated mineralizations, and thus probably with the spreading activity [Beurrier, 1988; Leblanc et al., 1991]. At Moho level, offsets induced by these faults can be shown to reach a few hundred meters, but are, on average, of lesser amplitude. The paleo-Moho, although locally tilted in the vicinity of these faults, has a regional south-eastward dip of about 10° (fig. 8b). Tilts are witnessed in the girdle around a NW-SE axis defined by the basal cumulates layering planes (fig. 8b).

Considering the asthenospheric flow structures, the Maqsad area can be divided schematically into a zone where foliations are sub-perpendicular to the paleo-Moho (average westward dip of 75°) and a broader zone of sub-horizontal foliations (fig. 6). In the zone of horizontal flow, shear senses indicate a clear divergence from a 2 km wide band parallel to the paleo-ridge axis (fig. 6). This band is characterized by an exceptionally well developed dunitic transition zone between the mantle harzburgites and the gabbroic cumulates (fig. 7): it reaches here a thickness of 300 meters compared to its usual thickness in ophiolites (ranging from a few meters to several tens of meters). The origin of these dunites is to be looked for in the incongruent melting of enstatite due to circulation of a basaltic melt equilibrated at greater depth [Berger and Vannier, 1984; Nicolas, 1986a; Kelemen, 1990]. All the undeformed chromite pods mapped in the Maqsad area crop out in this dunitic horizon. Further arguments for abundant melt circulation in that zone will be given in section 4. This evidence, together with the diverging shear sense pattern, led to equate the dunitic band with the paleo-spreading axis related to the Maqsad upwelling [Ceuleneer, 1991]. The azimuth of the vertical flow planes ranges from WNW-ESE to NNE-SSW with a NW-SE preferred orientation, sub-parallel to the paleo-ridge axis (fig. 8c). Flow lines are vertical in the southeastern part of the zone of vertical foliations (fig. 8d). There, the finite strain is moderate compared to the zones of diverging flow, as illustrated by the olivine fabrics (fig. 9a). The shear senses are very homogeneous throughout the zone of vertical flow and can be interpreted in terms of a relative increase of the upwelling velocity toward the band of dunites (fig. 6 and 7); this is further evidence for the equation of this band with the paleo-ridge axis. The dip of the flow lines decreases abruptly toward the northwestern extremity of the paleo-axial zone. There, the flow planes have a poorly defined dip. The finite strain is very strong and the strain facies strongly linear, pointing to channelling of the flow in a narrow horizontal conduit. The shear sense indicates northwestward divergence of the flow, i.e. away from the upwelling (fig. 6).

Beneath the paleo-axial zone, no clear rotation of the vertical flow is observed: the plastic deformation structures become more and more tenuous when penetrating into the dunites of the transition zone and finally vanish, as illustrated by the largely random olivine fabrics (fig. 9b) and by the undeformed nature of the chromite ore bodies cropping out there. It has to be stressed that, in spite of their lack of lattice fabric, individual olivine grains making up these dunites frequently present sub-grain boundaries diagnostic of hot working in asthenospheric conditions, which is consistent with a residual origin. Away from the axial zone, the rotation of the vertical flow structures to the horizontal becomes more and more progressive. This evolution is correlated with the progressive deformation of the dunitic horizon (fig. 7). At a distance of about 1 km from the axial zone, a sharp but progressive rotation of the foliation planes have been observed in the uppermost 50 m of the mantle section in a sense consistent with the divergence of the flow from the axial zone (fig. 7). At a distance of 6-7 km from the axial zone, this rotation initiates at a paleo-depth of about 2 km in a way illustrated on figure 7 and 8g,h. The rotation is rather progressive and in a sense opposite to the one observed in the section close to the axial zone; here, the distinction between moderately dipping flow planes related to the upwelling and the zone of diverging flow becomes somewhat arbitrary as it does not correspond to any shear sense inversion. The boundary has been fixed for a paleo-dip of about 30° because it corresponds to a marked transition in the strain facies: finite strain is moderate for higher dips and evokes the deformation style in the main upwelling, while for lesser dips, very intense finite strains are recorded, as illustrated by the fabric of figure 9c, by the reduced thickness of the dunitic transition zone (a few tens of meters, fig. 7) and by the extreme stretching of the chromite ore bodies. However, no sharp rotation, which should be related to the turn back of the flow, is observed (fig. 7). It might be revealed by a more detailed field study. It is also possible that this turn back occurs along a discontinuity and does not induce any rotation of the flow. This example illustrates the ambiguity in deducing absolute displacement fields from plastic strain structures. The interpretation in terms of flow trajectories proposed in figure 7, matches the observed plastic deformation field but is certainly not unique.

On the northern flank of the paleo-axial zone, no rotation of the flow has been put in evidence. This could be ascribed to late obliteration by the wide serpentinized fractures limiting this side of the paleo-axial zone. However, it cannot be excluded that it reflects some original asymmetry of the upwelling. Another asymmetric feature of the Maqsad diapir can be found in the flow lines trajectories in the zone of diverging flow. In the northeastern diverging zone, flow lines strike NNE-SSW, strictly perpendicular to the paleo-ridge axis (fig. 6); they can be followed a distance of about 12 km along the spreading direction; further away from the paleo-ridge, an E-W flow pattern is observed, apparently unrelated to the Maqsad upwelling (fig. 6). At a distance of a few kilometers away from the axial zone, shear sense inversion is observed about 100 m below the Moho (fig. 6). On the southwestern diverging zone, the flow lines define curved trajectories (fig. 6): they become

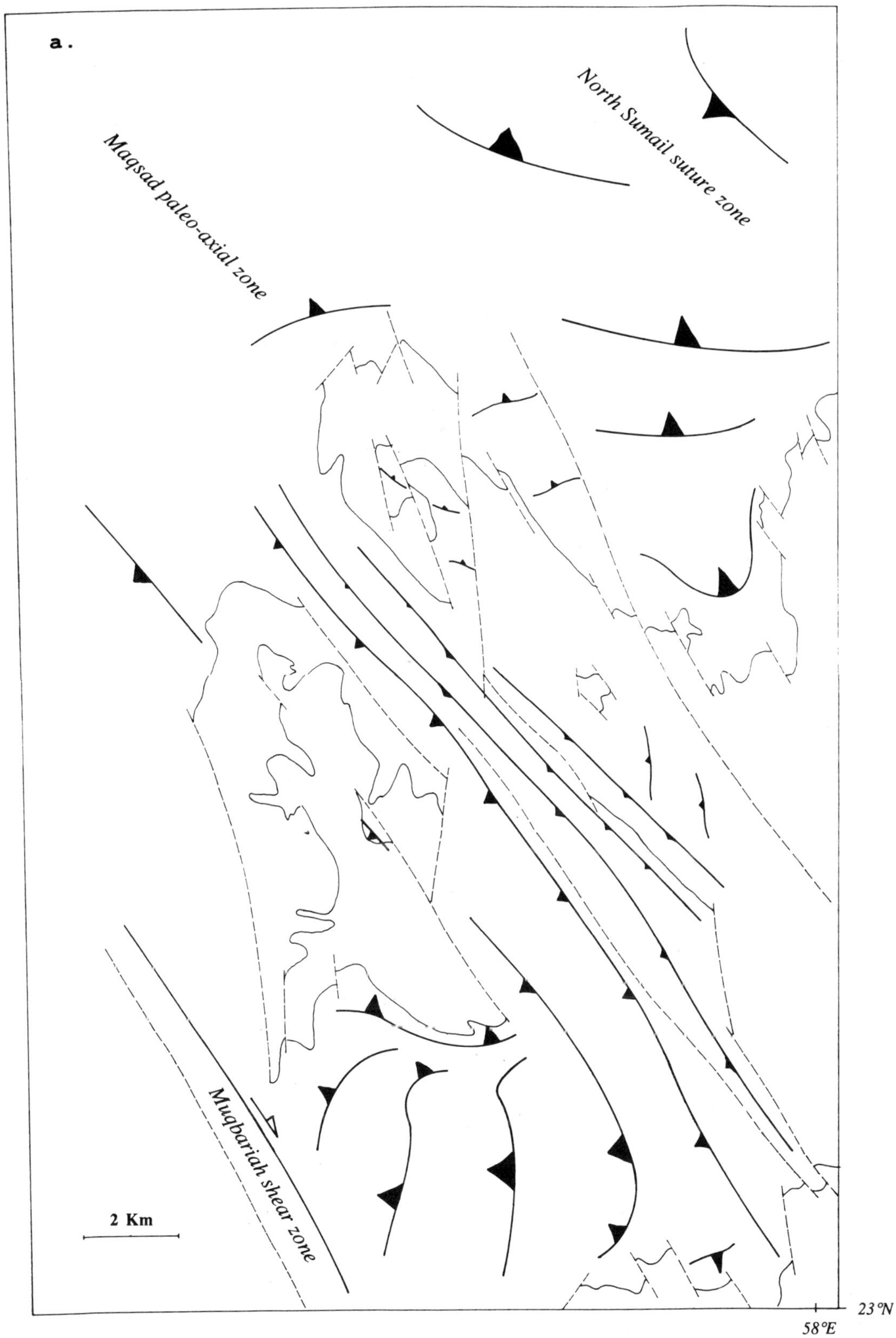

Fig. 6. Structural maps of the Maqsad area, in Oman. a) Flow planes trajectories; b) Flow line trajectories and shear senses; arrows: relative direction of movement of the upper block in area of weakly dipping foliations; stripes: trace of the foliation plane in area of sub-vertical foliations and lineations, the dot indicating the upwelling block.

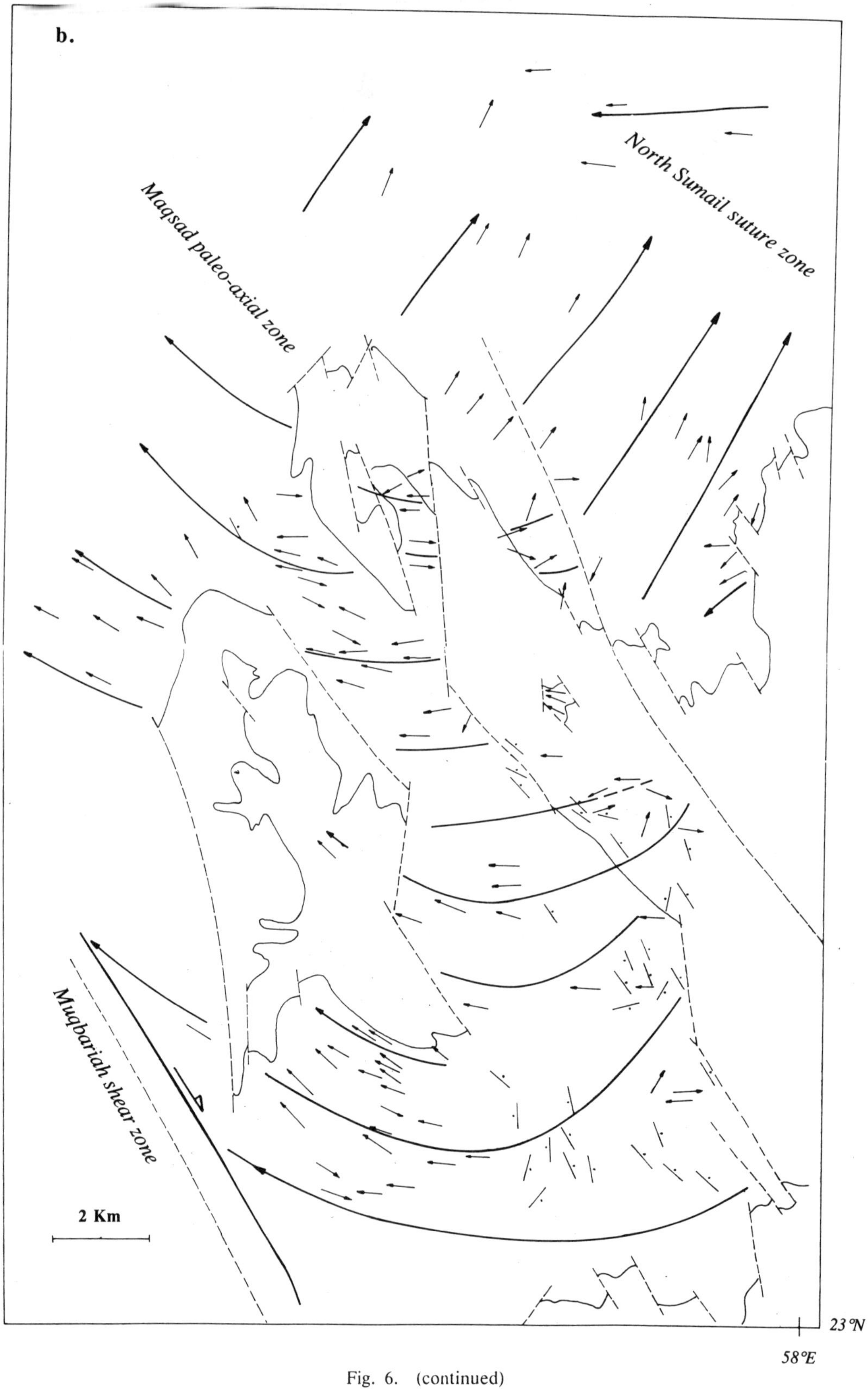

Fig. 6. (continued)

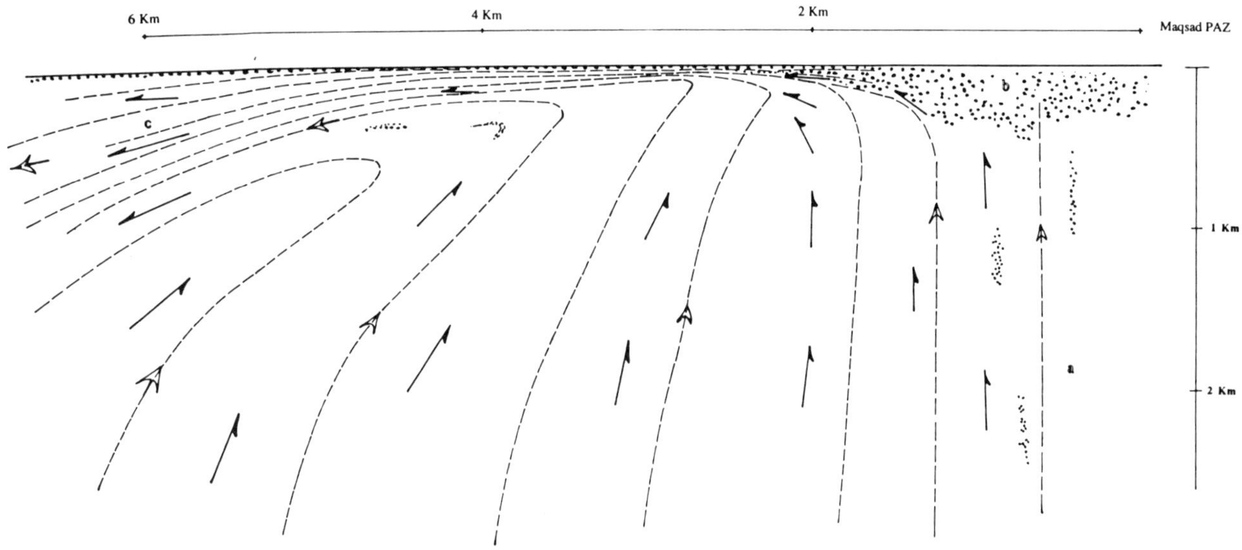

Fig. 7. Schematic cross section across the western part of the Maqsad upwelling. Trace of the flow planes and shear senses are shown (black arrows). Dotted lines and open arrows: interpretation in terms of flow trajectories. Dotted area: dunites. No vertical exageration. a, b and c refer to the olivine fabrics of figure 9.

oblique to the paleo-axial zone at a very short distance from it and turn progressively into parallelism with a 300 meters thick mylonitic shear zone (the Muqbariah shear zone) limiting the zone of diverging flow to the southwest, at about 12 km from the paleo-axial zone. The Muqbariah shear zone is likely responsible for this rotation, as its sense is consistent with the dextral play deduced from observations within the mylonites (fig. 6). The rotation occurred as the diverging mantle was still very ductile as no low-temperature microstructures overprint the peridotites in Maqsad, except in the immediate vicinity of the Muqbqriah mylonites.

3.2.3. Discussion: focusing and segmentation of mantle flow. The relationships between the Muqbariah shear zone and the asthenospheric structures in the Maqsad area suggest that the spreading activity recorded there was largely contemporaneous with the first detachment movements. This could explain why upwelling flow has been chilled there. Rapid cooling was certainly enhanced by vigorous hydrothermal circulation along the well-developed fault system, evidenced by alterations and mineralizations. The Maqsad case illustrates thus the interference of spreading and early emplacement tectonics, unavoidable in ophiolitic environment. In spite of these complications, some important consequences for asthenospheric processes can still be deduced from the mantle structures in Maqsad.

The most striking observation is the persistence of vertical asthenospheric flow up to Moho level. It reveals an unexpected geophysical process: the virtual disappearance of the dynamical boundary layer at the top of mantle upwellings. This phenomenon is localized in space to the spreading axis and appears genetically related to the occurrence of a thick summit horizon of undeformed residual mantle rocks. The dynamical boundary layer related to the upwelling progressively grows in the spreading direction, as revealed by the off-axis development of an horizon of intense horizontal plastic flow just beneath the Moho. These characteristics (upward increase of plastic strain, horizon of highly strained dunites, ...) are also those commonly observed in massifs where the "standard" flow pattern has been recorded (fig. 3). In Maqsad, it thus appears clearly that the structural and petrological character of the mantle on a vertical section away from the paleo-axial zone reflects the processes taking place upstream, at the top of the upwelling. Accordingly, Masqad could be considered as the "missing link" which allows one to interpret, in term of axial processes, the structures generally exposed in the ophiolitic outcrops and which correspond to an "off-axis" record. It suggests that the first order processes put in evidence in Maqsad could have a general application. Second order variations from one massif to the other, like the depth of the shear sense inversion, might reflect variations in such parameters as the spreading rate or the maturity of the diapir.

Another interesting property of the upwelling lies in its shape in cross section (open-downward fan, fig. 7). It can be interpreted in terms of focusing of the ascending flow toward the ridge axis. Both this focusing and the significant reduction of the upper boundary layer thickness can be attributed to a viscosity decrease of several orders of magnitude localized at the top of the upwelling [Rabinowicz et al., 1987]. The only geophysical process which seems to be able to induce such a weakening of the mantle is a catastrophic increase of the melt/rock ratio. As a matter of fact, evidence for a low viscosity crystal mush stage in the dunitic horizon can be found in the melt extraction structures frozen there (section 4.1.3). A possible mechanism for melt retention in the transition zone will be proposed later (section 4.3).

Morphological and petrological variations along the present-day ocean ridges reveal an axial segmentation at a scale of several tens of kilometers which is not necessarily related to the presence of transform faults [Macdonald et al., 1984; Langmuir et al., 1986; Batiza and Margolis, 1986,

Gente, 1987]. However, it is not clear whether or not this reflects a corresponding segmentation of the mantle upwelling. The question addressed now concerns evidence for 3-D complexity in mantle flow patterns recorded in ophiolites.

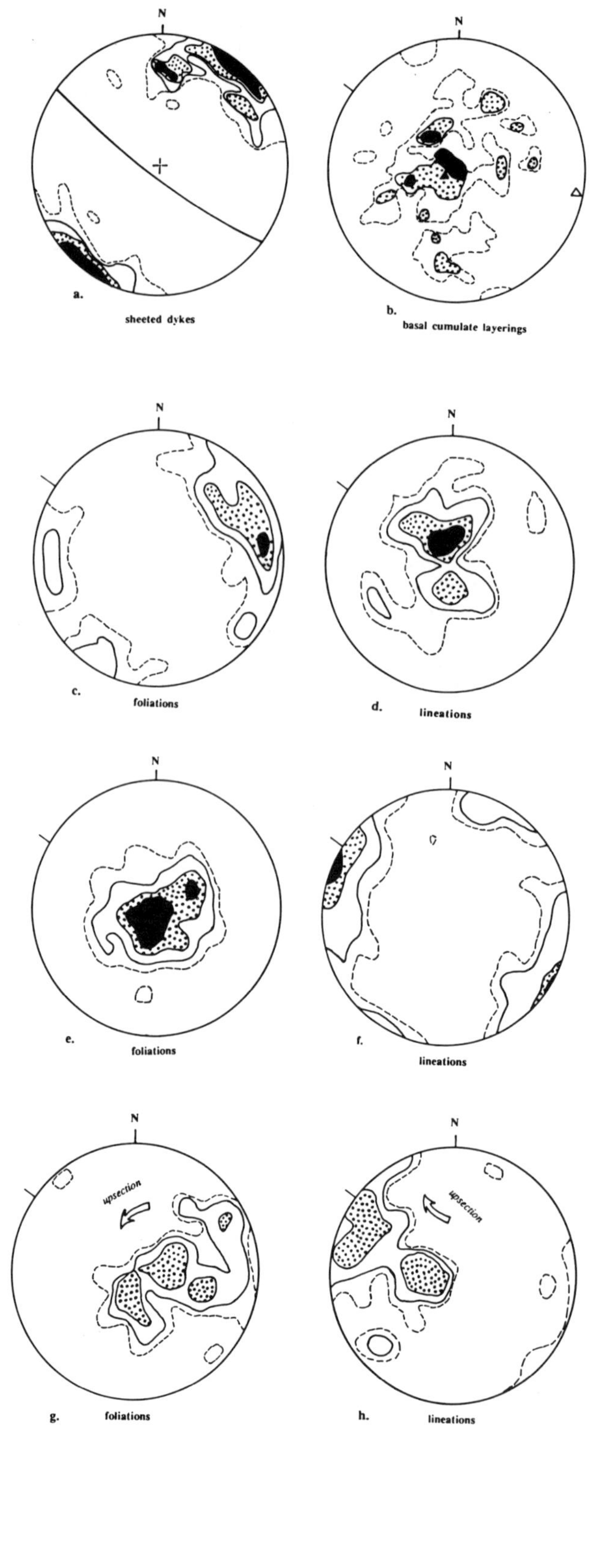

Far from any discontinuity (mylonitic shear zones) possibly related to paleo-transform faults, like those reported by Prinzhofer and Nicolas [1980], Karson [1984] and Reuber [1985], the three-dimensional nature of mantle upwelling would be demonstrated in the following ways:

In a zone of vertical flow, to observe closing - or at least curved - flow plane trajectories. In the Oman ophiolite, such a "pipe-like" geometry was proposed a few years ago, based on extrapolations of the data collected at that time [Ceuleneer et al., 1988; Nicolas et al., 1988]. This interpretation has not been corroborated by further mapping in the Maqsad area where the vertical flow planes define a narrow corridor parallel to the paleo-ridge, with no recognized along-strike closure [Ceuleneer, 1991]. The previous misinterpretation was due to two unrelated types of errors: (i) vertical planes departing from the general orientation were measured, as a rule, at the proximity of well developed faults not clearly identified at that time; (ii) moderately dipping planes were used in constructing the flow trajectories; in that case, the curvature reflects the progressive rotation of the foliation azimuth with decreasing dip into parallelism with the lithological contacts which strike roughly perpendicular to the paleo-ridge axis. This evolution is clearly illustrated in figure 8g,h. Considering the low density of field measurements in the few other areas of vertical flow and the difficulties put in evidence in Maqsad, we conclude that the existence of closing vertical flow plane trajectories has still not been demonstrated and cannot be used as evidence for the along axis segmentation of mantle upwellings.

In an area where the paleo-spreading axis has been accurately located, to catch the termination of the zone of vertical flow lines and their along-strike gradation into horizontal flow parallel to the ridge axis. This ideal situation might be realized in the Maqsad area of Oman. We have seen that in the northwestern part of the paleo-ridge segment cropping out there, vertical flow lines turn abruptly into the horizontal and keep an along axis orientation. However, it has to be noted that, in Maqsad, this particular information should be considered with some caution due to the local interference of the asthenospheric flow with shear motions related to early emplacement tectonics: it has been shown that the diverging flow on the western flank of the upwelling was progressively rotated into parallelism with a shear zone parallel to the ridge

Fig 8. Orientation of the internal structures in the Maqsad area. Equal area projection, lower hemisphere a) Sheeted dyke complex. 177 measurements; contours at 1.1, 2.2, 4.5, 8.4, 15%. b) Magmatic layering at the base of the crustal section; 80 measurements; contours at 1.25, 2.5, 3.75 and 5 %. c) High temperature foliations and d) lineations in the zone of upwelling flow. 91 measurements, contours at 2, 4, 6 and 8%. e) High temperature foliations and f) lineations in the zone of diverging flow. 135 measurements, contours 2, 4, 6 and 8%. g) and h) Progressive rotation of the plastic flow structures in the section 6 km away from the Maqsad paleo-axial zone (fig. 6); equal area projection, lower hemisphere. 50 measurements; contours: 2, 4, 8 and 12%. From Ceuleneer, 1986 and 1991).

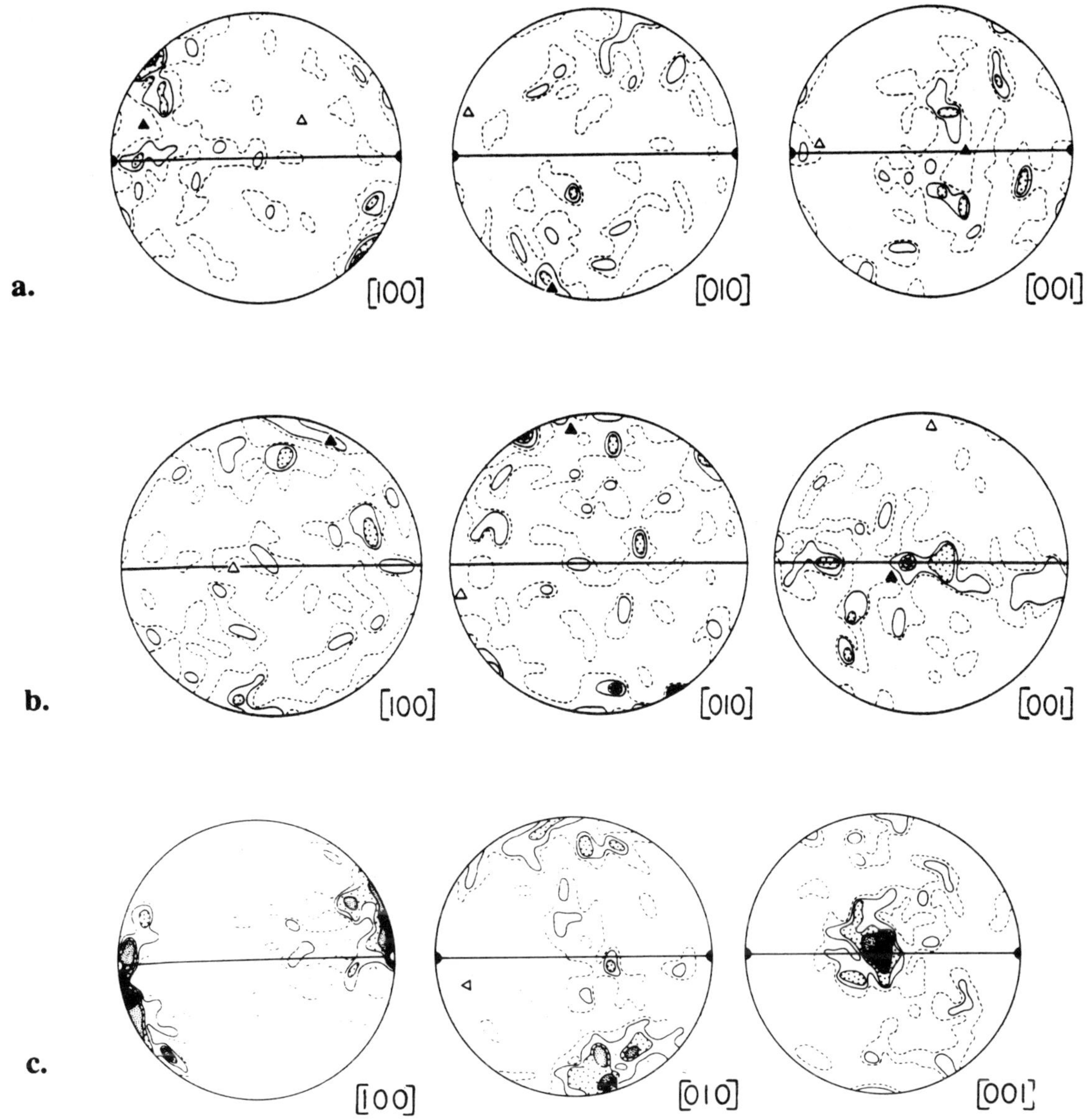

Fig. 9. Olivine preferred orientations in the Maqsad area (see figure 7 for location). Equal area projection on the lower hemisphere; 100 measurements: contours at 1, 2, 3, 4, and 10%. A. harzburgite in the paleo-upwellling. b) dunite at the top of the paleo-upwelling. c) harzburgite in the zone of diverging flow. From Ceuleneer (1986).

axis. Consequently, it can not be ruled out that the along axis flow lines mapped beneath the paleo-ridge axis in Maqsad are related to these complications. In spite of this restriction, it has to be noted that flow line trajectories with a strong along axis component have been frequently reported in ophiolites away from any recognized shear zone [Nicolas and Violette, 1982; Ceuleneer et al., 1988]. In ophiolites displaying a lherzolitic mantle section, it is even the dominant flow geometry [le Sueur et al., 1984; le Sueur and Boudier, 1986; Boudier and Nicolas, 1985]. Considering the relative abundance of this flow pattern (the most abundant after the "standard" flow pattern) it can be considered as an evidence for the existence of an along-strike component in the horizontal mantle flow beneath spreading centers, and as an indirect evidence for the three dimensional nature of the asthenospheric upwelling.

4. MELT MIGRATION STRUCTURES IN PERIDOTITES

Observations of magmatic structures in peridotites concern the textural and geometrical relationships between the residual matrix and the melt. Melt *migration paths* are decipherable macroscopically when: (i) crystallization products derived from this melt remained trapped in the mantle either as cumulate crystals interstitial within the residual phases or as mafic dykes; (ii) if the melt en route to the surface has caused some reaction which is witnessed in the modal composition of the residue. The possibility that abundant melt migration took place without leaving any macroscopic footprint of its passing cannot be ruled out. Some peculiar geochemical features, like the LREE-enrichment of most ophiolitic harzburgites, might be evidence for such process [e.g. Pallister and Knight, 1981; Frey, 1984; Bodinier et al., 1989], although it could also be ascribable to the melting mechanism [e.g. Prinzhofer and Allegre, 1985]. No direct evidence for the mechanisms *driving* melt migration can be found in the peridotites. However, the geometrical information gathered in the field and summarized below, coupled with theoretical arguments, offer serious constraints for models of dynamics of melt flow in the mantle.

4.1. Small-scale structures

"Small-scale" will refer here to the observations conducted on a single outcrop as opposed to those dealing with the distribution and orientation of the melt migration structures at the scale of a massif. Two opposite modes of occurrence of trapped melt are currently recognized in peridotites: either the melt relics ("cumulate" crystals) are scattered in intergranular position, or they are concentrated in pockets, veins, dykes or layerings whose thickness exceeds significantly the grain size of the residual matrix (a few millimeters).

4.1.1. Dykes and layerings. A first order observation definitely confirmed by generations of field geologists is the marked predominance of dykes and layerings compared to all other mode of occurrence of melt relics in mantle peridotites. Most of mantle peridotites display a more or less developed layering. The absence, rather than the occurrence of any layering is worth noting as a distinctive character of an outcrop. Layering, shown up by a regular variation in the modal composition of the peridotites (olivine/pyroxene ratio) is most often composed of deformed crystals and is parallel to the plastic foliation. A wide range of compositions has been reported, but pyroxenitic layerings (generally websteritic) are the most common. Cross-cutting dykes with such modal compositions are well documented and currently viewed as cumulates precipitated from a basaltic melt before the appearance of plagioclase at the solidus (i.e. at pressures greater than about 3 Kbar and/or at high temperature). As a matter of fact, clear zonation of the dyke composition has been put in evidence in some ophiolites: gabbro dykes are essentially restricted to the uppermost few kilometers of the mantle section and grade downwards to pyroxenitic dykes (e.g. Girardeau, 1979; Jackson, 1979; Boudier and Coleman, 1981; Reuber, 1982; Nicolas and Jackson, 1982; Ceuleneer, 1986]. The progressive transposition of pyroxenitic dykes into concordance with the high temperature (asthenospheric) foliation has been frequently reported. Layerings are thus likely derived from dykes emplaced beneath the spreading center before the cessation of plastic flow. It seems thus natural to attribute a similar origin and significance for melt extraction to layerings and dykes. A more pristine origin, not necessarily related to the *late* plastic flow and melt extraction event experienced by the massif, has been proposed for some layerings equilibrated at greater depth (spinel and garnet lherzolite facies), like those of Beni Bousera [Kornprobst, 1970; Polvé and Allègre, 1980; Allegre and Turcotte, 1986]. Dunitic layerings are also relatively frequent, although their occurrence is more occasional. Dunites can develop at the walls of pyroxenite and gabbro dykes [Boudier and Nicolas, 1977; Quick, 1981a,b]. As in the case of pyroxenites, dunitic layerings may be interpreted in terms of transposition by the plastic flow of an initially cross-cutting structure.

Cross-cutting dykes intrusive in mantle peridotites are almost as common as layerings, although their abundance (in terms of relative volume) is lesser. In spite of a rather heterogeneous distribution (both at a local scale and at the scale of a massif), their average abundance can be estimated to about one dyke cropping out every 10 to 100 meters [e.g. Reuber, 1982; Ceuleneer, 1986]. Their thickness ranges typically from a few centimeters to several decimeters. Dykes exceeding one meter in thickness are more occasional; they generally have a texture distinct from the thinner dykes which points to injection in a cooler environment. These are the pegmatitic dykes, most of them displaying a somewhat hydrated mineralogy (brown amphiboles rimming the pyroxene and concentrating locally in amphibolite veins), and the microgabbro to diabase dykes. Chromitite dykes, which are typically a few meters in thickness, are somewhat puzzling: the concentration of chromite points to the circulation of a high temperature magma, but the abundance of hydrated inclusions in the chromite crystals proves the importance of fluids in the crystallization process [e.g. Lorand and Ceuleneer, 1989; Leblanc and Ceuleneer, 1991]. However, most of the chromitite dykes have been transposed into concordance with the asthenospheric flow which proves their relation with the spreading event [Cassard et al., 1981; Ceuleneer and Nicolas, 1985]. The chromitite dykes are interesting in that they offer a unique opportunity to estimate the volume of melt which flowed through them [Lago et al., 1982]. Melt volumes between 300 and 400 times the present volume of the cavity (about 1500 m^3) have been found for a well preserved chromitite dyke in the Maqsad paleo-axial zone [Leblanc and Ceuleneer, 1991]. Accordingly, dykes might be avenues channeling most of the flow of melt produced in a mantle upwelling.

4.1.2. Porous flow and rooting of dykes . As melting begins at triple junctions between the different phases that makes a lherzolite, the melt is initially scattered in intergranular positions [e.g. Nicolas, 1986a; Sleep, 1988]. Consequently, whatever the importance of cracks in pooling the melt from the mantle, some kind of porous flow necessarily preceeds the formation of dykes. Evidence for porous flow in mantle peridotites is rather uncommon. The only places where it is relatively widespread is in the plagioclase lherzolite facies and in the dunitic transition zone of ophiolites. Assuming that the plagioclase and a certain fraction of clinopyroxene represent the frozen liquid, a sequence where the intergranular melt is progressively collected in small lenses which eventually coalesce into centimetric veins has been described in the Lanzo peridotites [Boudier and Nicolas, 1977]. These observations show the preponderant role played by plastic strain in concentrating small melt fractions and the structural control exerted by solid-state flow on melt migration geometry (veins are, as a rule, parallel to the shear plane in the deforming matrix). Similar observations have been conducted in other lherzolitic massifs. Nicolas [1986a] argues that they are relevant to the stage preceding the formation of thicker dykes with plurikilometric extension which pool the melt to the surface. However, the precise relationships between the large dykes and the small veins are not clearly documented in the field.

In most ophiolites, a complete section from the zone of initiation of partial melting to the crustal magma chambers is not accessible to observation: only the uppermost part of the melt drainage network is exposed (maximum depth of about 10 km beneath the Moho). However, this ideal situation is more closely approached in the rare ophiolites with a mantle section made of plagioclase lherzolite, due to the reduced depth interval of melting necessarily associated with this undepleted composition. The Trinity ophiolite is an example of such a cold environment where a small amount of melt was delivered to the surface, as shown by the poor and discontinuous development of the crustal section [le Sueur et al. , 1984; le Sueur and Boudier, 1986; Cannat and Lescuyer, 1990]. The following vertical zonation of petrological facies has been observed in Trinity (fig. 10). It contrasts with the one found in classical ophiolites [Ceuleneer and le Sueur, 1989 and unpublished data].

The maximum thickness of the mantle section in Trinity is of the order of 2-3 km. Most of its volume, up to a paleo-depth of a few hundred meters, is made up of a peculiarly homogeneous

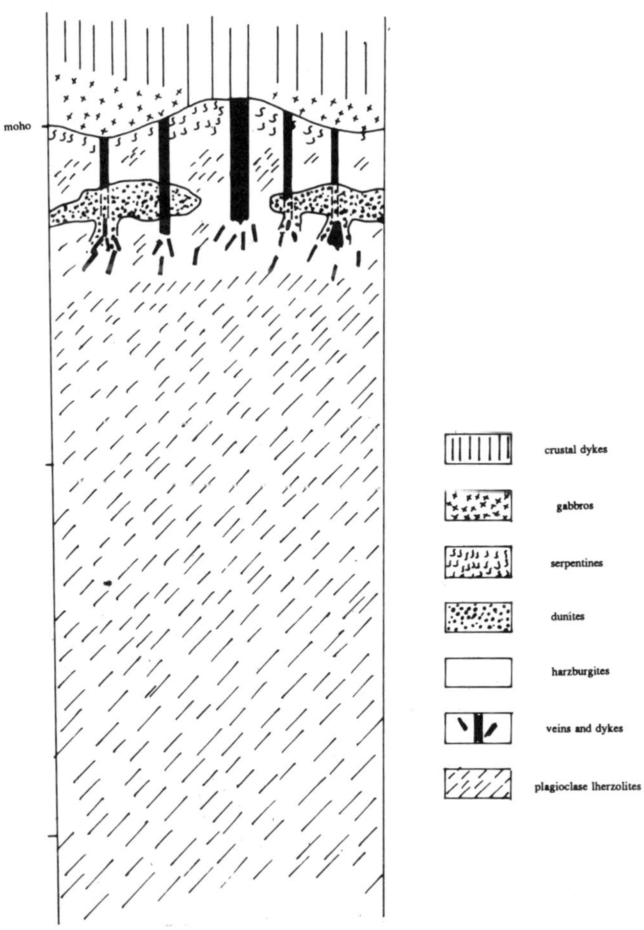

Fig. 10. Schematic lithological zonation within the Trinity peridotites.

plagioclase lherzolite. As the Moho is approached, melt segregation structures are observed in the form of plagioclase lenses parallel to the plastic flow plane (fig. 11a) similar to those described in Lanzo. A few tens of meters higher, the melt has been largely extracted from the matrix, and gabbroic veinlets are scattered in a residue of lherzolitic composition. Locally, the veins develop harzburgitic to dunitic walls. The dunites have their maximum development a few tens of meters higher, a level where the large hectometric dunitic bodies described by Quick [1981b] are scattered. These tabular dunites do not form a continuous horizon like in Oman but appear always at the same paleo-depth (200-300 m beneath the paleo-Moho). There is no evidence for a "crystal mush" stage in the Trinity dunites (no impregnation and/or fabric destruction at a scale exceeding a few decimeters, as observed in Oman); the dunitification process in Trinity generally postdates the acquisition of plastic flow structures (fig. 11b). The large dunitic bodies in Trinity were likely produced by repeated injection of an "aggressive" melt within closely spaced dykes rather than by the pervasive soaking and compaction of a large mantle horizon at the top of a diapir. Evidence for this process is provided by the rooting of the tabular dunites in the dunitic walls of pyroxenitic dykes cross-cutting the underlying lherzolites (fig. 11c) and by the relatively abundant sub-vertical lherzolitic and harzburgitic screens within the tabular dunites. The large dunitic bodies at Trinity are not located just beneath the mantle/crust transition but are separated from the Moho by a lherzolitic horizon 100-200 m thick displaying a homogeneous composition compared to the underlying peridotites. This horizon, where the plastic flow plane is sub-horizontal, is intruded by numerous pyroxenitic dykes (fig. 11d) perpendicular to the dominant orientation of the asthenospheric flow lines. In this horizon, the mafic dykes do not present reaction features with the host rocks (fig. 11d), which reflects injection in a cooler environment. Their abundance and thickness increase upsection; dykes a few meters thick are common at sub-Moho level where they form locally dyke "complexes" with screens of serpentinized lherzolite. This horizon acted probably as a cool and rigid lithospheric lid separating the asthenosphere from the crust.

4.1.3. Mush compaction. Structures in plagioclase lherzolites are relevant to situations where the melt/rock ratio is relatively low (<10%). Observations concerning melt migration mechanisms involving higher melt percentages have been conducted in the transition zone of ophiolites, mostly at the top of the Maqsad upwelling [Ceuleneer, 1991]. They are briefly summarized here. The transition zone of ophiolites (fig. 11e) corresponds to a very shallow environment (about 2 Kbar pressure, at most) where the enstatite has been incongruently melted through the action of a basaltic melt equilibrated at > 5 Kbar pressure [e.g. Nicolas, 1986a; Kelemen, 1990]. Incipient stages of this dissolution process have been frozen at rare places in the underlying harzburgites. Assuming that all the olivine crystals are residue of the solid mantle (which is not strictly true, as the incongruent melting of enstatite implies precipitation of secondary olivine) and that all the clinopyroxene and plagioclase crystallized from a melt (which is probably true, considering their undeformed microstructure), the melt/matrix relationships are very easy to read in the field.

We have seen (section 3.2.2.) that the largely random olivine petrofabric in the Maqsad transition zone bears witness to the

Fig. 11.a. Pooling of plagioclase in lenses parallel to the shear plane (slightly oblique to the foliation plane); Trinity ophiolite.

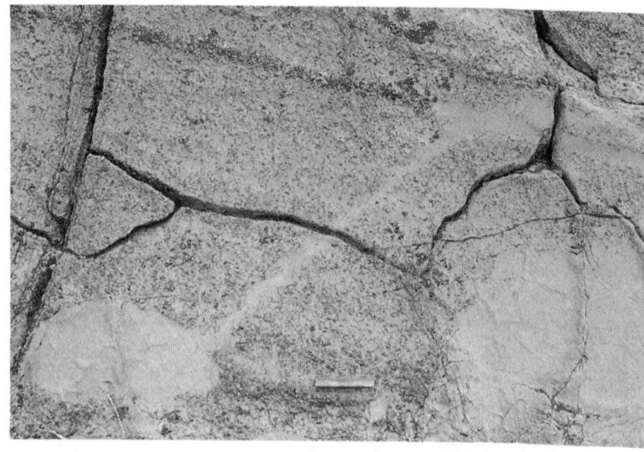

Fig. 11.b. Small discordant dunite pods and veins in the Trinity peridotites. The plastic deformation structure is underlined by a well-developed pyroxenitic layering. Note the channelling of the melt which induced the dunitification in a pyroxenite layer; Trinity ophiolite.

destruction of the mantle solid framework; this is attributable to a melt/rock ratio above the second percolation threshold. However, the transition zone, as presently exposed, has lost most of its interstitial melt. Fortunately, some melt relics remain trapped there and the various textural relationships they present with the solid matrix can be interpreted in terms of the various stages of compaction which led to the formation of the "barren" dunitic transition zone.

A stage of homogeneous impregnation can still be observed at some places in Maqsad (fig. 11f). However, this configuration seems to have been unstable: the inferred succession of events starting with the progressive constriction of these impregnated zones, likely due to lateral melt flow, and leading to the concentration of the melt in decimetric to metric spherical pockets is illustrated in figure 11f to 11j. Vertical gabbroic veins a few centimeters in thickness are still frequently rooted in these pockets (fig. 11i,j). Another frequently observed melt/matrix segregation structure is the lateral melt expulsion from magma pockets, which leads to the formation of sills (fig. 11k), a process which accounts for the layered feature of the transition zone. Other observations in the transition zone point to the local "mushy" behaviour of the melt/rock mixture there, like the injection of strongly impregnated dunites in decimetric cracks of large vertical extent (reaching the Moho, i.e. a few hundred meters in height); gravitational settling of the olivine crystals within these dykes is well documented. It is also frequent to observe the intrusion of crystal mush bodies, ranging in size from a few meters to a few hundred meters, in the barely crystallized crustal section [Benn et al., 1988; Reuber, 1988]. Such wehrlitic-troctolitic intrusions bear witness to an off-axis process which is enhanced by the very large plastic strain

Fig. 11.c. Rooting of dunites into underlying lherzolites at the wall of a pyroxenite dyke. Before the dunite formation, the pyroxenite dyke continued upward as evidenced by a diffuse trail of spinels in the dunites; Trinity ophiolite.

Fig. 11.d. Close-spaced pyroxenite dykes intrusive in the uppermost section of the Trinity peridotites. Note the total lack of dunitic margins; Trinity ophiolite.

Fig. 11.e. Thick dunitic transition zone at the top of the Maqsad paleo-upwelling. The 200 m-thick dunitic horizon (pale weathering colour) is sandwiched by the layered gabbros (top) and the harzburgites (foothill); Maqsad area of the Oman ophiolite.

Fig. 11.f. Zone of homogenous impregnation in the dunites of fig. 13b. Note the interstitial position of the clinopyroxene and plagioclase crystals (whitish grains) in the olivine matrix. Ruler: 10 cm; Maqsad area of the Oman ophiolite.

Fig. 11.g. Striction of an horizon of homogeneous impregnation in the same dunites; Maqsad area of the Oman ophiolite.

Fig. 11.h. Spherical pocket enriched in plagio- and cpx crystals in the same dunites; Maqsad area of the Oman ophiolite.

experienced by the mantle away from the axial zone. Interestingly, in conditions of low melt/matrix ratios and of incipient plastic deformation, which are realized away from the paleo-axial zone, the melt extraction structures mimic those observed in plagioclase lherzolites, i.e. the interstitial melt pools in small veins parallel to the shear plane (fig. 11l).

4.2. Large-scale migration patterns

4.2.1. Vertical zonation. As pointed out above, dykes, more or less affected by the plastic flow, constitute the dominant melt migration structure in ophiolites. A significant vertical evolution in the abundance, nature and orientation of the high-temperature dykes is currently observed in well-exposed sections [Reinhardt, 1969; Girardeau, 1979; Boudier and Coleman, 1981; Ceuleneer, 1986]: this classical zonation comprises a horizon, at great paleo-depth (roughly > 4 km, but this is only an approximate value), where the dykes are scarce and where dunites, both discordant and concordant may be very abundant, a horizon, at intermediate paleo-depth (from about 4 to 2 km), where the pyroxenitic dykes have a maximum abundance, and finally a shallow horizon (uppermost 2 km of

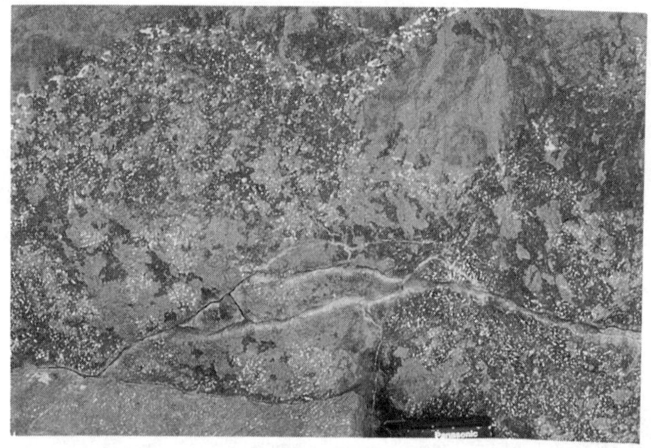

Fig. 11.i. Two adjacent pockets enriched in plagio- and cpx; Maqsad area of the Oman ophiolite.

Fig. 11.j. Rooting of a gabbro vein on an impregnated pocket; Maqsad area of the Oman ophiolite.

the mantle section) where gabbro dykes constitute the dominant facies. Away from the upwelling zones, where the transition zone is thin and highly deformed, the maximum

Fig. 11.k. Rooting of a gabbro sill on an impregnated pocket; Maqsad area of the Oman ophiolite.

Fig. 11.l. Pooling of plagio- and cpx in sub-horizontal veins parallel to the plastic flow plane. "Off-axis" transition zone; Maqsad area of the Oman ophiolite.

abundance of dykes is realized at Moho level. This zonation can be attributed either to an upward temperature decrease (i.e. it reflects the probability that the melt, or cumulate crystals derived from it, has to be trapped) or to an actual increase in the crack density.

In orogenic peridotites and in some ophiolite massifs, the dyke injection pattern indicates frequently a conjugate system of fracturation [Girardeau, 1979; Jackson, 1979; Reuber, 1982]. This has been interpreted in terms of shear fractures whose orientation is conditioned by the stress field induced by the plastic flow, the two injection planes being conjugate to the maximum compressive stress direction [Jackson, 1979; Nicolas and Jackson, 1982]. In the Oman ophiolite, most of the dykes are oriented parallel to the paleo-ridge axis, a direction which coincides, with the normal to the plastic flow direction. These dykes can be interpreted in terms of extension fractures. They are generally sub-vertical at shallow depth; downsection, their dip is less well defined and weakly dipping dykes coexist fequently with vertical ones on a same outcrop. This tendency appears clearly in the case of the Sarami massif. It is illustrated on stereonets of figure 12a,b where the poles of gabbro (i.e. shallow) dykes display a point maximum whereas the poles of pyroxenitic (i.e. deeper) dykes are distributed along a girdle, pointing to random paleo-dips at the massif scale. The dyke orientation pattern is not everywhere so simple. Dykes strongly oblique to the paleo-ridge axis can be dominant in some massifs and this is more frequent at great paleo-depths. This is illustrated in the case of the Hilti massif: on the stereonets of figure 12c,d it appears that the pyroxenite dykes are strongly oblique to the sheeted dyke complex, while the poles of gabbro dykes define a girdle pointing to their progressive rotation into parallelism with the sheeted dyke complex.

4.2.2. Horizontal zonation. Vertical zonation alone can not account for all the variability in dykes density and orientation observed in ophiolites [Ceuleneer, 1986; Nicolas et al., 1988]. Horizontal variations within a given horizon (i.e. paleo-depth) do exist. The question is to understand wheter or not they reflect along-ridge variability in the melt delivery, as assumed by Nicolas et al., 1988]. As a matter of fact, the mantle outcrops have integrated a successsion of events which ended in an off-axis environment with their incorporation into the lithophere. Consequently, horizontal zonations could also

Fig. 11.m. Vertical dunitic vein cross-cutting the harzburgites in the zone of upwelling; Maqsad area of the Oman ophiolite.

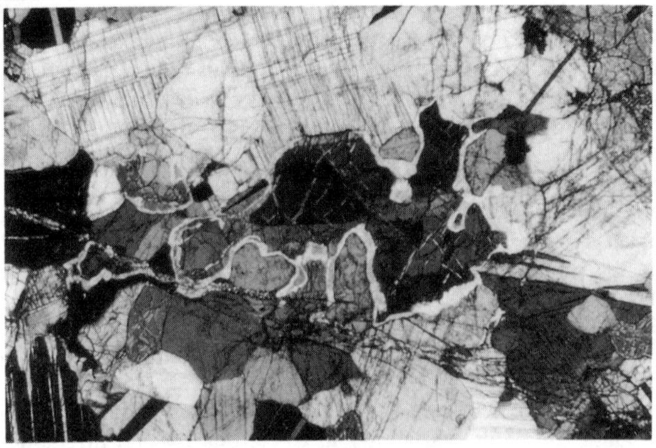

Fig. 11.n. Corrosion of a mantle xenocryst in a troctolitic vein cross-cutting the harzburgites in the zone of upwelling; Maqsad area of the Oman ophiolite.

Fig. 11.o. High density of intrusive gabbro dykes in the diverging zone of the Maqsad upwelling; Maqsad area of the Oman ophiolite.

reflect the position relative to the paleo-ridge axis, i.e. across-ridge variability. To analyse the dyke distribution and orientation in the Maqsad area of Oman is particularly relevant to that problem thanks to the accurate location of the paleo-spreading axis in that area.

In the zone of upwelling flow, contrasting with the classical zonation described above, the dyke abundance is peculiarly low up to the base of the transition zone. It is made of homogeneous peridotites which are relatively undepleted (opx-rich and containing a few percents of mantle diopside), compared to an average ophiolitic harzburgite. Rare dunitic veins (fig. 11m) point to melt circulation in channels of restricted extension. However, relics of the melt itself are very unusual there, an observation which demonstrates the great efficiency of the melt extraction process at greater, but unknown, depth in the upwelling. When present, they display particularly well-developed reaction structures with the residual crystals (fig. 11n). The dyke abundance increases suddenly at a distance of a few kilometers away from the paleo-axial zone, corresponding to the periphery of the upwelling (fig.11o). It is correlated with a marked increase in the plastic strain suffered by the peridotites. These dykes do not present reaction features with the peridotites. Some of them present slightly developed fine-grained margins pointing to injection in a significantly cooler mantle. The dykes tend to be oriented either parallel to the paleo-spreading axis, or perpendicular to the mantle flow line: in the northeastern diverging zone of Maqsad, these two directions coincide and only one family of dykes is found (fig. 13a); in the soutwestern diverging zone, where the flow direction is strongly oblique to the spreading direction, two preferred orientations of dykes are found (fig. 13b). It is worth

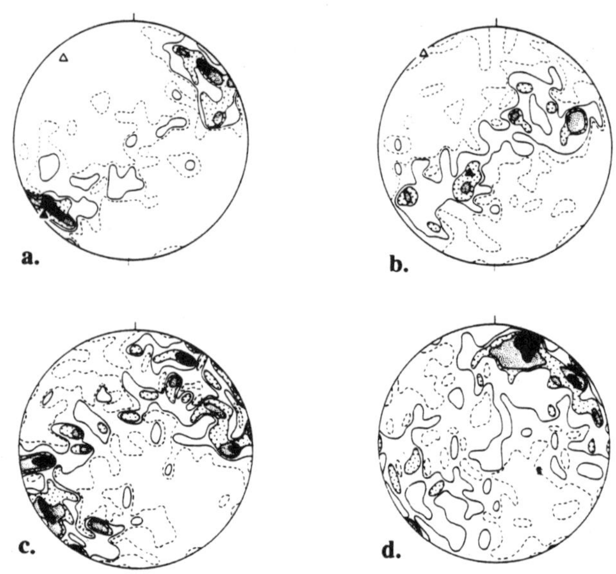

Fig. 12. Dykes orientations related to the "standard" flow pattern. Equal area projection, lower hemisphere; a) gabbro dykes in the Sarami massif of Oman (in that massif, the paleo-ridge has a N150°E orientation); 118 measurements; contours: .85, 1.7, 2.5, 3.4, 5%; b) pyroxentite dykes in the same massif; 196 measurements; contours. .5, 1.25, 2.25, 3.5%. c) gabbro dykes in the Hilti and Southern Fizh massif of Oman (in these massifs, the paleo-ridge has a N-S orientation); 227 measurements; contours: .5, 1, 1.5, 2, 2.5%; d) pyroxenite dykes in the same massifs; 516 measurements; contours: .4, 1, 2, 3, 4%; From Ceuleneer (1986).

noting that these dykes are absolutely similar in texture and modal composition, whatever their orientation; no clear relative chronology can be deduced from cross-cutting relationships. Accordingly, even if they are not strictly contemporaneous, they belong to the same magmatic event. The rare dunitic veins and gabbro veinlets cropping out in the core of the upwelling are, as a rule, perpendicular to the flow line azimuth (fig.13c). Veins and dykes rooted in the impregnated dunites of the transition zone have a largely random azimuth (fig. 13d). Late diabase and micro-gabbro dykes, which are relatively abundant in Maqsad and cross-cut all the mantle section, are parallel to the spreading axis (fig. 13e).

The dyke injection pattern observed in Maqsad illustrates the competition between lithospheric extension and asthenospheric flow in governing the stress orientation beneath a spreading axis. In the upwelling zone, the stress field is dominated by plastic flow as the vein orientation is independent of the paleo-ridge direction. In the mushy transition zone, the vein orientations, in a zone around a vertical axis, point to a stress field dominated by vertical compression with a poor azimuthal control. In a slightly off-axis environment, an equal proportion of cracks open in the directions parallel to the ridge and perpendicular to the flow lineation; this situation was probably frozen when the stress induced by the spreading superseded the one induced by the plastic flow. Further away from the ridge axis, the stress field orientation becomes definitely dominated by the spreading geometry as witnessed by the parallel to ridge injection pattern observed in Oman massifs located far from any recognized upwelling. Even when the lithosphere has cooled significantly, so that the melt is chilled within the mantle as diabase dikes, the stress field continues to be conditioned by the spreading geometry. The extremity of the northern diverging zone in Maqsad is characterized by a very uncommon melt extraction facies: the harzburgites there are abundantly (20% of the outcrop surface) intruded by thick (up to a few meters) pyroxenite and gabbro sills displaying a pegmatitic texture. These sills might represent the basaltic melts which were extracted at the tip of the Maqsad upwelling when it had not yet pierced through a previously accreted lithospheric segment. Another place where abundant pegmatitic dykes and sills have been recognized is the Batin area of Oman. Vertical flow lines and abundant dunites have been mapped there [Nicolas et al., 1988]. The Batin area can tentatively be interpreted as the eastward prolongation of the Maqsad paleo-axial zone, offset by the large Wadi-Tayin sinistral shear zone. The upwelling would have been frozen there at an earlier stage of its development than in the Maqsad area.

These observations and interpretations are summarized in figure 14. They show that strong variations in dykes facies and injection pattern are related to the stage of development of an upwelling and to the position relative to the paleo-axial zone, a conclusion we should keep in mind before ascribing such variations to along axis variability.

4.3. Discussion

The question which is addressed in the present section concerns the implications of ophiolite data for melt migration processes in the mantle. Two categories of melt migration models are generally considered: (i) one where the melt remains in intergranular position and migrates in response to the pressure gradients induced by matrix compaction, large scale solid-state flow, and/or melt buoyancy [e.g. Sleep, 1974; McKenzie, 1984; Scott and Stevenson, 1984, 1986, 1989; Spiegelman and McKenzie, 1987; Ribe, 1987]; (ii) one where melt migration is conditioned by crack formation and propagation [e.g. Shaw, 1980; Nicolas, 1986a; Phipps-Morgan, 1987; Sleep, 1988]. As the discontinuous and non linear nature of crack formation is difficult to integrate in the equations of fluid mechanics, most models do not treat explicitely the fracture mechanism and consider that melt migration occurs only by porous flow; this in spite of the fact that most authors agree that both mechanisms do exist in nature. Accordingly, an important constraint which should be searched for in mantle outcrops concerns the relative importance of porous flow and cracks, and their relationships.

Considering the abundance of dykes (or transposed ones) in the mantle section of ophiolites, we can state with confidence that cracks play an important role in draining magma through at least the few uppermost kilometers of the mantle. By the same way, the scarcity of intergranular melt relics there show that a very efficient melt segregation mechanism is operant at depths greater than about ten kilometers. The melt fraction which remains trapped within the solid network depends on surface properties of the melt and matrix [Waff, 1980; Toramaru and Fujii, 1986; von Bargen and Waff, 1988; Kohlstedt, this volume). Field evidence show that it is close to zero in the case of harzburgites and dunites. Such evidence led Nicolas [1986a] to propose that melt migration in the mantle is conditioned by the hydrofracturing mechanim, a single crack channelling, from time to time, a large volume of melt from the base of the melting column to the surface. However, as pointed out by Stevenson [1989], the propagation of large hydrocracks over vertical distances of several tens of kilometers, induced by a contrast of hydrostatic heads between the melt and matrix, is not achievable in the asthenosphere mainly in reason of its non-rigid behaviour and of the melt mobility.

If the dyking process is restricted to shallow depths, what observation do we have to constrain the earlier stages of melt segregation? Based on the data presented in section 3.2.1, we can distinguish schematically two contrasted natural situations relevant to that problem. One where the melt/matrix ratio is relatively weak (a few percent), illustrated mainly by plagioclase lherzolites, another one where the melt/matrix ratio is higher, close and locally above the second percolation treshold, represented by the transition zone of ophiolites at the top of paleo-upwellings. In the first situation, melt segregation is triggered by the shear flow; by the same way, the plastic flow geometry conditions the melt migration pattern: melt pools in small lenses parallel to the shear plane, which eventually coalesce into dykes perpendicular to the plastic flow direction. In the second situation, we have seen that melt percolation leads to very sharp lateral *porosity variations* at a scale of the order of one meter. Veins are preferentially rooted in the zones of high melt/rock ratio. The formation of veins from a compacting crystal mush has been accounted for by Stevenson [1989] in the frame of a model where viscosity is strongly dependent on the melt fraction. It is shown that very efficient melt migration takes place along the extension direction, which leads to the formation of restricted zones of very high melt/rock ratio. The characteristic spacing of veins predicted by Stevenson is a few orders of

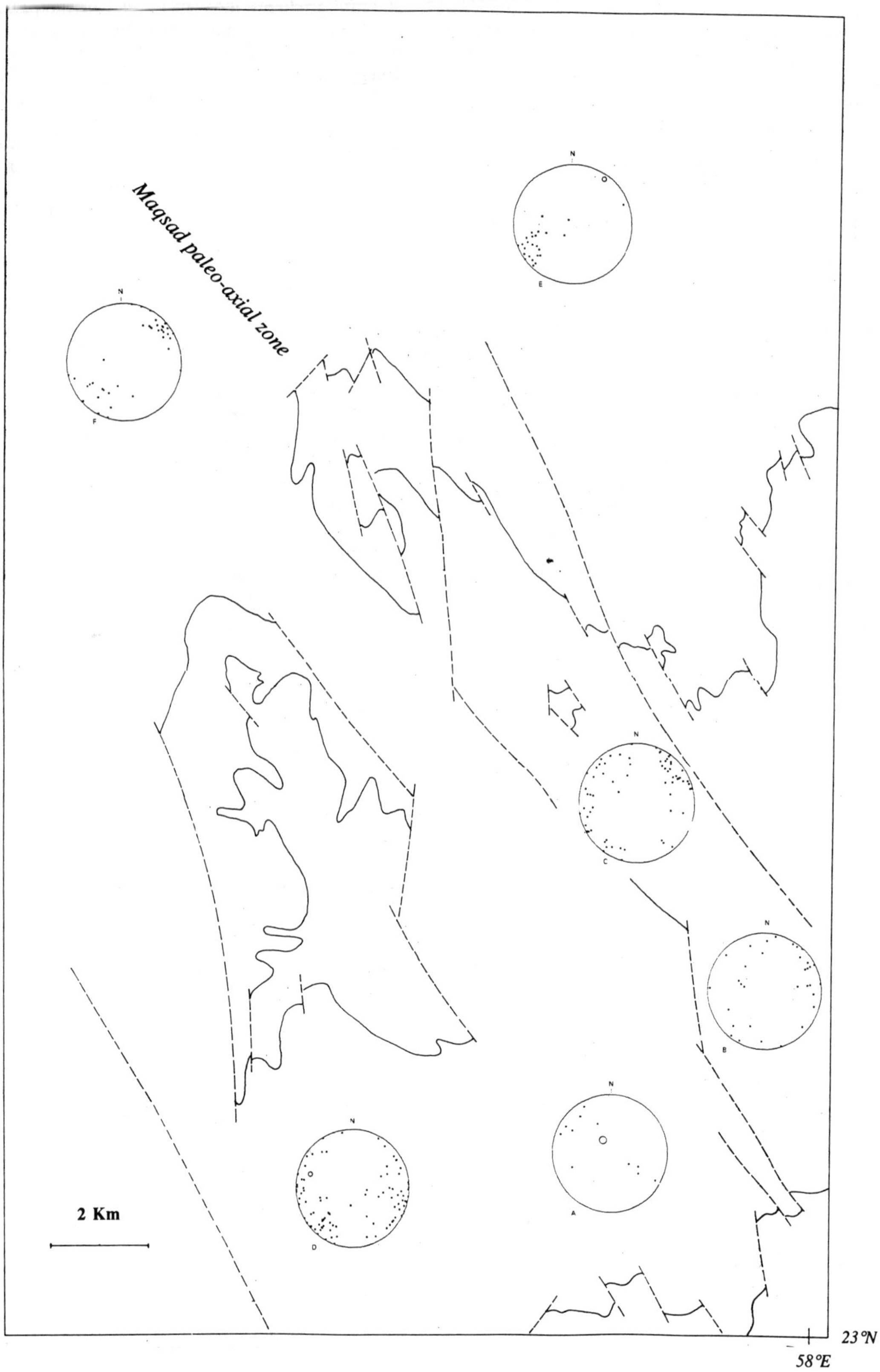

Maqsad paleo-axial zone
N
E
F
C
B
A
D
2 Km
23°N
58°E

magnitude lower than the compaction length, i.e. around one meter; it is remarkably similar to the one observed in the transition zone. This process issues in the formation of a network of close-spaced veins parallel to the direction of the maximum compressive stress inducing the compaction.

An important conclusion derived from these observations is that the melt, whatever the melt fraction and the importance of shear flow, does not travel by intergranular porous flow on distances exceeding a few meters. Further melt migration involves veins and eventually cracks. The main difference between the two situations lies precisely in the relative importance of veins and cracks. In the case of lherzolites, small melt fractions are efficiently extracted from the matrix as the opening of large cracks follows almost immediately the initiation of melting. In the example of the transition zone, where the shear strain is strongly reduced, melt migration tends to reorganize the melt in a network of close-spaced veins; massive melt migration over long distances through cracks happens later on in the segregation process.

Now the question is to decipher which one, if any, of these two situations presents the best analogy with melt migration in the asthenosphere. As the plagioclase lherzolite case might illustrate an actual context of initiation of melting, it is tempting to extrapolate the information gathered there to the melting zone beneath ridges. Accordingly, the mantle section of Trinity, where the melt segregation, and the formation of large dykes channeling the melt to the crust, takes place in a horizon a few tens of meters thick, could be a "shortened" image of what happens beneath oceanic ridges. However, melt extraction structures observed in plagioclase lherzolites formed in a shallow boundary layer where partial melting was very sluggish and where a small melt fraction was frozen shortly after it melted. In such a shallow and cooling environment, the influence of solid-state flow and of crack formation were necessarily enhanced. Conversely, in a mantle upwelling, where the melt is continuously produced and where the melt availability is not limited by thermal conditions, the migration process might depart strongly from the one recorded in lherzolites. At depths where partial melting initiates (about 80 km beneath ocean ridges, 150-200 km in hotspot plumes), a more rapid melt influx is expected for small excursions above the solidus, compared with melting in the plagioclase lherzolite facies [Takahashi, 1986; Herzberg et al., 1990]. In spite of the fact that the transition zone of ophiolites corresponds to a zone of melt accumulation, and not to a zone of partial melting, the physical conditions realized there might be closer to the ones realized at depth in the melting region of the mantle, than those realized in plagioclase lherzolite massifs. Rather than an immediate extraction from the melting zone to the surface, melt could distribute itself within a vein network whose orientation is conditioned by that of the maximum compressive stress. Melt could remain trapped in this "porous" reservoir as long as the dip of the veins is weak, which is probably the case at the base of the melting zone (fig. 15). An observation which supports this hypothesis is to be looked for in the dyke orientation pattern: even in the reduced paleo-depth interval exposed in ophiolites, it has been shown that the dyke orientation is conditioned by lithospheric stresses (vertical dykes parallel to the paleo-ridge axis) only at very shallow depths (about 2 km beneath the paleo-Moho). Deeper dykes can be strongly oblique to the ridge axis and moderately dipping, depending on the asthenospheric stress orientation.

This leads us to discuss the interaction between melt migration and plastic flow beneath the ridge axis. One of the first self consistent models of mantle flow and partial melting beneath ocean ridges are those of Buck and Su [1989] and Scott and Stevenson [1989]. They allowed the melt/rock ratio to increase up to high values, so that they got very low viscosities. This model accounts very well for the focusing of the flow but not for the asymmetry between the radius of the upwelling and the thickness of the upper boundary layer observed in the Maqsad diapir and consistent with the off-axis record ("standard" flow pattern). This implies a very local viscosity drop. In the model of Buck and Su, the melt fraction is mostly conditioned by the melt production. However, we have shown that an amount of melt exceeding a few percents is not stable in interstitial position and that the mechanism of pooling the melt in a vein network, and eventually into dykes, is particularly efficient. The effect of this redistribution of melt on the bulk viscosity of the matrix is presently not quantified. To understand the intimate coupling mechanism between solid-state flow and melt migration, and its effect on the viscosity, is thus particularly critical.

Field data, and theoretical considerations, have shown that the pressure and stress gradients lead to local melt concentrations and to strong porosity and viscosity variations at a local scale. This will modify the viscosity field and lead to a reorganization of the flow which will, in turn, affect the pressure and stress field. Such a coupling between the pressure field, melt migration, viscosity, and plastic flow geometry should exist in the very peculiar environment realized at the top of mantle upwellings in ophiolites. The upwelling, at least in the uppermost few kilometers of its total extension, is constituted of a rock almost entirely devoid of interstitial melt (with the exception of rare veins and dykes). At the very top of the upwelling, the facies changes suddenly into a crystal/melt mixture, where viscosities as low as 10^{15} Pa.s were likely realized [Rabinowicz et al., 1987]. The occurrence of this zone is correlated with the very sharp rotation of plastic flow lines.

Fig. 13. Orientation of some of the gabbro dykes in mantle peridotites from the Maqsad area. Dots: poles of dykes; open circle, local orientation of the plastic lineation in the peridotites. In that massif, the paleo-ridge axis has a N125°E orientation. a) dunitic veins and gabbro-troctolite veinlets in the paleo-upwelling. 11 measurements. b) gabbro veins rooted in the impregnated dunites of the transition zone; 34 measurements. c) gabbro veins and dykes in the transition zone for which no rooting relationships have been put in evidence; 80 measurements; d) gabbro dykes intrusive in the southwestern diverging zone; 116 measurements; e) gabbro dykes intrusive in the northeastern diverging zone; 38 measurements; f) diabase and microgabbro dykes cross-cutting all the Maqsad area; 42 measurements. From Ceuleneer (1991).

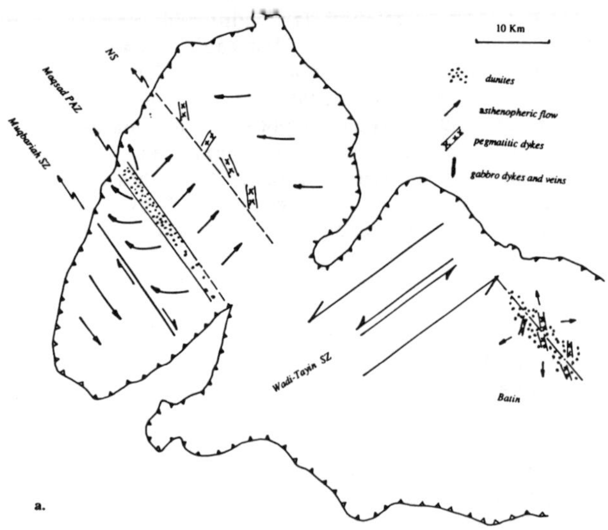

a.

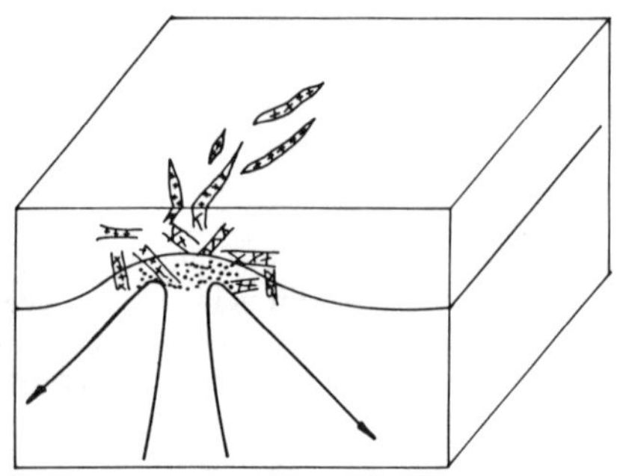

Batin stage

Away from the upwelling, where the melt has been largely extracted, the rotation of the flow has been shown to be more progressive. By the same way, the absence of evidence for a well-developed low viscosity zone at the top of the mantle section of Trinity could explain why vertical lineations have never been observed there, or in other ophiolites with a lherzolitic mantle section, the rotation of the vertical asthenospheric flow lines to the horizontal being too progressive to be caught in the shallow ophiolite record. This has been accounted for in the following way by Rabinowicz et al. [1987]. A viscosity drop of several orders of magnitude at the top of an upwelling current allows a considerable proportion of the flow to be channelled at shallow depth (fig. 16a). The dynamic pressure associated with the flow is found to increase suddenly by a few bars at the viscosity discontinuity (fig. 16b). Directly beneath the discontinuity, a negative pressure is found so that the pressure step will act against the rise of the buoyant magma. In the absence of a viscosity discontinuity, the pressure gradient is of the order of 0.05 $MPa.km^{-1}$. It is much weaker than the pressure gradient due to buoyancy of the melt (of the order of a few $MPa.km^{-1}$) so that the melt is free to move upward. However, in the case of a sudden increase by a few bars in the dynamic pressure, the melt has to form a horizon a few hundred meters thick to overcome this overpressure. Melt migration is probably at the origin of this feed-back: a local melt concentration will reduce the viscosity; the solid-state flow will be focused there so that the melt production will be enhanced in zones already enriched in melt. It eventually will lead to discontinuities in the plastic flow trajectories which will generate pressure gradients capable of trapping a large concentration of melt, which will be finally limited by the hydrofracturation mechanism. This sequence of

b.

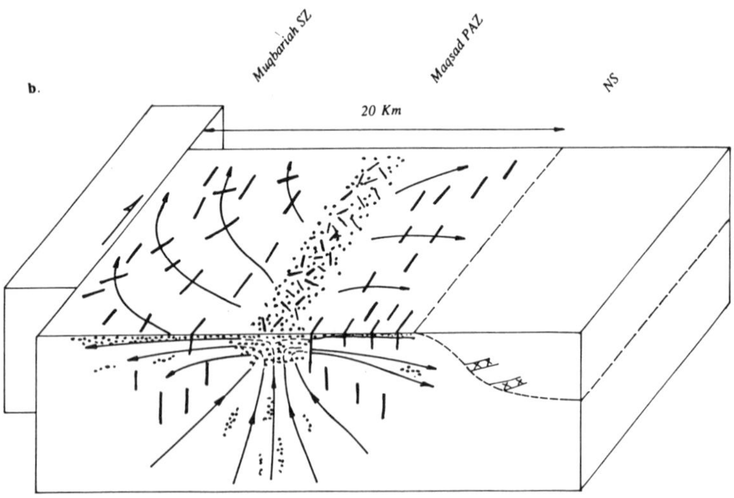

Maqsad stage

Fig. 14. a) Sketch of the mantle structures in the Sumail and Wadi-Tayin massifs of Oman. The paleo-ridge is supposed to be offset by the wide Wadi-Tayin sinistral asthenospheric shear zone. Another interpretation (Nicolas et al, 1988a) is to consider the Batin diapir as an off-axis structure and the Wadi-Tayin shear zone as a paleo-transform zone. b) Schematic cross-section across a mantle upwelling at two staes of its development, now exposed in Batin and Maqsad. Zonation and orientation of the melt migration structures are shown. SZ: shear zone; PAZ: paleo-axial zone; NS: North Sumail suture zone).

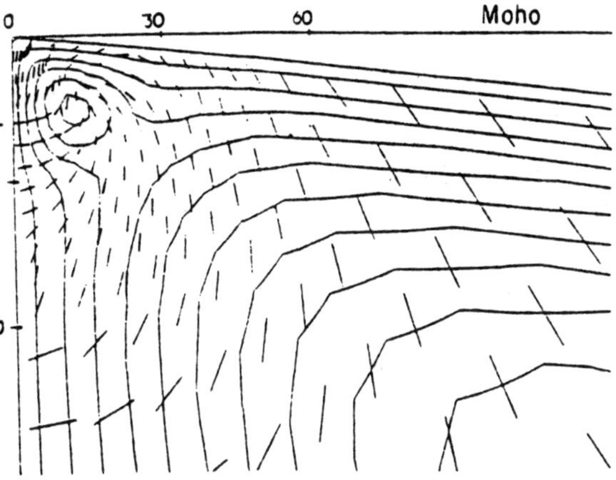

Fig. 15. Compressive stress orientation (lines) related to convective flow across a viscosity drop interface located at a depth of 30 km. From Rabinowicz et al. (1987).

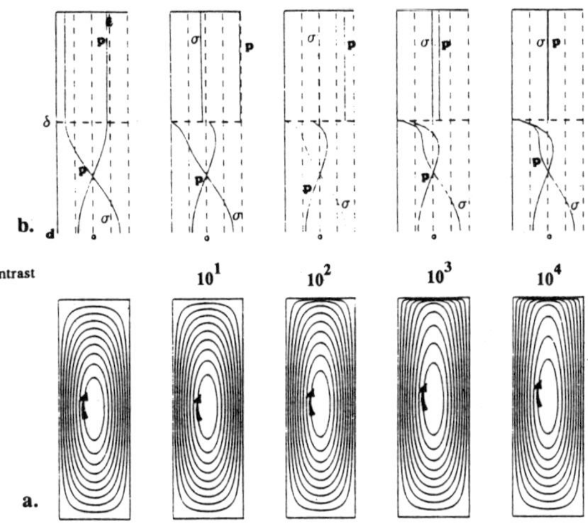

Fig. 16. a) Convective pattern in a box crossing a narrow viscosity drop interface at the very top of the box (1/25th of the total box thickness. Increasing viscosity contrasts from left to right (up to 4 order of magnitude). b) P pressure field and σ, stress field associated to these circulations. The low viscosity zone thickness has been enlarged for clarity. From Rabinowicz et al (1987).

events is, of course, highly speculative, but is consistent with the frozen structures observed in ophiolites. We suggest that similar feed-back processes might be operant at greater depth, among other, in the zone of initiation of partial melting where the sub-horizontal compressive stress field (fig. 15) is expected to induce the local accumulation of melt. Successive viscosity drops would contribute to focusing of the asthenospheric upwellings and the melt delivery beneath spreading ridges.

5. CONCLUSIONS

The main conclusions concerning mantle flow and melt migration beneath oceanic spreading centers which can be drawn from observations in peridotites are briefly summarized. Most of the mantle sections of ophiolites correspond to an *off-axis* record of sub-horizontal flow parallel to the spreading direction. This shows that the mantle flow geometry is largely conditioned by the spreading process. Plate drift is thus likely an important component of the force driving the asthenospheric flow beneath spreading axes. However, the vertical evolution of the finite strain and of the shear sense recorded in ophiolites cannot be explained by this "passive" component alone; an "active" component (i.e. flow driven by internal buoyancy forces) needs to be invoked to account for these gradients. If the vertical sections presently exposed in ophiolites are considered in the frame of progressive accretion of the spreading lithosphere, it can be estimated that the active flow is superseded by the passive flow at a distance of a few tens of kilometers away from the ridge axis. As a likely scenario, it can be proposed that the axial asthenospheric upwellings are primarily "forced" by the spreading plates but that the initiation of partial melting, and the resulting density contrasts, locally drive the flow.

An *axial* structure has been frozen in the Maqsad area of the Oman ophiolite, thanks to exceptional tectonic circumstances at the time of thrusting. It offers a unique opportunity to study the upwelling geometry and the melt delivery processes beneath a spreading axis. It is concluded that the current picture of narrow pipe-like axial upwellings has no observational support: the vertical flow planes define a structure elongated along the ridge axis with no observed along strike closure. Relatively broad (about 10 km) at depth, the upwelling narrows upsection to reach a width of about 2 km at Moho level. In spite of that, evidence for the 3-D nature of mantle circulation can be found in the evolution of the dip of the flow lines: vertical flow lines grade along strike into a sub-horizontal attitude, showing that a significant proportion of the upwelling asthenosphere is channelled along the ridge axis. However, no clear relation between this mantle flow pattern and crustal segmentation similar to the one observed in present-day oceans has been put in evidence in ophiolites.

The distribution and orientation of melt extraction features and of petrological facies in the Maqsad area are consistent with the picture drawn from plastic flow structures. Among other, dykes have been shown to be restricted to the periphery of the upwelling. Melt migration structures in the upwelling itself show that melt circulation was restricted to a network of close spaced veins. Reaction structures show that the last melt matrix equilibration was realized at a greater depth in the upwelling, not exposed in the shallow ophiolitic record (≤ 10 km beneath the Moho). The very top of the upwelling is a horizon of melt accumulation. It is correlated with the virtual destruction of the upper dynamical boundary layer. The progressive development of this boundary layer away from the spreading axis can be clearly followed. It confirms that the highly strained crust/mantle transition zone, and abundant cross-cutting dykes, observed in almost every ophiolitic massif, is an off-axis feature. It points to the coupling between melt migration and solid-state flow in the mantle. From field observations and theoretical arguments, it can be proposed that this coupling involves a feed-back mechanism between (i) the solid-state flow geometry, and the resulting pressure field, and (ii) the melt distribution, and the resulting viscosity field. The modalities of the *development* of such a feed-back are still largely unknown but are expected to be very sensitive to the small-scale melt migration mechanisms.

Some observations relevant to these mechanisms can be conducted in plagioclase lherzolites and in crystal mush horizons of ophiolites. It is shown that, whatever the melt fraction and the magnitude of the shear deformation, a melt phase which is initially distributed in intergranular positions is very efficiently concentrated in narrow zones of restricted extension. This compaction process induces some kind of self-organization of the melt, i.e. its repartition into a vein (i.e. zone of high melt/rock ratio) network whose geometry is conditioned by the local stress field. Veins are typically centimeter-thick and spaced by distances of the order of one meter. The coalescence of these veins will eventually induce crack formation (dykes) allowing rapid melt migration over greater distances, as observed in plagioclase lherzolites. However, this stage does not immediately follow the establishement of the vein network when the magnitude of the shear strain is moderate and the melt fraction important. If these conditions are realized in the melting region beneath mid ocean ridges, a vein network might be the dominant mode of melt migration to the surface. The ultimate stage of melt extraction through decimetric to metric dykes would be restricted to shallow depths and/or off-axis environments, a conclusion which is supported by observations in the Maqsad upwelling..

The existence of such a drainage network is expected to have various petrological consequences: it could trigger the mixing of incremental melt fractions produced in the melting column, which is compatible with a dynamic melting model [e.g. Navon and Stolper, 1987; Klein and Langmuir, 1987] and some isotopic data which require an efficient "averaging process" of the melt fractions produced from a heterogeneous mantle [e.g. Zindler and Hart, 1986]; it could favor the thermal homogenization of the mantle upwellings, leading to the buffering of melt composition in a restricted field and induce complex phenomena of differentiation (high pressure crystallization) which could account for some puzzling geochemical patterns of the erupted lavas [Ceuleneer et al., 1990].

Acknowledgments. We would like to thank Jason Phipps-Morgan and members of the RIDGE office who motivated this paper. It is mostly inspired from observations conducted in the Oman ophiolite. We are very indebted to Dr. Hilal Al Azri, Director of the Oman Geological Survey, who made field work possible and to professor A. Nicolas who introduced both of us to the study of ophiolites. Reviews from Mathilde Cannat, Donna Blackman and an anonymous reviewer have contributed to improve this paper. This is a contribution CNRS-INSU-DBT n° 411.

REFERENCES

Albarède, F., How deep do common basaltic magmas form and differentiate?, *J. Geophys. Res., in press*, 1992.

Allègre, C. and Turcotte, D.L., Implications of a two-component marble-cake mantle. *Nature, 323*, 123-127, 1986.

Auzende, J.-M., Bideau, D., Bonatti, E., et al.., Direct observation of a section through slow-spreading oceanic crust. *Nature, 37*, 726-729, 1989.

Auzende, J.-M., Ceuleneer, G., Cornen, G. et al.., Intraoceanic tectonism on the Gorringe Bank: observations from submersible, in *Ophiolites and oceanic lithosphere*, edited by J. Gass et al., pp. 113-120, Mem. Soc. Geol. London, 1984.

Batiza, R. and Margolis, S.H., Small non-overlapping offsets of the East Pacific Rise, *Nature, 320*, 439-441., 1986.

Benn, K., Nicolas, A. and Reuber, I., Mantle -crust transition zone and the origin of wehrlitic magmas: evidence from the Oman ophiolite, *Tectonophysics, 151*, 75-85, 1988.

Berger, E.T. and Vannier, M., Les dunites en enclaves dans les basaltes alcalins des îles océaniques: approche pétrologique, *Bull. Mineral., 107*, 649-663, 1984.

Berger, E.T. and Weber, B.C., Dunitification of mantle lithosphere below the Society archipelago: evidence for magma-mantle thermal and chemical transfers through xenoliths from the Papeno valley, Tahiti, *J. Geodyn.*, in press.

Beurrier, M., Géologie de la nappe ophiolitique de Samail dans les parties orientale et centrale des Montagnes d'Oman, *Documents du BRGM, 128*, 412 p., 1988.

Bodinier, J.-L., Guiraud, M., Dupuy, C. and Dostal, J., Geochemistry of basic dikes in the Lanzo massif (western Alps): petrogenetic and geodynamic implications, *Tectonophysics, 128*, 77-95., 1988.

Bodinier, J.-L., Vasseur, G., Vernières, J., Dupuy, C. and Fabries, J., Mechanims of mantle metasomatism: geochemical evidence from the Lherz orogenic peridotite, *J. Petrol.*, 1989.

Boudier, F., Structural study of three peridotite samples drilled at the western margin of the Mid Atlantic Ridge, in *Initial reports of the DSDP*, XLV, edited by Melson W.G. et al.,.pp603-608, 1978.

Boudier, F. and Coleman, R.G., Cross-section through the peridotite in the Samail ophiolite, Southeastern Oman mountains, *J. Geophys. Res., 86*, 2573-2592, 1981.

Boudier, F. and Nicolas, A., Structural control on partial melting in the Lanzo peridotites. Oregon, in *Dept. Geol. Mineral. Ind. , 96*, edited by H.J.B. Dick, pp 63-78, 1977.

Boullier, A.-M., Structure des péridotites en enclaves dans les kimberlites d'Afrique du Sud. Conséquences sur la constitution du manteau supérieur, *Bull. Soc. fr. Minéral. Cristallogr., 100*, 214-229, 1977.

Boullier, A.-M. and Gueguen, Y., SP-mylonites: origin of some mylonites by superplastic flow, *Contrib. Mineral. Petrol., 50*, 93-104, 1974.

Boullier, A.-M. and Nicolas, A., Texture and fabric of peridotite nodules from kimberlites at Mothae, Thaba Putosa and Kimberley, in *Lesotho kimberlites, Maseru, Lesotho*, edited by P.H. Nixon, 1973.

Buck, W.R. and Parmentier, E.M., Convection beneath young oceanic lithosphere: implication for thermal structure and gravity, *J. Geophys. Res., 91*, 161-174, 1986.

Buck, W.R. and Su, W., Focused mantle upwelling below mid-ocean ridges due to feedback between viscosity and melting, *Geophys. Res. Lett., 16*, 641-644, 1989.

Cann, J.R., A model for ocean crustal structure developed.

Geophys.J.R.Astron. Soc., *39*, 169-187, 1974.

Cannat, M. and Lescuyer, C., Ephemeral magma chambers in the Trinity peridotites, northern California, *Tectonophysics*, 1990.

Cannat, M., Juteau, T. and Berger, E.T.,. Petrostructural analysis of ODP Leg 109 serpentinized peridotites, in *Proceedings of the Ocean Drilling Program, Scientific Results, 106/109, College Station, Tex*, edited by R. Detrick et al.. ,pp. 47-56,1990a.

Cannat, M., Bideau, D. and Hébert, J., Plastic deformation and magmatic impregnation in serpentinized ultramafic rocks from the Garrett transform fault (East Pacific Rise), *Earth Planet. Sci. Lett.*, *101*, 216-232, 1990b.

Cassard, D., Structure et origine des gisements de chromite du massif du Sud (ophiolites de Nouvelle-Calédonie). *Thèse 3ème cycle*, Nantes, 239 p., 1980.

Cassard, D., Moutte, J., Nicolas, A., Leblanc, M., Rabinovitch, M., Prinzhofer, A. and Routhier, P., Structural classification of chromite pods from New Caledonia, *Econ. Geol.*, *76*, 805-831, 1981.

Ceuleneer, G., Structure des ophiolites d'Oman: flux mantellaire sous un centre d'expansion océanique et charriage à la dorsale, *Thèse Doc. Univ. Nantes*, 338 p., 1986.

Ceuleneer, G., Evidence for a paleo-spreading center in the Oman ophiolite: mantle structures in the Maqsad area, *Ophiolite genesis and evolution of oceanic lithosphere*, edited by T.J. Peters et al., pp.149-175, Kluwer Academic Press, 1991.

Ceuleneer, G. and le Sueur, E., Plastic flow and melt migration at the to of mantle diapirs: evidence from the Trinity ophiolite (California), *Terra*, *1*, 162, 1989.

Ceuleneer, G. and Nicolas, A., Structures in podiform chromite from the Maqsad district (Sumail ophiolite, Oman), *Mineralium Deposita*, *20*, 177-185, 1985.

Ceuleneer, G., Nicolas, A. and Boudier, F., Mantle flow patterns at an oceanic spreading centre: the Oman peridotites record, *Tectonophysics*, *151*, 1-26, 1988.

Ceuleneer, G., Rabinowicz, M. and Rosemberg, C., 3-D models of mantle plumes: thermal and petrological consequences, *EOS*, *71*, 43, 1625, 1990.

Coleman, R.G., Ophiolites - Ancient oceanic lithosphere? Springer Verlag, Minerals and Rocks, 12, 229 p, 1977.

Coleman, R.G., The diversity of ophiolites, *Geologie en Mijnbouw*, 141-159, 1984.

Cooper, R.F. and Kohlstedt, D.L., Rheology and structure of olivine-basalt partial melt, *J. Geophys. Res.*, *91*, 9315-9323, 1986.

Cordellier, F., Boudier, F.and Boullier, A.-M., Structural study of the Almklovdalen peridotite massif (Southern Norway), *Tectonophysics*, *77*, 257-281, 1981.

Crisp, J.A., Rates of magma emplacement and volcanic output, *J. Volc. Geother. Res.*, *20*, 177-211, 1984.

Darot, M. and Gueguen, Y., High-temperature creep of forsterite single crystals, *J. Geophys. Res.*, *86*, 6219-6234, 1981.

Dick, H.J.B., Partial melting in the Josephine peridotite I: the effect of mineral composition and its consequence for geobarometry and geothermometry, *Am. J. Sci.*, *277*, 801-832, 1977.

Dick, H.J.B., Abyssal peridotites, very slow spreading ridges and ocean ridge magmatism, in *Magmatism in the ocean basins*, edited by A.D. Saunders and M.J., pp. 71-105, Geological Society Spec. Pub., 42, 1989.

Dick, H.J.B., Fisher, R.L. and Bryan, W.B., Mineralogic variability of the uppermost mantle along mid-ocean ridges, *Earth Planet. Sci. Lett.*, *69*, 88-106, 1984.

Drury, M.R., Hoogerduijn Strating, E.H. and Vissers, R.L.M., Shear zone structures and microstructures in mantle peridotites from the Voltri massif, Ligurian Alps, N.W. Italy, *Geologie en Mijnbouw*, *69*, 3-17, 1990.

Ernst, T.G., Olivinknollen der basalte als bruchstücke alter olivinfelse, *Nachr. Akad. Wiss. Göttingen, Math. Phys.*, *1*, 147-154, 1935.

Etchecopar, A., Simulation par ordinateur de la déformation progressive d'un agregat polycristallin, *Thèse 3ème cycle, Nantes*, 135 p., 1974.

Etchecopar, A., A plane kinematic model of progressive deformation in a polycrystalline aggregate, *Tectonophysics*, *39*, 121-142, 1977.

Falloon, T.J., Green, D.H., Hatton, C.J. and Harris, K.L., Anhydrous partial melting of a fertile and depleted peridotite from 2 to 30 Kb and application to basalt petrogenesis, *J. Petrol.*, *29*, 1257-1282, 1988.

Forsyth, D.W., The evolution of the upper mantle beneath mid-ocean ridges, *Tectonophysics*, *38*, 89-118, 1977.

Francis, T.J.G., Generation of seismic anisotropy in the upper mantle along the mid-oceanic ridges, *Nature*, *221*, 162-165, 1969.

Francis, D., Mantle-melt interaction recorded in spinel lherzolite xenoliths from the Alligator Lake volcanic complex, Yukon, Canada, *J. Petrol.*, *28*, 569-597, 1987.

Frey, F.A., Rare earth element abundances in upper mantle rocks, in *Rare Earth Element geochemistry*, edited by P. Henderson, pp. 153-204, 1984.

Gente, P., Etude morphostructurale comparative de dorsales océaniques à taux d'expansion variés: *Thèse de l'Université de Brest*, 371 p., 1987.

Girardeau, J., Structure des ophiolites de l'Ouest de Terre Neuve et modèle de croûte océanique, *Thèse 3ème cycle, Nantes*, 154 p., 1979.

Girardeau, J. and Nicolas, A., The structure of two ophiolite massifs, Bay-of-Island, Newfoundland: a model for the oceanic crust and upper mantle, *Tectonophysics*, *77*, 1-34, 1981.

Goetze, C. and Evans, B., Stress and temperature in the bending lithosphere as constrained by experimental rock mechanics, *Geophys. J. R. astr. Soc.*, *59*, 463-478, 1979.

Green, H.W., Plasticity of olivine in peridotites, in *Electron microscopy in mineralogy*, edited by H.R. Wenk, pp. 443-464, Springer-Verlag, 1976.

Green, H.W. and Gueguen, Y., Origin of kimberlite pipes by diapiric upwelling in the upper mantle, *Nature Phys. Sci.*, *249*, 617-620, 1974.

Gueguen, Y. and Nicolas, A., Deformation of mantle rocks, *Ann. Rev. Earth Planet. Sci.*, *8*, 119-144, 1980.

Haggerty, S.E. and Sautter, V., Ultradeep (greater than 300 kilometers), ultramafic upper mantle xenoliths, *Science*, *248*, 993-996, 1990.

Helgason, J. and Zentilli, M., Field characteristics of laterally emplaced dikes: anatomy of an exhumed Miocene dike swarm in Reydarfjördur, Eastern Iceland, *Tectonophysics*, *115*, 247-274, 1985.

Herzberg, C., Gasparik, T. and Sawamoto, H., Origin of mantle peridotite: constraints from melting experiments to 16.5 GPa, *J. Geophys. Res., 95,* 15779-15804, 1990.

Jackson, E.D. and Thayer, T.P., Some criteria for distinguishing between stratiform, concentric, and alpine peridotite-gabbro complexes, *Int. Geol. Congr. Rep. Sess., 24th, 2,* 289-296, 1972.

Jackson, M., Structure des filons dans les massifs de peridotite: mécanismes d'injection et relations avec la déformation plastique, *Thèse, Université de Nantes,* 147 p., 1979.

Johnson, K.T.M., Dick, H.J.B. and Shimizu, N., Melting in the oceanic upper mantle: an ion microprobe study of diopside in abyssal peridotites, *J. Geophys. Res., 95,* 2661-2678, 1990.

Juteau T., Nicolas, A., Dubessy, J, Fruchard, J.-C. and Bouchez, J.-L., Structural relationships in the Antalaya ophiolite complex, Turkey: possible model for oceanic ridge, *Geol. Soc. Am. Bull., 88,* 1740-1748, 1977.

Karato, S.I., Grain-size distribution and rheology of the upper mantle, *Tectonophysics, 104,* 155-176, 1984.

Karson, J.A., Variations in structure and petrology in the Coastal Complex, Newfoundland: anatomy of an oceanic fracture zone, *Geol. Soc. London Spec. Publ., 13,* 131-144, 1984.

Kelemen, P.B., Reaction between ultramafic rock and fractionating basaltic magma; phase relations, the origin of calc-alkaline magma series, and the formation of discordant dunite, *J. Petrol., 31,* 1990.

Kirby, S.H., Rheology of the lithosphere, *Rev. Geophys. Space Phys., 21,* 1458-1487, 1983.

Klein, E.M. and Langmuir, C.H., Global correlations of ocean ridge basalt chemistry with axial depth and crustal thickness, *J. Geophys. Res., 92,* 8089-8115, 1987.

Kornprobst, J., Les péridotites et les pyroxenolites du massif ultrabasique de Beni-Bousera: une étude expérimentale entre 1100°C et 1550°C sous 15 à 30 Kb de pression sèche, *Contrib. Mineral. Petrol., 29,* 290-309, 1970.

Lagabrielle, Y. and Cannat, M., Ancient alpine jurassic ocean floor resembles the modern central Atlantic basement, *Geology, 18,* 319-322, 1990.

Lago, B., Rabinowicz, M. and Nicolas, A., Podiform chromite ore bodies: a genetic model, *J. Petrol, 23,* 103-125, 1982.

Langmuir, C.H., Bender, J.F. and Batiza, R., Petrological and tectonic segmentation of the East Pacific Rise, 5°30'-14°30'N, *Nature, 322,* 422-429, 1986.

Leblanc, M., Ceuleneer, G., Al Azri, H. et Jedwab, J., Concentration hydrothermale de Pd et de Pt dans les péridotites mantellaires du complexe ophiolitique d'Oman, *Compte Rendus Académie des Sciences, C.R.Acad.Sci., 312, II,* 1007-1012, 1991.

Leblanc, M. and Ceuleneer, G., Chilled chromitite dike from the Maqsad diapir (Oman ophiolite): evidence for chromite crystallization in a multicellular magma flow below an accreting oceanic crust, *Lithos,* in press, 1991.

le Sueur, E. et Boudier, F., Structures du complexe basique et ultrabasique de Trinity, Californie: genèse d'une ophiolite atypique, *Bull. Soc. Geol. France, 6,* 1007-1014, 1986.

le Sueur, E., Boudier, F., Cannat, M., Ceuleneer, G. and Nicolas, A., The Trinity mafic-ultramfic complex: first results of the structural study of an untypical ophiolite, *Ofioliti, 9,* 487-498, 1984..

Lorand, J.-P., Fe-Ni-Cu sulfides in tectonite peridotites from the Maqsad district, Sumail ophiolite, southern Oman: implications for the origin of the sulfide component in the oceanic upper mantle, *Tectonophysics, 151,* 57-73, 1988.

Lorand, J.-P. and Ceuleneer, G., Silicate and base-metal sulfide inclusions in chromites from the Maqsad area (Oman ophiolite, Gulf of Oman): a model for entrapment, *Lithos, 22,* 173-190, 1989.

Maaloe, S.and Printzlau, I., Natural partial melting of spinel lherzolite, *J. Petrol., 20,* 727-741, 1979.

Macdonald, K.C., Sempere, J.-C. and Fox, P.J., East-Pacific Rise from Siqueiros to Orozco fracture zones: along-strike continuity of axial neovolcanic zone and structure and evolution of overlapping spreading centers, *J. Geophys. Res., 89,* 6049-6069, 1984.

McKenzie, D.P., Finite deformation during fluid flow, *Geophys. J. R. astr. Soc., 58,* 689-715, 1979.

McKenzie, D.P., The generation and compaction of partially molten rocks, *J. Petrol.,25,* 713-765, 1984.

McKenzie, D.P. and Bickle, M.J., The volume and composition of melt generated by extension of the lithosphere, *J. Petrol., 29,* 625-679, 1988.

McKenzie, D.P., Roberts ,J.M. and Weiss, N.O., Convection in the Earth's mantle: toward a numerical simulation, *J. Fluid Mech., 62,* 465-538, 1974.

Mercier, J.-C., Olivine and pyroxenes, in *Preferred orientation in deformed metals and rocks: an introduction to modern texture analysis,* edited by H.R. Wenk, pp. 407-430, Academic Press, New York, 1985.

Mercier, J.-C. and Nicolas, A., Textures and fabrics of upper-mantle peridotites as illustrated by xenoliths from basalts, *J. Petrol., 16,* 454-487, 1975.

Mevel, C., Cannat, C., Gente, P., Marion, E., Auzende, J.-M. and Karson, J.A., Emplacement of deep crustal and mantle rocks on the west median valley wall of the MARK area (MAR, 23°N), *Tectonophysics, 190,* 31-54, 1991.

Michael, P.J. and Bonatti, E., Peridotite composition from the North Atlantic: regional and tectonic variations and implications for partial melting, *Earth Planet. Sci. Lett., 73,* 91-104, 1985.

Misseri, M., Structure des massifs ophiolitiques de Canyon Mountain (Oregon) et de Wadi Tayin (Oman): lithosphere d'arc insulaire, lithosphere oceanique, *Thèse 3ème cycle, Univ Nantes,* 166 p., 1982.

Misseri, M. and Boudier, F., Structures in the Canyon Mountain ophiolite indicate an island-arc intrusion, *Tectonophysics, 120,* 191-209, 1985.

Miyashiro, A., The Troodos ophiolite complex was probably formed in an island arc, *Earth Planet. Sci. Lett., 19,* 218-224, 1973.

Montigny, R., Le Mer, O., Thuizat, R. and Whitechurch, H., K-Ar and 40Ar/39Ar study of metamorphic rocks associated with the Oman ophiolite: tectonic implications, *Tectonophysics, 151,* 345-362, 1988.

Morton, J.L. and Sleep, N.H., A mid-oceanic ridge thermal model: constraints on the volume of axial hydrothermal heat flux, *J. Geophys. Res., 90,* 11345-11353, 1985.

Navon, O. and Stolper, E., Geochemical consequences of melt percolation: the upper mantle as a chromatographic column, *The Jounrnal of Geology, 95,* 285-307, 1987.

Nicolas, A., Melt extraction model based on structural studies in mantle peridotites, *J. Petrol., 27,* 999-1022, 1986a.

Nicolas, A, Structure and petrology of peridotites: clues to their geodynamic environment, *Rev. Geophys.*, *24*, 875-895, 1986b.

Nicolas, A. , Structure of ophiolites and dynamics of oceanic lithosphere. Kluwer, 367 p., 1989.

Nicolas, A. and Jackson, M., High-temperature dikes in peridotites: origin by hydraulic fracturing, *J. Petrol.*, *23*, 568-562, 1982.

Nicolas, A. and Poirier, J.-P., Crystalline plasticity and solid state flow in metamorphic rocks, Wiley, New-York, 444p, 1976.

Nicolas, A. and Prinzhofer, A., Cumulative or residual origin for the transition zone in ophiolites: structural evidence, *J. Petrol.*, *24*, 188-206, 1983.

Nicolas, A. and Violette, J.-F., Mantle flow beneath oceanic ridges: models derived from ophiolites, *Tectonophysics*, *81*, 319-339, 1982.

Nicolas, A., Boudier, F. and Bouchez, J.-L., Interpretation of peridotite structures from ophiolite and oceanic environments, *Am. J. Sci.*, *280(A)*, 192-210, 1980.

Nicolas, A. Lucazeau, F. and Bayer, R., Peridotite xenoliths in Massif Central basalts: textural and geophysical evidence for asthenospheric diapirism, in *Mantle xenoliths*, edited by P.H. Nixon, J. Wiley, 1986.

Nicolas, A., Ceuleneer, G., Boudier, F. and Misseri, M., Structural mapping in the Oman ophiolites: mantle diapirism along an oceanic ridge, *Tectonophysics*, *151*, 27-56, 1988.

O'Hara, M.J., The bearing of phase equilibria studies in synthetic and natural systems on the origin and evolution of basic and ultrabasic rocks, *Earth Sci. Rev.*, *4*, 69-133, 1968.

Ohtani, E.and Sawamoto, H., Melting experiment on a model chondritic mantle composition at 25 GPa, *Geophys. Res. Lett.*, *14*, 733-736, 1987.

Oliver, J., Solid Earth Science during the 21st Century, *EOS*, *72*, *11*, 121-126, 1991.

Oxburg, E.R. and Turcotte, D.L.,. Mechanisms of continental drift, *Rep. Prog. Phys.*, *41*, 1249-1312, 1978.

Pallister, J.S., Structure of the sheeted dyke complex of the Samail ophiolite near Ibra, Oman, J. *Geophys. Res.*, *86*, 2661-2672, 1981.

Pallister, J.S. and Knight, R.J., Rare-Earth element geochemistry of the Samail ophiolite near Ibra, Oman J. *Geophys. Res.*, *86*, 2673-2698, 1981.

Parker, R.L. and Oldenburg, D.W., Thermal model of ocean ridges, *Nature*, *242*, 137-139, 1973.

Pearce, J.A., Alabaster, T., Shelton, A.W. and Searle, M.P., The Oman ophiolite as a cretaceous arc-basin complex: evidence and implications, *Phil. Trans. r. Soc. London*, *300*, 299-317, 1981.

Phipps-Morgan, J., Melt migration beneath mid-ocean spreading centers, *Geophys. Res. Lett.*, *14*, 1238-1241, 1987.

Poirier, J.-P., Creep of crystals. Cambridge University Press, Cambridge, 1985.

Polvé, M. and Allègre, C., Orogenic lherzolite complex studied by 87Rb-87Sr: a clue to understand the mantle convection processes?, *Earth Planet. Sci. Lett.*, *51*, 71-93, 1980.

Presnall, D.C., Dixon, J.R., O'Donnel, T.H. and Dixon, S.A., Generation of mid-ocean ridge tholeiites, *J. Petrol. 20*, 3-35, 1979.

Prinzhofer, A.and Allegre, C., Residual peridotites and the mechanism of partial melting, *Earth Planet. Sci. Lett.*, *74*, 251-265, 1985.

Prinzhofer, A. and Nicolas, A., The Bogota peninsula, New Caledonia: a possible oceanic transform fault, *J. Geol.*, *88*, 387-398, 1980.

Prinzhofer, A., Nicolas, A., Cassard, D., Moutte, J., Leblanc, M., Paris, J.-P. and Rabinovitch, M., Structures in the New-Caledonia peridotites-gabbros: implications for oceanic mantle and crust, *Tectonophysics*, *69*, 85-112, 1980.

Quatrevaux, F., Girardeau, J. and Ceuleneer, G., Magma percolation related modal, textural and chemical heterogeneities in peridotites from the Maqsad diapir, Oman, EUG Abstract, Strasbourg, 1991.

Quick, J.E., Petrology and petrogenesis of the Trinity peridotite, an upper mantle diapir in the eastern Klamath mountains, northern California, *J. Geophys. Res.*, *86*, 11837-11863, 1981a.

Quick, J.E., The origin and significance of large, tabular dunite bodies in the Trinity peridotite, Northern California, *Contrib. Mineral. Petrol.*, *78*, 413-422, 1981b.

Rabinowicz, M., Ceuleneer, G. and Nicolas, A., Melt segregation and asthenospheric flow in diapirs below spreading centers: evidence from the Oman ophiolite, *J. Geophys. Res.*, *92*, 3475-3486, 1987.

Rabinowicz, M., Nicolas, A. and Vigneresse, J.-L., A rolling mill effect in asthenosphere beneath oceanic spreading centers, *Earth Planet. Sci. Lett.*, *67*, 97-108, 1984.

Raitt, R.W., Shor, G.G., Morris, G.B. and Kirk, H.K., Mantle anisotropy in the Pacific ocean, *Tectonophysics*, *12*, 173-186, 1971.

Reinhardt, B.M., On the genesis and emplacement of ophiolites in the Oman geosyncline, *Sweiz. Mineral. Petrogr. Mitt.*, *49*, 1-30, 1969.

Reuber, I., Générations successives de filons grenus dans le complexe ophiolitique d'Antalaya (Turquie): origine, évolution et mécanismes d'injection des liquides, *Thèse*, *Université Louis Pasteur, Strasbourg*, 246 p., 1982.

Reuber, I., Mylonitic ductile shear zones within tectonites and cumulates as evidence for an oceanic transform fault in the Antalaya ophiolite, SW Turkey, in *The geological evolution of the Eastern Mediterranean*, edited by J.E. Dixon and A.H.F. Robertson, pp. 319-334, *Spec. Pub. Geol. Soc.*, *17*, 1985.

Reuber, I., Complexity of the crustal sequence in northern Oman ophiolite (Fizh and southern Aswad block): the effect of early slicing?, *Tectonophysics*, *151*, 137-165, 1988.

Reuber, I., Nehlig, P. and Juteau, T., Axial segmentation at a fossil oceanic spreading centre in the Haylayn block (Semail nappe, Oman): off-axis mantle diapir and advancing ridge tip, *J. Geodyn.*, *13*, 1991.

Ribe, N.M., Theory of melt segregation - a review, *J. Volc. Geotherm. Res.*, *33*, 241-253, 1987.

Scambelluri, M., Hoogerduijn Strating, E.H., Piccardo, G.B., Vissers, R.L.M. and Rampone, E., Alpine olivine- and titanian clinohumite-bearing assemblages in the Erro-Tobbio peridotite (Voltri massif, NW Italy), *J. Metamorphic Geol.*, *9*, 79-91, 1991.

Schmeling, H., Numerical models on the influence of partial melt on elastic, anelastic and electric properties of rocks, part I, Elasticity and anelasticity, *Phys. Earth Planet. Int.*, *41*, 34-57, 1985.

Scott, D.R., The competition between percolation and

circulation in a deformable porous medium, *J. Geophys. Res.*, *93*, 6451-6462, 1988.

Scott, D.R. and Stevenson, D.J., Magma solitons, *Geophys. Res. Lett.*, *11*, 1161-1164, 1984.

Scott, D.R.and Stevenson, D.J., Magma ascent by porous flow, *J. Geophys. Res.*, *97*, 9283-9296, 1986.

Scott, D.R. and Stevenson, D.J., A self-consistent model of melting, magma migration and buoyancy-driven circulation beneath mid-ocean ridges, *J. Geophys. Res.*, *94*, 2973-2988, 1989.

Sen, G. and Presnall, D.C., Petrogenesis of dunite xenoliths from Koolau Volcano, Oahu, Hawaii: implications for Hawaiian volcanism, *J. Petrol.*, *27*, 197-217, 1986.

Seyler, M. and Mattson, P.H., Petrology and thermal evolution of the Tinaquillo peridotite (Venezuela), *J. Geophys. Res.*, *94*, 7629-7660, 1989.

Shaw, H.R., The fracture mechanism of magma transport from the mantle to the surface, *Physics of magmatic processes* edited by R.B. Hargraves, pp. 201-264, Princeton University Press, 1980.

Shearer, P. and Orcutt, J., Anisotropy in the oceanic lithosphee: theory and observations from the Ngendei seismic refraction experiment in the south-west Pacific, *Geophys. J.R. Astron. Soc.*, *80*, 493-526, 1985.

Sleep, N.H., Segregation of magma from a mostly crystalline mush, *Geol. Soc. Am. Bull.*, *85*, 1225-1232, 1974.

Sleep, N.H., Tapping of melt by veins and dykes, *J. Geophys. Res.*, *93*, 10255-10272, 1988.

Sleep, N.H. and Rosendahl, B.R., Topography and tectonics of mid-oceanic ridge axes, *J. Geophys. Res.*, *84*, 6831-6839, 1979.

Spiegelman, M. and McKenzie, D., Simple 2-D models for melt extraction at mid-ocean ridges and island arcs, *Earth Planet. Sci. Lett.*, *83*, 137-152, 1987.

Stevenson, D.J., Spontaneous small-scale melt segregation in partial melts undergoing deformation, *Geophys. Res. Lett.*, *16*, 1067-1070, 1989.

Takahashi, E., Melting of a dry peridotite KLB-1 up to 14 GPA: implications on the origin of peridotitic upper mantle, *J. Geophys. Res.*, *91*, 9367-9382, 1986.

Takahashi, E. and Kushiro, I., Melting of dry peridotite at high pressure and basalt magma genesis, *Am. Mineral.*, *68*, 859-879, 1983.

Toramaru, A. and Fujii, N., Connectivity of melt phase in a partially molten peridotite, *J. Geophys. Res.*, *91*, 9239-9252, 1986.

Violette, J.-F., Structure des ophiolites des philippines (Zambales et Palawan) et de Chypre: écoulement asthenospherique sous les zones d'expansion océaniques, *Thèse 3ème cycle, Nantes*, 161 p., 1980.

von Bargen, N. and Waff, H.S., Wetting of enstatite by basaltic melt at 1350° and 1.0 to 2.5 GPa pressure, *J. Geophys. Res.*, *93*, 1153-1158, 1988.

Waff, H.S., Effects of the gravitional field on liquid distribution in partial melts within the upper mantle, *J. Geophys. Res.*, *85*, 1815-1825, 1980.

Watson, S. and McKenzie, D.P., Melt generation by plumes: a study of Hawaiian volcanism. *J. Petrol.*, *32*, 1991.

Wei, K., Tronnes, R.G. and Scarfe, C., Phase relations of aluminium undepleted and aluminium depleted komatiites at pressure of 4-12 GPa, *J. Geophys. Res.*, *95*, 15817-15828, 1990.

Wilshire, H.G. and Kirby, S.H., Dikes, joints and faults in the upper mantle, *Tectonophysics*, *161*, 23-31, 1989.

Wyllie, P.J., Magma genesis, plate tectonics and chemical differentiation of the Earth, *Rev. Geophys.*, *26*, 370-404, 1988.

Yuen, D.A. and Fleitout, L., Thinning of the lithosphere by small-scale convective destabilization, *Nature*, *313*, 125-128, 1985.

Zindler, A. and Hart, S., Chemical geodynamics, *Ann. Rev. Earth Planet. Sci.*, *14*, 493-571, 1986.

G. Ceuleneer and M. Rabinowicz, UPR 234, GRGS, OMP, 14, avenue Ed. Belin, 31400 Toulouse.

The Physics of Magma Migration and Mantle Flow beneath a Mid-Ocean Ridge

DONALD L. TURCOTTE

Department of Geological Sciences, Cornell University, Ithaca, NY 14853

JASON PHIPPS MORGAN

Institute of Geophysics and Planetary Physics, UCSD, La Jolla CA 92093-0225

INTRODUCTION

A large fraction of the earth's volcanism occurs at mid-ocean ridges and results in the formation of the oceanic crust. The basic mechanism for how these magmas are generated is reasonably well known: As the sea floor spreads at an ocean ridge the mantle rock beneath must ascend to fill the gap. Hot mantle ascends along an adiabat beneath a spreading center which has a change in temperature with depth of about 0.5°K/km while the pressure dependence of the fertile mantle solidus beneath a ridge is about 4°K/km. Thus at a relatively shallow depth (~50-100km) ascending mantle crosses the solidus temperature and partial melting results. This mechanism for melting is commonly accepted and is supported by laboratory experiments which melt fertile mantle peridotites (cf. Hess, this volume). However, the amounts of melting beneath a mid-ocean ridge, the mechanisms of melt transport in the mantle (and lithosphere) and the interactions of melting and melt migration with the upwelling mantle flow at a spreading center are less well understood.

The goal of this review is to attempt to present the basic physics of flow, melting, and melt migration beneath a spreading center in a fairly simple and self-contained form so it can serve as an introduction as well as a guide to recent theoretical studies of flow and melting processes at a ridge. There have been two complementary theoretical approaches to investigating these problems. The first is to develop and examine analytical or quasi-analytical solutions which explore the physics of very idealized (often 1 - dimensional) systems. This analysis leads to important insights into the fundamental processes associated with melting, melt migration, and mantle flow. The second is to perform and analyze numerical experiments on more complex systems of equations. This approach allows the examination of more realistic geometries or non-linear interactions. In this review we will start by presenting analytical theories on melting and melt migration beneath a mid-ocean ridge. After this introduction we will summarize the implications of current 2-D and 3-D numerical experiments on the forces which shape mantle upwelling beneath a spreading center.

THE PHYSICS OF MELTING AND MELT MIGRATION

A detailed discussion of pressure release melting is given in McKenzie and Bickle [1988] and is discussed throughout this volume [see papers by Hess; Langmuir et al., and Grove et al.]. Geochemical and petrological studies roughly constrain the depths where most basaltic melt is generated. Small amounts of melting occur in the garnet lherzolite stability field (>65-80 km deep) [e.g. Salters and Hart, 1989]; however, most melting occurs in the 20-65 km depth range in the spinel peridotite stability field; and relatively little melting occurs in the <20 km depth plagioclase peridotite stability field. Petrological and geochemical data also imply that mid-ocean ridge basalts are generated by an average of 10-15% partial melting of the sub-ridge mantle (e.g. the maximum melt extraction is 20-30%). The generation of 6 km of oceanic crust from a mean amount of peridotite melting of 10% implies that a 60 km thick column of melted mantle should lie beneath this crust as it leaves the spreading center. Note that a column of this size also agrees with the petrological evidence that most melting occurs within the spinel peridotite stability field.

Mantle Flow and Melt Generation at Mid-Ocean Ridges
Geophysical Monograph 71
Copyright 1992 American Geophysical Union

Melt is originally produced along grain boundaries but segregates and ascends through the residual solid matrix due to its differential buoyancy to form the oceanic crust. A number of mechanisms have been suggested for melt migration. Initially the melt probably ascends by a porous flow mechanism. Thus melt ascent can be modelled using Darcy's law for flow in a porous medium [Frank, 1968; Sleep, 1974; Turcotte and Ahern, 1978; Walker et al., 1978; Ahern and Turcotte, 1979; Stolper et al., 1981].

POROUS FLOW MIGRATION

As melt accumulates on grain boundaries it will form a network of interconnected channels as illustrated in Figure 1a. We model these channels as a cubic network of circular channels (see Figure 1b). The side length of the cubic structure b is equivalent to the grain size. The diameter of the channels, δ, can be expressed in terms of the porosity $\phi - \phi_0$ where ϕ is the total porosity and ϕ_0 is the volume fraction of isolated pockets of liquid which do not contribute to the permeability:

$$\phi - \phi_0 = \frac{3\pi\delta^2}{4b^2} \tag{1}$$

The value of the unconnected porosity is a subject of considerable controversy [Watson, 1982]. In our calculations we will take $\phi_0 = 0.02$.

The permeability k of such a matrix of tubes is given by Bear [1972, p. 163]:

$$k = \frac{(\phi - \phi_0)\,\delta^2}{96} \tag{2}$$

Thus the permeability as a function of grain size and connected porosity is:

$$k = \frac{b^2(\phi - \phi_0)^2}{72\pi} \tag{3}$$

A more sophisticated geometrical model can be used but hardly seems justified considering the variability in grain size and magma viscosity. However, under quite general

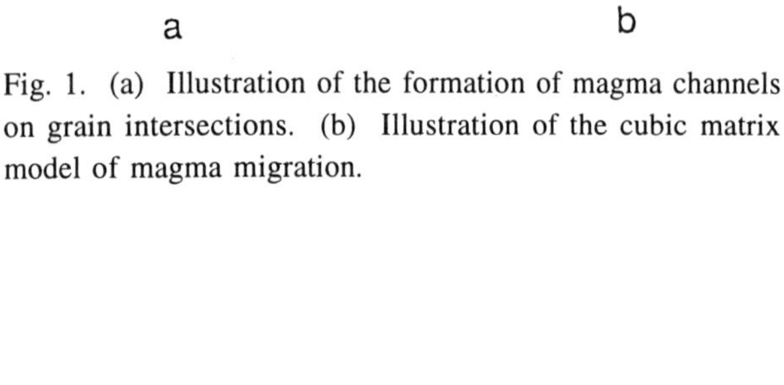

Fig. 1. (a) Illustration of the formation of magma channels on grain intersections. (b) Illustration of the cubic matrix model of magma migration.

conditions it can be shown that the permeability scales with the square of the interconnected melt fraction. We will now use the standard theory for flow in porous media to determine how this permeability allows the liquid to migrate.

We use Darcy's law to relate the relative velocity of the fluid through the solid matrix to the buoyancy pressure gradient:

$$\frac{1}{3}(\phi - \phi_0)(v - V) = \frac{k}{\eta_l}(\rho_s - \rho_l)\,g \tag{4}$$

where v is the liquid velocity (pore velocity), V the velocity of the solid matrix, ρ_l the liquid density, ρ_s the solid density, η_l the liquid viscosity, and g the acceleration of gravity. The left-hand side of (4) is the vertical fluid flux (Darcy velocity). In writing (4) we assume that the differential buoyancy between the liquid and the solid is balanced by the viscous pressure drop due to the flow of the liquid through the small channels. Implicit in this formulation of the problem is the ability of the matrix to deform as the liquid escapes upward (i.e. mantle compaction can occur on a time scale which is short compared to the time of melt segregation and ascent). We will show in a later section that this is a reasonable assumption.

The basic physics of mantle upwelling and pressure release melting is illustrated by solving the one-dimensional vertical flow beneath a mid-ocean ridge. For this case we can obtain analytical solutions to (1) to (4).

Within the melt zone the upward flux of mass must be equal to the flux of mass into the melt zone:

$$\rho_s V_0 = \frac{1}{3}(\phi - \phi_0)\,\rho_l v + \left(\frac{2}{3}\phi + \frac{1}{3}\phi_0\right)\rho_l V + (1 - \phi)\,\rho_s V \tag{5}$$

where V_0 is the mantle ascent velocity below the melt zone. The degree of partial melting, f, at any point in the melt zone is related to the liquid flux at that point by:

$$f\rho_s V_0 = \frac{1}{3}(\phi - \phi_0)\,\rho_l v + \left(\frac{2}{3}\phi + \frac{1}{3}\phi_0\right)\rho_l V \tag{6}$$

Combining (3) to (6) the liquid-filled porosity ϕ can be related to the degree of partial melting f by:

$$\phi^3 + C_1\phi^2 + C_2\phi + C_3 = 0 \tag{7}$$

$$C_1 = -(1 + 2\phi_0)$$

$$C_2 = 2\phi_0 + \phi_0^2 + V_0(f - 1)/\beta - \rho_s f V_0/\rho_l\beta$$

$$C_3 = \rho_s V_0 f/\rho_l\beta - \phi_0^2$$

and

$$\beta = -b^2(\rho_s - \rho_l)g/72\pi\eta_l$$

The temperature, T, within the melt zone must satisfy the steady-state energy equation:

$$\rho_s V_0 c_p \frac{dT}{dz} = \lambda \frac{d^2T}{dz^2} - Q_f \tag{8}$$

where λ is the thermal conductivity, c_p the specific heat at constant pressure (in writing (6) it is assumed that the specific heats of the liquid and solid are equal), and Q_f is the heat loss due to melting. The adiabatic temperature change has been neglected.

We assume that 25% of the mantle rock consists of a low melting temperature basaltic component. We also assume that the solidus temperature of the remaining 75% is never reached, thereby limiting partial melting to 25%. The partial melting of basalt in the mantle with 0.1% water as given by Ringwood [1975] is illustrated in Figure 2. We will approximate the equilibrium dependence of the partial melt fraction f on depth (pressure) and temperature with the equation:

$$f = A\{\exp[B(T - Cz - D)] - 1\} \tag{9}$$

This result is also given in Figure 2. The approximation for large degrees of partial melt is quite good but there is a considerable deviation for low degrees of partial melt between the depths of 75 and 150 km. If included in our model a small degree of melting would develop at depth but, as will be seen, no significant upward migration of magma would result. It should be emphasized that (9) is strictly an empirical result, alternative forms can certainly be used. Also, the data used in Figure 2 was based on experiments in

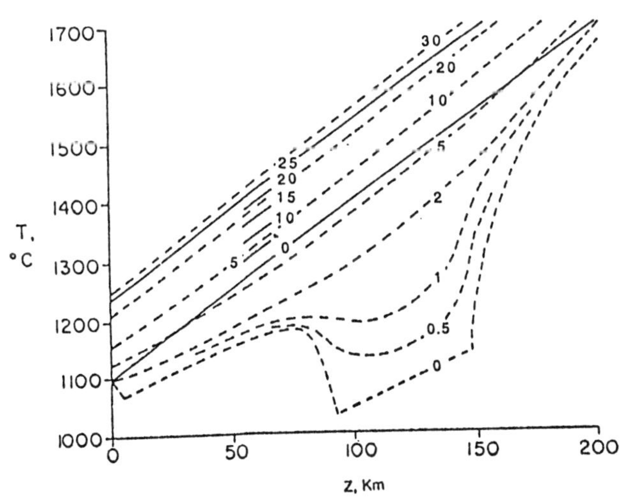

Fig. 2. Dependence of partial melt fraction of undepleted mantle (in percent) on temperature and depth. The dashed lines are values based on laboratory studies with 0.1% water [Ringwood, 1975] and the solid lines are from (9).

which the melt was not removed (batch melting), the removal of the melt may significantly influence the melt relations.

The heat loss due to partial melting Q_f is given by:

$$Q_f = L\rho_s V_0 \frac{df}{dz} \tag{10}$$

where L is the latent heat of fusion, since melting is caused by both changes in pressure and temperature:

$$\frac{df}{dz} = \frac{\partial f}{\partial z} + \frac{\partial f}{\partial T}\frac{dT}{dz} \tag{11}$$

Combining (9) to (11) we obtain:

$$Q_f = ABL\rho_s V_0 \exp[B(T - Cy - D)]\left[\frac{dT}{dz} - C\right] \tag{12}$$

Substitution of (12) into (8) yields:

$$\rho_s V_0 c_p \frac{dT}{dz} = \lambda \frac{d^2T}{dz^2} - ABL\rho_s V_0 \exp[B(T\text{-}Cz\text{-}D)]\left[\frac{dT}{dz} - C\right] \tag{13}$$

The solution within the melting zone requires the combined solution of (7), (9), and (13).

In the region of vertical flow beneath the melting zone f = 0. In this case (13) reduces to:

$$\rho_s V_0 c_p \frac{dT}{dz} = \lambda \frac{d^2T}{dz^2} \tag{14}$$

The temperature of the ascending rocks at great depth is T_m so that we require $T \rightarrow T_m$ as $z \rightarrow -\infty$. Also, the temperature and temperature gradient must be continuous at the base of the melt zone. The solution of (14) which satisfies continuity of temperature is:

$$T = T_m + (D + Cz - T_m)\exp\left[\frac{\rho_s c_p V_0}{\lambda}(z - z_0)\right] \tag{15}$$

where z_0 is the base of the melt zone. There is an upstream temperature precursor below the melt zone.

In the region above the melt zone the magma produced within the melt zone migrates upward. In general the solution for the migration velocity and magma filled porosity in this zone would require the solution of (7). However, in most cases of interest the magma migration velocity v is large compared with the mantle velocity V_0, that is $v >> V_0$. For this case (2) and (3) can be combined to give:

$$v = \frac{(\rho_s - \rho_l) g b^2 (\phi - \phi_0)}{24\pi\eta_l} \tag{16}$$

Also in this limit it follows from (6) that :

$$\frac{1}{3}(\phi - \phi_0)\rho_l v = f_0 \rho_s V_0 \tag{17}$$

TABLE 1: Variables in (69-75) and Parameter Values

Variable	Description	
u or u_i	velocity	
p	pressure	
τ_{ij}	deviatoric stress	
T	temperature	
ξ	mantle depletion	
$\dot{\psi}$	melt production rate	
ϕ	melt fraction (porosity)	
Parameter	**Value**	**Description**
λ	3.5 Wm^{-1} $°K^{-1}$	thermal conductivity
c_p	1 kJ kg^{-1} $°K^{-1}$	specific heat
L	400 kJ kg^{-1}	latent heat of fusion
A	0.4	constant in (9)
B	3.6×10^{-3} $°K^{-1}$	constant in (9)
C	3×10^{-3} °K m^{-1}	constant in (9)
D	1,100 °K	constant in (9)
ρ_s	3,300 kg m^{-3}	density of solid
ρ_l	2,700 kg m^{-3}	density of magma
b	2 mm	grain size
ϕ_0	0.02	unconnected porosity
η_l	1, 10 Pa s	viscosity of magma
κ	1 mm s^{-2}	thermal diffusivity
η	10^{18}-10^{21} Pa-s	mantle shear viscosity
χ	10^{18}-10^{21} Pa-s	mantle bulk viscosity
α	3×10^{-5} K^{-1}	thermal expansivity
β	2.4×10^{-2}	depletion buoyancy
γ	8×10^{-2}	melt buoyancy ratio

where f_0 is the total degree of partial melting in the melt zone. Combining (16) and (17) gives:

$$\phi - \phi_0 = \frac{6}{b}\left[\frac{2\pi\eta_l f_0 \rho_s V_0}{(\rho_s - \rho_l)\, g\rho_l}\right]^{1/2} \tag{18}$$

$$v = \frac{b}{2}\left[\frac{(\rho_s - \rho_l)\, g f_0 \rho_s V_0}{2\pi\eta_l\rho_l}\right]^{1/2} \tag{19}$$

for the connected porosity and magma migration velocity above the melt zone.

We now obtain numerical solutions for the structure of the melt zone. We solve (7), (9), and (13) for the variables ϕ, f, and T by stepwise integration from the base of the melt zone. The temperature and temperature gradient given by (14) must be continuous at the base of the melt zone. The unknown position of the base z_0 is obtained by iteration. The values of the parameters used in our calculations are given in Table 1. The temperature of the ascending mantle rock is T_m = 1300° C. Results are obtained for magma viscosities of 1 and 10 Pa s [Ryan and Blevins, 1987]. Mantle ascent velocities of 10 and 50 mm/yr are considered. The maximum ascent velocities beneath a ridge crest are of the order of the sea-floor spreading velocities, i.e. roughly within this range.

The temperature as a function of depth is shown in Figure 3. Also included is the dependence of the degree of partial melting on temperature and depth from (9). As the degree of partial melting increases, the temperature decreases because the latent heat of fusion has been extracted from the internal energy. To a first approximation this is the wet adiabat. If the temperature of the ascending mantle rock T_m were taken to be higher the melting zone would be displaced downward.

The degree of partial melting throughout the melt region is given in Figure 4. The degree of partial melting increases nearly linearly through the zone. The partial melt volume fraction (porosity) ϕ is given as a function of depth in the melt zone in Figure 5. Without melt migration, $\phi = f$, and the results in Figure 5 would be identical to the results given in Figure 4. When the pores become connected to form a permeable matrix, f exceeds ϕ meaning that the excess magma is squeezed out. In all cases the porosity does not exceed the assumed 2% unconnected porosity (ϕ_0 = 0.02) by more than 1%. The effects of varying viscosity and spreading rate are illustrated in Figure 5; lower viscosity magma flows more freely so that the connected porosity $\phi - \phi_0$ is less; the slower ascent velocity gives more time for magma migration so that the connected porosity is also less.

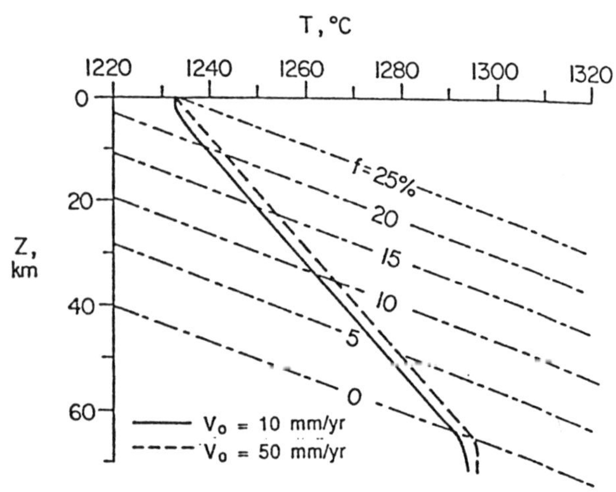

Fig. 3. Dependence of temperature on depth for mantle rock ascending beneath an ocean ridge at velocities of 10 and 50 mm/yr. The degree of partial melting f from (9) is also shown.

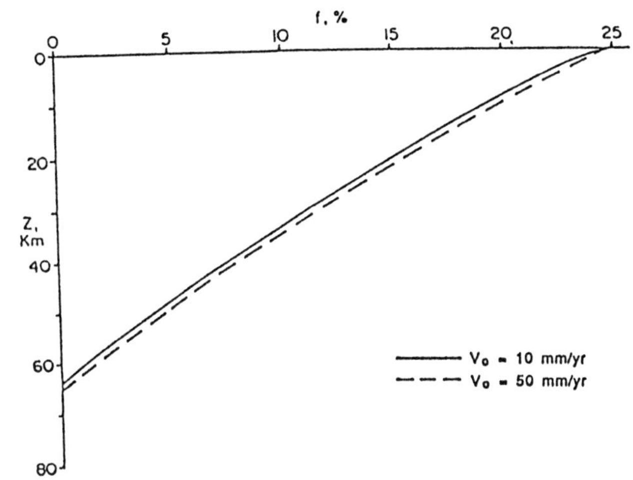

Fig. 4. Dependence of degree of partial melting f on depth z through the melt zone for two values of the magma viscosity η_l and two ascent velocities V_0.

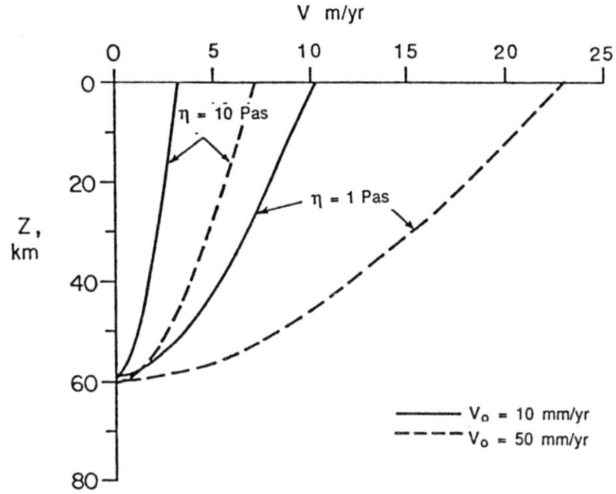

Fig. 6. Dependence of the magma migration velocity on depth through the melt zone for two values of the magma viscosity η_l and two ascent velocities V_0.

The dependence of the magma migration velocity v on depth in the melt zone is given in Figure 6. It is seen that the maximum migration velocity in each case is two to three orders of magnitude greater than the mantle ascent velocity. Higher mantle ascent velocities also result in larger magma migration velocities.

The general structure of an oceanic ridge is illustrated in Figure 7. Loss of heat to the surface creates a cold, rigid lithosphere. The cooling results in a freezing front which is defined by the solidus temperature of the magma. The degree of partial melting of the ascending mantle rock is also shown. In order to form the oceanic crust in the immediate vicinity of the oceanic ridge considerable lateral migration of magma must occur. Insufficient magma is produced immediately below the ridge crest.

The first partial melt fraction produced at the base of the melt zone migrates upward and mixes with the later partial melt fractions produced at shallower depths. The uniformity of this mixing will strongly influence the composition of the surface volcanic rocks. Heterogeneity of ridge basalts will result if off-axis, low melt-fraction magma reaches the surface and is erupted without mixing with near-axis, high melt-fraction magma. McKenzie and O'Nions [1991] have

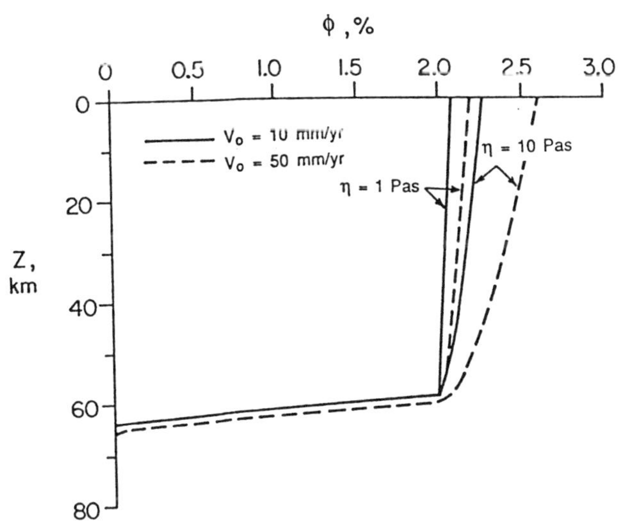

Fig. 5. Dependence of the partial melt volume fraction ϕ on depth z through the melt zone for two values of the magma viscosity η_l and two ascent velocities V_0.

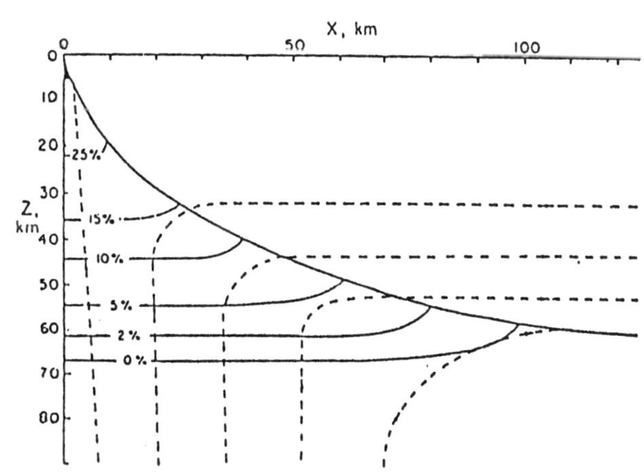

Fig. 7. Idealized ridge structure. Dashed lines show the motion of the matrix. Degrees of partial melting f are also shown. Solid line is base of lithosphere as determined by the solidus.

related partial melt distributions beneath ocean ridges to rare earth element concentrations in the surface magmas.

MATRIX COMPACTION

Although Ahern and Turcotte [1978] had used an order of magnitude calculation to show that matrix compaction would occur beneath an ocean ridge, McKenzie [1984] and Richter and McKenzie [1984] questioned the assumption. McKenzie [1984] developed the full equations for the viscous collapse of the matrix with Darcy flow of a low-viscosity, immiscible fluid. Matrix compaction was certainly important for the transient problems considered and led to the generation of solitary waves or "magmons" [Scott and Stevenson, 1984, 1986; Barcilon and Richter, 1986; Barcilon and Lovera 1989].

Ribe [1985a, b] studied compaction in detail for the ascending flow beneath ocean ridges and showed that free compaction would be expected and the results given above are valid. Ribe [1985a, b] considered a simple one dimensional model in which mantle material ascends at constant speed V_0 and reaches its solidus at $z = 0$ (in this analysis z is measured positive upwards), where a fixed volume fraction f_0 melts. Above $z = 0$, the melt occupies a volume fraction (porosity) ϕ, and migrates upward at a velocity v, and the unmelted residue (matrix) ascends at a velocity V. Although localized melting is unlikely to occur in the Earth's mantle, the model illustrates the essential physics of compaction.

The governing equations for the model [McKenzie, 1984] are conservation of mass for the fluid component

$$f\rho_s V_0 = \phi\rho_l v \tag{20}$$

conservation of mass for the matrix

$$\phi\rho_l v + (1 - \phi)\rho_s V = \rho_s V_0 \tag{21}$$

force balance for the melt

$$0 = -\phi\rho_l g - \frac{\eta_l\phi^2}{k}(v - V) - \phi\frac{dp}{dz} \tag{22}$$

and force balance for the matrix

$$0 = -(1 - \phi)\rho_l g + \frac{\eta_l\phi^2}{k}(v - V) - (1 - \phi)\frac{dp}{dz} + \left(\chi_s + \frac{4}{3}\eta_s\right)\frac{d^2V}{dz^2} \tag{23}$$

The dependent variables in (20) - (23) are the melt velocity v, the matrix velocity V, the porosity ϕ, and the melt pressure p; η_s is the shear viscosity, and χ_s is the bulk viscosity of the matrix. The assumptions made in deriving these equations are discussed in detail by McKenzie [1984]. Equation (22) is the strict form of Darcy's law, which states that the velocity of the melt relative to the matrix is proportional to the effective pressure gradient. The last term in (23) is the 'compaction' term, which describes the deformation of the matrix. For simplicity the viscosities χ_s and η_s have been assumed constant.

Equations (22) and (23) can be combined by eliminating dp/dz yielding

$$0 = -(1 - \phi)g\Delta\rho + \frac{\eta_l\phi}{k}(v - V) + \left(\chi_s + \frac{4}{3}\eta_s\right)\frac{d^2V}{dz^2} \tag{24}$$

where $\Delta\rho = \rho_s - \rho_l$. Equation (24) shows that melt segregation is governed by a balance of three forces: the differential buoyancy of the melt; the resistance to flow of melt through the matrix (Darcy resistance); and the resistance of the matrix to deformation (compaction term). If the compaction term can be ignored the flow is determined by a balance between differential buoyancy and Darcy resistance.

To complete the formulation we assume that the dependence of matrix permeability k on the porosity ϕ is given by (1). Substitution of (1), (20), and (21) into (24) gives

$$0 = -(1 - \phi)g\Delta\rho + \frac{72\pi\eta_l V_0}{\phi b^2}\left(\frac{\phi_0}{\phi}\frac{\rho_s}{\rho_l} - \frac{1 - \phi_0}{1 - \phi}\right) + (1 - \phi_0)V_0\left(\chi_s + \frac{4}{3}\eta_s\right)\frac{d^2}{dz^2}\left(\frac{1}{1 - \phi}\right) \tag{25}$$

In order to further simplify the problem we assume that $\phi_0 << 1$ (which implies $\phi << 1$), and that $\rho_s = \rho_l$ everywhere except in the first term of (25) (the Boussinesq approximation). Equation (25) then becomes

$$0 = -g\Delta\rho + \frac{72\pi\eta_l V_0}{b^2\phi^2}(\phi_0 - \phi) + V_0(\chi_s + 4\eta_s/3)\frac{d^2\phi}{dz^2} \tag{26}$$

This equation gives the dependence of ϕ on depth.

The problem can be further simplified by introducing the nondimensional quantities

$$\phi' = \frac{\phi}{\phi_0} \tag{27}$$

$$z' = \frac{z}{\delta_c} \tag{28}$$

where

$$\delta_c = \left[\frac{k_0\left(\chi_s + \frac{4\eta_s}{3}\right)}{\eta_l}\right]^{1/2} \tag{29}$$

is the compaction length with

$$k_0 = \frac{b^2 \phi_0^2}{72\pi} \tag{30}$$

Substitution of (27) to (30) into (26) gives (dropping primes)

$$\frac{d^2\phi}{dz^2} + \frac{1-\phi}{\phi^2} = R \tag{31}$$

where

$$R = \frac{k_0 g \Delta\rho}{\phi_0 V_0 \eta_l} \tag{32}$$

is a dimensionless buoyancy.

The appropriate boundary conditions for (31) can be determined as follows. The viscosity of the matrix is finite, so compaction of a material element of the matrix occurs gradually as it ascends beyond $z = 0$. At $z = 0$, no compaction has yet occurred, so $\phi = \phi_0$ there. As $z \to \infty$, however, compaction nears completion, and ϕ is determined by a balance of the first two terms of (24) (Darcy's law). The dimensionless boundary conditions for (31) are

$$\phi(0) = 1 \tag{33}$$

$$\frac{1-\phi(\infty)}{\phi^2(\infty)} = R. \tag{34}$$

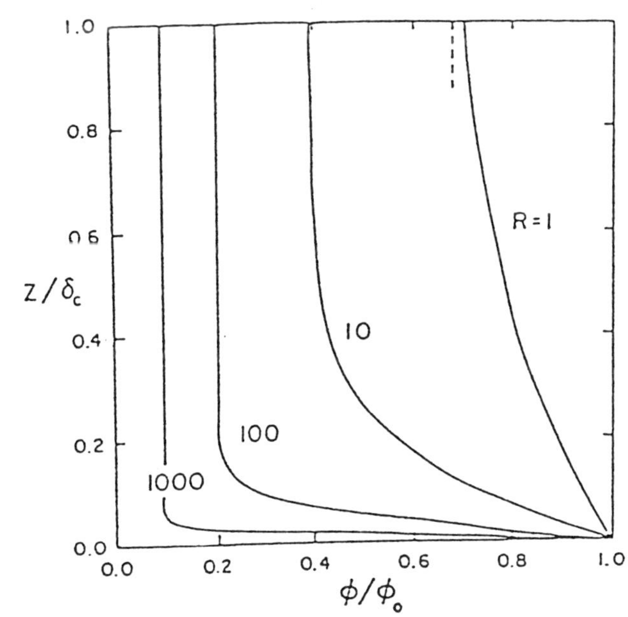

Fig. 8. Porosity profiles for the localized melting model. Compaction is confined to a boundary layer ('compaction layer') of thickness $\delta R' = R^{-1/2}\delta_c$. Dashed line shows $\phi(\infty)/\phi_0$ for R = 1.

The problem defined by (31) can be solved by defining $d\phi/dz = F$, whereupon it becomes

$$F\frac{dF}{d\phi} = R + \frac{\phi - 1}{\phi^2}. \tag{35}$$

Integration of (35) gives

$$F = \left\{2\left[R\left(\phi - \phi_\infty\right) + \ln\left(\phi/\phi_\infty\right) + \frac{1}{\phi} - \frac{1}{\phi_\infty}\right]\right\}^{1/2} \tag{36}$$

Substitution for F gives

$$z = \frac{1}{\sqrt{2}}\int_\phi^1 \left[R\left(\phi - \phi_\infty\right) + \ln\left(\phi/\phi_\infty\right) + \frac{1}{\phi} - \frac{1}{\phi_\infty}\right]^{-1/2} d\phi \tag{37}$$

Numerical evaluation of (37) gives z as a function of ϕ in the partially molten zone. The results are shown for several values of the buoyancy R in Figure 8. Compaction is confined to a boundary layer (compaction layer) whose thickness decreases with increasing R. Above this layer, the porosity ϕ has a constant value ϕ_∞ defined by (34). This state is known as uniform fluidization, and corresponds to a balance of buoyancy and Darcy resistance such that the porosity is constant. The (dimensional) thickness of the compaction layer is therefore given to order of magnitude by the reduced compaction length

$$\delta_R = R^{-1/2}\delta_c = \left[\frac{\phi_0 V_0\left(\chi_s + 4\eta_s/3\right)}{g\Delta\rho}\right]^{1/2}. \tag{38}$$

Substitution of the appropriate values from Table 1 shows that $\delta_R \sim 10 - 100$ m. This range of values is consistent with the estimate given by Ahern and Turcotte [1979] that a partially molten column exceeding about 80 m in height will collapse under its own weight. Note that the values used for χ_s and η_s (10^{15} - 10^{18} Pa s) are much less than the usual value of 10^{21} Pa s for the subsolidus viscosity of the mantle. This is because partial melting may reduce the effective viscosity of the crystalline matrix as a whole (not the individual grains) by orders of magnitude; for a discussion, see McKenzie [1984] or Kohlstedt [this vol.].

The upward drainage of magma through the solid matrix can have a range of implications. Magma erupting at the surface or forming intrusive bodies comes from a variety of depths and a variety of source regions [Kenyon and Turcotte, 1987]. There can be chemical interchange between the migrating fluid and the solid matrix. This can be particularly important in terms of incompatible trace elements such as the rare earths. The effect is similar to that in a chromatographic column [Ribe, 1985b; Richter, 1986; Navon and Stolper, 1987]. There is considerable evidence for non equilibrium behavior in the chemistry of midocean ridge basalts [Salters and Hart, 1989].

MAGMA FOCUSING

Beneath the crest of a fast spreading ridge the asthenosphere is likely to reach the base of the crust. Thus it is easy to envision how magma can migrate through the asthenosphere by the porous-flow mechanism to form a magma chamber. While this view may be essentially correct there are several problems with it. One of the major problems is the necessity of focusing the magma so that it reaches the immediate vicinity of the ridge axis. Perhaps the strongest observational constraint on melt migration is that the ridge neovolcanic zone is only 2-3 kilometers wide, independent of axial morphology or spreading rate [Macdonald, 1982]. Almost all crustal accretion occurs within the neovolcanic zone since the crustal thickness is essentially the same on and off axis [Detrick et al., 1987]. Melt emplacement near the surface to form oceanic crust is concentrated within an extremely narrow region indeed. In contrast Figure 7 and Figures 19-22 suggest that pressure release melting takes place in a much broader region on the order of 20-80 km on either side of the ridge axis.

Some process must focus this magma to be emplaced at the ridge axis. Gravity results in only a vertical migration of magma. The pressure drop associated with passive plate spreading (76) is of the proper sense to pull magma towards the ridge axis [Spiegelman and McKenzie, 1987; Phipps Morgan; 1987]. Spiegelman and McKenzie [1987] suggested that the viscous forces associated with passive upwelling are of the same order as the gravitational buoyancy force. The parameter that governs this effect is $\Gamma = 4\,\eta_s V_0/(\rho_s - \rho_l)gD^2$ where V_0 and D are a characteristic velocity and length associated with the mantle flow and η_s is the mantle viscosity. Taking $\eta_s = 2 \times 10^{19}$ Pa s, $V_0 = 50$ mm/yr, $\rho_s - \rho_l = 500$ kg/m^3, g = 10 m/s^2, and D = 100 km we find that $\Gamma = 3 \times 10^{-3}$ instead of ~1, which is needed for viscous forces to significantly shape local magma ascent. It does not appear that the mantle viscosity is sufficiently large for the viscous stresses to influence magma migration [Phipps Morgan, 1987].

We must then look for feasible mechanisms that can transport melt towards the ridge axis. This dilemma has led to two basic scenarios for the accumulation of melt to a narrow neovolcanic zone of crustal accretion: (1) Melting occurs within a broad region beneath a mid-ocean ridge and is focussed during melt migration to the ridge axis (see below for some possibilities); (2) Mantle upwelling and melting is strongly concentrated within a region directly beneath the ridge axis, and so mantle flow instead of melt migration transports the melts generated from a large region of the upwelling mantle below the melting region to the ridge axis. Two effects have been suggested to lead to this focussing of mantle upwelling: (1) the buoyancy force from large (10-20%) retained melt fractions within the upwelling mantle [Rabinowicz et al., 1987; Buck and Su, 1989; Scott and Stevenson, 1989]; and (2) an additional rheological feedback that the mantle viscosity is reduced when large melt fractions are present [Buck and Su, 1989]. Rabinowicz et al. [1987] and Buck and Su's [1989] numerical experiments show that mantle upwelling can be confined by this mechanism to a 10 km wide or smaller region beneath the ridge if retained melt fractions are large (>10%). Scott and Stevenson [1989] present numerical experiments that illustrate the buoyancy effects of a 6% melt fraction within the melting region, these experiments lead to a roughly 40 km wide melting region and still present the problem of how to transport this melt to the ridge axis.

There are several promising potential mechanisms for lateral melt migration beneath a MOR. Perhaps the best developed and most promising mechanism is that of Sparks and Parmentier (1991) who suggest that melt freezing will occur along a P-T solidus that slopes, like near-ridge isotherms, away from the ridge axis. Melt freezing leads to local pressure gradients that drive ascending melt away from the high pressure freezing region. This leads Sparks and Parmentier [1991] to propose that a large melt fraction will build up within roughly one viscous compaction length of the freezing boundary, and that melt will migrate up this 'channel' to the ridge axis. Another suggestion by Phipps Morgan [1987] is that melt, once generated, rapidly forms a network of veins within the melting region. He suggests that these veins will persist and be the natural channels for subsequent melt migration. Subsequent mantle deformation associated with the near-ridge turning from upwelling to horizontal motions will result in the passive deformation of this melt network along with its host rock into a strongly anisotropic fabric that is preferentially aligned towards the ridge axis and which will focus melt towards the neovolcanic zone. Ribe [1986, 1988] has also considered problems associated with magma focusing.

A second potential problem with the above porous flow models is that there is observational evidence that melt segregates in the asthenosphere to form veins [Nicolas, 1986]. The resulting magma flow within these veins may or may not be well described as a porous flow phenomena. Models for the formation and evolution of vein-like fractures were developed by Sleep [1988] and Stevenson [1989]. The formation of magma fractures within a deforming mantle matrix is an extremely complex phenomena which is currently difficult to treat.

MAGMA FRACTURE

We now turn to the problem of the migration of magma through the oceanic lithosphere at the ridge crest. There is ample observational evidence that magma collects in a magma chamber. Studies in Iceland and in ophiolites show that sheeted dike complexes are an essential feature of the generation of oceanic crust at spreading centers [Cann, 1974; Kidd, 1977]. A schematic representation of this process is illustrated in Figure 9. Magma is stored in a magma chamber at the base of the sheeted dike complex until the tensional stresses associated with sea-floor spreading are sufficiently large that an upward-propagating fluid-driven fracture initiates at the top of the magma chamber. This fracture creates a channel that allows the magma chamber to be fully or partially drained. Magma extruded at the surface forms pillow basalts. When flow ceases the magma in the dike solidifies forming the youngest member of the sheeted dike complex.

Magma driven fractures are the only mechanism that has been proposed for the transport of magma through the lithosphere. The basic formulation of fluid-driven fractures has been developed for hydraulically driven fractures by Geertsma and DeKlerk [1969] and Geertsma and Haafkens [1979]. Solutions must simultaneously consider the fluid flow through the crack, the elastic deformation of the surrounding rock, and the propagation of the fracture. Similarity solutions for laminar, pressure-driven, two-dimensional fluid fractures have been given by Spence and Turcotte [1985]. Mathematical details of the solutions have been given in Spence and Sharp [1985]. Similarity solutions for turbulent flow have been given by Emerman et al., [1986]. Applications of these solutions to the mechanics of dike injection have been given by Turcotte et al. [1988]. Spera [1987] has given an extended discussion of magma fracture.

Buoyancy-driven fractures as a mechanism for magma

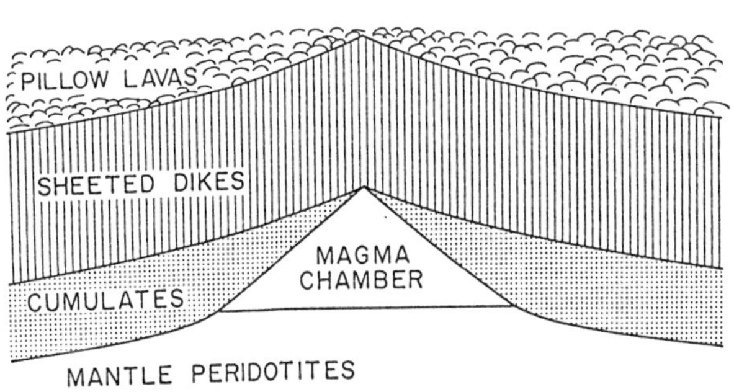

Fig. 9. Illustration (not to scale) of the oceanic crust at a spreading center.

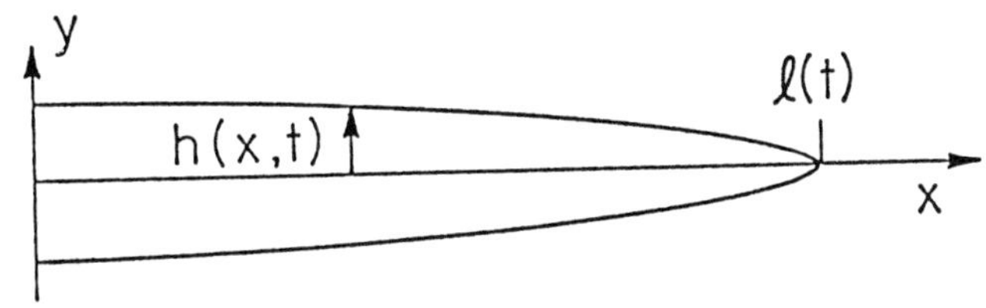

Fig. 10. Idealized geometry for two-dimensional fluid fracture.

migration have been studied by Weertman [1971], Anderson and Grew [1977] and Secor and Pollard [1975]. A solution for the steady upward propagation of a two-dimensional fluid fracture driven by buoyancy has been given by Spence et al., [1987]. They found a solution for a single value of the stress intensity factor. Lister [1990a, b; 1991] and Lister and Kerr [1990, 1991] have extended this to obtain solutions for any value of the stress intensity factor. In these solutions the elastic deformation is important only in the vicinity of the crack tip; below the crack tip the crack width and flow velocity are determined by a simple balance between the buoyancy force and the viscous force, and either laminar or turbulent flow results. Spence and Turcotte [1990] have obtained a solution for a transient buoyancy driven fracture neglecting the elastic tip effects.

It is instructive to solve a simplified set of equations in order to illustrate the interaction between the fluid flow and the crack propagation problems. A number of assumptions are made. The fracture is embedded in a uniform, impermeable medium with a shear modulus μ and Poisson's ratio ν. The magma is assumed to be a Newtonian fluid with a constant viscosity η_l. A solution is obtained for the two-dimensional case illustrated in Figure 10. Fluid is injected at $x = y = 0$ and the two-sided crack propagates in the +x and -x directions. Various conditions on the injection rate or injection pressure can be applied; in this paper we assume that the volumetric rate of injection per unit depth A is constant.

The length of the crack is $2l(t)$ and its width is $2h(x, t)$; $h \ll l$. The rate of supply is related to the width and mean velocity at the origin by

$$A = 2h(0)\bar{u}(0) \tag{39}$$

where $\bar{u}$ is the mean flow velocity. The volume of the crack is related to the volume of fluid injected by

$$At = 2\int_{-l}^{l} h(x,t)\,dx \tag{40}$$

where A is taken to be a constant. In most cases it is appropriate to make the lubrication approximation where the inertia terms are neglected and the applied pressure gradient is balanced by the shear stress. In laminar flow this balance gives [Hirs, 1974]

$$\bar{u} = \frac{1}{3} \frac{h^2}{\eta_l} \frac{\partial p}{\partial x} \tag{41}$$

and for turbulent flow

$$\bar{u} = \left[\frac{2^{5/4} 15 h^{5/4}}{\rho_l^{3/4} \eta_l^{1/4}} \left(-\frac{\partial p}{\partial x} \right) \right]^{4/7} \tag{42}$$

The Reynolds number at the entrance to the crack is given by

$$Re = \frac{\rho_l A}{\eta_l} \tag{43}$$

It is appropriate to assume that the flow is laminar if Re < 10^3 and turbulent if Re > 10^3. This transition occurs within the range of geologically plausible magma flow rates in a dike.

The elastic deformation of the medium is given by [Barenblatt, 1962]

$$p(x, t) = -\frac{\mu}{\pi(1 - \nu)} \int_{-l}^{l} \frac{\partial h(s, t)}{\partial s} \frac{ds}{(s - x)} \tag{44}$$

In the lubrication approximation it is appropriate to neglect the shear stress on the walls of the crack compared with the fluid pressure. In the elastic theory for the propagation of cracks a stress singularity occurs at the crack tip. The strength of this singularity is related to the fracture resistance of the elastic material. Thus it is appropriate to introduce the stress intensity factor K that is defined by [Irwin, 1957]

$$K = \frac{2 l^{1/2}}{\pi} \int_0^l \frac{p(x, t)\, dx}{(l^2 - x^2)^{1/2}} \tag{45}$$

It is necessary to solve (40), (44), (45), and either (41) or (42) for the variables h(x,t), $\bar{u}$(x,t), p(x,t), and l(t).

This is a particularly difficult set of equations to solve because of the singular behavior of the elastic deformation equation (44). However, reasonably accurate approximate solutions can be obtained by approximating some of the basic questions. A simplified elastic strain equation can be written

$$p(0) = \frac{\mu}{1 - \nu} \frac{h(0)}{l} \tag{46}$$

and this replaces (44). The conservation of mass equation (40) is replaced with

$$At = 2 l h(0) \tag{47}$$

The shear flow equations (41) and (42) are approximated by

$$\bar{u}(0) = \frac{1}{3} \frac{h(0)^2 p(0)}{\eta_l l} \tag{48}$$

for laminar flow and for turbulent flow.

$$\bar{u}(0) = \left[\frac{2^{5/4} 15 h(0)^{5/4} p(0)}{\rho_l^{3/4} \eta_l^{1/4} l} \right]^{4/7} \tag{49}$$

We first consider laminar flow in the limit that the fracture resistance is negligible compared with the viscous resistance to flow. Combining (39), (46), (47), and (48) gives

$$l = C_1 A^{1/2} \left[\frac{\mu}{(1 - \nu)\eta_l} \right]^{1/6} t^{2/3} \tag{50}$$

$$2h(0) = C_2 A^{1/2} \left[\frac{\eta_l (1 - \nu)}{\mu} \right]^{1/6} t^{1/3} \tag{51}$$

$$p(0) = C_3 \left(\frac{\mu}{1 - \nu} \right)^{2/3} \left(\frac{\eta_l}{t} \right)^{1/3} \tag{52}$$

with C_1 = 0.589, C_2 = 1.698, and C_3 = 1.44. Solving the complete equations numerically Spence and Sharp (1985) also obtained (50) to (52) with C_1 = 0.4636, C_2 = 1.49, and C_3 = 1.7279. Geertsma and Haafkens (1979) also using other approximations found (50) and (51) with C_1 = 0.68 and C_2 = 1.87.

We next consider turbulent flow in the limit that the fracture resistance is negligible compared with the viscous resistance to flow. Combining (39), (46), (47), and (49) gives

$$l = C_4 \frac{A^{3/8} \mu^{1/6} t^{2/3}}{\rho_l^{1/8} \eta_l^{1/24} (1 - \nu)^{1/6}} \tag{53}$$

$$2h(0) = C_5 \frac{A^{5/8} \rho_l^{1/8} \eta_l^{1/24} (1 - \nu)^{1/6} t^{1/3}}{\mu^{1/6}} \tag{54}$$

$$p(0) = C_6 \frac{A^{1/4} \rho_l^{1/4} \eta_l^{1/24} \mu^{2/3}}{(1 - \nu)^{2/3} t^{1/3}} \tag{55}$$

with C_4 = 1.40, C_5 = 0.714, and C_6 = 0.255. Solving the complete equations Emerman et al. [1986] also obtained (53) to (55) with C_4 = 1.4183, C_5 = 0.5121, and C_6 = 0.2593.

It is of interest to give the conditions under which the stress-intensity factor can be neglected. For laminar flow the condition is

$$\varepsilon = \frac{(1-\nu)^{3/4}K}{A^{1/4}\eta_l^{1/4}\mu^{3/4}} << 1 \qquad (56)$$

and for turbulent flow the condition is

$$\varepsilon = \frac{(1-\nu)^{3/4}K}{A^{7/16}\eta_l^{1/16}\rho_l^{3/16}\mu^{3/4}} << 1 \qquad (57)$$

If these inequalities are satisfied then the results given in (50) to (55) are valid. The shape of the laminar dike is

$$\frac{h}{h(0)} = \left[1 - \left(\frac{x}{l}\right)^2\right]^{2/3} \qquad (58)$$

and the shape of the turbulent dike is

$$\frac{h}{h(0)} = \left[1 - \left(\frac{x}{l}\right)^2\right]^{8/9} \qquad (59)$$

[Emerman et al., 1986].

We now give some examples for basaltic dikes. We first consider the role of the stress-intensity factor. Measured values of the critical stress-intensity factor have been summarized by Atkinson [1984]. As a typical value we take $K = 3$ MN/m$^{3/2}$. As typical parameters for dike injection we take $A = 1$ m^2/s, $\mu = 2 \times 10^{10}$ Pa, $\nu = 0.25$, $\eta_l = 10^2$ Pa s, and $\rho_l = 2700$ kg/m^3. From (56) we find $\varepsilon = 0.0109$ for laminar flow and from (57) we find $\varepsilon = 0.0192$ for turbulent flow. In most cases of dike injection it is appropriate to assume $\varepsilon << 1$. This is equivalent to neglecting the fracture resistance of the elastic medium compared to the flow resistance in the crack.

We first give the half-length l of a propagating dike as a function of time. The assumed properties of the magma and elastic medium are given above; also considered is a magma with a viscosity $\eta_l = 10^2$ Pa s and magma injection rates A =

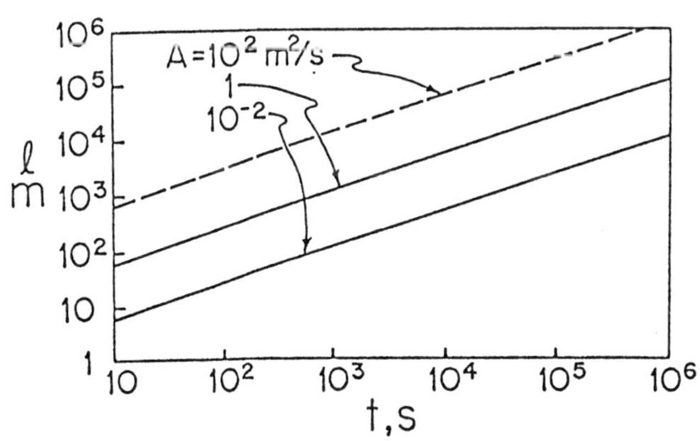

Fig. 11. The half-length l of two-dimensional fluid-driven dikes as a function of injection time t for several values of the injection rate A. The solid lines are laminar flow and the dashed line is turbulent flow, $\eta_l = 10^2$ Pa s.

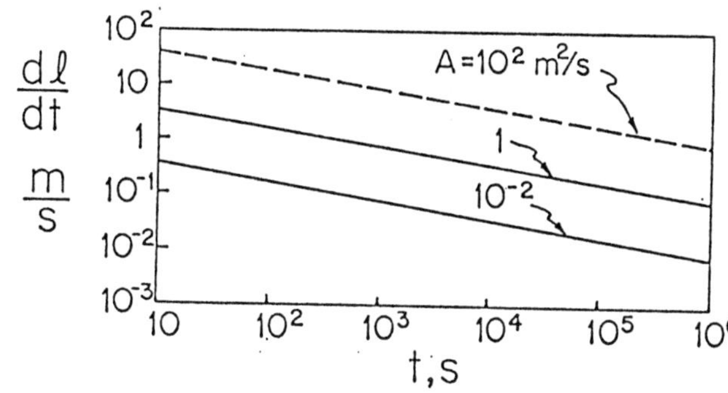

Fig. 12. Propagation velocities dl/dt as a function of the injection time t for several values of the injection rate A. The solid lines are laminar flow and the dashed line is turbulent flow, $\eta_l = 10^2$ Pa s.

10^{-2}, 1, and 10^2 m^2/s. The transition to turbulent flow is taken to be $Re = A\rho_l/\eta_l = 10^3$. The results from either (50) with $C_1 = 0.4636$ or (53) with $C_4 = 1.4183$ are given in Figure 11. The solid lines are laminar flow and the dashed lines are turbulent flow. The velocities of dike propagation are obtained by taking the time derivative of either (50) or (53) and the results are given in Figure 12. The maximum half-widths h(0) are obtained either from (51) with $C_2 = 1.49$ or from (54) with $C_5 = 0.5121$ and the results are given in Figure 13. Due to the increased resistance of turbulent flow, the lengths are shorter and the widths greater for turbulent flow than for laminar flow; also the required driving pressure is greater.

It is important to determine whether significant solidification will occur on the walls of the dike during magma injection. An estimate of the time required to solidify a crack of width 2h(0) has been given by Spence and Turcotte (1985). Taking the latent heat of fusion L = 400 kJ/kg, the specific heat $c_p = 1$ kJ/kg, the temperature difference between the wall rock and the magma $T_m - T_0 = 800$ °K, and the thermal diffusivity $\kappa = 0.5$ mm^2/s, the zone of solidification is shown by the dotted line SL in Figure 13. A dike would not be expected to propagate in conditions below this line.

We next consider the specific example of a dike in a sheeted dike complex at a mid-ocean ridge. Assuming $l = 7.5$ km, $\eta_l = 10^2$ Pa s, $A = 10^2$ m^2/s and other properties as above we find from Figure 11 that the required injection time is 400 s (~7 min.). The velocity of propagation from Figure 12 is 10 m/s and the maximum half-width from Figure 13 is h(0) = 2m. This is in excellent agreement with typical observed widths in Iceland and in ophiolites.

The model given above is appropriate for the dike as it propagates upwards from the magma chamber to the surface. Once it reaches the surface it behaves as an open channel for

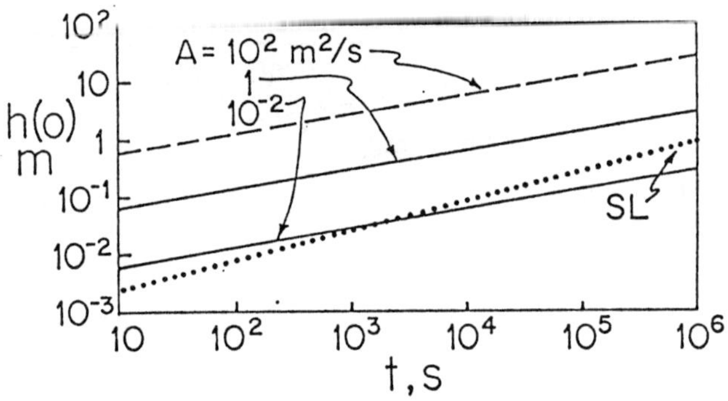

Fig. 13. Maximum half-widths h(0) as a function of the injection time t for several values of the injection rate A. The solid lines are laminar flow and dashed line is turbulent flow, $\eta_l = 10^2$ Pa s. Dikes would be expected to solidify beneath the dotted line denoted by SL.

the ascending magma until the magma chamber is drained. Under these circumstances it is a reasonable approximation to treat the buoyancy driven flow in a dike as a flow in a channel with buoyancy forces balanced by the viscous forces. Extensive studies of buoyancy driven flows of magma in pipes and channels have been given by Wilson and Head [1981]. The conditions under which viscous dissipation coupled with a strongly temperature-dependent viscosity can lead to a thermal instability have been given by Fujii and Uyeda [1974]. Further studies of the role of viscous dissipation have been given by Hardee and Larson [1977] and by Hardee [1986]. As discussed above, one-dimensional channel flow is often a good approximation for the flow in a magma-driven fracture. The model is illustrated in Figure 14. For laminar flow with a buoyancy drive the vertical flow rate Q in a channel of width 2h is [Hirs, 1974]

$$Q = \frac{2(\rho_s - \rho_l) g h^3}{3\eta_l} \tag{60}$$

The corresponding mean velocity, $\bar{u}$, in the channel is

$$\bar{u} = \frac{(\rho_s - \rho_l) g h^2}{3\eta_l} \tag{61}$$

For turbulent flow the vertical flow rate Q is given by [Hirs, 1974]

$$Q = \frac{2(30\,[\rho_s - \rho_l]\,g)^{4/7} h^{12/7}}{r_s^{3/7}} \left(\frac{2}{\eta_l}\right)^{1/7} \tag{62}$$

The corresponding mean velocity,

$$\bar{u} = \frac{(30\,[\rho_s - \rho_l]\,g)^{4/7} h^{5/7}}{\rho_s^{3/7}} \left(\frac{2}{\eta_l}\right)^{1/7} \tag{63}$$

As a specific example we consider the ascent of a basaltic magma through the mantle and take $\eta_l = 100$ Pa s, $\rho_l = 2800$ kg m^{-3}, $\rho_s = 3300$ kg m^{-3}, and $g = 10$ m s^{-2}. From (60) and (61) the transition from laminar to turbulent flow is found to occur for a dike width of 2h = 2 m and $\bar{u} = 16.7$ m s^{-1}. The dependence of the mean magma velocity on channel width is given in Figure 15.

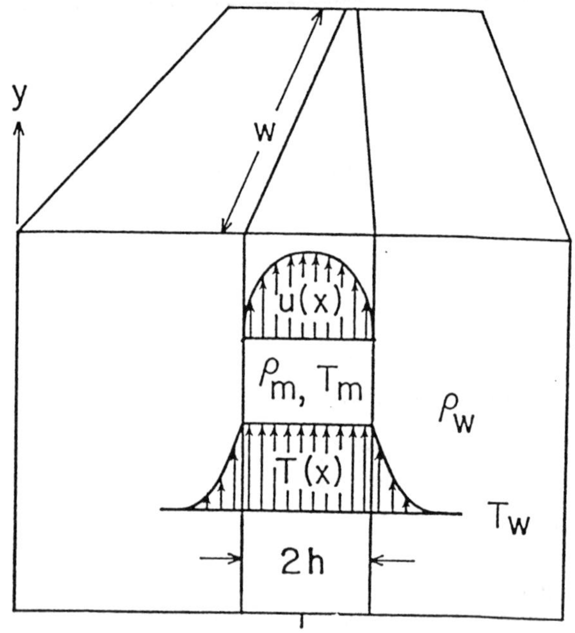

Fig. 14. Illustration of the channel-flow model.

If the crack is approximated by a constant width 2h and a horizontal depth w then the aspect ratio is $\alpha = w/2h$. In terms of the aspect ratio, the total flow volume V_m as a function of time is given by

$$V_m = \frac{4(\rho_s - \rho_l)\, g \alpha h^4 t}{3\eta_l} \tag{64}$$

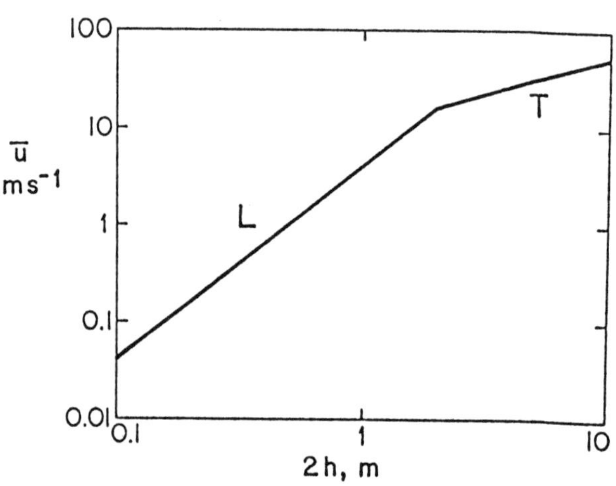

Fig. 15. Dependence of the mean velocity in the dike $\bar{u}$ on the dike width (2h) for laminar (L) and turbulent (T) flow.

for laminar flow and

$$V_m = \frac{4(30\,[\rho_s - \rho_l]\,g)^{4/7}\alpha h^{19/7} t}{\rho_s^{3/7}}\left(\frac{2}{\eta_l}\right)^{1/7} \tag{65}$$

for turbulent flow. The flow volume is given as a function of time in Figure 16 for several channel widths. The aspect ratio is taken to be $\alpha = 10^3$.

We now turn to thermal effects on the flow of magma through a channel. Because the wall rock has a temperature considerably below the solidus of the magma it would be expected that solidification of magma would occur on the boundaries of the flow channel. However, two effects tend to prevent solidification. The first is dissipative (frictional) heating in the magma and the second is the decrease in the melt temperature, T_m, of the magma as it ascends. For laminar flow this balance is given by [Turcotte and Schubert, 1982]

$$\rho_s L \frac{dh'}{dt} = \frac{(\rho_s-\rho_l)^2 g^2 h^3}{3\eta_l} - \frac{\rho_l(\rho_s-\rho_l) g h^3 c_p}{3\eta_l}\frac{dT_l}{dz} - \frac{\lambda(T_l-T_s)}{(\pi\kappa_s t)^{1/2}} \tag{66}$$

where dh'/dt is the rate of melting of the wall rock on one side of the dike (if dh'/dt is negative, it is the rate of solidification of the melt on one side). For simplicity we assume that the magma has a well-defined melt temperature, T_l , and the width of the channel 2h remains approximately constant. The first term on the right-hand side of (66) represents dissipative heating, the second term the decrease in melt temperature with depth, and the third term loss of heat to the wall rock.

Integration of (66) gives

$$h' = \frac{(\rho_s - \rho_l)^2 g^2 h^3 t}{3\eta_l \rho_s L}\left[1 - \frac{\rho_l c_p}{(\rho_s - \rho_l)\,g}\frac{dT_m}{dt}\right] - \frac{2k_s(T_l - T_s)}{\rho_s L}\left(\frac{t}{\pi\kappa_s}\right)^{1/2} \tag{67}$$

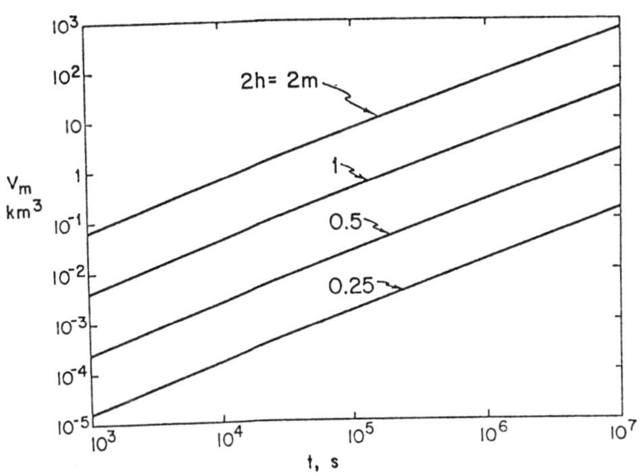

Fig. 16. Dependence of the flow volume V_m on time (t) for several dike widths (2h).

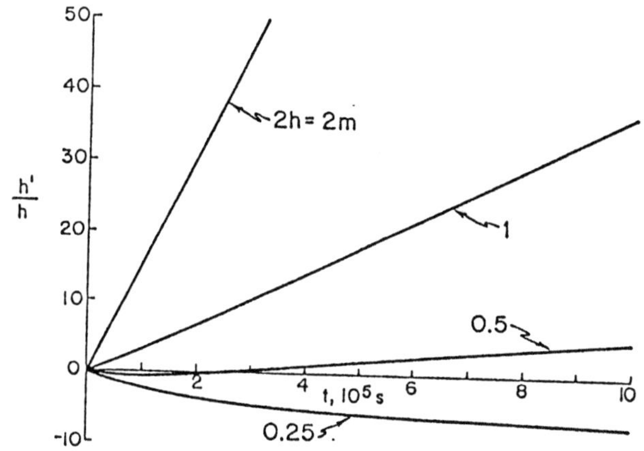

Fig. 17. Dependence of the ratio of the distance of solidification h' on the channel wall to the channel half-width (h) as a function of time t for several channel widths 2h.

The thermal balance for turbulent flow is

$$\rho_s L \frac{dh'}{dt} = \frac{(30)^{4/7}[(\rho_s - \rho_l)g]^{11/7} h^{12/7}}{\rho_l^{3/7}}\left(\frac{2}{\eta_l}\right)^{1/7} \times \left[1 - \frac{\rho_l c_p}{(\rho_s - \rho_l)\,g}\frac{dT_m}{dz}\right] - \frac{\lambda_s(T_l - T_s)}{(\pi\kappa_s t)^{1/2}} \tag{68a}$$

and integration gives

$$h' = \frac{(30)^{4/7}[(\rho_s - \rho_l)g]^{11/7} h^{12/7}}{\rho_l^{3/7}\rho_s}\left(\frac{2}{\eta_l}\right)^{1/7} \times t\left[1 - \frac{\rho_l c}{(\rho_s - \rho_l)\,g}\frac{dT_l}{dz}\right] - \frac{2\lambda_s(T_l - T_s)}{\rho_s L}\left(\frac{t}{\pi\kappa_s}\right)^{1/2}$$

We can now determine the width h' of the solidified or melted zone on the margins of a dike.

In addition to the parameter values given above, we take $\rho L = 1.12 \times 10^9$ J m^{-3}, $\lambda_s = 2$ W m^{-1} °K^{-1}, $T_l - T_s = 500$ °K, $\kappa_s = 0.7$ mm^{2}s^{-1}, $c_p = 10^3$ J kg^{-1} °K^{-1}, $dT_l/dz = -2$ °K km^{-1}. The dependence of h' on time is given in Figure 17 for several channel widths. It is seen that substantial melting of the wall rock can occur. This implies significant contamination of the primary magma.

MANTLE UPWELLING

The previous discussion of upwelling, melting, and melt migration treated mantle upwelling beneath a ridge with a simple 1-D approximation -- an ascending column of material. While this is reasonable for gaining insight into the physics of melting and melt migration, we need to consider both horizontal as well as vertical motions to gain insight into processes which enhance upwelling or lead to segmented upwelling along a spreading center. The basic reason for this is that the way material leaves a spreading center can play an important role in shaping ridge upwelling. It is likely that we now have a good physical description of the mechanisms which govern upwelling beneath a mid-ocean spreading center. At the temperatures and pressures beneath a spreading center we expect the mantle to deform by viscous creep. The mantle is known to deform by viscous flow in response to glacial loading and large-scale buoyancy forces indirectly sensed by seismic tomography.

Current difficulties in modelling ridge upwelling stem not from our inability to physically characterize viscous creep, but rather from our still incomplete knowledge of the dominant buoyancy forces shaping upwelling and a good description of the mantle viscosity structure at a ridge. (The modes in which melt migrates from the melting region to the surface may also play a fundamental role in shaping the flow field near a ridge.) Several recent studies of two dimensional mantle flow beneath spreading centers [Rabinowicz et al., 1987; Buck and Su, 1989; Scott and Stevenson, 1989; Sotin and Parmentier, 1989; Cordery and Phipps Morgan, 1992; Scott, this volume] have explored the possible influences on ridge axis upwelling structures from local ridge buoyancy forces associated with melt extraction (which lowers the density of the residual, more depleted mantle), with the presence of interstitial basaltic melt (which is roughly 15% less dense than neighboring mantle at 5-60 km depth), and with thermal buoyancy forces.

In order to model this complex, non-linear, system each model makes simplifying assumptions. For example, Scott and Stevenson [1989] include melt migration but introduce a thermal/melting model which neglects lateral variations in the sub-ridge thermal structure while Sotin and Parmentier [1989] ignore melt migration but include a more complete treatment of heat transport and melting. These studies also make different assumptions about the amount of melt that is present within and above the melting region which result in different predictions about the structure of mantle upwelling and melting beneath a spreading center. Rabinowicz et al. [1987] and Buck and Su [1989] assume that large melt fractions build up in the melting region and are effectively drained only directly below the ridge axis (at 6-10 km depth). The non-linear interactions between melting, melt extraction, melt migration, mantle viscosity, and mantle upwelling at a MOR are best studied through numerical experiments. With this approach we can explore ridge processes in a manner akin to an experimental petrologist who attempts to explain simple analogues to the polybaric melting of the mantle with a suite of laboratory experiments. In the following summary we will sketch the underlying physics involved in this modelling effort. Much of the numerical effort has relied on the rapid progress of the engineering community in developing numerical tools and algorithms that solve for fluid and energy transport as well as similar efforts to model mantle convection. See Zienkiewicz and Taylor [1991] and Fletcher [1988] for an introduction to current finite element and finite difference modelling techniques that form the backbone of most viscous flow and heat transport codes.

Until recently, numerical experiments like those discussed above were restricted to idealized 2-D ridge models with no flow in the along-axis direction. While this is a reasonable parameterization to explore the effects of plate spreading it does not allow the numerical experiment to capture an essential aspect of buoyant upwelling - that buoyancy forces alone can lead to plume-like upwelling instead of the sheet-like upwelling driven by plate separation. This problem is particularly critical when considering the effects of thermal buoyancy on near-ridge upwelling. First, lateral temperature gradients are likely to be small within the melting region because the temperature will be buffered to a partial melt solidus -- thus the upwelling itself has a relatively minor control on thermal buoyancy forces which will instead be more strongly shaped by the aging, cooling lithosphere. Second, shear flow away from the ridge axis will try to shape thermal downwelling to align in ridge perpendicular rolls in addition to a possible ridge parallel downwelling structure. Since a 2-D code cannot include possible ridge-perpendicular downwelling structures it will 'approximate' the heat transport of even steady-state ridge perpendicular rolls by the periodic or aperiodic formation of ridge parallel downwelling structures. Third, numerical experiments are usually constructed within a finite computational region, usually localized near the spreading axis. The length of this computational region often sets the length scale for thermal convection within the code - a length scale which may have little to do with the intrinsic length scale of the physical system.

For these reasons some numerical experimenters have explored models which do not include thermal buoyancy forces, an approximation based on the melting temperature buffer within the melting region which leads to small lateral temperature gradients within this region. In this review we will initially explore mantle flow physics through simple 2-D experiments which do not include thermal buoyancy. Later we will summarize the results of recent 3-D numerical experiments which properly treat the combined effects of thermal and compositional buoyancy.

Flow and energy transport within the mantle can be described by statements of mass conservation, momentum conservation, energy conservation, and a description of the melting process (phase change). This leads to the following set of non-linear equations which are the multidimensional time-dependent generalization of (8) and (20-23):

Mass Conservation

$$\nabla \cdot \mathbf{u} = -\dot{\psi} \quad (69)$$

$$\text{or } \nabla \cdot \mathbf{u} = 0 \quad (69a)$$

Mechanical Equilibrium

$$\left.\begin{aligned} 0 &= \tau_{ij,j} - \partial p/\partial x_i + g_i\rho\left(T,\xi,\phi\right) \\ \tau_{ij} &= \eta\left(\partial u_i/\partial x_j + \partial u_j/\partial x_i\right) + \chi\delta_{ij}\left(\nabla \cdot \mathbf{u}\right) \end{aligned}\right. \quad (70)$$

Energy Conservation

$$\partial T/\partial t + \nabla \cdot \left(\mathbf{u}T\right) = \kappa\nabla^2 T - \left(L/c_p\right)\dot{\psi} \quad (71)$$

Evolution of mantle depletion

$$\partial \xi/\partial t + \mathbf{u} \cdot \nabla \xi = \dot{\psi} \quad (72)$$

Description of mantle melting

$$\partial \psi/\partial t = F\left(T,\mathbf{u},\xi\right) \quad (73)$$

Description of melt migration

$$\phi = f\left(\text{melt migration processes, } \dot{\psi}\right) \quad (74)$$

Density equation of state

$$\rho = \rho\left(1 - \alpha T - \beta\xi - \gamma\phi\right) \quad (75)$$

with variables described in Table 1.

While the above description of energy conservation (71) is fairly straightforward, several buoyancy sources have been recently proposed as the 'dominant' forces shaping ridge upwelling. Initially, to partially isolate the effects of each term affecting viscous upwelling, we will introduce sequentially the terms in this equation.

The component of flow responding to plate separation is commonly called 'passive' mantle upwelling. In a local region about a ridge, surface plate motions can be treated as applied kinematic constraints on upwelling. A two-dimensional analytical solution exists for incompressible upwelling within a uniform mantle halfspace (a common idealization where the physical bottom is thought to lie well beyond the region of interest) due to two surface plates separating with a half-spreading rate U:

$$\left.\begin{aligned} u &= \frac{2U}{\pi}\left\{\tan^{-1}\left(\frac{x}{z}\right) - \frac{xz}{\left(x^2+z^2\right)}\right\} \\ v &= -\frac{2Uz^2}{\pi\left(x^2+z^2\right)} \\ p &= -\frac{4U\eta z}{\pi\left(x^2+z^2\right)} \end{aligned}\right\} \quad (76)$$

where u is the velocity component in the horizontal direction x and v is the velocity component in the vertical direction z. Directly beneath the ridge (x=0) the upwelling rate is $2U/\pi$, i.e. mantle ascending due to plate separation will rise at roughly the half-spreading rate. The magnitude of the mantle viscosity does not enter into this expression -- the magnitude of the sub-axial mantle viscosity *will* play an important role in determining whether local buoyancy forces significantly influence the fundamental passive upwelling pattern (or whether viscous pressure gradients significantly perturb vertical melt ascent due to buoyancy, see prior section on magma focusing).

Even for this simple flow the depth dependence of mantle viscosity can play a role in shaping how much lateral vs. vertical transport replenishes mantle that is incorporated into plate spreading. A more viscous mantle at depth will lead to a broader, slower deep upwelling region with shallower lateral flow towards the ridge. The influence of a low viscosity asthenosphere channel on ridge upwelling has not yet been investigated very much. Figure 18 shows the contrast between upwelling in a uniform viscosity halfspace vs. a system with a 100 km thick asthenosphere channel which is two orders of magnitude less viscous than the underlying mantle. This weak, thin low viscosity channel leads to significantly less total upwelling. This effect is intuitively understood because the asthenosphere acts as a decoupling zone between the surface plates and the slower lateral flow of the high viscosity halfspace beneath the asthenosphere. Since horizontal velocities are essentially zero at the base of the asthenosphere the net horizontal transport away from the spreading center is reduced, which in turn leads to less upwelling to supply material for plate spreading. The second effect is similarly characteristic but less intuitive -- upwelling is also more focussed when an asthenosphere channel is present as seen in Figure 18.

Thus a low viscosity channel leads to a broader, reduced upwelling, a region of lateral flow at the base of the channel towards the ridge axis, and a more focussed region of upwelling than passive plate spreading which incorporates a uniform mantle viscosity with depth. Note that boundary conditions on the edge of a computational region can also lead to 'uniform viscosity-like' upwelling even in the

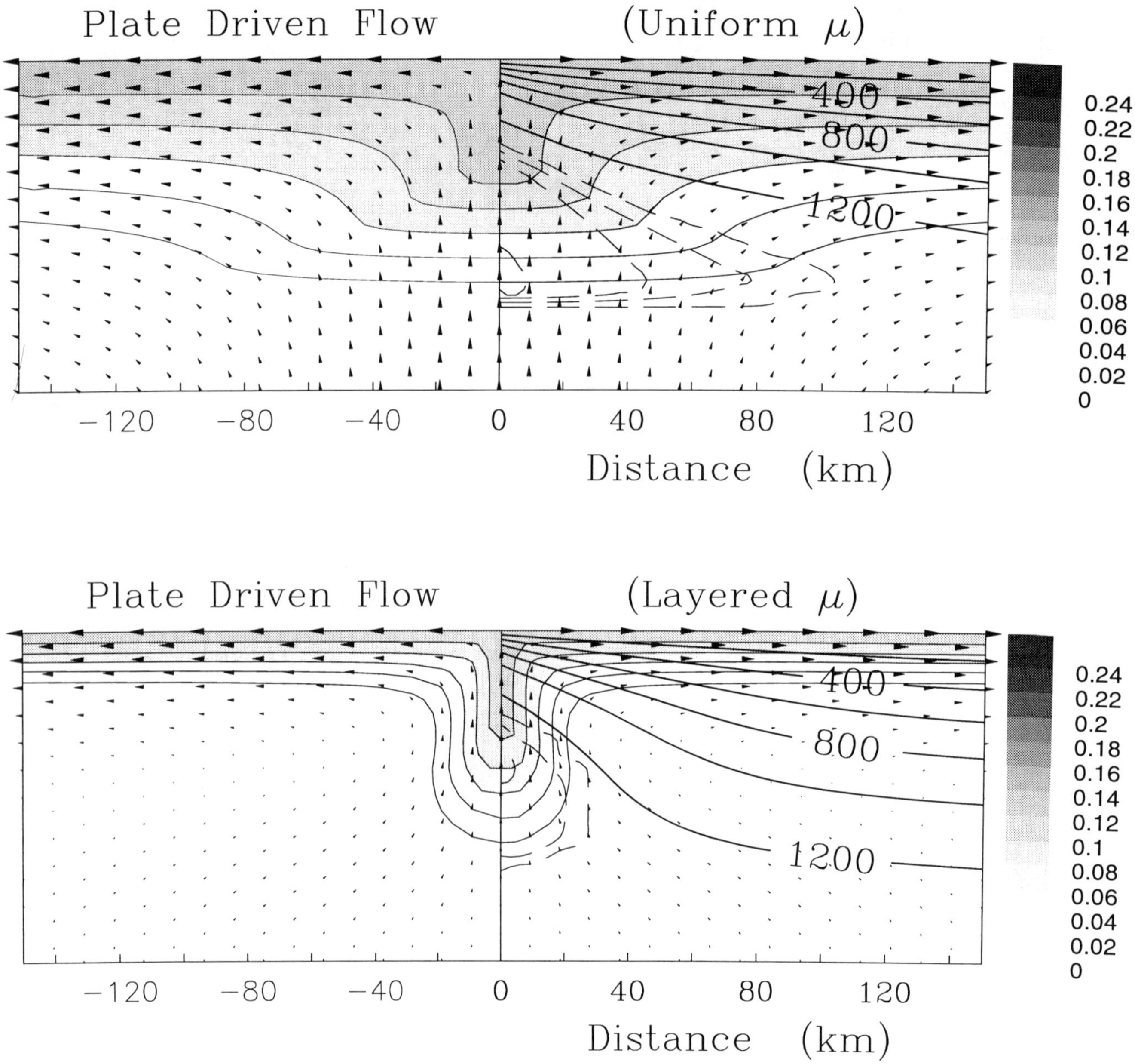

Fig. 18. Ridge upwelling due to plate spreading. (top) Plate-separation induced upwelling of a uniform viscosity halfspace. (bottom) Plate-separation induced upwelling for a simple layered viscosity structure with a 100 km thick asthsenosphere which lies above a mantle halfspace (100x more viscous than overlying asthenosphere). Temperature is shown by heavy 200° isotherms on the right panel of each figure and melt production rate (x 1000) is shown at 0.2 increment, dashed-line contours. (Melt production rate in km^2/Myr is normalized by the half-spreading rate). Mantle depletion due to melting and melt extraction is shown in the shaded field as the fraction of total melting which a given parcel of mantle has experienced. This also shows the effective buoyancy force due to compositional buoyancy associated with mantle depletion (cf. Table 2). The total crustal production is 2.6 km and 0.6 km, and the maximum upwelling rate 7 mm/yr and 6 mm/yr, for the top and bottom panels, respectively. (For scale, the top arrows show the imposed plate velocity of 10 mm/yr)

presence of a vertically stratified mantle rheology if a numerical experiment has either a passive halfspace solution as the outgoing velocity b.c. (e.g. Cordery and Phipps Morgan [1992]) or shear stress-free boundary conditions on the base and edge of the computational region. The fact that the boundary conditions can substantially alter the upwelling flow has just begun to be investigated and a good recipe for addressing this problem is still not known.

The reason why the effects of a low viscosity asthenosphere have yet to be studied in detail is both practical and pragmatic. First, codes which can effectively treat viscosity variations are just becoming available for 3-D numerical experiments. Second, the effects of local buoyancy are just beginning to be understood. Buoyancy effects which have been proposed to shape local ridge upwelling are: lateral changes in temperature (thermal buoyancy); the density change due to the presence of interstitial melt; and the effect of the extraction of a basaltic melt component from a mantle peridotite.

Let's first consider a system where melt, once generated, is extracted extremely efficiently and rapidly rises to form oceanic basaltic crust. Besides being easiest to study, this is not a bad first approximation since current geochemical measurements of abyssal peridotites [Johnson et al., 1990; Langmuir et al., this volume], laboratory studies of melt-rock systems [Kohlstedt, this volume] and geophysical measurements at ridges [Forsyth, this volume] are compatible with this scenario of a low residual melt fraction within upwelling, melting mantle. In this case we can neglect the buoyancy forces due to the presence of melt (i.e. γ=0 and α=0 in (75)). Melt depletion buoyancy forces tend to drive flow to a vertically stratified depletion structure with the lightest, most melted mantle overlying progressively less melted, denser mantle. Thus this force will enhance upwelling within the melting region but at the same time will inhibit a convective overturn. This flow enhancement can be seen by comparing Figures 18 and 19 (note that this effect is present, but more muted, for models where a low viscosity channel rheology already focusses upwelling as seen in Figures 18b and 19b). Sotin and Parmentier [1989] suggest that it is this feature of melt-extraction, enhanced upwelling, that is responsible for nearly uniform oceanic crust created at ridges varying in spreading rate from 10-180km/My (Figure 20).

While passive plate-driven upwelling alone predicts a decrease in crustal thickness at slower upwelling rates where the mantle is cooled to 15-25 km depths below the axis by heat conduction [Reid and Jackson, 1980], melt-extraction buoyancy forces will lead to melting which is limited not by the temperature structure within the melting range but instead by the rate at which depleted mantle can be carried away from the ridge axis by plate spreading -- thus predicting a uniform crustal thickness which depends only on the mean mantle temperature and composition beneath a spreading center. This result agrees with compilations of the spreading rate dependence of crustal thickness determined by seismic refraction and reflection experiments (Figure 20) which show essentially no dependence of mean crustal thickness on spreading rate [Chen, 1992].

While the density changes associated with 200-500°K lateral variations in temperature are similar to the density changes associated with 20% variations in the degree of melting and melt extraction (see Table 2) the influence of these effects are very different because melt extraction, unlike heat extraction, is an essentially irreversible process. Since thermal density changes are 'reversible' they will tend to drive a upwelling and downwelling cycle with cold material sinking, heating up, and rising at a later time. In contrast, melt extraction leads to a compositionally stratified flow, with the lightest (most melted) material nearer the surface. Sotin and Parmentier [1989] examined the relative importance of these two effects and found that the melt extraction effect tended to dominate near a spreading center, resulting in a mean crustal thickness which is insensistive to spreading rate. Thermal buoyancy contributed a time dependent upwelling oscillation which was superimposed on the compositional flow. Since thermal buoyancy often prefers the development of ridge-perpendicular (i.e. 3-D) rolls we will defer further discussion of the effects of thermal buoyancy until after an initial presentation of the characteristics of along-axis variations in upwelling.

The same basaltic melt extraction which leads to a more depleted, lighter, residual mantle also leads to a volume loss or sink (the departed melt) which must be replenished by additional inflowing mantle. This effect is usually ignored in studies of ridge upwelling which treat the mantle as incompressible but is easy to incorporate into any computational code [e.g. Phipps Morgan et al., 1987; Parmentier and Phipps Morgan, 1990; Phipps Morgan and Chen, 1992] if we replace a statement of incompressibility $\nabla \cdot \mathbf{u} = 0$ (69a) with a statement of mass conservation within the melting region $\nabla \cdot \mathbf{u} = \dot{\psi}$ (69), where $\dot{\psi}$ is the melt extraction rate. This effect is included in the experiments shown in Figure 21. The effects of the dilatational flow due to melt extraction are relatively minor and do not produce a qualitative change in the character of the flow as long as the melting region is relatively broad. The maximum upwelling rate and total crustal production are increased by a few percent and the melting region is slightly reduced in its across-axis extent.

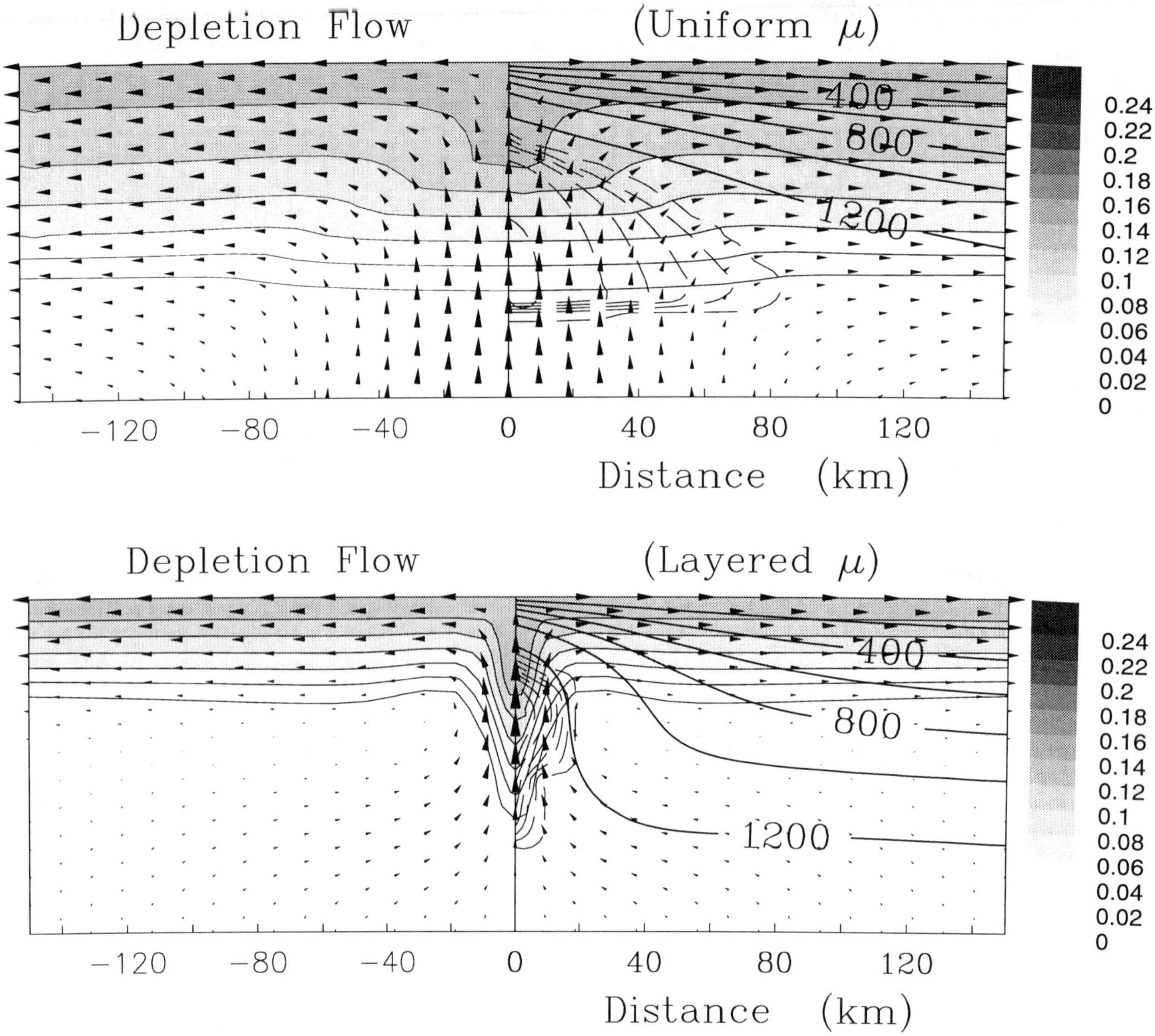

Fig. 19. Mantle upwelling due to plate spreading and compositional buoyancy forces due to melt extraction for a sub-ridge viscosity of 10^{19} Pa-s and a half-spreading rate of 10 mm/yr. Total density is shown by shaded fields. Annotations as in Figure 18. Top and bottom panels are for the same halfspace and layered viscosity structures as Figure 18. Note the significant enhancement of dynamic upwelling over passive flow alone. Annotations as in Figure 18. The total crustal production is 4.6 km and 1.3 km, and the maximum upwelling rate 11 mm/yr and 16 mm/yr, for the top and bottom panels, respectively.

The presence of a significant melt buoyancy force has the potential to qualitatively change the pattern of mantle upwelling at a MOR. Melt buoyancy, like thermal buoyancy, is a 'reversible' effect on mantle buoyancy. Once the melt is removed the overall mantle density will increase and this parcel of mantle has the potential to sink and be reincorporated into a recirculating cell beneath the ridge axis [Rabinowicz et al., 1987]. The melts produced by ridge upwelling would be considerably affected if mantle material passes multiple times through the melting and melt extraction zone. Because melt-retention buoyancy forces are most likely to be confined near the ridge axis, this

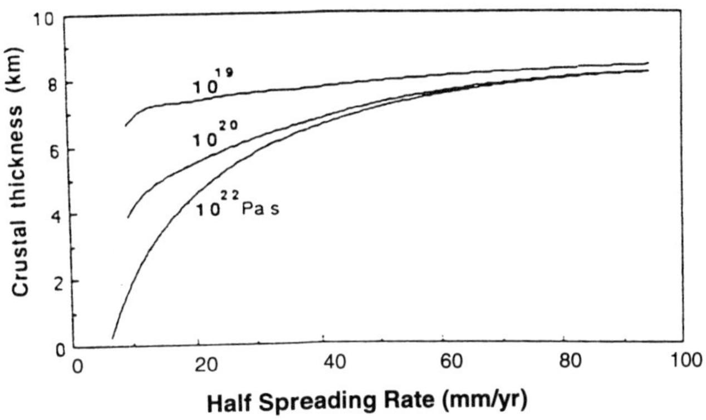

Fig. 20. Crustal thickness as a function of spreading rate for a range of mantle viscosities [from Sotin and Parmentier, 1989]. For a large enough spreading rate at a given viscosity, upwelling is dominated by plate spreading so that crustal thickness is independent of both viscosity and spreading rate. At slow spreading rates upwelling is increasingly dominated by compositional depletion buoyancy as viscosity is reduced. As a consequence, crustal thickness depends less on spreading rate but depends strongly on mantle viscosity. The observed weak dependence of crustal thickness on spreading rate implies that the sub-ridge asthenosphere viscosity is less than ~$5x10^{19}$ Pa-s -- i.e. in a regime where buoyancy effects can play a significant role.

source of buoyancy (unlike distributed thermal buoyancy) is relatively easy to include in a numerical experiment. The difficulties are not in treating this source of buoyancy, but rather in properly modelling the mechanisms of melting and melt migration.

Developing a proper quantification of chemical evolution during MORB petrogenesis is an active field of study [cf. Grove et al. and Langmuir et al. in this volume]. For petrological modelling purposes this level of description is

TABLE 2: Sources of Buoyancy Beneath a Mid-Ocean Ridge

Buoyancy Factor	Magnitude in (75)
thermal expansion (non-lithosphere part of thermal buoyancy)	α x ΔT (~200°C) = 0.6%
mantle depletion (compositional buoyancy due to melt extraction)	β x $\Delta\xi$ (~.2) = 0.5%
melt retention buoyancy (presence of melt in partially molten rock)	γ x $\Delta\phi$ (~.05) = 0.4%

essential. However, to explore the basic physical and thermal effects of melting it is often fruitful to adopt a much simpler picture of melting. Here we adopt the assumption that melting occurs with a depth-independent latent heat of fusion of 720 J/kg. This number is chosen in concert with a pressure-dependent solidus temperature of 1100°C + 3.25°C/km and inflowing 'asthenosphere' temperature of 1375°C to produce amounts of melting (~25%) which lead to typical ~6 km crustal thicknesses after assuming that most melt is extracted to form the crust. This captures much of the basic thermal and density evolution of the mantle as melts are generated which form oceanic crust.

Probably the next most important melting physics to include in numerical experiments to clarify the processes shaping mantle upwelling and melting is the physics and thermodynamics associated with the pressure-dependent spinel-garnet peridotite (at roughly 75 km depth) and plagioclase-spinel periditite (at roughly 25 km depth) phase transitions, both of which are likely to lie within the melting region. These phase transitions will have an impact on the density evolution due to melting and melt extraction. For example a plagioclase peridodite is thought to become denser instead of less dense with increasing melting and melt extraction. Density reductions due to melting are larger in the garnet than in the spinel stability field. We have explored possible mechanisms of melt migration earlier in this review. Here, we will present results which assume a porous flow treatment of vertical melt ascent due to melt buoyancy as in (4).

The process which we explore in the next experiments is the effect of melt retention in concert with the effects of buoyancy due to melting and melt extraction from the mantle matrix (i.e. α=0 in (75)). The principal result shown in Figure 22a is that melt retention above the melting region both localizes and strongly increases upwelling but does not lead to closed recirculating cells. Instead mantle rises greatly above its 'mantle depletion' level in the region where melt buoyancy is present and then sinks to a stable stratified level a relatively short distance off-axis before spreading laterally away from the ridge. When no viscosity stratification is present the off-axis thickness of the depleted layer is roughly equal to the depth of melting. (In general, the base of the on-axis density gradient which includes melt retention lies at the same depth as the off-axis depletion-density profile as seen in Figure 22b.) For models which incorporate a weak low-viscosity channel, upwelling is similarly enhanced (though total melting is still a factor of two less), but a low density root may still extend beneath the ridge axis (Figure 22b). Melt-retention-enhanced upwelling also leads to a greater preference for three-dimensional instead of two-dimensional upwelling for a given mantle viscosity and spreading rate.

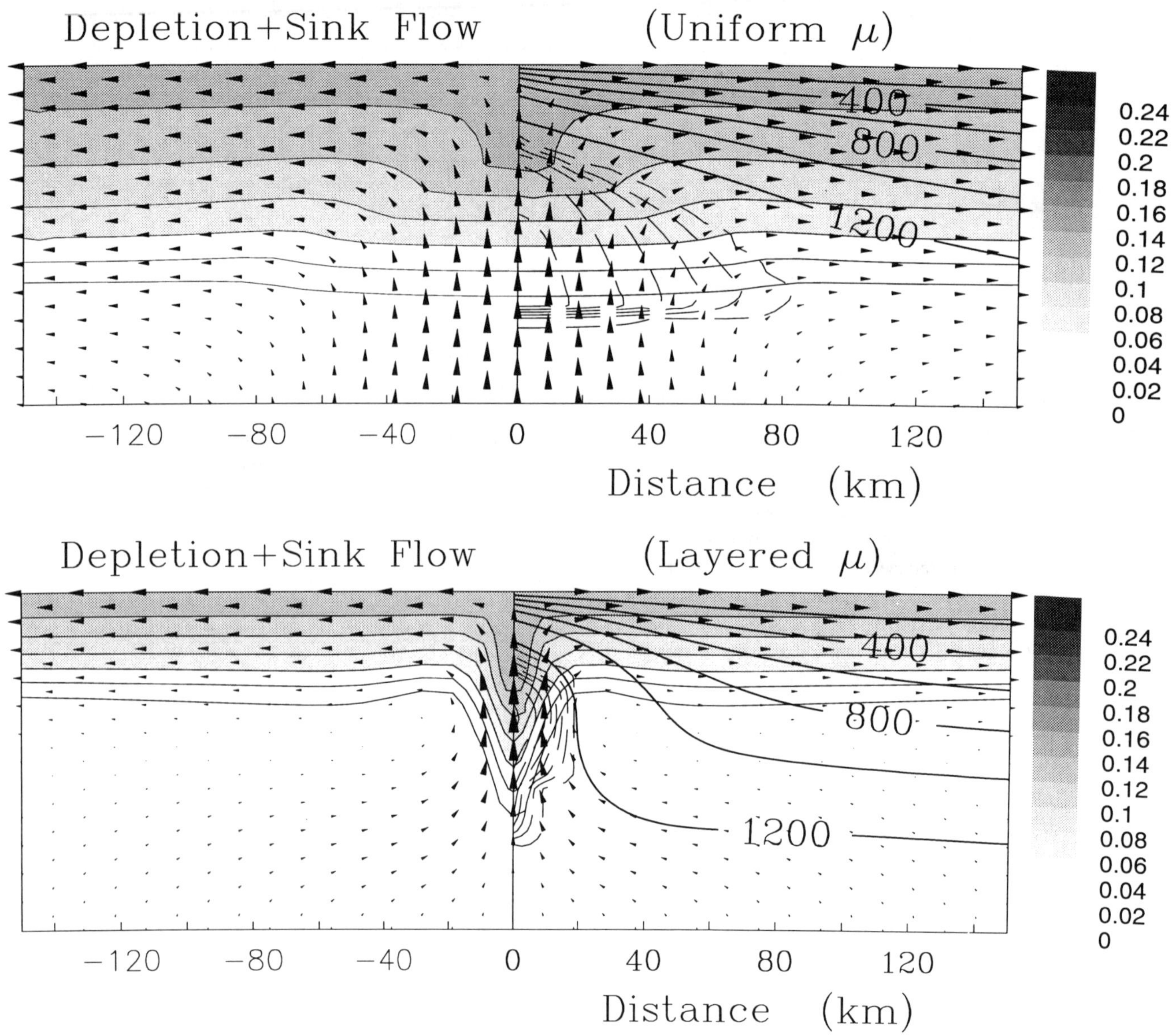

Fig. 21. Mantle upwelling due to plate spreading, sink-flow into the compacting region of melt production, and compositional buoyancy forces due to melt extraction for a sub-ridge viscosity of 10^{19} Pa-s and a half-spreading rate of 10 mm/yr. Top and bottom panels are for the same halfspace and layered viscosity structures as Figure 18. Annotations as in Figure 18. Mantle flow due to compaction in the melt production region somewhat enhances upwelling and melting over the incompressible treatment but does not qualitatively change the flow or upwelling regime (compare with Figure 19). The total crustal production is 4.6 km and 1.5 km, and the maximum upwelling rate 12 mm/yr and 16 mm/yr, for the top and bottom panels, respectively

3-D UPWELLING AND RIDGE SEGMENTATION

Along-axis variations in mantle upwelling have been proposed to explain along-axis variations in axial topography and gravity (see Forsyth, this volume, Phipps Morgan [1991], or Lin and Phipps Morgan [1992]). Clearly, the along axis variations in mantle temperatures and degree of partial melting associated with the finite length of spreading segments [Phipps Morgan and Forsyth, 1988] will promote local buoyant upwelling beneath the center of a segment. Whether this form of buoyant upwelling is a cause or merely a consequence of the

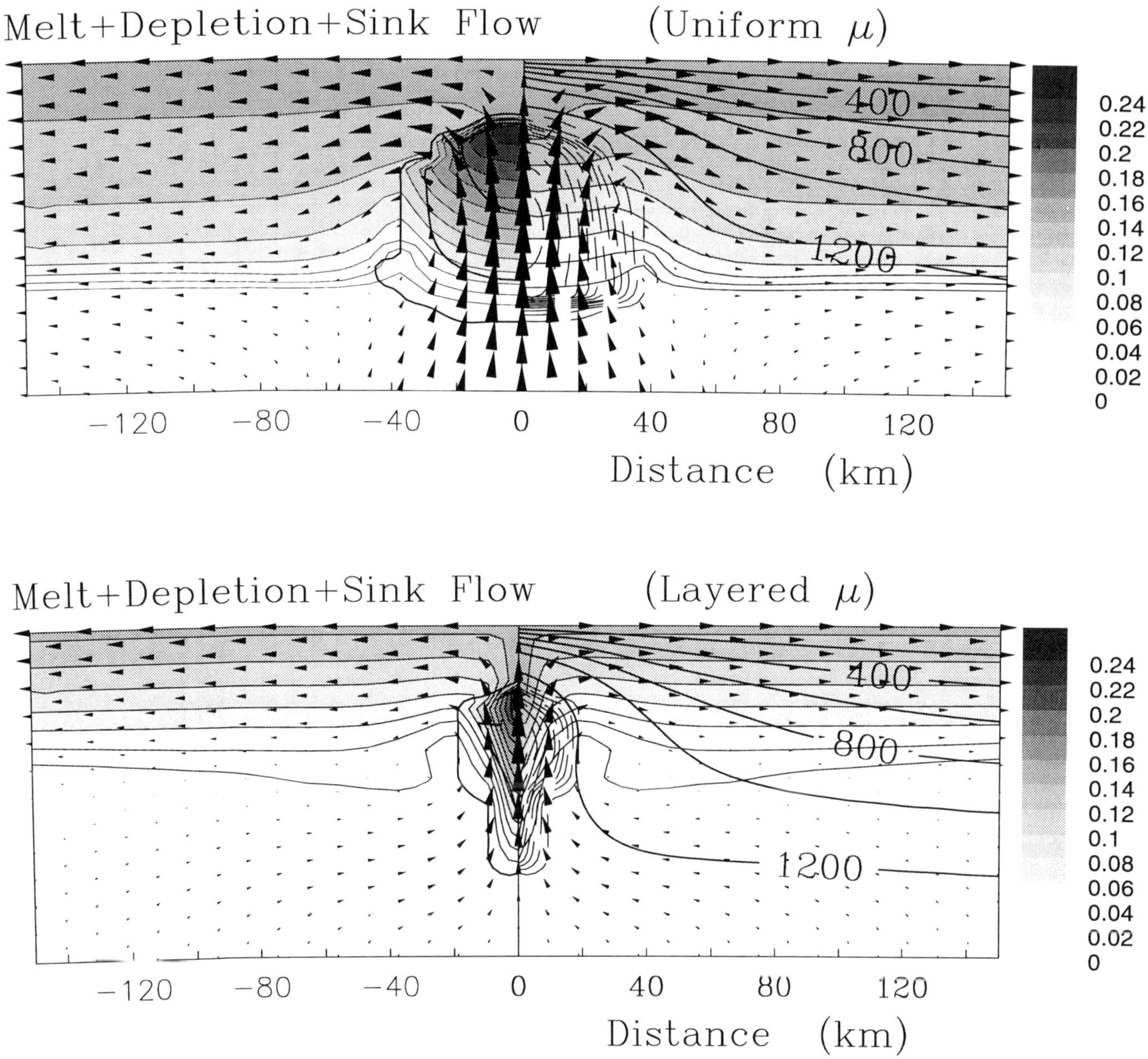

Fig. 22. Mantle upwelling due to plate spreading, the buoyancy of a retained melt phase (maximum melt fraction of 3%), sink-flow into the compacting region of melt production, and compositional buoyancy forces due to melt extraction for a sub-ridge viscosity of 10^{19} Pa-s and a half-spreading rate of 10 mm/yr. Top and bottom panels are for the same halfspace and layered viscosity structures as Figure 18. Annotations for the right side of each panel as in Figure 18. The left side of each panel shows the retained melt fraction in contours of 0.5% and the total (retained melt + compositional) buoyancy shaded in the same units as the depletion-buoyancy shown on the right side of each panel. The effects of melt buoyancy significantly enhance upwelling and melting over Figure 21. Compositional buoyancy still produces a stratified mantle composition (density) leaving the spreading center and restricts the formation of a recirculating cell directly around the melt migration The total crustal production is 6.5 km and 1.8 km, and the maximum upwelling rate 26 mm/yr and 22 mm/yr, for the top and bottom panels, respectively.

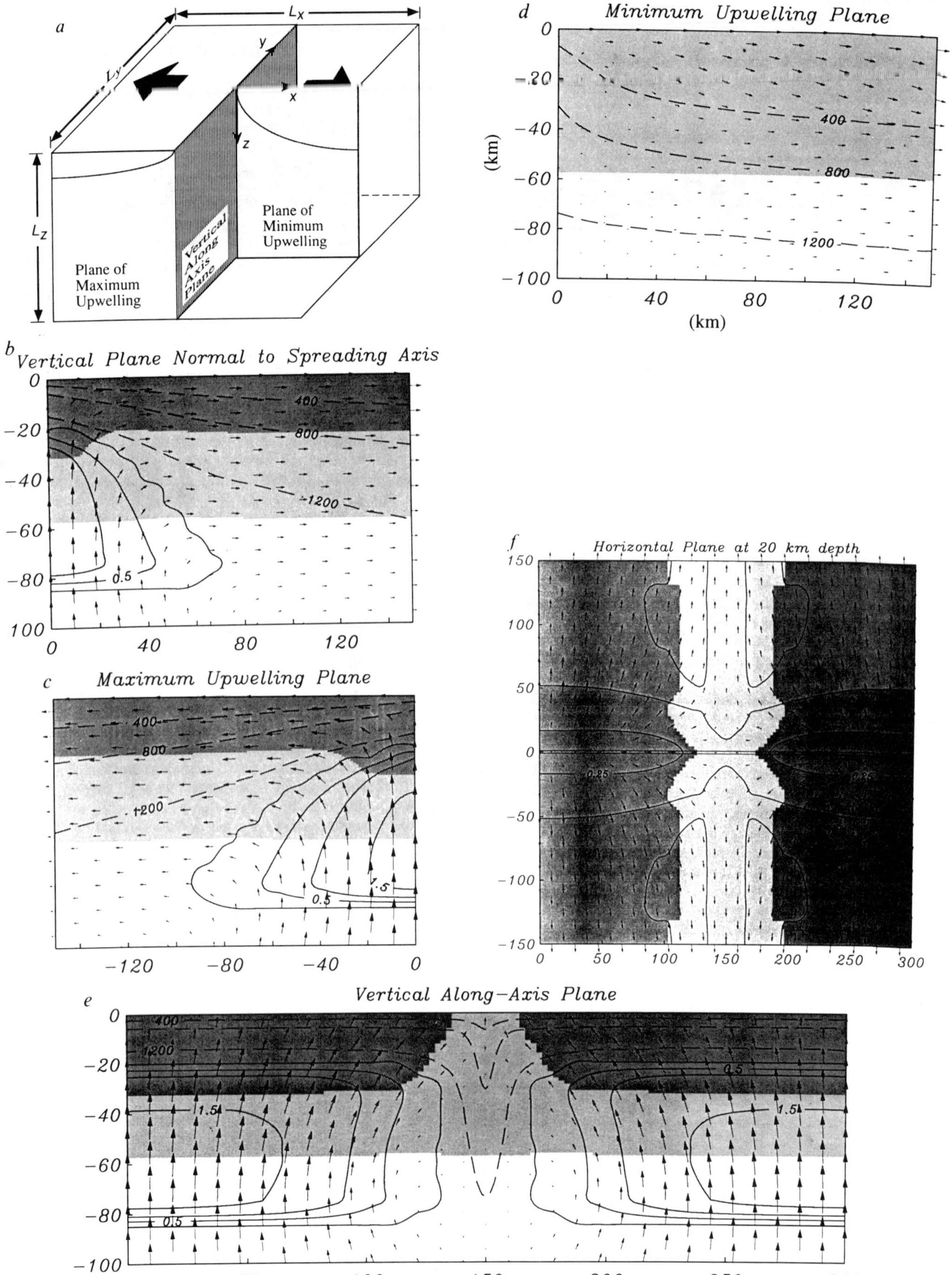
a
L_X
L_y
y
x
z
L_z
Plane of Maximum Upwelling
Vertical Along Axis Plane
Plane of Minimum Upwelling
b
Vertical Plane Normal to Spreading Axis
400
800
1200
0.5
c
Maximum Upwelling Plane
400
800
1200
0.5
1.5
d
Minimum Upwelling Plane
400
800
1200
(km)
(km)
e
Vertical Along-Axis Plane
400
1200
0.5
1.5
f
Horizontal Plane at 20 km depth
0.25

segmentation still remains to be determined. While the two-dimensional studies discussed above address the buoyancy forces which may influence the development of segmentation along a spreading center, they cannot directly explore segmentation since they are not three-dimensional numerical experiments. (A semi-numerical perturbation problem for the initiation of 3-D flow based on a 2-D numerical 'base state' has not yet been explored. This might be a fruitful way to reduce the computer resources needed to tackle the problem.)

So far there have been a few 3-D studies, all based on the 2-D Sotin and Parmentier [1989] scenario. Parmentier and Phipps Morgan [1990] explored the influence of buoyancy forces associated with mantle depletion on the formation of upwelling segmentation within a straight ridge segment. This approach involves creating a 3-D analogue to the experiment shown in Figure 19. A small (~1%) along-axis perturbation is added to an initial 2-D upwelling pattern. This initial along-axis perturbation is added to an initial 2-D upwelling pattern. This perturbation will either grow, if an intrinsically 3-D upwelling pattern is the preferred state, or will decay, if a 2-D pattern with no along-axis flow is the preferred state. A typical 3-D upwelling pattern is shown in Figure 23. Spreading rate and mantle viscosity are the two factors which can shape the intrinsic upwelling pattern for isoviscous flow at a particular value of β in (75). Figure 24 shows a 'phase diagram' for the preferred upwelling state as a function of these two variables. Faster plate spreading rates and higher sub-ridge mantle viscosities (both of which reduce the relative effect of local buoyancy forces) inhibit the formation of along-axis upwelling segmentation. The

Fig. 23. Structure of buoyant mantle flow beneath a spreading centre for a mantle viscosity of 10^{19} Pa-s and a half spreading rate of 10 mm/yr. *a.* Vertical and horizontal cross-sections through the model for which flow structure is depicted. *b.* Two-dimensional steady-state flow structure. Shading shows amount of melt depletion (light 1-7%, intermediate 7-14%, dark > 14%), dashed contours are isotherms (°C), solid contours show rate of melt production (x1,000) normalized by the half spreading rate (km^2), and arrows are flow velocity vectors. The spreading axis is at the top right. *c,d,e,f* Three-dimensional steady-state flow structure in the cross-sectional planes shown in *a*. Shading contours and arrows are as in *b* except that the solid contours in *f* show vertical velocity normalized by the half spreading rate. In *f* the double line shows the vertical plane containing the spreading axis. As a scale for velocity vectors in *b-f*, the maximum upwelling velocity in *c* is 18.5 mm/yr.

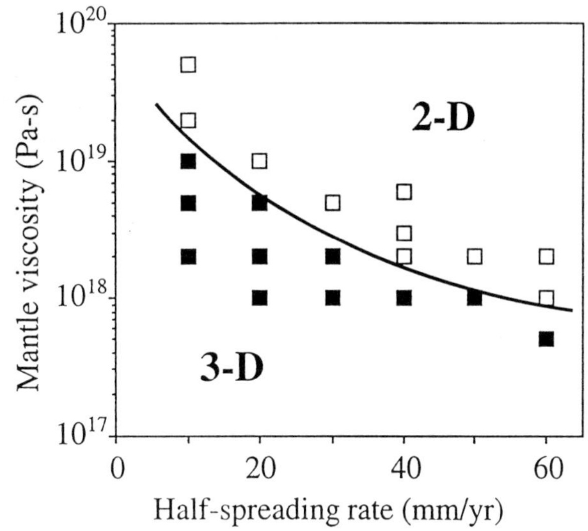

Fig. 24. Spreading rate dependence of upwelling structure in numerical experiments like that shown in Figure 23. Faster spreading and higher viscosity both tend to suppress the formation of a 3-D buoyant upwelling pattern.

results in Figure 24 suggest the intriguing possibility that if sub-ridge mantle temperatures and viscosities are relatively uniform about the global spreading center system, then the development of ridge segmentation may be strongly influenced by spreading rate. Fast-spreading ridges like the EPR could be intrinsically two-dimensional in upwelling structure except for the presence of transform faults and other ridge offsets while slow spreading ridges like the MAR could be intrinsically three-dimensional in upwelling structure with numerous quasi-stable zero-offset transforms which denote the boundaries of upwelling centers. This scenario is consistent with the striking spreading-rate dependence of along-axis variations in gravity, topography, and crustal structure shown by Lin and Phipps Morgan [1992].

Melt buoyancy appears to narrow the region of upwelling beneath a ridge segment which is seen in a comparison of Figures 23 and 25. However, melt buoyancy alone does not appear to significantly affect the wavelength or spacing of upwelling centers. Sparks and Parmentier [1992] developed a 3-D numerical code which includes thermal buoyancy beneath an accreting rigid lithosphere, but which does not include melt buoyancy. This is a more computationally intensive project which means that a much smaller parameter space has been explored to date. The striking feature of these experiments is the development of axis-perpendicular thermal 'rolls' at both slow and fast spreading ridges, as shown in Figure 26. This figure clearly shows that caution should be used in introducing thermal buoyancy

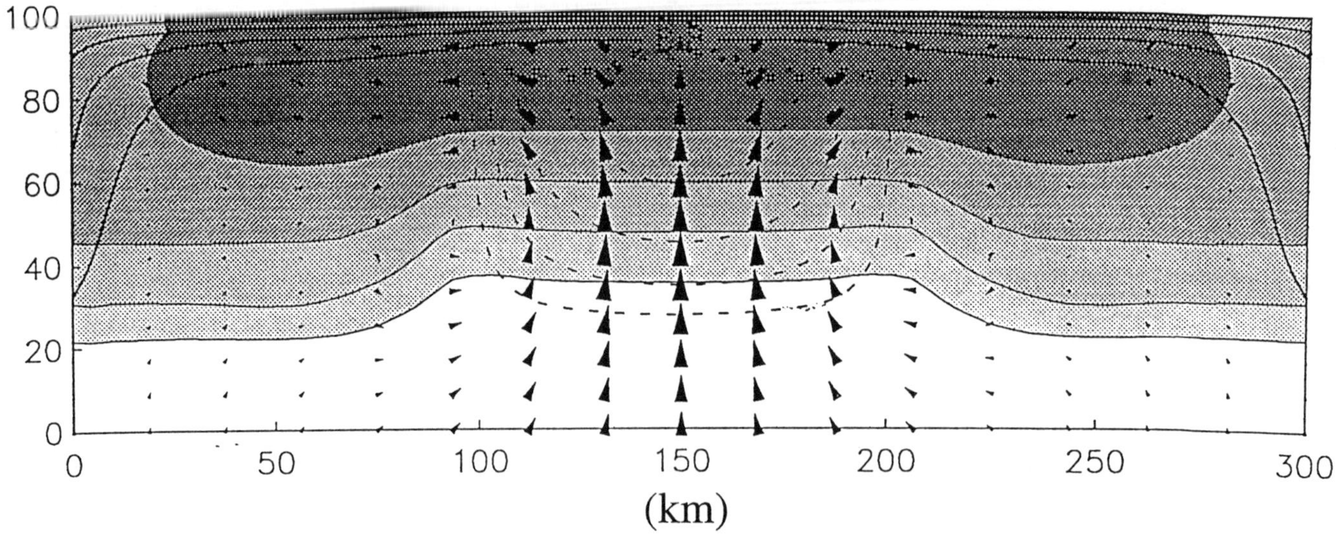

Fig. 25. Three-dimensional upwelling pattern due to plate spreading, the buoyancy of a retained melt phase (maximum melt fraction of 5%), sink-flow into the compacting region of melt production, and compositional buoyancy forces due to melt extraction for a sub-ridge viscosity of $2x10^{19}$ Pa-s and a half-spreading rate of 10 mm/yr. Same annotation conventions as in Figure 22 and 23. The presence of a significant buoyant melt phase appears to localize upwelling but not significantly change the spacing of ridge upwelling centers.

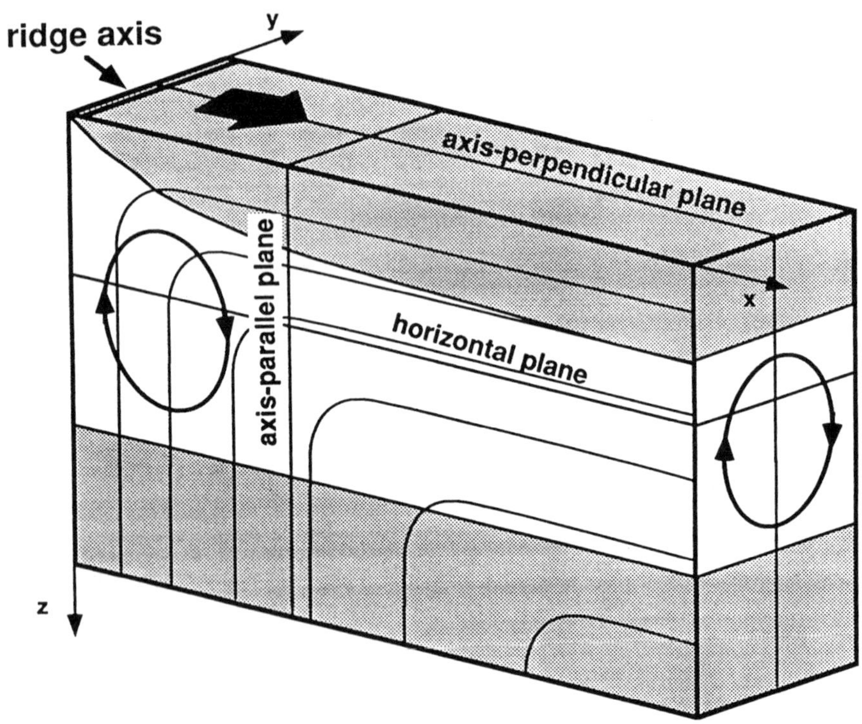

Fig. 26. Three dimensional upwelling pattern due to thermal buoyancy and depletion buoyancy beneath a rigid accreting lithosphere. Same notation as in Figures 22 and 23 compared to passive, plate driven flow a higher degree of melting is more localized beneath the ridge axis. A short distance away from the ridge axis, thermally driven convective motions carry depleted material to depth in ridge-perpendicular roll downwelling structures [Sparks and Parmentier, 1992].

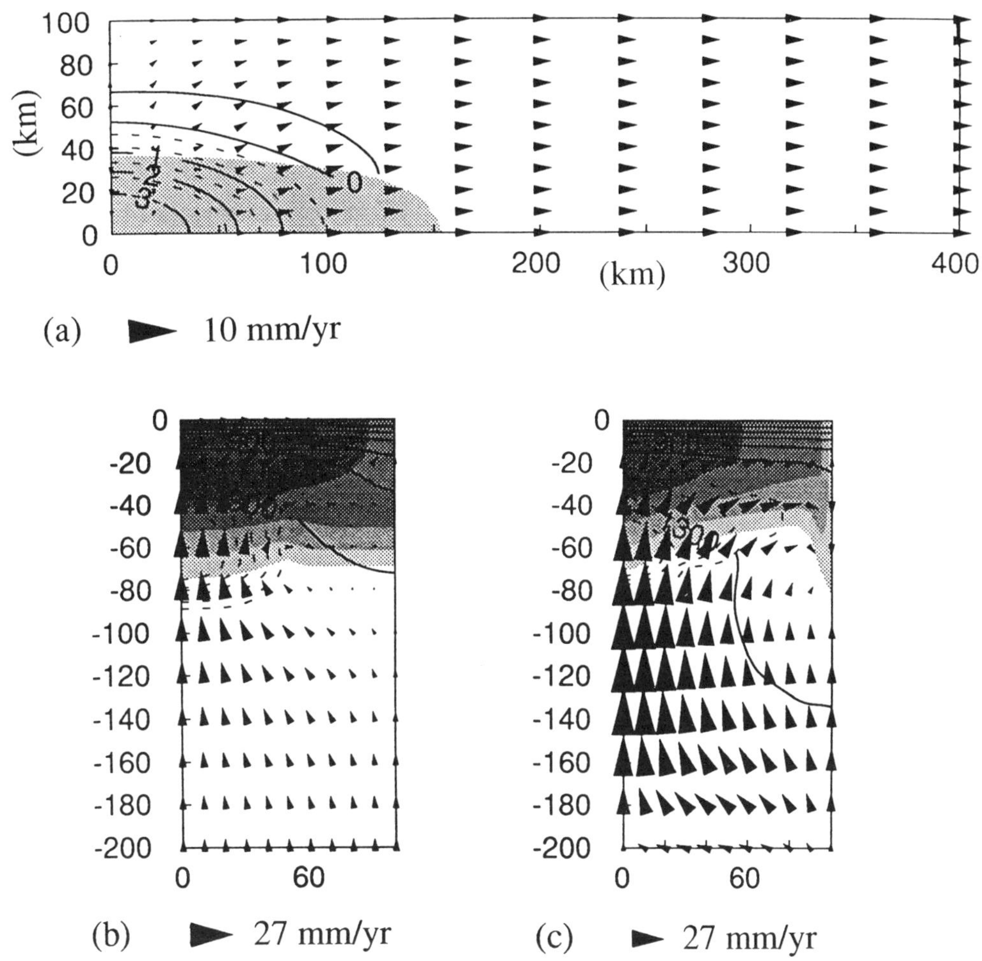

Fig. 26. (continued)

into a 2-D numerical experiment but raises several unresolved questions. In particular, these experiments suggest that at intermediate spreading rates an axis-parallel component of off-axis thermal downwelling can lead to enhanced upwelling and melting, suggesting that intermediate rate spreading ridges should have the largest crustal thicknesses along the global ridge system. This is not seen in current crustal thickness compilations -- suggesting caution is needed to avoid overinterpreting the results of these 3-D experiments.

In future 3-D studies we still need to obtain more insight into potential feed-back between the presence of a melt phase and its effect on locally reducing mantle viscosity and thereby shaping mantle upwelling as suggested by Buck and Su [1989] and Scott and Stevenson [1989]. A proper treatment of this problem in three dimensions is still beyond the capability of current numerical codes and computers but should become feasible within the next two or three years. The rapid increases in computing power mean that many new 3-D experiments will be performed in the next few years. We should soon have a much better understanding of the likely factors which lead to and shape along-axis segmentation.

SUMMARY

The initiation of melting beneath an ocean ridge occurs on grain boundaries at depths of 50-100 km. The resulting magmas form the 6 km thick oceanic crust. The basic driving force for the ascent of the magma is the differential buoyancy between the magma and the residual matrix. This review has emphasized two physical processes for the migration of the magma, porous flow migration along grain boundaries with a scale of millimeters and magma fracture with a scale of meters to kilometers. There is considerable observational evidence that both mechanisms are applicable to magma migration beneath ocean ridges. However, there is still a gap in theory between the millimeter scale and the meter scale.

Melt ascent appears to be focussed in the vicinity of the ridge crest. A number of possible mechanisms have been studied but uncertainties remain. The role of mantle upwelling in shaping melting and subsequent melt migration is still uncertain but we are now entering an era where a coordinated program of numerical experiments promises to yield significant new insights into this problem. Active mantle upwelling driven by compositional buoyancy due to melting and melt extraction, buoyancy due to the presence of a melt phase, and thermal buoyancy may all shape mantle upwelling in addition to the passive upwelling driven by plate spreading. These experiments also indicate that the basic flow beneath mid-ocean ridges may be fundementally three dimensional in nature.

Current theory presents a wide range of plausible and self-consistent physical models for upwelling and melt migration beneath a spreading center. Incorporating chemical predictions and observational constraints on the timescales of melt extraction and melt migration are likely to enhance our understanding of the depth, depth-range, and episodicity of melting processes, and potential depth-dependent transitions in the mechanisms of melt transport. However, future observational constraints on the width and three-dimensionality of the upwelling and melting region are critically needed to help sort out what physical processes, rheologies, and melt migration mechanisms lead to the strikingly narrow and continuous belt of mid-ocean ridge accretion that we see at the seafloor.

REFERENCES

Ahern, J. L., and D. L. Turcotte, Magma migration beneath an ocean ridge, *Earth Planet. Sci. Let,. 45*, 115-122, 1979.

Anderson, O. L., and P. C. Grew, Stress corrosion theory of crack propagation with applications to geophysics, *Rev. Geophys. Space Phys,. 15*, 77-104, 1977.

Atkinson, B. K., Subcritical crack growth in geological materials, *J. Geophys. Res., 89*, 4077-4114, 1984.

Barcilon, V., and O. M. Lovera, Solitary waves in magma dynamics, *J. Fluid Mech., 204*, 121-133, 1989.

Barcilon, V., and F. M. Richter, Nonlinear waves in compacting media, *J. Fluid Mech., 164*, 492-498, 1986.

Barenblatt, G. I., Mathematical theory of equilibrium cracks, *Adv. Ap. Mech., 7, 55-129*, 1962.

Bear, J., *Dynamics of Fluids in Porous Media*, Elsevier, New York, 1972.

Buck, W. R., and W. Su, Focused mantle upwelling below mid-ocean ridges due to feedback between viscosity and melting, *Geophys. Res. Let., 16*, 641-644, 1989.

Cann, J. R., A model for oceanic crustal structure developed, *Geophys. J. Roy. Astron. Soc., 39*, 169-187, 1974.

Chen,Y., Oceanic crustal thickness versus spreading rate, *Geophys. Res. Lett., 19*, 753-756, 1992.

Cordery, M.J., and J. Phipps Morgan, Melting and mantle flow beneath a mid-ocean spreading center, *Earth Planet. Sci. Let., 111*, 493-516, 1992.

Detrick, R.S., J.P. Madsen, P.E. Buhl, J. Vera, J. Mutter, J. Orcutt, and T. Brocker, Multichannel seismic imaging of an axial magma chamber along the East Pacific Rise between 4°N and 13°N, *Nature, 326*, 35-41, 1987.

Emerman, S. H., D. L. Turcotte, and D. H. Spence, Transport of magma and hydrothermal solutions by laminar and turbulent fluid fracture, *Phys. Earth Planet. Int., 44*, 249-259, 1986.

Fletcher, C.A.J., Computational techniques for Fluid Dynamics (2 volumes), Springer-Verlag, Berlin, 1988.

Frank, F. C., Two-component flow model for convection in the earth's upper mantle, *Nature, 220,*350-352, 1968.

Fujii, N., and S. Uyeda, Thermal instabilities during flow of magma in volcanic conduits, *J. Geophys. Res., 79*, 3367-3369, 1974.

Geertsma, J., and F. DeKlerk, A rapid method of predicting width and extent of hydraulically induced fracture, *J. Petrol. Technol., 21*, 1571-1581, 1969.

Geertsma, J., and R. Haafkens, A comparison of the theories for predicting width and extent of vertical hydraulically induced fractures, *J. Energy Res. Technol., 101*, 8-19,1979.

Hardee, H. C., Replenishment rates of crustal magma and their bearing on potential sources of geothermal energy, *J. Volcanol. Geotherm. Res., 28*, 275-296, 1986.

Hardee, H. C., and D. W. Larson, Viscous dissipation effects in magma conduits, *J. Volcanol. Geotherm. Res., 26*, 299-308, 1977.

Hirs, G. C., A systematic study of turbulent film flow, *J. Lubric. Technol., 96*, 118-126, 1974.

Irwin, G.R., Analysis of stresses and strains near the end of the crack traversing a plate, *J. Appl. Mech., 24*, 361-364, 1957.

Johnson, K.T.M., H.J.B. Dick and N. Shimizu, Melting in the oceanic upper mantle: An ion microprobe study of diopsides in abyssal peridotites, *J. Geophys. Res., 95*, 2661-2678, 1990.

Kenyon, P. M., and D. L. Turcotte, Along-strike magma mixing beneath mid-ocean ridges: Effects on isotopic ratios, *Earth Planet. Sci. Let., 84,* 393-405, 1987.

Kidd, R. G. W., A model for the process of formation of the upper oceanic crust, *Geophys. J. R. Astron. Soc., 50*, 149-183, 1977.

Lin, J., and J. Phipps Morgan, The spreading rate dependence of three-dimensional mid-ocean ridge gravity structure, *Geophys. Res. Let., 19*, 13-16, 1992.

Lister, J. R., Buoyancy-driven fluid fracture: The effects of material toughness and of low-viscosity precursors, *J. Fluid Mech., 210*, 263-280, 1990a.

Lister, J. R., Buoyancy-driven fluid fracture: similarity solutions for the horizontal and vertical propagation of cracks, *J. Fluid Mech., 217*, 213-239, 1990b.

Lister, J.R., Steady solutions for feeder dikes in a density-stratified lithosphere, *Earth Planet Sci. Let. 107*, 233-242, 1991.

Lister , J.R., and R.C. Kerr, Fluid-mechanical models of dike propagation and magma transport, in *Mafic Dikes and Emplacement Mechanisms,* A.J. Parker, P.C. Rickwood, and D.H. Tucker, eds., pp. 69-80, A.A. Balkema, Rotterdam, 1990.

Lister , J.R., and R.C. Kerr, Fluid-mechanical models of crack propagation and their application to magma transport in dikes, *J. Geophys. Res., 96*, 10,049-10,077, 1991.

Macdonald, K.C., Mid-ocean ridges: fine scale tectonic, volcanic, and hydrothermal processes within the plate boundary zone, *Ann. Rev. Earth, Planet, Sci., 10*, 155-190, 1982.

McKenzie, D., The generation and compaction of partially molten rock, *J. Petrol., 25,* 713-765, 1984.

McKenzie, D., and M. J. Bickle, The volume and composition of melt generated by extension of the lithosphere, *J. Petrol, 29*, 625-679, 1988.

McKenzie, D., and R.K. O'Nions, Partial melt distributions from inversion of rare earth element concentrations, *J. Petrol. 32*, 1021-1091, 1991.

Navon, O., and E. L. Stolper, Geochemical consequences of melt percolation: The upper mantle as a chromatographic column, *J. Geol., 95*, 285-307, 1987.

Nicolas, A., A melt extraction model based on structural studies in mantle peridotites, *J. Petrol., 27*, 999-1022, 1986.

Parmentier, E.M. and J. Phipps Morgan, Spreading rate dependence of three-dimensional structure in oceanic spreading centres, *Nature, 348*, 325-328, 1990.

Phipps Morgan, J., Melt migration beneath mid-ocean spreading centers, *Geophys. Res. Let., 14*, 1238-1241, 1987.

Phipps Morgan, J., Mid-ocean ridge dynamics: Observations and theory, *Rev. Geophys., U.S. National Report to IUGG, Supplement, 807-822*, 1991.

Phipps Morgan, J., and Y.J. Chen, The genesis of oceanic crust: Magma injection, hydrothermal circulation, and crustal flow, *submitted to J. Geophys. Res.*, 1992.

Phipps Morgan, J., and D.W. Forsyth, Three-dimensional flow and temperature perturbations due to a transform offset: Effects on oceanic crustal and upper mantle structure, *J. Geophys. Res., 93*, 2955-2966, 1988.

Phipps Morgan, J., E.M. Parmentier, and J. Lin, Mechanisms for the origin of mid-ocean ridge axial topography: Implications for the thermal and mechanical structure at accreting plate boundaries, *J. Geophys. Res., 92*, 12,823-12,836, 1987.

Rabinowicz, M., G. Ceuleneer, and A. Nicolas, Melt segregation and flow in mantle diapirs below spreading centers: Evidence from the Oman ophiolite, *J. Geophys. Res., 92*, 3475-3486, 1987.

Reid, I. and H.R. Jackson, Oceanic spreading rate and crustal thickness, *Mar. Geophys. Res., 5*, 165-172, 1981.

Ribe, N. M., The deformation and compaction of partial molten zones, *Geophys. J. Roy. Astron. Soc., 83*, 487-501, 1985a.

Ribe, N., The generation and composition of partial melts in Earth's mantle, *Earth Planet. Sci. Let., 73*, 361-372, 1985b.

Ribe, N. M., Melt segregation driven by dynamic forcing, Geophys. *Res. Let., 13*, 1462-1465, 1986.

Ribe, N. M., On the dynamics of mid-ocean ridges, *J. Geophys. Res., 93*, 429-436, 1988.

Richter, F. M., Simple models for trace elements fractionation during melt segregation, *Earth Planet. Sci. Let., 77*, 333-334, 1986.

Richter, F. M., and D. McKenzie, Dynamical models for melt segregation from a deformable matrix, *J. Geol., 92*, 729-740, 1984.

Ringwood, A. F., *Composition and Petrology of the Earth's Mantle*, McGraw-Hill, New York, 1975.

Ryan, M. P., and J. Y. K. Blevins, The viscosity of synthetic and natural silicate melts and glasses at high temperatures and 1 bar (10^5 pascals) pressure and at higher pressures, *U. S. Geol. Surv. Bull., 1764*, 563 p., 1987.

Salters, V. J. M., and S. R. Hart, The hafnium paradox and the role of garnet in the source of mid-ocean ridge basalts, *Nature, 342*, 420-422, 1989.

Scott, D. R., The competition between percolation and circulation in a deformable porous medium, *J. Geophys. Res., 93*, 6451-6462, 1988.

Scott, D. R., and D. J. Stevenson, Magma solitons, *Geophys. Res. Let.,11,* 1161-1164, 1984

Scott, D. R., and D. J. Stevenson, Magma ascent by porous flow, *J. Geophys. Res.*, 9283-9296, 1986.

Scott, D. R., and D. J. Stevenson, A self-consistent model of melting, magma-migration and buoyancy driven-circulation beneath mid-ocean ridges, *J. Geophys. Res., 94*, 2973-2988, 1989.

Secor, D.T. and D.D. Pollard, On the stability of open

hydraulic features in the earth's crust, *Geophys. Res. Let.*, *2*, 510-513, 1975.

Sleep, N. H., Segregation of magma from a mostly crystalline mush, *Geol. Soc. Am. Bull.*, *85*, 1225-1232, 1974.

Sleep, N. H., Tapping of melt by veins and dikes, *J. Geophys. Res.*, *93*, 10,255-10,272, 1988.

Sotin, C. and E.M. Parmentier, Dynamical consequences of compositional and thermal density stratification beneath spreading centers, *Geophys. Res. Let.*, *16*, 835-838, 1989.

Sparks, D.W., and E.M. Parmentier, Melt extraction from the mantle beneath spreading centers, *Earth Planet. Sci. Let.*, *105*, 368-377, 1991.

Spiegelman, M., and D. McKenzie, Simple 2-D models for melt extraction at mid-ocean ridges and island arcs, *Earth Planet. Sci. Let.*, *83*, 137-152, 1987.

Spence, D. A., P. W. Sharp, and D. L. Turcotte, Buoyancy-driven crack propagation; A mechanism for magma migration, *J. Fluid Mech.*, *174*, 135-153, 1987.

Spence, D. A., and P. Sharp, Self-similar solutions for elastohydrodynamic cavity flow, *Proc. Roy. Soc. London, Ser. A*, *400*, 289-313. 1985.

Spence, D. A., and D. L. Turcotte, Magma-driven propagation of cracks, *J. Geophys. Res.*, *90*, 575-580, 1985.

Spence, D. A., and D. L. Turcotte, Buoyancy-driven magma fracture: A mechanism for ascent through the lithosphere and the emplacement of diamonds, *J. Geophys. Res.*, *95*, 5133-5139, 1990.

Spera, F.J., Dynamics of translithospheric migration of metasomatic fluid and alkaline magma, in *Mantle Metasomatism*, M.A. Menzies and C.J. Hawkesworth, eds., pp. 1-20, Academic Press, London, 1987.

Stevenson, D. J., Spontaneous small-scale melt segregation in partial melts undergoing deformation, *Geophys. Res. Let.*, *16*, 1067-1070, 1989.

Stolper, E., D. Walker, B.H. Hager, and J.F. Hays, Melt segregation from partially molten source regions: The importance of melt density and source region size, *J. Geophys. Res.* *86*, 6261-6271, 1981.

Turcotte, D. L., and J. L. Ahern, A porous flow model for magma migration in the asthenosphere, *J. Geophys. Res.*, *83*, 767-772, 1978.

Turcotte, D. L., S. H. Emerman, and D. A. Spence, Mechanics of dike injection, in *Mafic Dike Swarms*, edited by H. C. Halls, and W. F. Fahrig, pp. 25-29, Geol. Assoc. Canada, Special paper 34, 1988.

Turcotte, D. L., and G. Schubert, *Geodynamics*, Wiley, New York, 1982.

Walker, D., E. M. Stolper, and J. F. Hays, A numerical treatment of melt/solid segregation: Size of the eucrite parent body and stability of the terrestrial low-velocity zone, *J. Geophys. Res.*, *83*, 6005-6013, 1978.

Watson, E.B., Melt infiltration and magma evolution, *Geology*, *10*, 236-240, 1982.

Weertman, J., Theory of water-filled crevasses in glaciers applied to vertical magma transport beneath oceanic ridges, *J. Geophys. Res.*, *76*, 1171-1183, 1971.

Wilson, L., and J. W. Head, Ascent and eruption of basaltic magma on the earth and moon, *J. Geophys. Res.*, *86*, 2971-3001, 1981.

Zienkiewicz, O.C., and R.L. Taylor, *The Finite Element Method* (4th ed. -- 2 volumes), McGraw-Hill, London, 1991.

Donald L. Turcotte, Dept. of Geological Sciences, Cornell University, Ithaca, New York 14853

Jason Phipps Morgan, Institute of Geophysics and Planetary Physics, UCSD, La Jolla, California 92093-0225

Petrological Systematics of Mid-Ocean Ridge Basalts: Constraints on Melt Generation Beneath Ocean Ridges

CHARLES H. LANGMUIR[1], EMILY M. KLEIN[2], AND TERRY PLANK[1]

1. Lamont-Doherty Geological Observatory of Columbia University, Palisades, NY 10964

2. Department of Geology, Duke University, Durham, NC 27706

Mid-ocean ridge basalts (MORB) are a consequence of pressure-release melting beneath ocean ridges, and contain much information concerning melt formation, melt migration and heterogeneity within the upper mantle. MORB major element chemical systematics can be divided into global and local aspects, once they have been corrected for low pressure fractionation and interlaboratory biases. Regional average compositions for ridges unaffected by hot spots ("normal" ridges) can be used to define the global correlations among normalized Na_2O, FeO, TiO_2 and SiO_2 contents, CaO/Al_2O_3 ratios, axial depth and crustal thickness. Back-arc basins show similar correlations, but are offset to lower FeO and TiO_2 contents. Some hot spots, such as the Azores and Galapagos, disrupt the systematics of nearby ridges and have the opposite relationships between FeO, Na_2O and depth over distances of 1000 km.

Local variations in basalt chemistry from slow- and fast-spreading ridges are distinct from one another. On slow-spreading ridges, correlations among the elements cross the global vector of variability at a high angle. On the fast-spreading East Pacific Rise (EPR), correlations among the elements are distinct from both global and slow-spreading compositional vectors, and involve two components of variation. Spreading rate does not control the global correlations, but influences the standard deviations of axial depth, crustal thickness, and MgO contents of basalts.

Global correlations are not found in very incompatible trace elements, even for samples far from hot spots. Moderately compatible trace elements for normal ridges, however, correlate with the major elements. Trace element systematics are significantly different for the EPR and the mid-Atlantic Ridge (MAR). Normal portions of the MAR are very depleted in REE, with little variability; hot spots cause large long wavelength variations in REE abundances. Normal EPR basalts are significantly more enriched than MAR basalts from normal ridges, and still more enriched basalts can erupt sporadically along the entire length of the EPR. This leads to very different histograms of distribution for the data sets as a whole, and a very different distribution of chemistry along strike for the two ridges. Despite these differences, the mean Ce/Sm ratios from the two ridges are identical.

Existing methods for calculating the major element compositions of mantle melts [Klein and Langmuir, 1987; McKenzie and Bickle, 1988; Niu and Batiza, 1991] are critically examined. New quantitative methods for mantle melting and high pressure fractionation are developed to evaluate the chemical consequences of melting and fractionation processes and mantle heterogeneity. The new methods rely on new equations for partition coefficients for the major elements between mantle minerals and melts. The melting calculations can be used to investigate the chemical compositions produced by small extents of melting or high pressures of melting that cannot yet be determined experimentally. Application of the new models to the observations described above leads to two major conclusions: (1) The global correlations for normal ridges are caused by variations in mantle temperature, as suggested by Klein and Langmuir [1987] and not by mantle heterogeneity. (2) Local variations are caused by melting processes, but are not yet quantitatively accounted for. On slower spreading ridges, local variations are controlled by the melting regime in the mantle. On the EPR, local variations are predominantly controlled by ubiquitous, small scale heterogeneites. Volatile content may be an important and as yet undetermined factor in affecting the observed variations in major elements.

We propose a hypothesis, similar to one proposed by Allegre et al [1984] for isotopic data, to explain the differences between the Atlantic and Pacific local trends, and the trace element systematics of the two ocean basins, as consequences of spreading rate and a different distribution of enriched components from hot spots in the two ocean basins. In the Atlantic, the hot spot influence is in discrete areas, and produces clear depth and chemical anomalies. Ridge segments far from hot spots do not contain enriched

Mantle Flow and Melt Generation at Mid-Ocean Ridges
Geophysical Monograph 71
Copyright 1992 American Geophysical Union

basalts. Melting processes associated with slow-spreading ridges vary substantially over short distances along strike and lead to the local trends discussed above, irrespective of hot spot influence. In the Pacific, enriched components appear to have been more thoroughly mixed into the mantle, leading to ubiquitous small scale heterogeneities. Melting processes do not vary appreciably along strike, so local chemical variations are dominated by the relative contribution of enriched component on short time and length scales. Thus the extent of mixing and distribution of enriched components influences strongly the contrasting local major element trends. Despite the difference in the distribution of enriched components, the mean compositions of each data set are equivalent. This suggests that the hot spot influence is similar in the two ocean basins, but its distribution in the upper mantle is different. . These contrasting relationships between hot spots and ridges may result from differences in both spreading rate and tectonic history. Unrecognized hot spots may play an important role in diverse aspects of EPR volcanism, and in the chemical systematics of the erupted basalts

The observations and successful models have consequences for melt formation and segregation. (1) The melting process must be closer to fractional melting than equilibrium melting. This result is in accord with inferences from abyssal peridotites [Johnson et al., 1990]. (2) Small melt fractions generated over a range of pressures must be extracted rapidly and efficiently from high pressures within the mantle without experiencing low pressure equilibration during ascent. This requires movement in large channels, and possibly more efficient extraction mechanisms than normally envisaged in porous flow models with small residual porosity. (3) Diverse melts from the melting regime produce variations in basalts that are observable at the surface. (4) Basalt data can be used to constrain the melting process (e.g. active vs. passive upwelling) and its relationship to segmentation. The data cannot be used to constrain the shape of the melting regime, however, for many shapes lead to similar chemical results. (5) Highly incompatible elements and U-series disequilibria results appear not yet to be explained by melting models, and may require additional processes not yet clearly envisaged.

1. INTRODUCTION

The basalts that form the upper portion of the ocean crust are the end product of melting of the mantle, migration of melt through the mantle towards the surface, and cooling and differentiation of magma near the surface. Because the basalts have experienced each of these processes, there is the possibility that they contain a chemical record of these events. Our challenge is to decipher their chemical codes to reveal their history.

Evidence to help us decipher the codes comes from observation, experiment, and theory. The observations show the chemical systematics that exist in mid-ocean ridge basalts (MORB), and the relationships between the chemical compositions and physical aspects of the ocean floor such as axial depth, crustal thickness and tectonic setting. Experiments provide the necessary data to model chemical changes that occur under different conditions of temperature, pressure and composition. Theory aids in imagining what processes could occur that cause chemical changes in basalts, and is the sole means of relating the observations and experiments to quantitative models of the entire crust formation process.

Currently, three major factors are considered to contribute to the diversity of MORB compositions. (1) Crystallization and differentiation processes at shallow levels in the crust and uppermost mantle modify the compositions of mantle melts. (2) Melting itself can lead to differences in composition, depending on the temperatures and pressures involved, and the modes of melt migration and segregation. (3) The mantle source may vary in its major element, trace element and isotopic composition, which can lead to inherited variations in basalt compositions.

Recognizing and isolating the different effects of melting and crystallization on the compositions of erupted basalts is a fundamental and ongoing problem in igneous petrology. Historically, the debate centered on whether basalt compositions are close to those of actual melts of the mantle [*Green and Ringwood*, 1967; *Green*, 1971; *Gast*, 1968], or whether their compositions have been so modified by differentiation in crustal magma chambers [*O'Hara*, 1965; *O'Hara*, 1977] or complex melting processes [*O'Hara*, 1985] that their melting histories are completely obscured. Although O'Hara's original argument that most basalts lie on low-pressure cotectics and thus are not "primary" is correct, it is now generally accepted that basalts still preserve aspects of their mantle melting history. The question of the relative importance of melting and differentiation processes is still not fully resolved, however, as will be clear from later discussion in this paper. Certainly one of the major questions is how well simple models of melting and differentiation approximate the more complex natural systems [e.g., *O'Hara*, 1977, 1985; *Allegre and Minster*, 1978; *Richter*, 1986; *Plank and Langmuir*, 1992].

It also has long been recognized that the mantle must be chemically heterogeneous on various scales. For example, hot spots are associated with large-scale enrichments of incompatible trace elements, such as K and Ba, and radiogenic isotope ratios, such as $^{87}Sr/^{86}Sr$ [e.g., *Schilling*, 1975; *Verma and Schilling*, 1982; *Schilling et al.*, 1983], compared to ocean ridges of average depth. The isotopic enrichments, in particular, require isolation of a mantle reservoir with enriched parent/daughter ratios for substantial portions of Earth history. On a smaller scale, there is isotopic variability in single seamounts [*Zindler et al.*, 1984] and over short time intervals for an individual ridge segment [e.g., *Shirey et al.*, 1987], and heterogeneities on the scale of meters are observed in ultramafic massifs such as the Ronda peridotite [*Reisberg and Zindler*, 1986/87). Although it is more difficult to distinguish major element source heterogeneity from the overprinting effects of melting and crystallization, some of the major element variability in MORB has been ascribed to major element variations in the mantle beneath ridges [e.g., *Langmuir and Hanson*, 1980; *Sigurdsson*, 1981; *Bryan and Dick*, 1982; *Klein and Langmuir*, 1987, hereafter referred to as *KL*87].

Thus, the effects of source composition, melting and crystallization form a complex web of chemical signatures. Our task is to understand what aspects of the chemical signature tell us about each process. Only then can one investigate the implications of basalt chemistry concerning melt formation and segregation.

We believe that developments in petrology in the past several years allow a much clearer separation among the effects of source composition, melting and crystallization, although there are still areas of dispute. Such developments include improvements in quantitative modeling of igneous processes, the acquisition of high-quality chemical data for basalts from diverse ridge segments, and more detailed experimental studies on peridotite melting and melt differentiation. It is now possible to better constrain those aspects of the data that may be attributed to each process, and to identify more clearly those aspects of the data for which interpretation remains ambiguous.

Our approach in this paper is first to present a working model for overall variability in MORB, which serves as a framework for subsequent presentation and discussion. We then present some basic concepts as necessary vocabulary. Some of this material is quite elementary, but is included to provide the necessary background to make these notes more accessible to non-petrologists. We then present observations of the major element chemistry of MORB, emphasizing aspects of the data that are critical to the problems of melting and melt migration. The basic major element observations are used to develop specific questions concerning their origin. We then consider modeling techniques, and present a new quantitative model for major elements that is able to treat fractional melting and mantle heterogeneity quantitatively. These modeling techniques are then applied to the questions raised by the data. Finally, we examine trace element data to test how they correspond to the major element models, and what new light they throw on the issues of melt formation and migration.

There are other important petrological approaches that we do not include in this paper. Notably, recent work on abyssal peridotites [*Dick et al.*, 1984; *Michael and Bonatti*, 1985; *Johnson et al.*, 1990; *Johnson and Dick*, 1992] has provided important evidence concerning melt formation and segregation, and has been able to address from a different point of view many of the same problems discussed in this paper. The abyssal peridotite literature is an important and necessary scientific complement to the work on basalts that is discussed here.

This paper includes a significant amount of background and review material for the non-specialist, in an effort to make the reasoning clear. However, it also includes new observations of the systematics of MORB compositions, new developments in modeling mantle melting and crystallization at high pressure, new constraints on the nature of mantle melting and crystallization at high pressure, new constraints on mantle melting chemistry and processes, and a new model for the origin of the local systematics of ocean ridges and the differences between MORB erupted on the Mid-Atlantic Ridge and East Pacific Rise.

1.1 A Working Model

It will help to understand the relevance of much of the material that follows if it can be placed in an overall framework for ocean crust creation.

It has been understood for many years that the ocean crust is formed by melting of the mantle as a result of decompression during upwelling beneath ocean ridges [e.g., *Carmichael et al.*, 1974; *Oxburgh*, 1980 and references therein]. Mantle ascending beneath the ridge will follow an adiabatic gradient of 1° to 2°C of temperature drop for each kilobar of ascent. The mantle solidus that controls where melting begins has a very different temperature gradient of about 13°/kbar, so that ascending mantle inevitably intersects the solidus and begins to melt (Figure 1). The pressure at which such melting begins is controlled by the overall temperature of the mantle: hotter mantle intersects the solidus deeper than cooler mantle [e.g., *Schilling*, 1973].

The breaking of crystal bonds that occurs during melting requires heat, and this heat of fusion is provided internally by the portion of mantle that is ascending. This causes the mantle to cool as it melts, and therefore its temperature decreases with decreasing pressure more rapidly than the adiabatic gradient (Figure 1). Because the energy available for melting depends on the temperature difference between the melting adiabat and the solidus at a given pressure, the more the parcel of mantle ascends, the more it melts. The amount of melting that results is about 1-2% per kbar of pressure release.

These considerations need to be combined with the overall geometry of the melting regime beneath ridges. Various possible geometries were discussed by *Oxburgh* [1980], but the triangular shape with a flat bottom that results from adiabatic upwelling of mantle with a laterally uniform temperature is now most accepted (see *McKenzie and Bickle*, [1988] for a useful discussion; Figure 2). The amount of melt in the melting regime is controlled by the amount of pressure release that each individual parcel of mantle within the melting regime has undergone (Figure 2). The higher the mantle temperature, the more melt that is produced, and hence the greater the thickness of the oceanic crust [e.g., *McKenzie*, 1984].

Because the melt is less dense than the mantle, it will move as the mantle melts. Therefore the many different melts present at any one time within the melting regime may erupt independently, mix, or react with the mantle through which they pass, in a collection of processes that have been called "dynamic melting" [*Langmuir et al.*, 1977]. The mean composition of the ocean crust, however, will be the mean composition of melts pooled from the melting regime.

The mean melt composition, and its relationship to mantle temperature and crustal thickness, can be envisioned simply by considering the "residual mantle column" that emerges outside the melting regime for each increment of spreading (Figure 2).

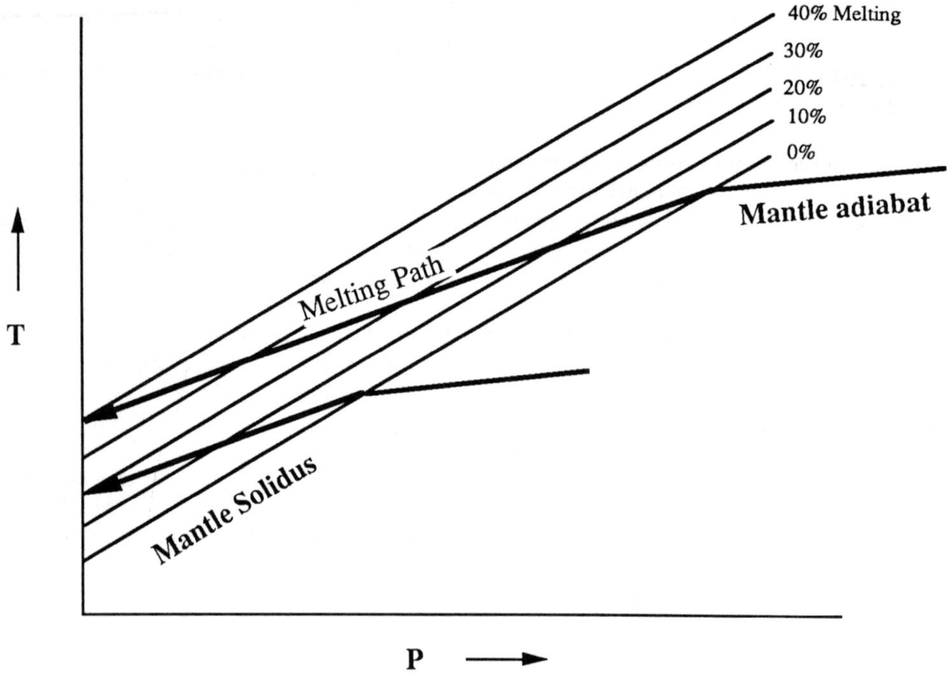

Fig. 1. Schematic pressure (P) vs. temperature (T) diagram showing melting paths for adiabatically ascending mantle. Melting contours are evenly spaced for illustration only. The two paths are for two different mantle temperatures. The further a mantle parcel ascends, the greater the extent of melting.

This column reflects all that has happened within the melting regime [*Plank and Langmuir,* 1992], and therefore provides a one dimensional summary of the melting regime. The deeper portions of the residual mantle column are produced at the outer corner of the melting regime, while the shallowest are produced at the center of the melting regime. The residual mantle column shows increasing depletion in melt upwards in the column, reflecting the increasing extent of melting upwards in the melting regime. The mean melt reflects the pooling together of all the various melts that were extracted to form the column. The mean pressure reflects the average pressure from which the melts are derived. Crustal thickness is the total amount of melt that is extracted, which is simply the mean extent of melting times the column height, if extraction is complete.

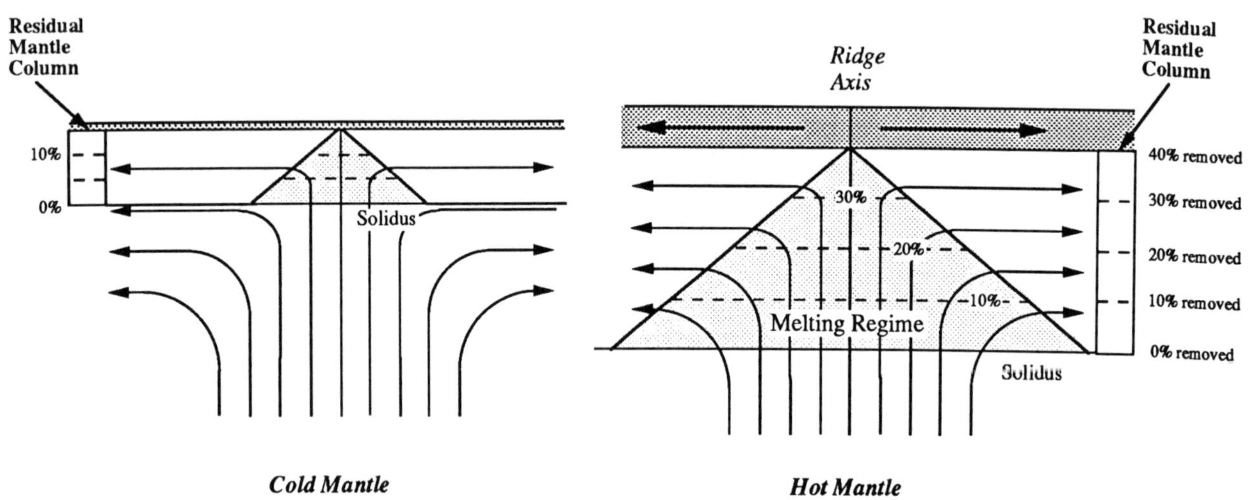

Fig. 2. Idealized steady-state melting regimes produced by spreading of the plates [after *Plank and Langmuir,* 1992]. Solid curves with arrows are mantle flow paths through the melting regime. Dashed lines are contours of the extent of melting in the melting regime or the extent of melt removed in the residual mantle column (RMC). The two melting regimes are for two mantle temperatures. Hotter mantle intersects the solidus deeper (see Figure 1), leading to greater extents of melting, a taller RMC, and thicker crust.

These considerations can be combined with simple chemical considerations for a general model of the composition and thickness of the ocean crust in response to lateral variations in mantle temperature [*KL*87]. The chemical considerations are that some elements, such as Na_2O, strongly prefer the melt to the solid mantle, and hence their concentrations in mantle melts are roughly inversely proportional to the extent of melting [*Jaques and Green*, 1980; *Dick et al.*, 1984]. Other elements, such as FeO, are sensitive to the pressure of melting, with higher pressures leading to higher FeO contents in the melt [*Langmuir and Hanson*, 1980]. The critical observation from regionally averaged MORB is that the extent of melting (inferred from Na_2O) and the pressure of melting (inferred from FeO) correlate positively [*KL*87]. This requires a model where higher mean extents of melting occur at higher mean pressures. To accomplish this, extraction must be very efficient, with melts from deep in the melting regime migrating toward the surface without re-equilibrating at shallow levels. Then higher mantle temperatures lead to taller melting columns, greater mean extents and pressures of melting, and greater crustal thickness. Ocean ridge depth is controlled largely by three factors: crustal thickness, and the temperature and composition of the underlying mantle. All of these factors lead to shallower depths for hotter mantle. Therefore relationships among crustal composition, crustal thickness and ridge depth can be envisioned as a response to variations in mantle temperature along the ocean ridge system.

At the same time, consideration of the melting process above indicates the likelihood of substantial compositional variability within the melting regime. Within any single melting regime created by a given mantle temperature, the smallest degree melts occur at the greatest pressures. This could lead to the opposite relationship between the extent of melting and pressure of melting for diverse melts from a single melting regime, as compared to average melts from diverse melting regimes created by variations in mantle temperature. Therefore, chemical variations can occur within a single melting regime that lead to diverse melts of the mantle in local regions. These chemical variations can have the opposite relationship between pressure and temperature of melting than that observed for mean melts among different melting regimes. Such variations could lead to local compositional variations that are distinct from global compositional variations.

The melts of the mantle do not arrive at the surface in an unadulterated state. They may pass through colder lithosphere, or react or undergo fractionation at high pressure. They certainly pass through magma fractionation systems at lower pressures, just beneath or within the crust, where substantial differentiation of their compositions may take place by a variety of processes. The simplest form of differentiation occurs in a closed system at low pressure, which we will refer to as "low pressure differentiation." All other processes of differentiation, such as high-pressure or *in situ* fractionation, we refer to as "complex differentiation processes."

This working model provides a qualitative picture of ocean crust formation: lateral variations in mantle temperature control the mean melt composition; variations in melt composition within a single melting regime control local chemical variability; and differentiation processes modify everything. We can see from the model the difficulty and complexity of addressing many important questions concerning melt formation and segregation. The model does not address the details of the melting process, such as whether melting occurs by fractional or equilibrium melting, or to what extent extraction is efficient. It assumes a homogeneous mantle when we know the mantle is heterogeneous in many aspects of its composition. It is a two-dimensional, across-strike model, when much of the excitement in the study of ridges in the past decade has concerned along-strike variations, and the next decade will reveal new discoveries about changes through time. It assumes passive upwelling when upwelling in some regions may be actively buoyant. It is spreading rate-independent when spreading rate has long been considered to be a primary control on the variability of ocean ridges.

The model serves, however, as a useful mental framework to aid in the digestion of what is presented on the following pages. It provides a structure within which to view the great complexity and variability of MORB, while the data from MORB can in turn be used to test the overall validity of the model. As we will see, the model requires elaboration and modification in order to account for aspects of the data that probably result from the complexities mentioned above.

2. Basic Theoretical Concepts

2.1 Phase Diagrams

The classical approach to petrologic problems is through phase diagrams. In a simple ternary system consisting of phases A, B and C (Figure 3*a*), the major element composition of a liquid produced by partial melting of the three-phase solid assemblage will be the eutectic composition, regardless of the initial phase proportions. The eutectic melt is invariant and conveys no information concerning source composition or extent of melting. Many experimental studies, in the interests of thermodynamic rigor and in the tradition of Bowen, have concentrated on such simple systems (e.g., the CaO-MgO-Al_2O_3-SiO_2 or CMAS system). If the earth were like these systems, and the magmas were on low-pressure cotectics, then the major element composition of the magma would contain no information about extent of melting or source composition.

Simple phase diagrams have provided the framework for petrologic studies throughout most of the century; they are the basis for how petrology is taught, and how petrologists think about petrologic problems. Because it is difficult to graphically represent even four components in a phase diagram, an aim of many petrologic studies in recent years has been to reduce the multi-component mantle system to a more tractable three or four component system [e.g., *Stolper*, 1980; *Elthon and Scarfe*, 1984]. An example of such a "pseudo-ternary" phase diagram is shown in Figure 3*b*. Pseudo-invariant points and pseudo-invariant curves on these diagrams are really regions in multi-component space [*Elthon and Scarfe*, 1984; *Stolper*, 1980; *Presnall and Hoover*, 1984; *Fujii*

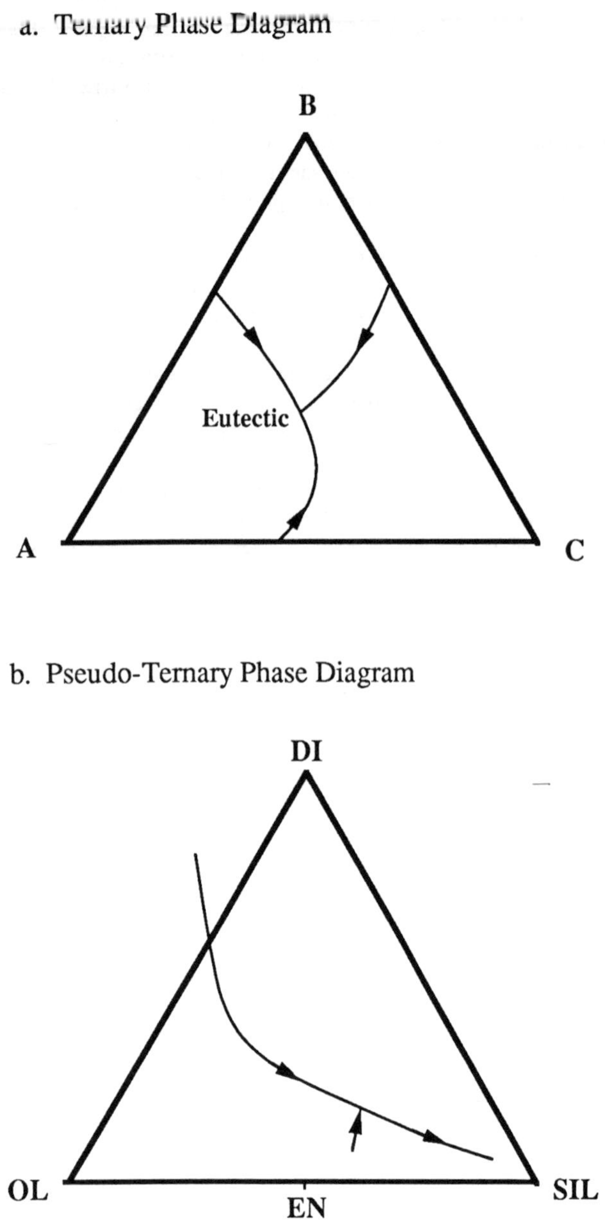

Fig. 3. *a*) A simple ternary phase diagram with a eutectic. Any bulk composition within the triangle would yield an initial melt composition at the ternary eutectic. *b*) Example of a pseudo-ternary phase diagram [after *Stolper*, 1980].

and Scarfe, 1985]. On such pseudo-ternary diagrams, however, one is naturally drawn to think of true invariant points, and by extension, one might conclude that mantle melts reflect neither source composition nor extent of melting. Indeed, earlier studies described mantle melting as "eutectic-like" [e.g., *Yoder*, 1976; *Mysen and Kushiro*, 1977], and many later studies have interpreted these diagrams as if melting were eutectic-like.

As many have recognized, however, there are three fundamental differences between melting in simple systems and in the mantle. First, there are eight important components (Si-Al-Fe-Mg-Ca-Na-Ti-Cr) and additional minor components (e.g., K, Fe^{3+}, P, CO_2, and H_2O) in the real system. For a given number of mineral and melt phases, the number of components governs the degrees of freedom: the large number of components and limited number of phases involved in mantle melting precludes invariant melting, which means that the melt composition can vary as melting proceeds. Second, all mantle phases show solid solution; for example, the Mg/Fe ratio of liquids in equilibrium with olivine increases as melting proceeds. Third, the mantle melts polybarically, so the constant pressure restriction of most phase diagrams does not hold. These theoretical considerations have been conclusively verified by recent experimental studies that show a wide range of melt compositions in equilibrium with lherzolite for a single bulk composition and a single pressure [e.g., *Jaques and Green*, 1980; *Fujii and Scarfe*, 1985; *Takahashi and Kushiro*, 1983; *Falloon and Green*, 1988; *Falloon et al.*, 1988]. It is nonetheless tempting to draw a pseudo-ternary diagram and use it in a classical manner to discuss mantle melting. Therefore, it is important to emphasize that:

1. There are no mantle eutectics, even on pseudo-ternary diagrams, and therefore mantle melting produces a range of compositions;
2. Melting is a polybaric process, and melts form and probably segregate over a range of pressures.

Phase diagrams provide a convenient graphical representation of complex compositions; they often can accurately predict what phases are stable; and they are the conventional vocabulary of petrologists. They can be used to illustrate principles and to demonstrate in a straightforward way the effects of complex processes. They can also be used with care to consider complex equilibria. These aspects make them a useful tool, as illustrated by other papers in this volume [*Hess*, this volume; *Grove et al.*, this volume]. For our purposes, however, the complexity and texture of the chemical signal is often reduced after projecting onto a pseudo-ternary diagram, and thus some of the information contained in the signal is lost. More subtly, the use of over-simplified phase diagrams can lead one who is not aware of the full complexity to think of multi-component equilibria incorrectly. For these reasons, in this paper we do not emphasize the phase diagram presentation of equilibria, and rely instead on the quantitative calculation of equilibria using individual elements or components.

2.2 *Partition Coefficients*

Almost all the chemical variability we will discuss reflects the fractionation of elements that occurs during the separation of liquid and crystal. A very simple concept that describes this fractionation is the partition coefficient, K_d, defined for a given single element or component as follows:

$$K_d = C_s / C_L \tag{1}$$

where C_s is the concentration of an element in a particular mineral, and C_L the concentration of the element in the liquid.

For a multiphase solid assemblage the K_d's for each individual mineral are combined into a "bulk distribution coefficient" (D) which is simply the concentration in the total solid (C_S) divided by the concentration in the liquid. Hence:

$$D = C_S / C_L \qquad (2)$$

= (fraction mineral 1)*(K_d for mineral 1) + (fraction mineral 2)*(K_d for mineral 2) + etc.

$$= \Sigma X_j K_{d\,i}^{\,j} \qquad (3)$$

where X is the fraction of the mineral and $K_{d\,i}^{\,j}$ is the K_d for element i between phase j and the liquid. (Symbols used throughout the paper are given in Table 1.)

The K_d is actually a synthesis of a great deal of thermodynamic information. Determining K_d's for real, non-ideal systems from thermodynamic data on pure components involves complex, empirical solution models. However, the K_d contains this information in a single value, and can be measured directly from experimental or natural mineral-melt pairs. Thus, if K_d's can be parameterized as a function of temperature, pressure, and composition, a thermodynamically valid and relatively simple description of equilibrium can result [*Langmuir and Hanson*, 1981; *Nielsen and Dungan*, 1983]. We will use this K_d approach to calculate both crystallization and melting. It is important to note that temperature, pressure and compositional variations affect the absolute value of the partition coefficient, as illustrated for the olivine K_{dMg} in Appendix B. Careful attention to the influence of these variables is particularly important for quantitative modeling of major element variations using a K_d approach, as shown in subsequent sections.

Another advantage of the K_d approach is that it unifies the treatment of major and trace elements. Trace elements have long been modeled with partition coefficients [*Gast*, 1968; *Shaw*, 1970; *Schilling*, 1971]. Elements with partition coefficients <1 are concentrated into the liquid and thus are "incompatible" in the solid phases. With respect to mantle minerals, such incompatible elements include the heavy alkali elements, the light rare earth elements, Ba, Th, U, Ta, Nb and Pb. Some major elements, such as Na and Ti, have small K_d's for most mantle minerals, and hence behave incompatibly. Elements that are preferentially incorporated into a particular mineral are "compatible" with respect to that mineral. For example, Sr is preferentially incorporated into plagioclase; it is thus incompatible if plagioclase is absent, but becomes compatible if plagioclase forms a significant fraction of the total solid assemblage. Trace elements can be more or less compatible than major elements. For example, Fe, Mg and Ni are all compatible in olivine, but Ni is the most compatible. Thus, the same principles can be applied to both major and trace elements, and each element becomes a part of a single spectrum of variability when viewed from the perspective of their K_d's. This perspective is exemplified by the use of "Onuma diagrams" [e.g., *Onuma et al.*, 1968; *Philpotts*, 1978] which show that different elements that occupy the same site in a mineral have partition coefficients that vary smoothly as a function of atomic radius, independent of their concentration in the liquid.

This idea, that each element is part of a continuum of variability, has been put to great use in the interpretation of trace element variations, and in particular in the interpretation of the rare earth elements (REE). The REE vary progressively in the extent of their incompatibility with atomic number, because all of them have a similar outer electron shell, and therefore the change in their chemical behavior results principally from the progressive change in ionic radius with atomic number. With respect to the primary mantle minerals, the light rare earth elements (LREE), such as La or Ce, are most incompatible, while the heavy rare earth elements (HREE), such as Yb or Lu, are only moderately incompatible and, if garnet is present, may become compatible. In order to make use of this regular variation in compatibility, it is common to plot the REE abundances in rocks as a function of increasing atomic number (called a Coryell-Masuda diagram) and normalize these abundances to a standard, such as chondritic meteorites, to eliminate the natural saw-toothed abundance pattern that results from the preferential creation of even numbered nuclides during nucleosynthesis. The resulting REE pattern can often reveal, at a glance, important features of the petrogenesis of the rock.

2.3 Mass Balance and Simple Equations for Melting and Crystallization

The simplest constraint for both major and trace elements is mass balance between solid and liquid:

$$C_o = (F * C_L) + [(1-F) * C_S] \qquad (4)$$

where C_o is the initial concentration of the element in the system, often called the "bulk composition", F is the fraction of liquid, and as noted above, C_S is the concentration of the element in the total solid assemblage. Mass balance applies strictly to closed systems, although there are mass balance constraints even in open systems.

It is useful to combine the simple concept of a distribution coefficient with mass balance. By substituting equation (2) into equation (4) we arrive at:

$$C_L / C_o = 1/[F + D*(1-F)] \qquad (5)$$

This simple equation was developed by *Schilling* [1966] and *Gast* [1968], and has formed the basis for most trace element modeling. It is called the "batch melting" or "equilibrium melting" equation when applied to melting, and the "equilibrium crystallization" equation when applied to crystallization.

Although this equation is simple, it is also general and powerful. First, it applies to all elements, major elements as well as trace elements. Second, provided there is enough time for liquid and solid to reach equilibrium, equation (5) describes any point in the melting or crystallization process. Phases may disappear and melt may move, but at any given point in time the concentrations in the solid and liquid define the distribution coefficient, and thus equation (5) holds.

TABLE 1. Definitions & Acronyms

C_L	concentration of an element (or component) in the liquid
C_o	initial concentration of an element (or component) in the system; bulk composition
C_S	concentration of an element (or component) in the total solid assemblage
C_s	concentration of an element (or component) in a particular mineral
C_p	heat capacity
D	bulk partition coefficient (or bulk distribution coefficient); see equation (2)
F	fraction of the system that is liquid
$\bar{F}$	mean extent of melting
F_{max}	maximum extent of melting
h_c	thickness of the ocean crust
H_f	heat of fusion
K_D	FeO-MgO partition coefficient between olivine and melt; see equations (12-14)
K_d	parition coefficient (or distribution coefficient); see equation (1)
$K_{d_i}^j$	partition coefficient for element (or component) i between phase j and the liquid
Mm	the mass of melt
M_s	the mass of solid prior to melting
$Na_{8.0}$	Na_2O normalized to 8 wt% MgO in order to correct for the effects of 1 atm crystal fractionation. See *Klein and Langmuir* [1987] for the general algorithm used to approximate the liquid line of descent for MORB; in order to explore more detailed relationships among samples from a particular region, the algorithm was often adjusted to better represent the liquid line of descent for the composition of samples from that region [see *Klein and Langmuir,* 1989]. Occasionally, it has been useful to normalize the data to a value other than 8 wt% MgO, e.g. $Na_{7.3}$ is normalized to 7.3% MgO.
$Fe_{8.0}$, $Si_{8.0}$, $Ti_{8.0}$, etc.	FeO, SiO_2, or TiO_2 normalized to 8 wt% MgO, as described above for $Na_{8.0}$. See *Klein and Langmuir* [1987; 1989] and herein for algorithms used to calculate these parameters
P	pressure
P_{22}	the pressure of the upwelling mantle parcel at 22% melting
$\bar{P}$	mean pressure of melting
P_f	the final pressure of melting
P_o	pressure at which ascending mantle intersects the solidus and begins to melt
T	temperature
X_j	fraction of the total solid assemblage that is mineral j.
XsTi	a measure of the amount of enriched component in MORB from the EPR, see Figure 26

AFM	Accumulated fractional melting
AMAR	Alvin at the MAR, region near 36°N on the MAR
EPR	East Pacific Rise
FAMOUS	French American Mid-Ocean Undersea Study, a region near 36°30'N on the MAR
JDF	Juan De Fuca Ridge
LLD	Liquid lines of descent
LREE	Light rare Earth elements
MAR	Mid-Atlantic Ridge
Mg#	Mg/(Mg+Fe), atomic ratio
MORB	Mid-ocean ridge basalt(s)
N-MORB	Normal mid-ocean ridge basalt
E-MORB	Enriched mid-ocean ridge basalt
T-MORB	Transitional mid-ocean ridge basalt
OFZ	Oceanographer Fracture Zone
PRMC	Periodically replenished magma chambers
REE	Rare earth elements
HREE	Heavy rare earth elements
LREE	Light rare earth elements
RMC	Residual mantle column

The simple mass balance equation illustrates several important principles concerning the relative effects of melt fraction and partition coefficient on liquid composition. Figure 4*a* shows the concentrations in the melt of elements with different D's as a function of F for batch melting. Note first that as F goes to 1 (all melt), C_L/C_o of all elements converge on 1, the initial starting concentration. Highly incompatible elements (D ~ 0) vary as 1/F, and thus can be sensitive indicators of the extent of melting (for a given C_o) or crystallization. For all values of F, compatible elements (D > 1) are preferentially partitioned into the solid, and therefore relative to the source concentration are in lower abundance in the melt. As F becomes smaller than D, the concentration of all elements approaches 1/D. This is true even for highly incompatible elements as F approaches 0. Thus, for incompatible elements, the maximum enrichment that can be achieved is 1/D, and the most any two elements in the liquid can be fractionated from one another is the ratio of their D's (Figure 4*b*). This 1/D constraint applies to a multitude of melting models, and thus is a very useful relationship.

Equation (5) and Figures 4*a* and 4*b* apply to both equilibrium melting and equilibrium crystallization. For melting the system begins as total solid (F = 0), whereas for crystallization the system begins as total melt (F = 1). The greatest incompatible element enrichments and fractionations occur at small values of F, and change little at large values of F. Compatible elements, on the other hand, change most at large values of F and are constrained by the 1/D limit over a wide range of F as F gets small. Thus small extents of partial melting affect incompatible elements substantially, but compatible elements less, while small amounts of crystallization affect compatible elements substantially, but incompatible elements and particularly incompatible element ratios (Figure 4*b*) very little. For these reasons, variations in the ratio of two incompatible elements are a more useful indicator of processes other than crystallization, such as variability in the extent of melting or in the source composition.

More complex processes affect these conclusions to a certain extent. For example, if the phases enter the melt in different proportions than they are present in the solid, then the D for any element changes as a function of F. This leads to:

$$C_L / C_o = 1/[F^*(1-p)+D_o)] \qquad (6)$$

where D_o is the bulk partition coefficient when F is zero [calculated using equation (3)], and p is a distribution coefficient weighted according to the phase proportions entering the liquid (calculated using an equation similar to equation (3) except that X_j would refer to the fraction of each mineral entering the melt). The difference in results using equation (6) as compared to equation (5) becomes apparent for elements that have high partition coefficients in the minerals that are melting preferentially (i.e., they have p's approaching or exceeding 1). For incompatible elements (both D_o and p near 0), there is little difference. Equation (6) rather than equation (5) is more applicable to mantle melting, since aluminous phases and clinopyroxene enter the liquid preferentially to olivine and orthopyroxene. Note that the 1/D limit applies equally to this equation as F becomes small relative to D_o, since D_o and p are usually similar in magnitude.

Chemical variations produced during fractional (in contrast to equilibrium) crystallization and melting, where solid and

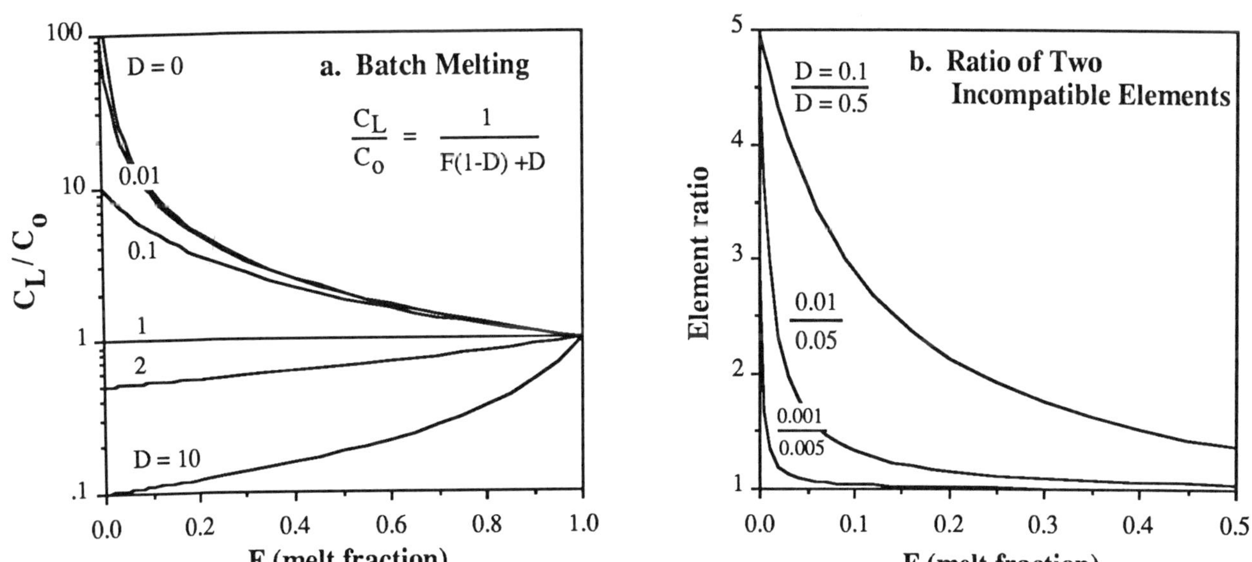

Fig. 4. Illustration of the effects of equilibrium melting on trace element abundances using equation (5) (batch melting equation). *a*) Variation in liquid concentration (C_L) (normalized to unit source concentration C_o = 1) as a function of melt fraction (F) for six "elements" with different bulk distribution coefficients (D). *b*) Change in the ratios of incompatible elements with different D's as a function of F. Each curve is for a different pair of "elements" that have the D's indicated. Note that for D<0.1, incompatible element ratios can be changed only at very low extents of melting.

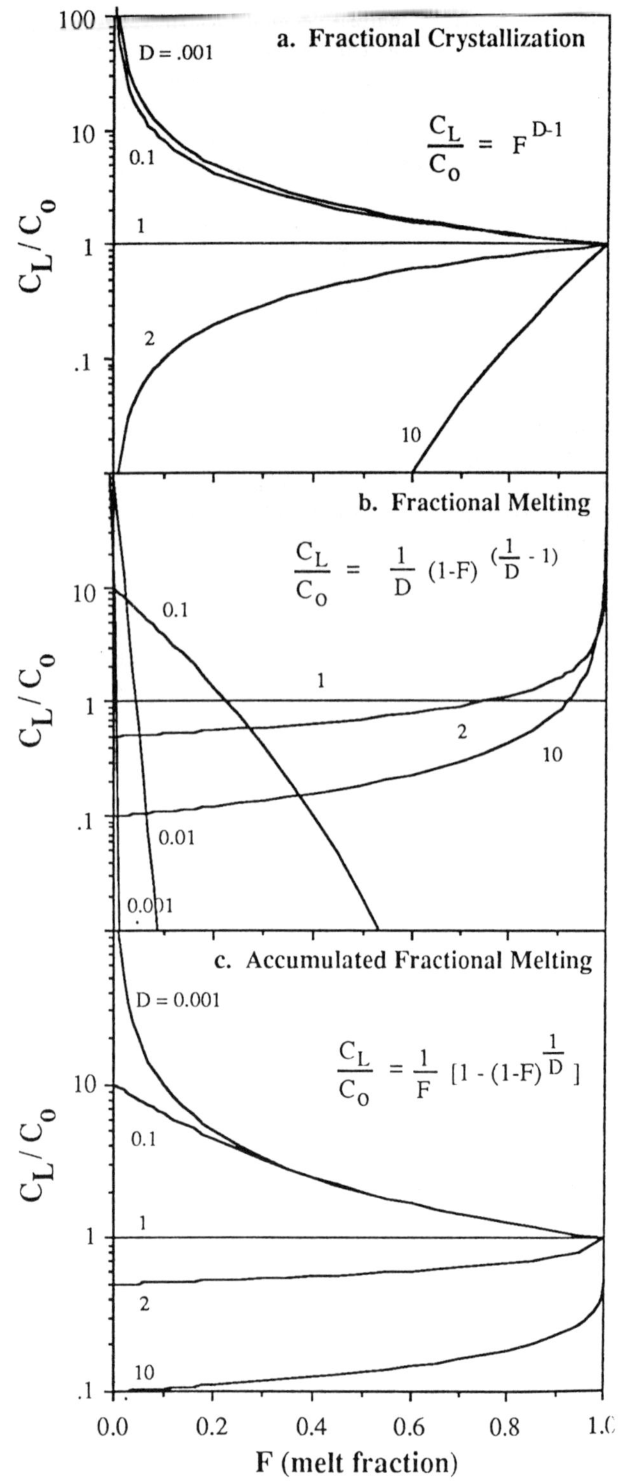

Fig. 5. Illustration of the effects of fractional crystallization and melting on trace element abundances. These results should be compared to Figure 4 — the notation is the same for the two figures. *a)* Fractional crystallization for five different values of D (equation 7). *b)* Fractional melting (equation 8). *c)* Accumulated fractional melting (equation 23). Note the very similar results produced by accumulated fractional melting to the equilibrium melting results of Figure 4.

liquid are separated from each other instantaneously throughout the process, are illustrated in Figures 5*a*-5*c*. The equations describing fractional crystallization and melting, assuming D remains constant during each process, are respectively [*Shaw*, 1970]:

$$C_L / C_o = F^{(D-1)} \tag{7}$$

and

$$C_L / C_o = (1/D)*(1-F)^{((1/D)-1)} \tag{8}$$

Like equilibrium crystallization, fractional crystallization has little effect on incompatible element ratios, but it causes compatible elements to change even more rapidly with decreasing F. Successive melts produced by fractional melting show strong depletions in incompatible trace elements, due to the rapid removal of these incompatible elements during the first increments of melting. On the other hand, if the melts produced during fractional melting are mixed prior to sampling, the pooled product of fractional melting is nearly identical to a melt produced by equilibrium melting to the same maximum F (Figure 5*c*).

2.3.1 Trace Element Patterns

The chemical systematics presented above lead to straightforward conclusions concerning the significance of trace element variations in MORB. For example, Figure 6 shows three REE patterns for MORB, one typical of "normal" MORB, one from near a hot spot that is highly "enriched", and one that is transitional. Since the LREE are incompatible, the

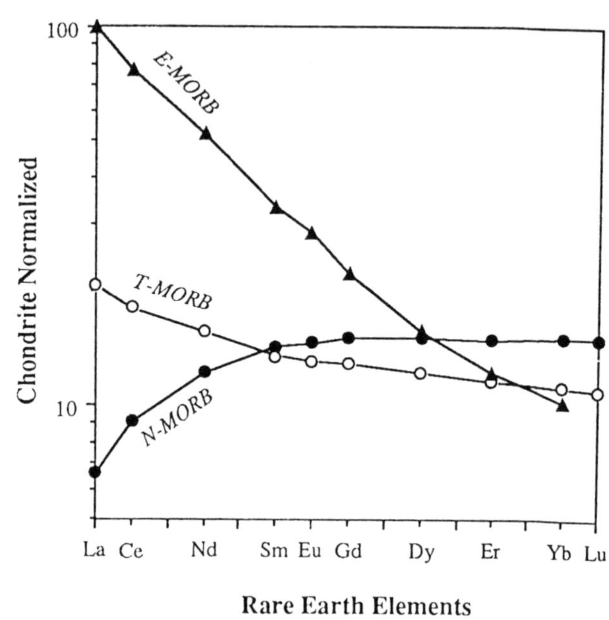

Fig. 6. Rare earth element patterns for three mid-ocean ridge basalts (MORB) from the Mid-Atlantic Ridge (MAR). Enriched E-MORB is from the Azores platform [*C.H. Langmuir*, unpublished data]; Transitional T-MORB is from the FAMOUS region [*Langmuir et al.*, 1977]; Normal N-MORB is from the MAR, between the Atlantis and Hayes Fracture Zones (MAR standard [LDGO, unpublished data]).

mantle source is depleted relative to the melts derived from it (in most cases — see *Langmuir et al.*, 1977). Therefore the LREE depletion of normal MORB requires a LREE depletion of its source. The elevated LREE of the "enriched" MORB requires a source that is enriched relative to that of normal MORB. A conventional trace element interpretation might also be that the enriched MORB is derived by lower extents of melting (leading to higher levels of the relatively incompatible lighter REE). The heavy REE depletion of the enriched MORB suggests that garnet, which preferentially retains the HREE, was a residual phase.

The smooth variations of the REE provide a context for explaining the variations of other trace elements. Two elements that have virtually identical D's maintain a constant ratio through all extents of melting, and may not be fractionated from one another during many of the processes that lead to mantle heterogeneity. For example, P and Nd, Zr and Sm, Y and the HREE, and Ti and Eu are all pairs of elements that show a relatively constant ratio in all ocean ridge basalts (Figures 7*a* and 7*b*, Table 2), which suggests that the D's for each pair of elements are very similar. P, Zr, Ti and Y are thus "analogues" for certain of the REE.

In principal, all of the trace elements could be plotted on a single diagram similar to a REE patterns, with the order determined by relative differences in their partition coefficients. These diagrams, commonly called "spidergrams," are often used to present an abundance of trace element data [e.g., *Wood*, 1979; *Sun*, 1980; *Thompson et al.*, 1983]. A spidergram is simply a REE diagram such as Figure 6, with other elements added to it. All elements more incompatible than La would be plotted to the left of La. These elements include Cs, Pb, Ba, Nb, Ta, Th, U, Rb and K. Analogue elements that behave similarly to the REE would plot in the middle of the REE pattern. Elements more compatible than Yb, such as Sc and V, would plot to the right of the heaviest REE. Spidergrams have the advantage of extending the range of D's that can be examined graphically in a single diagram, and of constructing a REE-like diagram in the absence of REE data. In practice, however, such diagrams often show irregular systematics from element to element, particularly for the most incompatible elements. Nonetheless, the principle behind these diagrams is based on the trace element equations: that a family of elements exhibit systematic relationships that depend primarily on their relative D's, provided the residual mineralogy is similar.

Trace elements provide a widely used method of classification for ocean ridge basalts. Schilling in particular has shown that basalts from near hot spots are relatively enriched in incompatible elements, and that this enrichment often decreased progressively with distance from the hot spot center. This variability in trace element enrichment has led to a simple three-fold descriptive classification for MORB. N-MORB (normal) have the characteristic depletion in the most incompatible elements, E-MORB (enriched) are enriched in incompatible elements, and are most prevalent on ridges influenced by hot spots, and T-MORB (transitional) are in between the two. On spidergrams, N-MORB have a positive slope, T-MORB have a flat slope, and E-MORB have a negative slope.

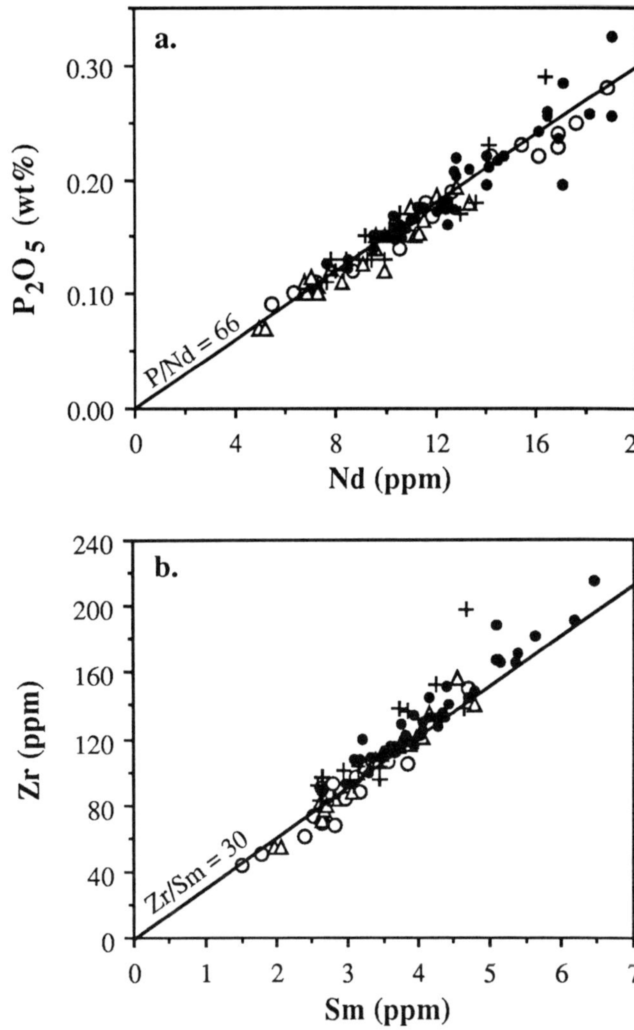

Fig. 7. Analogue element plots showing how some elements behave very similarly to certain REE during the melting and crystallization processes that occur on ocean ridges. *a*) P_2O_5 and Nd, *b*) Zr and Sm concentrations in MORBs. P and Zr are analogue elements to the rare earth elements Nd and Sm, respectively. Plusses for Australian Antarctic Discordance; open circles for Oceanographer Fracture Zone; triangles for Tamayo Fracture Zone; solid circles for CHEPR EPR data. Data sources in Table 2.

2.4 K_d's and Major Element Equilibria

Application of the simple batch melting equation to MORB data appears to be problematic because D can be a function of T, P and C_L. But at any point in a process, when T, P, and C_L are fixed, the equation applies. So if we are able to develop equations for D as a function of these variables, we can calculate compositional variations as a function of extent of melting (or crystallization).

Langmuir and Hanson [1980, 1981] showed how the combination of equations from K_d's, mass balance and mineral

TABLE 2. Analogue Element Ratios for MORB Suites

	N	Ref.	P/Nd Mean	P/Nd SD	Zr/Sm Mean	Zr/Sm SD	Ti/Eu* Mean	Ti/Eu* SD	Y/Yb Mean	Y/Yb SD
Oceanographer	19	1,2	65	5	29	2.7	6876	261	10.7	0.6
Tamayo	29	2,3	63	4	29.9	2	6939	224	10.2	0.4
CHEPR	48	4,2,5	65	5	32.1	1.7	6853	279	10.2	0.4
FAMOUS	12	4,2	69	3	30.5	3.9	6875	279	10.4	0.6
AAD	20	6	65	6	33	3.8	6695	493	10.2	0.7
Kolbeinsey	23	7	62	9	---	---	6662	447	---	---
Galapagos	10	8	59	22	28.9	2.8	6820	916	10.8	0.8
TOTAL	161		64.0	3.2	30.6	1.7	6817	101.61	10.4	.27
ID/DCP only	108		65.7	2.4	30.4	1.3	6886	26	10.4	0.24

References:
1 Shirey et al., 1987
2 Bender and Langmuir, unpublished data
3 Bender et al., 1984
4 Langmuir et al., 1977
5 Langmuir et al., 1986
6 Klein et al., 1991
7 Schilling et al., 1981
8 Perfit et al., 1983

stoichiometry can be used for the general calculation of multiphase equilibria. They showed that, given K_d's and a bulk composition, the stability, proportions and compositions of all phases in the system can be uniquely determined. Obviously, this approach allows calculation of liquid compositions during either crystallization or melting, if expressions for the K_d's are known. We present here a simple illustration of the approach, using olivine, because the constraints from olivine are important for understanding the chemical consequences of mantle melting.

The olivine formula $(Mg, Fe)_2SiO_4$ imposes a stoichiometric constraint in terms of cation fractions:

$$C_{olivine}^{MgO} + C_{olivine}^{FeO} = 0.6667 \quad (9)$$

Roeder and Emslie [1970] showed that for a limited compositional range the partition coefficients between olivine and liquid have a simple linear dependence on temperature:

$$\log K_{dMgO} = 3740/T - 1.87 \quad (10)$$

$$\log K_{dFeO} = 3911/T - 2.50 \quad (11)$$

(We will see below that the compositional dependencies become important for mantle melting. Variations of olivine K_d's with temperature, pressure, and liquid composition are presented in Appendix B). Since the temperature dependence of the two K_d's is very similar, the ratio of the two is nearly constant. This ratio, K_D, is about 0.3, and is relatively insensitive to changes in T, P or composition in the geological conditions of interest. K_D is defined as

$$K_D = [C_{ol}^{FeO}/C_L^{FeO}]/[C_{ol}^{MgO}/C_L^{MgO}] \quad (12)$$

From the definition of K_d in equation (1):

$$K_{dMgO}^{Ol/L} = C_s^{MgO}/C_L^{MgO} \quad (13)$$

$$K_{dFeO}^{Ol/L} = C_s^{FeO}/C_L^{FeO} \quad (14)$$

Substituting these expressions for C_s, and the generalized K_d expressions of equations (10) and (11) into equation (9) yields the following relation:

$$C_L^{MgO} * 10^{(A1/T)-B1} + C_L^{FeO} * 10^{(A2/T)-B2} = 0.6667 \quad (15)$$

where A1, A2, B1 and B2 are constants.

This is the equation for the olivine saturation surface first presented by *Roeder and Emslie* [1970], and provides the temperature at which any given liquid composition will be in equilibrium with olivine, and the composition of the olivine. The equation represents the olivine-liquid phase diagram, for real basaltic systems. A similar equation could also be written that would include complex compositional dependences — the K_d expressions simply become more complicated. Simultaneous solution for many phases gives phase compositions and proportions for any bulk composition at any P and T for which K_d's are known.

The olivine saturation surface is particularly useful for our purposes because olivine is the dominant mineral in the mantle, and all mantle melts are in equilibrium with olivine. Since the olivine composition in the mantle source is well known, all initial liquids must be in equilibrium with an olivine of this particular composition, and lie on the olivine saturation surface at the P and T of interest. The fact that K_D is constant requires (equation (12)) that the Fe/Mg ratio of the liquid is fixed. Figure 8 presents a simplified illustration of the olivine

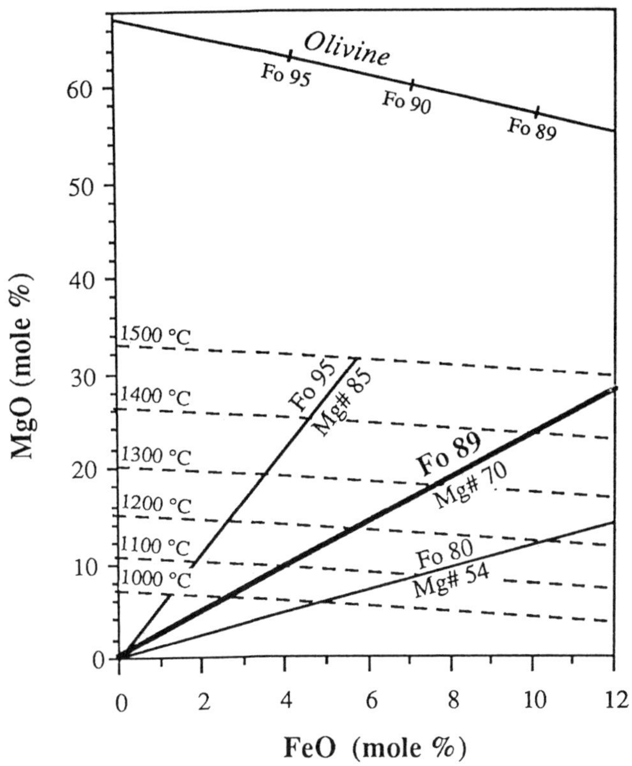

Fig. 8. Olivine saturation surface using the data of *Roeder and Emslie* [1970] and after *Langmuir and Hanson* [1981]. The forsterite (Fo) compositions above the lines radiating from the origin are olivine compositions that coexist with liquid compositions along the line (Mg# = MgO/[MgO+FeO] in cation mole%).

saturation surface. Melts in equilibrium with a mantle containing Fo_{89} olivine must plot on the line labelled Fo_{89}. Higher pressure melts of the mantle plot along the Fo_{89} melt line toward higher MgO and FeO contents. Crystallization of olivine from the melt leads to almost vertical trends on this diagram (until very high degrees of crystallization); therefore, provided olivine is the only mineral that has crystallized en route to the surface, liquids formed at higher pressures remain higher in FeO than those formed at lower pressures.

These simple principles of element partitioning provide a useful framework in which the observed variations of some elements in oceanic basalts can be understood. The behavior of the elements in more complex and open system models will be considered further in section 4.0. A more sophisticated approach that allows calculation of FeO and MgO in mantle melts for a variety of conditions and processes will be discussed at length in section 5.0.

3. OBSERVATIONS

3.1 Influence of Low Pressure Fractionation on MORB Compositions.

The chemical changes that occur during low-pressure crystallization are well documented by abundant experimental data [e.g., *Bender et al.*, 1978; *Walker et al.*, 1979; *Fisk et al.*, 1980; *Grove and Bryan*, 1983]. Computer programs are available that use the experimental data to calculate the changes in melt composition produced during crystallization, also called the liquid lines of descent [LLD; *Langmuir and Hanson*, 1981; *Ghiorso et al.*, 1983; *Nielsen*, 1985; 1990; *Weaver and Langmuir*, 1990; *Longhi*, 1991]. The programs generally are calibrated using a large experimental data base from many laboratories, and problems of interlaboratory biases commonly lead to errors in the temperatures of phase appearances of 10°-15°C. Such errors make the programs useful for general considerations of crystallization, but inadequate for detailed petrogenetic modeling of an individual rock suite. For this latter purpose, the program must be "tuned" to the data set of interest, preferably by having experiments on the same samples, analyzed in the same laboratory as the experiments themselves. Many of the programs have an advantage over experiments, however, in that bulk composition is preserved, and there are no kinetic effects. Thus LLDs calculated on the basis of experimental results may more accurately reflect crystallization processes than the experiments themselves.

Examples of 1 atm LLDs calculated for a high MgO sample from the East Pacific Rise near the Clipperton transform fault using the programs of *Weaver and Langmuir* [1990] and *Nielsen* [1985] are shown in Figure 9. The LLD's generally track the trends of the data, suggesting that these samples are related, to first order, by low-pressure fractionation.

The fit of both calculated LLDs in Figure 9 to the basalt data is particularly good for elements like Na_2O and TiO_2 that are incompatible with respect to the fractionating phases. Elements whose variations are strongly linked to a single mineral phase (such as Ca to clinopyroxene or Al to plagioclase) are more sensitive to the detailed fit of the calculated LLD. Note, for example, that both calculated LLDs in Figure 9 show a marked decrease in CaO with the arrival of clinopyroxene on the liquidus, producing a kink in each trend. The two LLDs differ in the precise position of the kink on this diagram due to differences in the clinopyroxene K_d relationships in the two programs. The offset, however, represents only a 12°C difference in the appearance of clinopyroxene on the liquidus, and this temperature difference is within the combined uncertainties in the experimental and analytical measurements that direct the programs. The fact that a mere 12° difference in the appearance of a phase can significantly affect the systematics of sensitive elements is the reason that detailed modeling of data from a particular region requires a program that is well calibrated by experimental data for those particular samples. This example underscores the importance of "tuning" the LLD program to the data set of interest for detailed petrogenetic modeling.

These diagrams illustrate some of the simplest and yet most important observations concerning MORB. The first is that samples from individual regions usually form rather smooth trends on plots of elemental (or oxide) abundances as a function of MgO. Petrologists often plot elemental variations as a function of MgO (or Mg# = Mg/(Mg+Fe) in cation mole %)

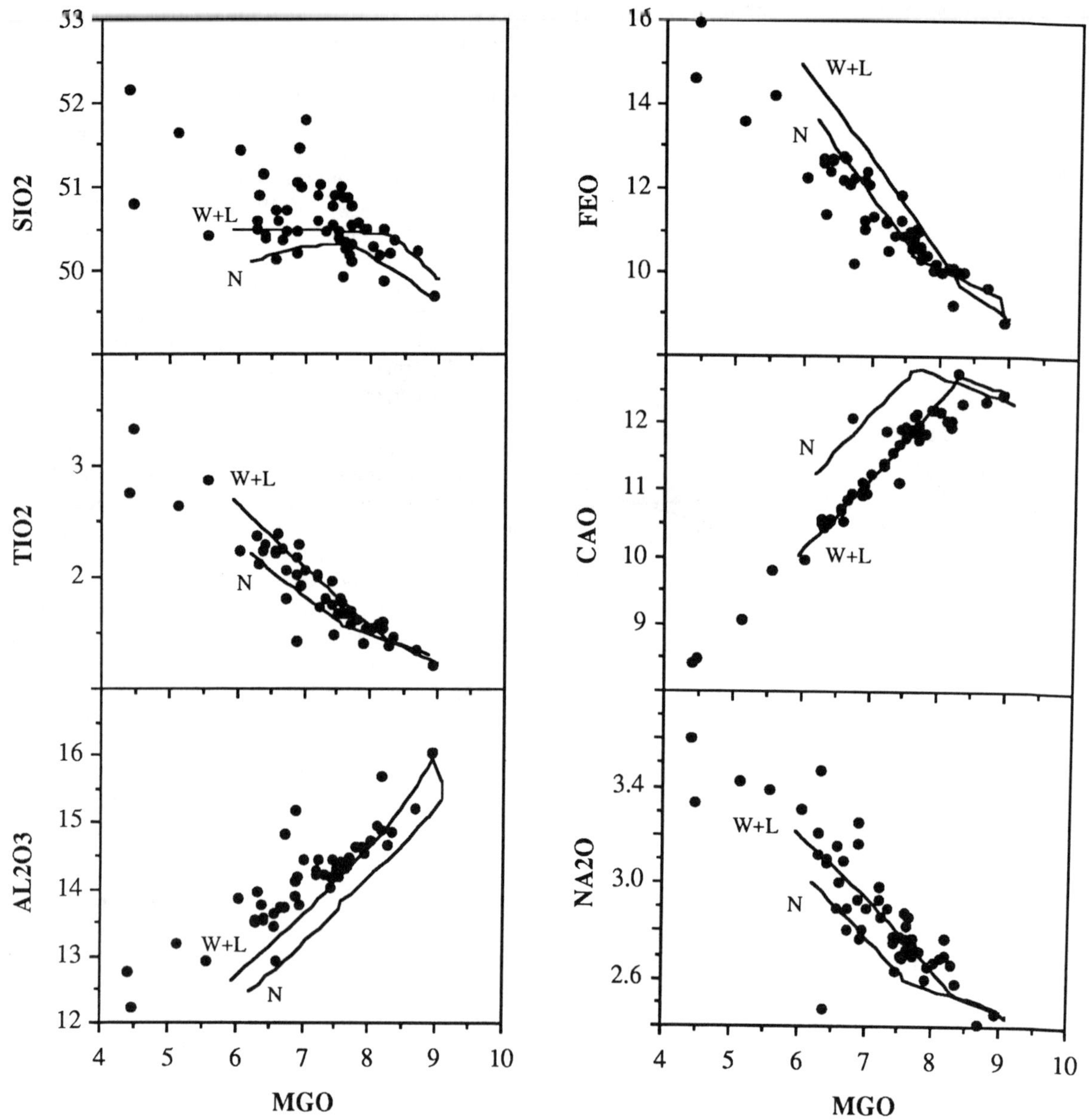

Fig. 9. MgO variation diagrams for microprobe data for basalt glasses from near the Clipperton Fracture Zone [*J. F. Bender and C. H. Langmuir*, unpublished data]. Lines are calculated liquid lines of descent (LLD) using the programs of *Weaver and Langmuir* [1990] (labeled W+L) and *Nielsen* [1985] (labeled N) for the primitive composition at around 9 wt% MgO. A more recent version of Nielsen's LLD program can be found in *Nielsen* [1990]. Oxides in wt%.

because MgO is compatible in olivine and therefore decreases during cooling and crystallization at low-pressure. Indeed, there is a simple relationship between MgO contents and magmatic temperature, as shown in Figure 10. Rocks with the highest MgO are the highest temperature magmas and so represent compositions least modified from primary mantle melts. However, as Figures 9 and 10 make clear, the vast majority of recovered basalts have experienced substantial cooling since their generation within the mantle, and this cooling has produced changes in the concentrations of all the elements. These chemical changes must be corrected for if we are to see through the substantial chemical variability caused by differentiation processes at low pressure.

Examination of data from many regions of the ocean floor shows the ubiquitous influence of the low-pressure fractionation signature on MORB compositions. Figures 11*a* and 11*b* shows MgO and Na_2O variations in samples from five regions that encompass most of the global range in Na_2O contents for MORBs. The data from the five regions form sub-parallel trends, and within each region, the Na_2O contents

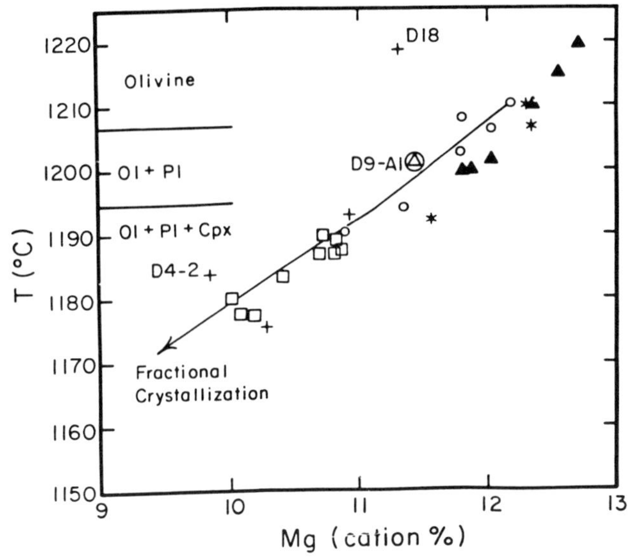

Fig. 10. Calculated liquidus temperatures for basalt glasses from the Tamayo region of the East Pacific Rise vs. MgO. The LLD for one of the samples is shown for reference. LLD, liquidus temperatures, and phase appearances are calculated from the method of *Weaver and Langmuir* [1990]. Figure is from *Bender et al.* [1984].

increase gradually with decreasing MgO contents. The trends of the data for each region are similar to calculated low-pressure LLDs.

In addition to the changes in melt composition produced during low-pressure crystallization, it is clear from the data presented in Figures 11*a* and 11*b* that there are at least two other aspects of chemical variability. First, the five regions form sub-parallel trends, so that at a given MgO content, each region shows a distinct and characteristic narrow range in Na_2O contents. Anticipating a topic that will be discussed in greater detail below, these differences in Na_2O contents (and other elements) at a given value of MgO from region to region along the mid-ocean ridge system appear to reflect global variations in parental magma composition. Second, it is apparent that even in the data from a single region, there is variation in Na_2O (and other elements) at a single value of MgO. While some component of this local variability may result from analytical error, there are often compositional systematics to this variability as well (shown below).

To summarize, we can divide the observed compositional variations into three categories. First, there is the variability that appears to result largely from low-pressure crystallization and varies systematically with MgO. Second, among samples collected from an individual ridge segment, there is *local* chemical variability that cannot be accounted for by low-pressure crystallization. Last, there is *global* chemical variability manifest as differences in average compositions (at a given MgO content) from region to region. Figure 12 illustrates the three levels of compositional variability: the lines show low pressure fractionation, the shaded boxes show the local variability at a given value of MgO, and the two open circles show the average difference between the two regions at a given value of MgO. Because the chemical changes produced by low-pressure crystallization obscure the *local* and *global* chemical systematics, it is necessary to correct the basalt data for the effects of differentiation. In essence, we want to determine the residual signal left after low-pressure differentiation. It is this residual signal that must then be accounted for either by some more complex fractionation process, by processes of melt generation and segregation, or by source heterogeneity.

3.2 Correction of Data for Low Pressure Fractionation

The existence of sub-parallel trends for the Na_2O LLDs (e.g., Figures 11*a* and 11*b*) provides a ready basis for "correcting" the data for the effects of low pressure fractionation. Figure 11*a* illustrates the method used by *KL*87 where Na_2O values are extrapolated along a constant slope to 8 wt% MgO (the calculated value of Na_2O at 8 wt% MgO is called $Na_{8.0}$). This constant slope is meant to approximate the slope of the olivine+plagioclase±clinopyroxene LLD. In this way, samples that lie along a perfectly linear fractionation trend of this slope would collapse to a single value of $Na_{8.0}$. Figure 11*a* shows the slope of the line used by *KL*87 to calculate $Na_{8.0}$, superimposed on five MORB suites. The constant slope method is clearly not appropriate for all data sets. For some it approximates the trend of the data and so a calculated value of $Na_{8.0}$ is similar to the real Na_2O contents of samples with 8 wt% MgO. For other regions, however, there are substantial differences between the slopes of the real data, and the slope used in the algorithm, so that calculated values of $Na_{8.0}$ give inaccurate estimates of Na_2O at 8 wt% MgO.

One problem with the constant slope used to normalize the data is that the slope of a fractionation trend varies naturally as a function of the composition of each MORB suite and is steeper with increasing Na_2O contents from region to region (Figure 11*b*). *KL*87 recognized this problem for the low Na_2O samples from the Kolbeinsey Ridge, which fall on a shallower trend than that assumed in the $Na_{8.0}$ calculation, but did not take this differences into account. Thus the $Na_{8.0}$ values calculated for fractionated samples from the Kolbeinsey Ridge are lower than the real Na_2O contents of samples with 8 wt% MgO.

Another problem with the constant slope used for fractionation correction is that an abrupt change in slope often occurs when a new mineral joins the fractionating assemblage. For example, there is a slight kink in Na_2O-MgO LLDs where clinopyroxene joins olivine and plagioclase as a fractionating phase (Figure 11*b*). This change in slope is apparent in some of the natural suites as well; samples above 8.0 wt% MgO from the ridges adjacent to the Kane Fracture Zone, for example, lie on a shallow slope reflecting only olivine and plagioclase crystallization (Figures 11*a* and 11*b*). Because the $Na_{8.0}$ equation from *KL*87 was based on a steeper olivine-plagioclase-clinopyroxene fractionation trend, the $Na_{8.0}$ values calculated from this equation will be erroneously high for the high MgO Kane samples.

Thus, the variations in slope predicted by both changes in starting composition and position along the LLD could lead to

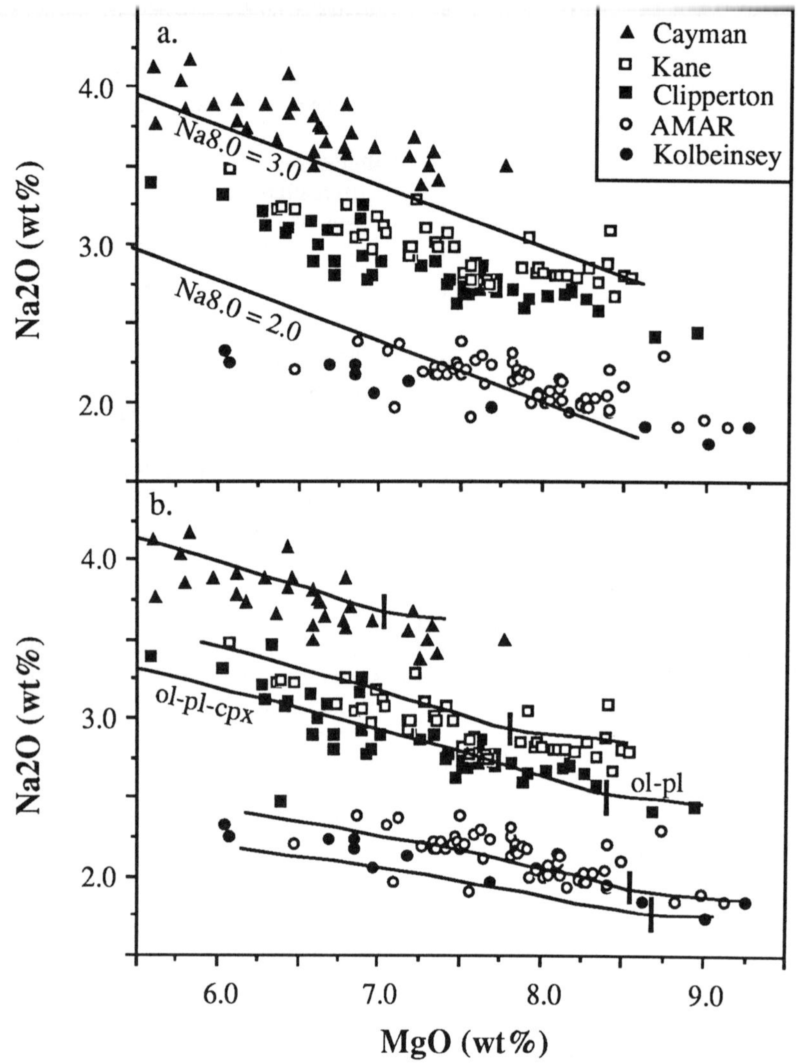

Fig. 11. *a)* MgO vs. Na_2O in MORBs from five regions. Samples are from the Mid-Cayman Rise, from near the Kane Fracture Zone at 23.22-23.67°N on the Mid-Atlantic Ridge (MAR), from the AMAR region at 36.42-36.79°N on the MAR, from near the Clipperton Fracture Zone on the EPR, and from the Kolbeinsey Ridge north of Iceland. Data from the Smithsonian Catalogue of Basalt Glasses [*Melson et al.*, 1977; and pers. comm.], except for Clipperton data (as in Figure 9), and Kolbeinsey data [*Schilling et al.*, 1983]. Lines show slope used by *Klein and Langmuir* [1987] to calculate $Na_{8.0}$ (Na_2O at 8% MgO). *b)* Same data with calculated LLDs for a primitive composition from each of the five suites using the program of *Weaver and Langmuir* [1990]. Vertical bar indicates the point at which clinopyroxene (cpx) joins olivine and plagioclase (ol-pl) as a fractionating phase, at which point the slope of the LLD steepens.

errors in $Na_{8.0}$ values calculated assuming a constant slope. It is difficult, however, to develop a single algorithm that takes into account these effects. One way to improve upon the constant slope method would be to use the LLD program to develop a $Na_{8.0}$ algorithm specific to the region of interest. For example, in modeling samples from the Kane region, different slopes could be adopted on either side of the clinopyroxene-in kink at 8% MgO. Samples with greater than 8% MgO can be projected to 8% MgO along the shallow olivine-plagioclase slope, and the samples with less than 8% MgO along the steeper olivine-plagioclase-clinopyroxene slope. Because local variability is defined as the deviation from a low pressure LLD, it is imperative that one pay particular attention to the details of the LLD in modeling local variations [*Klein and Langmuir*, 1989].

Will a more accurate $Na_{8.0}$ correction affect the global $Na_{8.0}$ variations as well? In order to assess any artifacts in the $Na_{8.0}$ global averaging that may have resulted from the constant slope method used by *KL*87, we have recalculated $Na_{8.0}$ averages for the five suites shown in Figures 11*a* and 11*b* using the slopes and kinks given in the LLDs calculated for those regions. These results are shown in Figure 13, along with the

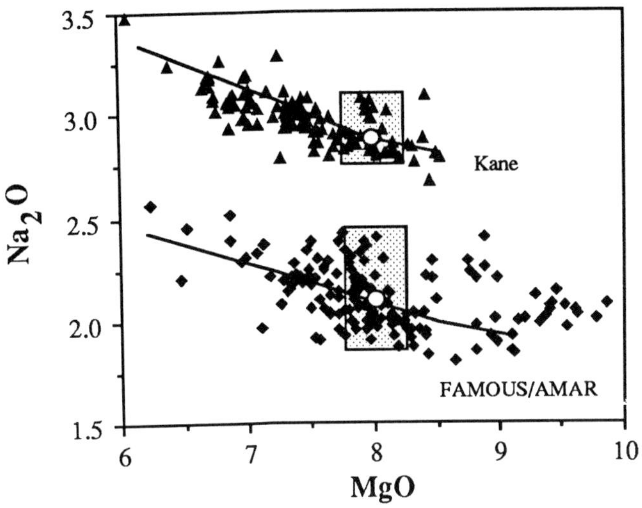

Fig. 12. MgO vs. Na_2O for Kane and FAMOUS/AMAR basalt glasses. Data from Smithsonian Catalogue of Basalt Glasses [*Melson et al.*, 1977; and pers. comm.]. Lines are LLD's for one composition from each of the suites. Shaded bar is variability in each of the suites at ~ 8 wt% MgO. Open circle is the average Na_2O content at 8 wt% MgO for each suite. Thus there are three levels of variability in the data: variability due to low-pressure crystal fractionation, local variability at a constant MgO value, and average differences among different regions that contribute to a global signal.

original determinations of $Na_{8.0}$ using the *KL*87 method. All of the new $Na_{8.0}$ averages are higher than the original ones. This is because the constant slope (0.373) used is often too steep for the data, and so overcorrects the Na_2O values of fractionated samples. For example, the constant slope of 0.373 is too steep for both the entire Kolbeinsey suite, and the region of the Cayman LLD above 7% MgO where only olivine and plagioclase fractionate. Taking into account these complexities for the Kolbeinsey and Cayman LLDs, leads to higher average $Na_{8.0}$ values for both regions.

The change in the $Na_{8.0}$ averages using the new and more accurate correction scheme, however, is typically only 0.2 wt% Na_2O, and most importantly, the relative changes among the regions are essentially the same. Because these effects are small and systematic, we have continued to use the *KL*87 correction scheme for global averages in this paper. The local systematics, however, have been investigated using the LLD appropriate for each particular region.

The method presented above for correcting Na_2O values for fractionation can be developed for any element that varies predictably with differentiation such as FeO and SiO_2 (see *KL*87 and *Klein and Langmuir*, 1989, for the algorithms used to calculate $Fe_{8.0}$ and $Si_{8.0}$). Since incompatible trace elements vary inversely with F, and are insensitive to the appearances of specific mineral phases, an equation can be derived simply from the relationship between F and MgO given by the LLD program to correct highly incompatible elements for fractionation. Such an equation for Ce, for example, is:

$$\log(Ce_{8.0}) = \log(Ce) - 0.11*(8-MgO) \qquad (16)$$

3.2.1 Interlaboratory Bias

Different laboratories calibrate their analyses to different standards, which often leads to a significant and systematic offset between analyses from different laboratories. Because these interlaboratory differences can be large, it is important to apply corrections to the various data in order to make them consistent with one another. The largest number of analyses of MORB have been determined by electron microprobe at the Smithsonian Institution in the laboratory of William Melson. Two other important sources of data for this paper are the analyses from Jean-Guy Schilling at the University of Rhode Island, and more recent analyses at Lamont.

Two independent approaches provide information on the necessary interlaboratory correction factors. There has been sufficient overlap in the samples analyzed by electron microprobe at the Smithsonian, and by microprobe and direct current plasma spectrometry at Lamont to determine systematic biases between these laboratories [e.g., *Klein et al.*, 1991]. This leads to correction factors for each element that can be applied to individual analyses. These factors cause $Na_{8.0}$ and $Fe_{8.0}$ calculated from Lamont data to be respectively 6-8% and 3-7% higher than the same samples measured at the Smithsonian.

It is also possible to compare regional averages for groups of samples determined by the different methods. This latter approach can also be used to evaluate the Rhode Island data. Figure 14 compares regionally averaged data from seven different ridge segments. Both the Lamont and Rhode Island data are similarly offset from the Smithsonian data. The offsets in the regional averages are similar to those calculated from the elemental correction factors for individual analyses (as above). In the following presentation of MORB data, we have applied

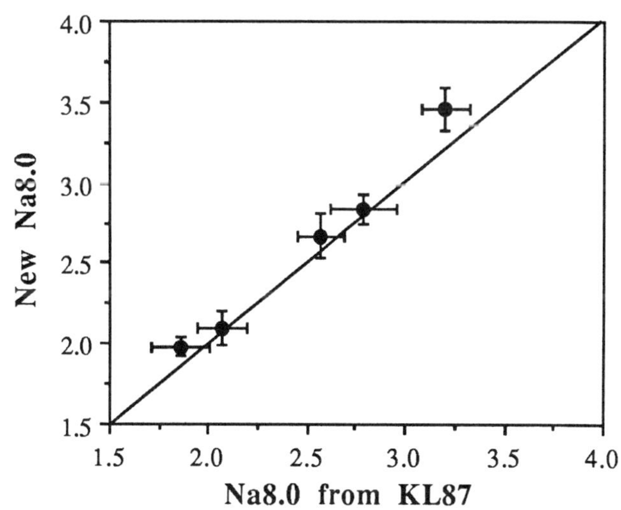

Fig. 13. $Na_{8.0}$ for the five suites shown in Figure 11. The original values given in *Klein and Langmuir* [1987] are compared to those calculated by taking into account the slopes of the individual LLDs shown in Figure 11*b*. Each point is an average of the $Na_{8.0}$ values for each region, and the error bars are one standard deviation of the $Na_{8.0}$ values.

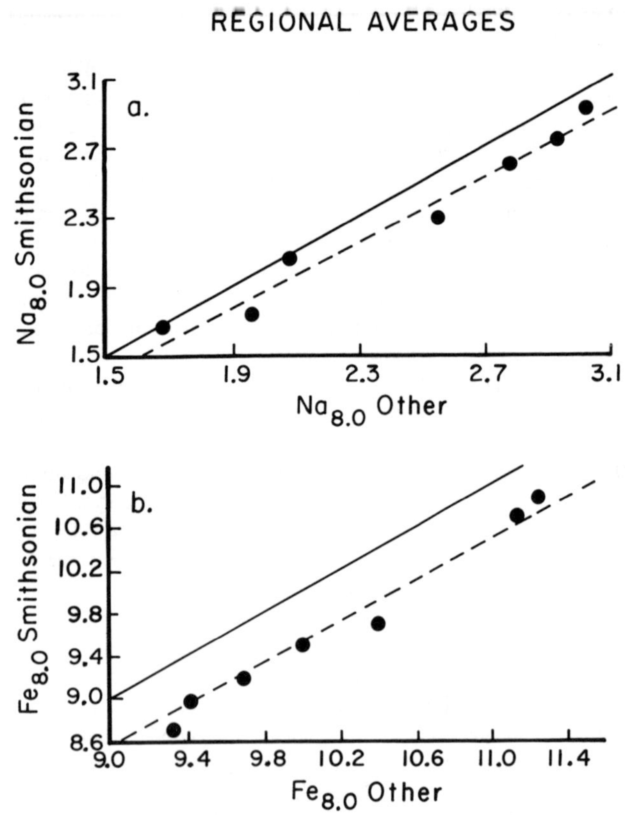

Fig. 14. A comparison of regional mean *a*) $Na_{8.0}$ and *b*) $Fe_{8.0}$ calculated from measurements by different laboratories. The individual samples are not identical in all cases, so the comparison includes sampling biases as well as calibration differences among the laboratories and analytical methods. "Smithsonian" data is by microprobe analysis [e.g., *Melson et al.*, 1977]. "Other" data is by microprobe or direct current plasma emission spectrometry at Lamont [e.g., *Klein et al.*, 1991] or by classical chemical methods at J.-G. Schilling's laboratory at the University of Rhode Island [e.g., *Schilling et al.*, 1983]. The data points, which span most of the global range, are from the following regions: Reykjanes Ridge (2), EPR near the Tamayo Transform, S. Kane, FAMOUS, and the Australian Antarctic Discordance (2). Solid line is the line for perfect agreement; dashed line through the data is a line for a constant percentage change. The interlaboratory correction factor is 7.5% for $Na_{8.0}$ and 5% for $Fe_{8.0}$.

corrections to analyses such that they are consistent with Lamont and Rhode Island analyses. Applying these corrections reveals systematics in the data that are not apparent in the uncorrected data.

3.3 Global Variability of Major Elements in MORB

Our examination of the major element systematics observed among mid-ocean ridge basalts begins with the systematics observed on a global scale, for regionally averaged data. For more than a decade, petrologists have noted systematic differences in MORB major element chemistry from region to region. Early, predominantly reconnaissance sampling led to the classification of MORB into broad petrographic provinces, with particular emphasis on compositional differences between MORB recovered near hot spots and those recovered far from hot spots, as well as differences among hot spots [e.g., *Miyashiro et al.*, 1969; *Melson and O'Hearn*, 1979; *Morel and Hekinian*, 1980; *Sigurdsson*, 1981; *Schilling et al.*, 1983; *Dmitriev et al.*, 1984]. With further sampling and study, differences in parental magma composition were recognized even among basalts recovered far from hot spots [e.g., *Langmuir and Hanson*, 1981; *Bryan and Dick*, 1982; *Langmuir and Bender*, 1984]. *Dick et al.* [1984] examined basalts from thirteen ridge segments in the Atlantic and Indian Oceans and suggested that they exhibited regional differences in parental magma composition as a result of variations in the extent of melting.

*KL*87 extended the investigation of global systematics of major element variations in basalts to include all ridge segments for which data were available as of about 1986. *KL*87 normalized the Na_2O and FeO data to a constant value of MgO in order to correct for low-pressure fractionation, as described above, and averaged the data from each region. They found that regional averages of $Na_{8.0}$, $Fe_{8.0}$ and CaO/Al_2O_3 correlated with the mean axial depth of the ridge, with the exception of data for samples collected very close to some hot spots. Application of interlaboratory corrections to the global data set and the availability of some new data clarifies important aspects of the global correlations (Figure 15). Each data point is an average of samples from approximately 100 km of ridge length. Although all data points together exhibit rough correlations between the chemical parameters and axial depth, it is clear from the figure that a useful separation can be made between (1) basalts from all normal ridges; (2) basalts from back-arc basins; and (3) basalts from ridge segments that have been influenced by certain hot spots (Azores, Jan Mayen, Bouvet, Tristan/Gough, Galapagos). For the normal ridges, $Na_{8.0}$, $Fe_{8.0}$ and axial depth correlate strongly. We call these relationships the "global correlations" since they apply to the global system of ridges that have not been substantially influenced by hot spots. Back-arc basins as a group also show correlations among $Fe_{8.0}$, $Na_{8.0}$ and depth, but the data are systematically offset to lower $Fe_{8.0}$ in comparison to normal ridges. Some ridges influenced by hot spots have anomalous major element compositions in comparison to normal ridges (i.e., low $Fe_{8.0}$ and sometime $Na_{8.0}$). For these ridges, there are also different relationships between chemistry and axial depth than normal ocean ridges, as discussed further below.

Careful examination of the regionally averaged data reveals finer distinctions within the global correlations. Basalts from the Indian Ocean have systematically lower $Fe_{8.0}$ (Figure 16*a*) and higher $Na_{8.0}$ than basalts from the Atlantic for a given depth. Normal sections of the EPR and the northern MAR are distinct-- basalts from the MAR between 23° and 32°N are high in $Fe_{8.0}$ for a given $Na_{8.0}$ (Figure 16*b*). It appears that substantial portions of the ocean ridge system have their own chemical signature, which is a cause of the width of the overall field of the global correlations.

The discussion above shows that the effects of hot spots, back-arc basins and regional variability lead to noise in the

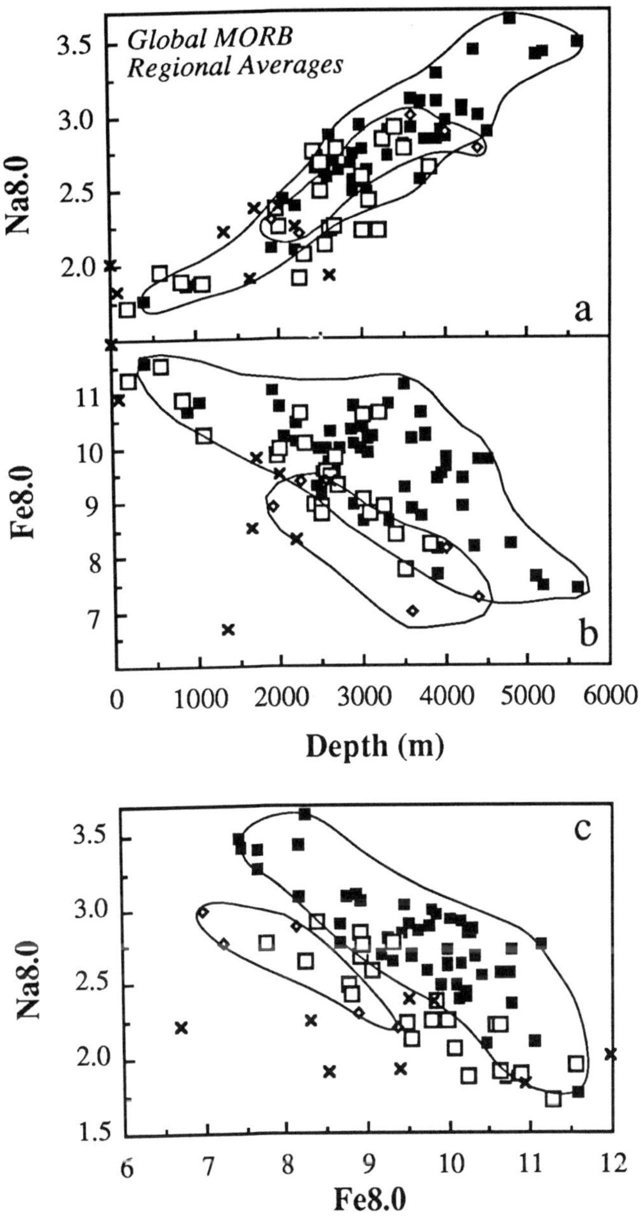

Fig. 15. Global correlations between regional averages of smoothed axial depth and the $Na_{8.0}$ and $Fe_{8.0}$ of MORBs. This figure is the same as Figure 2 in *KL*87, except all analyses have been recalculated to sum to 100% (all iron as FeO) and have been corrected for interlaboratory differences. Corrections were applied to all analyses determined at the Smithsonian Institution [*Melson et al.*, 1977] or that were reported to be consistent with the Smithsonian data [i.e., *Schilling et al.*, 1985], based on the factors reported in *KL*89 (their Figure 5), so that all data are consistent with those determined at Lamont and by wet chemistry at URI. Data sources as in *KL*87, with updates from the most recent Smithsonian Catalogue of Basalt Glasses [*Melson*, pers. comm.], *Batiza et al.* [1988], *Fryer et al.* [1990], and LDGO unpublished data from 23-32°N on the MAR. Different groups of MORB are distinguished: solid boxes are MORB from "normal" ridge segments; diamonds are from back-arc basins; open squares are from ridges influenced by the Galapagos, Azores, Jan Mayen, Tristan, Iceland, and Bouvet hot spots; Xs are from ridge segments immediately adjacent to these hot spots. Note that Iceland and adjacent ridges are also coincident with the normal MORB array.

global data set, but that global systematics emerge for different subsets of the data. For TiO_2, the regional effects are even more pronounced, so that the data set as a whole does not show obvious systematic relationships with other chemical parameters or with depth (Figure 17). However, separation of the data into different groups does clarify the $Ti_{8.0}$ observations. Regional averages of $Ti_{8.0}$ for normal ridges exhibit a positive correlation with $Na_{8.0}$, and hence with depth. Ridges influenced by hot spots show variable behavior. Basalts from Iceland are very high in TiO_2, which is characteristic of many hot spots world wide [e.g., *Feigenson et al.*, 1983; *Albarede and Tamagnan*, 1988]. Some other hot spot-influenced ridges (e.g. MAR near the Azores), however, have lower TiO_2 than normal ridges. Back-arc basins are all

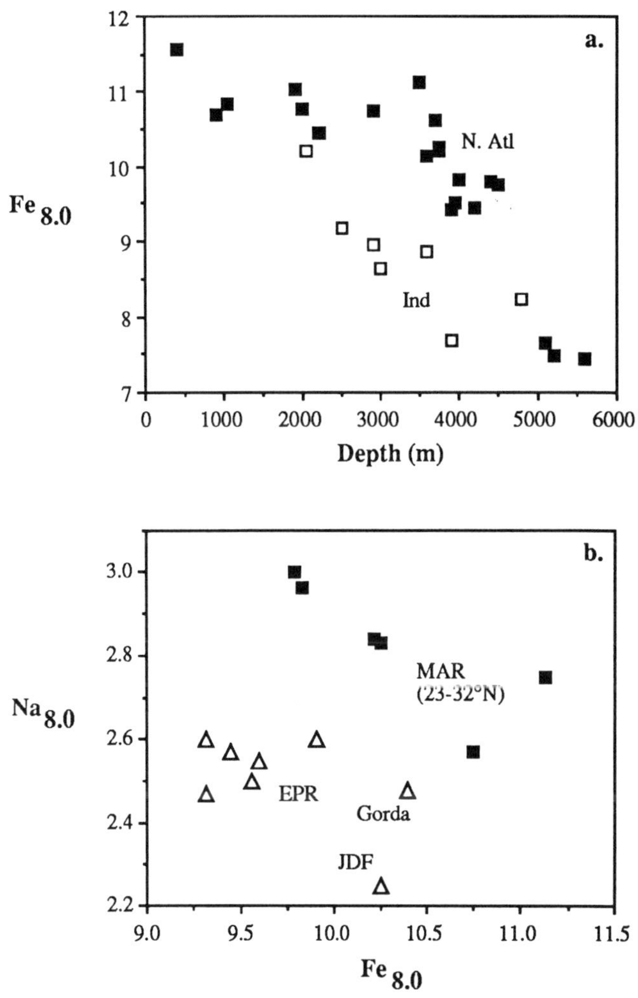

Fig. 16. Systematics within the "normal" ridge data for different ocean basins, as in Figure 15. *a*) Indian Ocean MORB are off-set to lower $Fe_{8.0}$ (shown) and higher $Na_{8.0}$ (not shown) than N. Atlantic MORB for the same axial depth. *b*) There is a systematic offset between N. Atlantic and Pacific MORB. These subtle inter-ocean differences contribute to the observed "width" of the global correlations for normal ridges of Figure 15.

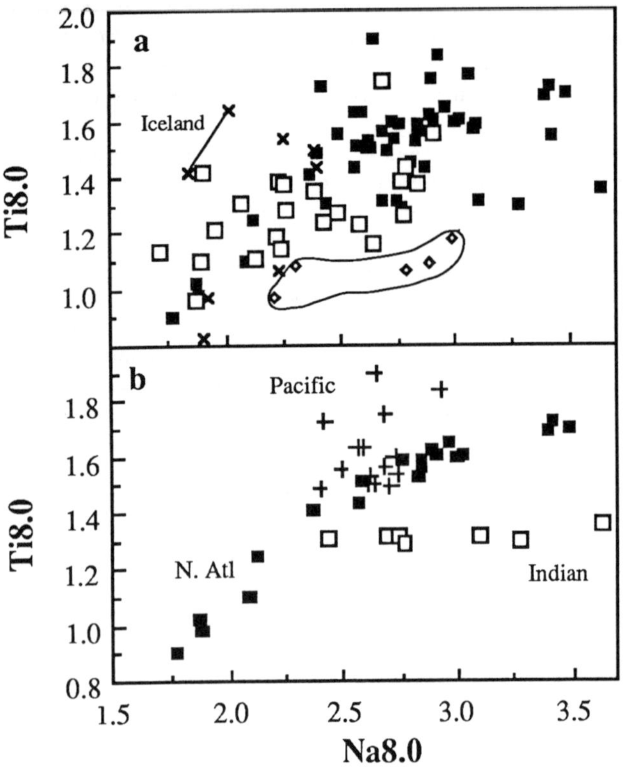

Fig. 17. Regional averages for $Ti_{8.0}$ in MORB. *a*) Global variations in MORB. Symbols as in Figure 15. *b*) Systematics within the "normal" MORB data for different ocean basins. Data sources as in Figure 15. $Ti_{8.0}$ calculated for samples with 5 - 8.5 wt% MgO from: $Ti_{8.0} = TiO_2 + 0.351(MgO-8)$.

low in TiO_2. Within the normal ridge population, MORB from Indian Ocean Ridges are low in TiO_2, and MORB from the Pacific are slightly high in TiO_2 (Figure 17*b*). $Ti_{8.0}$ in basalts from the North Atlantic correlates very well with both depth and $Na_{8.0}$. Thus $Ti_{8.0}$ appears to show similar global systematics to $Na_{8.0}$, but is strongly affected by the differences among different hot spots, by the back-arc environment, and by regional differences.

One of the most important aspects of the global correlations for normal ridges (Figure 15) is that they relate the composition of the ocean crust to a fundamental physical characteristic of the ridge: axial depth. In addition, *KL*87 showed that $Na_{8.0}$ correlates roughly with crustal thickness, a result that remains with the inclusion of more recent data (Figure 18). The relationship with crustal thickness shown in Figure 18 relies to some extent on Iceland and the Mid-Cayman rise (the thick and thin end-members), and the Mid-Cayman Rise crustal thickness value is inferred from geologic rather than seismic constraints [*Stroup and Fox*, 1981]. However, seismic data from the northern part of the Kolbeinsey Ridge [*Evans and Sacks*, 1979; unintentionally omitted from the compilation in *KL*87] suggest a crustal thickness of 15 km, confirming the association of low $Na_{8.0}$ values with anomalously thick crust, and data from the Arctic Ridge suggests the association of very thin crust with deep ridges.

In order to better fully characterize the relationship between crustal chemistry and crustal thickness, it is important that crustal thicknesses be determined in other regions, particularly those that are anomalously shallow or deep. One needs to bear in mind, however, the intrinsic limitations of seismic measurements of "crustal thickness." Ultramafic cumulates are petrologically part of the crust, and yet they may have high seismic velocities similar to mantle values. Crust with a significant thickness of such cumulates may thus appear anomalously thin seismically. There is also evidence in the Atlantic for substantial disruption of crust, and the presence of abundant serpentinized peridotite at high crustal levels [e.g., *Karson et al.*, 1987; *Bougault*, personal communication]. In this case, seismic "crustal thickness" may reflect tectonic disturbance or depth of hydrothermal alteration rather than the volume of magma derived from the mantle.

3.4 Regional Variability Around Hot Spots

Certain hot spots influence the chemical composition of ridges so that they do not show the same relationship between

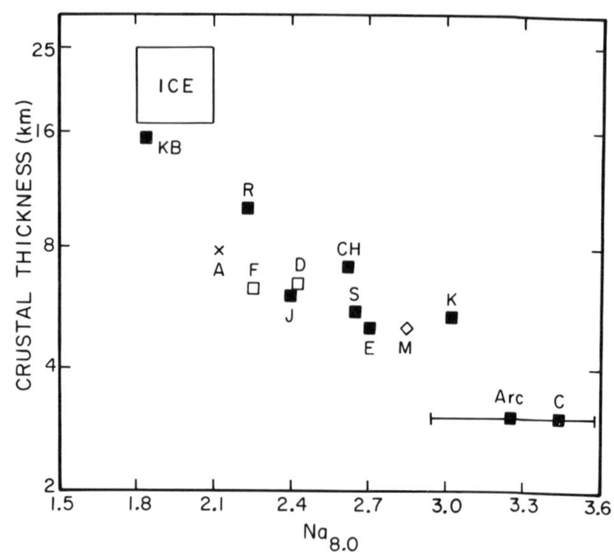

Fig. 18. Regional averages of $Na_{8.0}$ vs. crustal thickness. Data and references are the same as in *Klein and Langmuir* [1987] (their Figure 14) except Smithsonian data has been corrected for interlaboratory bias as in Figure 14 and new data points have been added. Kolbeinsey crustal thickness is from *Evans and Sacks* [1979]. The crustal thickness for the Reykjanes Ridge has been increased to 10 km based on considerations of the off-axis profile of *Bunch and Kennett* [1980], since this reflects fully mature crust. For the Arctic Ridge, there is a crustal thickness estimate [*Reid and Jackson*, 1981; *Jackson and Reid*, 1980] but no $Na_{8.0}$ data. $Na_{8.0}$ is estimated from the global $Na_{8.0}$-depth correlation (Figure 15). The error bar on the point represents the variation in $Na_{8.0}$ that would be possible across the entire global array for a mean depth for the Arctic Ridge of 4500 m [*IHO/IOC/CHS*, 1984]. Other regions shown are I (Iceland), R (Reykjanes Ridge), F (MAR at 37°N), A (Azores region), D (MAR at 45°N), J (Juan de Fuca), S (EPR near Siqueiros transform), E (East Pacific Rise), K (MAR, Kane transform), M (Mariana Trough), CH (5-14°N on EPR), KB (Kolbeinsey Ridge), Arc (Arctic ridge), C (Cayman Trough).

chemistry and axial depth as normal ocean ridges. This is apparent for MORB from the Galapagos platform (Figure 19). Note that these data are for individual locations rather than the regional averages discussed above. $Na_{8.0}$ increases, $Fe_{8.0}$ decreases, and depth shallows systematically towards the center of the hot spot. $Na_{8.0}$ and $Fe_{8.0}$ correlate negatively with a similiar slope to the global variability, but the relationship with depth is the opposite of that observed globally. The shallowest regions have the lowest $Fe_{8.0}$ and the highest $Na_{8.0}$ (Figure 19).

Not all hot spots, however, exhibit this anomalous relationship. For data from the Reykjanes Ridge, south of Iceland (Figure 20), $Na_{8.0}$ decreases and $Fe_{8.0}$ increases as the depth shallows. This is consistent with the overall global relationships among these parameters, although $Na_{8.0}$ may increase slightly over the Iceland hot spot.

As pointed out above, TiO_2 may also show a distinctive hot spot effect. The positive correlation between $Ti_{8.0}$ and $Na_{8.0}$ for normal ridges may breakdown near high-TiO_2 hot spots such as Iceland. Indeed, basalts from Iceland and the northern Reykjanes Ridge have among the highest TiO_2 and lowest $Na_{8.0}$ of any MORBs (Figure 17).

These observations from near hot spots indicate that major element chemical effects associated with some hot spots are

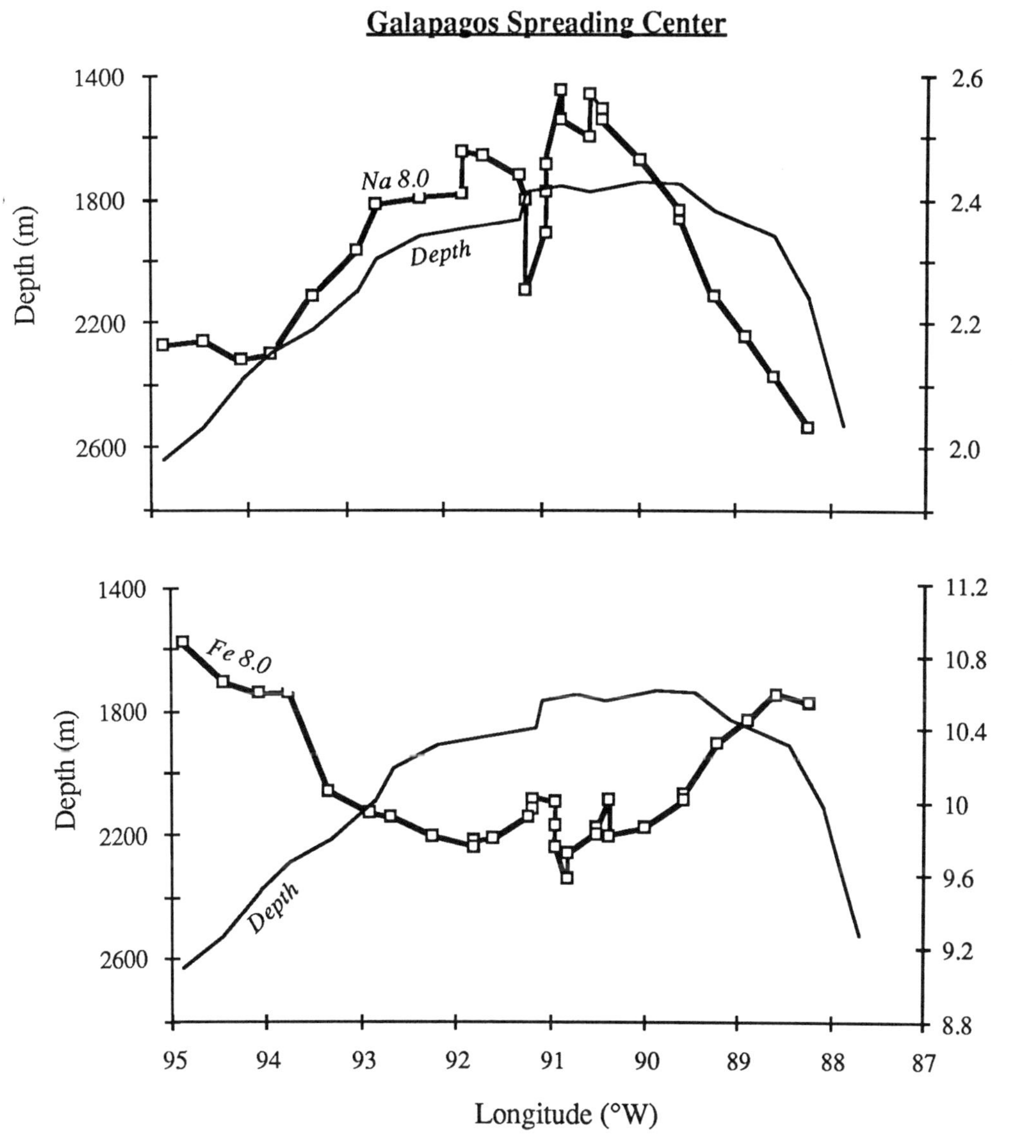

Fig. 19. Variations in basalt composition and axial depth across the Galapagos platform. Microprobe data from *Fisk et al.* [1982] and depths from *Schilling et al.* [1982]. Points plotted are five-point running averages of the data, in order to give a regional perspective and smooth out the smallest scale noise from individual dredges. The Galapagos Islands themselves are between 90°W and 92°W.

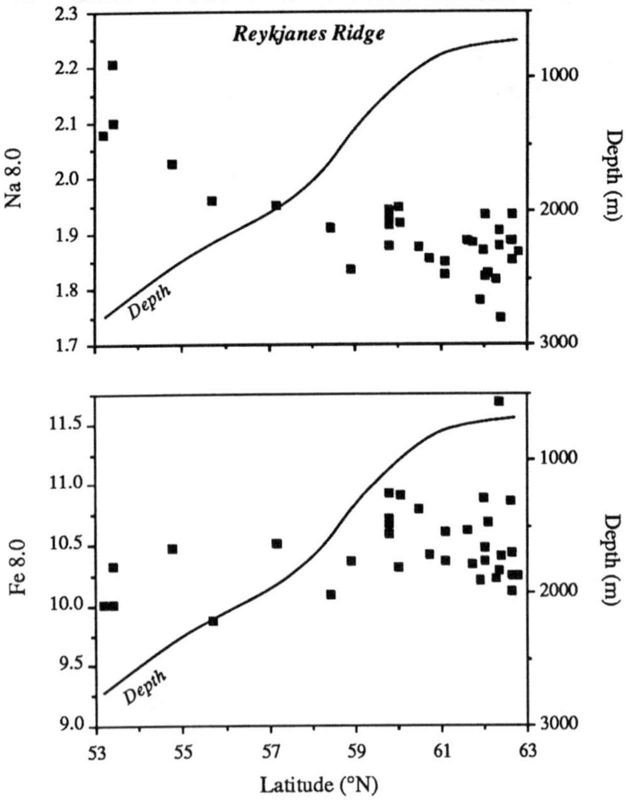

Fig. 20. Variations in basalt composition and axial depth along the Reykjanes Ridge. Each point is for a single sample group within a dredge. Microprobe data from *Melson et al.* [1977 and pers. comm.]. Depth curve is from dredge depths. The Reykjanes Ridge extends onto Iceland at about 63.8°N.

distinct from those exhibited in other regions, and therefore a single explanation cannot apply to the major element systematics of all MORB. It is also clear that different hot spots (e.g., Iceland and the Azores lead to distinct major element effects on MORB. This latter point has been emphasized previously by *Langmuir and Hanson* [1980] and *Schilling et al.* [1982].

3.5 Local Variability Of Major Elements in MORB

Above, we discussed the systematic global variability in average MORB chemistry over distances of hundreds to thousands of kilometers. Looking at the data from a global point of view is a comparatively recent development, because it has only been in recent years that enough high-quality data have been collected from diverse ridge segments. This global coverage is made up of a large number of investigations, representing more than two decades of studies of the local compositional variations among MORB recovered from individual ridge segments [e.g. *Muir et al.*, 1964]. Such studies have identified significant local chemical systematics for both major and trace elements. This chemical variability has been attributed to both magma chamber [*Bryan and Moore*, 1977; *Dungan and Rhodes*, 1978] and melting processes [*White and Bryan*, 1977; *Langmuir et al.*, 1977], although the evidence for variations in the extent of melting has in the past relied almost exclusively on trace element modeling.

Klein and Langmuir [1989; hereafter *KL*89], prompted by the work of *Brodholt and Batiza* [1989], explored the nature of local chemical variability and how it compares to global variability, using the fractionation-corrected parameters of *KL*87. Examining data from densely sampled ridge segments, *KL*89 found that there were systematic local chemical variations, and that these systematics can differ substantially from the global vector of variability. *KL*89 emphasized data from the Atlantic, but noted that the systematics for the East Pacific Rise (EPR) and the mid-Atlantic ridge (MAR) were very different. Here we present the nature of the local variability for the MAR and EPR, which may be representative of slow- and fast-spreading ridges, respectively.

3.5.1 Local Variability on Slow-Spreading Ridges

Figure 21 shows $Na_{8.0}$ vs. $Fe_{8.0}$ for densely sampled segments of slow-spreading ridges (mostly in the North Atlantic). Local trends from "normal" ridges are compared to the "normal" global trend (Figure 21*a*) and local trends from ridges near hot spots are compared to the global trend for hot spot-influenced regions (Figure 21*b*). The data from each individual region form short, sub-parallel trends stacked in the direction of the global vector, but the trends for each local region are oblique to the global trend on each plot. Particularly striking are the data from the section of the MAR between the Kane and Hayes Fracture Zones (Figure 21*a*). In this region, erupted basalts are normal MORB, yet the ridge depth varies by 1500 meters. Thus this region provides a natural experiment of the relationship between chemistry and depth independent of the anomalous effects produced by hot spots. Averages over long distances (100 km) from this region are consistent with the global correlations, and yet the local variability over short distances is of the opposite slope to the global variation on plots of $Na_{8.0}$ vs. $Fe_{8.0}$. Thus the local chemical variability has its own chemical systematics which are distinct from the global systematics among regional averages.

It is significant that a clear local vector also occurs for one of the most enriched portions of the ocean ridge system, near the Oceanographer Fracture Zone (Figure 21*b*). In this region, several mantle components have been identified on the basis of radiogenic isotope data, and very large variations exist in the REE patterns of erupted basalts [*Shirey et al.*, 1987]. Within each compositional group, the data exhibit clear positive correlations between $Na_{8.0}$ and $Fe_{8.0}$. This region, therefore, would potentially allow a clear distinction to be made between the processes causing the local vector and the effects of heterogeneous components in the mantle.

These observed chemical systematics appear to reflect processes that can occur within or beneath individual ridge segments, and will be referred to as the "local" correlations. Recall that in our discussion of the variations observed on $MgO-Na_2O$ diagrams (Figure 12), we referred to local variability

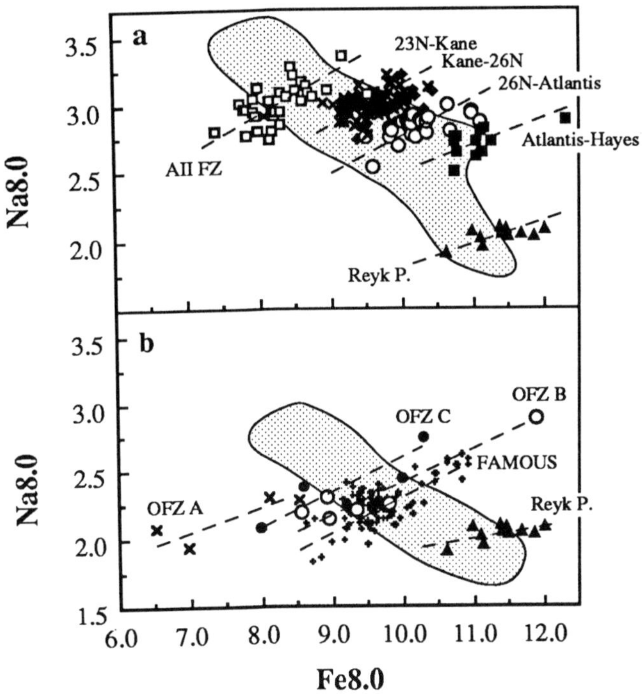

Fig. 21. $Fe_{8.0}$ vs. $Na_{8.0}$ for individual samples from various densely sampled ridge segments. *a*) Samples from "normal" ridge segments. Shaded field is for regional averages of data from "normal" ocean ridges from Figure 15. Solid triangles for Reykjanes Peninsula [*Jakobsson et al.*, 1978]; Xs for 23°N to the Kane Fracture Zone [*W. Melson*, pers. comm.]; filled diamonds from the Mid-Atlantic Ridge 23°N-26°N (*Langmuir*, unpublished data); open circles from 26°N-Atlantis FZ (*Langmuir*, unpublished data); filled squares from Atlantis FZ to Hayes FZ (*Langmuir*, unpublished data) ; open squares for Atlantis II FZ [*Johnson and Dick*, 1992]. Note how Atlantic data are offset to the high side of the normal field, while the Indian data are on the low side (see Figure 16). *b*) Samples from ridge segments influenced by hot spots, compared to shaded field for regional averages of hot spot-influenced ridges from Figure 15. Solid triangles for Reykjanes Peninsula as above; plusses for FAMOUS and AMAR regions of the MAR [*W. Melson*, pers. comm.]. Samples from near the Oceanographer Fracture Zone (OFZ) [*Bender and Langmuir*, unpublished data] are in three groups based on Ce/Yb as follows [*J. Bender*, unpublished isotope dilution data; *Shirey et al.*, 1987]: Ce/Yb = 9.7-10.7 for A, = 7.4-9.1 for B, = 4.2-6.5 for C. (Note that groupings A-C here are not the same as the A-B-C groups of *Shirey et al.* [1987]). All data have been renormalized to 100%, and corrected for interlaboratory biases as for Figure 15. The $Na_{8.0}$ and $Fe_{8.0}$ algorithms for samples south of Kane and the Reykjanes Peninsula have been adjusted as in *Klein and Langmuir* [1989], their Figure 5.

as the "scatter" in the data from an individual region at a constant value of MgO, that could not be explained by low-pressure crystal fractionation. Our examination of local variability here shows that there is fine-structure within the scatter of the data from individual regions that is revealed when the effects of fractionation are removed.

The local correlations are also not artifacts of the fractionation-correction applied to the data. This can be demonstrated by examining a region for which a large number of samples exist. The FAMOUS/AMAR region of the Mid-Atlantic Ridge is among the most densely sampled segments of the ocean ridge system, and spans less than 50 km of ridge length. In the Smithsonian data file, there are over 160 individual analyses from this region (Figure 22*a*). The 62 analyses that fall within the narrow range of 8.0+/-0.25 wt% MgO (Figure 22*b*) show no obvious correlations between Na_2O with MgO, yet Na_2O contents range from 1.85 wt% to 2.45 wt%. These samples do exhibit systematic variations between Na_2O and other major and minor elements (Figure 23). The trends in Na_2O, FeO and SiO_2 shown in Figure 23 are similar in slope and magnitude to the $Na_{8.0}$, $Fe_{8.0}$ and $Si_{8.0}$ variations for this region shown by *KL*89, although they do not span as great a range in values. Furthermore, although the local trends were originally recognized by *KL*89 on the basis of Na, Fe and Si co-variations, other significant correlations also exist between Na_2O and CaO, TiO_2 and K_2O, but not Al_2O_3 (Figure 23). Thus, it is clear from the FAMOUS/AMAR data that the Atlantic-type local trends occur not only in samples corrected to 8 wt% MgO but also in the raw data for samples with approximately 8 wt%

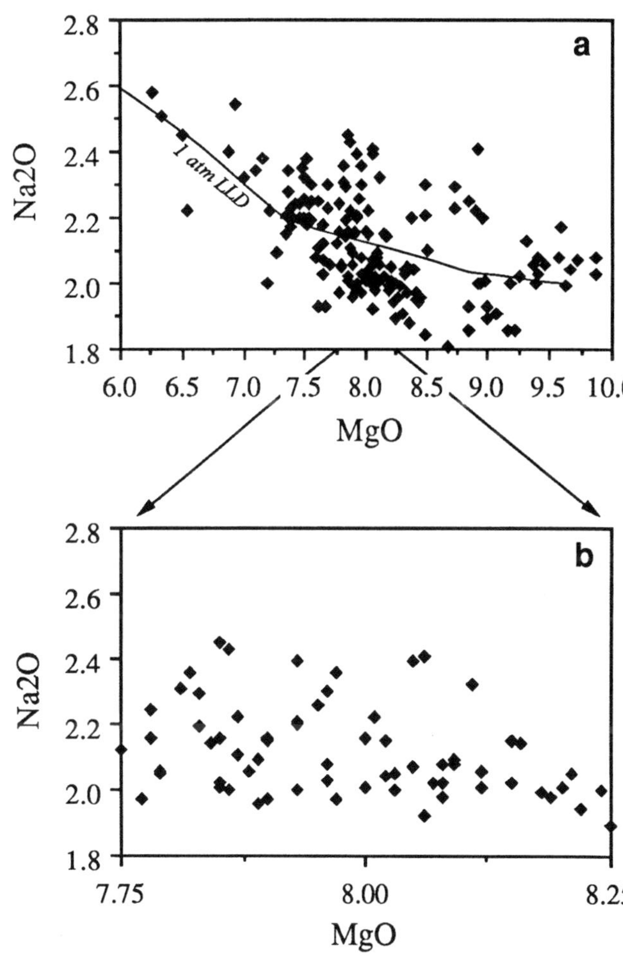

Fig. 22. Illustration of local variability for a well-sampled ridge segment. MgO vs. Na_2O for *a*) all samples from the FAMOUS/AMAR region of the MAR included in the Smithsonian Catalogue of Basalt Glasses [*Melson et al.*, 1977; and pers. comm.], and *b*) only those analyses with 7.75-8.25 wt% MgO. Data have been re-normalized to 100%, but interlaboratory correction factors have not been applied.

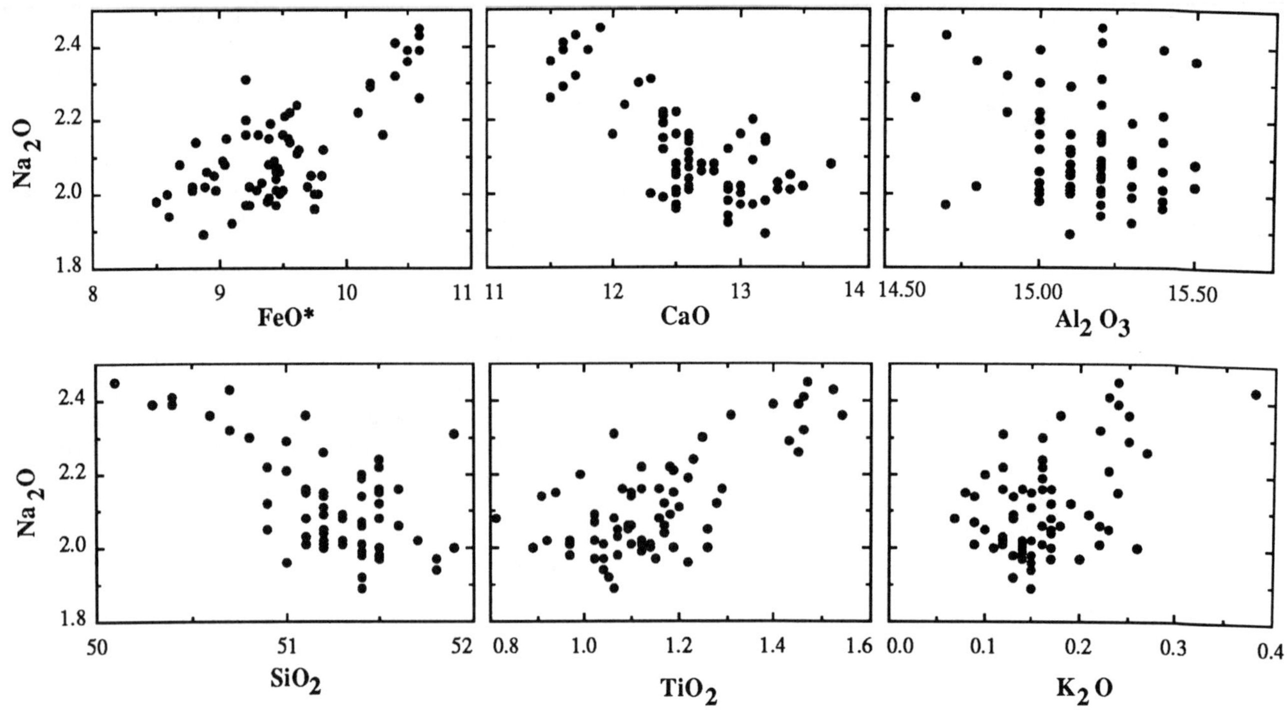

Fig. 23. Local variability for a narrow window of MgO, in order to remove all errors that could be associated with fractionation correction, using samples from FAMOUS/AMAR region with 7.75-8.25 wt% MgO, as in Figure 22*b*. Note that most elements show systematic correlations with one another even at constant MgO content.

MgO. This conceptually simplifies the problem of explaining the origin of these local trends; we simply need to identify processes that produce the observed variability at 8 wt% MgO.

The fact that clear local vectors occur for the FAMOUS/AMAR region, with transitional trace element chemistry characteristic of a hot spot margin, for the Oceanographer region, which is one of the most enriched parts of the ocean floor, and for the Kane to Hayes region, which is of normal MORB composition, shows that the local chemical systematics are not controlled by mantle heterogeneity or other effects that contribute to hot spots. Instead, it appears that Atlantic local variability is characteristic of the slow spreading ridge environment. Similar local vectors have recently also been found on the very deep and slow-spreading Southwest Indian Ridge by *Johnson and Dick* [1992] (Figure 21), so that these local trends occur over the entire range of axial depth and $Na_{8.0}$ values at slow spreading ridges.

3.5.2 Local Variability on the East Pacific Rise

The data from the fast-spreading East Pacific Rise (EPR) have very different systematics than those observed at slow-spreading ridges. In order to present the variability along the EPR, we have normalized the data to the mean MgO value for a large number of EPR samples (7.3 wt% MgO), because the EPR contains few samples with greater than 8 wt% MgO. Choosing this normalization value ensures that the fractionation corrections themselves have minimal influence on the distribution of the normalized data.

The EPR shows much less depth variation than the MAR, over both short and long distances along the axis. Nonetheless, individual EPR basalts have a large range in composition after correction for fractionation. The $Na_{7.3}$-$Fe_{7.3}$ variations for EPR data sets are parallel to the global vector, and the range for individual samples (Figure 24) is almost as large as the global range for regional averages! Note, however, that average EPR data plots in the middle of the global field which corresponds with its typical depth of about 2700 m (see Figure 15).

This variability for the EPR as a whole is in fact a form of local variability because it can occur between small ridge offsets along as little as 50 km of the axis. Figure 24*b* compares data from about 50 km of the EPR north of 12°N [*Langmuir et al.*, 1986; *Reynolds et al.*, 1992 in press] with the entire EPR from 5-14°N and shows that there is virtually complete overlap. Within this 50 km region there are no major ridge offsets and only a 100 m change in axial depth, so the chemical variations clearly occur on a local scale, even in the absence of significant disturbances of the crust or thermal perturbations of the mantle. The local EPR variability is similar to the total range of EPR data partly because the regional depth along the EPR varies little.

From the $Na_{8.0}$-$Fe_{8.0}$ perspective, there is striking overlap between the EPR local and global averaged data. But the

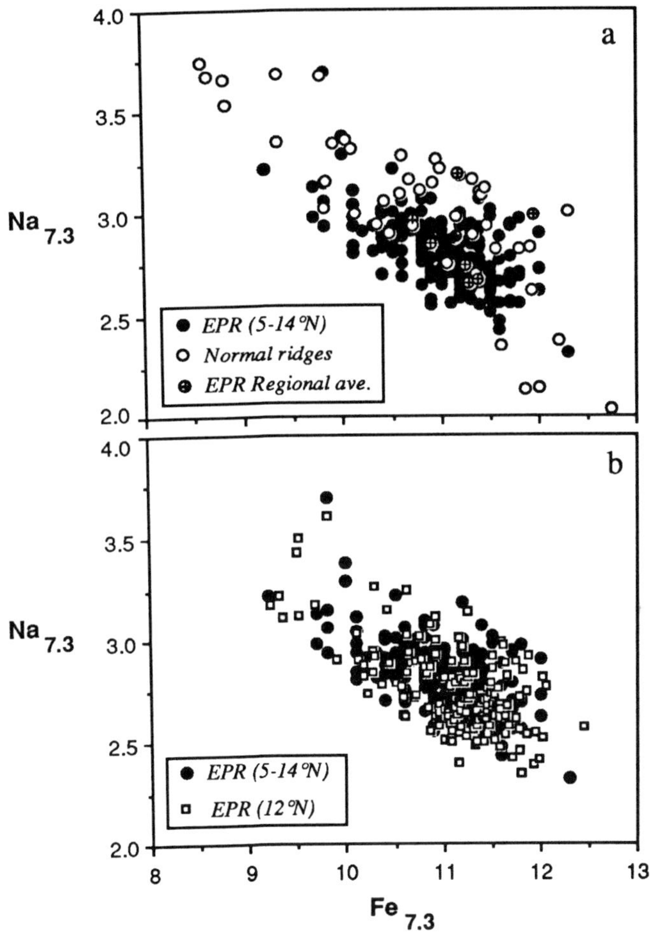

Fig. 24. $Na_{7.3}$ vs. $Fe_{7.3}$ for individual samples from 5-14°N on the East Pacific Rise (EPR) from the CHEPR expedition [microprobe analyses of glasses, *J. F. Bender and C. H. Langmuir*, unpublished data] compared to *a*) global regional averages for normal ocean ridges and EPR averages (from Figure 15), and *b*) individual samples from 12°N on the EPR from the VENTURE expedition [microprobe analyses of glasses, *Reynolds et al., 1992* and unpublished data]. Compositions are compared at 7.3 wt% MgO instead of 8 wt% MgO because 7.3% is the mean for all EPR basalts. Although the Na-Fe variations of EPR and the global vector are similar, this similarity does not extend to other elements.

similarity ends there. The systematics of the EPR data are not similar to the global vector for all elements. Plots of $Na_{7.3}$ vs. $Si_{7.3}$ and $Fe_{7.3}$ vs. $Si_{7.3}$ do not exhibit the characteristic slopes of the global data (Figure 25, see also *KL*89). In fact, some individual areas show weak positive correlations between $Fe_{7.3}$ and $Si_{7.3}$. There are a few samples from the EPR which have high $Na_{7.3}$ and low $Si_{7.3}$, creating one sparse limb of the EPR field that is parallel to the Atlantic data. Yet these same samples have low $Fe_{7.3}$, which is opposite to the MAR local vector. The characteristic inverse relationship between Fe and Si that is present for the global correlations and the slow-spreading local vector is completely absent in the EPR data. Therefore the EPR local systematics are distinct from both the global vector and the Atlantic local vector.

Two components of EPR local variability. Examination of the other major elements suggests there may be two components of local variability in the EPR data, one within "normal" MORB (N-MORB, with low K_2O/TiO_2), and one between N-MORB and "transitional" MORB (T-MORB, with higher K contents). In the sample group as a whole, plots of many of the major elements show scatter (Figure 26), until the N-MORB are separated from the T-MORB. *Reynolds et al.* [1992, in press] have pointed out that $Fe_{7.3}$ and $Ti_{7.3}$ correlate positively for the N-MORB alone in the 12°N region, and that temporal variations emerge much more clearly from consideration of the N-MORB alone.

This variation for the N-MORB, however, does not produce the striking negative $Na_{7.3}$-$Fe_{7.3}$ correlation for the EPR, because the N-MORB alone exhibit little variation in $Na_{7.3}$. Instead, the range in $Na_{7.3}$ results from a different compositional vector, that extends from N-MORB to T-MORB. The high K_2O, T-MORB component has low FeO and high Na_2O, and also high TiO_2, low SiO_2, and high Al_2O_3. These characteristics of the transitional component correlate quantitatively. For example, the excess TiO_2 (defined as the Ti in excess of the N-MORB Fe-Ti correlation line shown in Figure 26) correlates well with most of the major elements, as illustrated in Figure 27. The transitional component is slightly variable along the EPR, leading to local arrays that are slightly offset from one another for some elements.

Thus the Pacific local variability requires two compositional components. The component of variability that occurs within N-MORB has little variation in $Na_{7.3}$, and limited variation in many other elements. The substantial variability that occurs in EPR lavas results from a second chemical component that is related to the variable presence of "enriched" material. The major chemical variations correlate with the amount of this enriched component. This is apparent in Figure 28, where it can be seen that the spread in $Na_{7.3}$ and $Fe_{7.3}$ of EPR lavas relates to the presence and absence of the enriched component that creates T-MORB. It is the presence of this enriched component, therefore, that gives to the EPR data their unique geochemical signature, which contrasts with the slow-spreading local vector and the global vector. Although we have documented these local variations using data from the northern EPR, the superfast spreading region of the southern EPR [*Sinton et al.*, 1991] shows very similar chemical characteristics (Figure 29). As *Sinton et al.* [1991] point out, however, there is substantial variability in K_2O/TiO_2 along the southern EPR.

3.6 Summary of Observations

The observations presented above show that once the compositional effects of low-pressure fractionation have been removed, the major element compositions of ocean ridge basalts display certain fundamental systematics. It is our aim to explore the cause of these systematics with quantitative models.

First, on the largest scale of sampling, regional averages of basalt composition from most ridges form a global component

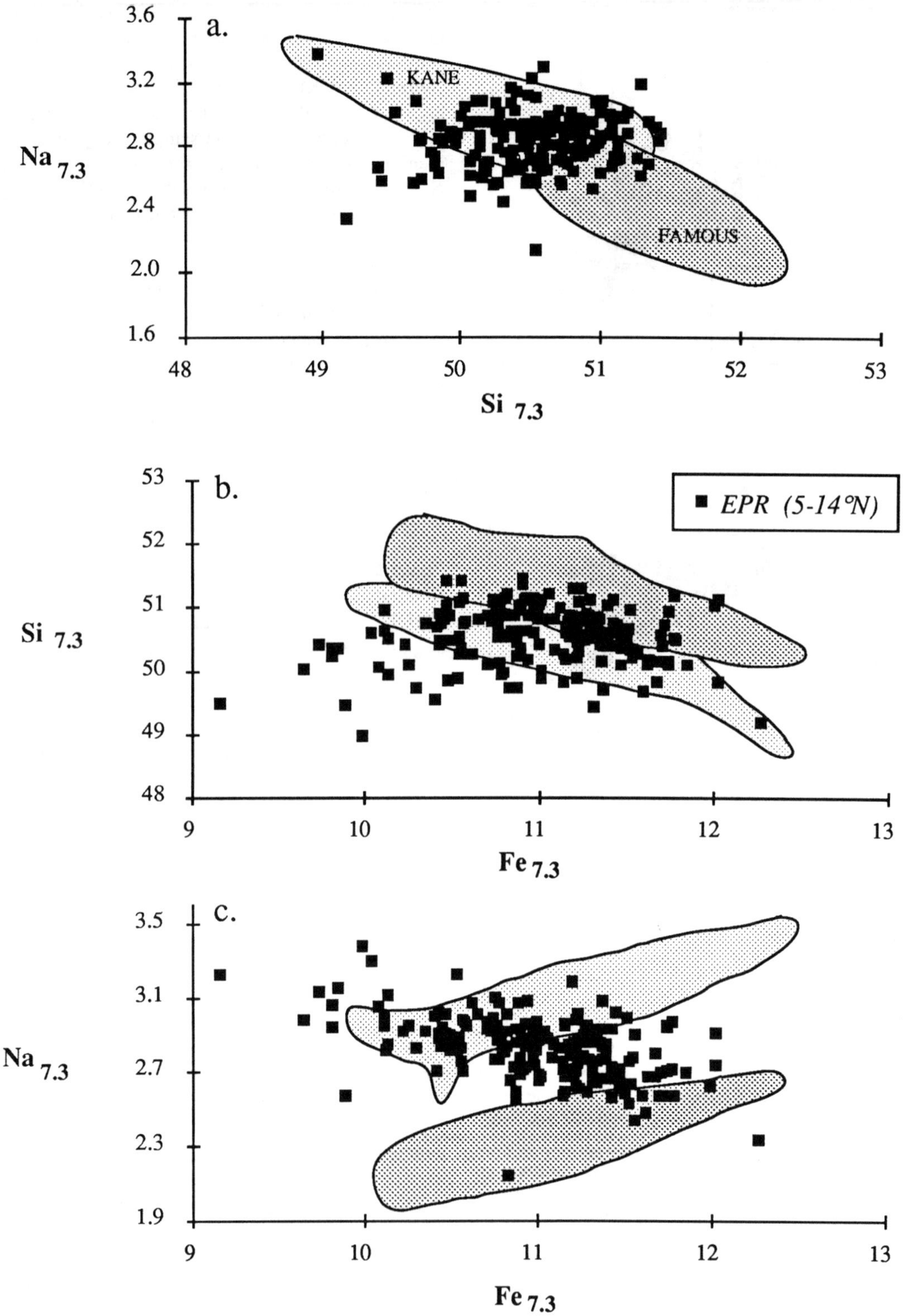

Fig. 25. *a)* $Si_{7.3}$ vs. $Na_{7.3}$, *b)* $Fe_{7.3}$ vs. $Si_{7.3}$, and *c)* $Fe_{7.3}$ vs. $Na_{7.3}$ for individual samples from 5-14°N on the EPR (CHEPR data as in Figure 24) compared to Atlantic local vectors. Fields for the Atlantic enclose individual samples from the MAR near the Kane Fracture Zone and the FAMOUS region, corrected for interlaboratory biases (data sources as in Figure 21; note that data have been normalized to 7.3 wt% MgO instead of 8 wt% MgO). EPR local variability is distinct from both the global and the local Atlantic vectors of variability (see discussion in text).

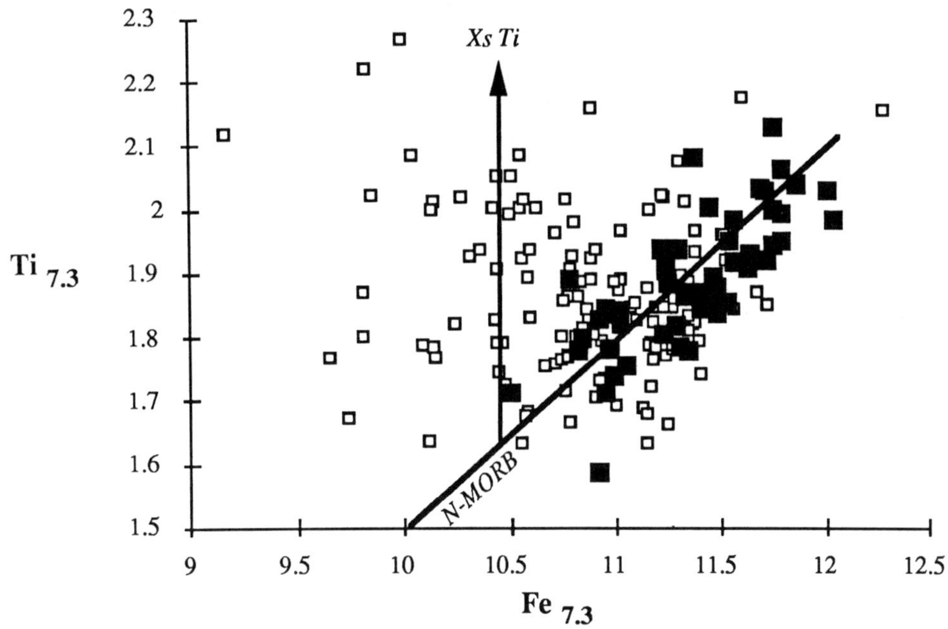

Fig. 26. $Fe_{7.3}$ vs. $Ti_{7.3}$ variations for individual samples from 5-14°N on the EPR (CHEPR data as in Figure 24). The most depleted CHEPR samples with K_2O/TiO_2 < 0.067 (solid squares) define a positive trend (N-MORB line), while the entire CHEPR data set is scattered. Xs Ti is defined as $Ti_{7.3}$ in excess of the N-MORB correlation line, as indicated by arrow.

of chemical variability (Figure 15). A key feature of these global chemical systematics is that they also correlate with physical characteristics of the ridge, namely axial depth and crustal thickness (Figures 15 and 18). Some hot spots, however, are anomalous in this global context; regional averages taken over the Galapagos platform, for example, show a relationship between chemistry and depth that is opposite to the global correlations (Figures 15 and 19). Back-arc basins exhibit the same correlations as normal ridges, but the data are offset to lower $Fe_{8.0}$. Iceland generally plots with the global correlations for most elements, but is slightly high in $Na_{8.0}$ and very high in $Ti_{8.0}$. Thus, although the global systematics represent one of the most prominent signals in the basalt data, there are exceptions, and no single explanation is likely to account for all characteristics of the data.

When we examine the local variability that is observed among individual samples from individual spreading segments, entirely different chemical systematics emerge. Among samples from slow-spreading ridges, chemical relationships are observed that have the opposite slope to the global $Na_{8.0}$, $Fe_{8.0}$ and $Si_{8.0}$ trends (Figure 21). These characteristics occur over the entire global range of depth and chemistry, and for ridge segments near and far from hot spot influence. On fast-spreading ridges, there are two important local chemical components that are distinct from both the global and the slow-spreading local compositional vectors (Figures 25 and 27). The amount of the enriched component appears to be a dominant control on the range of $Na_{8.0}$ and $Fe_{8.0}$ observed.

Our challenge, then, is to understand this complicated yet systematic fabric of compositional variability. Our approach is through quantitative models that include complex fractionation, mantle melting and melt segregation processes, and the effects of source heterogeneity. The comparison of these model results with the observations will then allow us to isolate that part of the chemical signal that is governed by the processes of melt formation and migration.

4.0 Modeling Of Mantle Melting: Background And Previous Work

The basic geochemical observations presented above need to be explained in terms of physical models of melting, differentiation or source heterogeneity. It is clear at the outset that each of the three factors will play some role: no crust can be created without melting; mantle heterogeneity is evident on all scales and in all types of analyses by which the mantle has been studied; complex processes of differentiation are evident in MORB and in all modern models of magma cooling. Our aim is to see how we can distinguish the chemical effects of the various processes in order to better constrain the melting process.

Constraints on the various processes arise from the comparison of MORB compositions to quantitative calculations of the chemical consequences of the possible differentiation and melting processes. The next sections, therefore, consider quantitative modeling approaches. Because we are ultimately concerned with melting processes, modeling of melting is considered in the main body of the text. Complex differentiation processes, however, can also contribute to the chemical systematics observed. Methodology for modeling these complex differentiation processes is presented in

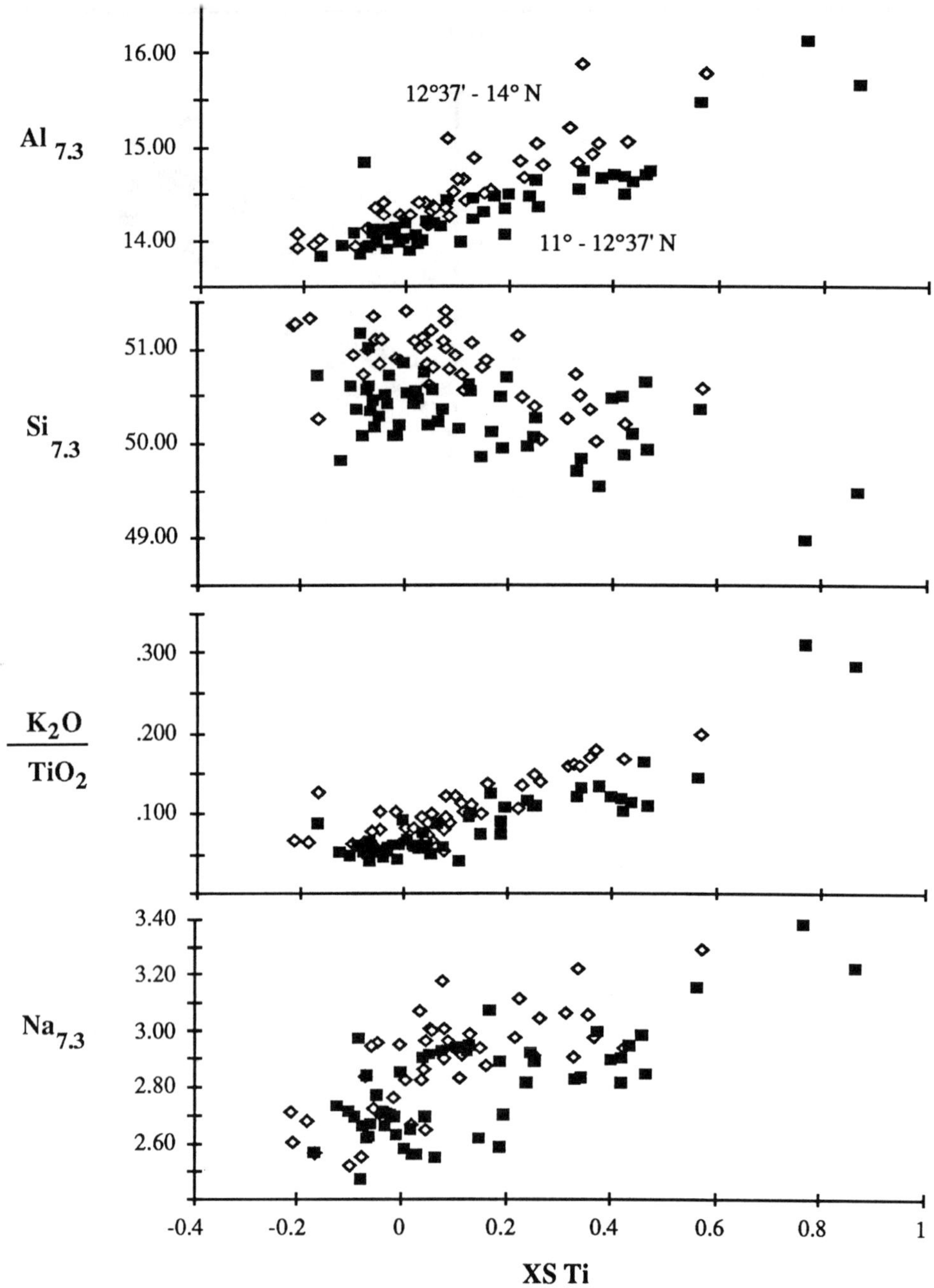

Fig. 27. XS Ti (defined in Figure 26) vs. other chemical parameters for individual analyses from two regions of the northern EPR (CHEPR data as in Figure 24). High XS Ti correlates with K_2O/TiO_2, and therefore reflects the T-MORB component in these samples. Note that samples with high $Na_{7.3}$ are most influenced by the T-MORB component, therefore it is this component that is controlling much of the EPR local variability.

Appendix C, and the results of these calculations are then incorporated into the discussion.

As we saw from the working model, modeling mantle melting requires consideration of three separate but related aspects of the problem. We need quantitative expressions for (a) a melting function, which gives the extent of melting as a function of decompression during adiabatic ascent; (b) a mixing function, which indicates how much of each particular melt composition is produced and pooled; and (c) a chemistry function, which gives the chemical compositions of the melts that are present throughout the melting regime. Each function has its own methodology and uncertainties. Additional

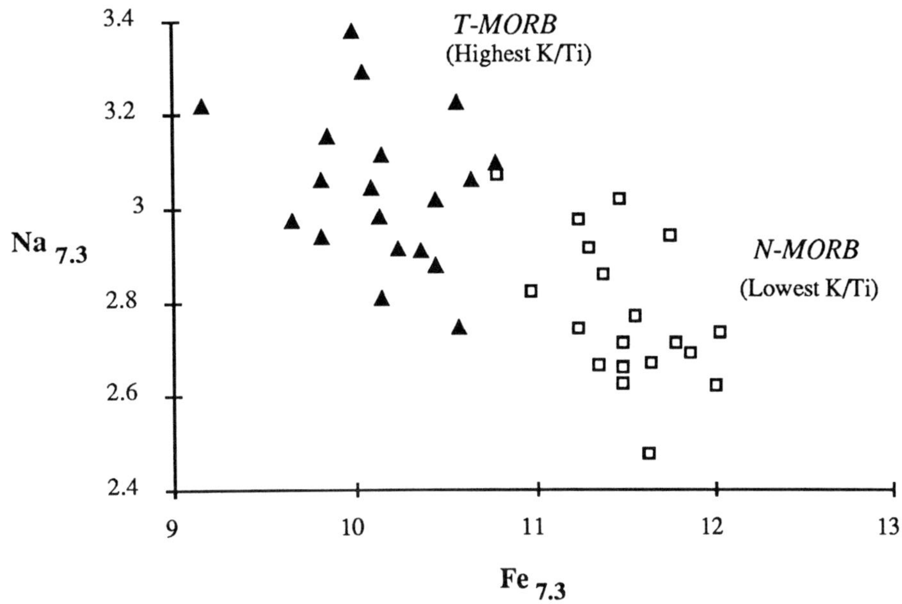

Fig. 28. The control on the EPR $Na_{7.3}$-$Fe_{7.3}$ local vector by enriched components. The most enriched samples (filled diamonds; K_2O/TiO_2 > 0.136) have the lowest $Fe_{7.3}$ and highest $Na_{7.3}$, while the most depleted samples (open squares; K_2O/TiO_2 < 0.055) have the highest $Fe_{7.3}$ and lowest $Na_{7.3}$. CHEPR data as in Figure 24.

complexities include how much melt is extracted from the solid, how it migrates, and how it chemically reacts on the way to the surface. Quantitative models for mantle melting must consider each of these aspects, for the final chemical composition of the basalt that is delivered to the magma reservoirs, where cooling and fractionation occur, depends on all of them.

4.1 The Importance of Mechanisms of Melt Extraction

Before turning to the three functions necessary to calculate melting, it is necessary to consider mechanisms of melt extraction. There are two end-member models that describe the relationship between melting and melt extraction: equilibrium melting, where the melt stays with the matrix as melting proceeds; or fractional melting, where the melt is instanteously separated from the solid as each increment of melt is generated (see equations (5) and (8)).

Whether mantle melting more closely approximates equilibrium or fractional melting will have profound effects on the melting, chemistry and mixing functions. If fractional melting occurs, the melting function will require a greater increase in temperature for a given melt fraction. Perfect fractional fusion could also lead to an effective termination of melting when clinopyroxene disappears from the residue [*Presnall*, 1969; *Hess*, this volume]. In all circumstances, fractional fusion leads to less melting and thinner crust for a given pressure of intersection of the solidus.

Fractional melting also influences the chemistry function because successive fractional melts are more depleted in the incompatible major and trace elements, and because the mantle will melt to a lesser overall extent. However, if fractional melts are pooled, the pooled melt composition closely resembles an equilibrium melt produced by melting to the same extent for most elements [*Shaw*, 1970]. The exception occurs where partition coefficients change significantly as a function of pressure. Then some of the incremental melts may have formed under conditions of very different partition coefficients.

The melting process influences the mixing function primarily through its influence on the physics of upwelling beneath the ridge [e.g., *Sleep*, 1974; *Ahern and Turcotte*, 1979; *McKenzie*, 1984]. Differential flow between melt and matrix depends strongly on the porosity, and therefore on the extraction model. Furthermore, if melting approaches equilibrium melting, then the greater porosity will lower the effective mantle viscosity, which can lead to the formation of instabilities or focussed flow in the melting regime [e.g., *Scott and Stevenson*, 1989; *Buck and Su*, 1989]. As we will see below, this can also affect the mixing function for pooled melts.

Neither pure equilibrium nor pure fractional melting are physically realistic; mantle melting is probably somewhere in between. A more realistic melting model would be that melt segregates once some threshold porosity has been reached. Melt is then continually being removed from the system, but some melt fraction always remains with the residue. This is the "continuous melting" model presented by *Langmuir et al.* [1977] as one aspect of the more complex "dynamic melting," which conceptually included reaction during ascent and a variety of melt extraction processes. Recent literature has used the term dynamic melting for the original continuous melting model [*McKenzie*, 1985*a*, 1985*b*; *Williams and Gill*, 1989;

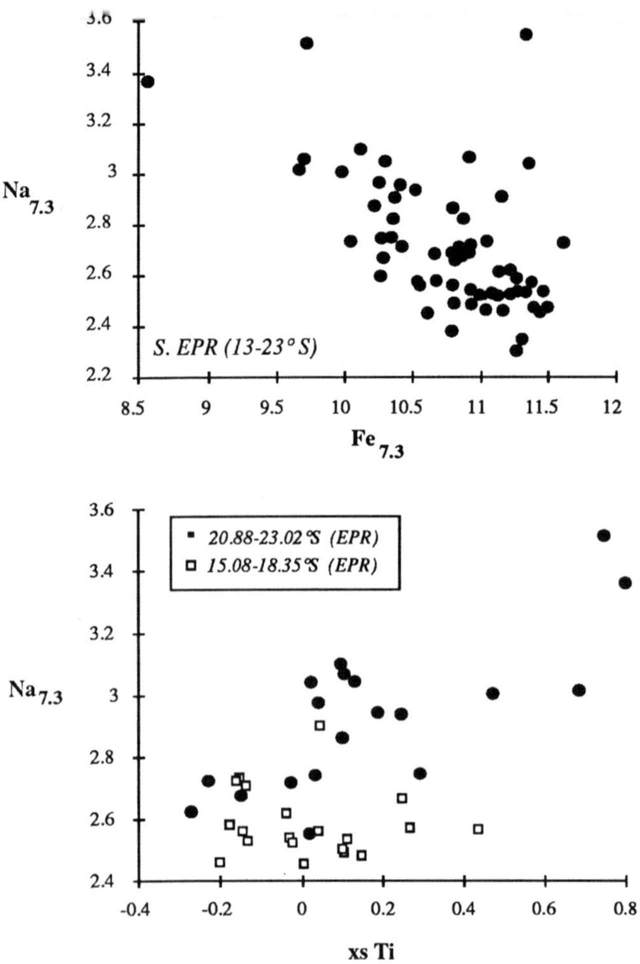

Fig. 29. Basalts from the superfast spreading region of the southern EPR [*Sinton et al.*, 1991] show local systematics similar to those of the northern EPR basalts (compare to Figures 24 and 27).

Elliott et al., 1991]. The important result for geochemical models is that continuous melting leads to effects that are intermediate between equilibrium and fractional melting, depending on the trapped melt fraction.

Even if melting locally approaches fractional melting, reaction of ascending melt with the surrounding solid matrix could lead to chemical effects similar to continuous melting. Measurements of physical properties suggest the possibility of complete extraction of melt, but the fact that melt is generated throughout a large vertical region would lead to continual replacement by melt transported from below. Further melting would then take place in the presence of a small melt fraction, which is the characteristic of the continuous melting model. The most striking contrast between equilibrium and fractional melting occurs when one solid phase disappears. At this point, the temperature needed for further fractional melting jumps discontinuously to a much higher value, while equilibrium melting continues progressively with continuously increasing temperature. However, if melt is added to a system that has melted fractionally, melting can continue without temperature discontinuities. The net result is that melting can proceed with a small instantaneous porosity. The flow of melt through the system allows a substantial amount of liquid to interact with the solid and produces chemical results similar to continuous melting with a significant porosity. The chemical effects in this case could be interpreted as resulting from a significant residual porosity, even if melt extraction were very efficient and the ultimate porosity in the residual rock was close to zero.

4.2 The Melting Function

Figure 1 showed schematically the two important reference slopes for an adiabatically ascending parcel of mantle. Because the solidus has a steeper dT/dP slope than the adiabat, mantle that ascends adiabatically will intersect its solidus and begin to melt. The amount of melt generated depends upon the total decompression experienced. The goal of the melting function calculation is to track the "melting path" that a parcel of mantle follows through the melting regime. The energy required to melt is the heat of fusion. Because the ascending mantle is in isolation from its surroundings, the heat of fusion must come from the internal energy of the melting mantle parcel, resulting in a temperature drop. Thus, the "melting path" follows a T-P slope between that of the solidus and the adiabat (Figure 30).

The important parameter in defining the melting path is $(\partial F/\partial P)_S$, or the change in melt fraction with change in pressure above the solidus during adiabatic decompression. Estimates of this parameter, which we will call γ, range from 1-2%/kbar [e.g., *Ahern and Turcotte*, 1979; *Cawthorn*, 1975; *McKenzie*, 1984]. *Cawthorn* [1975] in particular considered how various parameters affect the total amount of melt produced. This parameter (γ) provides a crucial tie between the physical and chemical models of melting. For example, if an ascending parcel of mantle intersects its solidus at 20 kbar (about 60 km depth) and melts 1%/kbar, then the maximum extent of melting would be 20%. If instead γ were 2%/kbar, then the maximum F would be 40%, twice as much total melt would be produced, and twice as much crust might be generated. These are substantial differences, and thus it is important to consider what variables contribute to γ, and how well we know them.

Because much of the information we need to calculate γ must come from laboratory experiments where results are given in units of temperature and not energy, it is more convenient to discuss thermal balances. The thermal balance for adiabatic melting is given by four terms: two for the thermal energy available (the slope of the solidus and the slope of the adiabat), and two for the thermal energy required (the heat of fusion, and the T increase above the solidus).

At a given pressure during melting, the thermal energy available for melting comes from the difference in temperature between the solid adiabat and the solidus (Figure 30). (There is an additional term involving the difference between the solid and liquid adiabats, but given the much larger uncertainties in the other parameters, this term can be ignored.)

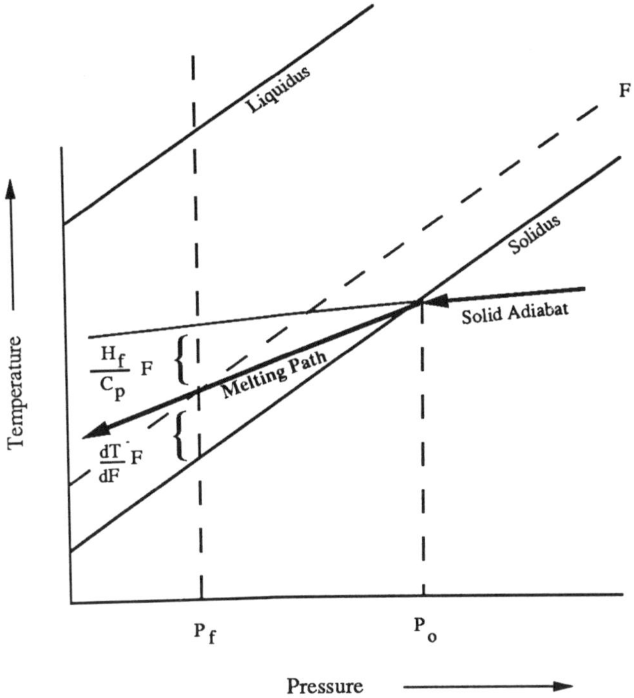

Fig. 30. Schematic pressure-temperature diagram showing the parameters considered in calculating the melting path for an adiabatically ascending mantle parcel, as discussed in section 4.2 in the text. As mantle upwells adiabatically, it intersects the solidus at some pressure (P_o) and begins melting; the melting path has a steeper slope than the solid adiabat because there is a temperature drop associated with the heat of fusion (H_f). At any given pressure (P_f) less than P_o, the energy available for melting can be expressed as the temperature difference between the solidus and the extension of the solid adiabat. This temperature difference is partitioned between the temperature drop associated with the heat of fusion, and the temperature increase above the solidus required to melt to an extent F. C_p is the heat capacity, and dT/dF is the spacing of the contours of F on this diagram.

For a one-component system, the solid phase melts at a single temperature at any one pressure. For a multi-component system like the mantle, however, melting occurs over a large temperature interval, from 0% melting at the solidus to 100% at the liquidus. Thus, at any given pressure, it takes a given temperature change above the solidus to melt a given amount, and this value for dT/dF determines the melting contours on a P-T diagram (e.g., Figure 1). The simplest way to contour F as a function of P and T is to assume a linear spacing of melting contours from solidus to liquidus, or a constant value of (dT/dF). The temperature increase required to melt a specific amount (F) would thus be F*(dT/dF).

The heat of fusion (H_f) is the amount of energy required to melt 100%. This heat is obtained by cooling the mantle, and the energy yielded by each degree of cooling is given by the heat capacity, C_p. If we assume that the heat of fusion is evenly distributed between liquidus and solidus, and that H_f and C_p remain constant over the melting interval, then the temperature drop upon melting is $F*(H_f/C_p)$.

For a given decrease in pressure from P_o (the pressure of intersection of the solidus) to P_f (the final pressure), the thermal balance between these four terms may be written as:

$$(P_f - P_o)*[(dT/dP)_{adb} - (dT/dP)_{sol}] = F*(H_f/C_p) + F*(dT/dF) \quad (17)$$

For incremental changes in pressure, the extent of melting as a function of pressure is:

$$\gamma = [(dT/dP)_{adb} - (dT/dP)_{sol}] / [H_f/C_p + dT/dF] \quad (18)$$

Thus we can calculate the amount of melting produced upon decompression by substituting into equation (18) values for the slope of the solidus, the slope of the adiabat, the heats of fusion and heat capacities of mantle materials, and a function relating temperature above the solidus to extent of melting. The uncertainty in $(\partial F/\partial P)$ thus relates to the uncertainties in these various parameters. Below we consider typical values for these parameters, as well as the uncertainties and complexities that might realistically be involved.

A typical value for $(dT/dP)_{adb}$ is 1°C/kbar, though it may be as high as 2°/kbar [e.g., *Cawthorn*, 1975]. These are small numbers compared to the slope of the solidus (typically 12°/kbar; see below), and so variations in this number do not lead to large variations in $(\partial F/\partial P)$.

In principle, the solidus should change slope as the mantle assemblage changes from plagioclase- to spinel- to garnet-lherzolite with increasing pressure [e.g., *Presnall et al.*, 1979]. These differences, however, are not obvious in the existing experimental data on natural systems. *McKenzie and Bickle* [1988] made an extensive compilation of data on the mantle solidus. Their compilation is reproduced in Figure 31*a* for the pressure interval from 0 to 40 kbar, which is relevant to magma genesis in the modern earth. The open symbols are those experiments where solid and liquid were present, while the closed symbols are those where there was no liquid; in principle, the mantle solidus should be the single line that separates the two symbols. It is clear from the figure, however, that there is significant overlap in the data. Much of this variability may reflect real differences among starting compositions and experimental techniques, rather than statistical error. The solid lines in the figure are drawn such that one is below most of the open symbols, while the other is above most of the closed symbols. These two lines illustrate the uncertainty in our knowledge of the solidus (about 60°C). The zero pressure intercept of the solidus could be anywhere from 1095°C to 1155°C. The slope is constrained to be between 11.5 and 13°/kbar. Thus there are considerable uncertainties in both the slope and intercept of the solidus. A regression of all the data in Figure 31*a* leads to the equation: [T = 1139 + 11.4*P]. If one considers only the experiments that bracket the solidus to better than 50°C, the regression yields: [T = 1118 + 12.7*P]. In our calculations we consider the effects of variations in both the intercept and slope of the solidus.

Common values for the heats of fusion and heat capacities of mantle minerals are taken to be 100 cal/gm and 0.3 cal/gm/°, respectively [e.g., *Carmichael et al.*, 1974]. However, the heat of fusion and heat capacity may vary significantly as a function

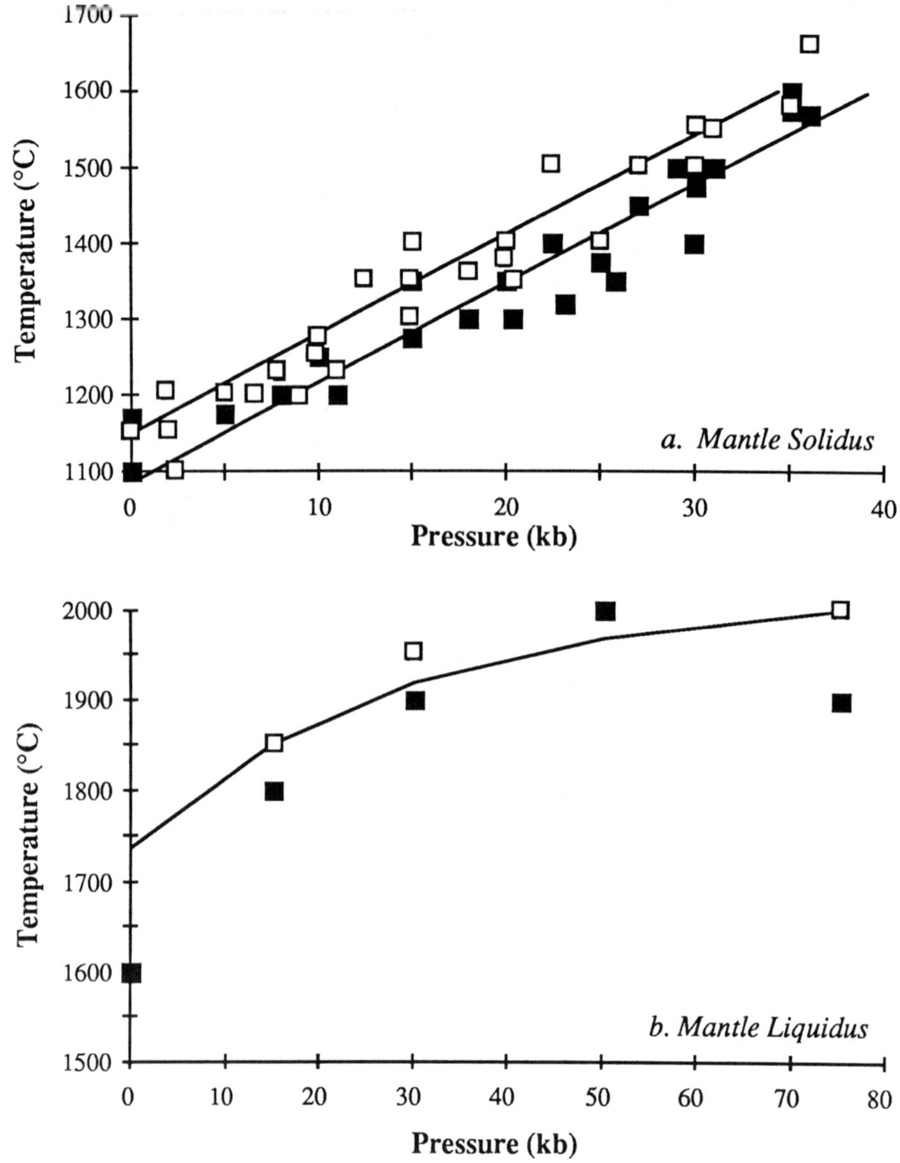

Fig. 31. *a*) Pressure-temperature diagram for estimating the slope and intercept for the mantle solidus. Data for both panels as compiled by *McKenzie and Bickle* [1988]. Filled squares show results from experiments in which only solid was present; open squares show results of experiments in which both solid and melt were present; lines are drawn so that the lower line is below most of the solid+melt data, and the upper line is drawn so that it is above most of the solid-only data. The offset in the lines suggest that the uncertainty in estimating the solidus is about 60°C at a given pressure. *b*) Pressure-temperature diagram for estimating mantle liquidus. Filled squares show experiments in which both solid and melt are present, and the open squares show experiments in which only melt is present. The curve shows the estimate of the liquidus of *McKenzie and Bickle* [1988].

of temperature, mineralogy and melt fraction [*Cawthorn*, 1975]. *Hess* [this volume] evaluates the most recent experimental data, and suggests an average value for the heat of fusion for mantle melting of 180 cal/gm. Varying the heat of fusion from 100 to 180 cal/gm/° has a large effect on γ; thus we consider the effects of both values in our calculations.

The way in which F is contoured in the melting interval must come directly from melting experiments. Existing data from peridotite melting experiments suggest that two important features must be taken into account. First, the melting interval narrows with pressure as the liquidus and solidus approach each other; and second, melting contours are closer together near the solidus.

The narrowing of the melting interval with pressure requires knowledge of both the mantle solidus and liquidus as a function of pressure. There are sparse data for the mantle liquidus because the high temperatures involved are difficult to achieve in the laboratory. The available data [*Takahashi,* 1986] are presented in Figure 31*b*. *McKenzie and Bickle* [1988] calculate an uncertainty in the liquidus of 7°C, but in practice the uncertainties in the liquidus are much greater than this value. For example, the minimum temperature interval over which the liquidus is bracketed is 50°C. Moreover, there are few data at low pressures, and almost any curve can be drawn for the liquidus below 15 kbar. Despite the lack of constraints on the slope of the liquidus, we can make a rough approximation of the narrowing of the melting interval with pressure. For the purposes of the melting calculations in this paper, we assume that the melting interval decreases by a factor of two in roughly 45 kbar (from experiments of *Takahashi*, 1986), and that the spacing between the melting contours decreases proportionally.

Some experiments suggest that substantial melt is generated over a small temperature interval near the solidus [see Figure 2 in *Jaques and Green,* 1980]. Knowledge of the melting relationships near the solidus is quite limited because there are few experimental data between 0 and 15% melting. During progressive melting of peridotite, it appears that clinopyroxene disappears as a residual phase after about 20% melting [e.g., *Jaques and Green*, 1980; *Hess*, this volume]. The subcalcic clinopyroxenes that exist at higher pressure might last to considerably higher melt fractions, however. In the absence of better experimental constraints, we assume that a change in melting behavior results from the disappearance of clinopyroxene at 22% melting, and that melting contours are more closely spaced for less than 22% melting, and more widely spaced for greater than 22% melting. We use 3.5°C/% for F<22% melting, and 6.8°C/% for F>22%. These contour intervals are appropriate for melting at 1 atm, and become compressed proportionally with the melting interval with increasing pressure. Figure 32 shows the melting contours resulting from these relationships. Note that we do permit melting to occur beyond 22% melting, where clinopyroxene may disappear, because residual porosity or melt fluxing from below will permit additional melting to take place, as discussed above in section 4.1.

Substituting into equation (18) the values of 1°/kbar for the slope of the adiabat, 13°/kbar for the slope of the solidus, 0.3 cal/gm/° for the heat capacity, 3.5°/% and 6.8°/% for dT/dF (at less than and greater than 22% melting, respectively), and (1-P/88) for the pressure correction on the melting contours, we arrive at:

$$\gamma = -12/[(H_f/0.3) + 350(1-P/88)] \quad \text{for } F < 0.22 \qquad (19)$$

and:

$$\gamma = -12/[(H_f/0.3) + 680(1-P/88)] \quad \text{for } F > 0.22 \qquad (20)$$

Figure 32 shows adiabatic melting paths for mantle that intersects the solidus at 30 kbar, calculated from equations (19) and (20) for two different values of the heat of fusion (100 and 180 cal/gm). Note that the total amount of melting differs considerably for the two cases. The estimates of γ given by equations (19) and (20) range from values similar to that used by *KL*87 (1.2%/kbar) to around 2%/kbar for the lower estimate

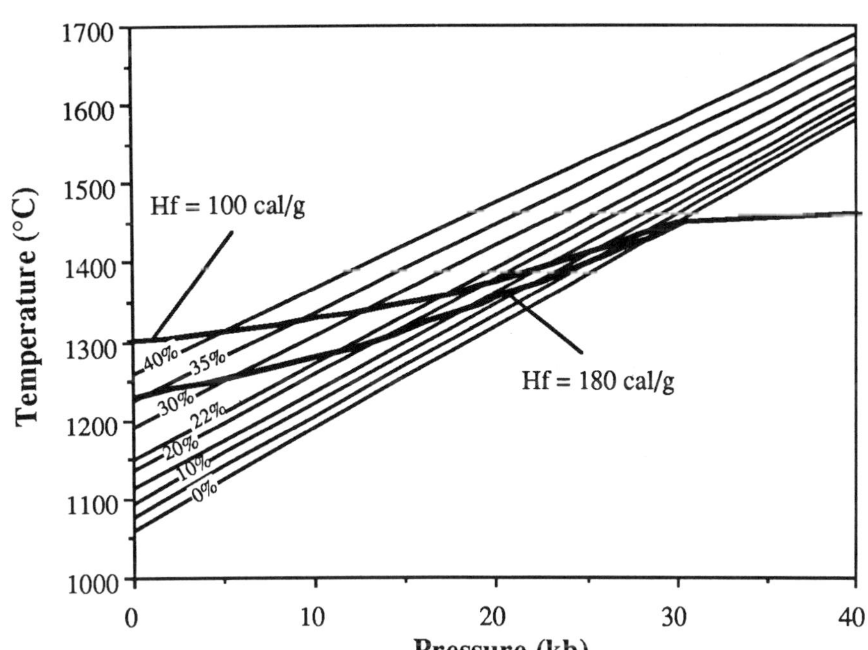

Fig. 32. Pressure-temperature diagram contoured for the extent of melting (%F) showing melting paths for mantle that intersects the solidus at 30 kbar. Melting paths and melting contours calculated from equations given in section 4.2 in the text. The two paths are for two different values for the heat of fusion (H_f), as indicated.

of the heat of fusion at high pressure. For comparison, the melting function used by *McKenzie and Bickle* [1988] averages 1%/kbar for P_o = 35 kbar, and 1.6%/kbar for P_o = 15 kbar.

The melting paths in Figure 32 are curved because the instantaneous slope depends on both the extent and pressure of melting. During the initial stages of adiabatic melting, the pressure is high, and the extent of melting low. Both of these factors lead to compressed melting contours and larger amounts of melting per unit of pressure release. The melting path during the initial stages of melting has a steeper dT/dP slope because more energy (temperature) is required for each unit of pressure drop.

In order to calculate the total amount of melting along a given trajectory, equations (19) and (20) can be integrated:

$$F_{max}(<22\%) = 3.015 \ln[(H_f/0.3 + 350 - 3.98P_f)/(H_f/0.3 + 350 - 3.98P_o)] \quad (21)$$

and for the portion above 22% melting:

$$F_{max}(>22) = 1.546 \ln[(H_f/0.3 + 683 - 7.76P_f)/(H_f/0.3 + 683 - 7.76P_{22})] \quad (22)$$

where P_{22} is the pressure of the mantle parcel at 22% melting and equals (1.0757 P_o - 18.1).

A melting path for fractional melting is less well constrained. The principal uncertainty is that the temperature of the mantle solidus continually rises as melt is removed, and this temperature increase cannot be determined from the equilibrium melting conditions of the experimental data. To approximate the effects of fractional melting, we use a dT/dF value of 6°/% melting for F<22%, and 8°/% melting for F>22%. Using these values, of course, leads to correspondingly less melting per unit of pressure release. For evaluation of mantle heterogeneity, we assume that the mantle solidus increases by 20° for each unit increase in the Mg# [Mg/(Mg+Fe), atomic ratio]. Both of these values need to be refined by further experimental work.

4.3 The Mixing Function and the Shape of the Melting Regime

The amount of melt generated under the ridge depends on the upwelling pattern as well as the ambient temperature of the underlying mantle. In addition, melts generated by fractional or equilibrium melting come from different depths in the melting regime, and hence there are chemical consequences associated with the melting process. In this section, we show how melts can be summed for different melting regimes and melting processes, and how the summing is relatively insensitive to the many variations in the "shape" of the melting regime beneath a ridge. A more thorough consideration of the effects of the shape of the melting regime is presented by *Plank and Langmuir* [1992].

4.3.1 An Illustrative Incremental Model

In general, because melting occurs wherever mantle upwells above its solidus, the physical form or "shape" of the melting regime is defined by the upwelling pattern. Consider a simple, two-dimensional, corner-flow pattern that results from passive upwelling in response to spreading [e.g., *Ahern and Turcotte*, 1979; *Phipps Morgan*, 1987; *McKenzie and Bickle*, 1988; Figure 33*a*]. The base of the melting regime is the mantle solidus, and the boundaries of the melting regime are defined by where mantle flows horizontally and no longer melts. Given a solidus, a melting function, and mantle temperature on the solidus, melting can be contoured in the melting regime. The end-points of the mantle flow paths (1-4) from the melting regime in Figure 33*a* are highlighted on the P-T diagram in Figure 33*b*.

Figure 34*a* illustrates schematically what happens in the simplified melting regime of Figure 33 during an increment of spreading. Boxes are used to represent units of mantle passing through the melting regime, with one box length equal to one increment of spreading. The shading reflects the amount of melt that is produced upon upwelling. Open boxes indicate no melting. During an increment of time, the plates spread, and all the white boxes as well as the boxes at the boundaries of the melting regime move one horizontal step. All the other boxes move up to replace the void and melt an additional increment. The total number of increments that each box has melted is proportional to the distance above the solidus.

Consider first an equilibrium melting model, where the melt remains in contact with the mantle all the way to the top of each column, and then is extracted. The melt fraction within each box increases upon decompression, and no melt leaves the matrix until the box reaches the top of its column. The melts are then tapped from the boxes at the top of each column in the melting regime, and these melts are pooled to form the oceanic crust. *KL*87 and *McKenzie and Bickle* [1988, hereafter *MB*88] presented pooled melt compositions produced by this kind of melting and mixing model, although neither study described the melting regime in this way.

In order to calculate the composition and volume of the melt mixture that forms the oceanic crust, it is useful to consider the steady-state *residue* of mantle melting that results from melt extraction within a given melting regime. At steady-state, the residue of mantle melting is the column of mantle that emerges outside the melting regime — the residual mantle column (RMC) (Figure 34*a*). The RMC is the net result of the integrated melting and extraction processes that occur within the melting regime. Each increment of spreading creates a new section of oceanic crust and a new RMC. Each box in the RMC, back-tracked along its flow path, represents a column in the melting regime. For example, the uppermost box in the RMC in Figure 34*a* originated from six increments of melting in the central column of the melting regime. Thus, the two-dimensional melting regime is reduced to a one-dimensional residual column.

Because the RMC is the complement to the steady-state pooled melt that forms the ocean crust, the RMC may be used to calculate important properties of the ocean crust, such as its composition and thickness, and the mean extents and pressures of melting by which it was derived. To calculate these properties, it is only necessary to consider what has happened to produce each box in the RMC. The final mean extent of

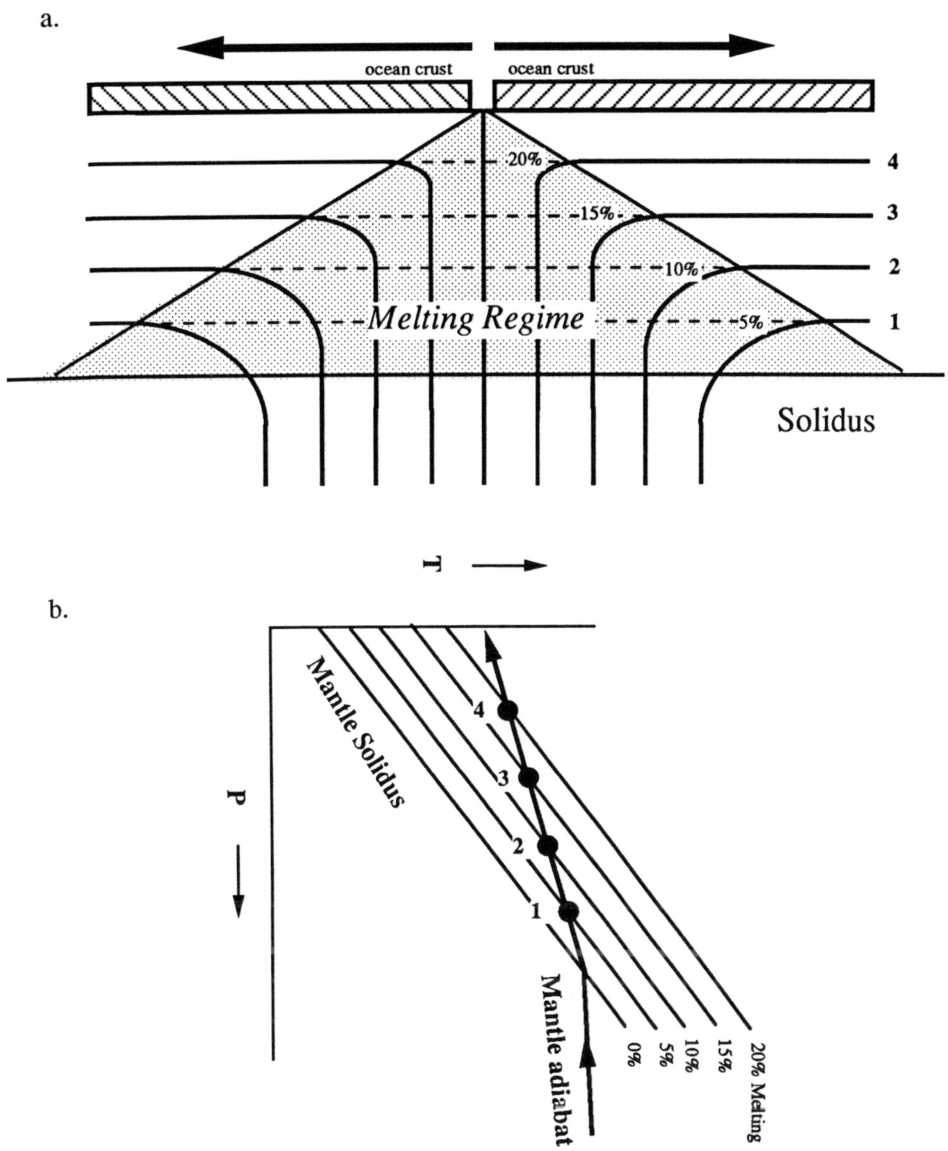

Fig. 33. *a*) Schematic view of the melting regime for simple corner flow circulation beneath the ridge resulting from passive upwelling as a result of spreading. Solid lines (without arrows) are mantle flow lines; the shaded region is the melting regime, contoured for the extent of melting (labelled dashed lines). *b*) Schematic pressure-temperature path for adiabatic melting resulting from upwelling and intersection of the mantle solidus. The region above the solidus is contoured for the extent of melting. The numbered dots (1-4) correspond to the maximum extent of melting that each numbered flow line in Panel (a) undergoes. Thus, the single melting path in (b) is a representation of the melting systematics that occur in the two-dimensional melting regime shown in Panel (a).

melting ($\bar{F}$), mean pressure of melting ($\bar{P}$), and composition of the pooled melt can be calculated simply by averaging the melt associated with each of the boxes in the RMC. The bookkeeping for this illustrative calculation is shown in Figure 34*a*, where the height of each box equals 1% melting and 1 unit pressure.

This simple incremental model can be used to illustrate how the mean properties of the oceanic crust can be calculated from the RMC.

1. The mean extent of melting, $\bar{F}$ is the total amount of melt produced, divided by the total amount of mantle that has melted during each increment of spreading. Since the RMC is the amount of mantle processed through the melting regime during each increment of spreading, $\bar{F}$ is the total melt divided by the total area of the RMC. $\bar{F}$ is calculated in Figure 34*a* by summing the melt associated with all the boxes in the RMC, and dividing by the total number of boxes in the RMC. Although the maximum amount of melting (F_{max}) for the

a. Equilibrium

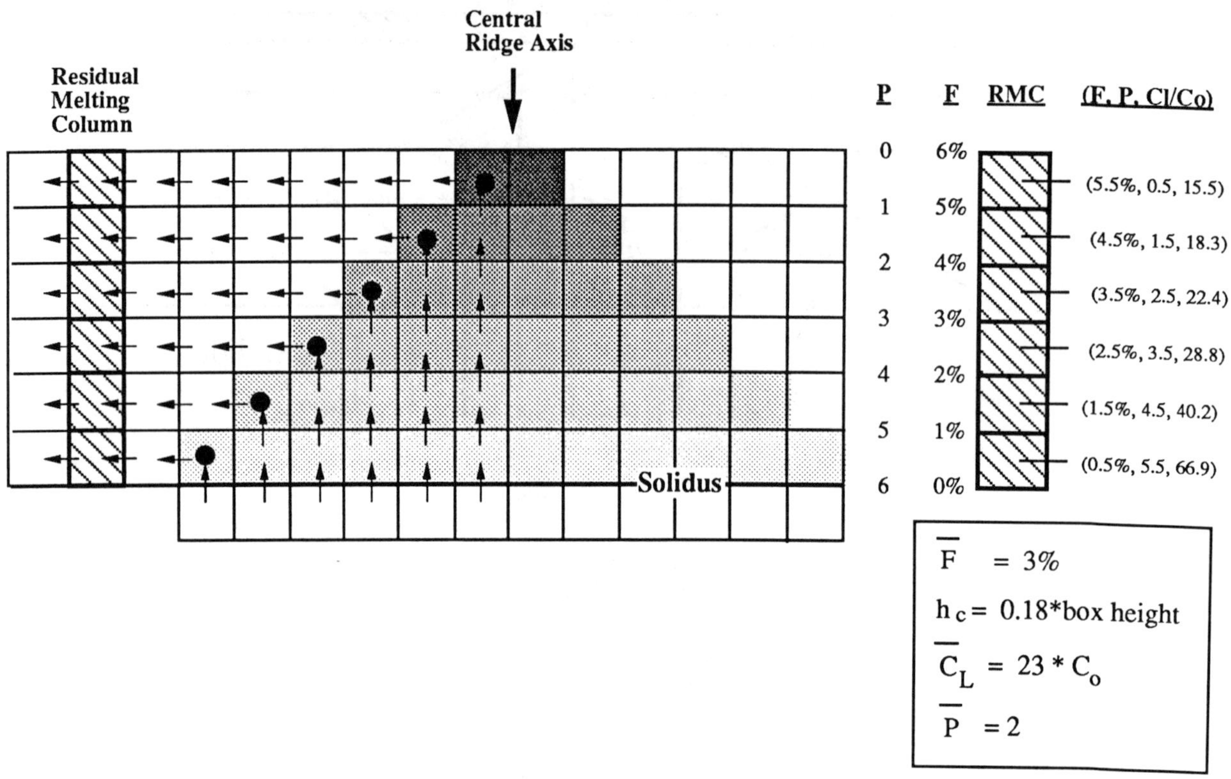

b. Fractional

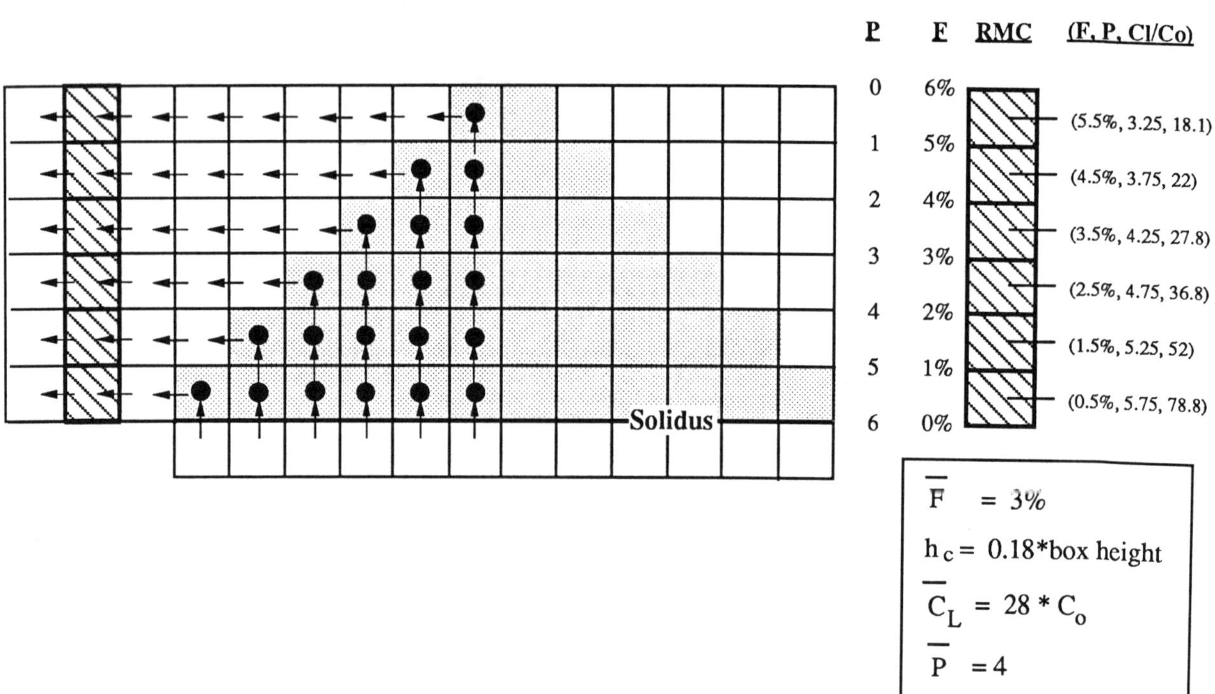

melting regime in Figure 34*a* is 6%, $\bar{F}$ is half the maximum, or 3%. [Note: In our figures, the mantle has not been compacted to account for the volume lost by melt extraction. Thus, the length of the RMC is the total pressure interval of melting, and not the true residual mantle column compacted for melt loss. This introduces small errors in the calculations of crustal thickness, but allows simpler illustrations of the relationship between the RMC, the melting regime, and crustal thickness.]

2. Because the melts that make up the ocean crust have been derived from various depths in the melting regime, the ocean crust reflects a mean of the pressures of equilibration of the individual melts. If the melts undergo no re-equilibration once separated from the solid residue of melting, then the mean pressure ($\bar{P}$) can be calculated as a weighted mean of the individual melts. In Figure 34*a*, $\bar{P}$ is 2 pressure units, which is less than half the pressure interval because more melt is generated in the shallow levels of the melting regime. [Note: *KL*87 defined a "mean pressure" as the pressure at which half the total melt formed shallower and half formed deeper. This is actually a median pressure, and differs slightly from the mean pressure defined here, which is a true mean. A true mean $\bar{P}$ is conceptually and computationally easier to compare to mean compositions.]

3. If all the melt pooled from the melting regime forms the ocean crust, then the total melt produced during each increment of spreading would equal the thickness of the ocean crust, h_c, once differences in densities (about 10%) are accounted for. Because the RMC is the total amount of mantle processed through the melting regime during each increment of spreading, if all the melt is extracted, the crustal thickness equals the height of the RMC multiplied by the mean extent of melting. If the boxes in Figure 34*a* were each 10 km in height, then (0.03*60)/0.9 or 2 km of crust would be generated during each unit of spreading.

4. For equilibrium melting, the liquid composition can be calculated anywhere in the melting regime given the source composition, the bulk distribution coefficient that applies to the appropriate extent of melting, and the batch melting equation (equation 5). The composition of the total aggregate melt from the entire melting regime can then be calculated by weighting the melt composition by the amount of melt with that composition (a weighted mean). For the example in Figure 34*a*, an element with D = 0.01 would be enriched about 23-fold in the final aggregate melt over the source concentration. Note that this is very similar to a batch melt calculated for the same $\bar{F}$ (using equation 5, a 3% batch melt = 25-fold enrichment for D = 0.01).

This latter point, that the composition of the pooled melt from the melting regime closely resembles a simple batch melt calculated from $\bar{F}$, is particularly important. It means that a MORB may be inverted to calculate an accurate $\bar{F}$ by assuming simple batch melting, even if the sample was derived by mixing of melts from throughout the melting regime.

4.3.2 *Effects of Fractional Melting*

Consider now the incremental model for fractional melting, where melt is extracted from the solid mantle as soon as it forms. This process is illustrated schematically in Figure 34*b*, where the solid circles represent points at which melt is extracted. During an increment of spreading, each box that upwells will melt one increment. These increments are then extracted from each box and pooled at the ridge axis. Note that in this model, melt is extracted from all depths, not just at the top of each of the melting columns, as for the equilibrium model presented above. Pooled melts derived by fractional melting thus reflect higher mean pressures. For the fractional melting example shown in Figure 34*b*, the mean pressure of melt extracted from the central melting column is half the total pressure interval (3.25 kbar), because equal increments of melt are extracted at each depth. This mean pressure contrasts with that for the equilbrium model, which is the shallowest pressure for the central column (0.5 kbar). The mean pressure for the fractional melts pooled from the entire melting regime in Figure 34*b* is twice that for the equilibrium case.

Although the most dramatic difference between pooled fractional and equilibrium melts is the mean pressure, there are other differences. The composition of the pooled melt will be slightly different for fractional vs. equilibrium melting. For the fractional case, the pooled melt derived from each vertical column has an equivalent composition to an accumulated fractional melt (AFM) melted to the maximum extent (e.g., 6% for the central column). Thus, the AFM equation of *Shaw* [1970] was used to calculate the concentrations in the melt associated with each column in the melting regime, or box in the RMC, in Figure 34*b*.

$$C_L = (C_o / F)*[1 - (1-F)^{(1/D)}] \qquad (23)$$

Fig. 34. A schematic incremental view of melting beneath the ridge and a calculation of the mean properties of the pooled melt from the melting regime. Arrows indicate solid mantle flow. The shape of the melting regime (shaded) is defined by the upward component of mantle flow and therefore by the boxes that move upward; the point where mantle flow is horizontal defines the upper boundary of the melting regime. Panel (a) is for equilibrium melting and Panel (b) for fractional melting. The shading increases in intensity with increasing amount of melt in each box as it moves upward; melt is pooled from only the top boxes of each of the columns (boxes with large dots at left side of melting regime). Also shown is the residual melting column (RMC) which is the steady-state residue of melt extraction. The extent of melting (F) and pressure of melting (P) in the melting regime are in arbitrary units. Values presented at the right of the diagram show calculations that result from a consideration of the melting systematics represented by the RMC. The numbers in the parentheses are the $\bar{F}$, $\bar{P}$ and mean concentration of an element with D = 0.01 (mean C_L/C_o) of the melt associated with each box in the RMC or column in the melting regime. $\bar{F}$, $\bar{P}$, C_L (mean C_L/C_o) and crustal thickness calculated for the entire melting regime is the average of the boxes in the RMC, and are shown at the lower right. After *Plank and Langmuir* [1992].

Because the AFM and batch equations give similar results, however, the concentration of the pooled fractional melt ($28{*}C_o$) is very similar to that of the pooled equilibrium melt ($23{*}C_o$).

Fractional melting should also lead to lower maximum (and mean) extents of melting for the same mantle temperature, because γ should be lower (as discussed above). For the sake of illustration, however, the fractional melting regime in Figure 34*b* was constructed for the same γ as for the equilibrium case. If a lower γ is adopted (as in some examples below), then the maximum extent of melting will be less than 6%, and the mean will be less than 3%, for the same total pressure interval. These effects also lead to less crust produced during each increment of spreading.

To summarize, fractional melting should lead to roughly twice the mean pressure of melting, lower mean extent of melting, greater incompatible element concentrations, and less crust than equilibrium melting for the same mantle temperature.

4.3.3 A Continuous Melting Regime

Although we presented a simple incremental model for the sake of illustration, the same principles apply to a continuous melting regime. Figure 35 shows a melting regime and RMC generated upon corner flow, where F is contoured linearly though the melting regime as 1.2%/kbar (a good approximation to equations 19 and 20). Also shown is the distribution of melts that are extracted from the melting regime at steady-state as a function of depth. This melt distribution is the complement to the RMC. In Figure 35, the mass of melt (M_m) created at each depth is linearly related to both F and P. For example, 20 times more melt is produced by 20% melting at 1.67 kbar than is produced by 1% melting at around 18 kbar. Mixing this linear distribution of melts together produces an oceanic crust with a characteristic composition and volume, as calculated by *KL*87 and *MB*88. This linear type of mixing results from linear melting contours with depth in the melting regime, and constant horizontal velocity outside the melting regime. One can imagine non-linear cases, however, that could result from more rapid flow through the shallow part of the melting regime, or non-linear melting contours [*Plank and Langmuir*, 1992].

In integral form, $\bar{F}$ and $\bar{P}$ are defined as:

$$\bar{F} = \frac{\int_0^{F_{max}} [M_m(F)]\, dF}{\int_0^{F_{max}} [M_s(F)]\, dF} \qquad (24)$$

$$\bar{P} = \frac{\int_{P_f}^{P_o} [P\, M_m(P)]\, dP}{\int_{P_f}^{P_o} [M_m(P)]\, dP} \qquad (25)$$

where M_s is the mass of solid prior to melting, P_o is the pressure of intersection of the solidus, and P_f is the final pressure. Because F is defined as M_m/M_s, equation 24 becomes:

$$\bar{F} = \frac{\int_0^{F_{max}} [M_m(F)]\, dF}{\int_0^{F_{max}} [M_m(F)/F]\, dF} \qquad (26)$$

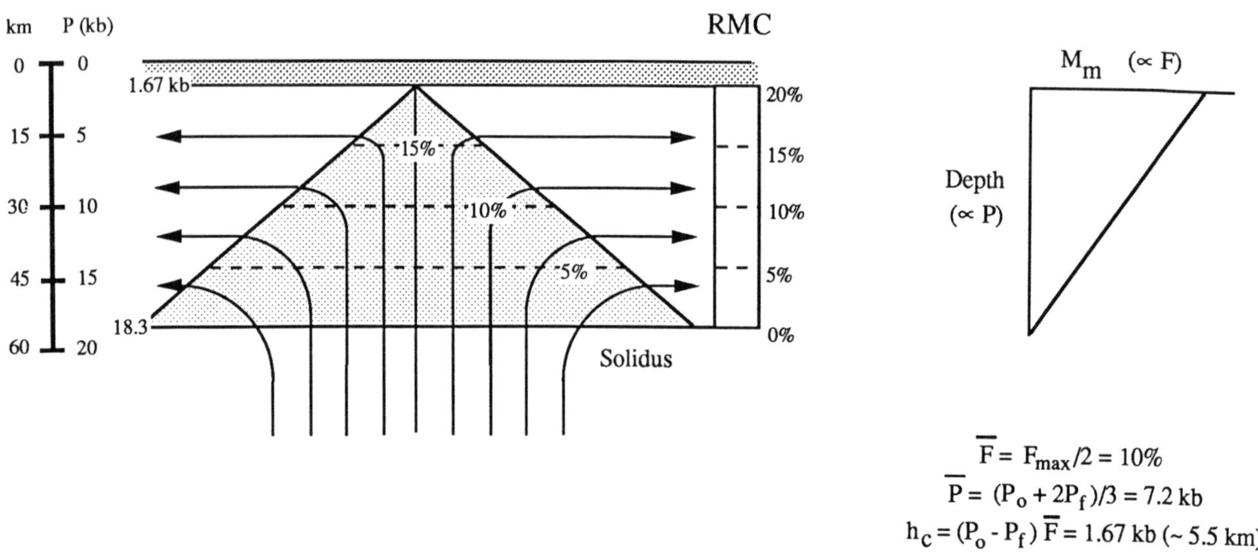

Fig. 35. As in Figure 34, but for a continuous melting regime where γ is 1.2%/kbar. Also shown is the distribution of melt as a function of depth. The mass of melt (M_m) produced anywhere in the melting regime is a linear function of both F and P because γ is constant in this model. The steady-state oceanic crust produced from this melting regime has a $\bar{F}$ of 10%, $\bar{P}$ of 7.2 kbar, and is about 5.5 km thick.

For linear mixing (Figure 35), $M_m = aF$ (a is a constant), and $\bar{F} = 1/2F_{max}$ (integrating equation (26)). This is a general result of any linear mixing model, and is the same as in the incremental model above, and the same as presented in *KL*87. Thus, although the maximum amount of melting beneath a ridge may be 20%, the mean amount of melting reflected in the composition of the oceanic crust will be 10%. If M_m and P are also linearly related such that $M_m = b - cP$ (where b and c are constant), then $\bar{P} = (P_o + 2P_f)/3$ (from equation 25). Thus $\bar{P}$ is shallower than half the pressure interval (it equals the upper third), reflecting the greater amount of melt generated at shallower pressures in the melting regime.

The chemical composition for the mean melts coming from the melting regime can be calculated by integrating the total liquid produced. This equation (from *KL*87) is:

$$\bar{C}_i = \frac{\int_0^{F_{max}} [C_i(F) * F]\, dF}{\int_0^{F_{max}} F\, dF} \qquad (27)$$

For a perfectly incompatible element, the liquid composition is simply $C_o/\bar{F}$. For most incompatible trace elements, the abundance calculated using a simple melting equation and $\bar{F}$ will be a very good approximation.

The integrals above are general, and can be used to calculate the mean properties of the ocean crust given any distribution of melt in a melting regime. Because the mean extent of melting and mean pressure of melting are reflected in the composition of the oceanic crust (e.g., in its Na_2O and FeO contents, respectively), we can use major element compositions to invert for melt distributions in the melting regime. Thus, coupling chemical data with measurements of the thickness of the oceanic crust (the total melt volume produced) provides powerful constraints on the melting processes beneath ridges.

4.3.4 Many Shapes Lead to the Same RMC at Steady-State

An important aspect of the mean extent and pressure of melting is that they are independent of the "shape" of the melting regime for a given pressure of intersection of the solidus, provided the melting contours are controlled by the amount of pressure release above the solidus, and there is no downward component of mantle flow within the melting regime (*Plank and Langmuir,* 1992). This can be simply illustrated by Figure 36, which shows two melting regimes with different shapes. The two shapes are produced by different mantle upwelling patterns, the curved shape by upwelling that decreases in velocity outward from the central axis, and the triangular shape by constant and rapid upwelling velocity. The horizontal velocity of the mantle that emerges out of both melting regimes is the same and equals the spreading rate. Because the flow of mantle out of the melting regime controls how much mantle may be processed through the melting regime, both shapes actually process the same amount of mantle during each increment of spreading. The flux of material (shaded) through both melting regimes is the same, and the different shapes are exactly balanced by the different upwelling patterns. Thus, although they have different shapes, both melting regimes lead to the same RMC, the same thickness of ocean crust, and the same mean extents of melting. This is a steady-state model; a new section of crust and residual mantle are produced during each increment of spreading.

The result that both shapes lead to the same RMC may seem counter-intuitive upon first glance at the problem. For example, the curved shape appears to contain a very large region of mantle that melts less than 5%, and one might think

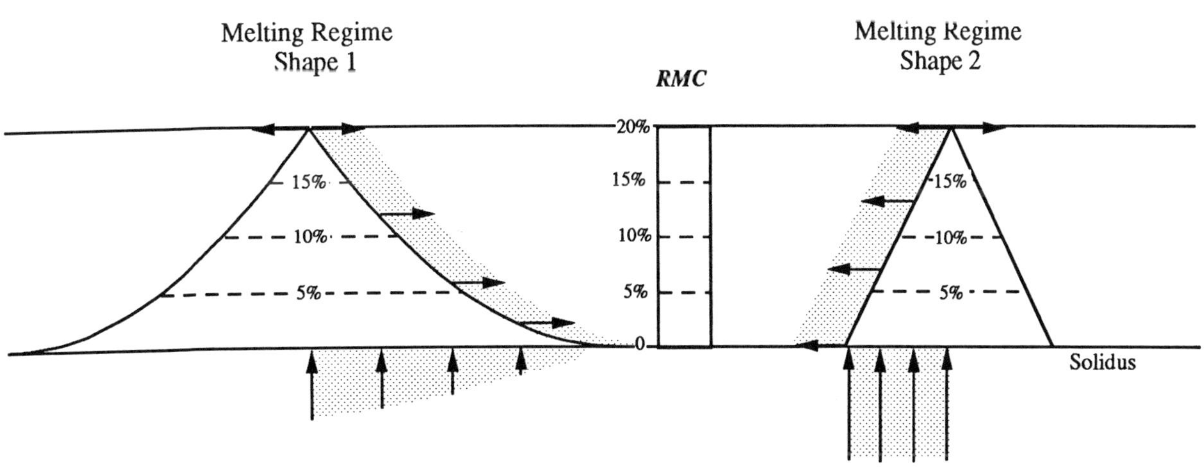

Fig. 36. The effect that two different shapes of the melting regime have on the residual melting column. Because the horizontal flow velocity is the same for both of these melting regimes, both shapes lead to the same residual melting column (RMC), and thus the same volume and composition for the pooled melt.

this would lead to lower mean extents of melting than for the triangular shape. But the current volume of the melting regime is not the same as the amount of mantle passed through each part of the melting regime per spreading increment. Upwelling velocities are controlled by the slope of the upper boundary of the melting regime, and the upwelling velocity compensates for the variations in shape. Since melting beneath ridges is ultimately linked to spreading, the ocean crust results from the amount of mantle that is processed through the melting regime for each increment of spreading, and not simply the total volume of mantle that makes up the melting regime.

Although many shapes lead to the same RMC, different mixing functions can result from complications such as active upwelling and differential melt extraction from different parts of the melting regime. These effects have chemical consequences that can lead to different relationships between chemistry and crustal thickness. These more complex mixing functions will be considered in section 6.1, when their theoretical consequences can be quantitatively compared to the observed data.

4.4 *The Chemistry Function*

The melting function gives the extent of melting with distance above the solidus, the mixing function gives the relative amounts of each extent of melting present in the melting regime. In order to calculate crustal compositions, we need to know the chemical compositions of the various melts produced within the melting regime.

4.4.1 *Experimental Constraints on the Compositional Variations Caused by Partial Melting of the Mantle*

A function that describes the liquid composition during partial melting of the mantle must ultimately come from experimental data. Conceptually, this is a simple problem. A peridotite sample is loaded into an experimental capsule, heated and compressed to the temperature and pressure of interest, then quenched and analyzed. In practice, technical difficulties limit the experimentalist to working only at moderate pressures and with large melt fractions in the charge. Existing data provide quite complete information on moderate extents of melting [e.g., *Jaques and Green*, 1980; *Stolper*, 1980; *Fujii and Scarfe*, 1985; *Falloon et al.*, 1988; *Takahashi and Kushiro*, 1983; *Takahashi*, 1986]. But there are virtually no data for the relatively small extents of melting that may be so important for understanding melting processes beneath the ridge. This gap in the experimental data is particularly troublesome in view of the evidence that at small extents of melting, volatile species may play an important role in the melting process, and volatile-bearing experiments are lacking. Nonetheless, the experimental data that do exist allow many important conclusions to be drawn.

It has been known for more than twenty years that basalts formed at greater pressures are more alkalic, silica undersaturated, and olivine normative than those formed at lower pressures [e.g., *Green and Ringwood*, 1967]. The interpretation of more recent experimental data over the years has been somewhat encumbered by the phase diagram approach, where the focus has been on finding isobaric pseudo-invariant points, rather than illustrating the actual compositional changes that are occurring as a function of pressure and extent of melting. When one does examine the experimental data element by element, some simple systematics emerge [e.g., *Jaques and Green*, 1980; *Fujii and Scarfe*, 1985; *KL*87].

Figures 37 and 38 present some of the compositional changes that occur during mantle melting, based on the peridotite melting experiments of *Jaques and Green* [1980], *Fujii and Scarfe* [1985], and *Falloon and Green* [1987]. Figure 37 shows the compositional variations with temperature at 10 kbar for four different peridotite compositions, which include different experimental configurations run in different laboratories. While there are offsets due to differences in the starting compositions, the qualitative trends for the various elements are consistent among the different sets of experiments. For example, although the absolute Na_2O and Al_2O_3 contents of the melts are sensitive to the starting material, all the melting trends for these elements decrease with increasing T. Figure 38 shows the melting trends as a function of F for a single peridotite composition at three different pressures [from *Jaques and Green*, 1980]. By combining the information given in Figures 37 and 38, and examining the results of other melting experiments, the systematics for each of the major elements as a function of extent and pressure of melting can be described.

Na_2O, as well as TiO_2 and K_2O (not shown), behave incompatibly during melting. Their concentrations in the melt are greatest at low F and decrease with increasing F. MgO and Al_2O_3 also show strong variations with F. Al_2O_3 decreases while MgO increases in the melt with increasing F. There are also small pressure dependencies on Al_2O_3 and MgO concentrations at a given F. SiO_2 and FeO show marked pressure dependences. Higher FeO and lower SiO_2 melts are produced at higher pressures. On the other hand, the effect of progressive melting at a single pressure is relatively small for both SiO_2 and FeO. SiO_2 increases slightly with melting. FeO changes little, although its trend shows a kink when clinopyroxene disappears as a residual phase (near 25% melting). CaO variations during melting are complex, but there is good agreement between different sets of experiments; CaO contents in the melt increase steadily upon melting, reach a maximum when clinopyoxene is consumed (near 25% melting), and then decrease upon continued melting.

Many of these qualitative trends in the major element contents of the melt can be understood by considering the mineralogical changes that occur during melting of peridotite. For example, CaO contents are strongly linked to the melting behavior of clinopyroxene. For a given pressure, the initial melt occurs at the lowest temperature, where partition coefficients for Ca-bearing components are highest. As the temperature increases, clinopyroxene dissolves into the melt whenever Ca decreases below its equilibrium value. At the same time, the K_d is decreasing with increased temperature, and

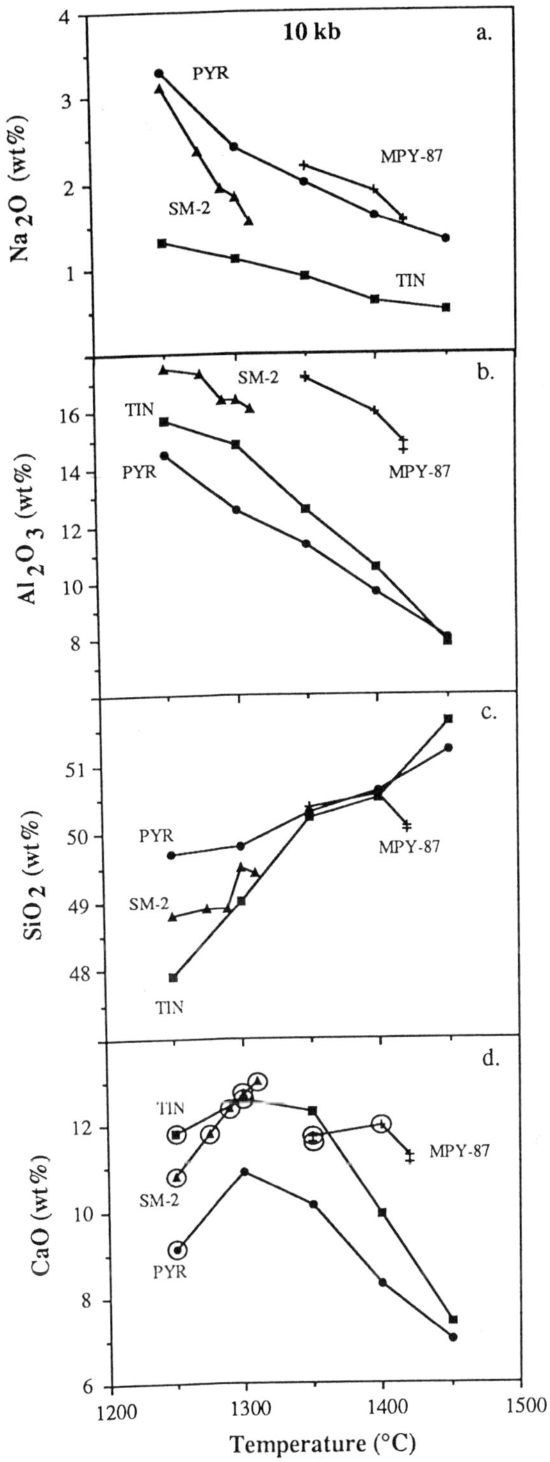

Fig. 37. Compositions of melts produced by melting of peridotite at 10 kbar as a function of temperature of the experiment for four different starting compositions. Solid circles are Hawaiian Pyrolite (PYR) from *Jaques and Green* [1980]; solid squares are depleted Tinaquillo lherzolite (TIN) from *Jaques and Green* [1980]; solid triangles are SM-2 from *Fujii and Scarfe* [1985]; and crosses are MORB Pyrolite sandwich experiments with about 10% basalt (MPY-87) from *Falloon and Green* [1987].

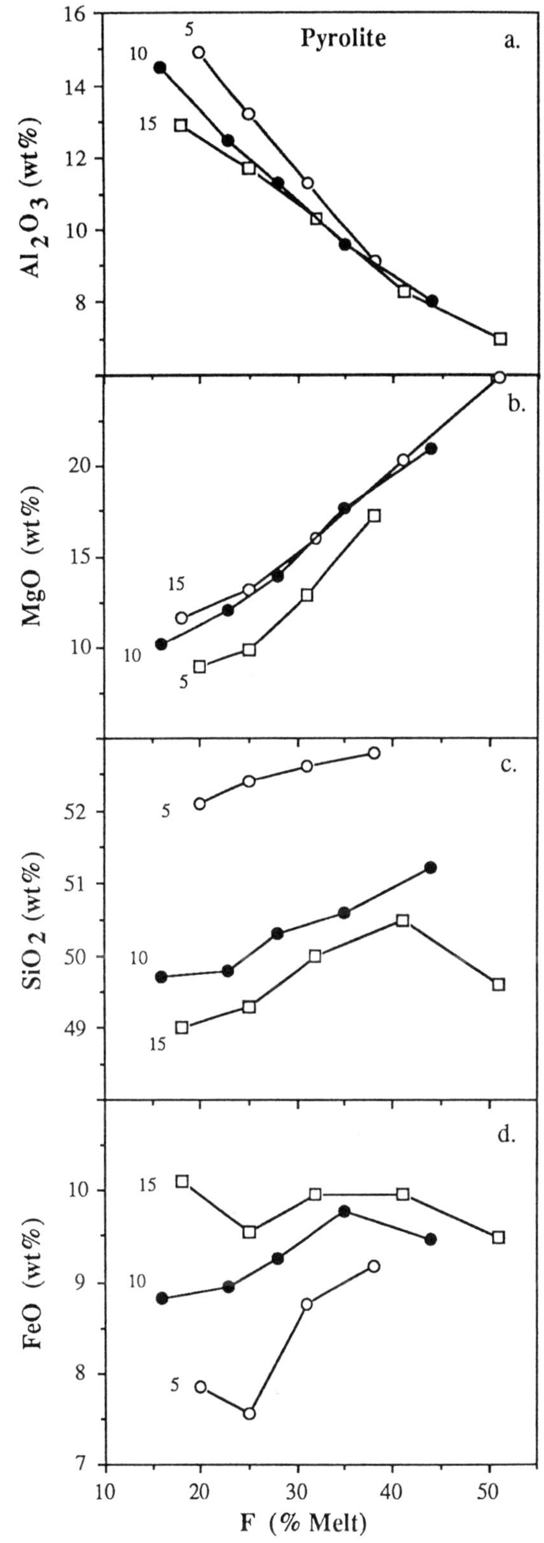

Fig. 38. Extent of melting (%F) vs. the compositions of melts produced by melting of peridotite of pyrolite composition at 5 kbar (open circles), 10 kbar (solid circles), and 15 kbar (open squares). Data from *Jaques and Green* [1980].

therefore the equilibrium content of CaO in the liquid increases. Thus CaO increases as long as clinopyroxene is available to contribute more CaO to the liquid. With the disappearance of clinopyroxene, CaO behaves incompatibly and is diluted by the addition of Ca-poor phases (olivine, orthopyroxene, spinel) to the melt.

Al_2O_3 decreases with melting because of the progressive melting of the aluminous phases, the progressive melting of clinopyroxene, and the progressive change in the spinel composition from Al-rich to Cr-rich. The small decrease in Al_2O_3 contents of the melt with increasing pressure is a consequence of more aluminous pyroxene compositions at higher pressures. The MgO, FeO and SiO_2 systematics are controlled largely by olivine partitioning relationships, as discussed in more detail below.

In addition to the qualitative trends indicated in the peridotite melting experiments, additional constraints on the melting behavior of the mantle come from combining experimental data with simple partitioning models. The partitioning models provide a means of extending the experimental data between and beyond the pressures, temperatures and compositions of the experiments themselves. For example, Na, Ti and K can be modeled as trace elements during batch or fractional melting, given information on their partition coefficients.

Since Na_2O plays a central role in our discussions, it is important to consider the constraints on its behavior. The Na_2O contents of olivine, orthopyroxene and spinel are almost zero, so the behavior of Na_2O during melting of spinel peridotite is controlled by the partition coefficient for Na between clinopyoxene and liquid, and by the relative abundance of clinopyroxene in the solid residue. Partitioning data for Na in clinopyroxene from various experimental studies at pressures up to 30 kbar are shown in Figure 39. The partition coefficient appears to increase with increasing temperature because of the increased jadeite component in residual clinopyroxene at higher pressures (and hence temperatures). The Na K_d is on the order of 0.2 for temperatures and pressures appropriate to most MORB. Since clinopyroxene makes up less than 20% of the mantle starting composition, the D for Na between solid and liquid begins at a value of less than 0.04, and continually decreases as melting proceeds and clinopyroxene is preferentially consumed by the melt. Thus Na behaves essentially as an incompatible trace element during melting of mantle peridotite. Similar approaches can be used to model Ti and K. Quantitative evaluation of Ti K_d's is given in Appendix A.

A second quantitative constraint comes from the very regular partitioning relationships for FeO and MgO between olivine and liquid, originally shown by *Roeder and Emslie* [1970]. The distribution coefficient for the exchange of Fe and Mg between olivine and melt [K_D; equation 12] is close to 0.3. Although there are slight temperature and pressure dependencies to this value [e.g., *Bender et al.*, 1978], the effects of increasing both temperature and pressure are compensatory, so that the value of K_D along the solidus remains close to 0.3. Olivine is always present during mantle melting, for it is the last phase to melt, and this fact, combined with the relatively constant K_D,

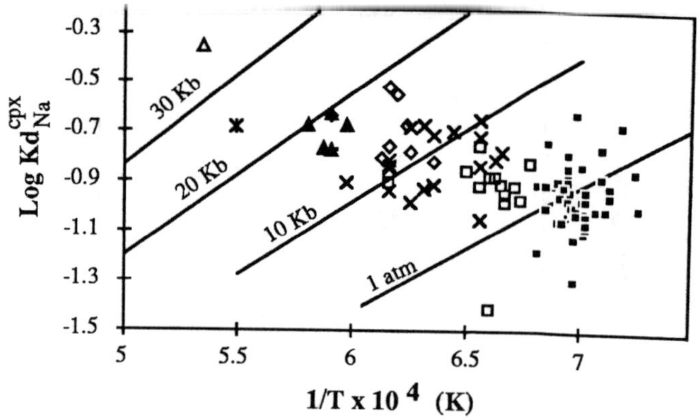

Fig. 39. Clinopyroxene melt K_d for Na at various pressures vs. inverse temperature for experiments on mafic systems. Data are from *Walker et al.*, [1979]; *Elthon and Scarfe* [1984]; *T. Falloon*, pers. comm. [1990]; *Grove and Bryan* [1983]; *Grove et al.*, this volume; *Tormey et al.* [1987]; some experiments reported in *Nielson and Drake* [1979]; *Bender et al.* [1978]. The different pressures are indicated by different symbols: small closed squares, 1 atm; open squares, 8 kbar; x's, 10 kbar; open diamonds, 15 kbar; solid triangles, 20 kbar; X with vertical slash, 25 kbar; and open triangle, 30 kbar. The curves were calculated from the equations given in Appendix A for the K_d's as a function of temperature and pressure. Each curve was calculated for different temperatures at a single pressure. Note the large pressure dependence for Na.

constrains the behavior of Fe and Mg during melting [*Hanson and Langmuir*, 1978; *Langmuir and Hanson*, 1980; Appendix B].

In fact, the strong pressure dependences of FeO and SiO_2 shown in the experiments can be predicted simply from the behavior of olivine partition coefficients as a function of temperature and pressure. The FeO and MgO K_d's for olivine increase with increasing pressure and with decreasing temperature. This produces competing effects on the olivine K_d's along the mantle solidus where pressure and temperature both increase. To determine whether the P or T effect dominates, the relative magnitude of their respective effects on the olivine K_d's must be considered. The pressure effect on the olivine K_d's is on the order of 5°/kbar, while the mantle solidus is on the order of 12°/kbar (see Appendix B for references). Thus, upon higher pressure melting of the mantle, the temperature effect on the olivine K_d's dominates, and the Mg and Fe K_d's are lowered. This means that the melt is richer in olivine components (higher Fe and Mg and lower Si) at higher pressures along the mantle solidus. This prediction based simply on the olivine K_d's provides independent confirmation of the trends in the Jaques and Green experiments, which suffered significant Fe loss [*Jaques and Green*, 1980; *Falloon and Green*, 1988].

Thus the qualitative compositional trends indicated by the peridotite melting experiments combined with quantitative partitioning relationships provide important constraints on the systematics of mantle melting. Any successful melting model must take into account the following three constraints:

1. Na, Ti and K can be modeled as incompatible trace elements, and their melting behavior must be consistent with

known clinopyroxene partition coefficients. The D for these elements decrease in the order Ti - Na - K.

2. FeO and SiO_2 contents of the melt are strongly controlled by the pressure of melting. FeO and MgO variations must be consistent with olivine partitioning, stoichiometry and mass balance. The FeO/MgO ratio of the melt must be consistent with a K_D of about 0.3 (equation 12).

3. CaO increases steadily with extent of melting as long as clinopyroxene remains in the solid residue, then decreases once clinopyroxene is consumed; this effect may vary with pressure, as the CaO contents of clinopyroxenes decrease with increasing pressure. Al_2O_3 contents of the melt decrease with increasing extent of melting. Thus the CaO/Al_2O_3 ratio increases with progressive melting as long as clinopyroxene is a residual phase.

4.4.2 A Critical Evaluation of Published Chemistry Functions

There have been a number of approaches to calculating mantle melt compositions during adiabatic ascent: *Langmuir and Hanson* [1980 (FeO and MgO only)], *KL*87, *McKenzie and Bickle*, [1988], and subsequent to the short course, *Niu and Batiza* [1991]. In the following section, we evaluate these melting models in light of the qualitative guidelines given above. We then present in section 5 a revised K_d approach for four elements, Fe, Mg, Na and Ti, which is fully quantitative and has several advantages over previous approaches.

*KL*87 calculated the concentrations of each oxide in the melt as a function of extent and pressure of melting, based on available experimental data and theoretical constraints. The abundances of different elements were calculated in different ways, depending on the information and theory available. For several elements (SiO_2, CaO, Al_2O_3, FeO), smoothed curves were fit to the experimental melt compositions as a function of extent and pressure of melting. An example is shown in Figure 40*a* for SiO_2. SiO_2 decreases with the pressure of melting and increases with the extent of melting. There is little experimental data to predict melt compositions at low extents of melting, and the data that do exist have experimental difficulties; nevertheless, the information available suggests that SiO_2 is low in melts produced by less than about 10% melting [c.f., *Falloon et al.*, 1988]. Using the general trends inferred from the experimental data as a guide, smoothed isobaric melting paths were drawn through the data (dashed lines in Figure 40*b*). The estimated SiO_2 content for any pressure and extent of melting can thus be read off the graph in Figure 40*b*.

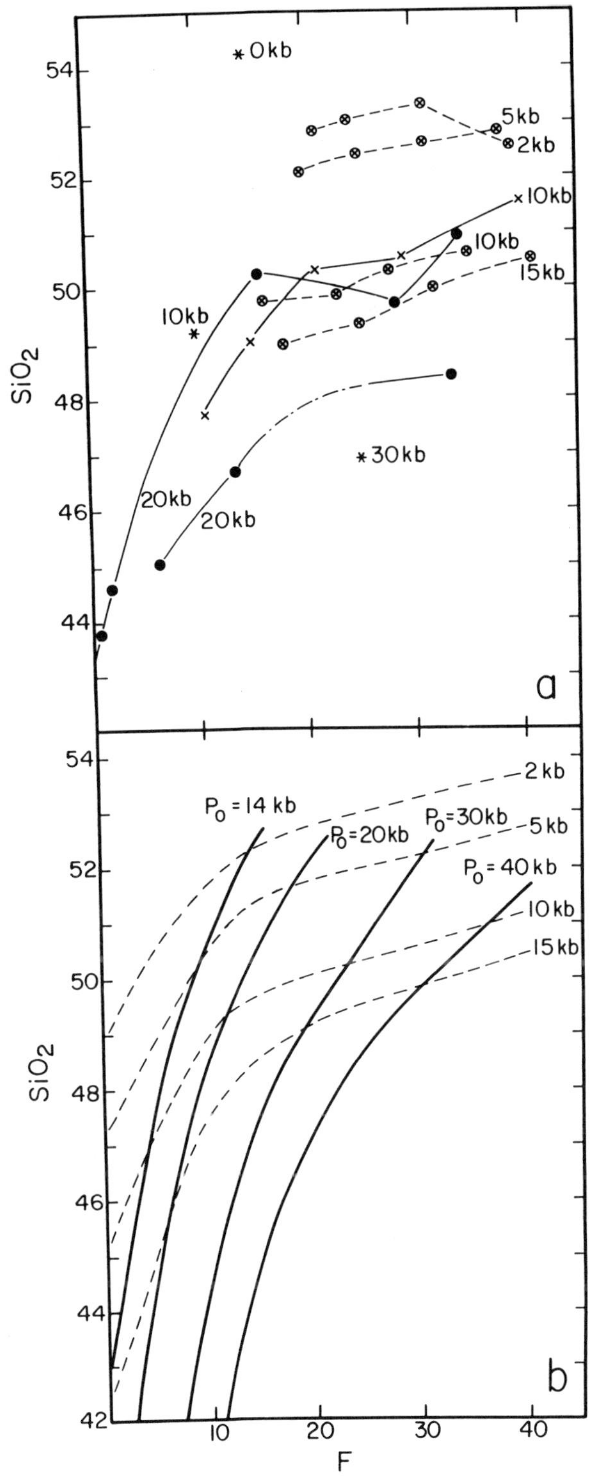

Fig. 40. SiO_2 (wt%) vs. F (extent of melting). *a*) Results of several experimental studies on the isobaric melting of various peridotitic compositions at the pressures indicated. Sources and starting compositions are: pyrolite (circled crosses), Tinaquillo lherzolite (crosses, only 10 kbar melts shown) [*Jaques and Green*, 1980]; garnet lherzolite +/-water (solid circles) [*Mysen and Kushiro*, 1977]; natural spinel lherzolite (asterisks; extents of melting estimated from reported Na_2O and TiO_2 abundances) [*Takahashi*, 1986]. *b*) From the trends of the experimental data shown in Panel (a), with particular emphasis on the pyrolite melting studies, isobaric melting curves (dashed) were estimated for 2, 5, 10, and 15 kbar. Assuming 1.2% melt/kbar pressure release, polybaric melting curves (solid) were estimated for P_o of 40, 30, 20, and 14 kbar, by fitting to the isobaric melting curves (e.g., for P_o = 40, 30% melting will be achieved by 15 kbar). The polybaric melting curves estimate the SiO_2 abundances in instantaneous melts present in the melting column at each pressure above the solidus. Note that SiO_2 contents at low extents of melting are unconstrained, and are probably too low for dry melting. From *Klein and Langmuir* [1987].

For those elements that behave as incompatible trace elements during mantle melting of spinel lherzolite (Na_2O, TiO_2), concentrations were calculated using partition coefficients consistent with peridotite melting experiments (e.g., Figure 39), and source compositions appropriate to the MORB mantle. MgO contents were constrained by the FeO contents using the method of *Langmuir and Hanson* [1980].

The *KL*87 chemistry function meets the three experimental constraints listed above, but suffers from being semi-quantitative and difficult for other investigators to reproduce.

McKenzie and Bickle [1988] presented a series of equations to parameterize experimental melt compositions. Instead of F, they defined a new parameter T', with which to parameterize the experimental data. T' is temperature normalized to the temperature interval between the solidus and liquidus of average peridotite, at the same pressure. By defining T', they were able to use virtually all experimental data, including experiments run on different starting compositions or experiments for which F was not determined. *MB*88 related T' to F by fitting a polynomial function to the limited number of experiments in which F had been determined.

Figure 41*a* shows experimental melt compositions plotted as a function of T', as well as the paramterization determined by *MB*88 to fit the data. For FeO and MgO, separate parameterizations were determined for 0, 10 and 20 kbar. SiO_2, calculated by difference from 100%, also shows a pressure dependence due to the pressure dependencies of the other parameters. These parameterizations express elemental variations as a function of inferred extent of melting at a given pressure, and are thus analogous, although by no means identical in form, to the isobaric melting paths estimated by *KL*87 (e.g., Figure 40*a*).

In spite of the large experimental data base used by *MB*88, and the quantitative nature of their parameterizations, their results do not produce improved representations of the chemical variations that occur during mantle melting. Indeed, one might argue that it is because *MB*88 used all experimental data and approached the data in a wholly statistical way that their isobaric melting curves have inherent problems. The problems are particularly apparent for CaO, FeO and Na_2O.

Figure 41 shows the experimental data for CaO and the preferred *MB*88 curve. Over a range of T' from -0.5 to -0.2 (corresponding to 0% to 34% melting), the original parameterization curve has a negative slope, which contradicts one of the constraints obtained from careful examination of the experimental data. The overlap of the various experiments prevents the data set as a whole from showing the positive slopes apparent for individual groups of experiments. Indeed, when they compared their results for CaO to the MORB data, the observed lack of agreement led them to add an arbitrary correction factor (a "pressure dependence") to the parameterization.

A second problem in the parameterization of *MB*88 concerns FeO. *MB*88 recognized that both FeO and MgO show strong pressure dependencies. However, their pressure dependence for FeO in the preferred parameterization is very small, changing by just over 1 wt% FeO over a pressure interval of 20 kbar at

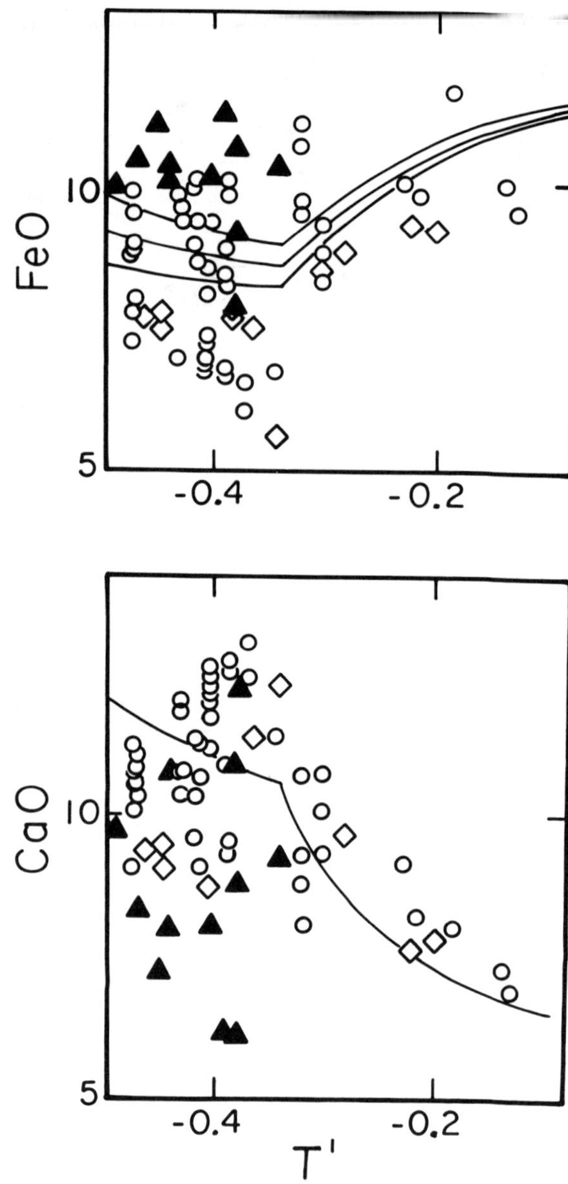

Fig. 41. Compositions of melts produced experimentally by melting of peridotite as a function of T' (which is related to extent of melting). Crosses mark compositions of experiments carried out at pressures < 0.5 GPa, circles at pressures of 0.5< P < 1.5 GPa, and triangles P > 1.5 GPa. Curves show parameterizations of the experimental melt compositions as a function of T'; where three curves are shown, they are the calculated compositions at pressures 0 (not labelled), 1 and 2 GPa. After *McKenzie and Bickle* [1988]; see this reference for experimental data sources and explanation of T'.

0% melting. Figure 41 shows that this change is too small to account for the data. The *MB*88 parameterizations also do not satisfy the K_D constraints for the partitioning of FeO and MgO between olivine and liquid (discussed above; equation 12). For example, their calculated normal MORB composition yields a K_D of 0.24 for equilibrium with an Fo_{90} olivine, which is far outside the experimental limits of 0.3±0.03 (but see below for an improvement by *Watson and McKenzie*, [1991].

A third problem concerns Na_2O. The bulk distribution coefficient (D) for Na_2O given by *MB*88 is 0.169, a value which exceeds by about a factor of four the D permitted by the experimental data (from Figure 39, and assuming a maximum of 20% cpx in the bulk mantle). The D also stays constant in the *MB*88 calculations, instead of decreasing as clinopyroxene progressively melts out of the residue and the pressure decreases. The source concentration of Na_2O used by *MB*88 is 0.44 wt%, while the source concentration for Na_2O in the mantle is well constrained to be about 0.3 (± 0.03) wt%, based on data from primitive mantle nodules [*Jagoutz et al.*, 1979], peridotite suites [*Frey et al.*, 1985] and chondritic considerations [*Hart and Zindler*, 1986]. The combination of these factors creates serious discrepancies between the *MB*88 Na_2O calculation and that expected from more appropriate values of D and source abundance, as illustrated in Figure 42. The *MB*88 values are too low at low extents of melting, and too high at higher extents of melting. At high values their calculations violate mass balance, with more Na present in the liquid than is present in the initial source. Thus, while the calculations consistent with the experiments have Na_2O vary by almost a factor of four from 0-20% melting, the *MB*88 calculations produce variations of only 60%.

Watson and McKenzie [1991] have recently presented a revised chemistry function. Their revised function has the same problems with respect to CaO and Na_2O, but has been made consistent with the olivine/liquid K_D for FeO and MgO.

The *MB*88 and *Watson and McKenzie,* [1991] approach has the advantage of being a quantitative parameterization that can be put into a variety of models, and it should be able to be modified to produce chemical variations more consistent with the experimental constraints given above. In their current forms, however, these chemistry functions are not adequate for detailed modeling of mantle melting.

Niu and Batiza [1991, hereafter *NB*91] have recently presented a third approach to the determination of a chemistry function for mantle melting. Their chemistry function is based largely on the experiments of *Jaques and Green* [1980], although they applied corrections to the data in an effort of account for the systematic errors identified by *Falloon and Green* [1987; 1988]. *NB*91 extract apparent D's from the experiments using a mass balance equation. This enables them, in turn, to reproduce a given set of experiments quite well. From one set of experiments to another, however, the apparent bulk distribution coefficients change, so one set of values cannot be applied to a variety of potential mantle compositions.

Because a complete set of experiments on their preferred mantle composition does not exist, *NB*91 manipulate the apparent D's in an effort to reproduce the limited experiments that do exist on this composition. This makes it difficult to evaluate the accuracy of the results outside the experimental range. The *NB*91 parameterization for equilibrium melting, however, does reproduce the general trends of the experimental data, such as the changes of Na_2O and CaO with F. Therefore, the *NB*91 approach should be able to effectively parameterize equilibrium melting where there is a complete set of experiments on a single bulk composition.

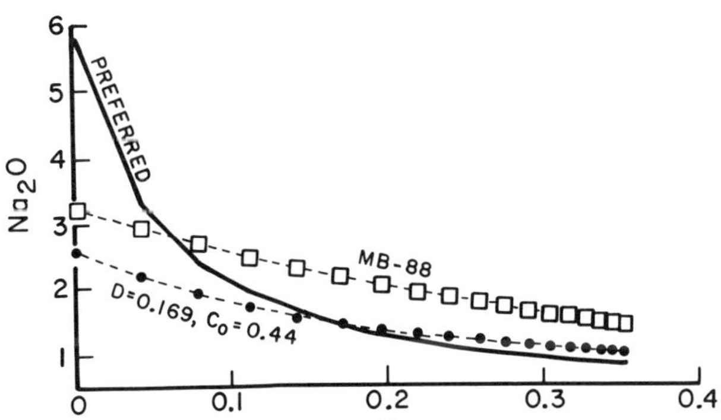

Fig. 42. Calculated wt% Na_2O in melts vs. extent of melting (F). Open squares show the results using the parameterizations of *McKenzie and Bickle* [1988]. Filled circles are a calculation for equilibrium melting using a bulk D of 0.169 and C_0=0.44 wt% Na_2O, the source abundance used by *McKenzie and Bickle* [1988]. The solid line shows a preferred equilibrium melting calculation that is consistent with the experimental data on Na partitioning, using a C_0 of 0.3 wt% Na_2O, a K_d for clinopyroxene of 0.15 (bulk D of about 0.03), and assuming that clinopyroxene progressively melts out of the residue and is no longer a residual phase by 22% melting (thus, D is not constant).

4.5 *Results of Previous Attempts to Combine Melting, Mixing and Chemistry Functions*

In sections 4.2, 4.3 and 4.4 above, we have described some of the reasoning and uncertainties in the development of the three functions necessary to calculate melt compositions produced during mantle upwelling beneath the ridge. In the following, we describe three previous attempts to combine melting, mixing and chemistry functions and compare these results to the MORB data set. We then present a new quantitative model for calculating mantle melt compositions.

Klein and Langmuir [1987] adopted a constant value for ($\partial F/\partial P$) of 1.2% melt/kbar, similar to the value used by *Ahern and Turcotte* [1979]. Their chemistry function, based on the results of equilibrium melting experiments at different pressures, required the assumption of batch melting during ascent. From the isobaric melting curves (e.g., dashed lines in Figure 40*b*), polybaric melting curves were calculated for different pressures of intersection of the mantle solidus (solid lines in Figure 40*b*). *KL*87 used a linear mixing function, where the polybaric melt compositions were weighted by F (their equation 9). This pooled melt is the complement to the residual mantle column depicted in Figures 2 and 34.

The pooled melt compositions calculated in this way for different pressures of intersection of the mantle solidus result from different mantle temperatures. If these pooled melt compositions then crystallize at low pressure to 8 wt% MgO, their compositions can be compared to the MORB data. The chemical variations produced by different pressures of intersection of the mantle solidus (taller or shorter RMCs) result in an inverse correlation between $Fe_{8.0}$ and $Na_{8.0}$. Mantle that intersects the solidus deep produces an aggregate melt of high $\overline{F}$ and $\overline{P}$ (low sodium, high iron), while mantle that intersects the solidus shallow produces an aggregate melt of low $\overline{F}$ and $\overline{P}$ (high sodium and low iron; Figure 43*a*). In addition, the resulting pooled melt compositions form an inverse correlation between $Na_{8.0}$ and CaO/Al_2O_3. Thus, the model adopted by *KL*87 reproduces the global vector of chemical variability in its overall trends. *KL*87 also modeled the expected correlation between $Na_{8.0}$ and crustal thickness (assuming two slightly different concentrations of Na_2O in the source), and these calculations agree with the trend of the observations (Figure 44).

In detail, however, the calculated chemical variations show discrepancies with the observed global chemical variations. Compared to the global vector, the *KL*87 model reproduces the observed global range in $Na_{8.0}$, but only half the observed range in $Fe_{8.0}$. Furthermore, the calculated melts are offset to lower $Fe_{8.0}$ contents for the same $Na_{8.0}$ (Figure 43*a*). Another important test is to see how well the *KL*87 model reproduces normal ocean crust of average depth and crustal thickness, such as the East Pacific Rise between 8°N and 14°N. The *KL*87 calculated melt composition for 6 km crustal thickness is compared to EPR data in Figure 43*b* where it can be seen that the calculated $Fe_{8.0}$ value is significantly lower than observed in the EPR data. In addition, the calculated $Si_{8.0}$ value is slightly high, despite the fact that *KL*87 used very low SiO_2 contents for low extents of melting (see Figure 43*a*), an aspect of their paper that has been criticized by *Falloon et al.* [1988].

In spite of these discrepancies, the general consistency of the calculated mean melt compositions and the observed regional averages suggests that the global vector results from regional variations in mantle temperature. The failure of the model to reproduce the observed variability in detail suggests that some aspect of the *KL*87 model (chemistry, melting and/or mixing function) needs modification.

MB88 combined their chemistry function (for equilibrium melting), with a non-linear melting function, and a linear mixing function. The *MB*88 results approximate the MORB data, but do not reproduce the absolute values or slopes of the observations in detail (Figure 43). Their total range in Na is 1.6-2.7, only two-thirds of the observed range in $Na_{8.0}$ values of 1.7-3.6. More importantly, all pooled melts with >2.4 wt% Na_2O produce less than 4 km of crust, while a pooled melt with 3 wt% Na_2O, a common value for MORB, produces zero crustal thickness! (Figure 44) These problems can be traced directly to their chemistry function. As discussed above, the inappropriately high D value for Na_2O results in too narrow a range in calculated Na_2O contents. Secondly, without the use of the arbitrary pressure correction for CaO noted above, there is very little range in the calculated CaO/Al_2O_3 ratios (their Figure 15*b*). Lastly, like the *KL*87 results, the *MB*88

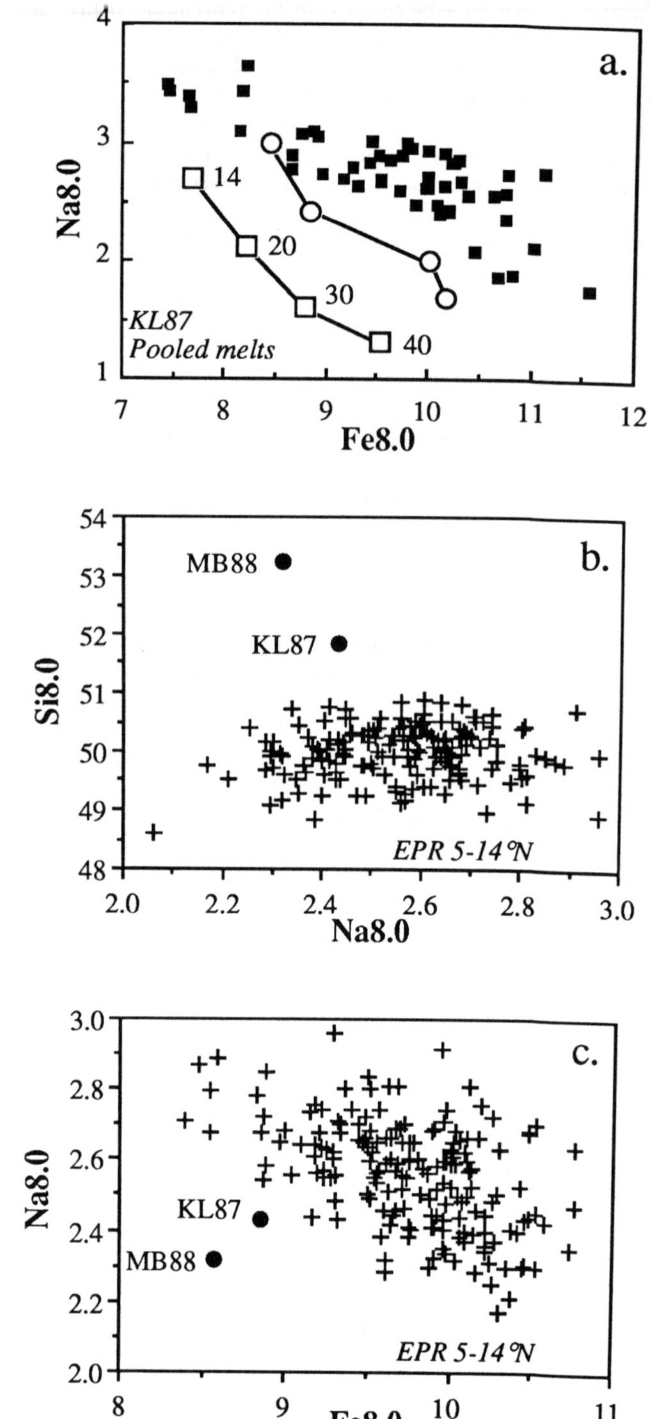

Fig. 43. $Na_{8.0}$ vs. *a*) $Si_{8.0}$ and *b*) $Fe_{8.0}$ for samples from the East Pacific Rise between 5°N and 14°N (data as in Figure 24), and the mean melt compositions for "normal" MORB, corrected for the effects of olivine fractionation to 8 wt% MgO, reported by *McKenzie and Bickle* [1988] and *Klein and Langmuir* [1987]. Note that both the *MB*88 and *KL*87 calculations are too high in $Si_{8.0}$ and too low in $Fe_{8.0}$ to account accurately for normal EPR compositions. This results from the fact that both of them calculated equilibrium melting rather than fractional melting (see discussion in text).

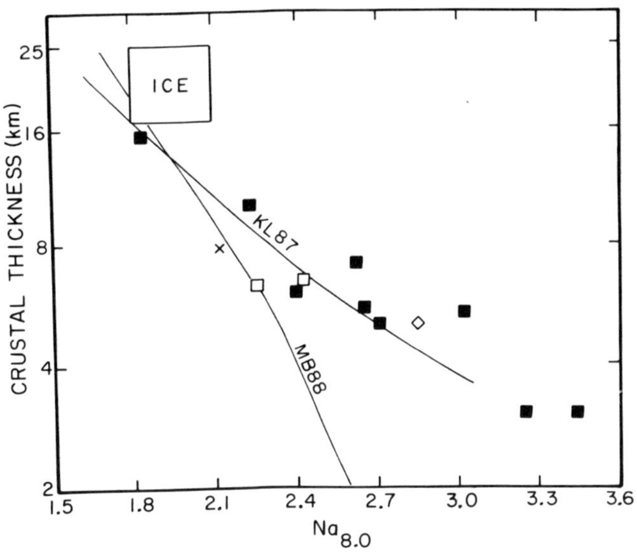

Fig. 44. $Na_{8.0}$ vs. crustal thickness for observed oceanic crust and calculated pooled melts from *KL*87 and *MB*88. The steepness of the *MB*88 results from the high effective partition coefficient for the Na_2O calculations in their model (see Figure 42). Crustal thickness data from Figure 18.

parameterization produces melts that have too little change in FeO, and generally too low FeO and too high SiO_2 contents (Figure 43).

In spite of our criticisms of the *MB*88 chemistry function, we should not lose sight of the fundamental agreement between the conclusions of *KL*87 and *MB*88. Both studies suggest that partial melting of the mantle exerts a major control on MORB chemistry, and particularly on Na_2O contents. Both suggest that the differences in partial melting are in response to mantle temperature, and correlate with crustal thickness. Both use equilibrium melting and a linear mixing function to calculate mean melt compositions. All of these factors indicate substantial agreement on the big issues concerning melt formation and migration. The agreement between the two studies also extends to an important discrepancy between the $Fe_{8.0}$ and $Si_{8.0}$ calculated from peridotite melting experiments and those observed in MORB.

Niu and Batiza [1991] used a different approach in the formulation of their chemistry function based on the empirical derivation of apparent D's for each oxide as a function of P and F for different bulk compositions. One of their goals was to improve upon the approach of *KL*87 and *MB*88 by modeling the effects of fractional, rather than equilibrium, melting. Their mixing function involves the pooling of melt compositions produced at 1% melt increments for different pressures of intersection of the solidus and final pressures of melting. Their melting function is based on an adiabatic gradient during melting of 6°/kbar from *McKenzie* [1984] and melting contours from the experiments of *Jaques and Green* [1980].

The *NB*91 approach reproduces the experiments of Jaques and Green more reliably than *MB*88, and captures the behavior of Na and Ca well. Thus the chemistry function for equilibrium melting is useful. There are difficulties with the application of the chemistry function to the general melting problem, however. First, it is not clear whether an apparent D determined for equilibrium melting at relatively large extents of melting can be used to model fractional melting. Second, for their melting function *NB*91 superimpose a melting adiabat from *McKenzie* [1984] onto melting contours determined by *Jaques and Green* [1980]. This juxtaposition does not take into account the thermodynamic link between temperature path and extent of melting, and causes their melting function to produce a large extent of melting near the solidus, without the requisite temperature drop that is required to supply the heat of fusion. This leads to an impossibly high value to 5% melt per kbar of pressure release near the solidus. Finally, *NB*91 do not use a suitable mixing function that is tied to physically realistic models of steady-state melt generation. Therefore although *NB*91 provide a useful parameterization of equilibrium melting experiments, their detailed conclusions concerning MORB generation need to be reconsidered in light of the need for a combination of appropriate chemistry, melting and mixing functions to model mantle melting beneath ridges.

5. New Quantitative Model for Mantle Melting

There are drawbacks to all three melting models discussed above. First, because equilibrium melting is ideally the type of melting that occurs in the experiments, any method based on empirical reproduction of the experimental results can be reliable only for equilibrium melting. Answering the question of how the mantle melts depends critically on being able to quantitatively model a variety of melting processes in addition to equilibrium melting, such as fractional melting and continuous melting. Second, previous model results are inflexible with respect to mantle source composition, in that they cannot reliably model mantle compositions that differ significantly from those of the experimental starting compositions. Thus, these models cannot be used to address questions of mantle heterogeneity.

These two difficulties can be overcome by an alternative approach to calculating the chemistry function for mantle melting. If partition coefficients between individual minerals and liquid can be parameterized as a function of temperature, pressure and liquid composition, then equilibria can be calculated rigorously. This approach has been used effectively for the calculation of fractional and equilibrium crystallization [*Langmuir and Hanson*, 1981; *Bender et al.*, 1984; *Nielsen and Dungan*, 1983; *Weaver and Langmuir*, 1990], for the evaluation of complex crystallization processes [e.g. *Nielsen*, 1989, 1990; *Langmuir*, 1989], and is the method used to evaluate high pressure crystallization in this paper for the results of section 6.3 and Appendix C. The equations apply equally to melting and crystallization, and can be applied to a diversity of compositions and processes.

The K_d approach to calculating chemical equilibria, like the other chemistry functions discussed in section 4.4, depends on an experimental data base. However, it has the advantage that the necessary data can be obtained from a variety of experiments, even those in which only one solid phase is present and the melt fraction is large. This contrasts with the approach that monitors change in the liquid composition and requires saturation of the liquid with multiple solid phases [e.g., *Stolper*, 1980; *Fujii and Scarfe*, 1985; *Kinzler and Grove,* 1992*a*]. Thus the K_d approach allows the use of experiments that have the fewest experimental difficulties and is able to make use of a much larger experimental data base.

The K_d approach we use is not similar to the approach of *NB*91, although they also use "distribution coefficients" for their calculations. The distribution coefficients used by *NB*91 are bulk distribution coefficients (D) for a multiphase assemblage. While K_d's can be parameterized as a function of T, P, and composition and can be applied to a variety of conditions, *NB*91 parameterize their D's as a function of melt fraction, which depends on the bulk composition and the melting process. Their results, therefore, apply rigorously only to a single bulk composition and a single process, equilibrium melting.

For the complete calculation of mantle melting, it would be necessary to consider all components and all possible phases. Then the proportions and compositions of all phases in the system could be determined for a given bulk composition. But such a complete understanding of partition coefficients for the mantle system is lacking. For melting of the mantle, however, there is an *a priori* constraint that allows MgO and FeO to be calculated even in the absence of complete characterization of the partition coefficients for all elements and all phases. The constraint is that olivine is always a residual phase, and hence the liquid must be in equilibrium with olivine at each pressure, temperature and extent of melting required by the melting function. This constraint allows the calculation of the FeO and MgO contents of the liquid as melting proceeds. The specific equations that allow this calculation are presented in Appendices A and B. We emphasize that the calculations require that olivine be present in the residue, but do NOT require that olivine be the only residual phase. The olivine-melt K_d's must be obeyed regardless of what other phases exist. The calculations apply from solidus to liquidus as melting proceeds.

For the accurate calculation of MgO and FeO contents during melting, it is necessary that the temperature, pressure and compositional controls on the olivine partition coefficients be taken into account. There have been many investigations of olivine/liquid equilibria, over a substantial range of P, T and X, and several hundred experiments are available to calibrate the K_d's (see references in Appendix B). We have re-examined these experiments in order to develop appropriate equations for the FeO and MgO K_d's for the calculations. In addition to T and P, the olivine K_d's are sensitive to melt composition, particularly to the concentration of the alkalis [e.g., *Roeder*, 1974; *Ford et al.*, 1983; *Kinzler and Grove*, 1992*a*; and Appendix B]. For a given P and T, the olivine K_d's increase substantially with increasing alkali content. The parameterization of the Fe and Mg K_d's in olivine is sufficiently precise that it should lead to relatively small errors in the melting calculations. The average error of recovery for the partition coefficients over a much wider range of pressure, temperature and composition than is relevant to mantle melting is 5%, which is on the order of 0.5 wt% FeO or MgO (Appendix B). The compositional dependence of the Fe and Mg K_d's is a major difference between these calculations and those carried out by *Langmuir and Hanson* [1980] and in the earlier version of this paper distributed in July 1990. The earlier calculations used only a temperature dependence on the olivine partition coefficients with no compositional dependence.

Partition coefficients also permit the calculation of major elements such as TiO_2, Na_2O and K_2O that behave similarly to trace elements. We treat these three elements as trace elements, but for TiO_2 and Na_2O the partition coefficients change with temperature, pressure, and the fraction of clinopyroxene in the residue. As discussed in Appendix B, these two elements seem to have different dependencies on pressure and temperature. The Na K_d for clinopyroxene has a particularly strong pressure dependence, presumably due to the increased stability of the jadeite component in clinopyroxene with increasing pressure. The effects of liquid composition on these partition coefficients should be added as more experimental data become available, but in general the values of the K_d's are small enough that compositional variation should not have a large effect on the computed results, once melting proceeds beyond a few percent.

To carry out the calculation, the source composition and mantle temperature at some depth below the solidus must be specified. Pressure and temperature for the initiation of melting are then determined by the mantle solidus (see Figure 31*a*). The subsequent relationships among temperature (T), pressure (P) and extent of melting (F) are given by the melting function discussed in section 4.2. T, P and F from the melting function are used to calculate the K_d's and liquid composition for Na_2O, K_2O and TiO_2, assuming non-modal melting with clinopyroxene decreasing by 0.7% in the residue for each percent of liquid produced. This liquid composition, along with T and P, allows calculation of the FeO and MgO K_d's between olivine and liquid. Then the solution to the quadratic equation in Appendix B gives the FeO and MgO contents of liquid and residual solid.

This calculation can be understood intuitively. Because the FeO-MgO exchange distribution coefficient, K_D, is about 0.3, the FeO/MgO of a liquid in equilibrium with any mantle olivine composition is fixed. Thus the liquid composition is constrained on an MgO-FeO diagram to a line radiating from the origin. The melting function gives temperature and extent of melting at a given pressure, for a given pressure of intersection of the solidus. The K_d equations then provide the K_d's for MgO and FeO, and these fix the point of liquid composition on the line. Combining these constraints with the mass balance relationships among melt, source and residue allows the calculation for all extents of melting.

The chemical calculations can be combined with diverse melting and mixing functions to quantify the chemical consequences of mantle melting processes. Specific advantages of the calculations are that they are accurate for the elements determined (Fe, Mg, Na, K and Ti); they permit a quantitative evaluation of different styles of melting, such as fractional or equilibrium melting; they can be used to quantitatively model mantle heterogeneity; and they can be used reliably for the small extents of melting that are difficult to address experimentally. Although the new calculations do not include all the major elements, the five elements considered are calculated rigorously, and independently of the other elements, so the results are valid. Most important for the topics considered here, we are able to calculate the two elements (Na and Fe) that are particulary diagnostic of MORB variability and thus to evaluate the origin of the local and global systematics of compositional variability.

5.1 Illustration of Calculated Results

In the following sections we begin with simple examples of melting to illustrate the quantitative results of the new melting model under simple melting conditions.

5.1.1 Isobaric Melting

The simplest case to consider is isobaric melting, where melting occurs as the temperature increases at constant pressure. F can be calculated from dT/dF (see section 4.2). Knowledge of T and F permits calculation of the Na_2O, K_2O and TiO_2 of the liquid, which permits the calculation of FeO and MgO K_d's, and finally the FeO and MgO contents of the liquid. Figure 45 shows results for MgO, and FeO and Na_2O for isobaric, equilibrium melting at several pressures. At the lowest extents of melting, the FeO and MgO contents are low because the K_d's are high due to both high alkali contents in the liquid and low temperatures. FeO and MgO both initially increase as the extent of melting increases, though of course the FeO/MgO ratio decreases. For a given MgO content or given extent of melting, FeO increases with increasing pressure (Figure 45).

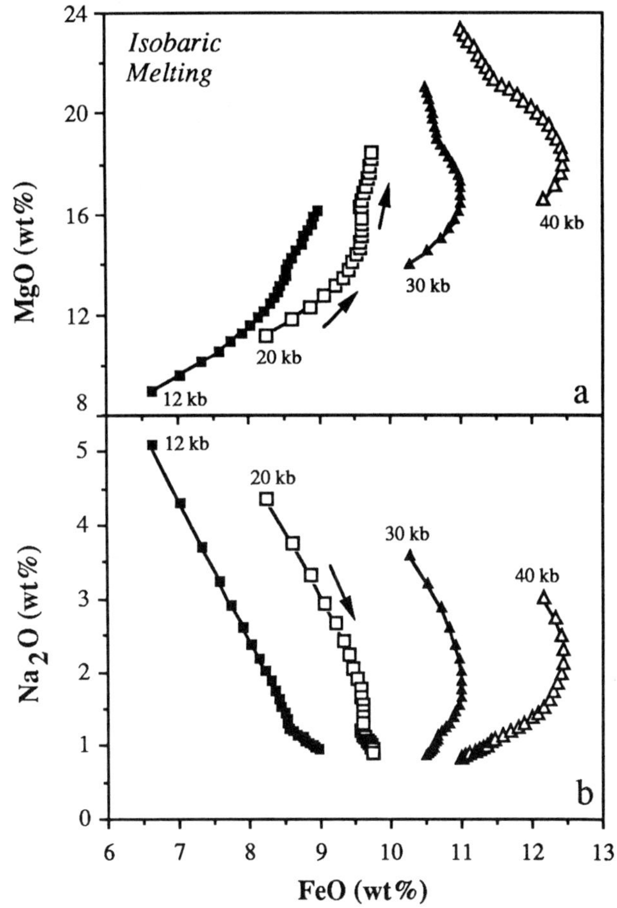

Fig. 45. Results of the new melting model for isobaric melting of the mantle. FeO vs. *a*) MgO and *b*) Na_2O for melting at 12, 20, 30, and 40 kbar. Arrows indicate increasing degrees of partial melting, from about 2% to 35% melting for each curve. The kinks in the paths correspond to where clinopyroxene is no longer a residual phase (at ~22% melting).

5.1.2 Equilibrium and Fractional Melting during Adiabatic Decompression from 20 kbar

In order to calculate melting during adiabatic ascent, a melting function must specify the amount of melting produced for a given pressure drop (section 4.2). The melting function differs for equilibrium and fractional melting, with fractional melting producing a smaller increment of melting for the same pressure decrease. Because the partition coefficients change with each pressure and temperature, all elements for fractional melting are calculated incrementally, with the liquid entirely removed at each 1 kbar pressure step.

We first present results that do not involve a mixing function. We simply follow the evolution of liquid compositions in a single unit of peridotite as it intersects its solidus and decompresses (Figure 46*a*). We call these trajectories "unit paths" since they apply to one unit of mantle as it passes through the melting regime. Note that this is distinct from the integrated melts that result from mixing of melts from many unit paths in a melting regime.

For equilibrium melting, the liquid stays with the residue as melting proceeds, and the curve in Figure 46*a* shows the progressive evolution of the liquid at the different pressures within the melting regime. For fractional melting, liquid is extracted at each pressure step. Figure 46*a* shows both the instantaneous melts produced at each pressure step, and the progressive change of the accumulated fractional melts, where each melt produced along the unit path is mixed with those produced previously. In all cases, the FeO contents first increase at low extents of melting, then decrease as melting proceeds. The early increase is due to the compositional effects on the partition coefficients. The Na_2O and K_2O concentrations are high in the first increments of melting, causing the K_d's to be relatively high, and FeO and MgO to be relatively low (see Figure 46*a*). As the extent of melting increases, Na_2O and K_2O decrease, causing a decrease in the K_d's and a corresponding increase in the FeO and MgO contents of the liquid. At still higher extents of melting, the alkali contents change relatively little, and the pressure and temperature effects

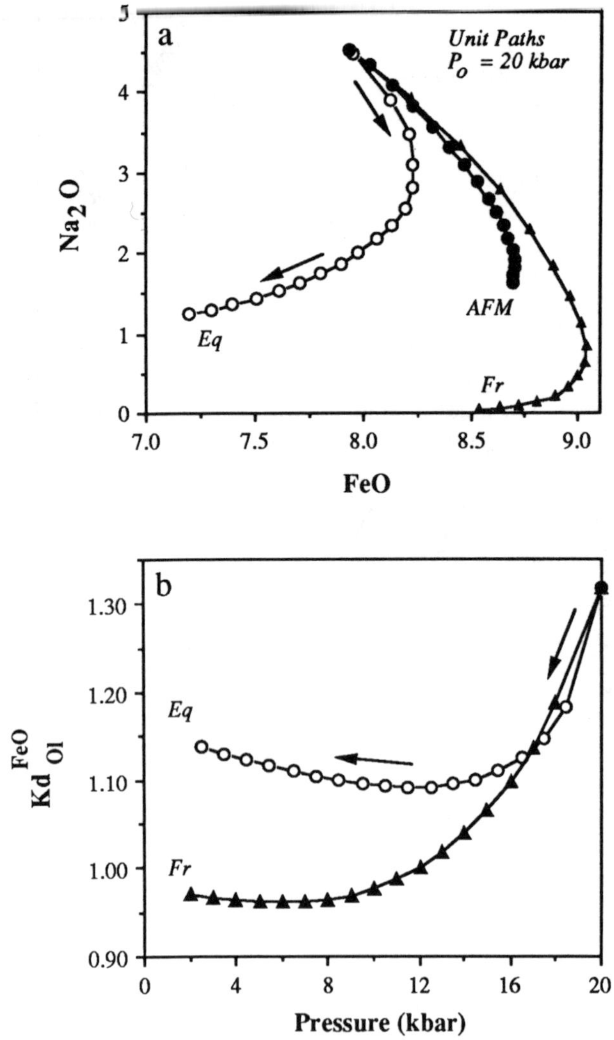

Fig. 46. *a)* Na_2O vs. FeO for melt compositions predicted from the new melting model for a single unit of mantle during progressive adiabatic melting. Three such "unit paths" are shown for equilibrium melting (Eq), fractional melting (Fr), and accumulated fractional melting (AFM) for mantle that intersects the solidus at 20 kbar, and melts to ~2 kbar (a pressure approximately equivalent to the base of the normal ocean crust). The AFM path results from summing the melts along the fractional unit path. Arrows show direction of increasing extent of melting (melting paths begin at 2% melting). Each point along the curve represents a 1 kbar drop in pressure. *b)* Variation in the olivine/melt K_d for FeO as a function of pressure along the unit path for equilibrium and fractional melting. K_d is calculated as described in Appendices A and B and section 5.0.

dominate. Since the FeO concentration in the melt is sensitive to pressure, FeO decreases substantially along the lower pressure portions of the unit paths.

The contrasts between the fractional and equilibrium unit paths can be understood as a consequence of the different temperatures and liquid compositions that exist at a given pressure. Fractional melting strips most of the alkalis from the residue at low extents of melting. This causes large decreases in the K_d's at low extents of melting and more rapid increase of FeO in the liquid (Figures 46*a* and 46*b*). The temperature of the peridotite for fractional melting is also higher at any given pressure and extent of melting. This causes the K_d's to be lower, and the FeO contents higher. The combination of these effects causes the fractional melts to be substantially richer in FeO than the equilibrium melts.

Kinzler and Grove (1992*a* and *b*) have recently and independently carried out calculations of mantle melting based on a different experimental data set and an entirely different method of calculation. They also found curved melting paths due primarily to compositional effects on olivine partitioning, similar to those presented here.

5.2 Importance of the Melting Process to Calculated Liquid Compositions

In Figure 46*a*, the three different melting processes lead to distinct paths of liquid compositions, which illustrates the importance of understanding the physical aspects of the melting process in evaluating models of basalt petrogenesis from experimental data. Contrasts become even more pronounced when diverse mixing functions are included. It is clear, therefore, that interpretation of MORB data cannot be made solely from equilibrium experiments alone, but depends critically on the melting process. Although this complicates the straightforward interpretation of experimental data, it is nonetheless advantageous. Because different melting processes lead to different effects on melt compositions, basalt compositions can be used to evaluate the melting processes that gave rise to them.

5.3 Calculation of Equilibrium and Fractional Melting Beneath Ocean Ridges

To apply the calculations to ocean ridges, we need to use appropriate mixing functions. As discussed in section 4.3, these functions differ for equilibrium and fractional melting. For equilibrium melting, we can simply sum the melts as indicated in the RMC. For fractional melting, each "box" of the RMC has provided fractional melts from the total range of pressures along the unit path, so there is an additional step of integration that is required to sum the instantaneous melts into an accumulated fractional melt. Therefore calculation of fractional melting requires summing of the melts along the unit path, to calculate each box of the RMC, and then summing the resulting AFM liquid from each box. Figures 47 and 48 illustrate both the unit paths within the melting regime and the integrated mean melts from the entire melting regime.

Calculated melt compositions for equilibrium and accumulated fractional melting are shown in Figures 47 and 48 for various pressures of intersection of the solidus (P_o). As P_o increases, so does the overall extent of melting, and the resulting mean melts have lower Na and higher Fe. Fractional melting leads to higher Fe than equilibrium melting for a given Na, due to a combination of four factors. (1) Above, we discussed the effects of fractional melting on the K_d's, which leads to higher Fe for the same unit path (Figures 46*a* and 46*b*).

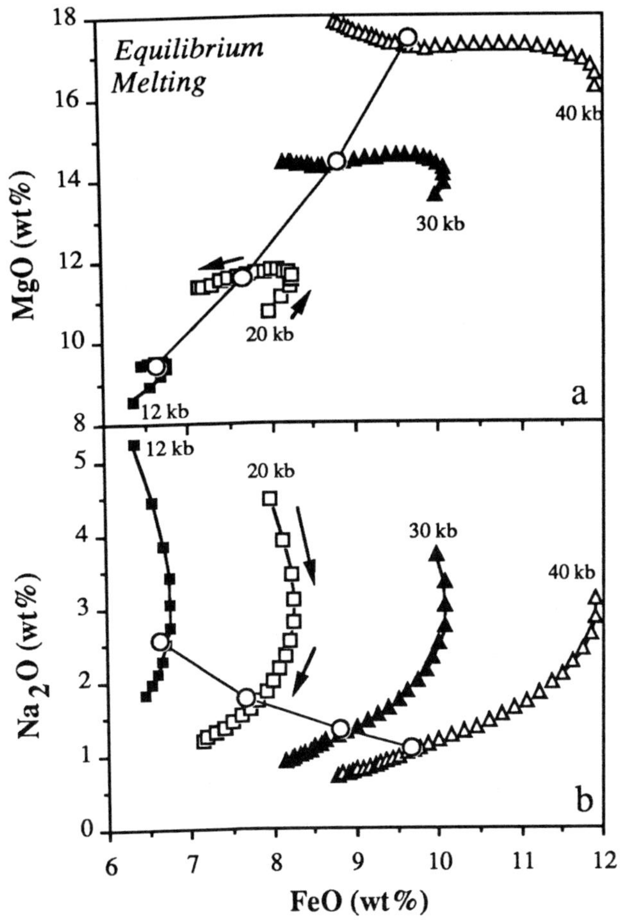

Fig. 47. FeO vs. *a*) MgO and *b*) Na_2O for equilibrium melts of the mantle, calculated from the new melting model. Each curve represents the family of melt compositions produced within the melting regime for a given pressure of intersection of the solidus (as labeled). Arrows show direction of increasing extent of melting (melting paths begin at 2% melting). Each point along the curve represents a 1 kbar drop in pressure. Open circles are melt compositions that result from pooling the melts along each curve.

(2) Pooling melts from throughout the melting regime leads to further increases in Fe. The mean pressure for any given column in the melting regime is twice as high for fractional melting (Figure 34 and section 4.3.2), leading to higher Fe melts. (3) The range of pressures for the various accumulated fractional melts from a single melting regime is lower because the central portion of the melting regime has a mean pressure of about half P_o, rather than a pressure corresponding to the base of the crust. The more limited range in pressures leads to a smaller decrease in FeO contents with greater extents of melting within any given melting regime. (4) Finally, the mean extent of melting is slightly lower, leading to slightly higher Na in the mean melts for any P_o.

One other important difference between equilibrium and fractional melting is the range in FeO for mean melts produced by a range in P_o. From P_o of 12 to 40 kbar, FeO varies from 7-11.5 wt% for fractional melting regimes, but only 6.5-9.5 wt% for equilibrium melting regimes (Figures 47 and 48). The range in mean pressure as a function of P_o is greater for fractional melting, and this leads to the greater range in FeO contents. For the evaluation of mean melts, therefore, fractional melting regimes produce higher overall Fe contents and a greater range in FeO contents.

It is important to consider the effects of uncertainties on the calculated results. Figure 49 shows the effects of changing various parameters in the melting function on melt compositions. The solid black line is the standard model we have been using, with a solidus slope and intercept of 13°/kbar and 1150°C, and a heat of fusion of 180cal/gm. Other curves show the effects of a 1° change in the solidus slope (which has an uncertainty of about 1°), the effect of changing the zero intercept of the solidus to 1100° from 1150°C, and the effect of changing the latent heat of fusion to 100 cal/gm from 180 cal/gm. The change in solidus intercept has the largest effect, causing a difference of about 1 wt% FeO. These uncertainties affect the absolute value of FeO, but not the relative changes in FeO for different pressures of intersection of the solidus. For example, in Figure 47 the effect of a change in the temperature

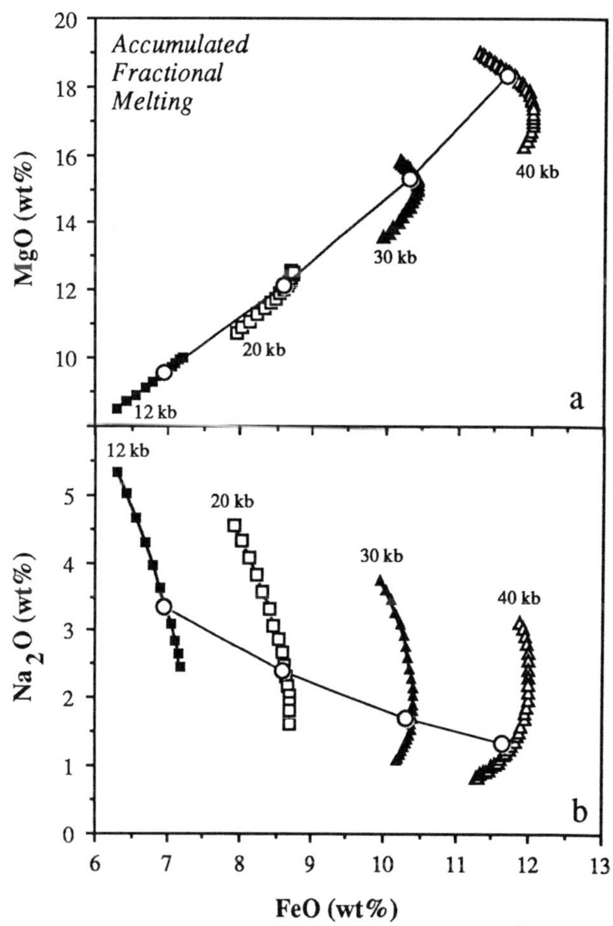

Fig. 48. As Figure 47, for accumulated fractional melting.

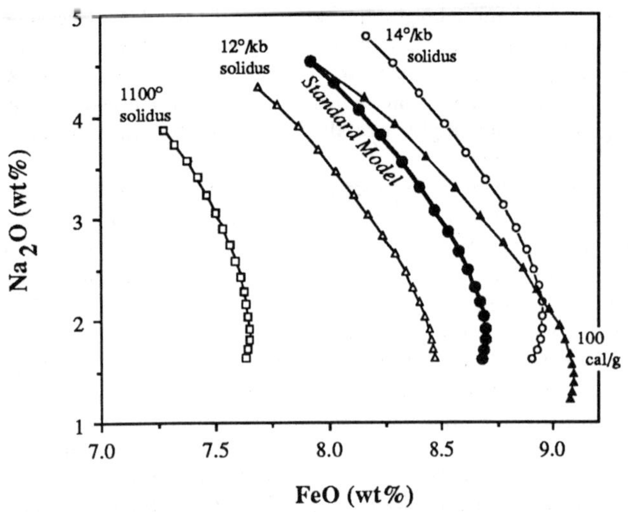

Fig. 49. FeO and Na_2O concentrations of calculated mantle melts for different melting parameters for a pressure of intersection of the solidus of 20 kbar. The "standard model" in this figure is for accumulated fractional melting, not equilibrium melting. Each curve shows the accumulated fractional melt composition. A lower temperature intercept for the solidus (1100°C) causes all K_d's to be higher, and hence FeO and Na_2O in calculated melts to be lower. A shallower solidus slope (e.g. 12°/kbar) causes all temperatures to be lower, producing a similar effect. A smaller heat of fusion allows more melting to occur, ultimately leading to lower Na_2O and higher FeO contents.

or slope of the mantle solidus would be to move all the curves to higher or lower FeO contents.

Uncertainties in Na_2O concentrations are less complicated because the Na K_d is much less sensitive to temperature and pressure than the K_d's for FeO and MgO in olivine. The principal uncertainty has to do with the source concentration of Na_2O. A primitive mantle value is 0.30+0.03 wt% [*KL87*, *Jagoutz et al*, 1979], but the mantle that gives rise to MORB may be slightly depleted in Na_2O. Furthermore, *Hofmann* [1988] has shown that the depletion in the MORB mantle is complementary to the enrichment in continents for a large number of elements. If the continents are subtracted from primitive mantle to form the depleted (MORB) mantle, then this depleted mantle would have a Na_2O concentration of around 0.26 wt%, or about 10% depletion relative to primitive mantle. A change in the source concentration of Na_2O translates directly to changes in the Na_2O concentrations in calculated melts, hence the various melting curves could move 10% relative in terms of Na_2O abundances as a result of permissible variations in source concentrations.

6. APPLICATIONS TO MORB DATA

We now have in place the systematics of the MORB data set and the quantitative tools to evaluate the effects of various melting styles, mixing functions, and mantle source compositions. In order to avoid an exhaustive evaluation of the chemical consequences of all possible variables, we begin by considering the qualitative effects of various models for melt formation beneath ridges. These qualitative considerations permit a rapid evaluation of many variables and enable us to select the most promising ones for a full quantitative treatment.

6.1 Mean Properties of the Ocean Crust Expected for Different Melting Regimes

The models discussed in the mixing function section (section 4.3) considered simple melting regimes for a single mantle potential temperature where melt extraction was complete, upwelling was passive, and there was no downward component of mantle flow near the ridge. We will call this the "standard model." In this section, we consider more complex models, such as active flow, incomplete melt focusing, and variations in mantle temperature and their effects on the mean properties of the oceanic crust that is produced. These results can then be used to select appropriate models that are likely to account for the observed data from the ocean crust. We first examine each of the variables in isolation. In Figure 50, each model is drawn to scale, with 1 kbar equivalent to 3 km of mantle or 3.3 km of crust and γ equal to 1.2%/kbar (a good approximation to equations 19 and 20), unless specified otherwise. For each melting regime, the maximum F (F_{max}), mean F ($\bar{F}$), crustal thickness (h_c) and mean P ($\bar{P}$) are calculated, as explained in section 4.3, and these parameters are plotted together in Figure 51. The parameters for the standard model have been chosen such that the resulting pooled melt produces average ocean crust (h_c = 5.5 km, $\bar{F}$ = 10%). It thus serves as a useful reference in discussing the results of the more complex models.

Mantle Temperature. Figure 2 showed two melting regimes that might result from upwelling mantle of different temperatures. Hot mantle intersects the solidus deeper than cold mantle, so P_o and mantle T correlate with one another. Because the melt fraction increases linearly with P for both melting regimes, the equations for linear mixing that were outlined above (section 4.2) can be used to calculate $\bar{F}$ (1/2 F_{max}) and $\bar{P}$ (P_o / 3 + P_f*2/3). Crustal thickness then equals the product of $\bar{F}$ and the height of the RMC [$\bar{F}$ *(P_f - P_o)]. Cold mantle intersects the solidus shallower, melts to a lower extent at shallower pressures, and creates less crust than normal mantle (Figures 50*a*-50*b*). Crustal thickness increases exponentially with $\bar{F}$ because both $\bar{F}$ and the height of the RMC increase with increasing mantle temperature (i.e., $h_c = \bar{F} * \bar{F} *$ 2/0.012, see Figure 51*a*). $\bar{P}$ and $\bar{F}$ are positively correlated because a greater amount of melting requires a larger pressure interval of melting, and thus higher mean pressures (Figure 51*b*).

Fractional Melting. Melting contours in the melting regime (and RMC) are more widely spaced for fractional melting because more decompression is required to compensate for the rise of the solidus due to melt loss. To account for this effect, we have used a lower value of γ (1.05%/kbar, a good approximation to equations (19) and (20) with the recommended changes suggested in section 4.2 for dT/dF). Because of this lower value of γ, fractional melting leads to a lower maximum and mean extent of melting and thus thinner

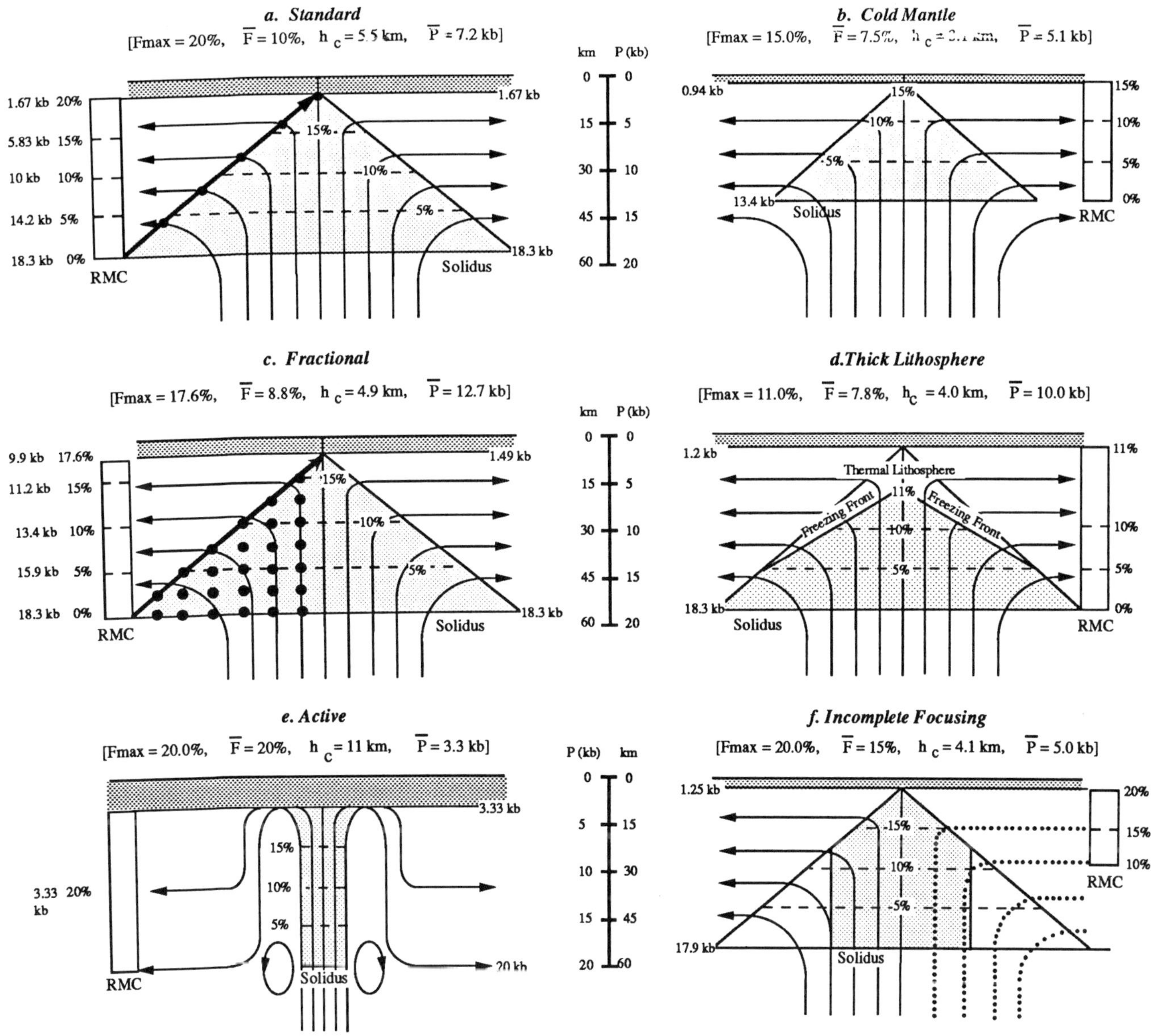

Fig. 50. Different models for the melting regimes. All melting regimes are drawn to scale, where 1 kbar is equal to 3 km of mantle or 3.3 km of crust, and γ is 1.2%/kbar, unless noted otherwise. The mean properties of the steady-state ocean crust generated by each melting regime are given, and may be calculated from the RMC. $\bar{F}$ is the mean extent of melting, $\bar{P}$ is the mean pressure of melting, h_c is crustal thickness, and F_{max} is the maximum amount of melting in the melting regime. *a*) The standard model is for equilibrium melting and compete melt focusing. This standard model generates average oceanic crust (e.g., h_c of 5.5 km, and $\bar{F}$ of 10%). *b*) Cold mantle intersects the solidus shallower leading to lower $\bar{F}$, $\bar{P}$ and h_c. *c*) Fractional melting is modeled here with a lower γ (1.05%/kbar) and continuous melt extraction. Crust produced by fractional melting is slightly thinner and has lower $\bar{F}$ than the standard model, and $\bar{P}$ is much higher. *d*) The melting regime beneath thick lithosphere is compressed by a freezing front. The melting contours are more widely spaced, because mantle may no longer upwell adiabatically. The resulting oceanic crust has lower $\bar{F}$ and higher $\bar{P}$ than the standard model. *e*) Active flow is exaggerated here, where all mantle melts to the base of the crust. Active flow leads to thicker crust, higher $\bar{F}$ and lower $\bar{P}$ than the standard model. *f*) If melting the "wings" of the melting regime is not focused into the ridge, crust formed at the ridge will be thinner, but derived by higher F and lower $\bar{P}$ than the standard model because the high P and low F melts in the corners of the melting do not contribute to the melt erupted at the ridge axis.

crust than the standard model for the same P_o (Figures 50*a*-50*c*). In practice, however, it would be difficult to distinguish on the basis of $\bar{F}$ and h_c alone between fractional melting and a slightly cooler mantle (see Figure 51*a*).

A major difference between fractional and equilibrium models, however, is the mean pressure of melting. For the same mantle temperature (P_o~18 kbar), fractional melting leads to higher mean pressures of melting ($\bar{P}$ = 12.7 kbar) than

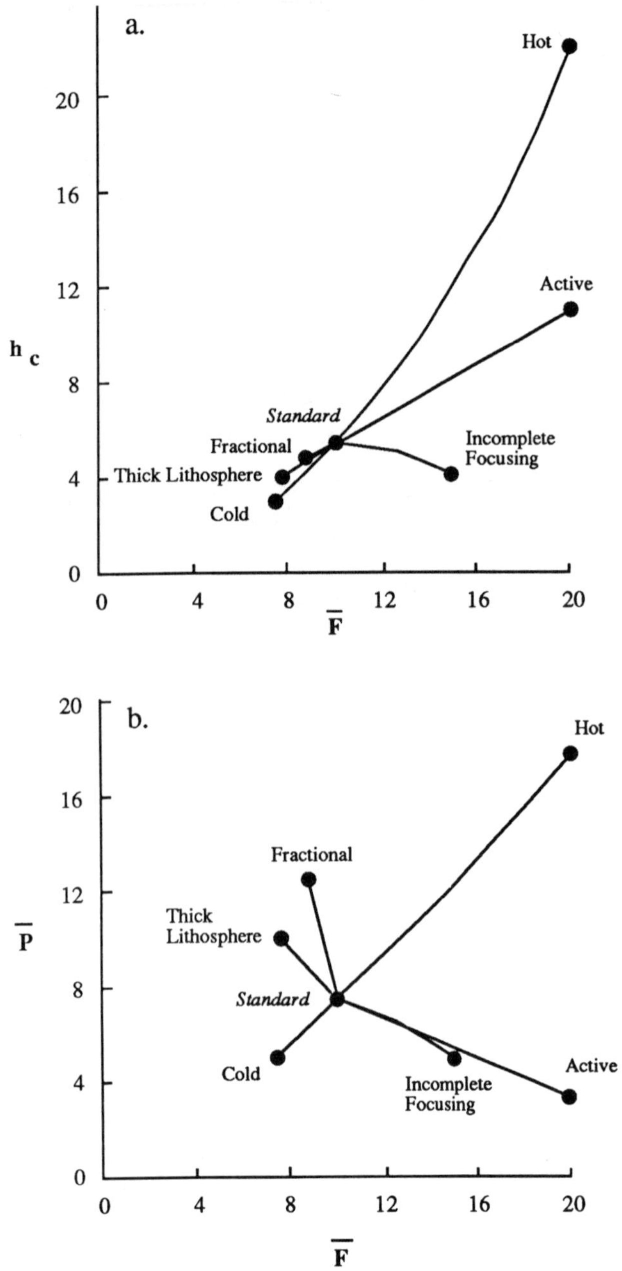

Fig. 51. Properties of the oceanic crust generated by the six melting regime models in Figure 50; mean extent of melting ($\bar{F}$); mean pressure of melting ($\bar{P}$); crustal thickness (h_c). This figure can be used to infer the progressive variations that would be produced by each process identified in the figure. For example, a small component of active upwelling would slightly lower $\bar{P}$ and raise F. A combination of active upwelling and fractional melting would cause little change from the equilibrium melting standard model.

equilibrium melting ($\bar{P}$ = 7.2 kbar) (Figures 50*a*-50*c*). This arises because the mean pressure for an equilibrium melt is the minimum pressure for any vertical column in the melting regime, while the mean pressure for an accumulated fractional melt is half the pressure interval of that column (Figures 34*a*-34*b* and Figures 50*a*-50*c*). Pooled fractional melts are thus distinct from pooled equilibrium melts in that they will reflect higher $\bar{P}$ for similar mean extents of melting (Figure 51*b*).

Thick Lithosphere. Figure 50*d* illustrates the possible effects on the melting regime of a thick thermal lithosphere, which may cause melting to stop at depths greater than the base of the crust. Such a situation might correspond to the melting regime adjacent to a fracture zone, or to a very slow-spreading ridge. The slowly upwelling mantle may no longer follow an adiabat and even cease to melt during upwelling at shallow levels. The upper boundary of the melting regime in this case coincides with a freezing front due to the deep penetration of the conductive thermal boundary layer. The melting regime does not extend to the base of the crust (as it does in the other models considered here). For the same mantle temperature (P_o ~ 18 kbar) and melting process (equilibrium), the thick conductive boundary layer leads to lower F_{max}, lower $\bar{F}$, higher $\bar{P}$ and thinner crust (Figure 50*d*). These effects are similar to those produced by fractional melting, but the effects on $\bar{F}$ and crustal thickness are greater for thick lithosphere.

Active Flow. Extra buoyancy due to melt retention or a decrease in mantle density upon melt depletion may lead to extra buoyancy and active upwelling beneath ridges [e.g., *Scott and Stevenson*, 1989]. Active upwelling may involve a narrowing of the zone of upwelling, more rapid rates of upwelling, and downward flow outside of the melting regime. These effects are exaggerated in Figure 50*e*, where flow is completely active. All mantle flows to the base of the crust, then "spills downward" to fill the RMC. In this end-member case, all the mantle melts to the maximum amount ($\bar{F} = F_{max}$) at the shallowest pressure ($\bar{P} = P_f$). Compared to the standard model, the totally active flow doubles $\bar{F}$ and substantially lowers $\bar{P}$ (Figure 51*a*). Crustal thickness is doubled as well due solely to the doubling of $\bar{F}$, since the height of the RMC is the same in the two cases. Because crustal thickness increases as a linear function of F in this model, the trajectory in Figure 51*a* between totally passive and totally active flow is linear, and differs substantially from the exponential relationship generated by mantle T. The relationship between $\bar{P}$ and $\bar{F}$ for totally active flow is of the opposite slope to the mantle T curve because $\bar{P}$ decreases dramatically as $\bar{F}$ increases (Figure 51*b*).

Incomplete Melt Focusing. Models thus far have assumed complete extraction of all melt at the ridge axis. It is likely, however, that some melt in the furthest corners of the melting regime never contributes to on-axis volcanism. Figure 50*f* illustrates incomplete focusing, where melt is focused and extracted only from the central shaded region. The RMC in Figure 51*f* reflects the incomplete focusing; it is truncated at the base and contains only residual mantle that has melted more than 10%. Mantle processed through the melting regime at greater depths than the 10% contour does not lose melt to the ridge axis and thus is not part of the RMC. The primary consequence of such incomplete focusing is to increase $\bar{F}$ and decrease $\bar{P}$ (Figure 51*b*), because the deep, low degree melts never contribute to ocean crust at the ridge axis. Although the effects on $\bar{F}$ and $\bar{P}$ are similar to active upwelling, much less crust is produced by incomplete melt focusing because much of the melt never makes it to the ridge (Figure 51*a*). Incomplete melt focusing thus leads to greater mean extents of melting and thinner crust compared to the standard model.

6.1.1 Variability of individual melts produced by different melting regimes

The discussion above is pertinent to the mean melts produced by different melting regimes and hence is most relevant to the global variability of ocean ridges, where compositions are averaged over substantial lengths of ridge. Local variability, on the other hand, is more likely a response to variations that occur within a single melting regime [*KL*89]. It is of interest, therefore, to consider how the different melting regimes affect the pressures and extents of melting for individual melt parcels

Conceptually, for equilibrium melting the individual melts are those that are produced at the tops of each vertical column in the melting regime (Figure 34*a*). Equilibrium melts are thus formed within the standard melting regime between ~2 and 18 kbar and 0-20% melting (Figure 50*a*). On a P-F diagram, these melts form a trajectory with a slope that corresponds to γ of 1.2%/kbar (Figure 52). The inverse relationship between F and P is simply a consequence of pressure release melting: low extents of melting occur deep in the melting regime, while high extents of melting occur closer to the surface. Since decompression melting is the mechanism of melt formation beneath ridges, all models will lead to an inverse relationship between F and P. Note that this inverse relationship for melts produced within the melting regime for a given mantle temperature (given P_o) is oblique to the trajectory of mean melts produced by different mantle temperatures (different P_o's).

The fact that pressure release melting occurs for all the models means that melts produced beneath a thick thermal lithosphere or by incomplete melt focusing lie on the same melting curve as the standard model (Figure 52). However, the effect of a thicker thermal lithosphere is to restrict melting to the lower part of the melting regime, while the effect of inefficient focusing is to sample only the central, shallower region of the melting regime. Thus, a single mantle temperature can produce quite different families of compositions, depending on the lithospheric thickness and efficiency of focusing.

Melts from a fractional melting regime are potentially more variable, since they are extracted from throughout the melting regime rather than just from the top of each melting column within the melting regime (see Figure 50*c*). The accumulated fractional melts generated from individual vertical columns within the melting regime have equilibrated at higher pressures than the equilibrium melts tapped from the tops of these columns, (Figures 34 and 50). Thus, for a given extent of melting the melt trajectory for accumulated fractional melts lies at higher pressures than for equilibrium melts (Figure 52). The different melting function (γ) for fractional melting also leads to a different P-F slope (Figure 52).

Similarly, the results for active upwelling depend on whether melt is extracted from throughout the melting regime, or just from the top of each melting column. If extracted from the top of the melting column, then there is no variability in the end-member case, since all mantle rises to and equilibrates at the base of the crust (Figure 50*e*). In this case, the melt produced by active upwelling will occupy a single point along the standard model melt trajectory shown in Figure 52. In contrast, if melts are sampled from throughout the melting column, then melting must approach fractional melting, and the instantaneous melts will lie along the trajectory for fractional melting, shown in Figure 52.

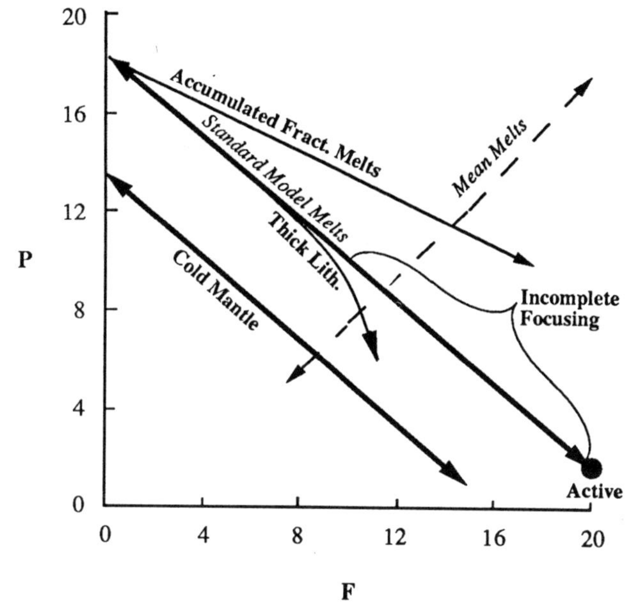

Fig. 52. The properties (F and P) of the individual melts (from the solidus to F_{max}) within the various melting regimes shown in Figure 50. The dashed line shows the mean melts produced by a range in mantle temperature. The other lines trace the $\bar{P}$ and $\bar{F}$ characteristics of extractable melts within a single melting regime beneath the ridge. For example, incomplete focusing causes the low F high P melts in the melting regime not to be extracted, so all extracted melts lie along the lower portion of the standard model line. The thick line labeled "cold mantle" shows an equilibrium melting model for a colder mantle temperature.

6.1.2 Implications

The utility of the above exercise is to illustrate how simple predictions of the mean properties of the ocean crust and the P-F characteristics of melts within the melting regime can be calculated and inferred given a melting model and RMC. Each of the above models makes specific predictions of relationships among $\bar{F}$, $\bar{P}$ and crustal thickness for the global variability, and between P and F for local variability. We can estimate crustal thickness from seismic measurements, $\bar{F}$ from the concentration of moderately incompatible elements (like Na), and $\bar{P}$ from the concentration of elements with pressure-dependent partition coefficients (like Fe). The new chemistry function can be used to calculate Fe and Na contents for the diversity of melting regimes. Therefore, by combining the physical models described above with the new chemistry function, we can directly compare the models to the data. Note that it is the combination of inferences concerning extent of melting, pressure of melting and crustal thickness that provides the most useful constraints. For example, active upwelling can produce thicker crust and greater mean extents of melting (lower Na) relative to the standard model, but the mean pressures are lower than those that would be generated by increasing mantle temperature. Thus the use of all three parameters permits distinctions among models that could not be made on the basis of any two alone.

Several important results are clear at the outset. First, mantle temperature is the ONLY parameter investigated that produces positive relationships between $\bar{F}$, $\bar{P}$ and crustal thickness for mean crust, and therefore a predicted inverse relationship between $Na_{8.0}$-$Fe_{8.0}$ and between $Na_{8.0}$ and crustal thickness. These are the principal characteristics of the global vector for MORB (Figures 15 and 18). Thus the qualitative conclusions of *KL*87 and *MB*88 still seem reasonable, and more complex melting regimes do not provide alternative explanations for the observed data. Furthermore, complex differentiation processes do not produce the requisite relationships among Na, Fe and crustal thickness either (Appendix C).

Second, it is a natural consequence of pressure release melting that melts generated from within the melting regime have the opposite relationship between F and P compared to mean melts pooled from throughout the melting regime, as noted by *KL*89. As a result, a variety of melting regimes should produce diverse magmas that range from low degree melts formed at high pressures to high degree melts formed at low pressures. Qualitatively, therefore, the cross-cutting "local trends" that occur on slow-spreading ridges appear to be consistent with a range of melts produced within a single melting regime. That said, it is also clear that the various types of melting regimes can differ substantially in the detailed character of these melts. An exact fit to the data holds the potential for better constraining the nature of the melting regime and the melting process. Precise values for local variations in crustal thickness, which are currently lacking, would be an invaluable aid in constraining the possibilities.

It is also likely that real melting regimes may embody combinations of the various isolated parameters discussed above. Hot mantle may have a larger melting regime and hence less efficient focusing. Fractional melting may promote passive upwelling, while equilibrium melting may promote active upwelling because the presence of melt will lower the viscosity and enhance the buoyancy of the upwelling mantle. These possibilities will need to be considered as we attempt to quantitatively reproduce the chemical characteristics of MORB.

Finally, this entire discussion has tacitly assumed a two-dimensional ridge, and we will need to consider the relationship of the melting regime to ridge segmentation. Large transform offsets may have a significant thermal effect on the melting regime and the ascending mantle. Active upwelling could occur at segment centers, with a thick thermal lithosphere at segment edges. Thus different processes could affect different portions of a ridge segment and the chemical expression of local variability.

6.2 The origin of the global vector

Although mantle temperature qualitatively accounts for the global vector, the quantitative efforts by *KL*87 and *MB*88 did not reproduce the absolute values of MORB Na and Fe (see Figure 43). In addition, mantle heterogeneity has not been quantitatively evaluated as a possible explanation for the global vector. *Albarede* [1992] and *Natland* [1989] have suggested that much of the major element signal is associated with ridges affected by hot spots. Since hot spots are derived from an isotopically heterogeneous mantle, perhaps major element heterogeneity of the mantle is a possible cause of the global vector. The new melting model can be used to determine what combination of conditions can produce the global vector of composition and to test whether mantle heterogeneity might account for the global vector.

The calculated mantle melts all have significantly more than 8 wt% MgO, and the compositions formed near the solidus at the higher pressures have as much as 18 wt% MgO. Thus the calculated mantle melt compositions must be fractionated to 8% MgO in order to compare them to the MORB data. This is problematic, for our calculations do not provide CaO or Al_2O_3 concentrations. CaO and Al_2O_3, however, can be reasonably estimated from the existing experimental data and, moreover, the effects of different amounts of CaO and Al_2O_3 within the permissible range do not seriously affect the crystallization results. The potential diversity of fractionation processes is a more serious problem. The crystallization pressure may vary with both crustal thickness and spreading rate, and the crystallization process may vary between an equilibrium, fractional, or *in situ* process [*Langmuir*, 1989]. Our evaluation of crystallization, therefore, must take into account some of the diversity of mechanisms by which crystallization may take place.

6.2.1 Evaluation of Fractional and Equilibrium Melting Models

The results of the new melting model for equilibrium melting were described above (section 5) and shown in Figure 47. Figure 53 compares the MORB array for normal ridges to calculated $Na_{8.0}$ and $Fe_{8.0}$ variations for mean melts from melting regimes that result from different mantle temperatures (different P_o). The figure shows both the calculated melt compositions, and these compositions adjusted to 8 wt% MgO by fractional crystallization at 1 atm and at 4 kbar. As explored at length in Appendix C, for the same parental magma composition, high pressure crystallization leads to higher $Fe_{8.0}$ than low pressure crystallization, but does not cause a significant change in the $Na_{8.0}$. The calculated melt compositions produced by equilibrium melting have Na_2O and FeO contents that are too low to account for the global array (Figure 53*a*). After crystallization at 1 atm, the melt compositions still do not have sufficient range in FeO, and crystallization at 4 kbar leads to too flat a slope. If the mantle Na_2O abundance were increased by 30% (to 0.40 wt%), then the melt array would intersect the MORB field, but the range in FeO would still be insufficient. Figure 54 compares the calculated $Na_{8.0}$ and crustal thickness values to the MORB data. The $Na_{8.0}$

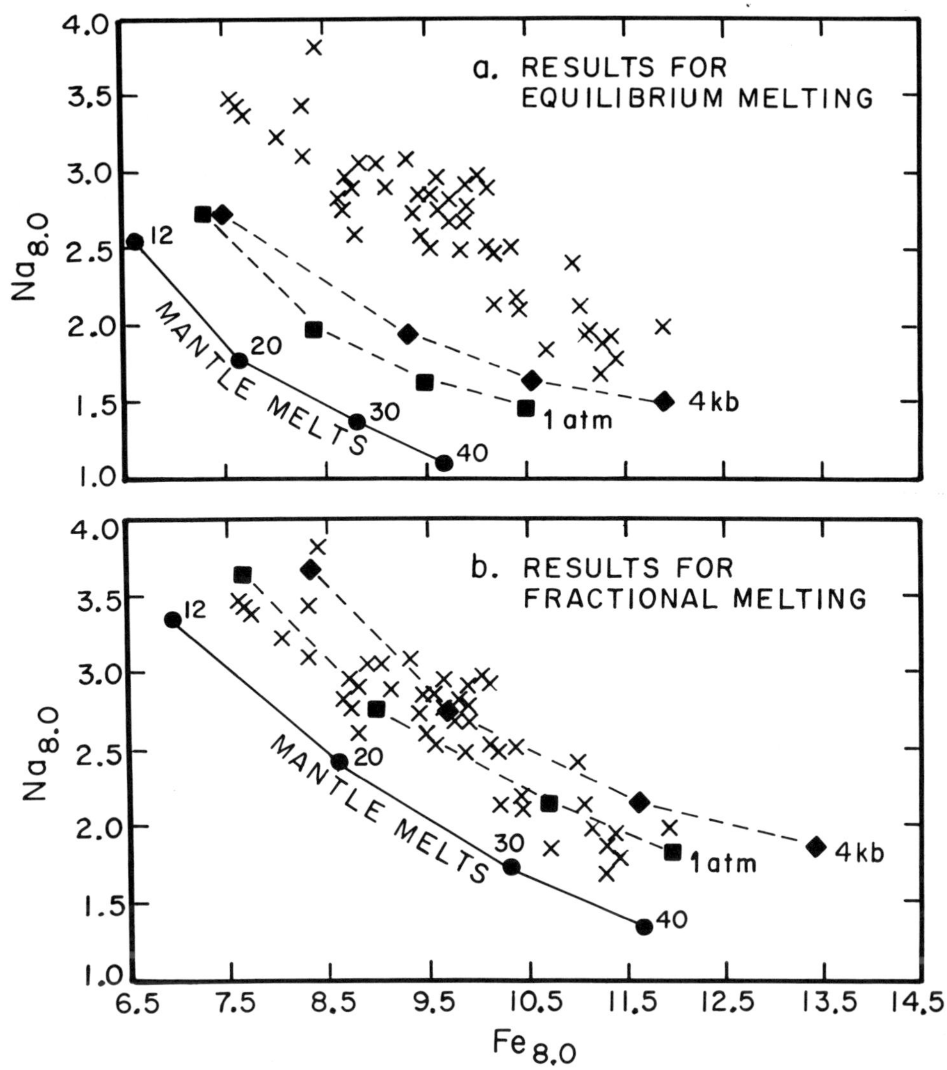

Fig. 53. $Na_{8.0}$ vs. $Fe_{8.0}$ for MORB, for calculated mantle melts and for the same melts fractionated at 8% MgO at pressures of 1 atm and 4 kbar. The X's are normal MORB (from Figure 15). The top Panel (a) shows the results for equilibrium melting and the bottom Panel (b) for fractional melting for a range in P_o from 12 - 40 kbar. The mantle melt curves are the same as those shown in Figures 47*b* and 48*b*. Each labeled dot along the mantle melt curve is the mean melt for a given melting regime. Note that equilibrium melting gives neither high enough $Na_{8.0}$ nor the right Na8-Fe8 slope to account for the date.

values for pooled equilibrium melts are too low for the observed range in crustal thickness, but again, an increase in the Na_2O concentration of the mantle source to 0.40 wt% would allow the calculations to overlap the data. Since this source abundance for Na_2O is unreasonably high for the depleted or primitive mantle (see discussion in section 5.3), we conclude that equilibrium melting does not adequately reproduce the global MORB array. The more rigorous and completely independent quantitative modeling reported here, therefore, results in the same discrepancies for equilibrium melting that were apparent in our evaluation of the *KL*87 and *MB*88 models.

As discussed above (section 5.3), accumulated fractional melting leads to higher mean pressures of melting, and a greater range in mean pressure for a range in P_o. The compositional consequences of fractional melting are higher $Fe_{8.0}$ contents for the same $Na_{8.0}$ contents and a greater range in $Fe_{8.0}$ for the same range of $Na_{8.0}$ (Figures 47 and 48). After fractionation correction to 8 wt% MgO, the observed pooled fractional melts calculated from the new melting model are coincident with the MORB array (Figure 53*b*). The results for fractional melting also straddle the observations for $Na_{8.0}$ and crustal thickness data (Figure 54). The detailed relationship between these calculated results and the observations depends in part on the parameters used for the calculation. Changing the mantle source concentration from 0.30 wt% to 0.26 wt% Na_2O results in a curve that skirts the low side of the global array. Reducing the 1 atm intercept of the solidus could also lower the $Fe_{8.0}$ values by about 1% (see Figure 49). Certainly,

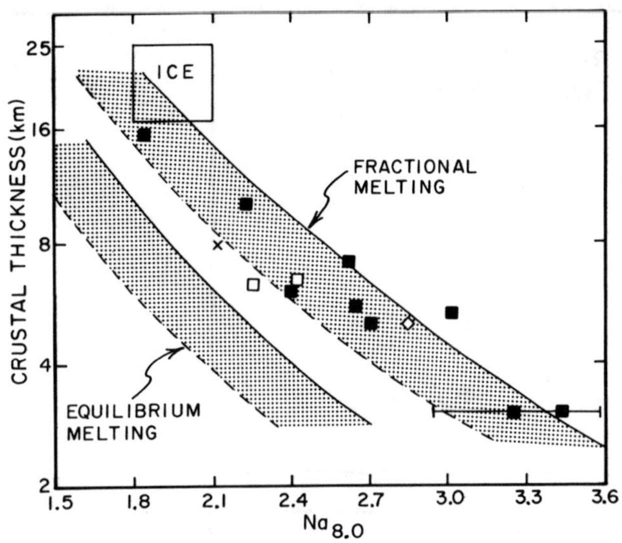

Fig. 54. $Na_{8.0}$ and crustal thickness for observed ocean crust (from Figure 18) and calculated mean melts (from Figure 53). Dashed lines are for C_0 of 0.26 wt% Na_2O, and solid lines are for C_0 of 0.30 wt% Na_2O. The equilibrium melting curves do not intersect the data field.

within the uncertainties of the parameters, the fractional melting calculations appear to reproduce the field of the observations more closely than the equilibrium melting calculations.

Another test of the consistency of these melting models is whether they reproduce the $Ti_{8.0}$-$Na_{8.0}$ MORB data. Figure 55 compares MORB data with a curve calculated for fractional melting. Although TiO_2 varied regionally (Figure 17), the calculated curve closely fits the data from normal North Atlantic ridges.

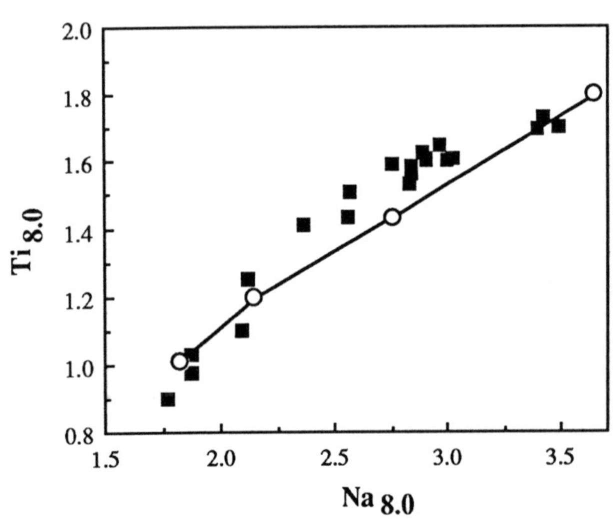

Fig. 55. Data from N. Atlantic ridges (from Figure 17*b*) are compared to calculated results for $Ti_{8.0}$ and $Na_{8.0}$. The model curve is for mean melts derived by fractional melting, with a source concentration for TiO_2 of 0.15 wt%. Melts have been normalized to 8 wt% MgO by fractional crystallization at 1 atm.

Fractional melting may also help to explain the discrepancy between the MORB data and the silica concentrations calculated by the *KL*87 and *MB*88 equilibrium models. Although the new quantitative model currently does not calculate SiO_2, the higher mean pressures for fractional melting should lead to lower silica concentrations, in better accord with the MORB data (see Figure 43). The lower SiO_2 contents obtained by fractional melting may also help to explain the lack of MORB compositions that are saturated with orthopyroxene under mantle conditions. Since the SiO_2 is lower than predicted from equilibrium melting, orthopyroxene, which is preferentially stabilized by higher SiO_2, may not appear as readily on the liquidus of accumulated melts produced by fractional melting.

In summary, these calculated results indicate that fractional melting models reproduce the global correlations very closely, while equilibrium melting models do not. These results are in remarkable agreement with arguments based on abyssal peridotites that have been made recently by *Johnson et al.* [1990]. A critical question, however, concerns the residual porosity that could be present during melting. Perfect fractional melting leads to 0% porosity, while equilibrium melting produces a porosity that increases with melt fraction during ascent of the mantle beneath the ridge, ultimately rising to values of 20-30%. *Langmuir et al.* [1977] argued for a few percent residual porosity based on trace element modeling, while *Johnson et al.* [1990] have argued for a residual porosity of less than 1%. The melting results presented here are not of sufficiently high resolution to constrain this problem well. Although equilibrium melting is ruled out for reasonable mantle source abundances of Na_2O, a continuous melting model with several percent residual porosity is permitted by the data. However, the perfect fractional melting model also is adequate.

6.2.2 Evaluation of Mantle Heterogeneity as an Explanation for the Global Compositional Variations

There are two aspects that need to be considered in order to evaluate the effects of heterogeneity in the mantle source. First, we need to consider reasonable ways in which source compositions might vary on a regional basis, for if all elements are allowed to vary randomly, the problem is intractable. Second, the effects of source variations on the melting and chemistry functions need to be quantified.

Constraints on the source variations come from direct observation of mantle materials. Data from mantle nodules and peridotite massifs tend to exhibit linear trends on MgO variation diagrams [e.g., *Frey et al.*, 1985; *Maaloe and Aoki*, 1977], an example of which is shown in Figure 56 for samples from the Ronda peridotite. Such trends are observed for all of the major elements that can be accurately determined, with the high Na_2O, low MgO samples showing high SiO_2, TiO_2, CaO and Al_2O_3 and slightly high FeO. These observations place constraints on how probable sources for MORB can vary, and on the Na_2O and FeO contents of undepleted mantle.

A curious feature of linear trends for MgO-Na_2O is that they are not consistent with an origin by different extents of partial melting. Partial melting leads to rapid depletion of Na_2O,

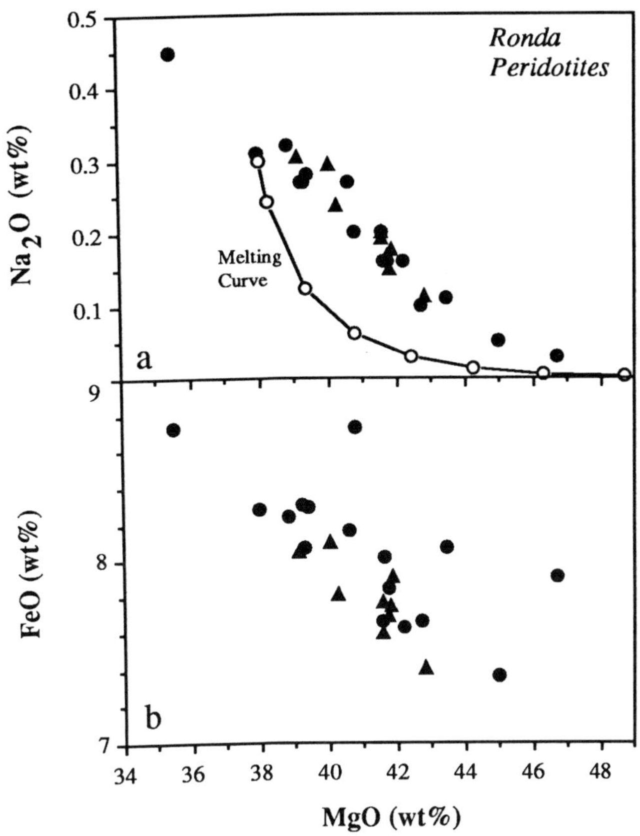

Fig. 56. MgO vs. Na_2O and FeO for Ronda peridotites. Closed circles are data from *Frey et al.* [1985]. Closed triangles are data from *Reisberg* [1988] determined at Lamont by DCP on Ronda spinel peridotites. Also shown for Na_2O is a calculated curve for changes in residual mantle composition as a function of increasing extents of melting. Note the peridotites are not explained as simple residues of equilibrium or fractional melting.

whether the melting process is equilibrium or fractional, and produces a concave upward curve on Figure 56*a*. Linear trends on variation diagrams are produced generally by mixing or unmixing. Therefore a simple interpretation of the observed trends is that they may result from either addition of basaltic material to a highly depleted mantle, perhaps in the form of subducted ocean crust, or by the extraction of a uniform melt composition to varying extents, as was suggested by *Frey et al.* [1985]. Despite these interesting complexities in the details of the depletion and enrichment process in the upper mantle, it is clear that the dominant control on major element variability is the relative amount of basaltic components that are in the mantle. Of course, another important aspect of mantle heterogeneity is the presence of ubiquitous veins in mantle peridotite. Such veins lead to a heterogeneous mantle on a small scale, and will be considered further in section 6.3.

The predominantly linear variations of the ultramafic data provide a range of compositions with which to evaluate major element heterogeneity. We use three compositions along the Ronda array as sources, one more depleted in basaltic constituents than the average mantle (41 wt% MgO), another more enriched (35.4 wt% MgO), and one of average mantle composition (38 wt% MgO). Values for all other elements can simply be taken from the linear data arrays.

Variations in mantle composition also produce a change in the temperature of the solidus: more fertile mantle has a lower solidus temperature at a given pressure. In order to take this effect into account, we have used a value of 20°C change in the solidus temperature per one percent change in the Mg# of the mantle. This value was estimated from existing experimental data, although it is not well constrained.

A shift in the solidus due to a shift in Mg# leads, further, to a change in the pressure of intersection of the solidus for an upwelling parcel of mantle (Figure 57). For example, a 1% decrease in the Mg# causes the solidus to decrease by 20°C, which leads to a 1.5 kbar increase in P_o (for a solidus slope of 13°C/kbar). These changes lead to variations in the extents and pressures of melting, which may counteract or accentuate the effects of varying source composition alone. For example, a lower Mg# leads to higher FeO contents for a given pressure (simply from the K_D relationship), and the lower solidus temperature leads to a greater P_o, which also contributes to higher FeO contents. Therefore the two effects enhance one another, and the FeO contents of melts from more fertile mantle will be higher. For Na_2O, however, the effects of source composition and change in extent of melting counterbalance. The higher Na_2O in the source enriches the melts, but this is partially offset by the increased amount of melting due to the greater P_o. For a 1% change in Mg#, the source changes substantially in Na_2O, yet the solidus intersection differs by only 1.5 kbar, which causes only a small change in the extent of melting. Thus the change in the source concentration for Na_2O dominates the dilution caused by greater melting.

These qualitative considerations show that source heterogeneity can be expected to lead to a positive correlation between FeO and Na_2O contents. This conclusion is rather insensitive to the uncertainties in the temperature-composition relations of the solidus. For example, if there were no change in solidus temperature with composition, FeO and Na_2O contents would still both increase, because the source concentrations increase. Therefore the slope of Na-Fe variations caused by mantle heterogeneity is well constrained in sign if not magnitude.

Quantitative Modeling. The new calculations permit a more rigorous evaluation of mantle heterogeneity, given appropriate input parameters. The calculations considered above showed the range of compositions produced by a range of mantle temperatures acting on a single source composition. Now we consider the consequences of a single mantle temperature acting upon a range of source compositions. Figure 58 shows the calculated variations in melt compositions produced by melting of the three different mantle compositions for a mantle temperature that would intersect a normal solidus at 20 kbar. In agreement with the qualitative discussion above, the more enriched mantle (lower Mg#) produces a pooled melt with higher Na_2O and somewhat higher FeO contents. The enrichment in FeO is limited, because the increase in alkalis

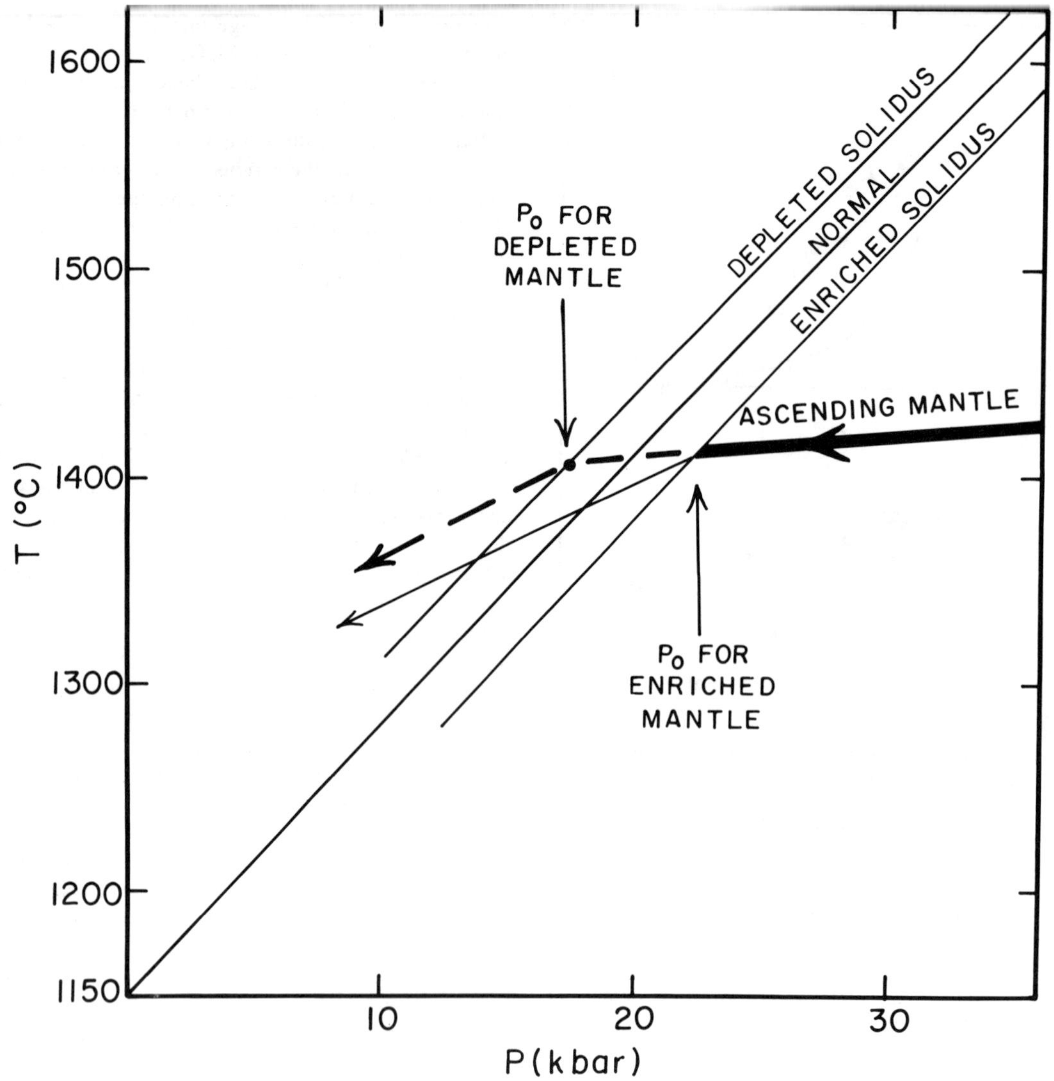

Fig. 57. P-T diagram illustrating schematically the consequences of changing mantle composition on melting, for a mantle temperature that would intersect the solidus at a pressure of 20 kbar for normal mantle. Enriched mantle has a lower solidus temperature which causes the pressure of intersection of the solidus (P_o) to be greater. Since melting starts deeper, more pressure release is possible, and the overall extent of melting is slightly higher. The combination of more fertile mantle and deeper melting leads to higher overall FeO contents. But the enriched mantle has substantially more Na_2O, which is not completely compensated for by the slightly greater melting extents. Therefore, for a given mantle temperature, melts of enriched mantle have higher FeO and higher Na_2O than melts of depleted mantle.

increases the FeO K_d's for olivine. Because the enriched mantle starts melting deeper and melts more, it also produces a greater quantity of melt, and hence a thicker section of oceanic crust (Figure 59). These results, although illustrated for a single mantle T, apply to the whole range of mantle temperatures. For example, temperature variations acting on relatively enriched mantle would lead to curves offset to higher Na_2O and FeO relative to normal mantle. These calculations demonstrate that mantle heterogeneity in the absence of mantle temperature variations leads to slopes that are opposite those of the global MORB data. Thus mantle heterogeneity, as observed in ultramafic nodules and peridotite massifs, cannot be the primary cause of the global correlations.

Heterogeneity could <u>contribute</u> to the observations, however. The width of the global correlations could result in part from mantle heterogeneity. This may be particularly important for some hot spots. For example, Iceland is slightly offset to the high $Na_{8.0}$-$Fe_{8.0}$ side of the global correlations, while the Azores and Jan Mayen are offset substantially to the low side (Figure 15). Iceland also has relatively thicker crust, while the Azores region has relatively thinner crust, in comparison to the melting calculations for normal mantle presented in Figure 54. Thus Iceland could be derived from a mantle that is slightly more fertile than normal mantle, leading to higher $Na_{8.0}$ and $Fe_{8.0}$ and thicker crust. The Azores region, on the other hand, could result from melting of depleted mantle.

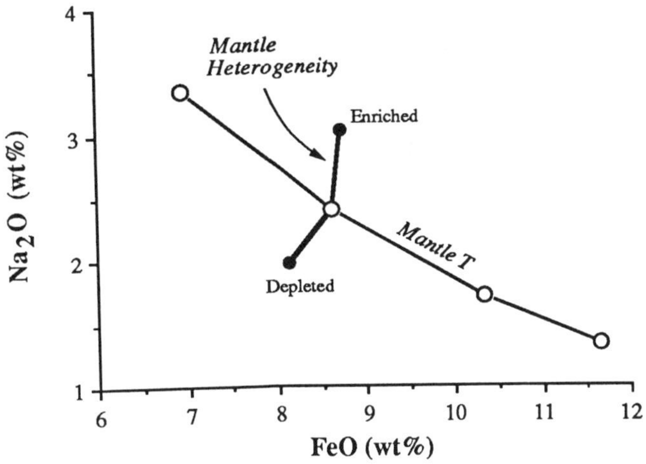

Fig. 58. Comparison of quantitative results for mantle temperature variations acting on a homogeneous mantle, with constant mantle temperature acting on a heterogeneous mantle. The mantle temperature curve is reproduced from Figure 53. The mantle heterogeneity curve is produced at the temperature that intersects the solidus of normal mantle at P_0 of 20 kbar. Enriched, normal and depleted mantle have, respectively: 36, 38, and 41 wt% MgO; 8.6, 8.4, and 7.9 wt% FeO; 0.4, 0.3, and 0.22 wt% Na_2O; and 0.02, 0.012 and 0.006 wt% K_2O. Note that constant temperature acting on a heterogeneous mantle produces arrays of melts that are roughly perpendicular to the global correlations observed in MORB.

Of course, the most incompatible elements are highly enriched in the Azores region. Thus, these calculations suggest trace element enrichment by a volatile(?) phase of a source that was previously depleted in its major element components. This explanation for the Azores might also account for the anomalous relationship between depth, $Na_{8.0}$ and $Fe_{8.0}$ that is observed approaching some hot spots (e.g. Figure 19). The lower $Fe_{8.0}$ would be the result of the depleted major element matrix of the mantle, and possibly the higher volatile contents during melting, while the higher $Na_{8.0}$ contents would result from the more recent enrichment event. This contrast between the Azores and Iceland is consistent with previous discussions of distinct origins for the sources of these two hot spots [*Langmuir and Hanson*, 1980; *Schilling et al.*, 1983].

6.2.3 Conclusions with Respect to the Global Variability of MORB Compositions

Based on the modeling above, we can conclude that:

1. The global correlations for normal ridges continue to be well explained by variations in mantle temperature acting on a relatively homogeneous mantle.

2. The melting process does not appear to be equilibrium melting. Pure fractional melting (0% residual porosity during melting) is consistent with the data, but there are significant uncertainties that would permit a melting model intermediate between fractional and equilibrium melting. For example, if the mantle is slightly more enriched than our normal mantle composition, the data would fall between the results for fractional and equilibrium melting. These uncertainties make it difficult to quantify the percent of residual porosity during melting from the basalt data alone.

3. Mantle heterogeneity, such as is observed in ultramafic nodules and peridotite massifs, cannot reproduce the global correlations.

4. The offset of some hot spots from the global correlations may result from major element heterogeneity in the mantle.

5. Back-arc basins show similar global systematics to other ocean ridges, but are systematically offset to lower $Fe_{8.0}$ and $Ti_{8.0}$. These offsets cannot currently be modeled quantitatively due to the lack of knowledge of the effects of small amounts of water on mantle and basalt phase equilibria. Suppression of plagioclase during crystallization could lead to lower calculated values of $Fe_{8.0}$ and $Ti_{8.0}$ [see *Fryer et al.*., 1990], but the effects of water on FeO during melting are not yet known.

6.3 The Origins of Local Major Element Variability

We turn next to explore the origin of the local vectors of chemical variability. This is a difficult undertaking due to the large number of factors that could play a role in the diversity of ridge environments. However, the regularity of the major element trends on slow- and fast-spreading ridges is striking, and should have a straightforward and consistent explanation. A key issue is whether the local variability (corrected for low pressure crystallization) results from melting processes, crystallization processes, or small scale mantle heterogeneity. In the following, we discuss how the data and quantitative modeling shed light on the problem.

*KL*89 focussed primarily on the origin of the Atlantic-type local chemical variations, and proposed that they were controlled by different extents of melting as individual or

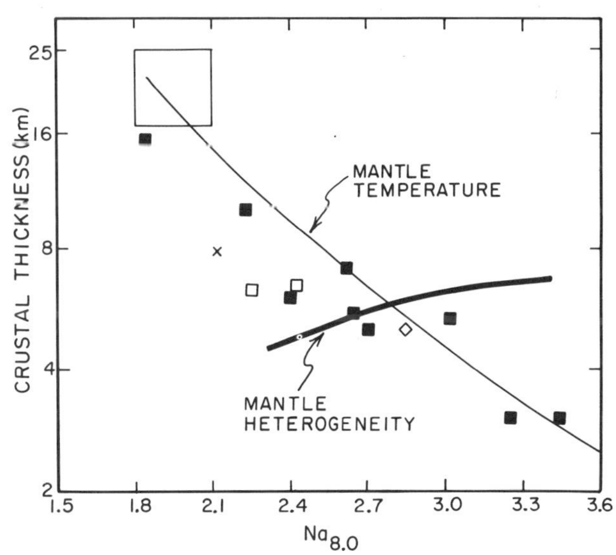

Fig. 59. Crustal thickness results for the same conditions described in Figure 58. Enriched mantle is more fertile and melts more, leading to both thicker crust and higher $Na_{8.0}$. As in Figure 58, the heterogeneity model produces results that are essentially perpendicular to the observations. Crustal thickness data from Figure 18.

partially pooled melts were sampled from the melting regime. Melts produced by smaller extents of melting would be sampled from just above the solidus and would show the greatest pressure of equilibration (hence, higher $Na_{8.0}$ and $Fe_{8.0}$ and lower $Si_{8.0}$). *KL*89, however, did not quantitatively explore the origin of the local vector, but rather showed a qualitative comparison between the Atlantic-type trends and the general trends for polybaric equilibrium melting inferred from the experimental data. They also noted that they could not rule out an origin by high-pressure fractionation, but this issue was not dealt with in depth. At the short course, Kinzler and Grove used data from new experiments and calculations to point out the potential importance of high-pressure fractionation in controlling the slow-spreading local vector [*Kinzler and Grove*, 1992*a* and *b*; *Grove et al.*, this volume]. Yet a third explanation for local variability comes from studies of trace elements and isotopes, and from the recognition of ubiquitous small-scale heterogeneities in mantle peridotites. Melting of a "veined mantle" as called upon for the trace element and isotope models [e.g., *Hanson*, 1977; *Wood*, 1979; *Zindler et al.*, 1984; *Prinzhofer et al.*, 1989] could lead to substantial local major element variability as well.

6.3.1 Quantitative Evaluation of Melting Models

In section 5, the new quantitative melting function was used to calculate the FeO and Na_2O contents of melts produced by adiabatic equilibrium and fractional melting within a given melting regime (Figures 47 and 48). We also showed above (section 6.2) that the global MORB array could be explained as an array of mean melt compositions produced by pooling fractional melts from throughout melting regimes of different solidus pressures (different mantle temperatures). Thus the success of the quantitative modeling in reproducing the global vector depends on the pooling of instantaneous melts. This, in turn, leads to predictions for the chemical characteristics of local variations, if those variations are caused by the eruption of melts from different parts of a single melting regime.

In Figure 60, the curved trajectories show the compositions of instantaneous melts produced within the melting regime for different pressures of intersection of the solidus, for both fractional and equilibrium melting corrected to 8% MgO. The paths of the individual melts cut across the global vector of mean melts at a high angle, as suggested by *KL*89. These calculated melting curves are compared to the local trends for slow-spreading ridge segments. It is clear that the calculated paths do not correspond with the shallow slopes of the MORB data. The fit is particularly poor for the fractional melting model, where the individual melts from the melting regime have substantial variation in Na_2O but very little variation in FeO. Thus while the global vector appears to be quantitatively consistent with the pooling of melts from different melting regimes, the slow spreading local vector is not well modeled by melts from within a melting regime. Of course, melting column effects are a completely inadequate explanation for the Pacific-type local vector, where $Fe_{8.0}$ and $Na_{8.0}$ correlate inversely (Figure 24), because calculated melts that occur within a melting regime form steep positive trends (Figure 60).

Modification of the mixing function to account for different upwelling mechanisms and patterns does not appear to aid in explaining the slow-spreading local trend. This conclusion results because different mixing functions control primarily how much of each melt along a given melting path, of given P_o, contributes to the final mean melt. For example, active upwelling causes all the melts to be made up of the high F (low Na_2O) portions of the path, while inefficient melt focusing causes melts from the low F (high Na_2O) part of the path to remain unsampled (Figure 52). Since all the melting paths in Figure 60 have little variation in FeO and much in Na_2O, no combination of melts along them can produce the shallow slope of the slow-spreading local trend. In the case of a thick thermal boundary layer, both the mixing and melting functions are affected. The highest F melts along the melting path are not produced, and the cooling effects of the boundary layer can cause less melting over a given pressure interval, leading to greater pressure changes for a given melting increment (Figure 52). This can produce fractional melting paths with positive slopes on a Na_2O-FeO diagram, but the slopes are still too steep to account for the slow-spreading trends. Thus, for a given P_o, none of the melting models we have investigated reproduces the observed Na-Fe slope of the local trends. Alternative explanations for the local vectors must be sought.

6.3.2 Complex Fractionation Models

One alternate explanation for the origin of the Atlantic-type local vector is that it results from either high-pressure crystallization [*KL*89; *Kinzler and Grove*, 1992*a* and *b*; *Grove et al.*, this volume], or some complex low-pressure crystallization process. These ideas have intuitive appeal, because crystallization processes should be strongly affected by spreading rate. For example, at slower spreading rates, cooling is more efficient, and crystallization might take place at deeper levels. Furthermore, the balances among frequency of injection, eruption and cooling in periodically replenished magma chambers (PRMC) could logically be expected to vary with spreading rate. Fortunately, these processes can be modeled quite well for basaltic systems, so we can evaluate their efficacy in creating the local vectors. The modeling methods and effects of high-pressure crystallization and complex low-pressure crystallization are discussed in detail in Appendix C. Here we present the more important results that forward the general line of discussion.

To evaluate the various models, it is useful to recognize that the local trends, originally defined by the correlations among $Na_{8.0}$, $Fe_{8.0}$ and $Si_{8.0}$, are also clearly observed in the Na_2O, FeO and SiO_2 abundances for the many samples from the FAMOUS area with approximately 8 wt% MgO (Figure 23), and that for these samples there are additional correlations among Na_2O and CaO, CaO/Al_2O_3, TiO_2 and K_2O. The existence of these systematics for many elements at a single value of MgO simplifies conceptually how we test the fractionation models: we seek processes that can produce the observed variability at 8 wt% MgO without worrying about subsequent low-pressure fractionation corrections.

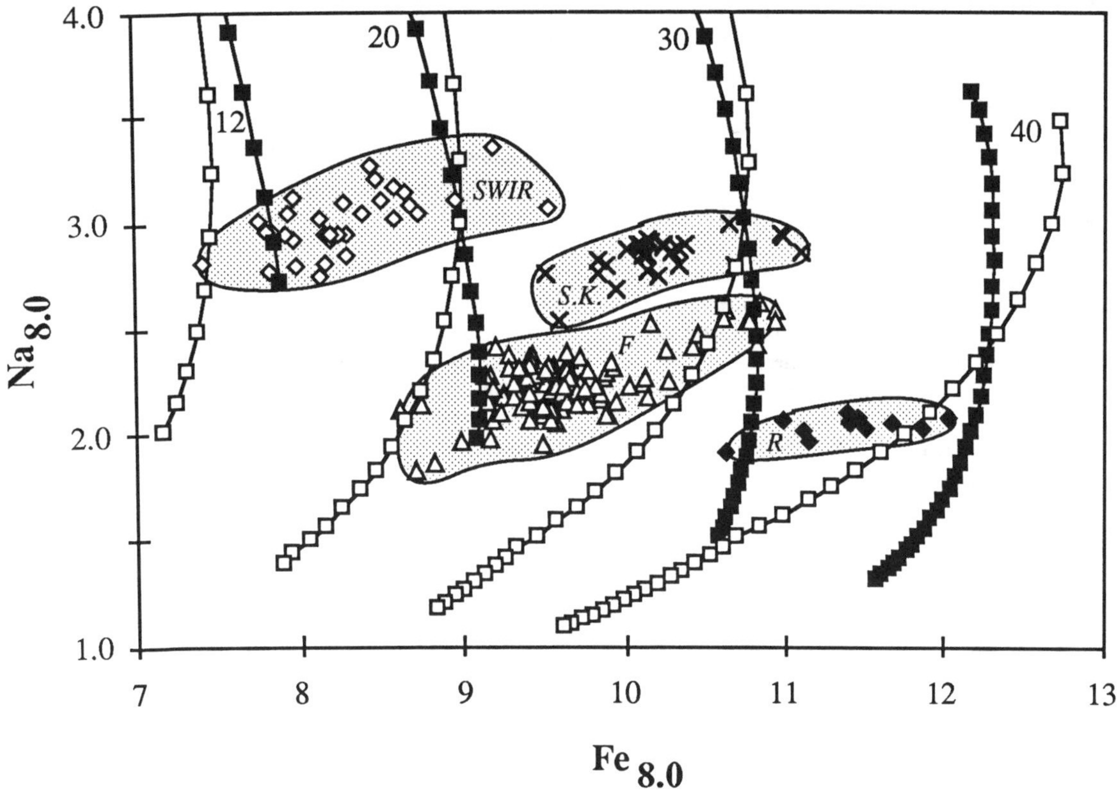

Fig. 60. Melting path curves for fractional (solid squares) and equilibrium (open squares) melting. Each path shows the liquid compositions in the melting regime, with the high Na_2O liquids forming at depth near the solidus, and the low Na_2O liquids forming near the base of the crust. Each point on the curves represents a pressure drop of 1 kbar. Curves are corrected for 1 atm fractionation based on the calculated correction for the mean melts (Figure 53). The fields on the diagram are for several regions which exhibit the known range in slow-spreading local vectors from Figure 21: R (Reykjanes Peninsula), F (FAMOUS area), SK (S. Kane Fracture Zone), SWIR (AII Fracture Zone).

Low-pressure Fractionation. As discussed in detail in Appendix C, mixing of evolved and primitive magmas by either *in situ* fractionation or in periodically replenished magma chambers can, under certain circumstances, lead to associated enrichments of both FeO and Na_2O at a given value of MgO (the two models differ substantially, however, in their effects on trace elements). Although these processes cannot produce the EPR local trend, they might account for the positive correlation between $Fe_{8.0}$ and $Na_{8.0}$ of the slow-spreading local data.

In general, the steady-state melts produced by periodically replenished magma chambers lie between fractional and equilibrium crystallization trends (Figure C1) as discussed in Appendix C, and thus do not produce large variations in either FeO or Na_2O at 8 wt% MgO. The one exception is if very high MgO liquids (>12 wt% MgO) mix with very fractionated liquids, such that the mixture lies outside the closed system crystallization field, producing large co-variations in FeO and Na_2O at 8 wt% MgO (Figure C2). However, for the FAMOUS samples, this would also produce a positive correlation between Na_2O and SiO_2 at 8 wt% MgO, as well as an inverse correlation between Na_2O and Al_2O_3. In contrast, one of the hallmarks of the MAR-type local vector is the inverse correlation between $Na_{8.0}$ and $Si_{8.0}$ (Figures 23 and 25). The FAMOUS data also show no correlation between Na_2O and Al_2O_3 (Figure 23) in contrast to the expected effects of such mixing processes.

Furthermore, as discussed in Appendix C, such mixed magmas have olivine alone on the liquidus just after mixing and evolve rapidly back to the multi-phase liquid line of descent of the parental magma. In order to observe the strong FeO-Na_2O enrichment, these magmas must be sampled just after mixing. Many of the FAMOUS samples, however, are saturated with both olivine and plagioclase at 8 wt% MgO, not olivine alone [e.g., *Bender et al.*, 1978]. This co-saturation is reflected by the consistent, positive slope that samples follow on a plot of MgO vs. Al_2O_3.

The difficulties in reproducing the slow-spreading local vector by periodically replenished magma chamber processes also apply to *in situ* differentiation. Crystallization of plagioclase and pyroxene, which produces an increase in FeO and Na_2O contents, is also expected to lead to an increase in SiO_2 and a decrease in Al_2O_3. These explanations, therefore, would produce systematics that are not observed in the MAR-type local vector and require special conditions that are not consistent with the widespread occurrence of the slow-spreading local trend. As discussed below, trace element systematics argue against an origin by low-pressure crystallization processes. This is not to say that PRMC or *in situ* crystallization processes do not occur, but rather that they do not appear to produce satisfactory explanations for the characteristics of either the MAR-type or EPR-type local vectors.

High Pressure Fractionation. At slower spreading centers such as the Mid-Atlantic Ridge, where the lithospheric cap may be thicker and where there is no evidence for crustal magma reservoirs, crystallization may occur at higher pressures. As discussed in Appendix C, crystallization at higher pressures leads to increased stability of clinopyroxene such that it may precede the arrival of plagioclase on the liquidus [e.g., *Bender et al.*, 1978]. Enhanced clinopyroxene crystallization leads to an inverse correlation between Na_2O and CaO/Al_2O_3, and early crystallization of the three-phase assemblage olivine-plagioclase-clinopyroxene leads to early enrichments in FeO, Na_2O and TiO_2. Thus, at least qualitatively, the observed chemical co-variations that characterize the MAR-type local vector are consistent with high-pressure fractionation. A potential difficulty, however, is that the expected enrichment in Na_2O is moderated by the more albitic plagioclase that crystallizes at higher pressures [e.g., *Bender et al*, 1978; *Fram and Longhi*, 1991].

In order to quantitatively evaluate whether high-pressure fractionation can account for the variability seen in the local trends, we modified the 1 atm liquid line of descent program of *Weaver and Langmuir* [1990] to include pressure dependencies for the distribution coefficients, as described in Appendix C and shown in Table C1. The program was calibrated to the available experimental data on FAMOUS compositions. Using a primitive composition similar to that of the FAMOUS region, calculations were carried out at 0, 4 and 8 kbar in order to compare the compositions calculated at 8 wt% MgO to the local trends seen in the FAMOUS/AMAR data (Figures 61, C5, and C6; see also results in Table C2.)

In qualitative terms, crystallization at 0, 4 and 8 kbar produces magma compositions at 8 wt% MgO that are similar to the local trends observed among the FAMOUS/AMAR samples with ~ 8 wt% MgO (see Figure 61 and Appendix C). In detail, however, there are a number of problems with the model. For example, the observed range in Na_2O in the FAMOUS/AMAR data is almost a factor of three greater than that produced by the high pressure modeling (Figure 61). This is in part a result of the pressure effect on plagioclase composition, but even if the crystallizing plagioclase at 8 kbar were less albitic, fractionation at 8 kbar could only produce about 50% of the total observed range in Na_2O. Similarly, high-pressure fractionation only produces about half of the total range in TiO_2, despite the fact that the total amount crystallized by 8 wt% MgO at 8 kbar is approximately 50%.

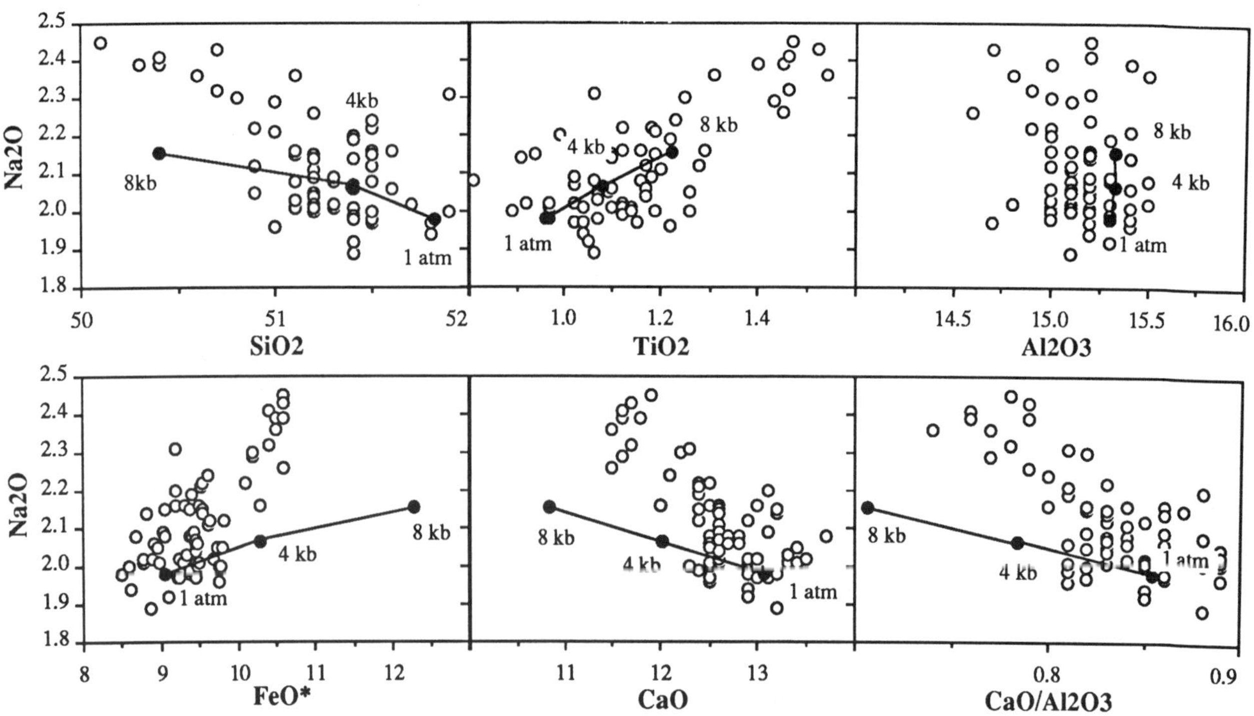

Fig. 61. Data from the FAMOUS/AMAR region with 8 wt% MgO (open symbols, reproduced from Figure 23) compared to calculated liquid compositions at 8 wt% MgO (solid symbols connected by lines) for crystallization at 1 atm, 4 kbar, and 8 kbar. These liquid compositions are calculated are from the LLDs shown in Figure C5 (see Appendix C for details).

Other high-pressure fractionation models (such as Model 2 considered in Appendix C) encounter similar difficulties in reproducing the local trends observed in the FAMOUS data. Thus, although it is logical that high-pressure fractionation may occur at slow-spreading ridges, and may contribute to the variability observed, it does not provide a quantitative explanation for the slow-spreading local vector. In general, it should also be recognized that high-pressure fractionation, like melting, has physical constraints associated with it. High-pressure fractionation will occur only if there is a thick lithosphere beneath the ridge, or if melts pass through the lithosphere above the melting regime far from the ridge prior to being focussed towards the ridge. It also seems likely that the melt would react with the cooler mantle rather than crystallize liquidus minerals in isolation from the mantle [e.g., *Keleman*, 1990]. Models of high pressure fractionation also need to be consistent with the melting process. For example, if fractionation occurs at 8 kbar, the melting column can extend only to a pressure of 8 kbar. High pressure fractionation, therefore, implies lower extents of melting, higher mean pressures of melting, and substantially thinner crust for a given mantle temperature.

6.3.3 Mantle Heterogeneity and the Local Vector

Ultramafic rocks show different types of heterogeneity on a variety of scales. One type of heterogeneity results from different amounts of removal or addition of mafic compositions, as modeled above for Ronda peridotite compositions (section 6.2.2) A second type of heterogeneity is veining, which is common in peridotites over distances of cm to tens of meters. These veins could represent simple modal variations of the same mineral composition as the host peridotite, or they could be of a different composition than the host peridotite. Indeed, both types of mantle heterogeneity, bulk compositional and veining, could be present over the hundred km distances in which we find the local vectors of variability, so we discuss each type of heterogeneity in turn.

Variations in bulk mantle composition were considered as an explanation for the global vector and illustrated in Figure 58. For a given mantle temperature, the variations in mantle composition considered produce a trend of mean melts that cross-cuts the global vector. The melting column paths for each of the mantle compositions would be a group of curves, each of which would have large changes in Na_2O for little change in FeO. Melts mixed from these various melting paths could produce a diversity of compositions, but again Na_2O varies too much and FeO varies not enough to explain the slow-spreading type local variations. If a greater change in Mg# for a given Na_2O content is assumed, it would be possible to produce slopes that correspond better to the slow-spreading local vector, but there is no independent evidence for this type of heterogeneity.

Melting of a veined mantle is more difficult to treat quantitatively, because the heterogeneities may occur on a small scale, there is a wide range of possible compositions, and there is little experimental data on non-standard mantle compositions. The two types of veins (mineralogical concentrations versus distinctly enriched material) are also likely to have very different effects.

Veins of pyroxene with the same composition as the host peridotite will serve as foci of melting. The solidus temperature will be the same as the surrounding peridotite, but substantially more melting will take place in the vicinity of the veins over a given temperature interval, because clinopyroxene is the dominant constituent of the melt. This principal is clear from consideration of any phase diagram: the closer the melt composition and the solid composition, the more melting that occurs for a given temperature change simply because of the lever rule. Now consider elements such as Na and Ti, which in the mantle are contained almost exclusively in pyroxene. At low extents of melting, the bulk partition coefficient is the same in the veins as elsewhere, because the liquid composition is simply C_o/D. Both C_o and D are proportional to the amount of pyroxene present, so the ratio is independent of the amount of pyroxene. Thus the initial liquid composition is essentially independent of modal variations. As the extent of melting increases, however, there are large differences, and the abundances of Na and other elements contained primarily in clinopyroxene will be much greater in the liquid for a given extent of melting (Figure 62). Thus, in the melting of mantle veins there are two counterbalancing effects: higher Na_2O for a given extent of melting, balanced by greater overall extents of melting. In terms of FeO variations, melting of clinopyroxene-rich veins can be expected to lead to lower concentrations of iron in the melt due to the lower concentrations of Fe and Mg in clinopyroxene relative to olivine. The higher Na_2O in the melt also tends to decrease the FeO contents of the melt low because of the K_d effect. If we choose arbitrary melting functions, we can arrive at liquid compositions that vary by about 0.5-1% in FeO with small changes in Na_2O, but the exact slope of the trends is not well constrained. Therefore reasonable evaluation of this model is still premature.

Enriched Veins: A Cause of the EPR Local Trends? A second type of vein is one that may have quite different mineralogy and chemical composition from the surrounding mantle. Such veins have often been called upon as an explanation for variable degrees of trace element and isotope enrichment in local provinces [e.g. *Prinzhofer et al.*, 1989; *Karsten, 1990*; *Michael et al.*, 1989]. Quantitative modeling of such material is not yet possible, but there are some indications that these veins may be important in controlling certain aspects of local variability. Current evidence suggests to us that such veins may cause the EPR local systematics identified in section 3.5.2, but probably are not important factors in the origin of the slow-spreading local trends.

One important characteristic of the EPR local trend is the association between major element composition and enrichment in incompatible trace elements (e.g., K/Ti). The variations in incompatible trace elements are far greater than can be explained by melting alone. In addition, *Klein et al.* [1986] found correlations between $Na_{8.0}$ and $Fe_{8.0}$ and the $^{87}Sr/^{86}Sr$ isotopic composition of basalts from some segments

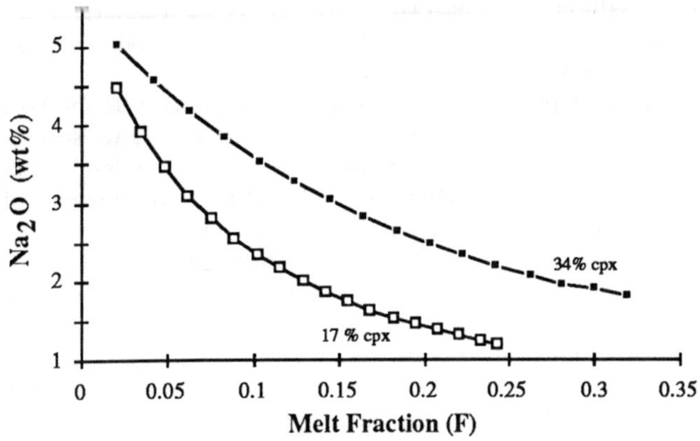

Fig. 62. Melting paths illustrating a possible effect of mineralogical heterogeneity during melting. The solid squares are for mantle with twice the amount of clinopyroxene (and twice the source concentration of Na_2O). The two curves are similar at low F because the ratio of C_o/D is similar. At higher F, the mantle with twice the amount of clinopyroxene produces melts richer in Na_2O.

segments of the northern EPR (Figure 63). Thus, we have evidence for heterogeneity in trace elements and isotopes, and clear correlations between indices of this heterogeneity and the major element signal. Furthermore, the presence of diverse lava types over very short distances and times [e.g., *Thompson et al.*, 1985; *Langmuir et al.*, 1986; *Sinton et al.*, 1990; *Hekinian et al*, 1989; *Reynolds et al.*, 1992] suggests that a range of mantle source compositions is present at any one time beneath the EPR axis. This observation seems most consistent with intermixed sources of material, such as a network of enriched veins, and suggests that the EPR local trend could result from such small-scale mantle heterogeneity. This would be consistent with arguments made previously on the basis of trace elements and isotopes [e.g. Prinzhofer et al, 1989].

If so, the correlation of trace element or isotopic enrichment with major element variations leads to interesting but perplexing implications regarding the melting behavior of these veins. The major element characteristics of these enriched melts are the association of lower $Fe_{8.0}$, lower $Si_{8.0}$, with higher $Na_{8.0}$ and Al_2O_3. These characteristics are not easily explained by the existing experimental data. It is important to recognize, however, that the enriched melts also tend to have higher volatile contents, particularly higher H_2O contents [*Michael*, 1988]. Thus, the major element characteristics of this enriched end-member may result from melting under conditions of high volatile abundances. Quantitative testing of the various aspects of this hypothesis for the EPR local trend must await further experimental data on the appropriate bulk compositions in the presence of water.

While the link between trace elements, isotopes and major elements suggests that the EPR local trend results from source heterogeneity, the lack of such a link for the local trends in the Atlantic suggests that the slow-spreading local trend does NOT arise from enriched veins. In part, the contrast between the EPR and slow-spreading local trends alone indicates that a similar explanation will not account for both of them.

Shirey et al. [1987] analyzed an extensive suite of samples from just north of the Oceanographer Fracture Zone. These samples were recovered from a small area, and hence reflect local variability. The data exhibit large variations in isotopic composition and incompatible element abundances, so there is source heterogeneity. If source heterogeneity were controlling the local major element systematics, there should be a correlation between trace element enrichment and major element variations.

Major element data for these samples (Figure 21) have clear local vectors of the slow-spreading type. The local vectors exist for the samples that have similar Ce/Yb ratios, that is, for subsets of the data that could be produced by different extents of melting. There is no correlation between trace element enrichment and position along a local vector for the data as a whole, as one might expect if the major element vector were controlled by small-scale heterogeneity. Therefore there is not the relationship between major and trace elements that is apparent on the EPR, and mantle heterogeneity does not appear to produce the local trend. The local trend can be present for subsets of the data where no heterogenetiy is required. Therefore mantle heterogeneity is an important aspect of this data set, but it is not the control on the local major element variability.

6.3.4 The Slow-Spreading Local Vector Problem

We have a reasonable hypothesis for the origin of the EPR local trend, although no quantitative model to account for its

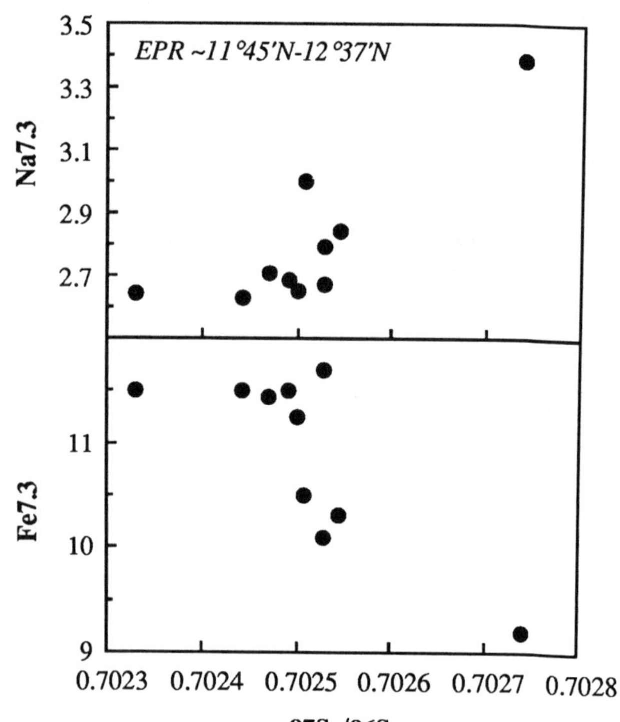

Fig. 63. $^{87}Sr/^{86}Sr$ vs. $Na_{7.3}$ and $Fe_{7.3}$ for EPR basalts from ~11°45'N-12°37'N. These correlations suggest that the high $Na_{7.3}$, low $Fe_{7.3}$ basalts are associated with a more enriched source. These high $Na_{7.3}$, low $Fe_{7.3}$ basalts are of the type that controls the EPR local vector. Data are from *Langmuir et al.* [1986, and unpublished], *Klein et al.*, [1986], and *Castillo et al.* [1992, in preparation].

diverse aspects. We do not yet, however, have a satisfying explanation for the slow-spreading local vector. Although some melting and fractionation processes produce qualitatively similar results that may contribute to the chemical variations, none of them fully accounts for the observations. We are thus faced with a puzzle: the pooled melts of the melting regime account remarkably well for the global systematics of regional averages, yet the local variations that go into the regional average compositions do not have a quantitative model to account for them. Furthermore, these local systematics are a remarkably robust feature of the data, for they appear along ridge segments representing the total range of axial depth, in different ocean basins, and in the presence or absence of hot spot influences. Therefore, whatever their origin, it must be a robust aspect of the crust formation process.

Although we do not have a solution to the slow-spreading local vector, there are other aspects of the data that are relevant, and hint at the direction of a solution. In the following, we examine other lines of evidence that appear to shed light on the problem.

6.3.5 Trace Element Evidence

Fractionation up to pressures of about 10 kbar involves the crystallization of plagioclase as well as pyroxenes and olivine. Plagioclase has a high partition coefficient for the element Sr. When plagioclase is a stable crystallizing phase, the bulk partition coefficient for Sr is about 1, so Sr concentrations change very little as crystallization proceeds. However, during mantle melting Sr behaves as an incompatible element because plagioclase is not likely to be a stable mantle phase over most of the melting interval. Sr thus provides a key to evaluating whether the local vector is created by melting or crystallization processes. Sr is not the only important trace element in this evaluation, however. Melting in general is better able to produce large variations in incompatible elements. Therefore, trace elements as a group potentially can constrain the origin of the local vector.

Carrying out such tests requires abundant major and trace element data on samples, preferably glasses, from individual ridge segments where a local vector is well-developed. Unfortunately, there are few ridge segments for which we can explore this relationship. In the following, we examine data from three regions on the Mid-Atlantic Ridge to see if the local vectors observed in these regions are likely to be caused by melting or crystallization processes.

Langmuir et al. [1977] presented major and trace element data on hand-picked glasses from the FAMOUS region that spanned a substantial compositional range. Among them are two samples with 8 wt% MgO that lie at the two ends of the local vector for the FAMOUS area discussed at length above. These two samples have substantially different abundances of all incompatible trace elements (varying by more than a factor of two), and Sr shows the same variability as other elements of similar incompatibility during mantle melting. This general behavior of Sr as an incompatible element is present for other FAMOUS data as well, as seen in Figure 64*a*. These large Sr

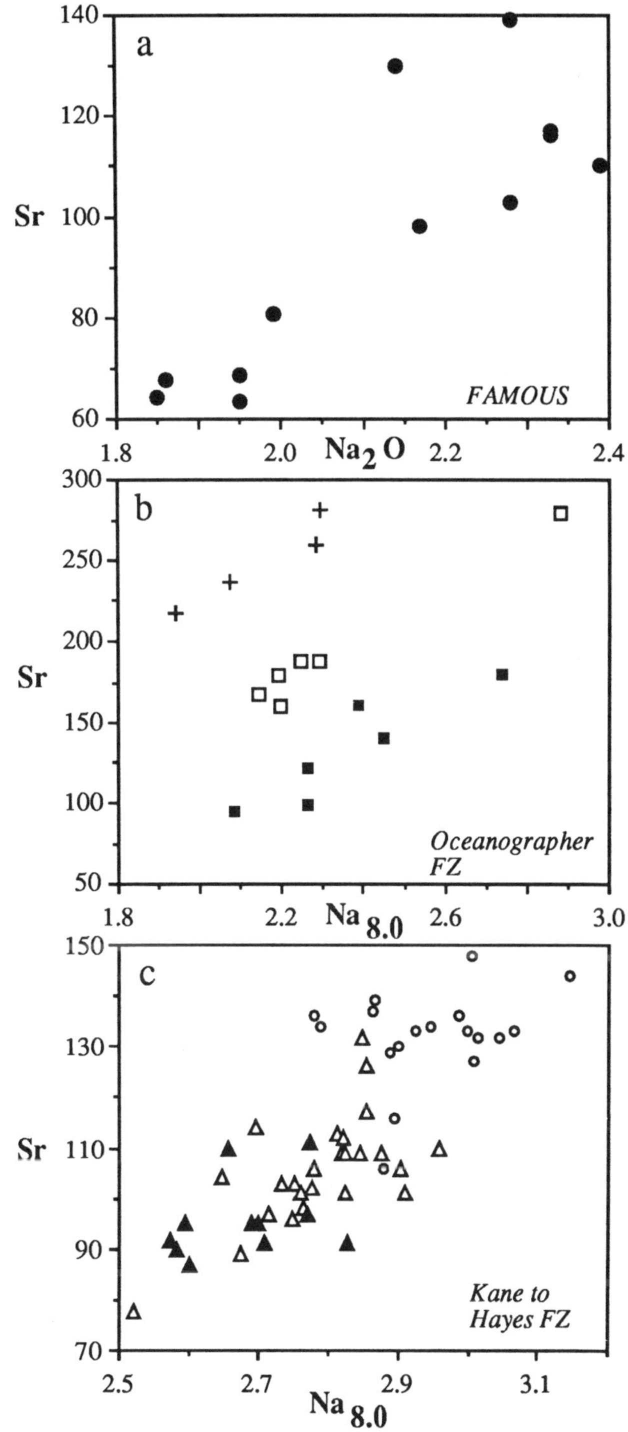

Fig. 64. $Na_{8.0}$ vs. Sr for MORBs from various regions on the Mid-Atlantic Ridge. *a*) High-MgO (> 8.5 wt%) samples from the FAMOUS region, obtained by plasma emission spectrometry [*Langmuir*, unpublished]. *b*) Samples from near the Oceanographer Fracture Zone. Plusses for Ce/Sm = 9.7-10.7; open squares for Ce/Sm = 7.4-9.1; solid squares for Ce/Sm = 4.2-6.5. Within each group, Na and Sr correlate positively. [*Langmuir and Bender*, unpublished data.] *c*) Samples collected between the Kane and Hayes FZs on the Mid-Atlantic Ridge. Open circles (23°N-24°N); open triangles (27°N to the Atlantis FZ); filled triangles (Atlantis FZ to Hayes FZ). Analyses by plasma emission spectrometry [*Langmuir*, unpublished].

variations preclude a process in which plagioclase is a major fractionating phase, and suggest control by mantle melting.

A second slow-spreading local-trend data set comes from a region of normal ocean ridge basalts, where there is no hot spot trace element signature, between the Kane and Hayes Fracture Zones on the Mid-Atlantic Ridge. This region is of particular interest, because there is a substantial gradient in axial depth, from more than 4000 meters near the Kane Fracture Zone, to about 2500 meters south of the Hayes Fracture Zone. The data for $Na_{8.0}$-$Fe_{8.0}$ were shown in Figure 21*a*; Figure 64*c* shows the $Na_{8.0}$-Sr relationships. The $Na_{8.0}$-$Fe_{8.0}$ diagram shows that the chemical variability is reflected in both global and Atlantic-type local vectors. The shallowest region has the highest $Fe_{8.0}$ and lowest $Na_{8.0}$, but the data for each local region exhibit a positive correlation between $Na_{8.0}$ and $Fe_{8.0}$.

$Na_{8.0}$ and Sr correlate positively for this region (Figure 64*c*). The correlation is obvious for the data as a whole, but this reflects in part the global component of variability. For each local region, there is still a positive correlation between the two parameters, but there is also significant variability in $Na_{8.0}$ for a given Sr content. This is particularly true for the southernmost region, near the Kane Fracture Zone. This data set as a whole may therefore require a two-fold explanation, one to account for the positive correlation between $Na_{8.0}$ and Sr, and another to explain the diversity of $Na_{8.0}$ at a single small range of Sr contents.

The third region is near the Oceanographer Fracture Zone. Figure 64*b* shows the $Na_{8.0}$-Sr variations for this region, with the samples divided into three groups on the basis of their Ce/Yb ratios. Within each group, there is the positive correlation that would be expected from melting processes. The component of heterogeneity appears to move sample groups to increasing Sr abundances for a given $Na_{8.0}$. Thus, in this diagram there is clear evidence for a separation of the local trend from control by mantle heterogeneity, and also the positive correlation between $Na_{8.0}$ and Sr that is characteristic of slow-spreading local trends as a whole.

The positive correlation between Sr and $Na_{8.0}$ in each of these regions suggests that *the slow-spreading local trend is produced by melting processes and not by crystallization processes.* Neither low-pressure crystallization, no matter how complex, nor high-pressure crystallization (to about 10 kbar) can be the dominant control on the slow-spreading local vector of variability, based on this correlation. This conclusion is possible because of the ubiquitous occurrence of plagioclase as a crystallizing phase from MORB at low to moderate pressures. The data also suggest that the slow-spreading local trend is not produced by local mantle heterogeneity. Therefore this robust signature of slow-spreading ridges is telling us some important characteristics of the melting process, even though we do not yet have a quantitative model to account for it.

This is not to say that high-pressure fractionation plays no contributing role. For the region near the Kane Fracture Zone, Sr does stay approximately constant, which supports high-pressure fractionation models that have been suggested for this region [*Tormey et al.*, 1987; *Grove et al.*, 1990; *Reynolds and Langmuir*, 1992 in press]. But high-pressure fractionation cannot be the root cause of the slow-spreading local vector.

6.3.6 *Evidence from Ocean Islands: Toward a Solution to the Local Vector Problem?*

One of the characteristics of the compositional dependencies of the partition coefficients incorporated in the quantitative melting model is that high alkali contents lead to low FeO contents. Indeed, small-degree melts from deep in the melting column do have very low FeO contents in our calculated models. There is another rock series that has often been thought to result from low extents of melting at high pressures — the alkaline rocks found on ocean islands and continents, regions where there is often a hot spot and thicker lithosphere that causes melting to take place at deeper levels. These rocks, some of which contain ultramafic nodules and must have ascended rapidly from the mantle, can have 7 wt% total alkalis at high MgO contents. Therefore, if the calculations are pertinent to the melting process, one would expect that these very alkaline rocks, particularly those with both high K_2O and high Na_2O, would have relatively low FeO contents.

But in contrast to the calculations, these alkaline rocks are often exceptionally high rather than low in FeO. Not only are alkaline rocks generally high in FeO, but the more alkaline the rock the higher its FeO contents. This is illustrated for rocks from the Hawaiian islands in Figure 65. In general, ocean island data for fresh volcanics exhibit a positive correlation between FeO and Na_2O, like the slow-spreading local vector, suggesting there is a "local vector" for ocean island volcanics as well as ocean ridge volcanics. For the ocean island rocks, the case for an origin by partial melting is also very strong, particularly since ultramafic xenoliths can occur in the alkaline rocks. The rocks also show substantial variability in incompatible element abundances. As Na_2O decreases, all the incompatible elements also decrease substantially, even when all data are normalized to a constant and high MgO content.

The high alkali and FeO contents, combined with the idea that ocean island volcanics are derived by low extents of melting from high pressures suggests that there is some additional factor influencing FeO and Na_2O contents during partial melting that is not part of our melting calculations — which means that it is not in the existing experimental data set. Possible candidates are certainly volatiles, which, because of their high solubility in magmas at high pressures, can play a particularly important role at low extents of melting at high pressures. A possible candidate for alkaline rocks is CO_2. Although the detailed effects of CO_2 on mantle melting are not well known, it is known that melting of CO_2-rich peridotite results in more alkaline rocks. Indeed, the creation of highly alkaline rocks can probably only occur in the presence of appreciable CO_2. The ultimate in alkalinity occurs in carbonatites [e.g., *Wyllie*, 1987; *Nelson et al.*, 1988].

There is increasing evidence that CO_2 is an important volatile constituent in the MORB source region as well. Dixon and Stolper [e.g., *Dixon et al.*, 1988] have shown that the CO_2 contents of erupted MORB are mostly controlled by the

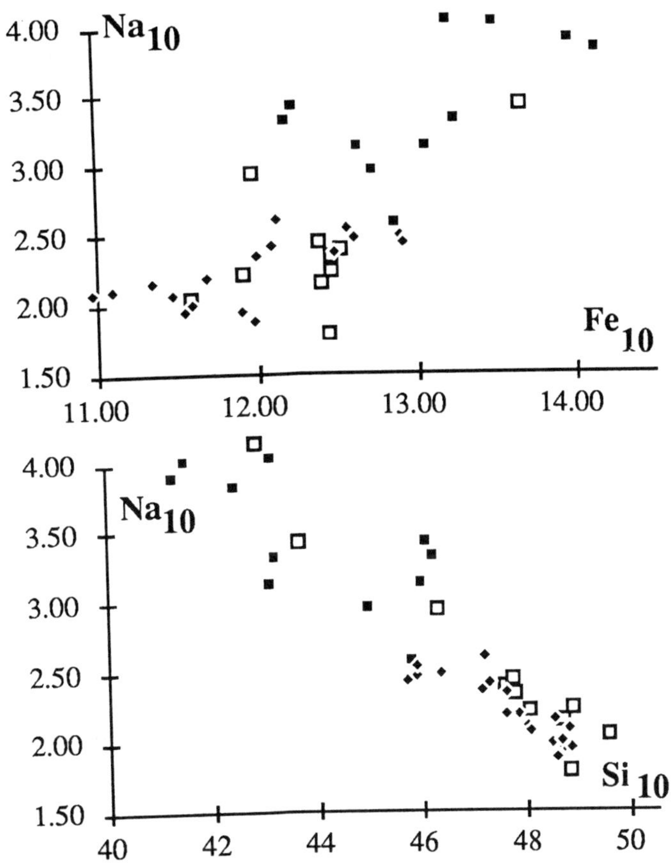

Fig. 65. Na_{10} vs. Fe_{10} and Si_{10} for basalts from Hawaii. Na_{10} is analogous to $Na_{8.0}$, where Na_{10} is the Na content projected to 10% MgO. Diamonds are samples from Mauna Kea [*Frey et al.*, 1990]; small solid squares from the Honolulu volcanics [*Clague and Frey*, 1981]; open squares from Loihi seamount. Samples were screened to have 7.5-12% MgO, > 41% SiO_2, and a sum of > 98% for the major element oxides. See *Langmuir* [1991] for more details.

solubility of CO_2, which is very sensitive to pressure. The erupted CO_2 contents simply reflect the pressure of degassing, and hence are minimum magmatic values, not indicative of the CO_2 contents at high pressures or in the source regions. In addition, there are special samples known as "popping rocks" that are occasionally recovered from the sea floor and have exceptionally high CO_2 contents. *Sarda and Graham* [1990] have suggested that the high inferred CO_2 contents of these rocks are more indicative of the actual abundance of CO_2 in MORB prior to degassing. Thus, it is possible that there is substantial CO_2 in the MORB mantle source, and that, as for ocean islands, the deep parts of the melting column are alkaline rocks with low SiO_2 and high FeO. Then the melting column vector would have much higher FeO contents with high Na_2O than we calculate based on the results of dry experiments. This might lead to melting column trends that are more similar to the slope observed for the Atlantic-type local vector. Whether CO_2 is the explanation or not, it is clear that there is some aspect of the melting process that can create high FeO contents with high alkali contents, for we see this effect in almost every primitive alkaline rock.

Such an alkaline composition might be an appropriate end-member for the creation of the slow-spreading local vector, for it would contain high Na_2O, high FeO, low SiO_2, and high concentrations of Sr and other incompatible elements. Thus, it may provide a solution to the two criticalaspects of the slow-spreading local vector that we have emphasized— the positive correlations of $Na_{8.0}$ with $Fe_{8.0}$ and Sr.

6.3.7 *Local Variability and Ridge Segmentation*

The discussion of local variability above relates to the chemical systematics without reference to the spatial location of the samples within a particular ridge segment. However, the spatial distribution of MORB compositions has been a major focus of research, and ultimately detailed three-dimensional models of melt formation and segregation may be constrained by the along- and across-strike chemical systematics of erupted lavas. In addition, the spatial systematics of the local variations may help to guide our thinking concerning their origins.

Early work in the FAMOUS area [*Bryan and Moore*, 1977; variations across the rift valley over a limited distance along strike. Such across-strike variations now seem best interpreted as short-term temporal variability. This classic data set therefore now provides strong evidence for substantial temporal variability in the chemistry of eruptives over short time intervals (thousands of years). The amplitude and period of temporal variability, should ultimately provide very useful constraints on melting models, but this information is just beginning to be available [e.g., *Hekinian and Walker*, 1987; *Reynolds et al.*, 1992], and cannot yet be used to constrain processes of melt formation and segregation.

There are sufficient data along strike, however, to begin to consider some of the spatial systematics of local chemical variability. For example, in the late 1970s and early 1980s attention focussed on chemical variations as a function of proximity to tectonic offsets along axis such as ridge offsets and propagating rifts [*Hekinian and Thompson*, 1976; *Schilling and Sigurdsson*, 1979; *Natland and Melson*, 1980; *Christie and Sinton*, 1981; *Sinton et al.*, 1983; *Bender et al.*, 1984; *Langmuir and Bender*, 1984]. In these and subsequent studies, evidence was found for local variations in the extent of melting and degree of fractionation. *Bender et al.* [1984], for example, noted chemical changes indicative of lower extents of melting adjacent to some transform faults, consistent with the prediction of such edge effects by Fox and coworkers [e.g., *Fox and Gallo*, 1984]. Subsequent work has shown that "magmatic edge effects" can occur even at some minute offsets of the East Pacific Rise [*Langmuir et al.*, 1986; *Sinton et al.*, 1991], but are not present at some of the largest offset transform faults [*LeRoex et al.*, 1989; *Klein et al.*, 1991; *Reynolds and Langmuir*, 1992 in press].

Recent work has also focussed on detailed sampling of entire ridge segments in order to explore the nature of along-strike chemical variability on the scale of a single spreading cell. In the Pacific, *Langmuir et al.* [1986] and *Sinton et al.* [1991] found substantial chemical variability over short distances, but also that several tectonic offsets often had significant chemical discontinuities across them, as first noted by *Thompson et al.* [1985]. In the Pacific, there is little variation in axial depth, gravity anomalies, and probably crustal thickness, and in the chemical data it does not currently appear that there are systematic chemical variations within offset-bounded ridge segments. *Batiza et al.* [1990] have suggested that regular variations across-axis in the Pacific can be studied using samples from seamounts.

For slow-spreading ridges, there have been many suggestions of a relationship between melt generation and position within a spreading segment [*Whitehead et al.*, 1984; *Lin et al.*, 1990]. Certainly, one can imagine a number of possible relationships between the melting regime and offsets of the ridge axis, and many of these relationships should have predictable petrological consequences. For example, preferential active upwelling in the centers of segments should lead to greater extents of melting at lower pressures, and greater crustal thicknesses. The combination of petrological data on carefully located samples with measurements of crustal thickness, geophysical properties such as gravity and theoretical modeling of melt segregation should lead to coherent and testable models that relate these various aspects of ocean crust formation.

Currently, however, there are few data that can be used to move in this promising direction, since combined detailed studies of ridge segments are just beginning. *Batiza et al.* [1988] sampled a 100 km-long, transform-bounded segment of the southern MAR, and found a pattern of chemical variability that they suggested reflected spatial systematics in the melting regime beneath the ridge (Figure 66*a*). They found a complex along-axis pattern where samples from the shallow center of the ridge segment had chemical characteristics similar to those at the deep ends of the segment. *Reynolds and Langmuir* [1992, in revision] examined the ridge segment south of the Kane transform in the N. Atlantic, and found a rather different pattern (Figure 66*b*). It is intriguing, however, that over the same length scale, independent of the ridge offsets, the pattern of chemical variation looks rather similar. The data from these two regions (Figure 66) shows that we are still not able to generalize relationships between chemistry and spatial position within a ridge segment. This type of data, however, holds much future promise for elucidating both the origin of the slow spreading local variations, and the relationships among melt production, ridge tectonics and basalt chemistry. Systematic spatial relationships that relate basalt composition to physical characteristics of the ridge, such as spreading rate, transform spacing, axial depth, gravity anomalies or stage of a temporal cycle will provide a coherent data set that can be used to test quantitative models for ocean crust generation.

6.4 The Nature of MORB Primary Magmas

The results of the modeling and the discussion of the origins of the global and local compositional vectors cast a different light on the long-standing question of MORB primary magmas [e.g., *O'Hara*, 1968*a*; *Elthon*, 1989; *Stolper*, 1980; *Fujii and Scarfe*, 1985; *Presnall and Hoover*, 1984; *Kinzler and Grove*, 1992*a* and *b*; *Falloon and Green*, 1987; *Elthon*, 1989]. The classical approach to this problem emphasized one or a small range in primary magmas, and that such magmas can be understood in terms of a pressure of segregation from the mantle. The correlation between pressure of melting and extent of melting, however, requires that melts of varying extents can be sampled from throughout the melting regime, as suggested by *Langmuir et al.* [1977], and that melts sampled at the surface result from pooling of diverse compositions from different pressures, as emphasized also by *O'Hara* [1985]. This means that it is more accurate to speak of mean pressures and extents of melting, and that the primary magmas that may arrive at the surface are not compositions that were themselves in equilibrium with the mantle at a single specified temperature and pressure. This complicates the classical approach to the primary magma problem of using multiple saturation experiments to test whether a particular composition is a primary magma in equilibrium with the mantle at some specific temperature and pressure. Melts at the surface are mean melts derived from ranges of T and P.

Another important corollary is the likelihood of a wide diversity of primary magma compositions present beneath all ridge axes, since the melting regime contains a variety of melt compositions within it. The variety is produced both by the different extents and pressures of melting within the melting regime, and as a result of melting a mantle that is heterogeneous on a range of scales. Therefore, there are multiple MORB primary magmas, and the challenge is to understand the systematic differences rather than to identify which of them is the true primary magma.

The primary magmas can also have a rather wide range of MgO contents. The exact value depends in part on the melting process — fractional melting causes significantly higher MgO contents than does equilibrium melting. MgO contents are controlled primarily by the pressure of intersection of the solidus. Mean melts from hot spot regions where melting begins at greater depth may have greater than 15% MgO, and hence be picritic. The deepest ridges over the coldest mantle may produce magmas with as little as 10% MgO. Thus proponents of both high and low MgO primary magmas were partially correct, for both exist in different regions. Average ocean ridge basalt, however, is now quite well constrained to be generated by an intersection of the solidus at about 20 kbar, a mean extent of melting of about 10%, at mean pressures of about 10 kbar, and to contain approximately 12 wt% MgO.

6.5 Interpretation of MORB Major Element Chemistry

We have now observed how major elements vary on many ocean ridges, and considered a number of theoretical approaches to account for these observations. If the working model applied generally, and both global and local trends corresponded with the calculated variations in mantle temperature and variations within the melting regime,

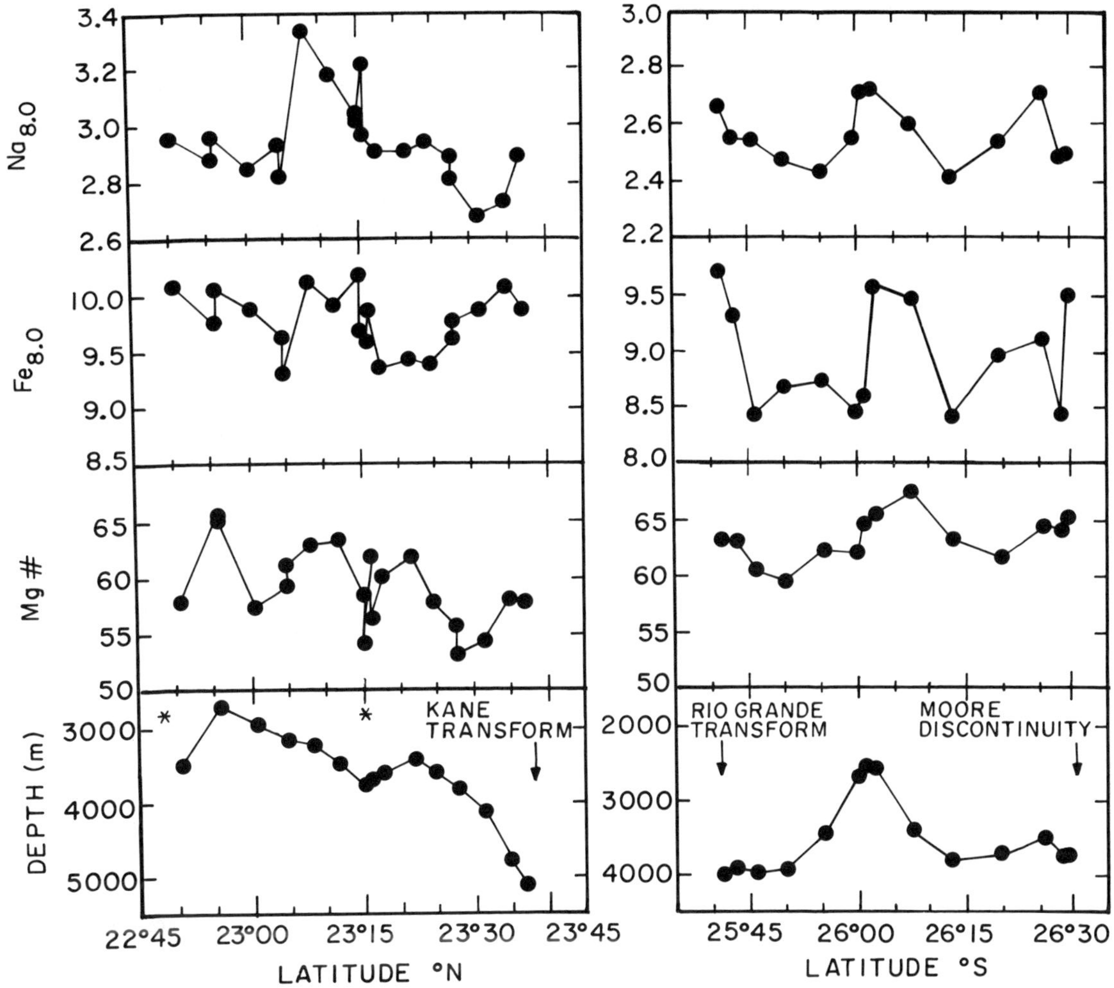

Fig. 66. Comparison of data for ridge segments along the Mid-Atlantic Ridge, one from south of the Kane Transform in the N. Atlantic [*Reynolds and Langmuir*, 1992], and the other from south of the Rio Grande transform in the S. Atlantic [*Batiza et al.*, 1988]. The S. Atlantic data, taken from a figure presented by *Batiza et al.* [1988], were not corrected for interlaboratory bias. The data points are for samples collected along strike of the ridge, with an effort made to sample the neovolcanic zone where possible. Thus, the patterns reflect as closely as possible the instantaneous variation in chemistry along a ridge segment. For each side of the figure, the scales are adjusted so both horizontal and vertical scales are comparable. The S. Atlantic data are for a single ridge segment. The N. Atlantic data span two ridge segments, separated by small non-transform discontinuities whose position is indicated by asterisks.

respectively, then interpretation of MORB data would be simple. We could write equations for mantle temperature and extent of melting that could be used for chemical analyses from any particular ridge segment. The Na_2O and FeO contents of individual basalts would specify both the mantle temperature and the depth within the melting regime from which that particular basalt was derived. This approach has been utilized recently by Niu and Batiza [1991].

But the situation is not so straightforward. There is structure within the global correlations caused by hot spots and by subtle differences between ocean basins. And there are distinct local trends on the MAR and EPR, neither one of which corresponds with the melting calculations in their present state of advancement. How, then, at this stage of our understanding, can we interpret a particular data set? For the East Pacific Rise, it is our current view that most of the local variations are caused by small scale heterogeneities. Hence variations in $Na_{8.0}$ along strike are not directly indicative of variations in mantle temperature and may not even indicate extent of melting. Instead, there may be information concerning the nature of melting and melt segregation from a heterogeneous mantle — but the content of this information remains to be elucidated. For slow-spreading ridges, although we do not have a quantitative explanation for the local trend, it is at least clear that it is caused by melting processes. Therefore, even though we cannot yet quantitatively reproduce it, we interpret the local trends provisionally as indicative of position within a melting regime — possibly with additional complications of small scale temperature variations associated with segmentation.

If we accept these provisional interpretations of the slow-spreading local trends, then the intersection of these trends with the global correlations gives the regional mantle temperature. This is the concept behind the $Fe^G_{8.0}$ parameter of *KL*89, a parameter which simply indicates the $Fe_{8.0}$ contents at which a local trend intersects the global array. $Fe^G_{8.0}$ tracks variations in regional depth on slow-spreading ridges very well. The use of this parameter is still open to question, however, because we do not yet have the theory that adequately accounts for the slow-spreading local trends. Use of this parameter on the EPR is not appropriate, since the local variations there have a different cause. The best estimate of $Fe^G_{8.0}$ will be to have a suite of data from a particular ridge segment, and to determine where those data intersect the global correlations. Then the new melting model allows us to calibrate $Fe^G_{8.0}$ to temperature. First we need an equation for the global correlations for normal ridges, which is:

$$Na_{8.0} = -0.36Fe_{8.0} + 6.26 \quad (28)$$

$Fe^G_{8.0}$ is the intersection of the local trend with this global correlation line. (Note that the specific equations for $Fe^G_{8.0}$ from *KL*89 do not apply here, because they were based on Smithsonian data, while the data in this paper and the calculations are based on Lamont data. $Fe^G_{8.0}$ should be determined for each data set). Then we have the following relationship between P_o and $Fe^G_{8.0}$, determined for the fractional melting model, which turns out to be almost linear:

$$P_o = 6.11Fe^G_{8.0} - 34.5 \quad (29)$$

P_o then gives the temperature of intersection of the solidus ($1150 + 13*P_o$). Application of these relationships is uncertain near hot spots, because we do not yet know the cause of the offset of many hot spot points from the $Na_{8.0}$-$Fe_{8.0}$ relationships for normal ridges.

7. TRACE ELEMENT CONSTRAINTS

Our approach thus far has emphasized the major element compositions of MORB. Approaching melting through major elements, however, is a recent development. The common approach for many years was instead to approach melting problems through the trace element compositions of the samples [e.g. *Schilling*, 1975; *White and Bryan*, 1977; *Langmuir et al.*, 1977; *Perfit et al.*, 1983; *Clague et al.*, 1981], and to use the major elements mostly as a sample screen. Trace elements appear to be an ideal tool to investigate melting processes, because the ratio of two moderately incompatible trace elements should change little during crystallization, but markedly during melting [*Gast*, 1968]. In practice, however, incompatible trace elements often vary more than one would predict, and these variations have not yet been shown to have straightforward relationships with the major element variations just presented. On the other hand, moderately incompatible and compatible elements do seem to show similar systematics to the major elements. In this discussion, we focus on the contrasting global and local systematics of trace element abundances in MORB, and also on the important distinction between the highly incompatible elements as a group, and the other less incompatible trace elements.

7.1 Global Variability of Trace Elements

The major element model for the global correlations carries with it specific predictions for the abundances of trace elements, if the source were homogeneous. The source is likely to be quite homogeneous for moderately incompatible elements [e.g., *Hoffmann*, 1988], and indeed, for samples far from hot spots, some compatible and moderately incompatible trace elements correlate well with $Na_{8.0}$ in a way that can be predicted by melting. For example, Sc behaves similarly to CaO/Al_2O_3, and forms an inverse trend with $Na_{8.0}$ for regionally averaged samples (Figure 67*a*; *KL*87). *KL*87 also showed that Sm/Yb correlates positively with $Na_{8.0}$, and that Ni also shows systematic relationships with $Na_{8.0}$ when the effects of olivine fractionation have been taken into account. Therefore for samples not dominated by hot spot effects, the global trace element variations are generally consistent with the inferences from the major element global vector, and hence with different extents of melting of the mantle on a regional basis.

As one examines progressively more incompatible elements, the relationships become more scattered. Figure 67*b* shows averages of $Na_{8.0}$ and $Ce_{8.0}$, (see equation 16) for the calculation of $Ce_{8.0}$, for the same regions as on the Sc plot. There is still a reasonable correlation which corresponds with melting models, but some regions have $Ce_{8.0}$ contents that are slightly high for a given $Na_{8.0}$. The EPR, for example, is higher than regions with comparable $Na_{8.0}$ from the Mid-Atlantic Ridge. The lower bound of the data, however, corresponds well with a melting model based on the major elements. *Plank and Langmuir* [1992, hereafter *PL*92] have called this line, which serves as a useful reference, the "baseline". Once one moves to elements more incompatible than Ce, the data become very scattered. For example, there is no simple relationship, between $Na_{8.0}$ and $Ba_{8.0}$, a highly incompatible element (Figure 67). Thus, coherence with major element global variations seems to be a function of element incompatibility. Highly incompatible elements seem to vary independently of the major elements. Of course, it is for just these elements that mantle heteroegeneity becomes increasingly important, as is well known. Note that the relationships shown in Figure 67 are for MORB samples far from hot spots. Including the samples near hot spots only increases the scatter on these diagrams.

Although hot spots are not part of any global correlations for trace element abundances, there are important spatial relationships associated with hot spots near ridges. These relationships have been extensively documented by Schilling and coworkers over the last twenty years [e.g., *Schilling*, 1973; *Schilling*, 1975; *Schilling et al.*, 1976; *Schilling et al.*, 1983]. Figure 68 shows the ratio of two light rare earth elements (REE), La and Sm, plotted against latitude and depth for the northern Mid-Atlantic Ridge axis. There is an obvious relationship between the occurrence of the hot spots (e.g., Azores, Iceland, and Jan Mayen), shallow ridge axes and higher La/Sm ratios. These observations led *Schilling* [e.g., 1972] to postulate mixing between a deep, incompatible element enriched plume source, and a shallow, depleted "normal" ocean

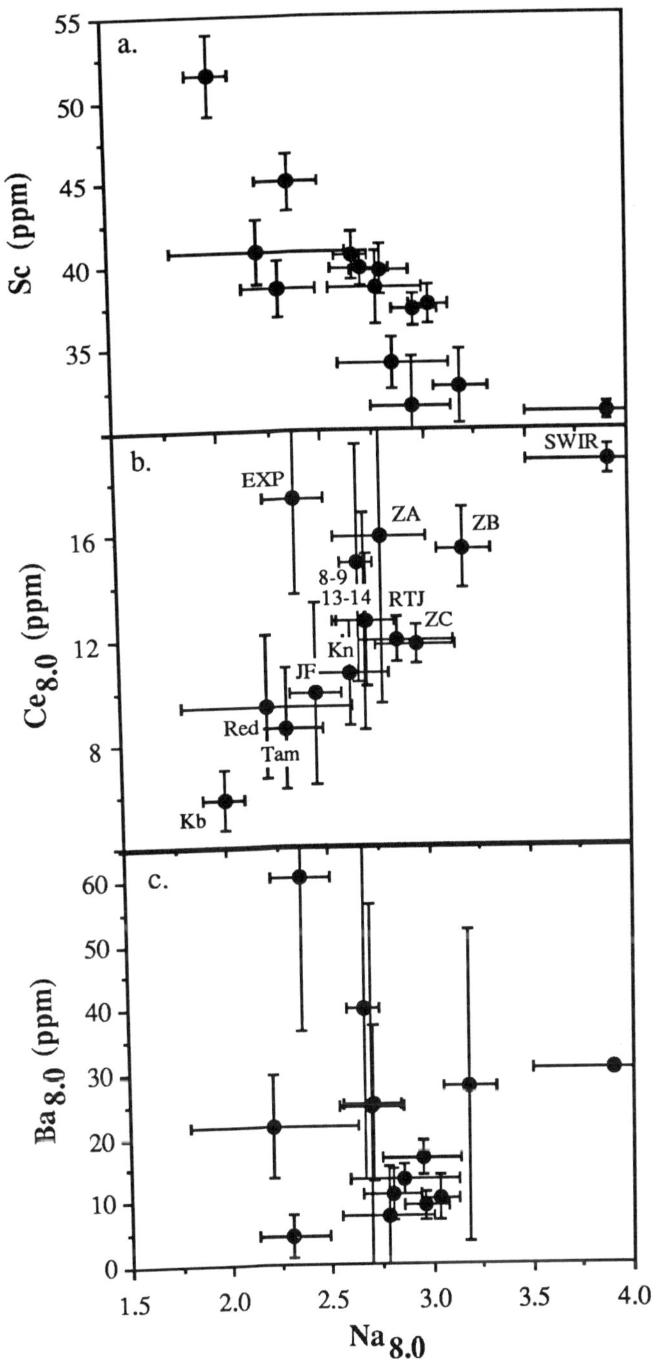

Fig. 67. $Na_{8.0}$ vs. *a)* Sc, *b)* $Ce_{8.0}$, *c)* $Ba_{8.0}$ for regionally averaged data. Each point is the average of samples from an individual region, and the error bars are one standard deviation of the values. $Ce_{8.0}$ and $Ba_{8.0}$ calculated using equation given in the text. Regions are: Kolbeinsey Ridge (Kb), 17°N in the Red Sea (Red), near Tamayo transform of the EPR (Tam), Juan Fernandez microplate (JF), near Kane of the MAR (Kn), 13-14°N on EPR (13-14), 8-9°N on EPR (8-9), south Explorer Ridge (Exp), Zones A, B and C of the Australian Antarctic Discordance (ZA, ZB, ZC), near the Rodriguez or Indian Ocean Triple Junction (RTJ), and around 30°S on the Southwest Indian Ridge (SWIR). For references, see *Plank and Langmuir* [1992].

ridge basalt source. Although some workers have questioned the depths in the mantle of the differing source compositions, it is nevertheless clear that a dominant signal in incompatible trace element variability is mantle heterogeneity associated with hot spots, a conclusion also indisputably supported by radiogenic isotope data [e.g., *Zindler and Hart*, 1986].

Observations regarding the width of the chemical anomalies as well as the different chemical signatures associated with different plumes have led to complex and quite detailed models for plume/ridge interactions [e.g., *Schilling*, 1985; *Hanan et al.*, 1986]. *Schilling* [1985], for example, has shown a correlation between the width of the chemical anomaly on the ridge, and the distance between the plume and the ridge. Thus, the sharp spikes in La/Sm that occur at latitudes 71°N and 35°N might be attributed to plumes that are presently far from the ridge axis. Although Schilling's models explain a large number of observations remarkably well, there are aspects of the long wavelength signal that are difficult to understand. For example, the gradients around both Iceland and the Azores are highly asymmetric. There are also spikes in incompatible element abundances at 43°N and 45°N along the Mid-Atlantic Ridge that are not currently well explained by off-axis hot spots.

It is important to note as well that the relationships for hot spots between incompatible element enrichment and depth are

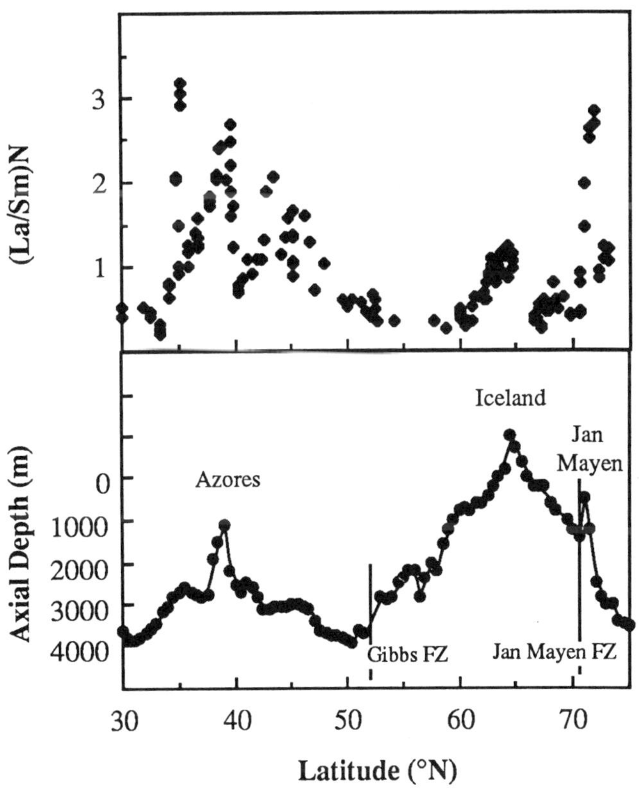

Fig. 68. Latitude (°N) vs. smoothed axial depth and (La/Sm)N for the northern Mid-Atlantic Ridge. Latitudinal depth variations, taken at 5° intervals, are from *LeDouaran and Francheteau* [1981] and *Vogt* [1986]; La/Sm data are from *Schilling et al.* [1983].

not "global," because the magnitude of incompatible element enrichment does not diminish systematically from the shallowest to the deepest ridge segments. This is apparent from Figure 68, where it can be seen that Iceland is a much shallower feature than the Azores, and yet has a substantially lower La/Sm ratio. Thus in trace elements as well as major elements and depth, the Azores and Iceland are very different hot spots.

These diverse aspects of the trace element data can be understood in part as the result of two contrasting effects. Some of the deepest parts of the ocean ridge system, such as the Mid-Cayman Rise [*Thompson et al.*, 1979] and Australian Antarctic Discordance [*Klein et al.*, 1991] are moderately enriched in incompatible elements compared to shallower ridges because of the lower extents of melting that gave rise to them. Hotter temperatures for normal mantle cause greater extents of melting and relative depletion in incompatibles. But the hottest regions, usually associated with hot spots, have a large trace element excess associated with mantle heterogeneity. Therefore both the deepest and shallowest ocean ridges are often more enriched in incompatible elements than normal depleted ridges that lie at intermediate depths. The contrasting trace element behavior leads to some of the great chemical boundaries in the oceans, such as the Hayes Fracture Zone, where mantle enriched by the Azores province is bounded on the northern side of the fracture zone, while relatively hot normal MORB mantle lies on the southern side [*Bougault and Treuil*, 1980]. The temperature may change progressively across the fracture zone, but the trace elements change abruptly because of the hot spot trace element influence on one side and not on the other.

Another important aspect of the global variability of trace elements is the contrast between the EPR and MAR. The EPR has a single hot spot, associated with Easter Island, that has a very limited effect on depth and chemistry. In contrast, much of the depth and chemistry of the Atlantic are strongly influenced by hot spots. The EPR, however, does seem to have some long wavelength gradients in trace element chemistry not obviously associated with hot spots. *Sinton et al.* [1991] have shown that there is a smooth gradient in incompatible element enrichment along several hundred kilometers of the southern EPR. The EPR also has apparently random occurrences of enriched MORB, in contrast to the highly localized occurrences associated with hot spots on the MAR. Thus the MAR and EPR, and possibly fast- and slow-spreading ridges in general, appear to have a rather different relationship with hot spot activity.

7.2 *Local Variability in Trace Elements*

Substantial variations in trace element abundances and ratios are well established for many small-scale regions at a variety of spreading rates [e.g., *White and Bryan*, 1977; *Langmuir et al.*, 1977; *Hekinian and Walker*, 1987; *Shibata et al.*, 1979; *Thompson et al.*, 1985; *Batiza and Vanko*, 1984; *Michael et al.*, 1989; *Karsten et al.*, 1990]. In some cases, the variability appears to be associated with the segmentation of the ridge [e.g., *Bender et al.*, 1984], while in others there are simply substantial differences among closely spaced samples. The trace element variations are often interpreted as evidence for different extents of melting of the mantle or for source heterogeneity, but these interpretations are seldom connected to the major element data.

The local systematics observed in the major elements provide a new perspective from which to view the trace element variations. We now can compare trace element variations, which have often been interpreted by quite specific partial melting models, with major element variations that are also considered to provide information concerning the melting process. We have seen above that there is a useful "baseline" based on regional averages that relates major and trace element compositions. From some trace elements, there is also evidence for pooling of individual melts from throughout the melting regime. For example, *Salters and Hart* [1989] found a relationship between Lu/Hf ratios and Hf isotopes that has implications similar to those that can be inferred from the relationship between $Na_{8.0}$ and $Fe_{8.0}$ — that melts are pooled from throughout the melting regime. The Lu-Hf relationships support the inference from major elements that there is a local vector that reflects processes within the melting regime.

Upon examining individual samples from any specific region in detail, however, an important contrast emerges between the major and trace element models — the trace element models almost always suggest a smaller inferred extent of melting than we have found for the major elements. For example, *Bender et al.* [1984] found evidence for extents of melting of around 5% for the Tamayo Region on the EPR, yet evidence from major elements and crustal thickness suggests instead a mean extent of melting of 10% or more. Therefore we see an overall consistency between average trace element composition and major element models for samples far from hot spots, but models based on separate analyses from an individual region lead to contradictory results.

The source of this contrast between major and trace element models can be seen in an examination of the unaveraged data. Although the averaged $Ce_{8.0}$ data for samples far from hot spots showed a reasonable correlation with $Na_{8.0}$, examination of the unaveraged data shows a large degree of scatter (Figure 69*a*). When individual samples are identified separately (Figure 69*b*), the data often form steep yet well-defined trends, where $Ce_{8.0}$ may vary by a factor of three or more, while $Na_{8.0}$ varies by only 20% [*PL*92]. These steep trends, for example, are apparent in the data from the Red Sea near 17°N, the EPR near 9°N, and the FAMOUS area. It is these steep trends that require small extents of melting for models based on trace elements. On the $Ce_{8.0}$-$Na_{8.0}$ diagram (Figure 69), the melting curve has a steep slope only at small extents of melting, because only when F is small can there be significant separation between elements with small D's. Therefore attempting to fit a steep trend with a trace element model, one is driven to small F's to arrive at the necessary fractionation among incompatible elements.

These steep trends provide another view of the long-standing observation in MORB studies of a "trace element excess" that

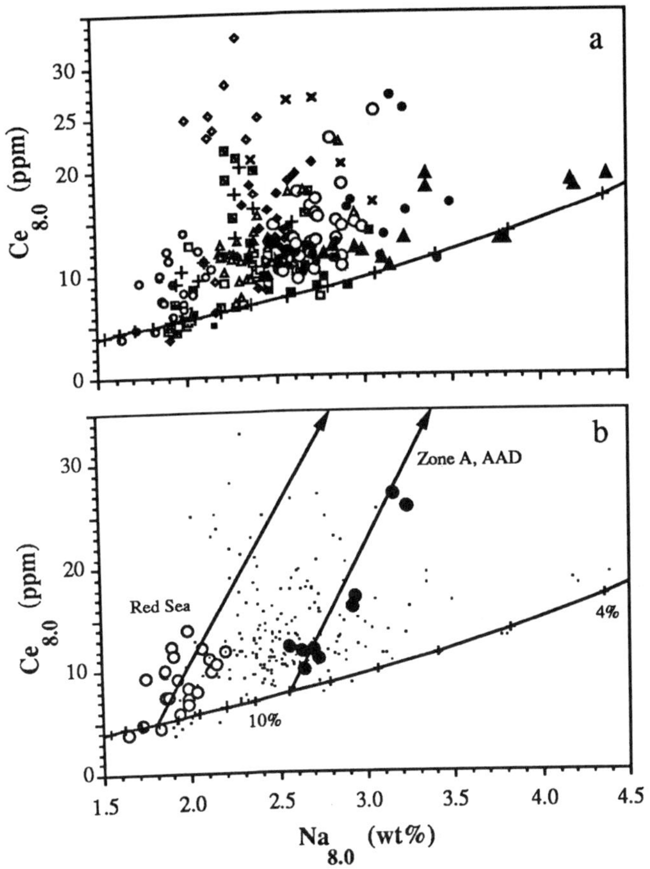

Fig. 69. $Na_{8.0}$ vs. $Ce_{8.0}$ for MORB, with a calculated melting "baseline". *a*) Basalt data points for individual samples (same as in Figure 25). Melting curve is fit through the bottom of the data array for an equilibrium melting model, where C_0 for Na_2O = 0.3 wt%, C_0 for Ce = 0.7 ppm, D_{Ce} = .001 and D_{Na2O} = 0.03. Extent of melting (4-17%) indicated as tick marks along the melting curve. *b*) Steep trends shown by the data from the Red Sea and Zone A of the AAD, and mixing lines between melts on the baseline and a 0.2% melt of the same mantle source. Figure from *Plank and Langmuir* [1992].

is present after correction for low pressure differentiation. When least squares calculations have been carried out to evaluate low pressure fractionation, the incompatible trace elements often show too much variation. These variations have been interpreted in terms of complex magma chamber processes [e.g., *Bryan and Moore*, 1977; *Dungan and Rhodes*, 1978; *Hekinian and Walker*, 1987] or as manifestations of distinct parent magmas derived by different extents of melting of the mantle [e.g., *White and Bryan*, 1977; *Langmuir et al.*, 1977]. This latter argument was based on studies of high MgO samples that could not have undergone substantially different amounts of fractionation. The steep $Na_{8.0}$-$Ce_{8.0}$ correlations create difficulties, however, even for a partial melting explanation, for the variations in trace element ratios exceed those predicted by major element melting models.

Another aspect of trace element data that poses difficulties for melting models is the Th-U-Ra disequilbria studies that are beginning to appear [e.g. *McKenzie*, 1985*a*; *Williams and Gill*, 1989; *Goldstein et al.*, 1989; 1991; *Reinitz and Turekian*, 1989; *Rubin and Macdougall*, 1988; 1990]. Taken at face value, these new data would seem to require very low extents of melting beneath ridges in order to fractionate these highly incompatible elements from one another during melting. However, the Ra excesses with respect to Th are so large that the melt fractions required are smaller than seems physically reasonable. In addition, new experimental data [*Latourette and Burnett*, 1992] show that U is more incompatible than Th in clinopyroxene, which suggests that equilibrium melting processes will not fractionate the elements from one another in the sense that is observed. These new data are particularly puzzling and very important, for the fractionations of the U-series disequibria must take place on very short time scales (10^3-10^5yrs) prior to eruption. It appears that some process other than equilibrium melting will be required to account for the data.

The trace element data thus clearly provide another indication of our incomplete understanding of the melting process. The bright side is that local variability of trace elements potentially contains much information concerning the melting process that we do not receive from major elements alone.

One idea that might aid in explaining this effect was raised by *O'Hara* [1985], who suggested that trace element abundances and ratios are influenced by the shape of the melting regime. If true, then the trace element signature could be used to constrain the shape, and hence the mixing function, that is applicable to ridges under different conditions. We therefore need to consider from a theoretical point of view what the local trace element variations may be revealing concerning the nature of the melting regime beneath the ridge. We can then return to potential interpretations of the trace element data.

7.3 Trace Elements and Melting Regime Shape

O'Hara [1985] suggested that incompatible elements may be differentially enriched by pooling melts from melting regimes of complex shapes. *PL92* argue that this conclusion is based on a misunderstanding of the difference between the $\bar{F}$ of the melting regime, as defined in section 4.3.1, and the maximum F at the top of the central column of the melting regime. This misunderstanding has propagated in the literature, with many papers comparing the results of pooled melt calculations with batch melting to the maximum F of the melting regime rather than the $\bar{F}$ of the melting regime.

The maximum F, of course, is applicable only to the top of the central column of the melting regime, where the extent of melting is greatest and the incompatible elements most diluted. The $\bar{F}$ refers to the melts pooled from throughout the melting regime, and it is this $\bar{F}$ that is the appropriate reference point for trace elements as well as major elements. The $\bar{F}$, as is clear from the residual melting column concept, is simply the total amount of melt divided by the total volume that has melted. A perfectly incompatible element will be exclusively in the liquid, and its concentration will be simply $C_o/\bar{F}$. This is identical to the concentration produced by batch melting to the same extent. In fact, the $\bar{F}$ may always be calculated from the perfectly incompatible element (D = 0), which will increase with C_o/F, regardless of the mixing or melting functions involved. It is not identical for elements that are not perfectly incompatible, but the discrepancies with batch melting are small.

This is illustrated in Figure 70*a*, where melts up to 5% are pooled, with a mean extent of melting of 3%. The solid line shows the compositions of the individual melts, and the triangle the composition of the pooled melt, which lies exactly on the line calculated by the simple model. Figure 70*b* shows

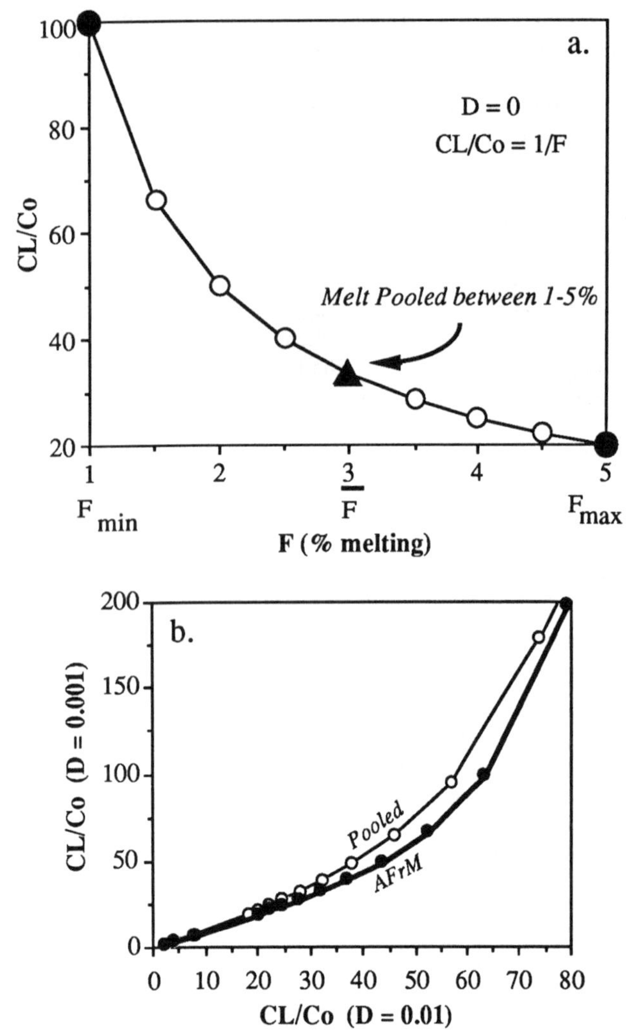

Fig. 70. Comparison of the chemical consequences of pooling melts vs. a simple batch melt to the same mean extent. *a*) Curve is concentration of a perfectly incompatible element vs. extent of melting (F) for equilibrium melting between 1 and 5%. Large triangle is mixture of melts pooled between 1 and 5%, which lies on the melting curve itself at the same mean $\bar{F}$. *b*) The concentration of two incompatible elements (D = 0.001 and 0.01) for simple accumulated fractional melting (AFM) and pooled AF melts for the same mean $\bar{F}$ using the equations given in *Plank and Langmuir* [1992].

the results of the same calculation for two elements whose D's differ by a factor of ten. In this case the pooled melt differs very slightly from the batch melting curve, but the differences in fact are indistinguishable, if one considers simply the analytical and geological uncertainties in any chemical analysis. The same result applies for a very wide range of shape models, so long as the melting function is continuous, and melts are mixed in proportion to F. *All reasonable shapes for melting regimes beneath ridges reduce to this kind of pooling, so the shape of the melting regime has little effect on the composition of the pooled melt, in comparison to a batch melt to the same* $\bar{F}$. Of course, as was the case with Fe, this conclusion does not apply for elements that show a strong pressure dependence for their distribution coefficients.

This result can be understood intuitively from Figure 70*b*. Mixing produces straight lines on such a diagram. If the melting function is continuous, there are compositions in the melting regime all along the batch melting curve, so there is a mixing end-member all along the curve. Mixtures of all the points along the curve never lead to a composition that deviates significantly from the curve.

These considerations show that the shapes considered by *O'Hara* [1985] do not produce trace element variations that are significantly different from batch melting to the same $\bar{F}$. The positive aspect is that simple trace element models can be accurate despite a lack of knowledge of the details of the shape. The negative aspect is that trace elements also then provide no constraints on the shape.

These results also have important implications for recent "inverse" models of melting beneath ridges proposed by *McKenzie and O'Nions* [1991]. They attempt to invert the distribution of melt in the melting regime from rare earth element analyses, using a single averaged analysis of the REE to arrive at a solution. Their K_d's and source are set *a priori*. Because most MORB are derived by melting from within the spinel peridotite stability field, the appropriate models to evaluate the results of *McKenzie and O'Nions* [1991] are those of constant K_d's, and variable extents of melting. In this case, a batch melt of the mean extent of melting within the melting regime will be essentially indistinguishable from a mixture of melts from throughout the melting regime. The surprising implication is that NO unique solution of melt distribution can be obtained from the REE average, for any volume of melt from any depth within the spinel stability field that has melted to that mean extent will yield the same REE pattern.

These simple considerations also show what process could produce a result that differs substantially from batch melting [*PL92*]. If mixing took place between two distal portions of the curve in Figure 70*b*, with no intermediate compositions participating, then mixtures would lie along a mixing line that could deviate substantially from the batch melting curve. This situation might be applicable to ridges, for the presence of small amounts of volatiles in the mantle source causes there to be a deep, volatile saturated solidus far below the dry solidus that we have used in our melting models, as illustrated in Figure 71, and emphasized previously by *McKenzie* [1985*b*] and *Galer and O'Nions* [1986]. In this region between the volatile-

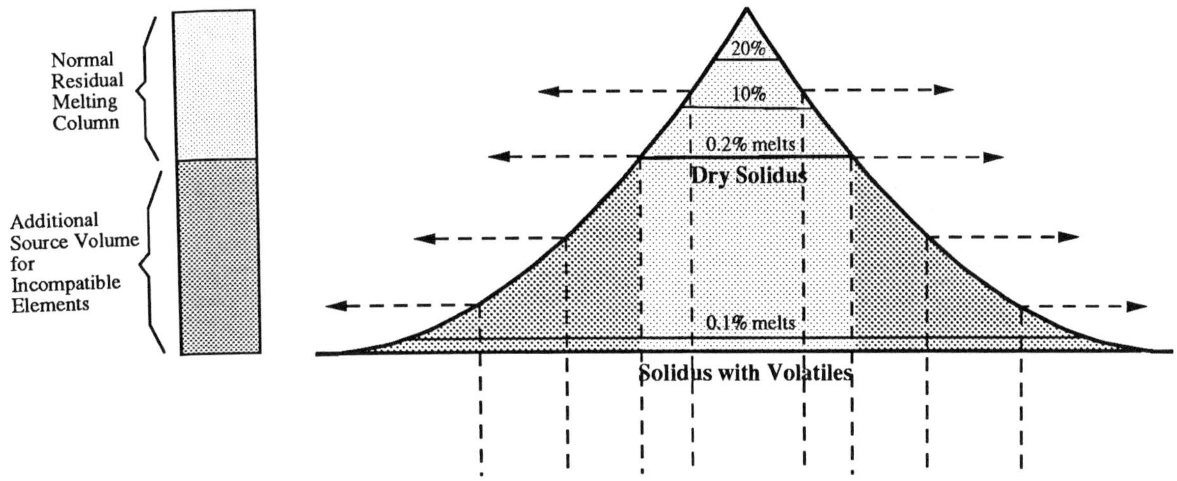

Fig. 70. Comparison of the chemical consequences of pooling melts vs. a simple batch melt to the same mean extent. *a*) Curve is concentration of a perfectly incompatible element vs. extent of melting (F) for equilibrium melting between 1 and 5%. Large triangle is mixture of melts pooled between 1 and 5%, which lies on the melting curve itself at the same mean $\bar{F}$. *b*) The concentration of two incompatible elements (D = 0.001 and 0.01) for simple accumulated fractional melting (AFM) and pooled AF melts for the same mean $\bar{F}$ using the equations given in *Plank and Langmuir* [1992].

saturated and dry solidi, the extent of melting will be very small, controlled by the small amount of volatiles contained within the mantle. These volatile-rich melts will be very low in viscosity and density, enabling them to segregate and contribute to the large volume of melt produced above the "dry solidus." These low degree melts would have high levels of enrichment of the most incompatible elements and could produce the steep $Na_{8.0}$ — incompatible element trends mentioned above. The extra low degree melts would have to come from out in the wings of the melting regime (see Figure 71), and so *PL*92 point out that an important and probably prohibitive physical aspect to this model is that there would have to be a mechanism that focused these extra melts from the most distant parts of the melting regime into the central axis.

A second possible model for these aspects of local trace element variability, however, stems from another aspect of the occurrence of the steep trends. The best examples occur either in the vicinity of hot spots, or on the EPR. In both of these environments there is evidence for enriched veins — though possibly on different scales. We saw from the melting models with Na_2O that the increase in source abundance in such veins overwhelms any dilution effect from increased extents of melting, so that melting of such veins should lead to anomalously enriched compositions. Variable mixing of melts from the veins and from the surrounding more depleted mantle might lead to the steep trends in Figure 69*b*. This undoubtedly accounts for some of the data, particularly where there is a correlation between incompatible element abundances and the radiogenic isotope ratios (e.g., $^{87}Sr/^{86}Sr$) such as we observe for portions of the EPR.

Therefore local trace element variations for the most incompatible elements vary more than one would expect from the major element models. Two possible explanations are mixing of low degree melts from the wings of the melting regime, or melting of small-scale heterogeneities. Evidence from Th-Ra-U, however, suggests there may be other processes exerting important effects of which we are not yet cognizant. Steep trends have not yet been discovered on normal sections of the Atlantic, but are common in the Pacific and where there are hot spot influences. Although these various major and trace element aspects of local variability are puzzling, they begin to make some sense when one considers the overall importance of spreading rate in influencing local variability.

8. Significance of Spreading Rate: A Key to Understanding the Local Trends?

Many of the important aspects of ocean crust formation correlate or seem to be affected by spreading rate, such as the striking associations of rugged topography and rift valleys with slow-spreading ridges, and gentle topography and axial swells with fast-spreading ridges. Physical models predict spreading rate-dependencies on the pattern of melt and mantle flow [e.g., *Scott and Stevenson*, 1989; *Spiegelman and McKenzie*, 1987], which may influence the composition of melt formed and the volume of melt segregated to form the oceanic crust. It has been suggested that spreading rate correlates with crustal thickness for ridges far from hot spots [*Reid and Jackson*, 1981], with basalt major element composition [*Bass*, 1971; *Morel and Hekinian*, 1980], with isotopic heterogeneity [*Cohen and O'Nions*, 1982; *Batiza*, 1984], with gravity characteristics [*Small and Sandwell*, 1989], and with certain aspects of tectonic and magmatic segmentation [*Schouten et al.*, 1985; *Dick*, 1989; *Sinton et al.*, 1991]. We have found that the local vectors of compositional variability exhibit a striking contrast between slow- and fast-spreading ridges. Other aspects of MORB chemistry, however, such as the global correlations, appear to

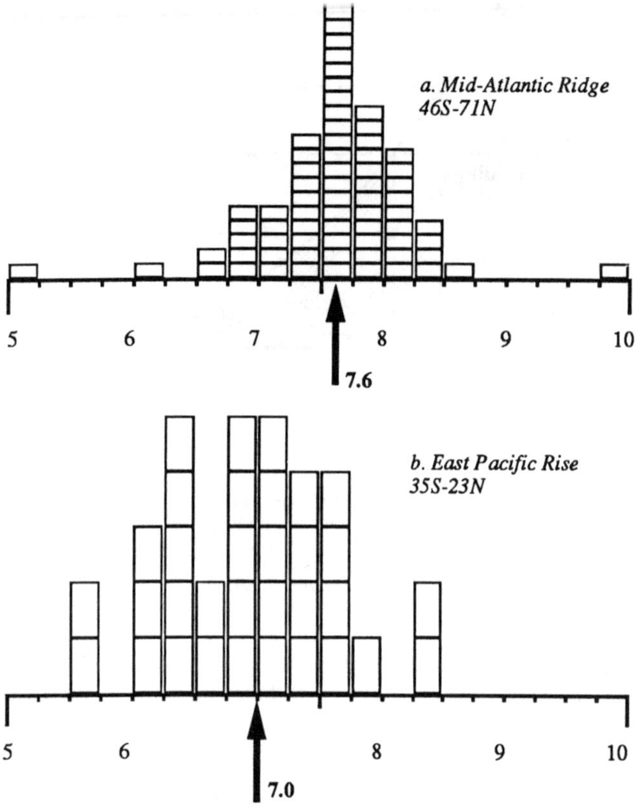

Fig. 72. *a*) Histograms of MgO wt% in samples from the Mid-Atlantic Ridge (46°S-71°N; 527 analyses) and East Pacific Rise (35°S-23°N; 400 analyses). *b*) Histograms of MgO wt% averaged per degree latitude for the individual analyses shown in Panel (a). Data from the Smithsonian Institution Catalog of Basaltic Glasses [*W. Melson*, pers. comm.], *Sinton et al.* [1991] and references in Figures 21 and 24.

be independent of spreading rate. In the following, we examine which aspects of chemical variability correlate with spreading rate and which do not, and consider the implications. The implications include a possible model to account for the difference in chemical systematics between the MAR and EPR.

8.1 Spreading Rate and Global Chemical Variability

The first major observation is that the global correlations (Figure 15) are independent of spreading rate. Virtually the entire range of $Na_{8.0}$, for example, occurs along the slow-spreading northern Mid-Atlantic Ridge [see Figure 2 *KL87*]. In addition, basalts with high $Na_{8.0}$ are recovered from the the slow-spreading Mid-Cayman Rise and the Southwest Indian Ridge, as well as the medium-spreading Australian Antarctic Discordance. Furthermore, the superfast spreading southern EPR erupts basalts with a limited and intermediate range of $Na_{8.0}$ values [*Sinton et al.*, 1991]. Thus, there is no apparent correlation between $Na_{8.0}$ and spreading rate.

There are important aspects of the global major element data set, however, that do show systematic relationships with spreading rate: average MgO contents of basalts and overall compositional diversity. Figure 72 shows histograms for the MgO contents of basalts from the EPR and MAR. The histograms are presented in two ways: as raw data, and as averages by degree of latitude so that densely sampled regions are not over-emphasized. In both treatments of the data, EPR basalts are significantly displaced toward lower MgO (lower temperature) than Atlantic basalts. The lower mean MgO contents of the EPR probably reflects the presence of crustal magma reservoirs. These reservoirs serve as the locations of mixing of diverse magma batches that are emplaced in the crust, and also as locations for cooling. Thus it makes sense that regions of faster spreading give rise to both more homogeneous and more differentiated basalt compositions.

The second important aspect is that the total range of variability increases substantially as spreading rate decreases. When the data are averaged region-by-region and examined as a function of spreading rate, there is an increase in the total range of mean MgO contents with decreasing spreading rate (Figure 73). Thus, slow-spreading ridges include basalts that are very primitive (e.g. FAMOUS, Kane to Hayes, Iceland), moderately differentiated (e.g. Reykjanes Ridge, Azores platform), or very differentiated (e.g. Mid-Cayman rise). On the other hand, all the data sets from fast-spreading EPR segments show similar mean MgO contents and similar standard deviations. The exception to this generalization is the occurrence of very low MgO (high SiO_2) volcanics such as andesites and dacites at fast-spreading ridges. Such rocks have not yet been observed from slow-spreading ridges, but are quite common on the Galapagos Rise [*Byerly et al.*, 1976; *Christie and Sinton*, 1981; *Fornari et al.*, 1983], and also along the East Pacific Rise. Occurrences on these ridges are usually associated with propagating rifts or other small non-transform offsets. Sinton and Detrick [1992], however, have recently independently considered the spreading rate dependence of MgO, and suggest that intermediate spreading rates may have

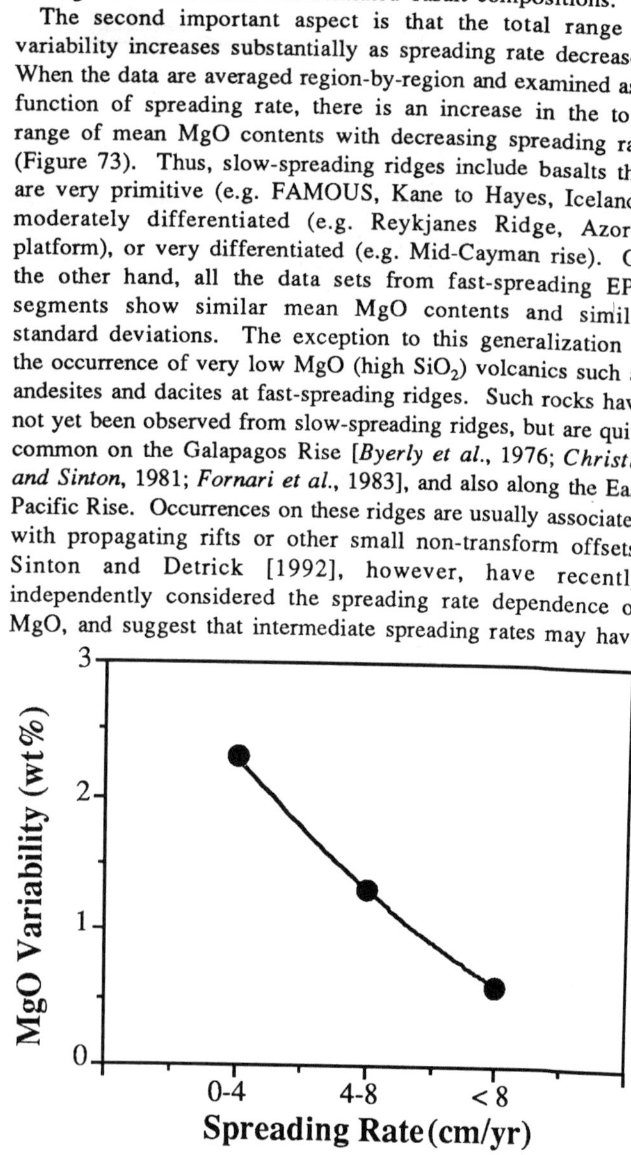

Fig. 73. Variability in MgO contents of basalts vs. spreading rate for the same data set as Figure 72 with the addition of data from intermediate spreading rate ridges. The variability is the total range in regional averages of MgO for more than 100 km segments within each of the three spreading rate intervals. For example, all lengths of the EPR have very similar mean MgO contents, while slow-spreading ridges range from very primitive to very evolved compositions. This variability does not include the high SiO_2 samples that occur periodically on intermediate and fast-spreading ridges.

the maximum variability, being in a transitional regime between the contrasting crustal fractionation systems of slow and fast-spreading ridges.

Many other chemical indices in addition to MgO exhibit an inverse relationship between diversity and spreading rate. For example, as noted above, slow-spreading ridges encompass the entire range in $Na_{8.0}$ values (Iceland vs. Mid-Cayman rise), while fast-spreading ridges from the EPR show only intermediate values of $Na_{8.0}$. *Batiza* [1984] suggested that variability in $^{87}Sr/^{86}Sr$ correlates inversely with spreading rate, (although a more recent study did not support this finding [*White et al.*, 1987]).

This inverse relationship is not restricted to chemical composition. It is true, for example, of variations in depth: slow-spreading ridges include the entire depth range, while fast-spreading ridges are almost always between 2500 and 3000 meters. It is also true of ridge morphology. Fast-spreading ridges look alike, while slow-spreading ones can vary from deep rift valleys, such as near the Kane Fracture Zone, to very small rift valleys, such as near the Azores, to none at all on the Rekykjanes Ridge or Kolbeinsey Rise. Thus one of the primary observations of spreading rate dependence is that on a regional basis, slow spreading ridges are highly variable in chemistry, morphology and depth, while fast-spreading ridges show litle variation, with depth and regional chemistry at mean levels with respect to ridges as a whole.

8.2 Spreading Rate and Local Variability

As discussed at length above, there are striking differences between major and trace element variability on slow- and fast-spreading ridges. The local trends of composition are completely distinguishable, and the expression of trace element heterogeneities is strikingly different as well. On well-studied slow-spreading ridges such as the northern MAR, the trace element signature is bimodal. Some regions are dominated by hot spot effects and erupt exclusively enriched basalts, while other long lengths of ridge can erupt exclusively depleted normal MORB compositions. On the EPR, in contrast, transitional and enriched MORB occur frequently, with no attribution to hot spot effects. The local trends in the Atlantic occur independent of the local mantle heterogeneity, while on the EPR the major element and mantle heterogeneity signals seem to be closely linked. These are such striking contrasts that they demand some sort of simple explanation.

A direction for an explanation comes from considering more carefully the spatial distribution of trace element patterns that occur along the MAR and EPR. We begin with consideration of two "normal segments" along each ridge, one between 8°and 14°N on the EPR [*Langmuir et al.*, 1986], and the other between the Kane and Hayes Fracture Zones on the MAR. Both these regions are well sampled, are not influenced by known hot spots, and are at average axial depths for ridges, with similar $Na_{8.0}$ contents. The sole obvious difference between these two normal segments of ridge is that their spreading rates differ by a factor of three.

Figure 74 presents the contrasting histograms of Ce/Sm ratios for about 50 samples that have been analyzed for REE from each region. The Kane-Hayes samples are all light REE depleted, with Ce/Sm ratios of 2.71±0.27. The EPR samples are much more variable, and have a much higher mean Ce/Sm for the entire population (3.74±0.67). These are striking contrasts for two reasons. First, both these ridge segments should be on the "baseline" for normal MORB removed from hot spots, since they have similar values of $Na_{8.0}$. Second, as just pointed out, standard deviations for most parameters are much higher for slow-spreading than fast-spreading ridges, yet here, for almost a thousand km of ridge, the MAR data set is much more homogeneous. This contrast is apparent in Figure 75, which shows how Ce/Sm varies with distance along each ridge. The higher and more variable values for the EPR are clear.

The extensive data set of Jean-Guy Schilling for much of the N. Atlantic can be used to consider whether these relationships are unusual [*Schilling et al.*, 1983]. To make this comparison, it is necessary to convert Schilling's data for La/Sm ratios to Ce/Sm ratios, since he usually does not analyze Ce, and the Kane-Hayes and EPR data sets frequently have no La. In Schilling's data set, Ce/Sm and La/Sm correlate very well (R = .988), so that a conversion factor can be applied to the La/Sm

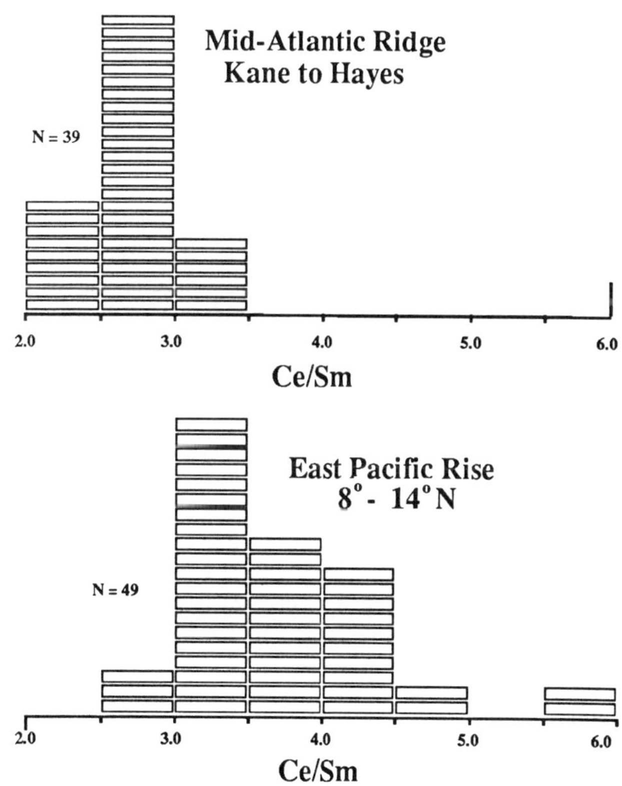

Fig. 74. Comparison of the distribution of REE data from normal sections of the Mid-Atlantic Ridge and the East Pacific Rise. The MAR data are from more than 1000 km of ridge between 24 and 34°N, extending from the Kane to the Hayes Fracture Zones [*Langmuir*, unpublished data by plasma spectrometry]. The EPR data are from ~700 km of ridge from 8°N-14°N [*Bender and Langmuir*, unpublished data by isotope dilution]. Over these distances, the MAR data for a normal section of ridge are both more depleted and more homogeneous than the EPR data.

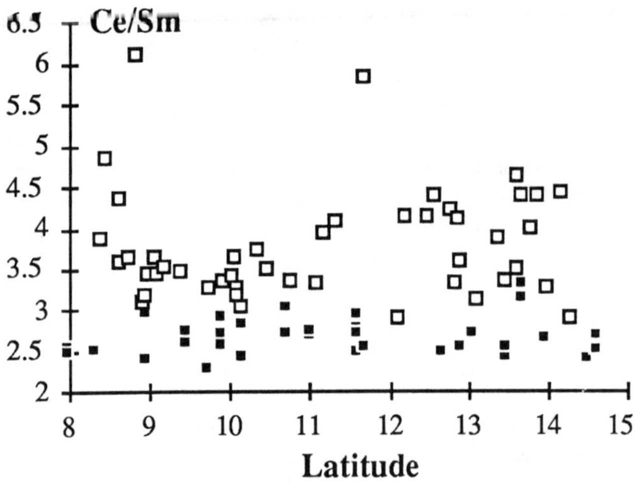

Fig. 75. Distribution of Ce/Sm ratios with distance along the EPR and MAR, using the same data set as Figure 74. Latitude is for the EPR data set. MAR data are at the same scale, but latitudes are not shown. Open squares are for the CHEPR EPR data set, closed squares are from Kane to Hayes. Note that the high Ce/Sm samples from the EPR are widely distributed along the ridge. The MAR does contain even more enriched samples than the EPR, but these are localized in specific hot spot regions (see Figure 68). No enriched samples were recovered on axis in the Kane to Hayes region.

data to arrive at Ce/Sm values. The conversion is Ce/Sm= 2.025*La/Sm+0.9.

Schilling's data set shows that for a much greater length of ridge than considered above, normal Atlantic segments are both more depleted and more homogeneous than the EPR. There are six normal segments of three degrees of latitude in Schilling's data set. Each of these segments has a mean Ce/Sm less than 2.9, much lower than the EPR data, and does not contain enriched basalts — hence the standard deviations are low. Other comparable data sets from the EPR are not available, but the distribution of data from the southern EPR reported by *Sinton et al.* [1991] appear to have a similar appearance to the northern EPR data. The random distribution of relatively enriched rocks is a striking characteristic of the southern as well as northern EPR data sets. Therefore the contrasting characteristics observed in the 8-14°N and Kane to Hayes regions appear to be general phenomena associated with the EPR and MAR.

Of course, the MAR also contains extensive lengths of ridge near hot spots that consist of enriched MORB. The population is bimodal, with a clear separation between enriched and depleted lengths of ridge (Figure 68). Where enriched basalts occur, the enrichment often appears gradually, over long gradients, such as that south of the Azores, where Ce/Sm increases from 2.5 to 9 progressively over seven degrees of latitude. Therefore the relative homogeneity of normal sections of the MAR contrasts with the high standard deviation for the MAR as a whole, which comes from the presence of both hot spot-influenced and "normal" sections of ocean ridge.

In view of these contrasts between the MAR and EPR, a most striking observation is that the MEAN Ce/Sm for the two ridges is identical , if both enriched and depleted portions of the MAR are included. This is apparent in Figure 76, which illustrates the dispersion of the entire N. Atlantic data set, and that mean MAR and EPR values are identical. The means are the same whether one takes the entire Atlantic data set from 22°N to 72°N as individual samples, or divides it into equal latitude intervals to remove all sampling bias. To summarize: for ridge lengths of several thousand kilometers, the means from both EPR and MAR are the same. The Atlantic is more dispersed, with more enriched samples than occur anywhere on the EPR. However, for normal sections of ridge, the EPR is more variable in incompartible elements and contains more enriched compositions. Therefore in trace elements as in major elements, there is a striking contrast between these two best-studied ocean ridges.

8.3 A Model to Account for Contrasts between EPR and MAR Major and Trace Element Data

The major and trace element contrasts between MAR and EPR might be explained by the interaction of a constant plume flux with variable spreading rate. Consider that the plume flux into both ocean basins may be comparable. For the Atlantic, the basin is bounded by continents on all sides, is slow spreading, and has recently opened. The Pacific, on the other hand, has a

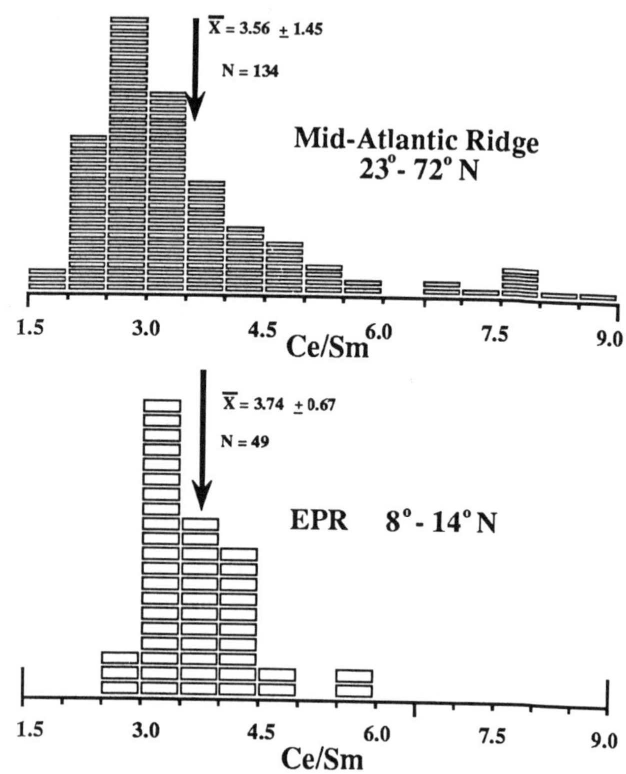

Fig. 76. Comparison of the EPR data set from Figure 74 with data from the more than 5000 km of the MAR, which includes several N. Atlantic hot spots. Data are from Kane to Hayes and also from *Schilling et al.* [1983]. The distribution of the individual samples for the two ridges is very different, but the mean is the same. Compared to the northern MAR as a whole, the EPR data are more homogeneous. Mixing of the N. Atlantic data could give rise to the EPR distribution. The same contrast is present in the data whether the data are examined as individual samples or by window of latitude, so the result does not depend on sampling bias.

very long history of spreading, and is a large basin with very fast spreading rates. Processing and hence mixing of the mantle will be much more rapid and efficient in the Pacific than the Atlantic. These factors could lead to plumes becoming mixed on a fine scale into the Pacific mantle, and remaining clearly separated in the Atlantic mantle. Faster spreading would dilute and mix the plume flux more efficiently into the upper mantle prior to melting at the spreading center.

At the same time, the melting process is strongly affected by spreading rate. On slow-spreading ridges, there are large gradients in depth along strike, suggesting a complex and diverse melting regime that varies substantially over intermediate distances [e.g. *Lin et al* 1990]. Therefore along a spreading segment, melts are erupted from diverse portions of the melting regime. On fast-spreading ridges, the melting regime is two dimensional and does not vary nearly as much along strike. Melts undergo a similar history and mixing all along the ridge. Small scale heterogeneities can be preserved because of rapid melt processing and vertical transport along and possibly across the ridge as well.

These two concepts explain qualitatively the major observations from the MAR and EPR. The MAR would have some very depleted and homogeneous portions, which are not influenced by hot spots. Other portions of the MAR would have relatively pure hot spot signature, leading to large depth and chemical anomalies. Temperature anomalies, variations in depth, crustal thickness and extent of melting would all be preserved. The slow-spreading rate and frequent ridge offsets associated with it leads to substantial variations along strike, and much variability within the melting regime. This gives rise to the major element local trends as a result of variable melting processes, strongly influenced by the pronounced tectonic fabric of the ridge. Because of the relative isolation of mantle influenced by the hot spots, large scale heterogeneities would be incompletely mixed, so that large blocks of mantle could have their own heterogeneous composition, and their own local melting trend. This may be the case for the Oceanographer data. Of course, there is evidence for some small scale heterogeneity around hot spot regions in the Atlantic. This could result from the partial mixing that takes place as the plume material interacts with the rest of the mantle.

For the Pacific, the plume component has become distributed on a small scale through the mantle. As *Allegre and Turcotte* [1986] pointed out in a different context, mantle mixing often leads to thin layers of material that have been folded and refolded, giving rise to ubiquitous small scale heterogeneities. This material could then give rise to random enriched basalt. But this component is not present everywhere. In a region where it does occur, such as the Pacific, the basalt is never as enriched as the most enriched basalts from the Atlantic, because the plume component residing in the enriched layers is always diluted by the surrounding depleted mantle matrix.

At the same time as small scale heterogeneity contributes to compositional diversity of erupted basalts, melting processes alone beneath the EPR seem to have limited effect. The temperature of the mantle appears to be uniform over large distances along strike, possibly due to more efficient mantle processing or tectonic history. And ridge segmentation is subdued in comparison to slow-spreading ridges. Thus, long and short wavelength effects on the EPR are muted in comparison to the Atlantic, and there is no local trend caused by melting processes. Instead, the local trend is a consequence of small scale heterogeneity.

This model explains many aspects of EPR volcanism — the lack of a slow-spreading major element vector, the random distribution of enriched MORB, the lack of very enriched MORB. It also might explain the gradual gradients in chemistry observed in the southern EPR [*Sinton et al.*, 1991]. Such long and smooth gradients might be the vestiges of incompletely mixed plume material. It could also explain why there is such a variety of compositions on many Pacific seamounts [e.g. *Batiza and Vanko*, 1984; *Zindler et al.*, 1984]. If the sub-Pacific mantle is veined, seamounts would sample all this variety without any of the homogenizing influence of sub-axial magma chambers.

There are also predictions for the MAR. For example, the lack of this veined material would suggest that off-axis seamounts along normal portions of the MAR removed from hot spots will not contain enriched MORB — in contrast to the common experience in the Pacific. One might also expect preservation of older heterogeneities in the Atlantic, since mixing is less efficient.

This model also appears to account for some of the differences in the details of the global correlations between ocean basins. Figure 16*b* showed that Pacific ridges were offset to lower $Fe_{8.0}$ for the same $Na_{8.0}$ in comparison to the Kane to Hayes region in the Atlantic. The Atlantic as a whole, however, is characterized by normal MORB regions with relatively high $Fe_{8.0}$, and hot spots with relatively low $Fe_{8.0}$. The combination of these two would lead to intermediate levels of $Fe_{8.0}$ — just what is observed for the Pacific data. Therefore the same model that accounts for the gross features of the data, explains some of the detailed complexities of the global correlations.

Although this model is consistent with many aspects of the major and trace element data, and the spatial occurrence of the data along the ridges, there are important additional issues. First, the differences between the EPR and MAR may have multiple causes. Some may result from spreading rate, but others may be due to the differing history and tectonic environment of the two ocean basins. For example, the ubiquitous small scale heterogeneity of the EPR is also apparent on the slower-spreading Juan de Fuca, Endeavour and Explorer Ridges in the northern Pacific [e. g. *Michael et al.*, 1989; *Karsten et al.*, 1990], so it appears this characteristic may apply to other parts of the Pacific basin, though apparently not the Galapagos Rise [*Schilling et al.*, 1981; *Sinton et al*, 1991]. Second, hot spots themselves are variable, and have different major element and incompatible element signatures. Ce/Sm is a neutral trace element parameter to use for it is a ratio that is enriched in all hot spots. Trace elements that are more incompatible are likely to show more complex relationships between the ocean basins. An encouraging aspect of the differences between the MAR and EPR is that they may provide a fingerprint with which spreading rates and histories of ancient crust that no longer makes up the sea floor can be inferred.

9. CONCLUSIONS

One of the aims of this paper was to develop further the systematics and framework of the chemical variations of MORB on various scales, for the clear statement of the systematics provides the relationships that quantitative models need to explain. Correction of the global data set for interlaboratory biases reveals a significant amount of structure in the global correlations of *KL*87. Back-arc basins and many hot spots separate cleanly from the normal MORB field, and the normal MORB form a coherent array that is useful for constraining melting models. There are significant if subtle differences among the different ocean basins, and striking differences in the distribution of the data as a function of spreading rate. On a smaller, local scale, the contrasts between the local trends in major and trace elements for the EPR and slow-spreading ridges are very clear. The relationships among these local trends, spatial position along ridge segments, and temporal variability, however, are still not well known. The relationship between major and trace elements for the slow-spreading local trends also needs much better definition. These areas of investigation are likely to lead to important new developments over the next decade.

A second aim of the paper was the development of quantitative models for mantle melting and fractionation. Although much remains to be done in this direction, the models provide a useful framework in which to carry out detailed evaluation of global and local data and the processes that give rise to them.

Interaction between the observations and the models leads to the following conclusions concerning melt formation and segregation beneath ocean ridges, and the generation of the diversity of ocean ridge basalts.

I. Mantle temperature variations are important.

The global vector of variability can be explained quantitatively by isentropic melting of a homogeneous mantle that intersects the mantle solidus at different pressures as a result of temperature variations. Mantle temperature differences exert a major control on MORB chemistry, and must be considered as one of the important variables of melting and melt segregation. Major element heterogeneity is ruled out as a cause of the global correlations for major elements.

II. Continuous melting with low residual porosity is indicated by both the basalt and abyssal peridotite data.

The high FeO contents of the global correlations and the large FeO differences among regions preclude equilibrium melting. Some of the most depleted peridotite compositions also are difficult to reconcile with purely equilibrium melting [*Johnson et al.*, 1990]. Existing data, however, cannot place firm constraints on values of the residual porosity, because of the uncertainties in the calculations and the details of the melting process.

III. Melts can be extracted from a range of pressures within the mantle without substantial low pressure re-equilibration prior to emplacement in the crust. They must be extracted rapidly and efficiently, and yet mix prior to eruption.

The high mean pressures of formation indicated in MORB major element compositions require the extraction of melts from deep in the melting regime without substantial modification. An even stronger constraint is given by the positive correlation between pressure of melting and extent of melting that is required to model the global vector. This relationship can come about only if melts are delivered from great depth to the surface without substantial re-equilibration during ascent, and suggests that melts form and pool over a range of pressures extending from near the solidus to near the top of the melting regime.

Conclusions II and III provide rigorous constraints for porous flow models with low porosity, because slow travel along small channels may inevitably lead to reaction upon ascent and re-equilibration at low pressures. Transport must be efficient, with minimal reaction during ascent. This suggests channelized flow and rapid ascent.

IV. Petrology does not highly constrain the physical shape of the melting regime because many different shapes lead to identical petrological results.

Active upwelling vs. passive upwelling, however, should have different petrological consequences. Detailed petrological studies of individual ridge segments coupled with seismic studies of crustal thickness should provide quantitative tests of melting models as they apply to ridge segments and to ridge segmentation.

V. The contrasting local vectors on slow-spreading ridges and the EPR are very regular in their occurrence, and result from mantle processes.

The melting column model of *KL*89 does not quantitatively account for the local trends observed on slow-spreading ridges or the EPR. Fractionation models, including high pressure fractionation, may contribute to the local vector in certain regions, but also fail to quantitatively account for the local vector. All fractionation models are inconsistent with the large variations in Sr that usually occur along with the major element variations. Therefore the local trends have to do with melting processes in the mantle. Alkali basalts from ocean islands also exhibit a local trend similar to that observed on slow-spreading ridges. The melting models fail completely to produce alkali basalts by small extents of melting at high pressure. Therefore we know there are aspects of the melting problem that are not adequately treated with current models. Certainly one of these aspects is volatile content. CO_2 may play an important role in melting beneath ocean ridges.

If the local vector is produced by different extents of melting within the melting regime, then it may become a tool to "map" the melting regime, showing from where in the mantle melts are delivered to the crust, and the spatial relationship between position on the ridge axis and position within the melting regime. In addition, the magnitude of the local vector may constrain the extent of mixing of melts in the melting regime or in the crust.

The local variability of chemical data from the Pacific is as systematic as that from the slow-spreading ridges, but define a different array in compositional space. We propose that this array is produced by melting of a mantle that is heterogeneous on a small scale. Such heterogeneity is not observed on normal ridges in the Atlantic.

VI. There is no systematic global correlation between mean extent of melting and spreading rate. Spreading rate controls the standard deviation of variability of many parameters, however.

In the limiting case of extremely slow spreading, however, basalts may be derived by the lowest extents of melting. The clearest correlation with spreading rate is with the standard deviation of chemical variability, both for melting and fractionation processes. Faster spreading ridges appear to be associated with limited mantle temperature variations, possibly due to more efficient mantle processing or tectonic history.

VII. Hot spots and small scale heterogeneities play an important role in controlling the chemical observations.

Heterogeneity in the Atlantic is on a large scale and relates to established hot spots. Heterogeneity on the EPR is mostly dispersed and ubiquitous. The mean Ce/Sm ratios of basalts from the two ridges are identical. The contrasts between the MAR and EPR may result from the sub-Pacific mantle having mixed in hot spot material more efficiently. This idea also helps to explain the different local trends of the MAR and EPR. The EPR is of constant depth, and on axis exhibits few variations that relate directly to the melting regime. Small scale heterogeneities control most of the major and trace element characteristics of the Pacific local vector. On the MAR, hot spot material is separated, and creates long wavelength gradients. The complexity and diversity of melting environments of a slow-spreading ridge causes the major element characteristics of the slow-spreading local vector.

Small scale veins may also play an important role in controlling the way the mantle melts, the distribution of melt within the melting regime, the relationship between pressure and extent of melting, the efficiency of extraction, and melt focusing. Extraction from a heterogeneous medium could have important physical and chemical effects that influence basalt compositions and the creation of the crust in general.

VIII. There is an unknown process that influences highly incompatible elements.

The fractionation of Th from U coupled with the Th isotopic disequilibrium requires fractionation of these highly incompatible elements on a time scale of less than 100,000 years. The even larger Ra fractionations, coupled with new experimental data suggesting U is more incompatible than Th, show that equilibrium (or fractional) melting cannot be the cause of these fractionations. Therefore there must be some additional diffusive or disequilibrium process that is influencing the distribution of these elements in the melting regime. Surprises for melting models may be in store.

Finally, we note that the various chemical constraints seem to require a combination of characteristics that are not fit by a single model of the melting regime. Melting that is close to fractional suggests movement of small degree melts at low porosity. One would think these would interact with their surroundings. On the other hand, the major element evidence shows that some melts reach the surface without interacting with the mantle during ascent. The global systematics require extensive melt pooling. The local variability suggests melts can often ascend in small batches without being thoroughly homogenized.

Aspects of these diverse observations might be reconciled with a model in which the style of segregation changes during ascent. Porous flow with low porosity may apply deep in the system. Progressively larger conduits may channel the melt at shallower levels. Ultimately, large channels may allow rapid movement of the melt to the surface without further interaction with the surrounding mantle. This hierarchy of melt delivery systems may require different sorts of physical mechanisms for different levels of the system. It is also possible that the degree of communication and separation of the melt delivery systems along strike may vary with depth in the mantle and segregation style, which may themselves be influenced by the thickness, distribution and composition of heterogeneous veins. A hierarchy of mantle flow and melt delivery systems may aid in the explanation of the hierarchical segmentation that may exist along spreading ridges [*Langmuir et al.*, 1986; *Macdonald et al.*, 1988].

APPENDIX A: CALCULATION OF FeO, MgO, Na_2O AND TiO_2 DURING ADIABATIC MELTING OF THE MANTLE

The contents of FeO and MgO in mantle melts are constrained by the fact that olivine is always a residual phase. *Hanson and Langmuir* [1978] and *Langmuir and Hanson* [1980] used this fact to develop equations to calculate the possible concentrations of FeO and MgO that could be produced during mantle melting. Actual melting paths could not be calculated, but the field of possible compositions still provided useful constraints. The knowledge that olivine must be residual is quite powerful, for it means that the FeO and MgO contents of mantle melts will be constrained by the partition coefficients for these two elements between olivine and liquid. If these partition coefficients can be expressed as a function of temperature, pressure and composition, and the pressure, temperature and relevant other compositional variables are known, then the FeO and MgO content of the melt is fully constrained.

Two additional developments allow this method of calculation to be developed further: equations for K_d's for FeO and MgO as a function of P, T and liquid composition (see Appendix B), and the existence of a melting function, which provides T, P and the extent of melting for any pressure of intersection of the mantle solidus. These developments now allow a rigorous calculation of FeO and MgO contents in mantle melts for different pressures of intersection of the mantle solidus. Different melting functions and melting processes can be easily incorporated into the model.

The calculation works as follows. For any mantle potential temperature, the melting function provides T, P and F as the mantle ascends, as detailed in section 4.2 We wish to determine six parameters: the FeO and MgO contents of the residue, olivine and liquid. We have as given the bulk mantle composition, MgO_o, and FeO_o. We have two equations for the K_d's for MgO and FeO between olivine and liquid as a function of pressure, temperature and liquid composition.

$$K_{dMgO} = MgO_{OL}/MgO_L = f(T,P,x) \quad (A1)$$

$$K_{dFeO} = FeO_{OL}/FeO_L = g(T,P,x) \quad (A2)$$

There is also one mass balance equation for each element:

$$MgO_o = F*MgO_L + (1-F)*MgO_s \quad (A3)$$

$$FeO_o = F*FeO_L + (1-F)*FeO_s \quad (A4)$$

and the equation required by olivine stoichiometry:

$$MgO_{OL} + FeO_{OL} = 66.67 \text{ mol\%} \quad (A5)$$

We need finally a relationship between the Fe/Mg ratio of the olivine and the residue composition. In most cases, the Fe/Mg ratios of the dominant mantle phases (olivine and two pyroxenes) are similar, and we take the Mg/Fe ratio of the olivine and the total residual solid to be the same.

$$MgO_{OL}/FeO_{OL} = MgO_s/FeO_s \quad (A6)$$

We therefore have six equations that relate the six unknowns. Solution of this set of equations leads to a quadratic with one element in the liquid as the unknown variable:

$$0 = A*FeO_L^2 + B*FeO_L + C \quad (A7)$$

$$A = F\,K_{dFeO}(1-K_{dFeO}/K_{dMgO}) \quad (A8)$$

$$B = K_{dFeO}(66.67F/K_{dMgO} - MgO_o - FeO_o) - 66.67F \quad (A9)$$

$$C = 66.67FeO_o \quad (A10)$$

All other solid and liquid compositions can then be determined from equations A1 - A6. These equations thus permit the straightforward calculation of liquid compositions for a variety of melting functions, melting mechanisms (e.g. fractional or equilibrium melting), source compositions and mantle temperatures. The one additional key piece of information that is required to carry out the calculation is the dependence of the partition coefficients on temperature, pressure and composition (these equations are developed in Appendix B).

It is important to recognize that these equations are not calculating liquid compositions when olivine is the ONLY residual phase, they solve the general problem as long as olivine is present as ONE of the residual phases. Equilibrium relationships between olivine and liquid apply independently of whatever other phases may or may not be present. A source of uncertainty in the calculation, however, is in equation (A6). This relationship applies well for plagioclase or spinel peridotite, where essentially all of the FeO and MgO is in olivine and pyroxenes, which have similar MgO/FeO ratios. Garnet, however, is much more FeO rich, and a garnet-rich residuum would lead to a different equation in place of A6. Since garnet melts out of the residue rapidly, this is not a serious problem for the conclusions of the present paper, but it may be important for small degrees of melting at higher pressures where garnet is an important residual phase. The effect of residual garnet will be to make the olivine more magnesian (less FeO) than the total residue. This will cause FeO contents of the liquid to be slightly lower, as one would expect when an Fe-rich phase such as garnet is left in the residuum.

It is possible with present data to calculate three other elements, Na, Ti, and K, using a K_d approach. K has very low partition coefficients in all mantle phases, and hence can be calculated as an incompatible element with very low D's. Clinopyroxene is the major host for Na and Ti in the mantle, so we have examined the experimental data set for clinopyroxene/liquid partitioning. These elements can also be treated as trace elements, and all that is required are equations for the dependence of the K_d's on temperature and pressure. Compositional effects are also likely to be important, but because these two elements are moderately incompatible, their D's are low in the mantle/melt system, and small errors in the K_d's do not affect the liquid composition substantially after the extent of melting is greater than a few percent. The behavior of the partition coefficients is very important, however, for calculating residual mineral compositions. Whereas a variation in the D for TiO_2 of 0.003 to 0.01 has little effect on the calculated melt composition at 10% melting, it causes variation of more than a factor of three in the residual solid composition. Because the emerging data on compositional variations in residual ultramafics may offer important constraints on melting processes [*Johnson et al.*, 1990], it is important that the partition coefficients be determined accurately. Equations for these two elements have been determined based on over 100 experimental data from the literature (see references in the caption to Figure A1). Dependence on both temperature and pressure are important. The experimental data are presented in Figure A1, along with curves that show the equations used to calculate the appropriate K_d's as a function of T and P. The equations are:

$$\text{Log}_{10}\, K_d(\text{Ti, Cpx/L}) = (10000 + 90\,P(\text{kb}))/T^{o}\text{K} - 7.5 \quad (A11)$$

$$\text{Log}_{10}\, K_d\,(\text{Na, Cpx/L}) = (5100 + 73\,P(\text{kb}) / T^{o}\text{K} - 4.48 \quad (A12)$$

In the calculation of liquid compositions, the contribution of orthopyroxene to the bulk partition coefficient is taken as a constant (.008 for Na and .01 for Ti), and the modal clinopyroxene is decreased as melting proceeds assuming that it makes up 60% of the melt fraction.

The advantage of this K_d approach to calculating mantle melt compositions is that many of the problems associated with performing mantle melting experiments, such as analyzing small melt pockets and equilibrating multiple phases, can be circumvented. For example, K_d's can be determined from experimental run products that consist of large quantities of melt and a small number of phases. This makes a wider variety of experiments available to characterize the K_d's. In Appendix B, several hundred olivine/liquid pairs are considered, which

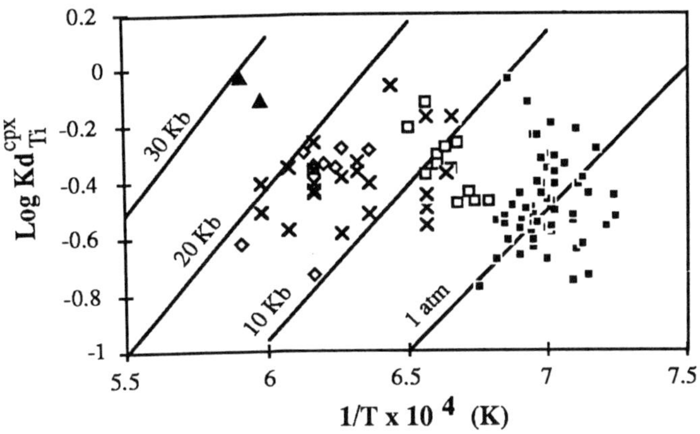

Fig. A1. Clinopyoxene/melt K_d for Ti vs. inverse temperature for experiments on mafic systems. Data sources and symbols as in Figure 39. Although the temperature dependence is not well constrained, the absolute errors involved are small, because the range of temperature of the calculations is similar to that of the experimental data points, and the K_d's vary over a small range. Since Ti is moderately incompatible in the mantle system, errors in liquid composition for moderate extents of melting are not strongly dependent on errors in the K_d's.

constitutes an order of magnitude more data than is available for other approaches to mantle melting calculations. Because these calculations are rigorous for any particular set of K_d's and melting function, they provide a useful test of other melting models that are based on empirical or qualitative reproduction of the relatively small experimental data set on multiply saturated liquids.

APPENDIX B: OLIVINE LIQUID PARTITIONING AS A FUNCTION OF TEMPERATURE, PRESSURE AND LIQUID COMPOSITION

The dependence of the partition coefficients for FeO and MgO in olivine on pressure, temperature and composition is of major importance for the calculations presented in Appendix A. The behavior of the olivine K_d's for MORB compositions is fairly well known, at least up to 15 kbar [e.g., *Bender et al.*, 1978]. However, the mantle beneath ridges may begin to melt at higher pressures. In addition, these initial melts will have high alkali contents, which can have a significant effect on olivine partition coefficients [*Roeder*, 1974].

Ford et al. [1983] developed equations for the pressure-, temperature- and compositional-dependencies of olivine K_d's. *Beatty* [1992] recently investigated the temperatures of liquids in equilibrium with olivine, and developed equations for K_d's that are independent of temperature. These latter equations are strongly constrained by olivine stoichiometry [*Langmuir and Hanson*, 1981], and are not particularly useful for calculations of liquid compositions. However, the wealth of recent experimental data on a range of melt compositions can be used to refine the K_d expressions developed by *Ford et al.* [1983].

In order to develop expressions for the olivine partition coefficients that are calibrated over the range of pressures and compositions likely for mantle melting, we have investigated over 250 experiments from the literature. These experiments span a range of pressures of up to 60 kbar, and a range of compositions including MORBs [*Grove and Bryan*, 1983; *Grove et al.*, 1990; *Bender et al.*, 1978; *Walker et al.*, 1979; *Tormey et al.*, 1987], highly alkaline basalts [*Sack et al.*, 1987], lunar basalts [*Grover et al.*, 1980] andesites [*Grove et al.*, 1982], komatiites [*Agee and Walker*, 1990] and the iron- and alkali-free CMAS system [*Presnall et al.*, 1979]. Thus we have included an even wider range of compositions and pressures than are expected to occur in the melting regime.

Roeder and Emslie [1970] originally showed a linear relationship between log K_d and inverse temperature, which is what would be predicted from the thermodynamics of ideal solutions. A rough correlation between log K_d and inverse temperature is present in the large experimental data set, but use of this simple linear relationship between the log (K_d) and 1/T leads to substantial errors in some regions of compositional space (Figure B1). The average error for a fit based only on the temperature dependence for the 1 atm data is 15%.

To investigate the compositional effects, we examined the data in small temperature and pressure windows to see if any obvious relationships were apparent. The spread in the data was clearly related to pressure and to variations in the alkali, Si and Al contents of the liquids. This relationship is apparent if one fits the 1 atm data with a linear 1/T dependence, and then examines relationships between the residuals from the 1/T fit and the liquid compositions. The residuals vary linearly with a compositional term, which includes a sum of the SiO_2, Al_2O_3, Na_2O and K_2O contents of the liquid composition (Figure B2). The constants in the term were determined by fitting the residuals to the compositional variables. No other compositional terms improve the fit to the residuals.

We then took the fit to the 1 atm data including the compositional dependencies and determined the residuals of the high pressure data from the 1 atm regression. A simple correlation exists between the residual from the 1 atm fit and the pressure of the experiment for geological compositions (Figure B3*a*). Thus, in addition to the temperature and compositional dependencies, a pressure dependent term must be included. Some of the experiments were in simple systems,

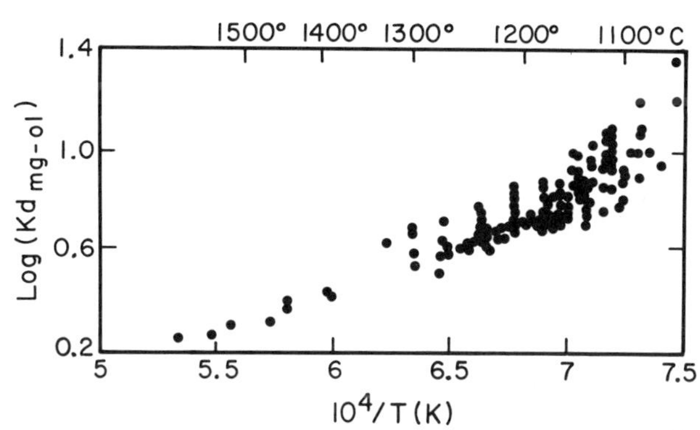

Fig. B1. The entire experimental data set for olivine/melt K_dMgO plotted against inverse temperature. Although there is a rough correlation, the scatter is large.

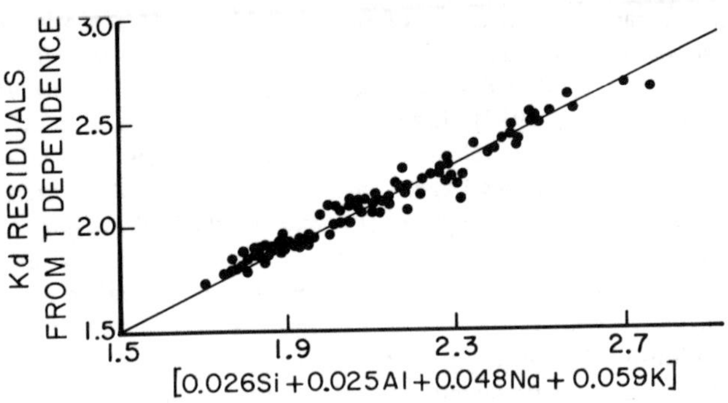

Fig. B2. Residuals of a linear fit between the K_d's for MgO and inverse temperature plotted against a linear combination of four elements in the liquid, for the 1 atm data alone. Compositional data are in cation mole percent.

however, with a much larger range of composition than occurs in the earth. From these experiments, it was clear that there was also a compositional term to the pressure dependence, due primarily to the amount of Al_2O_3, which probably changes its coordination in the liquid as pressure increases (Figure B3*b*). Thermodynamic considerations suggest that the log (K_d) should depend on P/T rather than pressure alone. These aspects of the data suggest a multiple linear regression of the data with log K_d as the dependent variable and 1/T, P/T, and Na_2O, K_2O, SiO_2 and Al_2O_3 as the independent variables. Regression of the high pressure data shows that the Al dependence decreases as pressure increases. All terms in the equation are significant, and there are no additional compositional terms that improve the fit. The final equation is:

$$\ln K_d Mg = 4400/T + 13P/T + [.026Si + .025Al + .048Na + .059K] - .011Al(P-.001)/P + .136(P-.001)/P - 3.231 \qquad \text{(B1)}$$

where the compositional terms are in cation mole percent and pressure is in kilobars. Average error of recovery of the K_d's is 4% for the entire data set. The error is evenly distributed among compositional groups; errors for the high pressure data alone are 4.2%, for komatiites 4.5%, and for high-SiO_2 rocks, 5.5%. Data from certain specific studies have systematically higher errors.

Although this equation covers a very broad range of compositions, it is not ideal for our purposes because it includes terms for Si and Al, and we are unable at this time to calculate reliably the Si and Al contents of mantle melts. Much of the Si and Al dependence in equation B1 results from the data from simple systems such as CMAS, where the Si and Al variations are much larger than in natural data sets from real rocks. We carried out an additional regression of the data, therefore, leaving out the CMAS data. For this smaller yet geological applicable data set, Na and K alone account for most of the compositional dependence. Residuals from the 1 atm 1/T dependence correlate linearly with Na and K (Figure B4*a*). For

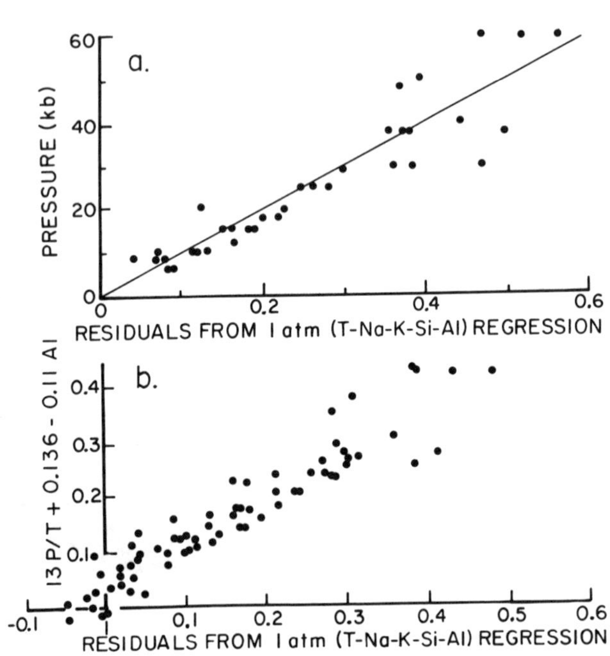

Fig. B3. *a*) The 1 atm expression for temperature and compositional dependence was applied to the high pressure data. Residuals from the 1 atm expression correlate linearly with the pressure of the experiment, except for those experiments in simple systems where there are large variations in Al content. *b*) Taking into account the Al effect and making the pressure term P/T, which is consistent with thermodynamic considerations, leads to a linear relationship with the residuals for the entire high pressure data set.

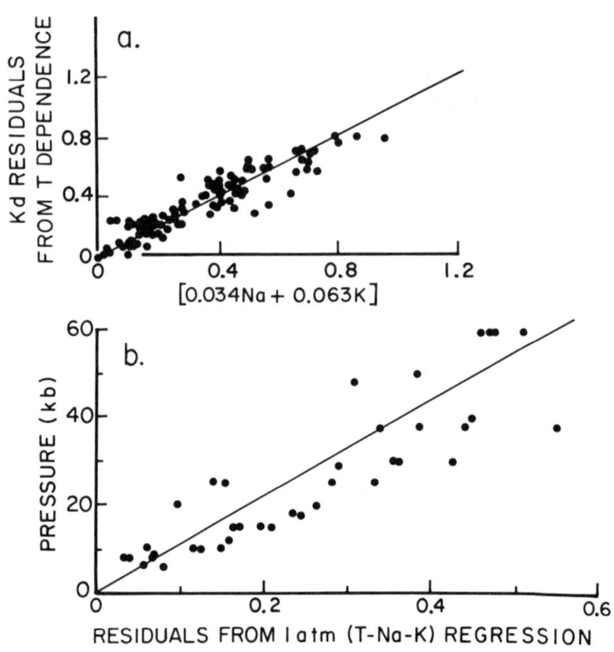

Fig. B4. *a*) Residuals of a linear fit between the K_d's for MgO and inverse temperature plotted against a linear combination of Na and K in the liquid (in cation mole percent). This relationship applies to geological compositions only, and not to the entire compositional range observed in simple systems. *b*) Linear relationship between residuals from the 1 atm temperature, Na and K fit and the pressure of the experiment.

the high pressure data, residuals from the temperature and composition fit to the 1 atm data correlate linearly with pressure (Figure B4*b*). The K_d equation with only Na and K as the compositional variables is:

$$LnK_dMg = 6921/T + .034Na + .063K + .0114P - 3.27 \qquad (B2)$$

This equation, which is used for the Fe and Mg modeling reported in the text, recovers the data with an average error of 5.3%.

The magnitude of these errors needs to be considered in the context of interlaboratory bias, experimental errors, and the variability observed in the natural data that we are trying to explain. Consideration of analyses of similar samples by different microprobes indicates that interlaboratory biases can be significant. For example, the MgO contents determined by the Smithsonian microprobe [*Melson et al.*, 1977] are systematically 6% lower than values determined at Lamont [*Klein*, 1989]. Thus errors of 5% in the K_d's may be as accurate as can be expected given that several different microprobes were used to analyze the experimental run products, and that there are additional uncertainties in the reported temperatures and pressures of the experiments. The total range in $Fe_{8.0}$ that we are trying to understand is 50% of the mean $Fe_{8.0}$ value, and thus is large relative to the errors for the K_d expression. Furthermore, the relative variations calculated for the FeO and MgO contents of mantle melts will be more precise than the absolute values, and it is the relative differences that largely constrain the models considered.

Finally, it should be noted that there is almost no experimental partitioning data on volatile-bearing systems. Since carbon and hydrogen are potentially important components in the mantle, particularly at low extents of melting, the effects of volatile species on partitioning at high pressures for the appropriate compositions is important future work.

APPENDIX C: MODELING COMPLEX DIFFERENTIATION PROCESSES

Much of the compositional variability observed in MORB can be attributed simply to low-pressure crystallization processes (see section 3.1). In addition, some aspects of the residual variability may result from more complex fractionation processes. These may include (a) magma evolution in periodically replenished magma chambers [*O'Hara*, 1977], where a magma chamber can be modeled simply as a holding tank with inputs and outputs; (b) *in situ* fractionation [*Langmuir*, 1989], where solidification takes place along the margins of a magma chamber in a temperature gradient that extends from solidus to liquidus; and (c) high-pressure fractionation, which has been suggested to be an important factor in MORB evolution [e.g., *Bender et al.*, 1977; *Elthon et al.*, 1982]. To evaluate the importance of these fractionation processes in contributing to the diversity of MORB compositions, it is necessary to quantitatively examine their chemical effects.

C.1 Periodically Replenished Magma Chambers (PRMC's)

There is abundant petrographic and geochemical evidence for magma mixing in MORB [e.g., *Kuo and Kirkpatrick*, 1982; *Dungan and Rhodes*, 1978] and common sense argues for replenishment of magma chambers when they exist. In the model of periodically replenished magma chambers [*O'Hara*, 1977; *O'Hara and Matthews*, 1981], magmas undergo constant recharge, mixing, crystal fractionation and eruption, and melts evolve to steady state compositions if input and output parameters remain constant. The steady state is a cycle through two end-member compositions — one that occurs after replenishment with a new pulse of magma and mixing with the magma already in the chamber, and the other that occurs after crystallization in the chamber, just before the addition of the new magma pulse. O'Hara's original treatment dealt with compositions after crystallization and before mixing, but compositions after mixing and before crystallization may also be important.

The net chemical effects of periodically replenished magma chambers can be related simply to the liquid lines of descent for equilibrium and fractional crystallization in a closed system [*Langmuir*, 1980; 1989]. Steady-state liquid compositions after crystallization lie between equilibrium and fractional liquid lines of descent (Figure C1), while the steady state compositions after mixing lie on mixing lines between the end point on the liquid line of descent and the composition of the primitive magma entering the chamber (Figure C2). This is true for both major and trace elements.

This relationship to fractional and equilibrium liquid lines of descent means that closed system crystallization paths can be used to evaluate the range of compositions produced in a PRMC at steady-state. For example, Figure C1 shows the total calculated range in FeO and Na_2O expected for liquids produced during crystallization in a steady-state periodically replenished magma chamber. The narrow width of this field down to low values of MgO shows that while this process may lead to complexity in the details of the fractionation trends, the effect is relatively small, and virtually negligible for high MgO compositions.

On the other hand, mixing along the liquid lines of descent can lead to compositions that differ significantly from the closed system path, particularly if there is mixing of a very high MgO magma with a very differentiated magma (Figure C2). This is possible in the "after mixing" step of the PRMC cycle. Although such mixtures have chemical compositions that differ substantially from the closed system LLD's, they also have recognizable chemical properties that are diagnostic. The mixed magma has higher SiO_2, FeO, Na_2O, TiO_2 and incompatible trace elements, and lower CaO, Al_2O_3, CaO/Al_2O_3 and compatible trace elements than magma produced by closed system differentiation process. In order to create significant variability, however, mixing must occur between a very primitive and a very differentiated magma. Mixing between only a moderately evolved MORB and a typical MORB with 7-8% MgO, for example, will not produce large effects.

It is also important to note two characteristics of such mixed magmas. They have a single liquidus phase, olivine, and when

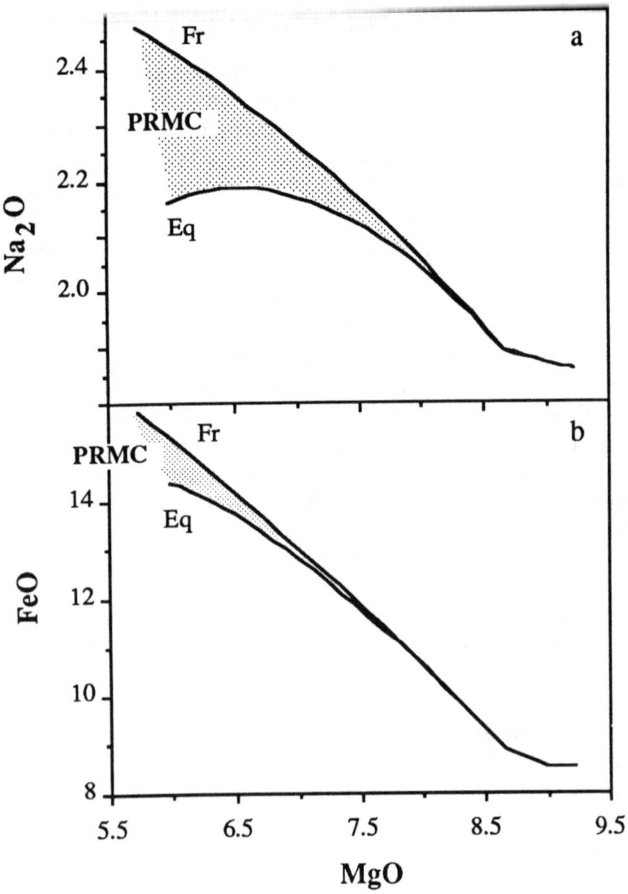

Fig. C1. MgO vs. *a*) Na_2O and *b*) FeO for equilibrium and fractional crystallization paths (LLDs) from *Weaver and Langmuir* [1990] for a single starting composition. It can be demonstrated [*Langmuir*, 1989] that the steady-state products of a periodically replenished magma chamber (PRMC) after the crystallization stage of the cycle lie between fractional and equilibrium crystallization, as indicated by the shading.

they cool they evolve quickly back towards the liquid line of descent of the parental magma. This is illustrated by LLD calculations in Figure C2. Therefore, to differ substantially from closed system evolution, these magmas must be caught immediately after mixing, prior to any subsequent cooling and differentiation. These characteristic properties are occasionally present in some MORB [see, e.g., *Bender et al.*, 1984].

In conclusion, PRMC's lead to liquids that continually return towards the closed system liquid line of descent. It is our intuition, based on the theory and calculations concerning PRMC's, that replenished magma chambers are not a dominant influence on the MORB chemical signal after fractionation correction. In the general case, the variations are quite similar to those produced by closed system crystallization. The one special case that can create significant variations produces specific compositional correlations and mineralogical characteristics that are recognizable, and not particularly common in our experience.

This is not at all to say, however, that mixing of magmas is not important. In fact, mixing of diverse magmas appears to be required to create the pooled melts that account for the global vector of compositional variation (see section 6.2).

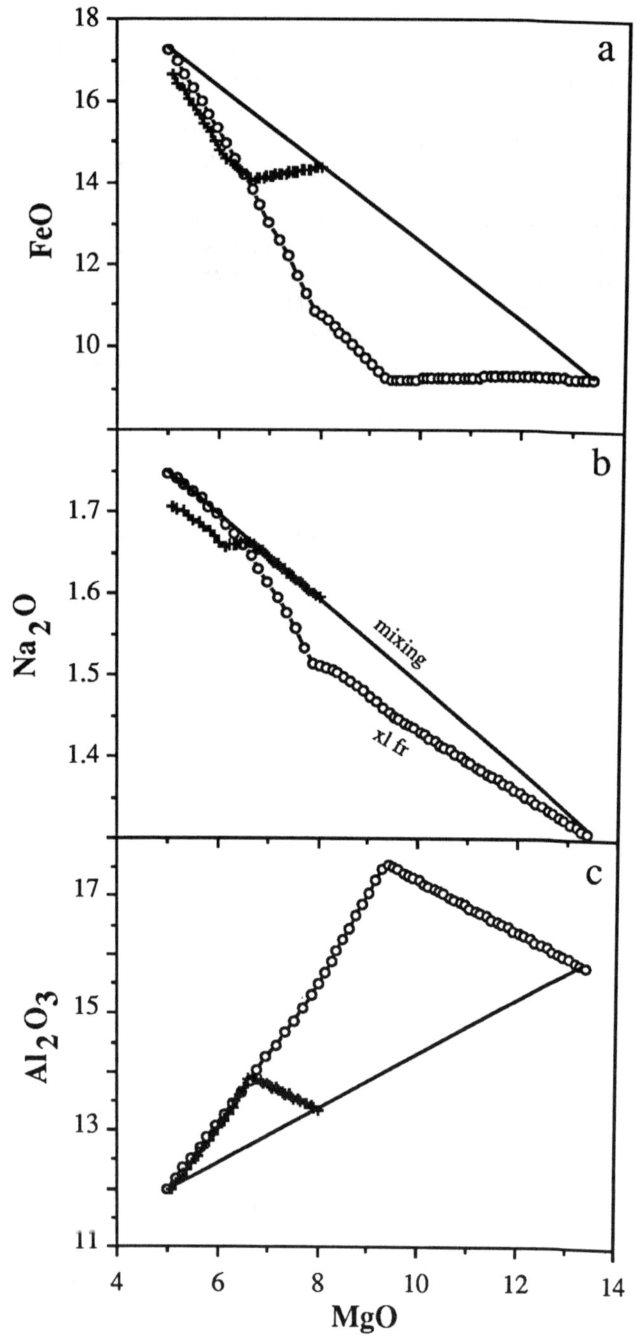

Fig. C2. MgO vs. FeO, Na_2O and Al_2O_3 for crystallization and mixing. The line formed of open circles (xl fr) is a 1 atm LLD for a primitive basalt; the thin straight line is a mixing line between the primitive basalt and a derivative basalt with 5 wt% MgO; the line formed of the crosses shows the LLD for a mixed melt composition; all LLD's calculated using the method of *Weaver and Langmuir* [1990].

C.2 In Situ Fractionation

Another magma chamber process that may affect differentiation paths is *in situ* crystallization at the margins of a magma chamber. Magma chambers lose heat at their margins, and therefore solidify by crystallization in boundary layers along their margins. The magma in the interior of the chamber will evolve if residual melt from the boundary layer returns to the interior of the chamber and mixes with the magma that has not yet become part of the boundary solidification zone, a process that has become known as compositional convection [see, e.g., *Sparks et al.*, 1984; *Sparks,* 1989]. The geochemical and petrological effects of this process have been considered by *Langmuir* [1989]. This form of differentiation differs from classical fractional crystallization because the evolution of the liquid in the chamber is controlled by addition of residual liquids from near the solidus rather than by the classical fractional crystallization which involves subtraction of minerals on the liquidus. The process of *in situ* crystallization differs from periodic replenishment because both mixing end-members evolve and change in composition, while for PRMC's, one end-member is both fixed and relatively primitive.

One of the characteristics of *in situ* crystallization is the removal of minerals that are not on the liquidus of the interior magma in the chamber. The composition of the returned liquid is controlled by minerals crystallizing at all levels in the boundary layer, ranging from liquidus to solidus. This means, for example, that a melt may show signs of clinopyroxene fractionation (decreasing CaO and CaO/Al_2O_3), before clinopyroxene appears on the liquidus as a crystallizing phase.

Figure C3 illustrates this effect with an *in situ* fractionation trend where liquid is returned after 50% crystallization in a boundary layer. The normal trend for fractional crystallization

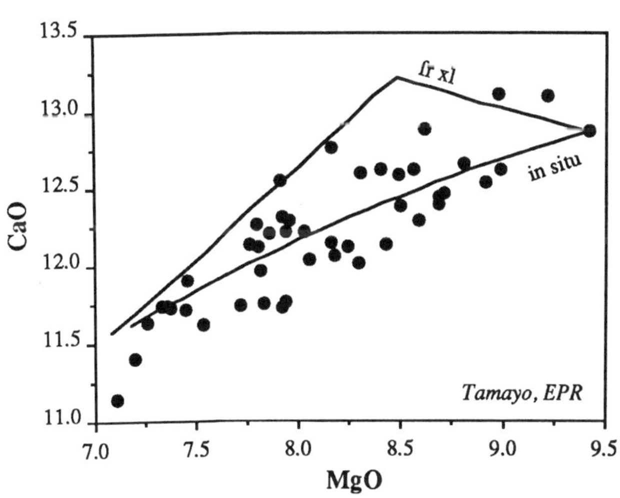

Fig. C3. CaO vs. MgO for data from the Tamayo region of the East Pacific Rise (filled circles; *Bender et al.*, 1984). Lines are calculated LLDs for simple fractional crystallization [*Weaver and Langmuir*, 1990], and *in situ* crystallization assuming return of liquid after 50% crystallization in the boundary layer.

does not show clinopyroxene crystallizing until the kink in the liquid line of descent at an MgO content of 8.5 wt%. The *in situ* crystallization path, however, shows CaO decreasing immediately, due to return of liquid from the boundary layer where clinopyroxene has crystallized. The *in situ* path appears to better reproduce the Tamayo data trend on this diagram.

The key parameter for the *in situ* model is the mean fraction of liquid that is returned to the chamber. For low mean melt fractions, the liquid line of descent may differ substantially from that of fractional crystallization. This can be understood from the simple reasoning using partition coefficients in section 2.2. When the melt fraction in the boundary layer becomes small, most elements reach their 1/D limit. Therefore moderately incompatible to compatible elements no longer change appreciably in concentration. Highly incompatible elements, with D's that approach zero, continue to increase in abundance. This means that the liquid returned to the magma chamber can be very enriched in incompatibles, and yet not have a particularly extreme major element composition. Thus *in situ* crystallization has the potential of creating large enrichments in incompatible elements, provided a large extent of crystallization in the boundary layer has occurred prior to return of the liquid to the chamber.

During melting of the mantle, clinopyroxene is the major mineral entering the melt up to about 20% melting, and for this reason it is often difficult to distinguish between melting, involving primarily the addition of clinopyroxene, and removal of clinopyroxene by crystallization. Of course, the F's involved are different and the two processes are likely to occur at markedly different pressures, but both processes produce the same general trends, including increases in Na_2O and FeO and a decrease in CaO/Al_2O_3, and therefore in a qualitative sense may be difficult to tell apart.

The mineral plagioclase, however, plays a critical role in identifying the effects of *in situ* crystallization. At low pressures plagioclase is a stable phase and its crystallization will impart a distinctive chemical signature to the liquid. Plagioclase is particularly important with respect to trace elements, because the element Sr, incompatible during mantle melting processes, has a sufficiently high partition coefficient in plagioclase that its bulk distribution coefficient approaches 1 during crystallization. Therefore, any crystallization process involving plagioclase should be distinguishable from mantle melting processes. (Of course, plagioclase is also a possible mantle phase at very low pressures, but it melts out of the residue at low extents of melting, and appears not to play an important role as a residual mantle phase in MORB genesis.)

C.3 High Pressure Fractionation

While the evidence for low-pressure crystallization of MORB magmas is clear, some have argued that basaltic magmas also experience significant extents of high pressure fractionation prior to equilibration at low pressure. Proponents of ubiquitous high pressure fractionation cite as evidence the occurrence of highly magnesian pyroxenes in some ophiolite suites and occasionally in oceanic basalts [e.g., *Elthon et al.*,

1982; *Elthon and Scarfe*, 1984; c.f. *Elthon*, 1991], and the position of some MORB suites on pseudo-phase diagrams [e.g., *Tormey et al.*, 1987].

Liquid line of descent programs that model crystallization at 1 atm have been successful due to the abundance of high quality experimental data at low pressure (see section 3.1). In theory, such liquid line of descent programs can be modified to model the effects of crystallization at high pressure; it is necessary only to know the pressure dependencies of the partition coefficients. Although both quality and quantity of experimental data at high pressure are much poorer than at 1 atm, the available data are nonetheless sufficient to begin quantitative evaluation of high pressure crystallization processes. We have incorporated pressure-dependencies of the K_d's for olivine, plagioclase and clinopyroxene into the liquid line of descent program of *Weaver and Langmuir* [1990; see Table C1 for K_d functions used]. Due to the paucity of high-pressure data, these results, although quantitative, are subject to greater uncertainties than the well-constrained calculations at 1 atm.

Because of interlaboratory biases and the complexity of natural compositions, all LLD calculations are most reliable if they are tuned to the rocks under consideration. To evaluate the effects of high pressure crystallization, we have calibrated the LLD program to the FAMOUS compositions, which we will use as our test case for the high pressure crystallization model. The agreement of the calibration with high pressure experimental data on FAMOUS compositions is illustrated in Figure C4 where the phase appearances predicted by the high-pressure liquid line of descent program are compared with the experimentally determined phase appearances reported by *Bender et al.* [1978]. Note the overall consistency between the experimental and calculated appearances of olivine, plagioclase and clinopyroxene up to a pressure of 8 kbar. In detail, the calculated liquidus curves for olivine and plagioclase are 5-10° lower than the experimentally determined liquidus curves. Nonetheless, at any one pressure, the experimentally determined and calculated temperature intervals for olivine + liquid stability are virtually identical. Although the appearance of clinopyroxene on the liquidus is not well constrained at low pressures in Bender et al.'s experimental data due to nucleation problems, the calculated appearance of clinopyroxene at 1 atm is in agreement with other experimental data for a similar

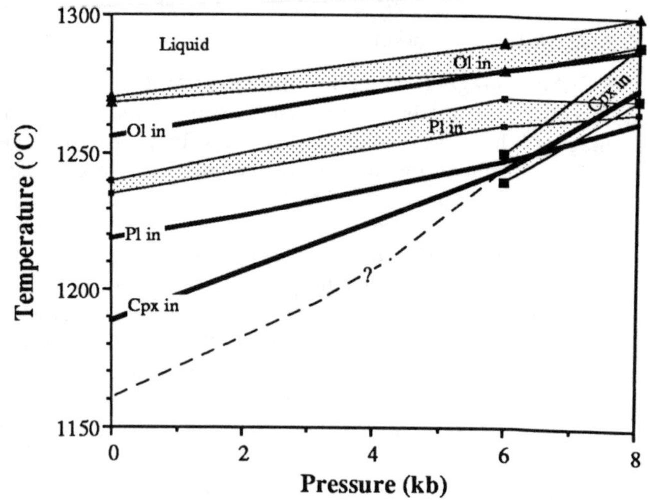

Fig. C4. Temperature vs. pressure diagram of phase appearances determined experimentally by *Bender et al.* [1978] for a sample from the FAMOUS region shown by the broad shaded bands representing the uncertainty in the experimental determinations. For example, the shaded band for plag-in lies between experiments in which there was no plagioclase, and experiments where plagioclase first appeared. The dashed line extending from the "clinopyroxene in" band at high pressure indicates great uncertainty in the position of this boundary due to nucleation problems in the low pressure experiments. Also shown (heavy solid lines) are the phase appearances calculated for the same melt composition using the liquid line of descent program of *Weaver and Langmuir* [1990], modified for crystallization at high pressure as described in the text.

TABLE C1. Distribution Coefficient Equations with Pressure

Phase	Component	Expression: $Log_{10}K_d$ =
Olivine	MgO	$[6921/T + 3.4Na_L + 6.3K_L + 0.00001154P - 3.27]/2.3$
	FeO	$Log_{10}K_d^{Mg} - 0.523$
Plagioclase	$CaAl_2O_4$	$2446/T - 1.122 + 0.2562An_L$
	$NaAlO_2$	$[3195 + 3283An_L + 0.0506P]/T - [1.885An_L + 2.3715]$
Clinopyroxene	MgO	$[3798 + 0.021P]/T - 2.28$
	FeO	$Log_{10}K_d^{Mg} - 0.6198$
	$CaSiO_3$	$[1783 + 0.0038P]/T - 0.753$
	$CaAl_2O_4$	$[2418 + 0.068P]/T - 2.3$
	$NaAlO_2$	$[5087 + 0.073P]/T - 4.48$
	TiO_2	$[1034 + 0.053P]/T - 1.27$

P in bars, T in kelvins, compositional terms in mole fraction starting composition [*Grove and Bryan*, 1983]. Lastly, both the experimental and calculated data show a cross-over in the appearances of plagioclase and clinopyroxene on the liquidus between 6 and 7 kbar.

It is important to note that inconsistencies between the experimental data and the calculated liquid lines of descent do not necessarily imply inaccuracies in the calculated liquid line of descent. The liquid line of descent calculations are based on estimates of the distribution coefficients derived from numerous experiments and therefore minimize experimental problems, such as disequilibrium, that may have occurred in any one experiment. Furthermore, due to problems such as sodium or iron loss, the experimentally determined liquid lines of descent may not accurately reflect closed system fractionation, although the distribution coefficient determined from the experiment may accurately represent the partitioning between solid and the modified melt composition. Thus, a less-than-perfect correspondence between experimental and calculated liquid lines of descent should not necessarily be attributed to problems with the calculation. On the other hand, the liquid line of descent program is as imperfect as its reproduction of the experimentally determined partition coefficients. For modeling of the subtle variations that exist within and between different magmatic suites, the bulk composition dependencies of the K_d's become important, and caution should be exercised in using a generic liquid line of descent program to interpret such subtle variations.

C.3.1 General Results of High Pressure Fractionation

As can be seen in the P-T diagram shown in Figure C4, the primary effect of crystallization at higher pressure is to increase the temperature of appearance of clinopyroxene

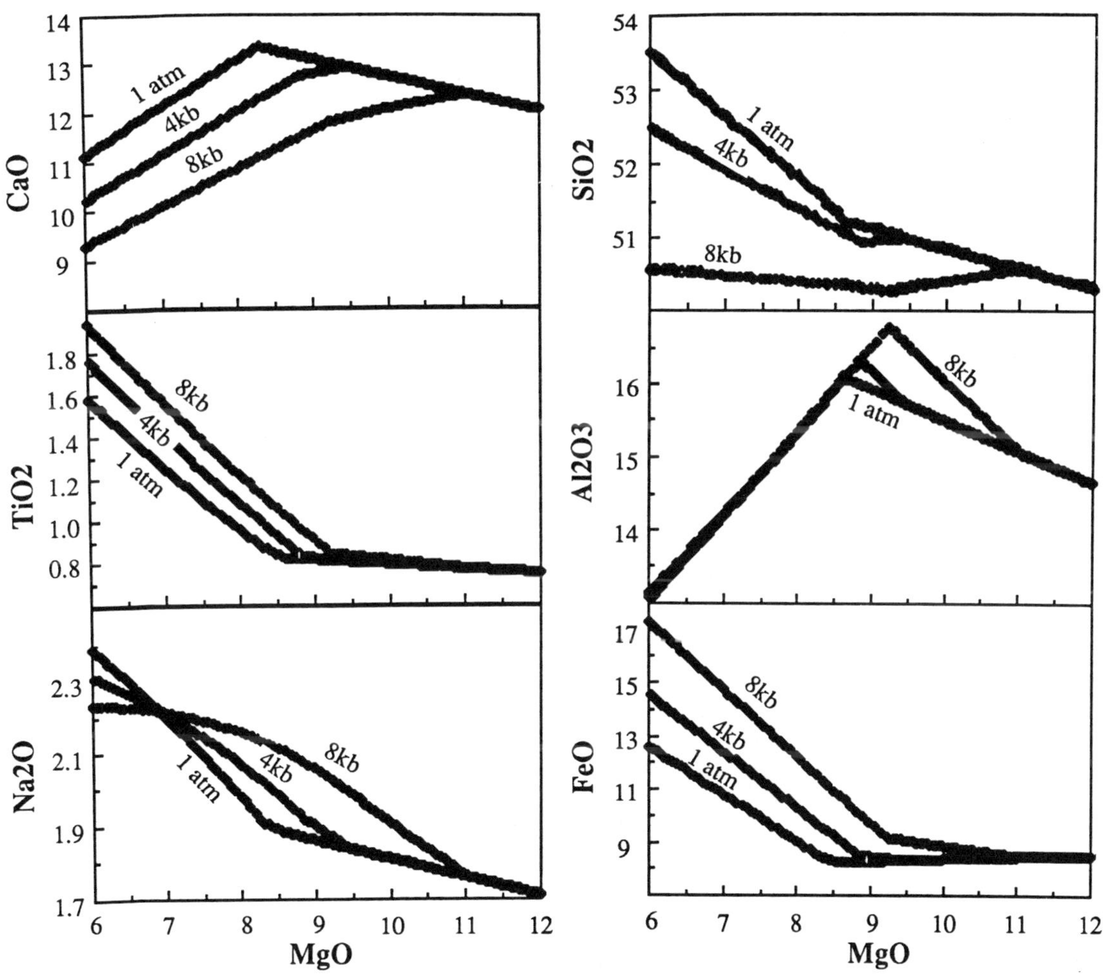

Fig. C5. Calculated liquid lines of descent for Model 1 crystallization at 1 atm, 4 kbar, and 8 kbar of the melt composition presented in Table C2, using the program of *Weaver and Langmuir* [1990] modified for high-pressure crystallization as described in the text and shown in Table C1. Compositions in figure are not exactly the same as those in the Table C2, because the compositions have been corrected for interlaboratory bias so as to be comparable with the Melson data for the FAMOUS/AMAR region presented in Figures 22-23. These results are also shown in Figure 61 and compared to the FAMOUS/AMAR data.

relative to olivine and plagioclase. In terms of compositional effects, liquid lines of descent on MgO variation diagrams show very different slopes when olivine crystallizes alone as compared to when plagioclase and clinopyroxene join the solid assemblage. Therefore the earlier appearance of these latter phases causes these changes of slope to occur at higher MgO contents, which in turn causes the abundance of some oxides at 8 wt% MgO to differ from those produced by differentiation at low pressure.

These effects are illustrated in Figure C5, where liquid lines of descent for a single high MgO parental composition are crystallized at different pressures. The starting composition, listed in Table C2 (starting "system" composition), was chosen by adding 2% olivine to a high-MgO, low-Na sample composition from the FAMOUS/AMAR region. This composition was crystallized fractionally at pressures of 1 atm, 4 kbar, and 8 kbar (Figure C5) and the phase appearances and compositions are presented in Table C2. Relative to differentiation at 1 atm, at 8 wt% MgO a higher pressure of crystallization usually causes Na_2O, FeO, TiO_2 and K_2O to be higher, and CaO, Al_2O_3 and SiO_2 to be lower. An important additional effect is that at higher pressure the crystallizing plagioclase is of a more sodic composition [e.g., *Bender et al.*, 1978; *Fram and Longhi*, 1991], and therefore the 8 kbar liquid line of descent shows a marked leveling-off of Na_2O enrichment as fractionation proceeds.

C.3.2 Models for High Pressure Fractionation

There are numerous possible high-pressure crystallization scenarios. Consider first a fractionation model in which there is a single parental magma composition, different batches of which segregate from the mantle and fractionate at different pressures, followed by immediate transport to the surface and eruption. This is Model 1, and would lead to the results described above and shown in Figure C5 and Table C2.

In another possible model (Model 2), different batches of the same parental magma composition fractionate at some high pressure, but the various batches fractionate to different extents prior to transport to the surface, where they fractionate further to 8 wt% MgO, and erupt. For Model 2, we used the 8 kbar liquid line of descent for the high pressure stage of crystallization, and at various stages of crystallization at 8 kbar, the residual liquid composition was "removed" and fractionated at 1 atm. The results of this exercise are shown in Figure C6. Differences in the fractionating assemblages between 1 atm and 8 kbar are evident. For example, on the plot of MgO vs. CaO, it is clear that at 8 kbar, clinopyroxene and olivine are stable at >10 wt% MgO, leading to decreasing CaO with decreasing MgO; in contrast, the same magma composition (>10 wt% MgO) at 1 atm shows increasing CaO with decreasing MgO due to the crystallization of olivine alone. The range in compositions produced by Model 2 is bracketed by the 1 atm and 8 kbar LLDs, as it was for Model 1,

TABLE C2. Model 1

Pressure = 1 atm	System	Liquid	Plag	Olivine	CPX
fraction:		0.9993		0.0007	
SiO2	50.00	50.01		40.86	
TiO2	0.75	0.75		0.00	
Al2O3	14.65	14.66		0.00	
FeO	8.30	8.30		9.84	
MgO	12.50	12.48		49.31	
CaO	12.05	12.06		0.00	
Na2O	1.75	1.75		0.00	
TEMP=	*1291*	*Frac. Liquid Remaining:*			*0.9993*
	System	*Liquid*	*Plag*	*Olivine*	*CPX*
fraction:		0.9919	0.0054	0.0027	
SiO2	50.93	50.97	47.68	40.32	
TiO2	0.82	0.83	0.00	0.00	
Al2O3	16.10	16.04	33.63	0.00	
FeO	8.02	8.05	0.00	12.72	
MgO	8.97	8.92	0.00	46.97	
CaO	13.24	13.25	16.61	0.00	
Na2O	1.92	1.93	2.08	0.00	
TEMP=	*1199*	*Frac. Liquid Remaining:*			*0.8929*
	System	*Liquid*	*Plag*	*Olivine*	*CPX*
fraction:		0.9813	0.0102	0.0028	0.0057
SiO2	51.25	51.30	47.92	40.16	53.39
TiO2	0.87	0.89	0.00	0.00	0.25
Al2O3	15.66	15.59	33.47	0.00	3.15
FeO	8.27	8.36	0.00	13.52	4.20
MgO	8.65	8.59	0.00	46.31	17.97
CaO	13.34	13.30	16.43	0.00	20.85
Na2O	1.95	1.96	2.19	0.00	0.20
TEMP=	*1192*	*Frac. Liquid Remaining:*			*0.8314*

Pressure = 4 kbar	System	Liquid	Plag	Olivine	CPX
fraction:		0.9999		0.0001	
SiO2	50.00	50.00		40.86	
TiO2	0.75	0.75		0.00	
Al2O3	14.65	14.65		0.00	
FeO	8.30	8.30		9.83	
MgO	12.50	12.50		49.32	
CaO	12.05	12.05		0.00	
Na2O	1.75	1.75		0.00	
TEMP=	*1308*	*Frac. Liquid Remaining:*			*0.9999*
	System	*Liquid*	*Plag*	*Olivine*	*CPX*
fraction:		0.9989		0.001	0.0001
SiO2	50.67	50.68		40.49	53.17
TiO2	0.80	0.81		0.00	0.29
Al2O3	15.71	15.73		0.00	4.25
FeO	8.12	8.12		11.80	3.84
MgO	9.89	9.85		47.71	19.42
CaO	12.92	12.93		0.00	18.79
Na2O	1.88	1.88		0.00	0.23
TEMP=	*1240*	*Frac. Liquid Remaining:*			*0.9238*
	System	*Liquid*	*Plag*	*Olivine*	*CPX*
fraction:		0.9745	0.0132	0.003	0.0092
SiO2	50.64	50.67	48.68	40.28	52.95
TiO2	0.83	0.85	0.00	0.00	0.32
Al2O3	16.31	16.24	32.96	0.00	4.62
FeO	8.28	8.41	0.00	12.92	4.26
MgO	9.22	9.14	0.00	46.80	19.31
CaO	12.76	12.70	15.83	0.00	18.28
Na2O	1.96	1.97	2.54	0.00	0.26
TEMP=	*1223*	*Frac. Liquid Remaining:*			*0.8562*

Pressure = 8 kbar	System	Liquid	Plag	Olivine	CPX
fraction:		0.9989		0.0011	
SiO2	50.00	50.01		40.85	
TiO2	0.75	0.75		0.00	
Al2O3	14.65	14.66		0.00	
FeO	8.30	8.30		9.85	
MgO	12.50	12.46		49.30	
CaO	12.05	12.06		0.00	
Na2O	1.75	1.75		0.00	
TEMP=	*1324*	*Frac. Liquid Remaining:*			*0.9989*
	System	*Liquid*	*Plag*	*Olivine*	*CPX*
fraction:		0.9963		0.0005	0.0032
SiO2	50.25	50.25		40.72	52.94
TiO2	0.77	0.77		0.00	0.36
Al2O3	15.05	15.09		0.00	5.16
FeO	8.25	8.26		10.54	3.54
MgO	11.50	11.46		48.73	20.48
CaO	12.38	12.37		0.00	17.27
Na2O	1.80	1.80		0.00	0.25
TEMP=	*1299*	*Frac. Liquid Remaining:*			*0.9668*
	System	*Liquid*	*Plag*	*Olivine*	*CPX*
fraction:		0.9948	0.0008	0.0002	0.0042
SiO2	49.96	49.95	49.66	40.25	52.20
TiO2	0.84	0.85	0.00	0.00	0.42
Al2O3	16.75	16.78	32.29	0.00	6.54
FeO	8.93	8.95	0.00	13.06	4.54
MgO	9.64	9.60	0.00	46.69	20.29
CaO	11.82	11.80	15.06	0.00	15.65
Na2O	2.07	2.07	2.99	0.00	0.36
TEMP=	*1256*	*Frac. Liquid Remaining:*			*0.8162*

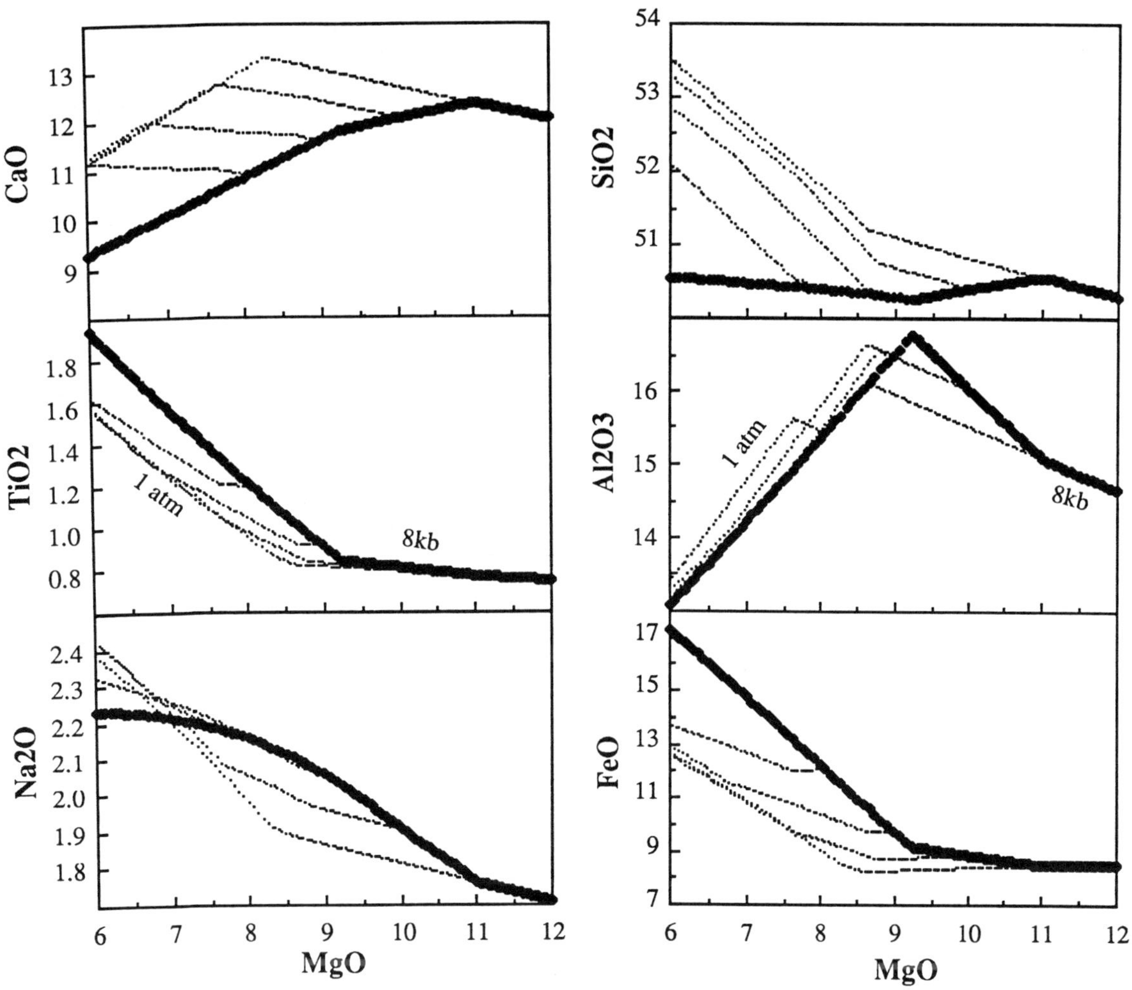

Fig. C6. Calculated liquid lines of descent for fractional crystallization of the parental magma composition shown in Table C2, using the liquid line of descent program of *Weaver and Langmuir* [1990], modified for high-pressure fractionation as described in the text and shown in Table C1. In this model (Model 2), crystallization begins at 8 kbar (solid curves), and at different stages of differentiation (different values of MgO) batches of this evolving magma are transported to low pressure where they fractionate at 1 atm (dotted curves).

and therefore the total range in any one oxide is virtually the same as that produced by Model 1 above.

Other fractionation models for high pressure differentiation can be envisioned. For example, one could consider the case of polybaric fractionation in which a particular parental magma composition begins crystallizing at high pressure, and as it migrates toward the surface it fractionates at successively lower pressures. This model, however, produces only a single composition at the surface, rather than producing the whole range of compositions at 8 wt% MgO such as would be necessary to account for the Atlantic-type local vector.

The results shown in Figure C5 and C6 are based on experiments that investigated basalts from the FAMOUS area, and are primarily applicable to examining the effects of high pressure fractionation in samples of this parental magma composition. We have also examined, however, the available experimental data for a starting composition from the Kane region, which differs from the FAMOUS composition in having higher Na_2O and SiO_2 and lower FeO. There are some differences in the pressure-dependencies derived from experiments for the two starting compositions, but there is agreement for the two different compositions in the overall trends of the effects of high pressure crystallization. Thus these calculations are able to give a reasonable estimate of the importance of high pressure fractionation for explaining the global and local MORB systematics (discussed in sections 6.2 and 6.3).

Acknowledgements. This paper results from over a decade of on-going work, to which dozens of people have contributed in important ways, particularly John Bender. The immediate manuscript has benefited from numerous interactions at the short course and reviews by or discussion with Rodey Batiza, Donna Blackman, Dana Desonie,

Bob Detrick, Kevin Johnson, Jason Phipps Morgan, Yao Ling Niu, Jennifer Reynolds and Katrin Wiese. Discussions with Ro Kinzler and Tim Grove provided impetus to investigate the role of high pressure fractionation thoroughly, and to address compositional dependence of the olivine partition coefficients. Thorough formal reviews by and discussion with Dave Christie, Jean-Guy Schilling and John Sinton aided very much in revising the manuscript. Charlene Straub worked long hours far beyond the call of duty in editing the manuscript and preparing it for publication. Joan Totten helped with the final preparation of the manuscript. Patty Catanzaro drafted many of the figures. This work has been supported by numerous grants from the National Science Foundation, particularly OCE 90-14925 (CHL) and OCE 90-16317 (EMK).

REFERENCES

Agee, C. B., and D. Walker, Aluminum partitioning between olivine and ultrabasic silicate liquid to 6 GPa, *Contrib. Mineral. Petrol., 105*, 243-254, 1990.

Ahern, J. L., and D. L. Turcotte, Magma migration beneath an ocean ridge, *Earth Planet. Sci. Lett., 45*, 115-122, 1979.

Albarede, F., How deep do common basaltic magmas form and differentiate?, *J. Geophys. Res., 97*, 10,997-11,009, 1992.

Albarede, F., and V. Tamagnan, Modelling the recent geochemical evolution of the Piton de la Fournaise volcano, Reunion Island, 1931-1986, *J. Petrol., 29*, 997-1030, 1988.

Allegre, C. J. and Y. Bottinga, Tholeiite, Alkali Basalt and Ascent Velocity, *Nature, 252*, 31-31, 1974.

Allegre, C. J., and J. F. Minster, Quantitative models of trace element behavior in magmatic processes, *Earth Planet. Sci. Lett., 38*, 1-25, 1978.

Allegre, C. J., and D. L. Turcotte, Implications of a two-component marble-cake mantle, *Nature, 323*, 123-127, 1986.

Allegre, C. J., B. Hamelin, and B. Dupre, Statistical analysis of isotopic ratios in MORB: the mantle blob cluster model and the convective regime of the mantle, *Earth Planet. Sci. Lett., 71*, 71-84, 1984.

Bass, M. N., Variable abyssal basalt populations and their relation to sea-floor spreading rates, *Earth Planet. Sci. Lett., 11*, 18-22, 1971.

Batiza, R., Inverse relationship between Sr isotope diversity and rate of oceanic volcanism has implications for mantle heterogeneity, *Nature, 309*, 440-441, 1984.

Batiza, R., and D. Vanko, Petrology of young Pacific seamounts, *J. Geophys. Res., 89*, 11,235-11,260, 1984.

Batiza, R., W. G. Melson, and T. O'Hearn, Simple magma supply geometry inferred beneath a segment of the Mid-Atlantic Ridge, *Nature, 335*, 428-431, 1988.

Batiza, R., Y. Niu, and W. C. Zayac, Chemistry of seamounts near the East-Pacific Rise: Implications for the geometry of sub-axial mantle flow, *Geology, 18*, 1122-1125, 1990.

Beatty, P., C. Ford, and D. Russel, Partition coefficients for olivine-melt and orthopyroxene-melt systems, *Contrib. Mineral. Petrol., 109*, 212-224, 1991.

Bender, J. F., F. N. Hodges, and A. E. Bence, Petrogenesis of basalts from the project FAMOUS area: experimental study from 0 to 15 kbars, *Earth Planet. Sci. Lett., 41*, 277-302, 1978.

Bender, J. F., C. H. Langmuir, and G. N. Hanson, Petrogenesis of basalt glasses from the Tamayo region, East Pacific Rise, *J. Petrol., 25*, 213-254, 1984.

Bottinga, Y. and C. J. Allegre, Thermal aspects of sea-floor spreading and the nature of the oceanic crust, *Tectonophys. 18*, 1-17, 1973.

Bougault, H., and M. Treuil, Mid-Atlantic Ridge: Zero-age geochemical variations between Azores and 22°N, *Nature, 286*, 209-212, 1980

Brodholt, J. P., and R. Batiza, Global systematics of unaveraged MORB compositions: Comment on "Global correlations of ocean ridge basalt chemistry with axial depth and crustal thickness" by E. M. Klein and C. H. Langmuir, *J. Geophys. Res., 94*, 4231-4240, 1989.

Bryan, W. B., and J. G. Moore, Compositional variations of young basalts in the Mid-Atlantic Ridge rift valley near lat. 36°49'N, *Geol. Soc. Am. Bull., 88*, 556-570, 1977.

Bryan, W. B., and H. J. B. Dick, Contrasted abyssal basalt liquidus trends: Evidence for mantle major element heterogeneity, *Earth Planet. Sci. Lett., 58*, 15-26, 1982.

Buck, W. R. and W. Su, Focussed mantle upwelling below mid-ocean ridges due to feedback between viscosity and melting, *Geophys. Res. Lett., 16*, 641-644, 1989.

Bunch, A. W. H., and B. L. N. Kennett, The crustal structure of the Reykjanes Ridge at 59°30'N, *Geophys. J. R. Astron. Soc., 61*, 141-166, 1980.

Byerly, G. R., W. G. Melson, and P. R. Vogt, Rhyodacites, andesites, ferro-basalts and ocean tholeiites from the Galapagos spreading center, *Earth Planet. Sci. Lett., 30*, 215-221, 1976.

Carmichael, I. S. E., F. J. Turner, and J. Verhoogen, *Igneous Petrology*, 739 pp., McGraw-Hill, New York, 1974.

Castillo, P. R., E. Klein, S. B. Shirey, J. F. Bender, R. Batiza, and C. Langmuir, Relationship of Sr-Nd-Pb isotopes of EPR and MAR basalt glasses to bathymetry and ridge segmentation, *EOS, 69*, 1477, 1988.

Cawthorn, R. G., Degrees of melting in mantle diapirs and the origin of ultrabasic liquids, *Earth Planet. Sci. Lett., 27*, 113-120, 1975.

Clague, D. A., F. A. Frey, G. Thompson, and S. Rindge, Minor and trace element geochemistry of volcanic rocks dredged from the Galapagos Spreading Center: Role of crystal fractionation and mantle heterogeneity, *J. Geophys. Res., 86*, 9469-9482, 1981.

Cohen, R. S., and R. K. O'Nions, The lead, neodymium and strontium isotopic structure of ocean ridge basalts, *J. Petrol., 23*, 229-324, 1982.

Condomines, M., C. H. Hemond, and C. J. Allegre, U-Th-Ra radioactive disequilibria and magmatic processes, *Earth Planet. Sci. Lett., 90*, 243-262, 1988.

Dick, H. J. B., Abyssal peridotites, very slow spreading ridges, and ocean ridge magmatism, in *Magmatism in the Ocean Basins*, edited by A. D. Saunders and M. J. Norry, pp. 71-105, Geol. Soc. London Spec. Pub. No. 42, 1989.

Dick, H. J. B., R. L. Fisher, and W. B. Bryan, Mineralogic variability of the uppermost mantle along mid-ocean ridges, *Earth Planet. Sci. Lett., 69*, 88-106, 1984.

Dixon, J. E., E. Stolper, and J. R. Delaney, Infrared spectroscopic measurements of CO_2 and H_2O in Juan de Fuca basaltic glasses, *Earth Planet. Sci. Lett., 90*, 87-104, 1988.

Dmitriev, L.V., A. V. Sobolev, A. V. Uchanov, T. V. Malysheva, and W. G. Melson, Primary differences in oxygen fugacity and depth of melting in the mantle source regions for oceanic basalts, *Earth Planet. Sci. Lett., 70*, 303-310, 1984.

Dungan, M. A., and J. M. Rhodes, Residual glasses and melt inclusions in basalts from DSDP legs 45 and 46: Evidence for magma mixing, *Contrib. Mineral. Petrol., 67*, 417-431, 1978.

Dupre, B., B. Lambret, D. Rousseau, and C. J. Allegre, Limitations on the scale of mantle heterogeneities under ocean ridges, *Nature, 294*, 552-554, 1981.

Elliott, T. R., C. J. Hawkesworth, and K. Gronvold, Dynamic melting of the Iceland plume, *Nature, 351*, 201-206, 1991.

Elthon, D., Pressure of origin of primary mid-ocean ridge basalts, *Magmatism in the Ocean Basins*, edited by A. D. Sounders and M. J. Norry, pp. 125, Geol. Soc. Spec. Publ. London 42, 1989.

Elthon, D., and C. M. Scarfe, High-pressure phase equilibria of a high-magnesia basalt and the genesis of primary oceanic basalts, *Am. Mineral., 69*, 1-15, 1984.

Elthon, D., J. F. Casey, and S. Komor, Mineral chemistry of ultramafic cumulates for the North Arm Mountain massif of the Bay of Islands ophiolite: evidence for high pressure crystal fractionation of oceanic basalts, *J. Geophys. Res., 87*, 8717-8734, 1982.

Etoubleau, J., O. Corre, J. L. Joron, H. Bougault, and M. Treuil, Costa Rica Rift: variably depleted basalts in the same hole. In *Init. Repts. DSDP 69* (eds. J. R. Cann et al.) pp. 765-773, 1983.

Evans, J. R., and Sacks, I. S., Deep structure of the Iceland plateau, *J. Geophys. Res., 84*, 6859-6866, 1979.

Falloon, T. J., and D. H. Green, Anhydrous partial melting of MORB pyrolite and other peridotite compositions at 10 kb: Implications for the origin of primitive MORB glasses, *Mineral. and Petrol.*, *37*, 181-219, 1987.

Falloon, T. J. and D. H. Green, Anhydrous partial melting of peridotite from 8 to 35kb and the petrogenesis of MORB, *J. Petrol. Sp. Lithosphere Issue*, 379-414, 1988.

Falloon, T. J., D. H. Green, C. J. Hatton, and K. L. Harris, Anhydrous partial melting of a fertile and depleted peridotite from 2 to 30 kb and application to basalt petrogenesis, *J. Petrol.*, *29*, 1257-1282, 1988.

Feigenson, M. D., A. W. Hofmann, and F. J. Spera, Case studies on the origin of basalt II. The transition from tholeiitic to alkalic volcanism on Kohala volcano, Hawaii, *Contrib. Mineral. Petrol.*, *84*, 390-405, 1983.

Fisk, M. R., J. G. Schilling, and H. Sigurdsson, An experimental investigation of Iceland and Reykjanes Ridge tholeiites, I. Phase relations, *Contrib. Mineral. Petrol.*, *74*, 316-374, 1980.

Fisk, M. R., A. E. Bence, and J.-G. Schilling, Major element chemistry of Galapagos Rift Zone magmas and their phenocrysts, *Earth Planet. Sci. Lett.*, *61*, 171-189, 1982.

Ford, C. E., D. G. Russell, J. A. Craven, and M. R. Fisk, Olivine-liquid equilibria: temperature, pressure and composition dependence of the crystal/liquid cation partition coefficients for Mg, Fe2+, Ca and Mn, *J. Petrol.*, *24*, 256-265, 1983.

Fornari, D. J., M. R. Perfit, A. Malahoff, and R. Embley, Geochemical studies of abyssal lavas recovered by DSRV Alvin from Eastern Galapagos Rift, Inca Transform and Ecuador Rift. 1. Major element variations in natural glasses and spatial distribution of lavas, *J. Geophys. Res.*, *88*, 10,519-10,529, 1983.

Fox, P. J., and D. G. Gallo, A tectonic model for ridge-transform-ridge plate boundaries: implications for the structure of oceanic lithosphere, *Tectonophysics*, *104*, 205-242, 1984.

Fram, M. S., and J. Longhi, Plagioclase/melt partitioning as a function of pressure, *EOS*, *72*, 304, 1991.

Frey, F. A., C.-Y. Suen, and H. W. Stockman, The Ronda high temperature peridotite: Geochemistry and petrogenesis, *Geochim. Cosmochim. Acta*, *49*, 2469-2491, 1985.

Frey, F. A., W. S. Wise, M. O. Garcia, H. West, S.-T. Kwon, and A. Kennedy, Evolution of Mauna Kea volcano, Hawaii: Petrologic and geochemical constraints on postshield volcanism, *J. Geophys. Res.*, *95*, 1271-1300, 1990.

Fryer, P., J. M. Sinton, and J. A. Philpotts, Basaltic glasses from the Mariana Trough, *Initial Rep. Deep Sea Drill. Proj.*, *60*, 601-609, 1981.

Fryer, P., B. Taylor, C. H. Langmuir, and A. G. Hochstaedter, Petrology and geochemistry of lavas from the Sumisu and Torishima backarc rifts, *Earth and Planet. Sci. Lett.*, *100*, 161-178, 1990.

Fujii, T., and C. M. Scarfe, Composition of liquids coexisting with spinel lherzolite at 10 kbar and the genesis of MORBs, *Contrib. Mineral. Petrol.*, *90*, 18-28, 1985.

Galer, S. J. G., and R. K. O'Nions, Magmagenesis and the mapping of chemical and isotopic variations in the mantle, *Chem. Geol.*, *56*, 45-61, 1986.

Gast, P. W., Trace element fractionation and the origin of tholeiitic and alkaline magma types, *Geochim. Cosmochim. Acta*, *32*, 1057-1086, 1968.

Ghiorso, M., I. S. Carmichael, M. L. Rivers, and R. O. Sack, The Gibbs Free Energy of mixing of natural silicate liquids; an expanded regular solution approximation for the calculation of magmatic intensive variables, *Contrib. Mineral. Petrol.*, *84*, 107-145, 1983.

Goldstein, S. J., M. T. Murrell, and D. R. Janecky, Th and U isotopic systematics of basalts from the Juan de Fuca and Gorda Ridges by mass spectrometry, *Earth Planet. Sci. Lett.*, *96*, 134-146, 1989.

Goldstein, S. J., M. T. Murrell, D. R. Janecky, J. R. Delaney, and D. A. Clague, Geochronology and petrogenesis of MORB from the Juan de Fuca and Gorda Ridges by ^{238}U-^{230}Th disequilibrium, *Earth Plant. Sci. Lett.*, *107*, 25-41, 1991. (Erratum, 109, 255-272, 1992)

Green, D. H., Composition of basaltic magmas as indicators of conditions of origin: application to oceanic volcanism, *Phil. Trans. Roy. Soc. Lond.*, *268*, 707-725, 1971.

Green, D. H., and A. E. Ringwood, The genesis of basaltic magmas, *Contrib. Mineral. Petrol.*, *15*, 103-190, 1967.

Grove, T. L., and W. B. Bryan, Fractionation of pyroxene-phyric MORB at low pressure: an experimental study, *Contrib. Mineral. Petrol.*, *84*, 293-309, 1983.

Grove, T. L., D. C. Gerlach, and T. W. Sando, Origin of calc-alkaline series lavas at Medicine Lake volcano by fractionation, assimilation and mixing, *Contrib. Mineral. Petrol.*, *80*, 160-182, 1982.

Grove, T. L., R. J. Kinzler, and W. B. Bryan, Natural and experimental phase relations of lavas from Serocki volcano, *Proc. ODP, Sci. Results*, *106/109*, 9-17, 1990.

Grover, J. E., D. H. Lindsley, and A. E. Bence, Experimental phase relations of olivine vitrophyres from breccia 14321: the temperature- and pressure-dependence of Fe-Mg partitioning for olivine and liquid in a highlands melt-rock, *Proc. Lunar Plant. Sci. Conf.*, *11*, 179-196, 1980.

Hamelin, B., B. Dupre, and C. J. Allegre, Lead-strontium isotopic variations along the East Pacific Rise and the Mid-Atlantic Ridge: a comparative study, *Earth Planet. Sci. Lett.*, *67*, 340-350, 1984.

Hanan, B. B., R. H. Kingsley, and J. G. Schilling, Pb isotope evidence in the south Atlantic for migrating ridge-hotspot interactions, *Nature*, *322*, 137-144, 1986.

Hanson, G. N., Evolution of the suboceanic mantle, *Geol. Soc. London*, *134*, 235-253, 1977.

Hanson, G. N., and C. H. Langmuir, Modelling of major elements in mantle systems using trace element approaches, *Geochim. Cosmochim. Acta*, *42*, 725-741, 1978.

Hart, S. R., and A. Zindler, In search of bulk Earth composition, *Chem. Geol.*, *57*, 247-267, 1986.

Hekinian, R., and G. Thompson, Comparative geochemistry of volcanics from rift valleys, transforms and aseismic ridges, *Contrib. Mineral. Petrol.*, *57*, 145-62, 1976.

Hekinian, R., and D. Walker, Diversity and spatial zonation of volcanic rocks from the East Pacific Rise near 21°N, *Contrib. Mineral. Petrol.*, *96*, 265, 1987.

Hekinian, R., G. Thompson, and D. Bideau, Axial and off-axial heterogeneity of basaltic rocks from the East Pacific Rise at 12°35'N-12°51'N and 11°26'N-11°30'N, *J. Geophys. Res.*, *94*, 17,437-17,463, 1989.

Hofmann, A. W., Chemical differentiation of the Earth: the relationship between mantle, continental crust, and oceanic crust, *Earth and Planet. Sci. Lett.*, *90*, 297-314, 1988.

Humphris, S. E., G. Thompson, J.-G. Schilling, and R. H. Kingsley, Petrological and geochemical variations along the Mid-Altantic Ridge between 46°S and 32°S: Influence of the Tristan de Cunha mantle plume, *Geochim. Cosmochim. Acta*, *49*, 1445-1464, 1985.

IHO/IOC/CHS. GEBCO - General Bathymetric Chart of the Ocean (5th Edition). Internat'l Hydrographic Organization/Intergovernmental Oceanographic Commission, Canadian Hydrographic Service, Ottawa, Canada, 74 p., 1984.

Irving, A. J., A review of experimental studies of crystal/liquid trace element partitioning, *Geochim. Cosmochim. Acta*, *42*, 743-770, 1978.

Jackson, H. R., and I. Reid, Crustal refraction and seismic reflection results from FRAM 1, *EOS*, *61*, 276, 1980.

Jakobsson, S. P., J. Jonsson, and F. Shido, Petrology of the western Reykjanes Peninsula, Iceland, *J. Petrol.*, *19*, 669-705, 1978.

Jagoutz, E., H. Palme, H. Baddenhausen, K. Blum, M. Cendales, G. Dreibus, S. Spettel, V. Lorenz, and H. Wanke, The abundances of major, minor and trace elements in the Earth's primitive mantle as derived from primitive ultramafic nodules, *Proc. Lunar Planet. Sci. Conf.*, *10th*, 2031-2050, 1979.

Jaques, A. L., and D. H. Green, Anhydrous melting of peridotite at 0-15 kb pressure and genesis of tholeiitic basalts, *Contrib. Mineral. Petrol.*, *73*, 287-310, 1980.

Jochum, K. P., A. W. Hofmann, E. Ito, H. M. Seufert, and W. M. White, K, U and Th in mid-ocean ridge basalt glasses and heat production, K/U and K/Rb in the mantle, *Nature*, *306*, 431-436, 1983.

Johnson, K. T. M., H. J. B. Dick, and N. Shimizu, Melting in the oceanic upper mantle: an ion microprobe study of diopsides in abyssal peridotites, *J. Geophys. Res.*, *95*, 2661-2678, 1990.

Johnson, K. T. M., and H. J. B. Dick, Open system melting and temporal and spatial variation of peridotite and basalt at the Atlantis II Fracture Zone, *J. Geophys. Res. 97*, 9219-9241, 1992.

Karson, J. A., et al., Along-axis variations in seafloor spreading in the MARK area, *Nature, 328*, 681-685, 1987.

Karsten, J. L., J. R. Delaney, J. M. Rhodes, and R. A. Liias, Spatial and temporal evolution of magmatic systems beneath the endeavor segment, Juan de Fuca Ridge: tectonic and petrologic constraints, *J. Geophys. Res., 95*, 19235-19256, 1990.

Keleman, P. B., Reaction between ultramafic rock and fractionating basaltic magma I. Phase relations, the origin of calc-alkaline magma series, and the formation of discordant dunite, *J. Petrol., 31*, 51-98, 1990.

Kinzler, R. J., and T. L. Grove, Primary magmas of mid-ocean ridge basalts. I. Experiments and methods, *J. Geophys. Res.*, in press, 1992*a*.

Kinzler, R. J., and T. L. Grove, Primary magmas of mid-ocean ridge basalts. II. Applications, *J. Geophys. Res.*, in press, 1992*b*.

Klein, E. M., Geochemistry of ocean ridge basalts: Mantle processes revealed by major element, trace element and isotopic variations, Ph.D. Thesis, Columbia University, New York, 1989.

Klein, E. M., and C. H. Langmuir, Global correlations of ocean ridge basalt chemistry with axial depth and crustal thickness, *J. Geophys. Res., 92*, 8089-8115, 1987.

Klein, E. M., and C. H. Langmuir, Local versus global variations in ocean ridge basalt composition: a reply, *J. Geophys. Res., 94*, 4241-4252, 1989.

Klein, E. M., C. H. Langmuir, and H. Staudigel, Geochemistry of basalts from the Southeast Indian Ridge, 115°N-138°E, *J. Geophys. Res., 96*, 2089-2108, 1991.

Klein, E. M., C. H. Langmuir, J. Bender, R. Batiza, and A. Zindler, Nd and Sr isotopic composition of volcanic glasses from the East Pacific Rise (8°N-12°N), *EOS, 67*, 1255, 1986.

Kuo, L. C., and R. J. Kirkpatrick, Pre-eruption history of phyric basalts from DSDP Legs 45 and 46: evidence from morphology and zoning patterns in plagioclase, *Contrib. Mineral. Petrol. 79*, 13-27, 1982.

Langmuir, C. H., A major element approach to basalts, PhD Thesis, SUNY, Stony Brook, NY, 1980.

Langmuir, C. H., Petrology data base, vol II and III, in *East Pacific Rise Data Synthesis Final Report*, edited by S. Tighe, Joint Oceanographic Institutes Inc., Washington, D.C., 1988.

Langmuir, C. H, Geochemical consequences of *in situ* crystallization, *Nature, 340*, 199-205, 1989.

Langmuir, C. H., Major element systematics of ocean island basalts, Caltech Plume Symposium, Pasadena, CA, 1991.

Langmuir, C. H., and J. F. Bender, The geochemistry of oceanic basalts in the vicinity of transform faults: Observations and implications, *Earth Planet. Sci. Lett., 69*, 107-127, 1984.

Langmuir, C. H., and G. N. Hanson, An evaluation of major element heterogeneity in the mantle sources of basalts, *Phil. Trans. R. Soc. Lond., 297*, 383-407, 1980.

Langmuir, C. H., and G. N. Hanson, Calculating mineral-melt equilibria with stoichiometry, mass balance, and single-component distribution coefficients, in *Thermodynamics of Minerals and Melts, Advances in Physical Geochemistry, Vol. 1*, edited by R. C. Newton, A. Navrotsky, and B. J. Wood, pp. 247-271, Springer-Verlag, New York, 1981.

Langmuir, C. H., J. F. Bender, and R. Batiza, Petrological and tectonic segmentation of the East Pacific Rise, 5°30'-14°30'N, *Nature, 332*, 422-429, 1986.

Langmuir, C. H., J. F. Bender, A. E. Bence, G. N. Hanson, and S. R.Taylor, Petrogenesis of basalts from the FAMOUS area: Mid-Atlantic Ridge, *Earth Planet. Sci. Lett., 36*, 133-156, 1977.

LaTourrett, T. Z., and D. S. Burnett, Experimental determination of U and Th partitioning between clinopyroxene and natural basaltic silicate liquid, *Earth Planet. Sci. Lett., 100*, 227-244, 1992.

LeDouaran, S. L., and J. Francheteau, Axial depth anomalies from 10 to 50° north along the Mid-Atlantic Ridge: correlation with other mantle properties, *Earth Planet. Sci. Lett., 54*, 29-47, 1981.

LeRoex, A. P., H. J. B Dick, and R. L. Fisher, Petrology and Geochemistry of MORB from 25°E to 46°E along the Southwest Indian Ridge: Evidence for contrasting styles of mantle enrichment, *J. Petrol., 30*, 947-986, 1989.

Lin, J., G. M. Purdy, H. Schouten, J.-C. Sempere, and C. Zervas, Evidence from gravity data for focused magmatic accretion along the Mid-Atlantic Ridge, *Nature, 344*, 627-632, 1990.

Longhi, J., Comparative liquidus equilibria of hypersthene-normative basalts at low pressure, *Am. Mineral., 76*, 785-800, 1991.

Longhi, J., Liquidus equilibria and solid solution in the system $CaAl_2Si_2O_8$-Mg_2SiO_4-$CaSiO_3$-SiO_2 at low pressure, *Am. J. Sci., 287*, 265-331, 1987.

Maaloe, S., and K. Aoki, The major element composition of the upper mantle estimated from the composition of lherzolites, *Contrib. Mineral. Petrol., 63*, 161-173, 1977.

Macdonald, K. C., P. J. Fox, L. J. Perram, M. F. Eisen, R. M. Haymon, S. P. Miller, S. M. Carbotte, M.-H. Cormier, and A. N. Shor, A new view of the mid-ocean ridge from the behavior of ridge-axis discontinuities, *Nature, 335*, 217-225, 1988.

McKenzie, D., The generation and compaction of partially molten rock, *J. Petrol., 25*, 713-765, 1984.

McKenzie, D., ^{230}Th-^{238}U disequilibrium and the melting processes beneath ridge axes, *Earth Planet. Sci. Lett., 72*, 149-157, 1985*a*.

McKenzie, D., The extraction of magma from the crust and mantle, *Earth Planet. Sci. Lett., 74*, 81-91, 1985*b*.

McKenzie, D., and M. J. Bickle, The volume and composition of melt generated by extension of the lithosphere, *J. Petrol., 29*, 625-679, 1988.

McKenzie, D., and R. K. O'Nions, Partial melt distributions from inversion of rare earth element concentrations, *J. Petrol., 32*, 1021-1091, 1991.

Melson, W. G., and T. O'Hearn, Basaltic glass erupted along the Mid-Atlantic Ridge between 0-37°N: Relationships between composition and latitude, in *Deep Sea Drilling Results in the Atlantic Ocean: Ocean Crust, Maurice Ewing Ser., vol. 2*, edited by M. Talwani, C. G. Harrison, and D. E. Hayes, pp. 249-261, AGU, Washington, D.C., 1979.

Melson, W. G., G. R. Byerly, J. A. Nelson, T. O'Hearn, T. L. Wright, and T. Vallier, A catalogue of the major element chemistry of abyssal volcanic glasses, *Smithsonian Contrib. Earth Sci., 19*, 31-60, 1977.

Michael, P. J., The concentration, behavior and storage of H_2O in the suboceanic upper mantle: Implications for mantle metasomatism, *Geochim. Cosmochim. Acta, 52*, 555-566, 1988.

Michael, P. J., and E. Bonatti, Peridotite composition from the North Atlantic: Regional and tectonic variations and implications for partial melting, *Earth Planet. Sci. Lett., 73*, 91-104, 1985.

Michael, P. J., R. L. Chase, and J. F. Allen, Petrologic and geologic variations along the southern Explorer Ridge, northeast Pacific Ocean, *J. Geophys. Res., 94*, 13,895-13,918, 1989.

Miyashiro, A., F. Shido, and M. Ewing, Diversity and origin of abyssal tholeiite from the Mid-Atlantic Ridge near 24° and 30°N, *Contrib. Mineral. Petrol., 23*, 38-52, 1969.

Morel, J. M., and R. Hekinian, Compositional variations of volcanics along segments of recent spreading ridges, *Contrib. Mineral. Petrol., 72*, 425-436, 1980.

Muir, I. D., C. E. Tilley, and J. H. Scoon, Basalts from the northern part of the rift zone of the Mid-Atlantic Ridge, *J. Petrol., 5*, 269-301, 1964.

Mysen, B. O., and I. Kushiro, Compositional variations of coexisting phases with degree of melting of peridotite in the upper mantle, *Am. Mineral., 62*, 843-856, 1977.

Natland, J. G., and W. G. Melson, Comparisons of basaltic glasses from the East Pacific Rise and Siqueiros Fracture Zone, near 9°N, *Init. Reports, DSDP, 54*, 705-23, 1980.

Natland, J. H., Partial melting of a lithologically heterogeneous mantle: inferences from crystallization histories of magnesian abyssal tholeiites from the Siqueiros Fracture Zone, in *Magmatism in the Ocean Basins*, edited by A. D. Saunders and M. J. Norry, pp. 41-70, Geol. Soc. Sp. Publ. No. 42, 1989.

Nelson, D. R., A. R. Chivas, B. W. Chappell, and M. T. McCulloch, Geochemical and isotopic systematics in carbonatites and implications for the evloution of ocean-island sources, *Geochim. Cosmochim. Acta, 52*, 1-17, 1988.

Neumann, E.-R., and J.-G. Schilling, Petrology of basalts from the Mohns-Knipovich Ridge: The Norwegian-Greenland Sea, *Contrib. Mineral. Petrol., 85*, 209-223, 1984.

Nielsen, R. L., EQUIL: A program for the modeling of low pressure differentiation processes in natural mafic magma bodies, *Computers and Geosci., 11*, 531-546, 1985.

Nielsen, R. L., Phase equilibria constraints on liquid lines of descent generated by paired assimilation and fractional crystallization: trace elements and Sr and Nd isotopes, *J. Geophys. Res., 94*, 787-794, 1989.

Nielsen, R. L., The theory and application of a model of open magma system processes, in *Modern Methods of Petrology: Reviews in Mineralogy, 24*, edited by J. Nicholls and J. K. Russel, pp. 65-106, Mineralogical Society of America, 1990.

Nielsen, R. L., and M. J. Drake, Pyroxene-melt equilibria, *Geochim. Cosmochim. Acta, 43*, 1259-1272, 1979.

Nielsen, R. L., and M. A. Dungan, Low pressure mineral-melt equilibria in natural anhydrous mafic systems, *Contrib. Mineral. Petrol., 84*, 310-326, 1983.

Niu, Y., and R. Batiza, An empirical method for calculating melt compositions produced beneath mid-ocean ridges: Application for axis and off-axis (seamounts) melting, *J. Geophys. Res., 96*, 21,753-21,777, 1991.

O'Hara, M. J. and R. E. Matthews, Geochemical evolution in an advancing periodically replenished, periodically tapped, continuously fractionated magma chamber, *J. Geol. Soc. London, 138*, 237-277, 1981.

O'Hara, M. J., Are any ocean floor basalts primary magma?, *Nature, 220*, 683-686, 1968*a*.

O'Hara, M. J., The bearing of phase equilibria studies in synthetic and natural systems on the origin and evolution of basic and ultrabasic rocks, *Earth Sci. Rev., 4*, 69-133, 1968*b*.

O'Hara, M. J., Geochemical evolution during fractional crystallization of a periodically refilled magma chamber, *Nature, 266*, 503-507, 1977.

O'Hara, M. J., Importance of the "shape" of the melting regime during partial melting of the mantle, *Nature, 314*, 58-62, 1985.

O'Hara, M. J., Primary magmas and the origin of basalts, *Scot. J. Geol., 1*, 19-40, 1965.

Onuma, N., H. Higuchi, H. Wakita, and H. Nagasawa, Trace element partition between two pyroxenes and the host lava, *Earth Planet. Sci. Lett., 5*, 47-51, 1968.

Oxburgh, E. R., Heat flow and magma genesis, in *Physics of Magmatic Processes*, edited by R. B. Hargraves, pp. 161-199, Princeton University Press, Princeton, NJ, 1980.

Perfit, M. R., D. J. Fornari, A. Malahoff, and R. W. Embley, Geochemical studies of abyssal lavas recovered by DSRV *Alvin* from Eastern Galapagos Rift, Inca Transform, and Ecuador Rift 3. Trace element abundances and petrogenesis, *J. Geophys. Res., 88*, 10,551-10,572, 1983.

Philpotts, J. A., The law of constant rejection, *Geochim. Cosmochim. Acta, 42*, 909-920, 1978.

Phipps Morgan, J., Melt migration beneath mid-ocean spreading centers, *Geophys. Res. Lett., 14*, 1238-1241, 1987.

Plank, T., and C. H. Langmuir, Effects of the melting regime on the composition of the oceanic crust, *J. Geophys. Res., 97*, 19,749-19770, 1992.

Presnall, D. C., The geometric analysis of partial fusion, *Am. J. Sci., 267*, 1178-1194, 1969.

Presnall, D. C., J. R. Dixon, T. H. O'Donnell, and S. A. Dixon, Generation of mid-ocean ridge tholeiites, *J. Petrol., 20*, 3-35, 1979.

Presnall, D. C. and J. D. Hoover, Composition and depth of origin of primary mid-ocean ridge basalts, *Contrib. Mineral. Petrol., 87*, 170-178, 1984.

Price, R. C., A. K. Kennedy, M. Riggs-Sneeringer, and F.A. Frey, Geochemistry of basalts from the Indian Ocean triple junction: Implications for the generation and evolution of Indian Ocean ridge basalts, *Earth Planet. Sci. Lett., 78*, 379-396, 1986.

Prinzhofer, A., E. Lewin, and C. J. Allégre, Stochastic melting of the marble cake mantle: evidence from local study of the East Pacific Rise at 12°50'N, *Earth Planet. Science Lett., 92*, 189-106, 1989.

Reid, I., and H. R. Jackson, Oceanic spreading rate and crustal thickness, *Marine Geophys. Res., 5*, 165-172, 1981.

Reinitz, I., and K. K. Turekian, $^{230}Th/^{238}U$ and $^{226}Ra/^{230}Th$ fractionation in young basaltic glasses from the East Pacific Rise, *Earth and Planet. Sci. Lett., 94*, 199-207, 1989.

Reisberg, L. C., The isotopic and geochemical systematics of the Ronda Ultramafic Complex of southern Spain, PhD. Thesis, Columbia University, N. Y., N. Y., 1988.

Reisberg, L., and A. Zindler, Extreme isotopic variations in the upper mantle: evidence from Ronda, *Earth Planet. Sci. Lett., 81*, 29-45, 1986/87.

Reynolds, J. R., C. H. Langmuir, J. F. Bender, K. A. Kastens, and W. B. F. Ryan, Spatial and temporal variability in the geochemistry of basalts from the East Pacific Rise, *Nature*, 359, 493-499, 1992.

Reynolds, J. R., and C. H. Langmuir, Petrological systematics of the spreading cell south of the Kane Transform: implications for models of ocean crust formation, *J. Geophys. Res.*, in revision, 1992.

Richter, F. M., Simple models for trace element fractionation during melt segregation, *Earth Planet. Sci. Lett., 77*, 333-344, 1986.

Roeder, P. L., Activity of iron and olivine solubility in basaltic melts, *Earth Planet. Sci. Lett., 23*, 397-410, 1974.

Roeder, P. L., and R. F. Emslie, Olivine-liquid equilibrium, *Contrib. Mineral. Petrol., 29*, 275-289, 1970.

Rubin, K. H., and J. D. Macdougall, ^{226}Ra excesses in mid-ocean ridge basalts and mantle melting, *Nature, 336*, 158-161, 1988.

Rubin, K. H., and J. D. Macdougall, Dating of neovolcanic MORB using ($^{226}Ra/^{230}Th$) disequilibrium, *Earth and Planet. Sci. Lett., 101*, 313-322, 1990.

Sack, R. O., D. Walker, and I. S. E. Carmichael, Experimental petrology of alkalic lavas: constraints on cotectics of multiple saturation in natural basic liquids, *Contrib. Mineral. Petrol., 96*, 1-23, 1987.

Salters, V. J. M., and N. Shimizu, World-wide occurrence of HFSE-depleted mantle, *Geochim. Cosmochim. Acta, 52*, 2177-2182, 1988.

Salters, V. J. M., and S. R. Hart, The hafnium paradox and the role of garnet in the source of mid-ocean ridge basalts, *Nature, 342*, 420-422, 1989.

Sarda, P., and D. Graham, Mid-ocean ridge popping rocks: implication for degassing at ridges, *Earth Planet. Sci. Lett., 97*, 268, 1990.

Schilling, J-G., Rare earth fractionation in Hawaii volcanic rocks, Ph.D. Thesis, Mass. Inst. Technology, 1966.

Schilling, J-G., Sea-floor evolution: rare-earth evidence, *Phil. Trans Roy. Soc. London, A268*, 663-706, 1971.

Schilling, J-G., Iceland mantle plume: Geochemical study of Reykjanes Ridge, *Nature, 242*, 565-571, 1973.

Schilling, J-G., Azores mantle blob: Rare earth evidence, *Earth Planet. Sci. Lett., 25*, 103-115, 1975.

Schilling, J-G., Upper mantle heterogeneities and dynamics, *Nature, 314*, 62-67, 1985.

Schilling, J-G., Geochemical and isotopic variation along the Mid-Atlantic Ridge axis from 79°N to 0°N, in *The Geology of North America, Vol. M, The Western North Atlantic Region*, edited by P. R. Vogt and B. E. Tucholke, Geological Society of America, 1986.

Schilling, J-G., and H. Sigurdsson, Thermal minima along the axis of Mid-Atlantic Ridge, *Nature, 282*, 370-5, 1979.

Schilling, J-G, R. N. Anderson, and P. Vogt, Rare earth, Fe and Ti variations along the Galapagos spreading centre, and their relationship to the Galapagos mantle plume, *Nature, 261*, 108-113, 1976.

Schilling, J-G., M. B. Bergeron, and R. Evans, Halogens in the mantle beneath the N. Atlantic, *Phil. Trans. Royal Soc. London, A297*, 147-178, 1980.

Schilling, J-G., R. H. Kingsley and J. D. Devine, Galapagos hot spot-spreading center system 1. Spatial petrological and geochemical variations (83°W-101°W), *J. Geophys. Res.*, *87*, 5593-5610, 1982.

Schilling, J.-G., H. Sigurdsson, A. N. Davis, and R. N. Hey, Easter microplate evolution, *Nature*, *317*, 325-331, 1985.

Schilling, J-G, M. Zajak, R. Evans, T. Johnson, W. White, J. D. Devine and R. Kingsley, Petrologic and geochemical variations along the Mid-Atlantic Ridge from 29°N to 73°N, *Am. J. Sci.*, *283*, 510-586, 1983.

Schouten, H., K. D. Klitgord, and J. A. Whitehead, Segmentation of mid-ocean ridges, *Nature*, *317*, 225-229, 1985.

Scott, D. R., and D. J. Stevenson, A self consistent model of melting, magma migration and buoyancy driven circulation beneath mid-ocean ridges, *J. Geophys. Res.*, *94*, 2973-2988, 1989.

Shaw, D. M., Trace element fractionation during anatexis, *Geochim. Cosmochim. Acta*, *34*, 237-243, 1970.

Shibata, T., G. Thompson, and F. A. Frey, Tholeiitic and alkali basalts from the mid-Atlantic Ridge at 43°N, *Contrib. Mineral. Petrol.*, *70*, 127-141, 1979.

Shirey, S. B., J. F. Bender, and C. H. Langmuir, Three-component isotopic heterogeneity near the Oceanographer transform, Mid-Atlantic Ridge, *Nature*, *325*, 217-223, 1987.

Sigurdsson, H., First-order major element variation in basaltic glasses from the Mid-Atlantic Ridge: 29°N to 73°N, *J. Geophys. Res.*, *86*, 9483-9502, 1981.

Sinton, J. M. and R. S. Detrick, Mid-ocean ridge magma chambers, *J. Geophys, Res.*, *97*, 197-216, 1992.

Sinton, J. M., D. S. Wilson, D. M. Christie, R. N. Hey, and J. R. Delaney, Petrologic consequences of rift propagation on oceanic spreading centers, *Earth Planet. Sci. Lett.*, *62*, 193-207, 1983.

Sinton, J. M., S. M. Smaglik, and J. J. Mahoney, Magmatic processes at superfast spreading mid-ocean ridges: glass compositional variations along the East Pacific Rise 13°-23°S, *J. Geophys. Res.*, *96*, 6133-6155, 1991.

Sleep, N. H., Segregation of magma from a mostly crystalline mush, *Geol. Soc. Am. Bull.*, *85*, 1225-1232, 1974.

Small, C., and D. T. Sandwell, An abrupt change in ridge axis gravity with spreading rate, *J. Geophys. Res.*, *94*, 17383-17392, 1989.

Sparks, R. S. J., *In situ* differentiation, *Nature*, *340*, 187, 1989.

Sparks, R. S. J., H. E. Huppert, and J. S. Turner, Fluid dynamics of involved magma chambers, *Phil. Trans. R. Soc., London*, *A310*, 511-534, 1984.

Spiegelman, M., and D. McKenzie, Simple 2-D models for melt extraction at mid-ocean ridges and island arcs, *Earth Planet. Sci. Lett.*, *83*, 137-152, 1987.

Stakes, D. S., J. W. Shervais, and C. A. Hopson, The volcanic-tectonic cycle of the FAMOUS and AMAR Valleys, Mid-Atlantic Ridge (36°47'N): Evidence from basalt glass and phenocryst compositional variations for a steady state magma chamber beneath the valley midsections, AMAR 3, *J. Geophys. Res.*, *89*, 6995-7028, 1984.

Stolper, E., A phase diagram for mid-ocean ridge basalts: Preliminary results and implications for petrogenesis, *Contrib. Mineral. Petrol.*, *74*, 13-27, 1980.

Stroup, J. B., and P. J. Fox, Geologic investigations in the Cayman Trough: Evidence for thin oceanic crust along Mid-Cayman Rise, *J. Geol.*, *89*, 395-420, 1981.

Sun, S.-S., Lead isotopic study of young volcanic rocks from mid-ocean ridges, ocean islands and island arcs, *Phil. Trans. R. Soc. London*, *A297*, 409-446, 1980.

Sun, S.-S. and G. N. Hanson, Origin of Ross Island basanitoids and limitations upon the heterogeneity of mantle sources for alkali basalts and nephelinites, *Contrib. Mineral. Petrol.* *52*, 77-106, 1975.

Sun, S.-S., R. W. Nesbitt, and A. Y. Sharaskin, Geochemical characteristics of mid-ocean ridge basalts, *Earth Planet. Lett.*, *44*, 119-138, 1979.

Sun, S.-S., M. Tatsumoto, and J.-G. Schilling, Mantle plume mixing along the Reykjanes Ridge axis: Lead isotope evidence, *Science*, *190*, 143-147, 1975.

Takahashi, E., and I. Kushiro, Melting of a dry peridotite at high pressures and basalt magma genesis, *Am. Mineral.*, *68*, 859-879, 1983.

Takahashi, E., Melting of a dry peridotite KLB-1 up to 14 GPa: implication on the origin of peridotitic upper mantle, *J. Geophys. Res.*, *91*, 9367-9382, 1986.

Thompson, G., W. B. Bryan, and W .G. Melson, Geological and geophysical investigation of the Mid-Cayman Rise spreading center: Geochemical variation and petrogenesis of basalt glasses, *J. Geol.*, *88*, 41-55, 1980.

Thompson, G., W. B. Bryan, R. Ballard, K. Hamuro, and W. G. Melson, Axial processes along a segment of the East Pacific Rise, 10°-12°N, *Nature*, *318*, 429-433, 1985.

Thompson, R. N., M. A. Morrison, A. P. Dickin, and G. L. Hendry, Continental flood basalts Arachnids rule OK?, in *Continental Flood Basalts and Mantle Xenoliths*, edited by C. J. Hawkesworth and M. J. Norry, pp. 158-185, Shiva, Cheshire, 1983.

Tormey, D. R., T. L. Grove, and W. B. Bryan, Experimental petrology of normal MORB near the Kane Fracture Zone, 22°-25°N, mid-Atlantic ridge, *Contrib. Mineral. Petrol.*, *94*, 121-139, 1987.

Verma, S. P., and J.-G. Schilling, Galapagos hot spot center system, 2, $^{87}Sr/^{86}Sr$ and large ion lithophile element variations (85°W-101°W), *J. Geophys. Res.*, *87*, 10,838-10,856, 1982.

Vogt, P. R., Portrait of a plate boundary: the Mid-Atlantic Ridge axis from the equator to Siberia, in *The Western North Atlantic Region, The Geology of North America, vol. M*, edited by P. R. Vogt and B. E. Tucholke, plate 8A, Geol. Soc. Am., Boulder, CO., 1986.

Watson, S., and D. McKenzie, Melt generation by plumes: A study of Hawaiian volcanism, *J. Petrol.*, *32*, 501-537, 1991.

Walker, D., T. Shibata, and S. E. DeLong, Abyssal tholeiites from the Oceanographer Fraction Zone. II. Phase equilibria and mixing, *Contrib. Mineral. Petrol.*, *70*, 111-125, 1979.

Weaver, J., and C. Langmuir, Calculation of phase equilibrium in mineral-melt systems, *Computers and Geosciences*, *16*, 1-19, 1990.

White, W. M., and W. B. Bryan, Sr-isotope, K, Rb, Cs, Sr, Ba, and rare-earth geochemistry of basalts from the FAMOUS area, *Geol. Soc. Am. Bull.*, *88*, 571-576, 1977.

White, W. M., A. W. Hofmann, and H. Puchelt, Isotope geochemistry of Pacific mid-ocean ridge basalt, *J. Geophys. Res.*, *92*, 4881-4893, 1987.

Whitehead, J. A., H. J. A. Dick, and H. Schouten, A mechanism for magmatic accretion under spreading centres, *Nature*, *312*, 146-148, 1984.

Williams, R. W., and J. B. Gill, Effects of partial melting on the uranium decay series, *Geochim. Cosmochim. Acta*, *53*, 1607-1919, 1989.

Wood, D. A., A variably veined suboceanic upper mantle -- Genetic significance for mid-ocean ridge basalts from geochemical evidence, *Geology*, *7*, 499-503, 1979.

Wyllie, P. J., Discussion of recent papers on carbonated peridotite, bearing on mantle metasomatism and magmatism, *Earth Planet. Sci. Lett.*, *82*, 391-397, 1987.

Yoder, H. S., *Generation of Basaltic Magma*, pp. 209, National Academy of Sciences, Washington, 1976.

Zindler, A., and S. Hart, Chemical Geodynamics, *Annu. Rev. Earth Planet. Sci.*, *14*, 493-571, 1986.

Zindler, A., H. Staudigel, and R. Batiza, Isotope and trace element geochemistry of young Pacific seamounts: Implications for the scale of upper mantle heterogeneity, *Earth Planet. Sci. Lett.*, *70*, 175-195, 1984.

C. Langmuir and T. Plank, Lamont-Doherty Geological Observatory of Columbia University, Palisades, NY 10964

E. Klein, Department of Geology, Duke University, Durham, NC 27706

Fractionation of Mid-Ocean Ridge Basalt (MORB)

TIMOTHY L. GROVE AND ROSAMOND J. KINZLER

Department of Earth, Atmospheric and Planetary Sciences, Massachusetts Institute of Technology, Cambridge

WILFRED B. BRYAN

Department of Geology and Geophysics, Woods Hole Oceanographic Institution, Woods Hole, MA

This paper presents new experimental data on MORB lavas obtained over the pressure range 0.001 to 10 kbar, discusses the evidence for the operation of fractional crystallization in MORB suites and offers a method for calculating the fractional crystallization history experienced by a MORB. Two important saturation boundaries that control MORB compositional evolution during fractional crystallization are the olivine (oliv)+plagioclase (plag) and oliv+plag+high-Ca pyroxene (augite) boundaries. A method is presented for locating these two saturation boundaries as functions of liquid composition and pressure of crystallization. Expressions also are presented for predicting the compositions of oliv, plag and augite. With these tools the effect of differentiation processes on MORB magma evolution can be deduced over the pressure range of 0.001 to 10 kbar. The compositional variations exhibited by suites of MORB from the mid-Atlantic near the Kane Fracture Zone at 23°N, the AMAR region at 36°47'N, the Reykjanes Peninsula in Iceland and the mid-Cayman rise in the Caribbean are consistent with fractional crystallization at pressures of 3 to 6 kbar (oceanic mantle fractionation). At 10-12°N on the East Pacific Rise MORB compositional variation is consistent with fractional crystallization at pressures of 0.001 to 2 kbar (ocean crustal fractionation). The occurrence of fractional crystallization in the oceanic upper mantle indicates cooler upper mantle temperatures beneath some spreading centers, which allow magmas derived by partial melting of mantle peridotite to crystallize and differentiate as they rise. A part of the positive correlation between Na_2O and FeO at 8 wt % MgO observed in spatially associated MORB suites [Klein and Langmuir, 1989] can be ascribed to fractional crystallization of MORB magmas over a depth range of 0 to 18 km. The compositional variation in MORB suites and the mineralogical evidence preserved in plagioclase phenocrysts indicate that MORB magma production and evolution involves fractionation of some mantle-derived partial melts prior to aggregation (or mixing).

INTRODUCTION

Fractionation of MORB magma is caused by crystallization (or freezing) induced when basaltic liquid is cooled by the mantle and crust encountered during its ascent from the source region to the active spreading center. Fractionation is one of several processes that may influence the compositional characteristics of a MORB magma during its evolutionary history. A simple and often-adopted thought experiment is to divide the history of MORB production and evolution into two separate sets of processes: 1) melt generation in the mantle source and 2) melt modification after separation from source. The concept is illustrated in Figure 1*a*. Magmas are produced in the MORB source mantle during adiabatic decompression by a near-fractional melting process. Melts are removed from their source after small amounts of melting (<1%) and aggregated (or mixed) with other partial melts at shallower depths [Ahern and Turcotte, 1979; McKenzie, 1984; McKenzie, 1985; Johnson et al., 1990]. The mantle-derived aggregated magma is then modified at even shallower depths by processes which include fractional crystallization, assimilation of, or reaction with, adjacent wallrock, Soret diffusion and magma mixing. Note that the box representing fractionation and mixing in Figure 1*a* represents a depth range over which these processes occur, and is not meant to represent a magma chamber.

Mantle Flow and Melt Generation at Mid-Ocean Ridges
Geophysical Monograph 71
Copyright 1992 American Geophysical Union

Melt Generation Processes

MORB magmas generated by partial melting of mantle peridotite during adiabatic decompression that are separated from their source and not further modified upon ascent to their site of eruption have been called primary magmas. Such magmas are almost never sampled at mid-ocean ridges. The operation of fractional melting in the mantle source of MORB implies that any hypothetical primary magma is likely to be a mix of many melt parcels produced over a range of depths from mantle peridotite source regions that change continuously in composition as fractional melting depletes the mantle source residue. Therefore, a hypothetical primary magma might be viewed as a mix (or aggregate) of magmas produced over a range of depths from a range of source compositions. Petrologists have used a number of criteria to identify primary magmas. A common measure is to select the lavas with the highest MgO content. These lavas, called primitive, are closest in composition to the primary magmas in equilibrium with the MgO-rich minerals in the mantle source. Elthon [1990] and Presnall and Hoover [1987] have assembled subsets of the most primitive compositions from the thousands of available chemical analyses of MORB. These compilations eliminate almost all the sampled MORBs. Only a few 100 remain in these selected data sets, and most of these show evidence of fractional crystallization (*Mg #* [Mg/(Mg+Fe2+)] < 0.70). Therefore, virtually all MORB magmas erupted on the ocean floor have experienced modification subsequent to separation from their source region.

Melt Modification Processes

After primary magmas are blended in the zone of aggregation the second set of processes in the thought experiment operates to modify these partial melts. Petrologists and geochemists traditionally view the process of modification as one in which a lava composition (or some subset of lava compositions) may be identified as a parent magma. The parent magma is one that has been least affected by melt modification processes and may be closest to a primary magma. Crystallization induced by cooler surroundings modifies the composition of this parent magma to produce derivative magmas. If the derivative magmas are sampled sequentially during the crystallization process by a series of eruptions from the magma reservoir, the resulting suite of compositions records the modification process. If fractional crystallization is the only modification process that operates on a parent magma, the compositional variation of the derivative suite provides a record of the conditions. The mixing (or aggregation) process discussed above for primary magmas may also play a role during the process of modification by crystallization. Dungan and Rhodes [1978], Walker et al. [1979] and others have identified evidence preserved in mineral assemblages and lava compositional variations that indicates the operation of mixing processes. O'Hara and Mathews [1981], O'Hara [1985] and Langmuir [1989] have constructed models that anticipate the chemical evolution of MORB magmas when mixing and fractional crystallization processes combine. Other processes have been put forth as potentially important modifiers of MORB magma chemistry. The Soret process [Walker and DeLong, 1982] involves diffusive exchange of silicate melt species and requires a sustained thermal gradient, since silicate melt diffusion is significantly slower than thermal diffusion. Assimilation of country rock by crystallizing MORB magma is another potential modifying process. Keleman et al. [1990] discuss the geochemical signatures expected from assimilation of peridotite country rock by MORB magma and conclude that extensive dissolution of peridotite by rapidly ascending

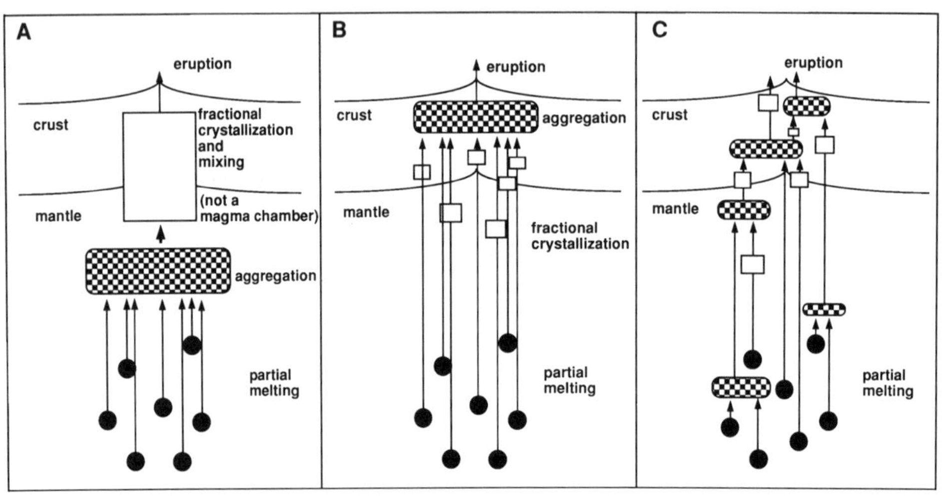

Fig. 1. Possible processes for MORB magma production and evolution. *a*) A view of MORB genesis in which melting and differentiation events are separated by an aggregation step. *b*) An extreme view in which fractional melts are produced during adiabatic ascent, the melts are segregated, moved to shallower, cooler mantle, where they are modified by differentiation before aggregation/mixing beneath the spreading center. *c*) Our preferred view of MORB genesis in which fractional melting, aggregation and differentiation occur throughout the upper mantle and the depth ranges of aggregation/mixing and differentiation overlap.

MORB magma is not likely to occur. Despite the potential for melt modification by mixing, Soret or assimilative processes, fractional crystallization is a dominant process, and is the focus of discussion in this paper.

Alternative Models for Melt Production and Modification

The separation of melt production and melt modification processes as suggested by Figure 1*a* may represent an artificial oversimplification of a more complex process. In part, the view of magma evolution expressed in Figure 1*a* arises from an older assumption about the process that led to MORB production in mantle source regions. In this older view of melt generation, primary MORB magma was produced by batch melting processes, and the debate concerning MORB genesis was whether MORB magma originated as a large percentage batch melt (>30%) at depths from 60 to 80 km [O'Hara, 1968; Stolper, 1980] or as a smaller degree melt (15 to 20 %) at shallower depths from 27 to 33 km [Presnall et al., 1978; Fujii and Scarfe, 1985]. If the melt production mechanism is a batch process, then it is logical to separate it from the melt modification process. However, if melting is fractional, then there is the possibility that melt production and modification are not separated by a single zone of aggregation and that modification processes may occur before aggregation. Figure 1*b* shows an alternative thought experiment for the production and modification of MORB magma beneath a spreading ridge. This model may be viewed as an endmember case in which fractional melts experience modification processes by fractionation prior to the aggregation process. The mantle thermal structure will determine whether melting or crystallization occurs. In the convecting mantle beneath ocean spreading centers there are regions where thermal conditions result in melting, and regions in which freezing or crystallization may occur. The aggregated magma erupted at the ocean spreading center in this model (Figure 1*b*) is composed of a collection of many parcels of melt that have experienced individual histories of melting and modification prior to aggregation. In this model fractional melting occurs over a range of depths by adiabatic decompression and melt modification occurs over a range of shallower depths. The transition from melting to fractionation occurs in the mantle, where the temperature of convecting mantle drops below the solidus, and the zone of fractionation extends into the oceanic crust.

A final model (Figure 1*c*) allows for additional complexity. In this model aggregation occurs at many levels during the melt production and modification process. Melt aggregation occurs in the region of melt production, and in the region of melt modification. As in the previous model melt modification by fractional crystallization occurs when mantle ambient temperature drops below the melt's liquidus.

In this paper we examine the compositional variations and the phenocryst assemblages of MORB lavas for evidence of the operations of fractionation and aggregation. We will show that compositional variability present in MORB suites indicates that fractional crystallization occurred over a range of depths from ≈18 to 0 km beneath the spreading center. We also show that compositional variability in plagioclase in some MORB suites is most easily reconciled, if melt production and modification occur before aggregation. We conclude that the third model discussed above most closely approximates the way in which melt modification by fractional crystallization fits into the overall evolution of MORB magmas.

EXPERIMENTAL AND ANALYTICAL METHODS

Experimental Procedure

The experiments used natural, glassy MORBs, or mixtures of MORB+oliv, as starting materials (see Grove and Bryan [1983], Tormey et al. [1987], and Grove et al. [1990], for descriptions of materials). Chips of the freshest portion of the MORB sample were selected, and ground to a powder in a SPEX shatterbox. Table 1 summarizes the compositions chosen for the experiments. Experimental conditions are reported in Table 2. Melting experiments at 0.001 kbar were carried out in a Deltech DT31VT quenching furnace at controlled oxygen fugacity conditions using the techniques described by Tormey et al. [1987].

Experiments at 2 kbar were carried out in an Ar gas-medium using a rapid quench cold seal gas pressure vessel constructed from a high-strength molybdenum alloy: zirconium-hafnium carbide-molybdenum (ZHM). The design of this vessel is

TABLE 1. Starting Compositions Used in Experiments Reported in Tables 2 and 3

	KN-54-2 -2-1[a]	ALV-1690-20[b]	AII-96 -18-1[a]	ALV-528 -1-1[a]	70-002[b]	ALV[b]-2004-3-1
n	4	106	4	8	368	7
SiO_2	51.7(6)[c]	49.9(3)	50.5(2)	48.8(1)	49.6(1)	49.6(2)
TiO_2	1.62(3)	1.71(3)	1.45(7)	0.80(2)	1.20(2)	1.20(2)
Al_2O_3	16.3(1)	15.5(1)	15.3(2)	15.7(1)	15.8(2)	16.3(1)
FeO	8.70(16)	9.92(8)	9.26(3)	9.26(16)	8.98(7)	8.65(9)
MgO	7.86(6)	7.68(9)	8.56(3)	10.2(7)	8.66(6)	9.13(5)
MnO	0.19(2)	0.17(5)	0.17(3)	0.19(3)	0.17(3)	0.09(5)
CaO	10.2(1)	10.8(2)	11.2(5)	11.9(1)	11.9(1)	11.7(2)
K_2O	0.25(2)	0.14(1)	0.12(1)	0.10(1)	0.10(1)	0.10(1)
Na_2O	3.55(5)	3.13(6)	3.13(16)	2.09(8)	2.67(6)	2.66(14)
P_2O_5	-[d]	0.18(7)	-	0.03(2)	0.12(4)	0.04(1)

References: KN-54-2-2-1, *Thompson et al.* [1981]; ALV-1690-20, *Grove et al.* [1990]; AII-96-18-1, *Tormey et al.* [1987]; ALV-528-1-1, *Grove and Bryan* [1983]; 70-002, *Thompson et al.* [1990]; and ALV-2004-3-1, R. J. Kinzler, W. B. Bryan,and T. L. Grove, manuscript in preparation. For analyses of San Carlos olivine see *Grove and Juster* [1989].

[a] Electron microprobe analysis of fused whole rock, KN-54-2-2-1 and AII96-18-1 were analysed with MIT MAC-5; ALV-528-1-1 was analysed with MIT JEOL 733.

[b] Electron microprobe analysis of natural glass; all analyses were carried out on MIT JEOL 733.

[c] Units in parentheses indicate standard deviation of element analyzed. Thus 51.7(6) should be read 50.7 ±0.6 wt.% SiO_2.

[d] Dash signifies that element was not analyzed.

similar to that pictured in Williams [1968]. The vessel is 1" in outer diameter by 12" long. A 0.25" by 11" inner hole, bored smooth and rounded at its base contains the high pressure gas. A cold seal closure at the top of the vessel serves as a pressure seal and is cooled by a recirculating water bath. The vessel is enclosed in an inconel 600 outer sheath and bathed in Ar gas at a small positive pressure. The end of the vessel containing the experiment is positioned in a Deltech furnace so that a temperature gradient of less than 2 ^{o}C is present across the sample. Temperature is monitored by a $Pt-Pt_{90}Rh_{10}$ thermocouple placed at the position of the sample on the outside of the pressure vessel. Thermocouples were calibrated using the melting points of NaCl, Au and Pd on the IPTS 1968 temperature scale. The temperature difference between the outside and inside was determined by direct measurement, and experimental temperatures are corrected for this difference. Experiments were terminated by removing the vessel from the furnace and inverting it so that the experimental charge fell to the water-cooled end of the vessel and quenched. Samples were placed in 0.188" diameter by 0.375" long by 0.007" wall Pt capsules that were welded shut. We developed the following procedure to ensure that the gas pressure was transmitted to the sample. A thin-walled (0.002") graphite capsule was loaded with about 5 mg of starting material and powdered graphite was packed in the open end. A second "getter" capsule loaded with wustite ensured a reducing atmosphere. Both were placed in the Pt capsule and surrounded by powdered graphite. The oxidation state of the experimental assembly was unknown, but was probably between the magnetite-wustite and iron-wustite buffers, because no Fe metal blebs were observed in the silicate charge, and wustite was the only phase present in the "getter" capsule at the end of the experiment. At the end of the experiment the thin-walled graphite capsules were always flattened, indicating that pressure was transmitted to the sample in the container.

Experiments at 8 and 10 kbar were carried out using a 1/2" solid-medium piston cylinder apparatus [Boyd and England, 1960]. The temperature was monitored and controlled using $Pt-Pt_{90}Rh_{10}$ thermocouples. The pressure medium consisted of an outer cell of NaCl and an inner cell of Pyrex. This pressure cell was calibrated against the melting point of gold at 10 kbar [Akella and Kennedy, 1971] and the observed pressure correction was 0-1 kbar. This pressure correction lies within the uncertainties of the calibration and the reported pressures do not include a friction correction. Approximately 10 mg of starting material was packed into a graphite container, closed with a graphite lid and welded into an outer Pt capsule. The graphite isolates the sample from the Pt outer capsule and prevents Fe loss. The Pt capsule keeps the redox conditions constant near the $C-CO_2-CO$ buffer. The 3 mm sample assembly was placed in a ceramic sleeve and centered in the hot spot of a graphite furnace using crushable alumina (An900) spacers. The thermocouple was located 3.5 mm above the center of the furnace, or 1.75 mm above the top of the sample capsule, and separated from the capsule by a 1.75 mm wafer of An900 ceramic. The temperature difference between the thermocouple position and the center of the sample was measured to be 45 ^{o}C. Temperatures reported in Table 2 are corrected for this temperature difference. Failure to reproduce the sample geometry from experiment to experiment results in inconsistent temperature measurements. Temperatures were reproduced in the assembly described above to within +/- 15 ^{o}C.

Analytical Methods

Experimental products were analyzed with the MIT 4-spectrometer JEOL 733 Superprobe, using an accelerating voltage of 15 keV and a beam current of 10 nA for all phases. Data were reduced using Bence and Albee [1968] matrix corrections with the modifications of Albee and Ray [1970]. A beam spot size of 10 microns was used for glass analyses, a 1 micron beam spot size was used for the crystalline products. Major element compositions of all the experimental products are presented in Table 3.

Attainment of Equilibrium and Conservation of Mass

The experiments described here were all synthesis experiments. The starting material was brought to the temperature and pressure of the experiment, held there for a specified time and terminated by quenching. We did not carry out reversal experiments for all possible changes in composition, and therefore can not claim to have proven equilibrium with respect to all possible exchange components. Our goal was to produce liquids and coexisting solid phases that are close approximations of equilibrium phase assemblages and that can be used to understand how MORB melts crystallize and differentiate in the mantle and crust. Several criteria indicate that our experiments provide a sufficiently close approach to equilibrium to be relevant to MORB crystallization. These criteria are: 1) maintenance of constant silicate sample composition, 2) homogeneity of product phases (with the exception of augite, see below), 3) achievement of consistent exchange of major elements between minerals and melts and 4) generation of experimental products that did not change in composition with continued time.

A materials balance calculation relating the compositions of phases (Table 3) to the initial starting bulk composition (using the method of unweighted least squares; Bryan et al. [1969]) (reported in Table 2) was used to evaluate whether the bulk composition of the sample remained constant during the experiment. The most common change in sample composition is the exchange of Fe between the charge and the container. For an experiment to be successful, the composition estimated by the mass balance of the phases present must approximate closely the bulk composition of the starting basalt or basalt-mineral mix. Only experiments with < 5 % relative Fe loss were considered acceptable.

The homogeneity of experimental products can be estimated from the reported standard deviations on the phase compositions (Table 3). In the case of the elevated pressure experiments homogeneous glass and olivine were generally

TABLE 2. Experimental Conditions and Phase Assemblages (Compositions are Reported in Table 3)

Experiment[a]	P (kbar)	T (°C)	duration (hours)	phases present[b]	phase proportions[c]	ΣR^2 [d]	%Fe loss[e]
KN-54-2-2-1							
KN-6	0.001	1210	3	gl			
KN-9	0.001	1207	4	ol,sp,gl			
KN-8	0.001	1205	7	ol,pl,gl			
KN-2	0.001	1196	9	ol,pl,gl			
KN-5	0.001	1185	7	ol,pl,gl			
KN-3	0.001	1177	5	ol,pl,gl			
KN-7	0.001	1170	13	ol,pl,gl			
KN-4	0.001	1161	22	ol,pl,cpx,gl			
ALV-1690-20							
ALV2-4	2	1176	9	ol,pl,gl	0.03:0.11:0.87	0.13	-
ALV2-5	2	1163	22	ol,pl,cpx,gl	0.05:0.21:0.06:0.69	0.08	-
ALV2-8	2	1152	36	ol,pl,cpx,gl	0.06:0.20:0.05:0.70	0.04	1.0
ALV2-10	2	1148	26	ol,pl,cpx,gl	0.05:0.19:0.04:0.71	0.09	-
ALV2-7	2	1145	36	ol,pl,cpx,gl	0.08:0.30:0.08:0.53	0.28	4.0
ALV2-9	2	1126	44	ol,pl,cpx,gl	0.07:0.33:0.12:0.48	0.27	2.3
AII-96-18-1							
H51	8	1240	5	ol,pl,cpx,gl	tr[f]:tr:0.06:0.93	0.05	-
H50	8	1230	14	ol,pl,cpx,gl	tr:0.12:0.18:0.71	0.02	-
H60	8	1215	10	ol,pl,cpx,gl	tr:0.15:0.19:0.66	0.01	-
ALV-528-1-1							
H62	8	1265	5	ol,pl,cpx,gl	0.03:0.04:0.06:0.87	0.25	1.0
H57	8	1250	6	ol,pl,cpx,gl	0.06:0.09:0.05:0.82	0.15	1.0
H32	8	1235	3	ol,pl,cpx,gl	0.05:0.14:0.11:0.70	0.01	-
H35	8	1225	8	ol,pl,cpx,gl	0.07:0.21:0.17:0.55		
70-002							
H59	8	1240	6	ol,pl,cpx,gl	tr:tr:tr:0.99		
H38	8	1225	10	ol,pl,cpx,gl	0.04:0.24:0.14:0.58	0.80	4.5
H58	8	1210	12	ol,pl,cpx,gl	0.07:0.36:0.21:0.34	0.01	-
90 % ALV-2004-3-1 + 10 % San Carlos Olivine (by wt.)							
H258	10	1250	11	ol,pl,cpx,gl	0.10:0.04:0.17:0.69	0.68	3.2
H137	10	1235	7	ol,pl,cpx,gl	0.09:0.09:0.20:0.62	0.00	-

[a] Starting compositions used for experiment. See Table 1 for bulk compositions.
[b] Abbreviations: ol = olivine, pl = plagioclase, cpx = clinopyroxene, gl = glass, sp = spinel.
[c] Experimental phase proportions estimated in wt.% by materials balance calculation.
[d] Sum of squared residuals for the materials balance calculation.
[e] Percent relative iron loss estimated with the materials balance calculation. - indicates that iron loss was < 0.5% relative.
[f] Phase present in trace proportion, not resolvable by materials balance calculation.

produced. Plagioclase grew as homogeneous un-zoned crystals in experiments that contained > 70 % melt. In highly crystalline experiments the interiors of plagioclase crystals were compositionally heterogeneous, but a population of relatively homogeneous rim compositions could be characterized. Augites were compositionally the most heterogeneous, as reflected in the higher standard deviations in reported averages of compositions. The exchange distribution coefficient (K_D) for Fe-Mg exchange ($K_D^{Fe\text{-}Mg}$ = $[X_{Fe}^{xtal}][X_{Mg}^{liq}]/\{[X_{Mg}^{xtal}][X_{Fe}^{liq}]\}$) between mineral and melt are comparable to those observed in other similar experimental studies. The average oliv $K_D^{Fe\text{-}Mg}$ from the 12 high pressure assemblages reported here is 0.29 +/- 0.02 (one sigma standard error). The average augite $K_D^{Fe\text{-}Mg}$ from the 12 high pressure assemblages reported here is 0.27 +/- 0.02 (one sigma standard error). Plag / liquid Ca-Na exchange distribution coefficients ($K_D^{Ca\text{-}Na}$ = $[X_{An}^{xtal}][X_{SiO2}^{liq}][X_{NaAlO2}^{liq}]/\{[X_{Ab}^{xtal}][X_{CaAl2O4}^{liq}]\}$,

TABLE 3. Electron Microprobe Analyses of Products of Experiments Reported in Table 2

	number	SiO_2	TiO_2	Al_2O_3	Cr_2O_3	FeO	MgO	MnO	CaO	K_2O	Na_2O	P_2O_5	Total
KN-6 gl	5	51.4(6)[a]	1.55(7)	16.1(3)	0.06(2)	8.55(8)	7.96(10)	0.15(4)	10.2(1)	0.28(2)	3.46(11)	-[b]	99.7
KN-9 gl	4	51.9(3)	1.50(7)	17.0(2)	0.06(2)	7.82(35)	8.06(9)	0.23(1)	9.93(9)	0.28(1)	3.71(6)	-	100.5
KN-8 gl	7	51.2(4)	1.66(7)	16.2(4)	0.07(5)	9.91(46)	7.50(20)	0.23(4)	10.1(1)	0.30(1)	3.65(21)	-	100.8
KN-2 gl	4	52.5(3)	1.74(5)	16.1(1)	0.05(3)	8.63(15)	7.47(9)	0.15(3)	10.3(8)	0.29(2)	3.47(7)	-	100.7
KN-5 gl	7	51.5(4)	1.90(8)	14.9(3)	0.06(2)	9.09(18)	6.80(18)	0.18(8)	10.4(2)	0.33(2)	3.60(7)	-	98.8
KN-3 gl	6	52.5(3)	2.09(8)	14.9(3)	0.07(1)	9.59(22)	6.67(24)	0.17(3)	10.3(1)	0.32(3)	3.47(14)	-	100.1
KN-7 gl	5	52.6(5)	2.26(7)	14.2(6)	0.04(4)	9.56(35)	6.26(20)	0.19(2)	10.5(8)	0.38(1)	3.56(16)	-	99.6
KN-4 gl	5	51.9(3)	2.38(6)	13.6(2)	0.06(3)	11.3(1)	6.06(4)	0.23(4)	10.4(1)	0.34(4)	3.27(9)	-	99.5
ALV2-4													
gl	7	49.7(3)	1.93(6)	14.3(2)	0.0[c]	10.7(2)	7.19(8)	0.22(5)	10.6(7)	0.18(1)	3.17(8)	0.25(2)	98.2
pl	3	52.7(6)	-	29.1(4)	-	0.58(3)	0.30(1)	-	12.0(4)	0.05(1)	4.17(26)	-	98.9
ol	5	38.3(3)	0.06(1)	0.0	0.0	18.7(1)	41.2(3)	0.22(5)	0.52(8)	-	-	-	99.1
ALV2-5													
gl	8	49.8(2)	2.35(3)	13.6(2)	0.0	12.2(2)	6.70(6)	0.25(4)	10.5(2)	0.19(2)	3.23(12)	0.32(3)	99.2
pl	6	52.0(4)	-	29.3(2)	-	0.58(11)	0.25(7)	-	13.3(3)	0.05(1)	3.90(24)	-	99.5
cpx	11	52.2(5)	0.87(19)	3.05(70)	0.28(6)	7.48(61)	17.3(9)	0.22(3)	18.2(1.2)	-	0.37(9)	-	100.0
ol	3	38.2(3)	0.04(2)	0.09(5)	0.04(1)	21.7(2)	39.6(3)	0.32(3)	0.49(1)	-	-	-	100.4
ALV2-8													
gl	10	50.0(3)	2.47(5)	13.7(1)	0.0	11.7(3)	6.60(8)	0.19(5)	10.5(2)	0.22(1)	3.36(11)	0.36(4)	99.0
pl	5	53.2(4)	-	29.2(1)	-	0.57(9)	0.26(9)	-	12.7(3)	0.05(1)	4.28(2)	-	100.2
cpx	10	52.8(5)	0.73(9)	2.47(33)	0.31(6)	7.21(21)	17.9(4)	0.19(2)	18.6(5)	-	0.33(6)	-	100.6
ol	3	38.1(3)	0.07(2)	0.06(2)	0.0	22.1(2)	39.5(3)	0.32(2)	0.41(5)	-	-	-	100.6
ALV2-10													
gl	10	49.8(3)	2.30(4)	13.7(1)	0.0	12.0(2)	6.68(6)	0.22(4)	10.7(1)	0.19(1)	3.18(18)	0.32(3)	99.0
pl	6	52.1(3)	-	29.5(2)	-	0.56(19)	0.26(8)	-	13.2(2)	0.04(1)	4.00(1)	-	99.6
cpx	9	52.0(5)	0.85(2)	3.05(51)	0.41(4)	7.15(59)	17.0(6)	0.18(4)	19.2(4)	-	0.34(7)	-	100.1
ol	6	38.3(3)	0.06(2)	0.0	0.0	21.0(4)	40.3(7)	0.31(2)	0.41(3)	-	-	-	100.6
ALV2-7													
gl	12	49.8(5)	3.32(8)	12.9(2)	0.0	12.9(3)	6.08(9)	0.27(4)	9.86(8)	0.25(2)	3.11(16)	0.43(6)	98.9
pl	5	53.6(3)	-	28.4(1)	-	0.54(5)	0.18(4)	-	12.0(1)	0.06(1)	4.51(10)	-	99.3
cpx	10	52.1(4)	0.87(15)	2.24(54)	0.18(4)	8.67(43)	17.5(5)	0.25(3)	17.7(8)	-	0.29(4)	-	99.8
ol	3	37.3(5)	0.11(2)	0.08(3)	0.05(1)	24.1(2)	37.1(3)	0.37(4)	0.47(3)	-	-	-	99.7
ALV2-9													
gl	12	49.0(4)	3.70(12)	12.5(1)	0.0	13.6(3)	5.91(7)	0.25(4)	9.66(15)	0.28(2)	3.29(10)	0.53(4)	98.8
pl	6	53.9(3)	-	27.8(6)	-	0.52(14)	0.30(22)	-	11.6(4)	0.08(1)	4.88(18)	-	99.1
cpx	11	52.4(9)	0.95(20)	2.13(54)	0.17(4)	9.35(70)	18.0(1.0)	0.27(3)	16.7(1.5)	-	0.30(3)	-	100.3
ol	3	37.2(1)	0.07(2)	0.05(1)	0.03(1)	25.9(3)	36.0(4)	0.42(5)	0.40(2)	-	-	-	100.0
H51													
gl	9	50.2(2)	1.40(6)	15.8(1)	0.05(3)	9.48(11)	8.01(13)	0.14(6)	10.6(1)	0.12(1)	3.13(10)	0.15(3)	99.1
pl	5	52.6(3)	-	29.0(5)	-	-	0.53(13)	0.56(30)	12.4(1)	-	0.04(1)	4.22(6)	99.9
cpx	13	51.0(6)	0.69(8)	5.83(35)	0.36(5)	6.04(22)	18.3(5)	0.17(5)	16.1(7)	-	0.41(3)	-	98.9
ol	5	39.6(2)	0.06(3)	0.07(5)	0.08(2)	14.9(4)	44.4(5)	0.23(1)	0.37(2)	-	-	-	99.7
H50													
gl	10	49.1(3)	1.82(2)	15.6(2)	0.03(2)	11.2(3)	7.64(27)	0.19(3)	9.77(13)	0.14(1)	3.57(14)	0.12(1)	99.2
pl	6	53.3(2)	-	28.8(4))	-	0.53(8)	0.33(17)	-	12.3(7)	0.04(1)	4.54(11)	-	99.8
cpx	11	52.1(4)	0.80(12)	4.28(6)	0.17(3)	7.88(43)	19.2(1.0)	0.21(3)	14.8(1.3)	-	0.44(6)	-	99.9
ol	3	38.6(7)	0.06(3)	0.11(3)	0.08(2)	18.0(3)	42.2(2)	0.25(2)	0.41(3)	-	-	-	99.7
H60													
gl	9	48.7(4)	1.90(11)	15.3(2)	0.08(4)	11.5(3)	7.27(3)	0.09(18)	9.62(18)	0.15(2)	3.61(11)	0.13(4)	98.4
pl	5	53.2(2)	-	28.5(3)	-	0.74(17)	0.47(16)	-	11.8(12)	0.04(2)	4.67(12)	-	99.4
cpx	6	52.0(5)	0.69(12)	4.04(74)	0.21(7)	7.34(34)	18.7(9)	0.17(9)	15.9(8)	-	0.43(8)	-	99.5
ol	4	38.6(5)	0.04(3)	0.12(2)	0.05(3)	18.5(3)	41.9(4)	0.21(2)	0.43(2)	-	-	-	99.9
H62													
gl	8	49.0(2)	0.90(4)	16.3(1)	0.11(3)	9.79(20)	9.12(9)	0.13(7)	11.5(13)	0.11(2)	2.33(14)	0.06(4)	99.4
pl	5	50.0(3)	-	30.9(4)	-	0.58(16)	0.44(15)	-	14.8(2)	0.03(2)	3.00(10)	-	99.8
cpx	11	51.4(8)	0.55(10)	5.47(1.1)	0.28(5)	5.76(38)	18.2(1.2)	0.21(4)	17.4(1.1)	-	0.32(6)	-	99.6
ol	4	39.2(4)	0.0	0.11(1)	0.09(2)	15.1(2)	44.0(5)	0.10(6)	0.44(3)	-	-	-	99.0
H57													
gl	8	49.3(1)	0.89(2)	15.9(1)	0.08(2)	10.1(1)	8.46(8)	0.15(4)	11.9(1)	0.12(1)	2.23(6)	-	99.1

TABLE 3. (continued)

	number	SiO_2	TiO_2	Al_2O_3	Cr_2O_3	FeO	MgO	MnO	CaO	K_2O	Na_2O	P_2O_5	Total
pl	5	50.3(4)	-	29.9(7)	-	0.74(11)	0.78(17)	-	14.5(2)	0.03(2)	2.99(10)	-	98.2
cpx	10	51.9(1.2)	0.37(17)	5.33(1.7)	0.44(11)	6.19(75)	19.2(2.2)	0.13(5)	16.2(2.1)	-	0.27(8)	-	100.0
ol	4	39.6(4)	0.0	0.16(6)	0.09(3)	15.4(4)	44.7(2)	0.16(3)	0.51(2)	-	-	-	100.6
H32													
gl	7	49.2(2)	1.02(8)	15.6(1)	0.11(2)	11.1(1)	8.32(11)	0.27(12)	11.6(2)	0.14(1)	2.28(6)	0.16(3)	99.8
pl	5	50.8(5)	-	31.1(4)	-	0.35(13)	0.38(11)	-	14.6(3)	0.03(2)	3.10(11)	-	100.4
cpx	10	51.9(9)	0.53(14)	5.56(86)	0.37(10)	6.19(44)	18.5(1.0)	0.08(11)	17.2(1.2)	-	0.30(3)	-	100.6
ol	3	40.0(3)	0.0	0.09(3)	0.10(1)	15.4(5)	44.9(1.0)	0.42(12)	0.42(2)	-	-	-	100.9
H35													
gl	7	48.9(3)	1.72(13)	15.4(2)	0.08(1)	11.9(4)	7.88(11)	0.22(6)	10.4(7)	0.16(1)	3.12(14)	0.26(3)	99.9
pl	6	51.9(3)	-	29.9(6)	-	0.35(13)	0.31(24)	-	13.3(2)	0.05(1)	3.90(1)	-	99.7
cpx	15	52.0(1.0)	0.92(30)	4.40(92)	0.20(4)	6.49(80)	18.1(1.0)	0.0	18.1(8)	-	0.35(8)	-	100.6
ol	3	39.1(3)	0.04(1)	0.06(1)	0.09(2)	18.9(6)	41.8(5)	0.43(18)	0.39(2)	-	-	-	100.8
H59													
gl	8	49.2(1)	1.11(4)	16.0(7)	0.11(2)	8.63(17)	9.50(15)	0.05(3)	11.8(1)	0.08(3)	2.61(5)	0.08(2)	99.2
pl	3	47.2(3)	-	32.9(5)	-	0.54(12)	0.98(21)	-	17.0(2)	0.0	1.65(10)	-	100.3
cpx	7	51.4(9)	0.50(13)	6.03(1.2)	0.55(9)	4.78(14)	18.7(8)	0.09(5)	17.7(6)	-	0.10(4)	-	99.9
ol	3	39.7(3)	0.02(3)	0.0	0.08(4)	12.7(3)	46.6(5)	0.15(2)	0.40(2)	-	-	-	99.7
H38													
gl	9	48.9(4)	2.80(19)	14.5(1.3)	0.08(2)	11.6(7)	7.9(1.8)	0.30(9)	10.2(7)	0.27(5)	3.39(36)	0.37(3)	100.3
pl	6	53.6(4)	-	28.1(9)	-	0.48(16)	0.54(28)	-	12.4(5)	0.07(2)	4.27(25)	-	99.5
cpx	17	52.6(1.0)	0.92(41)	3.63(32)	0.23(6)	7.25(99)	17.7(1.4)	0.08(8)	18.2(9)	-	0.35(5)	-	101.0
ol	3	39.7(2)	0.0	0.05(1)	0.05(3)	19.2(3)	42.1(5)	0.21(6)	0.37(9)	-	-	-	101.7
H58													
gl	3	49.0(1.2)	3.06(7)	14.0(2)	0.06(2)	14.1(5)	6.47(22)	0.23(4)	9.58(12)	0.26(1)	3.22(22)	0.40(1)	100.4
pl	5	52.7(6)	-	29.1(4)	-	0.63(14)	0.59(20)	-	12.9(4)	0.07(1)	4.10(27)	-	100.1
cpx	10	51.9(4)	1.02(17)	3.35(46)	0.20(3)	7.98(96)	16.8(6)	0.16(4)	18.0(9)	-	0.34(5)	-	99.8
ol	4	37.7(6)	0.06(3)	0.11(4)	0.08(4)	23.5(7)	37.2(9)	0.29(5)	0.48(2)	-	-	-	99.4
H258													
gl	8	48.7(2)	1.21(2)	17.2(2)	0.0	8.99(25)	8.10(26)	0.25(5)	10.0(1)	0.13(2)	3.84(13)	0.13(2)	98.6
pl	5	53.1(6)	-	28.6(6)	-	0.44(17)	0.49(30)	-	12.3(4)	0.03(1)	4.26(21)	-	99.1
cpx	42	51.5(8)	0.81(19)	6.89(1.1)	0.21(4)	6.14(32)	18.3(6)	0.17(4)	16.1(7)	-	0.56(11)	-	100.5
ol	5	40.0(1)	0.03(1)	0.05(1)	0.0	14.0(4)	45.7(1)	0.11(7)	0.31(1)	-	-	-	100.3
H137													
gl	8	48.6(3)	1.59(7)	17.2(1)	0.06(4)	10.2(2)	8.52(15)	0.15(6)	10.2(1)	0.16(1)	3.19(11)	0.16(3)	100.0
pl	6	52.6(4)	-	30.1(4)	-	0.40(8)	0.27(16)	-	13.1(4)	0.07(1)	4.15(18)	-	100.7
cpx	6	51.1(4)	0.68(12)	7.26(94)	0.24(12)	6.20(1.1)	19.4(4)	0.16(9)	15.3(1.9)	-	0.49(14)	-	100.8
ol	3	39.9(1)	0.05(1)	0.11(8)	0.03(1)	15.8(3)	44.8(5)	0.17(1)	0.40(7)	-	-	-	101.3

[a] Units in parentheses indicate standard deviation of element analyzed. Thus, 51.4(6) should be read 51.4 ± 0.6 wt.% SiO_2.
[b] Dash signifies that element was not analyzed.
[c] 0.0 indicates that element was analyzed, but found to be below detectability limit.

molar quantities) show a pressure dependence and a liquid compositional dependence. Plagioclase becomes more Na-rich as pressure increases [Lindsley, 1970], but MORB magmas also move toward silica undersaturation by crystallization at high pressures. The effect of this SiO_2 decrease is to make the crystallizing plag more Ca-rich. The values of $K_D{}^{Ca\text{-}Na}$ in our experimental data varied over a large range. This range in K_D values is comparable to that obtained in other natural basalt experimental studies [Thompson, 1974, 1975; Mahood and Baker, 1986; Grove and Juster, 1989]. We present a model that treats the effects of pressure and liquid composition on plag Ca-Na exchange (see Appendix 2).

Bartels et al. [1991] and Kinzler and Grove [1992a] have carried out experiments of variable duration and at the same conditions of temperature and pressure to determine the time necessary to achieve homogeneous experimental products. For high temperature, near-liquidus experiments with > 80 wt % liquid, experimental durations of six hours resulted in consistent phase assemblages and compositions. For highly crystalline, lower temperature experiments longer experimental durations were required. The augites in short-duration, low temperature experiments are heterogeneous in composition, and reaction rims are present on olivine and plagioclase. Longer duration experiments (>14 hours) produced more homogeneous augites, but compositional heterogeneity persists in even the longest duration experiments (up to 72 hours). The amount of homogeneity in the 72 hour experiment does not differ significantly from that achieved in the 14 hour experiment. The average composition of the experimentally produced augites appears to become constant for experiments with durations between 6 and 72 hours. The presence of compositional heterogeneity in augite is inconsistent with complete attainment of equilibrium. Similar heterogeneity is present in all previous experimental studies, and in many cases the experimental durations of the previous studies were significantly shorter than our experimental durations. The consistent and reproducible phase appearances and mineral-liquid exchange K_D's in the present study suggest that the deviations from equilibrium are not large, and we conclude that the experiments are sufficiently close to equilibrium that they can be used in the interpretation of MORB magma evolution. Experiments of far greater duration than are presently feasible would be required to produce true equilibrium phase compositions.

RESULTS

Experiments at 0.001 Kbar on a Lava From the Mid-Cayman Rise

These experiments were carried out on the most primitive MORB (KN-54-2-2, Thompson et al. [1980]) from the mid-Cayman rise at 0.001 kbar to provide information on the near-ocean-floor differentiation path followed by this parental magma. At 0.001 kbar, and 1200 °C KN-54-2-2 has oliv as a liquidus phase. After a short interval of oliv crystallization, plag joins the crystallization assemblage. The oliv+plag assemblage crystallizes over a 50 °C temperature interval, and at 1150 °C augite appears as a crystallizing phase.

Experiments at 2 Kbar on a Lava From Serocki Volcano

These experiments were carried out on a "normal" MORB from the MARK area, an active spreading center on the mid-Atlantic ridge south of the Kane Fracture Zone (KFZ) at 23 °N. The sample, ALV-1690-20 was collected by the ALVIN submersible from Serocki volcano [Humphris et al., 1990]. Grove et al. [1990] carried out 0.001 and 8 kbar melting experiments on ALV-1690-20 and located the oliv+plag+augite saturation boundary for this composition at these pressures. The experiments presented here locate the liq+oliv+plag+augite boundary for ALV-1690-20 at 2 kbar. The experiments at 0.001, 2 and 8 kbar on ALV-1690-20 provide information on the effect of pressure on this 4-phase boundary in temperature-composition space, as well as the effect of pressure and temperature on the compositional variation of the saturating mineral phases.

Experiments at 8 and 10 Kbar on Parental MORBs

Several compositions that have been the subject of previous 0.001 kbar experimental studies were chosen for experiments at 9 and 11 kbar. These samples were chosen because they are candidate parental magmas for individual MORB suites. By parental we mean a least-evolved magma from which other more evolved compositions can be derived by fractional crystallization. Chosen for this study were: sample AII-96-18-1, a parental group MORB from the MARK area [Tormey et al., 1987], ALV-528-1-1, a parental group MORB from the FAMOUS area [Grove and Bryan, 1983]; and samples 70-002 and ALV-2004-3-1 from the East Pacific Rise (EPR) at 11°45'N (Thompson et al. [1989], R. J. Kinzler, W. B. Bryan, and T. L. Grove, manuscript in preparation). These samples are all near multiple saturation with oliv+plag+augite on their 8-kbar liquidi.

DISCUSSION

Estimation of Fractionation Paths

The method described below allows us to reconstruct differentiation process(es) and test whether or not such processes could lead to the compositional variability observed in a suite of related MORB magmas. In order to predict the paths followed by differentiating magmas one needs to know the locations of primary phase volumes and multiple saturation boundaries, and how changes in magma composition and depth of crystallization alter the positions of these saturation boundaries. Furthermore, one needs to predict the compositions and proportions of phases that crystallize along the saturation boundaries, as functions of pressure and changing magma composition.

Constraints From Simple System Experiments

Controls of bulk composition. The system forsterite (Fo)-diopside (Di)-anorthite (An)-albite (Ab) at 0.001 kbar [Yoder and Tilley, 1962; Biggar and Humphries, 1981] serves as an excellent analog system for examining bulk composition controls on differentiation. Figure 2 (after Biggar and Humphries [1981]) shows ternary diagrams for Fo-Di-plag where the plagioclase composition varies from anorthite to albite. There is a dramatic change in composition of liquids that are multiply saturated with Fo+Di+plag as the Ab content of the system changes. The composition of the liquid saturated with Fo, Di and plag contains nearly equal proportions of Di and An in the Na-free systems, but becomes highly enriched in the plag component in the pure Ab system. Similarly, the Fo+plag+melt boundary moves toward plag, as the Ab content of the system increases. The effects of bulk composition evident in Figure 2 are not introduced by the projection scheme, because Biggar and Humphries [1981] used the projection scheme of O'Hara [1976] to ensure that albite and anorthite plot as molar equivalents. The change in the positions of the Fo + plag boundary can be expressed as a change in the proportions of crystallizing phases, from 0.40 Fo:0.60 An (weight proportions) on the Fo-An boundary to 0.05 Fo:0.95 Ab on the Fo-Ab boundary.

Control of pressure. The effect of pressure on the oliv+plag and oliv+plag+augite saturation boundaries has been investigated by Presnall et al. [1978] in the Fo-Di-An system. The position of the oliv+plag saturation boundary can be inferred by comparing the 0.001 and 7 kbar phase diagrams.

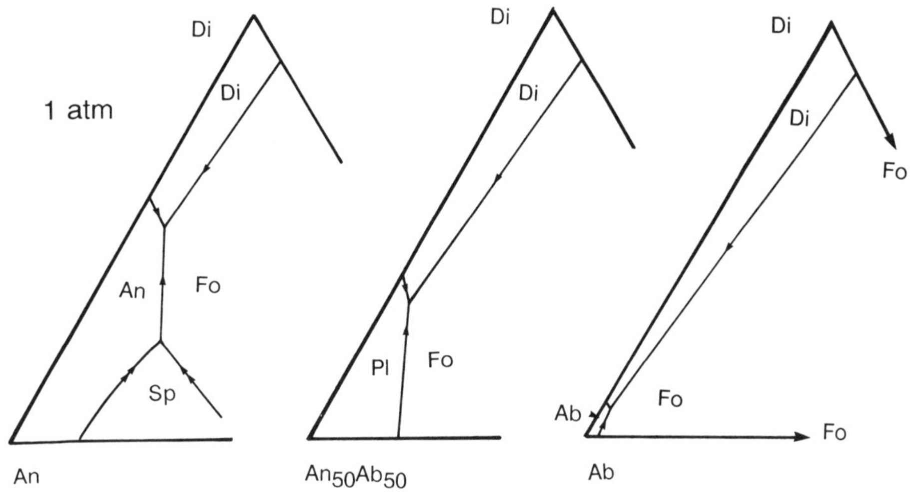

Fig. 2. Phase diagrams at 0.001 kbar for the systems Fo - Di - $Plag_{ss}$ from Biggar and Humphries [1981]. The ternary diagrams show the effect of varying Anorthite (An, $CaAl_2Si_2O_8$) and Albite (Ab, $NaAlSi_3O_8$) components on the composition of a liquid multiply saturated with Fo - Di - Plag. The effect of increased Ab content in the system is to move the 4 - phase boundary toward Plag.

At 7 kbar, Fo+An do not coexist with liquid. The positions of the boundaries have not been determined at pressures between 0.001 and 7 kbar, but one can infer the effect of pressure on the phase boundaries on the Fo-Di-An diagram (Figure 3). The major effect of pressure is to move the Fo+Di+An saturation boundary toward the Fo-An join. At about 5 kbar the reaction Fo+An = spinel+liquid terminates the coexistence of Fo and An, and a diopside + spinel phase volume separates the Fo and An primary phase volumes at 7 kbar pressure. If one infers the position of the Fo+An+diopside+spinel+liquid point at 5 kbar and compares it to the position of the Fo+An+melt boundary at 0.001 kbar, the Fo+An+melt boundary and the point are nearly coincident. This coincidence indicates that pressure variation from 0.001 to 5 kbar has little effect on the position of the Fo+An saturation boundary. Therefore, the effect of increased pressure is to move the Fo+An+diopside boundary down toward the Fo-An binary at nearly constant proportions of Fo:An.

Constraints from Experiments on MORB and MORB-Analog Compositions

Changes in bulk composition and pressure of crystallization also affect the characteristics of liquids saturated with oliv+plag or oliv+plag+augite in natural MORB compositions. As an example we show the 0.001-kbar saturation boundaries (Figure 4) for a high-Na_2O MORB from the mid-Cayman rise (3.55 wt % Na_2O, KN-54-2-2, Table 1,2 and 3) and a FAMOUS MORB with lower Na_2O (2.09 wt % Na_2O, ALV-528-1-1, Table 1 and Grove and Bryan, 1983). The projection scheme is described in Appendix 1 and is similar to the method of projection employed by O'Hara [1976], except that oxygen units are used for the projected components. Note that the oliv+plag saturation boundary for KN-54-2-2 is displaced toward plag, and the oliv+plag+augite saturation boundary for KN-54-2-2 is enriched in plag component relative to the FAMOUS 3- and 4-phase boundaries. The oliv+plag+augite saturated liquids in Figure 4 are located at the high-Cpx ends of the trends exhibited by the two sets of experiments and are distinguished by different symbols. These shifts in natural system saturation boundaries are similar to those observed in the simple Fo-Di-An-Ab system. Again, the shifts reflect differing proportions of oliv + plag that crystallize along the oliv - plag boundary. In the

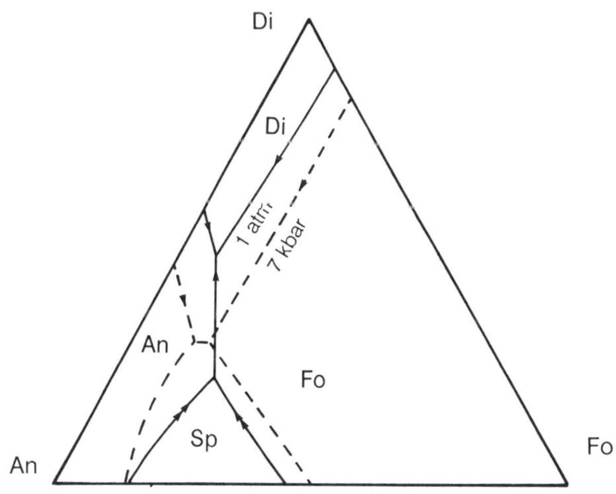

Fig. 3. The effect of pressure in the system Fo - Di - An on the composition of a liquid saturated with Fo, Di and An is shown in this ternary diagram from Presnall et al. [1978]. Phase relations are shown for 0.001 kbar and 7 kbar. The assemblage Fo - Di - An - Liq is not stable above about 5 kbar, so the influence of pressure must be inferred from the 7-kbar relations. The effect of increased pressure is to move the boundary toward the Fo-An side line.

Cayman example the proportions are 0.23 oliv:0.77 plag. In the FAMOUS example the proportions are 0.32 oliv:0.68 plag. Note that this change in proportions moves the position of the oliv + plag boundary in the projection scheme significantly, and therefore the projection scheme provides a sensitive indicator of the changes in crystallizing proportions. Moreover, the spread in the Oliv:Plag of the projected experimental data points defining the FAMOUS boundary in Figure 4 would translate to about a 0.02 variation in the fractional phase proportions for each area.

The next step is to provide a quantitative description of the influence of changing bulk composition on the locations of multiple saturation boundaries. Once such a quantification is available, it is possible to predict crystallization paths for any MORB composition. Several methods for estimating the crystallization path of a lava are available. Ghiorso [1987] reviewed computational methods that use thermodynamic or empirical expressions to compute the conditions under which a liquid becomes saturated with a mineral phase. Weaver and Langmuir [1990] and Nielsen [1988] present empirical methods based on single element mineral-melt partition coefficients. Longhi and Pan [1988], Grove and Juster [1989], Longhi [1991], and Kinzler and Grove [1992a] have presented methods that use information from experimentally determined phase equilibria to estimate mineral saturation sequence and crystallizing phase proportions. These methods rely on manipulations that can be performed with phase diagrams to estimate crystallizing phase proportions (as outlined in Morse, 1980, Chapters 8 and 9). Grove and Juster [1989] and Kinzler and Grove [1992a] devised a method that predicts the influence of compositional variables and pressure on the

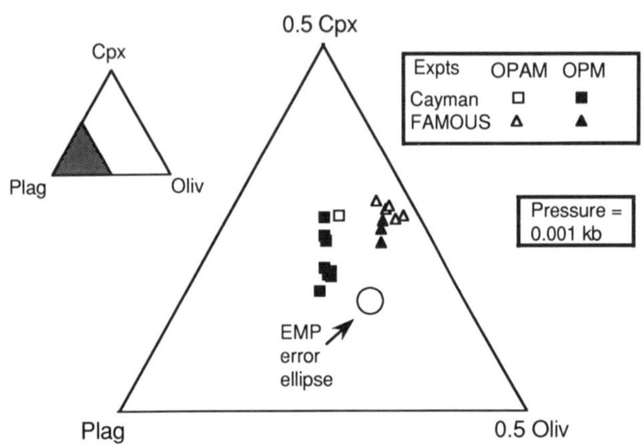

Fig. 4. The effect of variation in bulk composition on phase boundaries for natural MORB lavas is shown by the contrasting position of the OPM and OPAM boundaries defined by experiments on a high-Na_2O Mid-Cayman Rise lava (KN54-2-2), and a low-Na_2O FAMOUS lava (528-1-1),. The OPM boundaries (solid symbols) are defined by the compositions of experimentally produced melts that coexist with oliv+plag. The OPAM (open symbols) boundaries are defined by the compositions of experimentally produced melts that coexist with oliv+plag+augite. The 2 sigma error ellipse (EMP error) is the calculated standard deviation of the mean for replicate electron microprobe analyses of the 528-1-1 starting material. The projection scheme is described in Appendix 1.

compositions of liquids saturated with oliv, plag, augite and a low-Ca pyroxene (low-Ca pyx). The equations are presented in Appendix 2. The key compositional parameters that control the position of the 5-phase saturation boundary are: *Mg #*, *A #* (alkali #, $[Na_2O+K_2O]/[Na_2O+K_2O+CaO]$) and TiO_2 content. The justification of *A #* as an important variable is clearly shown in Figure 2 in the simple system where variations in Ab/An are equivalent to changes in *A #*. Longhi and Pan [1988] and Shi and Libourel [1991] have shown the systematic effects of variations in *Mg #* on the positions of saturation boundaries in projection schemes, and TiO_2 was found to be a significant compositional variable in MORB systems [Kushiro, 1975; Grove and Juster, 1989; Kinzler and Grove, 1992a]. Thus, using the method of Kinzler and Grove [1992a] we calculate the boundary defined by liquids saturated with oliv+plag+augite+low-Ca pyroxene (the OPALM boundary) as a function of pressure and *Mg #*, *A #* and TiO_2 content in the melt.

At fixed pressure, and for given values of *Mg #*, *A #*, and TiO_2 content, the 5-phase OPALM boundary projects as a point that lies at the termination of the 4-phase oliv+plag+augite+silicate melt boundary (OPAM). The OPAM boundary projects as a line terminated by the OPALM point in the Oliv-Cpx-Qtz pseudo-ternary. In the Oliv-Cpx-Plag pseudoternary the OPAM line collapses on top of the OPALM point and both project to a point. Figure 5 shows OPAM boundary lines and OPALM points at their ends in a perspective drawing for different pressures, and in pseudoternary projections. Most MORB magmas will follow the crystallization sequence: oliv, followed by oliv + plag, followed by oliv +plag + augite. Therefore, if one can predict the position of the olivine primary phase boundary, the OPM plane and the OPAM boundary line, this information will be sufficient to model crystallization. The simple geometric relations among the OPALM point, OPAM line and olivine+plag+silicate melt surface (OPM surface) can be used to predict this crystallization sequence. We use the expressions of Kinzler and Grove [1992a] to predict the OPALM boundary as a function of pressure and system composition. We therefore need to estimate the shape of the OPAM boundary line that extends from the OPALM point and the OPM plane.

As discussed above for the simple CMAS system, the effect of increasing pressure is to move the OPAM boundary toward the Oliv-Plag-Qtz plane, tracking along the oliv+plag+silicate melt boundary (OPM). Variations in pressure do not significantly affect the position of the OPM boundary relative to oliv and plag. The ternary perspective drawing (Figure 5) shows the pressure effect schematically. The OPAM boundary is the dark line defined by the intersection of the three planes; the OPM, the oliv+augite+silicate melt and the augite+plag+silicate melt boundaries. As pressure increases, the augite primary phase volume expands, causing the intersection as defined by the OPAM boundary line to move down toward the oliv-plag join. The OPM boundary shortens, and the oliv+augite+silicate melt and augite+plag+silicate melt boundaries shift along with the 4-phase OPAM boundary.

We have extensive coverage of the 0.001-kbar OPAM boundary in the MORB suite from the MARK area (Tormey

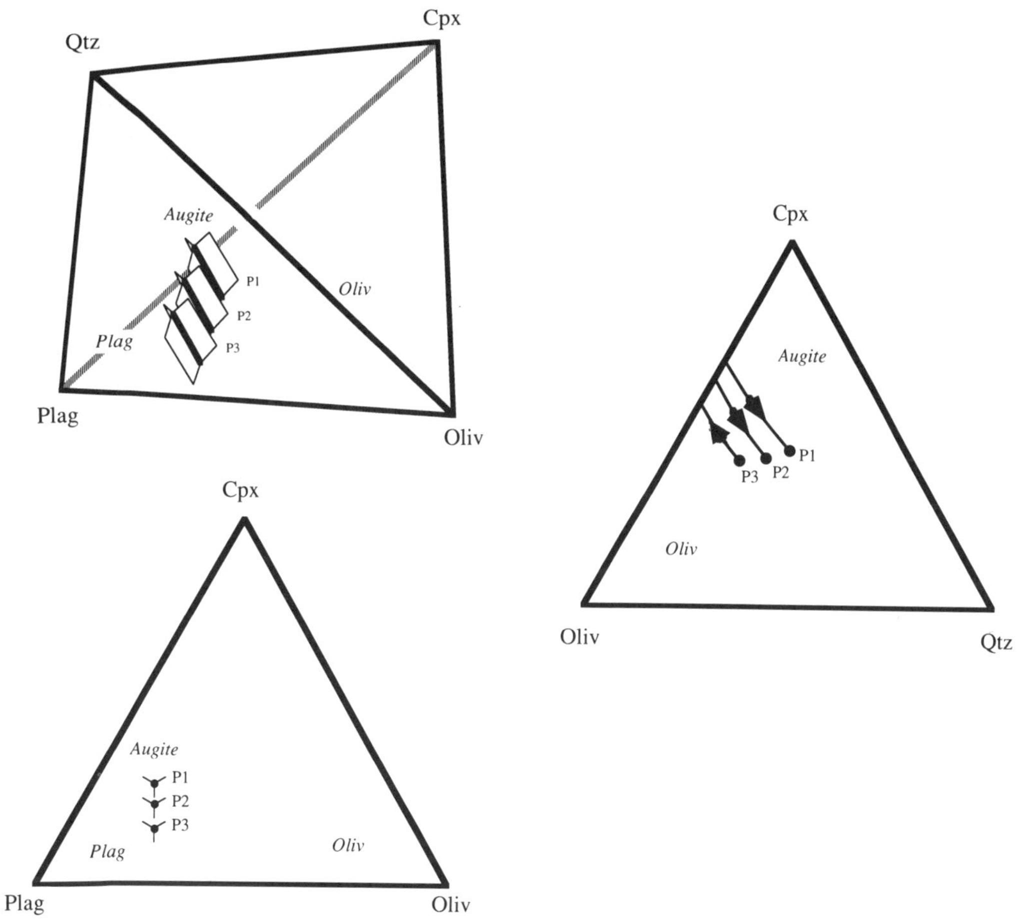

Fig. 5. *a*) The pseudoquaternary tetrahedron Oliv-Plag-Cpx-Qtz viewed in 3-D perspective shows the effect of pressure on the positions of the oliv+plag+augite+melt (OPAM) boundary (thick, dark line) and the oliv+plag+melt (OPM), oliv+augite+melt and plag+augite+melt boundaries (planes that intersect to form the dark line). These planes separate the primary phase volumes of oliv, plag and augite. P1 = experimentally determined 0.001-kbar OPAM boundary for KFZ lavas from Tormey et al. [1987]. P2 and P3 show schematically the 4- and 8-kbar relations. *b*) Projection onto the pseudoternary Oliv-Plag-Cpx from Qtz shows that the OPAM boundary "line" collapses to a point, and the OPM, oliv+augite+melt and plag+augite+melt "planes" collapse to lines. *c*) The Oliv-Cpx-Qtz pseudoternary shows the effect of increased pressure on the position of the OPAM boundary (solid curves with arrows). The solid circles indicate the inferred positions of the OPALM "points".

et al. [1987], Grove et al. [1990], see their Figures 4 and 2, respectively). The experimentally determined 4-phase OPAM boundary projects as a line in Oliv-Cpx-Qtz and Oliv-Plag-Qtz pseudoternaries and collapses to a point in the Oliv-Plag-Cpx pseudoternary that is coincident with the OPALM point. The OPAM assemblages experimentally produced at 0.001, 2 and 8 kbar on the MARK area lava ALV-1690-20 further substantiate the simple geometric relations between OPALM points and OPAM lines. Liquids saturated with oliv+plag+augite (the OPAM assemblage) have been produced experimentally for this sample at 0.001 and 8 kbar [Grove et al., 1990] and at 2 kbar (Tables 2 and 3). These liquids are plotted on the Oliv-Cpx-Qtz and Oliv-Plag-Cpx pseudoternaries in Figure 6. Of the two 8-kbar liquids shown on the projection, one is also saturated with a low-Ca pyx, and was one of 51 such liquids used by Kinzler and Grove [1992a] to constrain the expression used to predict the OPALM 5-phase points. The other OPAM assemblages shown in Figure 6 were not part of the data set used to predict the variation of the OPALM points with pressure and compositional variables. The predicted OPALM 5-phase points are plotted in Figure 6 for the ALV-1690-20 composition at pressures of 0.001, 2 and 8 kbar. OPAM lines were extended from these boundaries in the Oliv-Cpx-Qtz pseudo-ternary projection toward the Oliv-Cpx sideline. These OPAM lines were constructed in the following manner: an OPAM line was drawn by passing a line from the OPALM point through the 0.001-kbar oliv+plag+augite saturated liquids. OPAM lines with parallel slopes were extended from the 2-kbar and 8-kbar OPALM points to define inferred positions for the OPAM boundaries at these pressures. The arrows on the OPAM lines show the path followed by liquids with decreasing temperature. The

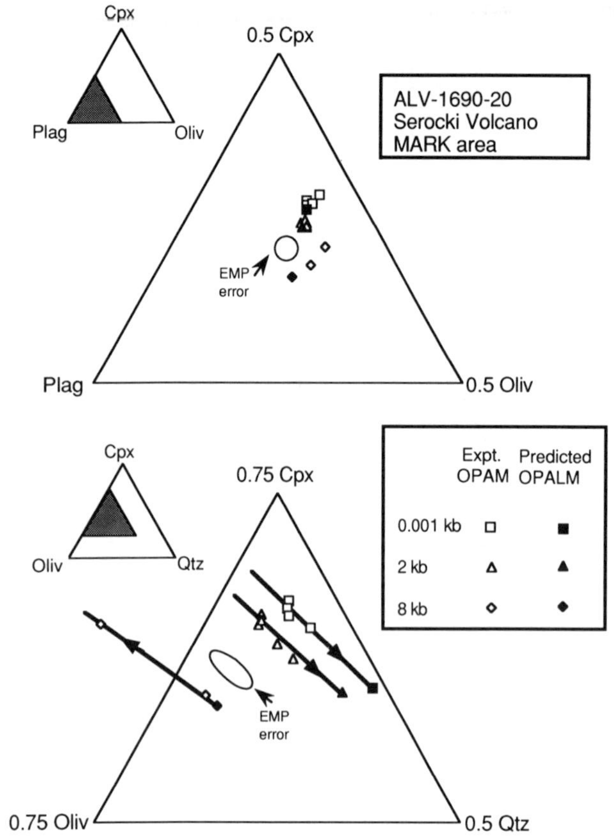

Fig. 6. Pseudoternary projections showing experimentally produced liquids saturated with oliv, plag and augite at 0.001 kbar (open squares), 2 kbar (open triangles) and 8 kbar (open diamonds) on composition ALV-1690-20 from Serocki volcano at the MARK area, south of the KFZ. The positions of the OPALM boundaries predicted at 0.001 kbar (solid square), 2 kbar (solid triangle) and 8 kbar (solid diamond) using the procedure outlined in the text and in Appendix 2 also are shown. The 2 sigma error ellipses (EMP error) are calculated as the standard deviation of the mean of replicate electron microprobe analyses of the ALV-1690-20 starting composition. Projection scheme is described in Appendix 1.

change in direction at 8 kbar is a result of the change in composition of the saturating augite. At 8 kbar the augite is sub-calcic, and projects at a higher Qtz component than the OPALM point. Therefore, liquids at this pressure evolve to SiO_2-undersaturated compositions. Notice the close match between the inferred OPAM boundaries and the experimental data in the Oliv-Cpx-Qtz pseudoternary. The predicted OPALM points are close in position to the experimentally produced oliv+plag+augite (OPAM) saturated liquids in the Oliv-Cpx-Plag pseudoternary. This near-coincidence is consistent with the observation that the OPALM point and the OPAM boundary collapse to a point in this projection. As discussed above, the symbol closest to the OPALM point in Figure 6 is labelled wrongly as OPAM and is actually saturated with a low-Ca pyx. This experimental point and the predicted OPALM point lie within electron microprobe analytical uncertainty. The other experimental point on the 8 kbar OPAM boundary is saturated with the OPAM phases. It projects to SiO_2-undersaturated composition in the Oliv-Cpx-Qtz pseudoternary, and shows the largest separation from the predicted 8 kbar OPALM point in the Oliv-Plag-Cpx pseudoternary. We have no explanation for this discrepancy, but it may be related to SiO_2 undersaturation.

Figure 7 shows the same comparison in the two pseudoternary projection schemes using the 8-kbar experimental data presented in this study. Oliv+plag+augite saturated liquids from samples 528-1-1, AII-96-18-1 and 70-002 (Tables 2 and 3) are shown along with the predicted 5-phase boundaries for these compositions. The correspondence between these experimentally produced liquids and the OPAM boundaries inferred from the predicted OPALM points is sufficiently close for the purposes of developing a model for elevated pressure crystallization. The expected analytical uncertainty based on 2 sigma errors for a typical electron microprobe analysis of these experimental glass compositions is also plotted in Figures 6 and 7, and shows

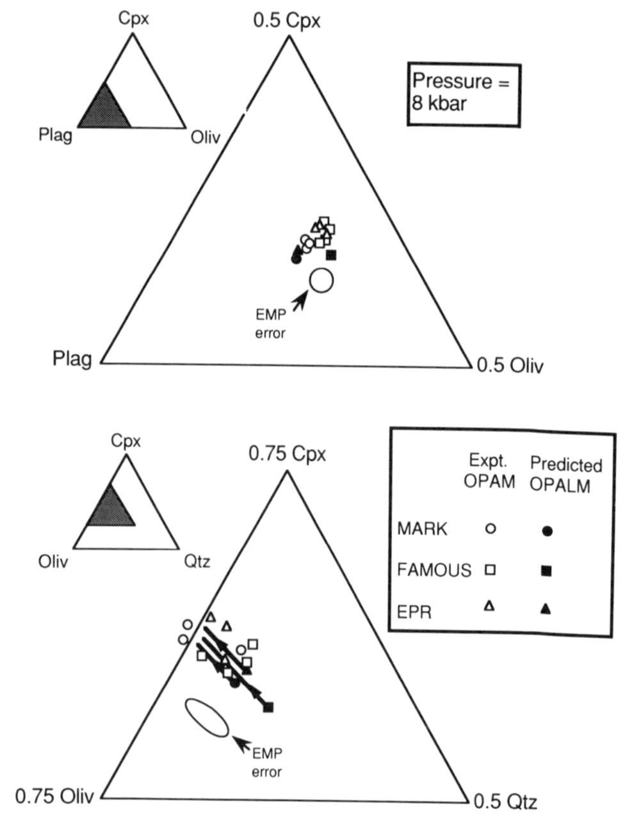

Fig. 7. Pseudoternary projections showing experimentally produced liquids saturated with oliv+plag+augite at 8 kbar (open symbols). Also shown are the predicted OPALM boundaries for the same compositions (solid symbols) and the inferred positions of the OPAM boundaries. The 2 sigma error ellipses (EMP error) are calculated for the standard deviation of the mean on replicate electron microprobe analyses of the AII-96-18-1 starting composition (see Tormey et al. [1987]). Projection scheme is described in Appendix 1.

that our predicted OPAM line recovers the positions of the oliv+plag+augite saturated liquids to within analytical uncertainty. We model fractionating liquid on an OPAM boundary by estimating phase proportions on the OPAM boundary that keep the liquid at a constant position in the Oliv-Plag-Cpx pseudoternary (see Appendix 2). This technique is sensitive to small changes in phase proportions and a change of 0.02 in phase proportions is sufficient to send a fractionating liquid off on a trajectory that is inconsistent with the inferred OPAM boundary.

Estimating the Fractionation History of MORB Suites

Fractional crystallization in the oceanic mantle at the Kane Fracture Zone (KFZ). Bryan et al. [1981] identified KFZ lavas that were representative of liquid compositions. Some of these are aphyric, and some are phyric lavas in which the phenocrysts and residual liquids have been retained in equilibrium proportions. The lavas in these two subsets, assumed to represent "liquids", are shown in Figure 8 in the Oliv-Cpx-Qtz and Oliv-Plag-Cpx pseudoternary projections. The samples were collected from rift floors and walls near the ridge-transform intersection south of the KFZ, to the south of the KFZ for 80 km, and to the north of the fracture zone for about 20 km. The transform offset is 130 km. Systematic compositional differences between lavas of the north and south rifts indicate differences in mantle sources [Bryan et al., 1981; Machado et al., 1983; Langmuir and Bender, 1984]. Therefore, generalizations about this data set must be made with care, and in this and subsequent discussions we will restrict specific models of petrogenesis to restricted geographic areas, and to sample groups that show coherent geochemical trends.

There are some notable characteristics of this subset of the KFZ regional whole-rock data. The first is the apparent absence of lava bulk compositions that have been modified by oliv+plag+augite differentiation at near ocean-floor (0.001-kbar) conditions. This is in contrast to the East Pacific Rise, north of the Clipperton Fracture Zone, an example discussed below. A signature of near-sea floor differentiation is observed in the glasses of the phyric subset of KFZ lavas, which record within sample differentiation upon emplacement [Bryan et al., 1981, Tormey et al., 1987]. If near ocean-floor fractionation had modified the whole rock compositions, we would expect a clustering of lava compositions near the 0.001-kbar OPAM boundary in Figure 8. We observe instead that the lava compositions mainly project to a cluster near the OPAM boundaries predicted for the pressure range from 2 to >4 kbar in the Oliv-Plag-Cpx pseudoternary. In the Oliv-Cpx-Qtz projection there is a trend of the lavas toward the Oliv-Qtz sideline. Part of this elongation is a result of analytical uncertainty. Examination of major element variations in this data set indicates two groups, the one that projects near the >4 kbar OPAM boundary has slightly higher normative Qtz and the group that projects near the 2 kbar OPAM boundary has lower normative Qtz. In order to further explore whether the compositional variations exhibited by the lavas are generally consistent with fractional crystallization the data are plotted in Figure 9 in terms of wt % TiO_2 versus wt % MgO. The KFZ data set shows a general increase in TiO_2 abundance with decreasing MgO abundance. Superimposed on the diagram are two models that show the results of oliv+plag+augite fractionation from 2 parental lava compositions, calculated at 5 kbar. The lavas are generally dispersed along the modelled trends. Therefore, the compositional variations displayed by this subset of the KFZ MORB are generally consistent with fractionation of oliv+plag+augite below the base of the oceanic crust. This fractionation would have occurred in the upper oceanic mantle at pressures of 3 kbar or more, equivalent to depths of > 10 km. The inference that evolution of these lavas occurred by fractional crystallization at 2<P<6 kbar is consistent with recent seismic evidence for absence of a seismically detectable crustal magma chamber along this portion of the mid-Atlantic ridge [Detrick et al., 1990].

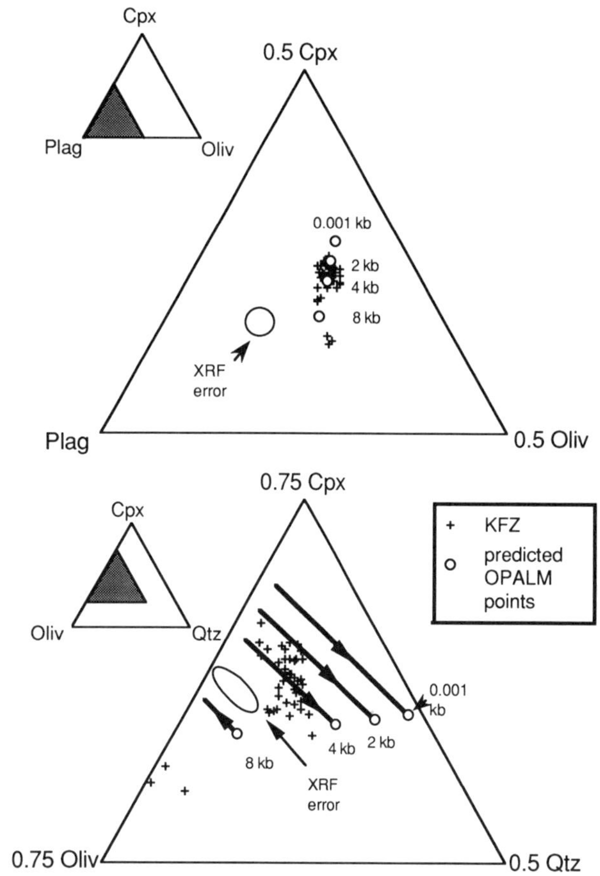

Fig. 8. Pseudoternary projections of lavas (pluses) from the Kane Fracture Zone (KFZ) determined by Bryan et al. [1981] to be representative of liquid compositions. Also shown are predicted OPAM boundaries for a representative KFZ parental magma (AII96-18-1) at 0.001, 2, 4, and 8 kbar. The 2 sigma error ellipses (XRF error) are shown for a representative XRF analysis. The XRF error ellipses are large, because each point represents only a single analysis (see Tormey et al. [1987]) and errors are proportional to counting statistics. Projection scheme is described in Appendix 1.

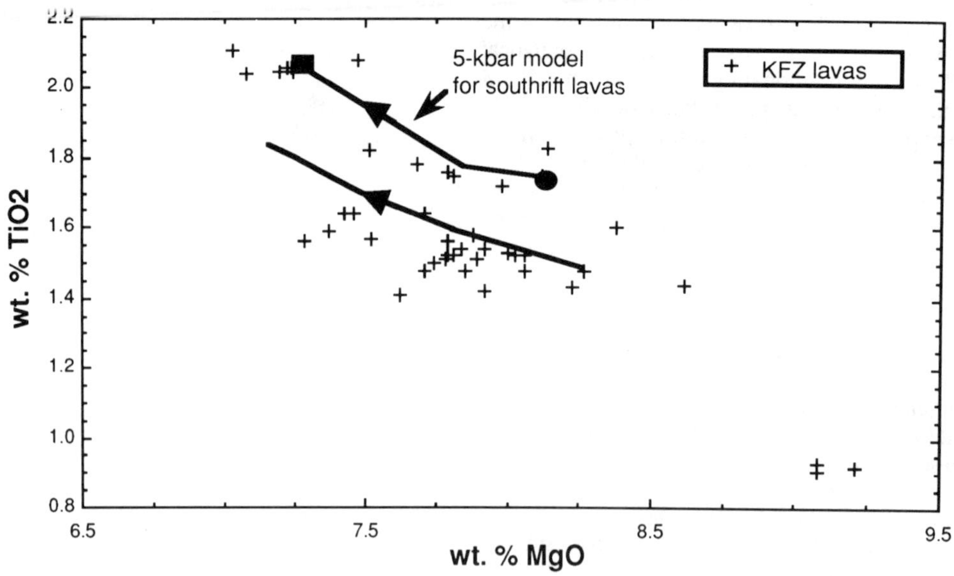

Fig. 9. Wt % TiO_2 versus wt % MgO diagram showing the KFZ data set (pluses) shown in Figure 8. Solid curves are model paths that show the effects of 5-kbar fractionation on two parental lavas from the KFZ data set. The upper curve relates parent and derivative lavas sampled from the MARK area (shown again in the lower pseudoternary projection in Figure 9 and detailed in Table 4).

Within suites of lavas collected from restricted geographical areas, we can examine compositional variations in more detail for evidence of fractionation at elevated pressure. A region located just south of the Kane Fracture Zone (the MARK area) was sampled in dredges GS104-14, -15, -17, -19, -20, -25, -29 and AII96-18 [Bryan et al., 1981]. For these samples Bryan et al. [1981] obtained a comprehensive set of major, trace and rare earth element abundances. Machado et al. [1983] report isotopic information. Tormey et al. [1987] modelled the compositional trend displayed by these lavas by fractional crystallization of oliv+plag+augite at elevated pressures. The projected positions of these lavas in the Oliv-Cpx-Qtz and Oliv-Plag-Cpx pseudoternaries (Figure 10) indicate that they are related by oliv+plag+augite fractionation at pressures between 4 and 6 kbar. A model of 1% oliv, followed by 8% oliv+plag+augite fractionation generally matches the major element compositional variations between the parental lava GS104-17-37 and the evolved lavas sampled in dredge GS104-25. This model is shown in Figures 9 and 10 and described in Table 4. Further details of the model (how phase compositions and proportions were calculated) are discussed in Appendix 2. The additional spread in the projected positions of the lavas relative to the calculated model in Figure 10 is largely due to scatter related to analytical uncertainty. Grove et al. [1990] discuss another example of elevated pressure fractionation in the vicinity of Serocki volcano at the MARK area, south of the Kane Fracture Zone.

Fractional crystallization at the mid-Cayman Rise. Thompson et al. [1980] noted that compositional variations exhibited in lavas collected from the mid-Cayman Rise (MCR) were consistent with augite as a fractionating phase along with oliv and plag. They also noted that augite was rare or absent as a phenocryst phase, and oliv+plag were the common phenocryst minerals preserved in MCR basalts. Experiments at 0.001 kbar show that sample KN54-2-2 (Tables 2 and 3), the most primitive of the MCR samples, has a 100 ^{o}C interval of oliv+plag crystallization, before augite appears (Figure 4). The liquids produced by crystallization of this potential parent at near ocean-floor conditions do not resemble the more evolved MCR lavas. The MCR lavas are displaced from the predicted 0.001-kbar OPAM boundary in the Oliv-Cpx-Qtz and Oliv-Plag-Cpx pseudoternary projections (Figure 11). In the Oliv-Plag-Cpx projection the MCR lavas cluster near the predicted 4-kbar OPAM boundary. This same data set is shown in a wt % TiO_2 versus wt % MgO diagram in Figure 12. As was observed for the KFZ data set, the MCR lavas generally show an increase in TiO_2 abundance with decreasing MgO abundance. Calculated models of oliv+plag+augite fractionation at 5 kbar from 2 distinct parent compositions are superimposed on the diagrams. The lavas are generally dispersed along the calculated trends. Therefore, the compositional variations observed in the MCR lava suite also are consistent with a model involving differentiation in the upper oceanic mantle (5 kbar = ~15 km).

Fractionation within the oceanic crust at the East Pacific Rise. Thompson et al. [1989] identified a suite of MORB lavas from the East Pacific Rise (EPR) that display effects of fractional crystallization within the oceanic crust. These lavas were sampled at the active spreading center between 10.63 and 10.93 ^{o}N, north of the Clipperton Transform (10.25 ^{o}N), where the neovolcanic zone shallows progressively from 3100 m at the ridge-transform intersection to a high of 2500 m at 10.93 ^{o}N. A group of samples from dredges 51, 52, 53, 54 and 57 show coherent compositional relations that Thompson et al. [1989] ascribe to near ocean-floor fractional crystallization.

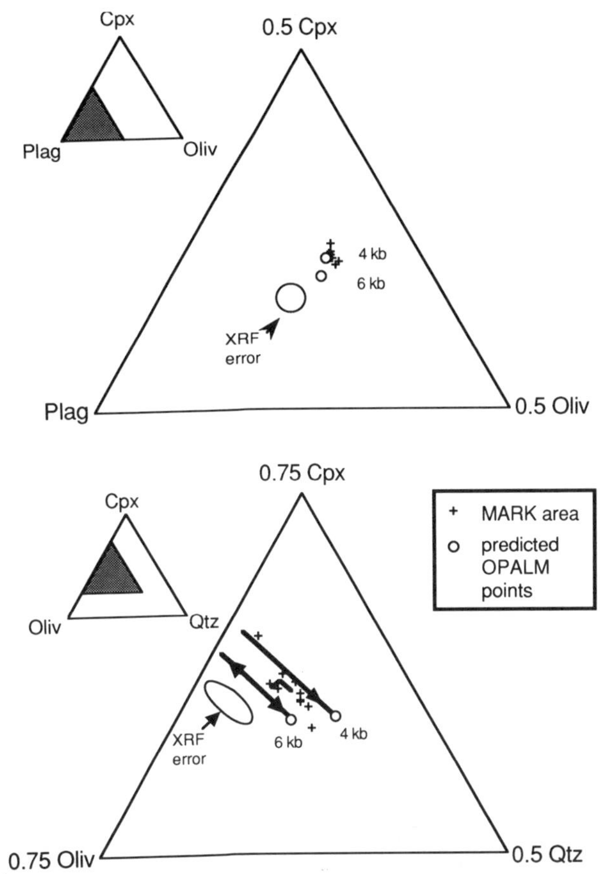

Fig. 10. Pseudoternary projections of lavas (pluses) from the MARK area, an active spreading center south of the KFZ [Bryan et al., 1981], that form a geochemically coherent and spatially associated group. These lavas are a subset of the lavas shown in Figures 8 and 9. The thin, solid, curved line in the lower projection shows the fractional crystallization model shown in Figure 9 and detailed in Table 4. Also shown are predicted OPAM boundaries at 4 and 6 kbar for AII96-18-1, a member of this suite. The 2 sigma error ellipses (XRF error) are shown for a representative XRF analysis. Projection scheme is described in Appendix 1.

Both major and trace element variations are consistent with oliv+plag+augite fractionation. These samples contain microphenocrysts of oliv, plag and augite, strengthening the case for fractionation of these phases. In Figure 13 we show the data of Thompson et al. [1989] for these EPR MORBs on the Oliv-Cpx-Qtz and Oliv-Plag-Qtz pseudoternaries. Also shown are the predicted OPAM boundaries for these compositions at 0.001 kbar and 2 kbar. Note that a subset of the lavas cluster near the predicted 0.001-kbar OPAM boundary. The more evolved lavas (dredges 52 and 53, distinguished as samples with the highest Qtz component in the Oliv-Cpx-Qtz pseudo-ternary) cluster along the predicted OPAM boundary at 2 kbar. The EPR lavas show a more systematic increase in TiO_2 abundance with decreasing MgO abundance (Figure 14), relative to the data sets from the KFZ and the MCR, consistent with fractional crystallization from a more limited range of parent magma compositions over a more limited range of pressures. The solid circles in Figure 14 show the average compositions for each of the 6 dredges of Thompson et al. [1989]. One could draw a single trend of oliv+plag+augite fractionation through these data. However, although the less evolved compositions are consistent with augite saturation at 0.001 kbar, the more evolved compositions are not (Figure 13). If the entire data set were related by a single trend of oliv+plag+augite fractionation from a common parent then the fractionation process must have moved to deeper levels in the crust as it proceeded. In fact, two distinct fractionation trends can be recognized in the data (shown in Figure 14). The lavas on the 2-kbar OPAM boundary are offset towards slightly higher TiO_2 abundances at similar MgO abundance, relative to those on the 0.001-kbar OPAM boundary. As discussed below in the section on the development of local trends, such elevations in minor components (TiO_2 and Na_2O) and FeO are an expected consequence of fractional crystallization at elevated pressures. The lavas that have compositional variations consistent with crystallization at pressures near 2 kbar were sampled from the southern, deeper portion of this active spreading center, near the ridge-transform intersection. The observation that these lavas are more evolved than those from the northern, shallower portion of this active spreading center may reflect a combination of a more limited magma supply and a cooler differentiation environment caused by their proximity to the ridge-transform intersection [Sinton et al., 1983; Langmuir and Bender, 1984]. Furthermore, the seismic study of Detrick et al. [1987] identified a reflector associated with a magma chamber within the oceanic crust beneath the shallow portion of this spreading center, at 10.93°N. However, the reflector disappears just to the south of the high and is not identified over the length of the ridge approaching the Clipperton Transform. The inferred presence of a magma chamber beneath the shallow, northern portion of this spreading segment, and inferred absence of a seismically detectable magma chamber within the oceanic crust at the southern, deeper portion of this spreading segment are consistent with the inferences of depths of evolution from the compositions of the lavas.

The Pyroxene Paradox

Many discussions of MORB petrogenesis conclude that fractionation of augite (along with oliv and plag) is required to relate suites of temporally and spatially associated MORB lavas. It often is found that models of the major and trace element variations observed in a MORB suite require the participation of augite as a crystallizing phase. However augite is often absent as a phenocryst phase or is only a minor component. Among the writers who have called upon "phantom" augite as a crystallizing phase are: Dungan and Rhodes [1978], Fisk et al. [1980], O'Donnell and Presnall [1980] and Fujii and Bougault [1983]. This apparently contradictory role for clinopyroxene has been referred to as the "pyroxene paradox" by Dungan and Rhodes [1978], Fisk et al. [1980] and Francis [1986]. Our prediction of phase relations for MORB provides one explanation for the pyroxene paradox. Figure 5 can be used to illustrate the point.

TABLE 4. Fractional Crystallization Model for MARK Area South Rift Lavas

SiO_2	Al_2O_3	TiO_2	FeO	MgO	CaO	Na_2O	K_2O	P_2O_5	Oliv	X_{Fo}	Plag	X_{An}	Cpx	X_{En}	INC	F
Starting Composition																
50.34	15.47	1.47	9.63	8.46	11.38	2.73	0.10	0.15								
50.58	15.67	1.49	9.60	8.11	11.53	2.77	0.10	0.15	1.0	0.84					0.01	0.99
50.59	15.61	1.51	9.73	8.04	11.48	2.78	0.10	0.15	0.12	0.84	0.54	0.78	0.34	0.55	0.02	0.97
50.65	15.32	1.65	10.35	7.65	11.25	2.86	0.11	0.17	0.12	0.82	0.54	0.77	0.34	0.54	0.02	0.88
50.71	14.94	1.83	11.15	7.16	10.95	2.95	0.13	0.19	0.12	0.80	0.54	0.75	0.34	0.53	0.02	0.77
Desired End Point Composition																
50.37	15.06	2.05	11.23	7.20	10.89	2.76	0.13	0.21								
(22)	(22)	(1)	(2)	(6)	(4)	(9)	(2)	(2)								

Starting composition is an average of AII96-18-1 and -2. Desired end point is an average of GS104-25, -7, -8, -41, -3, -9 and -5 (parenthesized units represent 2 sigma values for average). Compositions from Bryan et al. [1981] are normalized to 100% excluding MnO. Phase proportions (wt. fraction) removed in fractionation step(s) to yield derivative liquid composition are indicated on same line as derivative liquid. Oliv = olivine, Plag = plagioclase and Cpx = augite. Compositions of phases calculated to be in equilibrium with the residual liquid at each step shown are provided in terms of: X_{Fo} = Forsterite content in oliv, X_{An} = Anorthite content in plag, X_{En} = Enstatite content in cpx. See Appendix II for method of calculation and discussion of Fe-Mg and Ca-Na exchange equilibria used. INC = size in mass units of steps used, F = fraction of liquid remaining.

Consider a magma that is saturated with augite at elevated pressures and that lies on boundary P2 in Figure 5. If this magma cools at P2, it will crystallize oliv, plag and augite. If this magma is extracted from its reservoir and delivered to the ocean floor, it will no longer be on the oliv+plag+augite saturation boundary. The oliv+plag+augite boundary at ocean floor conditions will plot at higher values of Cpx (at P1), and the lava will find itself on the oliv+plag saturation boundary at P1, and will crystallize only these two phases upon cooling. Therefore, it may have experienced elevated pressure fractionation at P2 involving augite, but augite will not crystallize upon eruption of the lava. Thompson et al. [1980] and O'Donnell and Presnall [1980] have proposed such a mechanism for explaining the paradox. Other complex factors can produce an imprint of augite fractionation. Magma mixing [Dungan and Rhodes, 1978; Walker et al., 1979] of a primitive basalt with one that has evolved to augite saturation and experienced significant augite crystallization can produce a signature of pyroxene fractionation in the resulting mixed MORB magma. Langmuir [1989] has proposed in situ crystallization as a process to explain a signature of augite fractionation. In this process, mixing of cooler, augite-saturated magmas that evolve near the side of the chamber and hotter, less-evolved magmas from the interior of the same magma chamber, produces a signature of augite crystallization in the mixed magma. The evidence for magma mixing in MORB is clear, and well documented in many studies. There is no easy way to use compositional variations to uniquely distinguish a process of fractional crystallization from a process of fractional crystallization+mixing. The arguments we present for fractional crystallization at elevated pressures are based on the similarity between the observed trends and those inferred for fractional crystallization of oliv+plag+augite based on experimental evidence.

Calcic Plagioclase in MORB Lavas

Some MORB lavas contain Ca-rich plagioclase that shows textural evidence for resorption and disequilibrium with the surrounding host magma. These disequilibrium textures have been repeatedly described in the ocean-floor literature. Meyer and Shibata [1990] described An_{85}-An_{80} plag xenocrysts in basalts from Serocki volcano near 23 °N at the MARK area, south of the KFZ. Flower et al. [1978] described An_{88} plag in lavas sampled in Hole 396B from this same region, south of the KFZ. Equilibrium plag compositions can be estimated (see Appendix 2) for the host lavas using a plag-melt Ca-Na exchange distribution coefficient ($K_D{}^{Ca\text{-}Na}$). The equilibrium plag composition for the Serocki lava containing the $An_{85\text{-}80}$ plag varies from An_{72} to An_{65} over the pressure range of 0.001 kbar to 8 kbar (Figure 15, labelled *a*). The equilibrium plag composition for the lava containing the An_{88} plag ranges from An_{70} to An_{64} over the same pressure range (Figure 15, labelled *b*). Therefore, the host lava compositions will not crystallize the observed An-rich plag compositions under anhydrous conditions.

One process that might explain the presence of An-rich plagioclase is elevated pressure fractionation in which augite

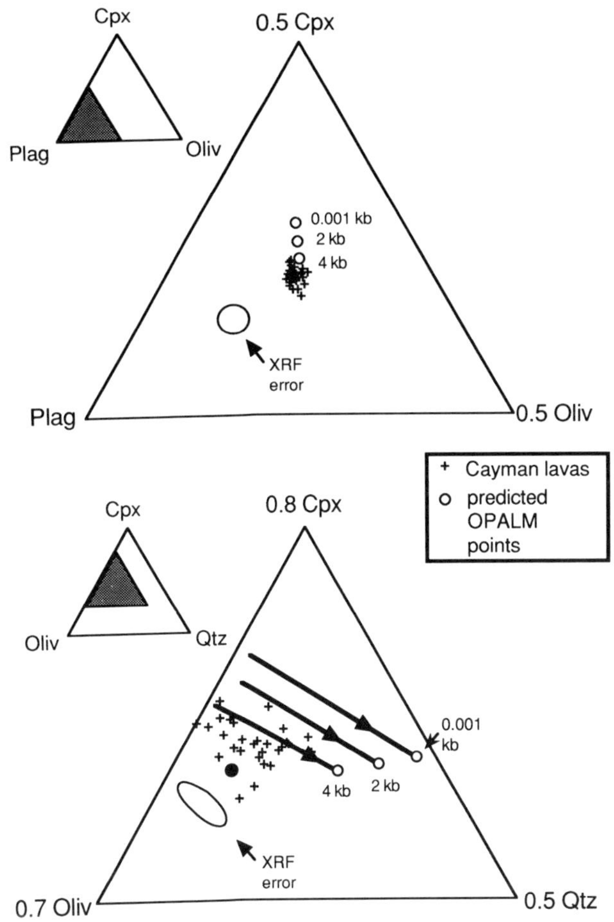

Fig. 11. Pseudoternary projections of lavas (pluses) from the mid-Cayman Rise (MCR) from Thompson et al. [1980]. Also shown are the predicted OPAM boundaries for a parental MCR composition (KN54-2-2, indicated by solid circle in lower projection) at 0.001 kbar, 2 and 4 kbar. The 2 sigma error ellipses (XRF error) are shown for a representative XRF analysis. Projection scheme is described in Appendix 1.

has participated as one of the crystallizing phases. We show below that this is not a viable mechanism. If a magma with a high Ca/Na ratio initially crystallizes oliv+plag at elevated pressure, precipitating an An-rich plagioclase, reaches the OPAM boundary and then crystallizes the 3-phase assemblage, the result will be to reduce the Ca/Na ratio of the residual liquid. If the liquid derived from this fractionation process is sampled along with the An-rich accumulated crystals from the earlier oliv+plag fractionation event, cognate xenocrysts of An-rich plagioclase could be preserved in a liquid with a low Ca/Na ratio. For example, in the next section we present fractionation models calculated at 5 and 0.001 kbar for an estimated primary MORB magma from Kinzler and Grove (1992b). Both models are calculated to MgO abundances of 8 wt %. The 0.001-kbar fractionation calculation yields plagioclase from An_{75}, when plag first appears in the crystallizing assemblage, to An_{72}, the plag in equilibrium with the 0.001-kbar derivative magma at 8 wt % MgO. The 5-kbar fractionation calculation also yields a range of plagioclase compositions, from An_{72}-An_{68}. The observed An-rich plag cannot be produced by these fractionation processes from an aggregate MORB primary magma (Figure 15, labelled *c*).

A viable mechanism for producing such a crystal-liquid association is to mix two magmas, one with a high Ca/Na ratio and An-rich plagioclase phenocrysts, and the other with a low Ca/Na ratio. These two magmas might be fractional melts extracted from different mantle sources, one that has seen extensive melt extraction (high Ca/Na, i.e. from the shallow part of the melting region) and one that is more fertile (low Ca/Na, i.e. from a deep part of the melting region) [Dick et al., 1984; Klein and Langmuir, 1987, 1989]. The range of plag compositions that would crystallize from fractional melts calculated after Kinzler and Grove [1992b] is An_{96}-An_{75} (Figure 15, labelled *d*). In calculation *d* presented in Figure 15, the magmas ascend to 2 kbar and then crystallize oliv until they just saturate with plag. The magma derived from the most fertile mantle, deepest in the melting region, crystallizes An_{74} plag; the magma derived from the most depleted mantle, shallow in the melting region, crystallizes An_{96} plag. If fractionation had occurred at greater depths, the An content of the plag would have been lower. The implication of the presence of An-rich xenocrystic plagioclase is that the fractionation event that produced these crystals occurred before the aggregation of the fractional melts. This evidence supports the scenario for melt generation and evolution where fractionation occurs prior to aggregation (Figure 1*b*), or the more complex scenario where aggregation/mixing occurs over the depth ranges of fractionation and melting (Figure 1*c*). Sobolev and Shimizu [1991] found melt inclusions trapped in olivine phenocrysts in MORB to contain very low rare earth and incompatible element abundances that indicate derivation from a highly depleted mantle source. In this case olivine (rather than plagioclase) crystallized from the melt prior to aggregation, but the evidence supports the proposal made above for the operation of fractionation prior to aggregation.

Residues of Elevated Pressure Fractionation in Mantle Peridotite

Augite- and plag-bearing cumulate assemblages preserved in dikes and veins in peridotites provide evidence that elevated pressure fractional crystallization occurs in the mantle. Dispersed plag- and augite-rich zones and dikes are found below the mafic cumulates of ophiolites to depths up to 1 km. These features have been variously interpreted as melt impregnation or cumulates of basalt crystallization [Nicolas and Prinzhofer, 1983; Nicolas, 1986]. Some of these zones are up to 2-3 km wide and consist of plagioclase-rich, augite-rich and dunite veins [Nicolas, 1986]. The zones grade horizontally into homogeneous lherzolite. The crystallization products of basaltic magmas also are common as cm-sized dikes and fracture fillings, sometimes surrounded by dunite-rich reaction zones, as in the Trinity peridotite [Quick, 1981], and in the Lanzo [Bodinier, 1988] and Oman [Nicolas, 1986] ophiolites.

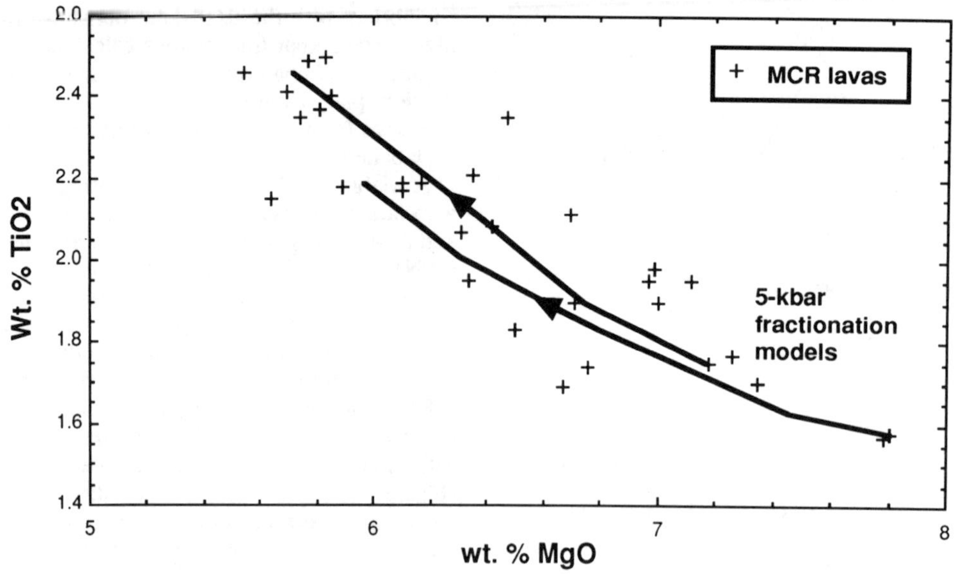

Fig. 12. Wt % TiO_2 versus wt % MgO diagram showing the MCR data set (pluses) shown in Figure 11. Curves with arrows show fractional crystallization models calculated at 5 kbar from two potential parental compositions in the MCR data set.

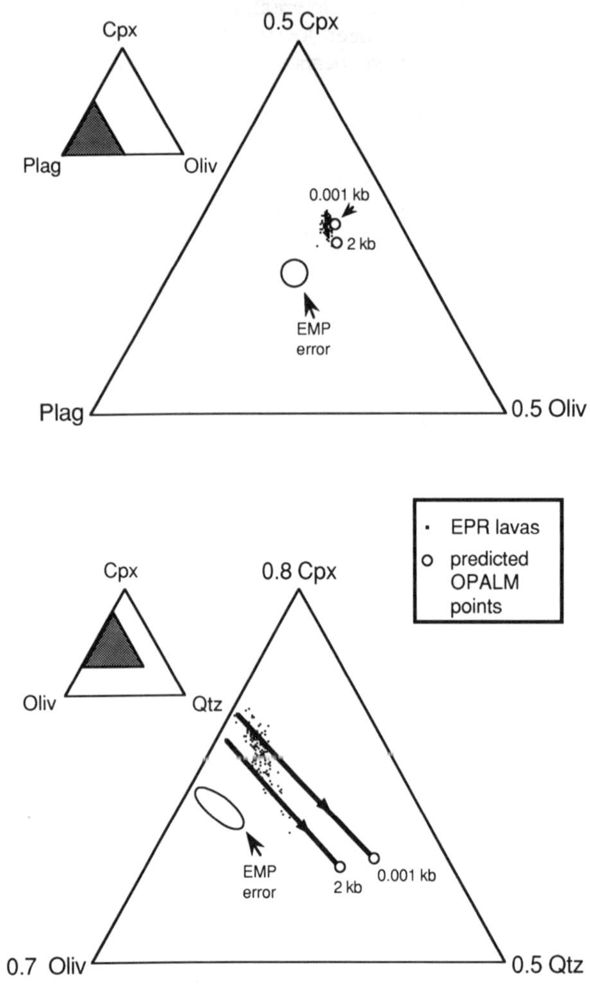

Plagioclase peridotites represent 20% of the peridotites dredged from the ocean floor [Dick, 1989]. One of the distinctive compositional characteristics of plagioclase peridotites is that they contain nearly 10 times the Na and Ti abundances of normal abyssal peridotites. Dick [1989] noted that the compositional variations in plagioclase peridotites lay along a peridotite-basalt mixing line, and did not follow a trend expected if the plagioclase peridotites were residues variably depleted by a melting process. He proposed that plagioclase peridotites were normal abyssal peridotites impregnated with varying amounts of trapped basaltic melt. We suggest that rather than being impregnated with melt, some plagioclase peridotites contain the cumulates of oliv+plag+augite fractionation that took place in the upper oceanic mantle. Such assemblages would consist of the minerals residual from the oceanic mantle fractional crystallization processes proposed on the basis of compositional variations in some erupted basalt compositions. Therefore the presence of plag- and augite-bearing cumulate assemblages in upper mantle peridotite supports the inference drawn from compositional variations in some MORB suites that fractional crystallization occurred at upper mantle pressures.

Role of Fractional Crystallization in the Development of Local Trends

Klein and Langmuir [1989] examined the compositional variations at several well studied locations along active

Fig. 13. Glasses from the East Pacific Rise between 10.6 and 10.9 °N projected onto the Oliv - Cpx - Qtz and Oliv - Plag - Cpx pseudoternaries. Thompson et al. [1989] report averages of these glass data for dredges. Also shown are the predicted positions of the 0.001 kbar and 2 kbar oliv - plag - augite boundaries. Error ellipses show representative 2-sigma uncertainties for electron microprobe analyses of a glass.

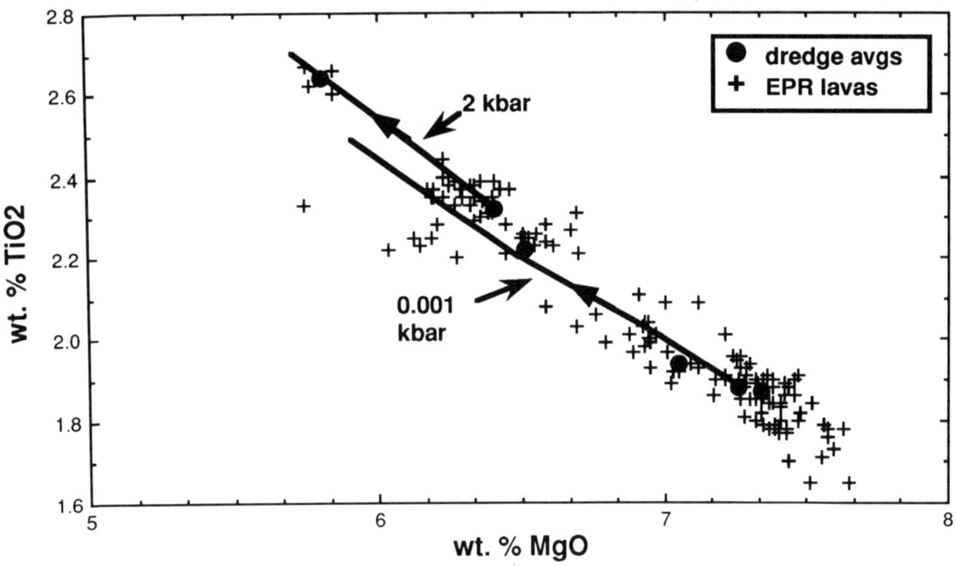

Fig. 14. Wt % TiO_2 versus wt % MgO diagram showing the EPR data set (pluses) shown in Figure 13. Curves with arrows show fractional crystallization models calculated at 0.001 and 2 kbar from two potential parental compositions in the EPR data set. Solid circles are averages of glass data from single dredges [Thompson et al., 1989].

spreading centers. To correct their data for the effects of fractional crystallization, Klein and Langmuir plotted compositions of MORB adjusted to 8 wt % MgO. One of their projections [Klein and Langmuir, 1989; Figure 5b] is redrawn in Figure 16 and shows schematically the global trend and the local trends that they recognized. The local trends are coherent trends of compositional variation at 8 wt % MgO shown by MORB from restricted geographic areas. The global trend is the variation of regionally averaged MORB data from the world's oceans. In Figure 16 we have outlined local trends from the mid-Cayman Rise, the Kane Fracture Zone (KFZ), the AMAR area and the Reykjanes Peninsula. These outlines enclose the fractionation-corrected analyses of MORB from the designated areas. Note that the compositional trends on

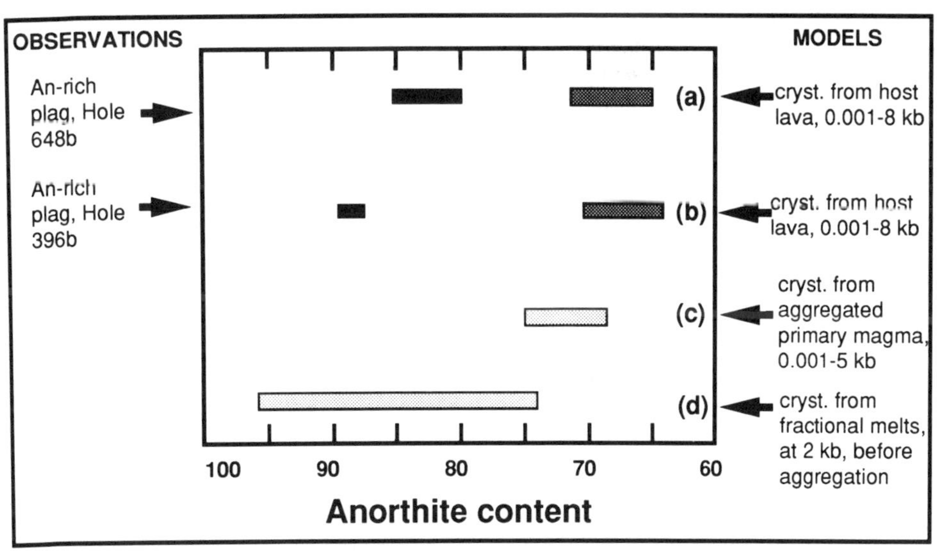

Fig. 15. Observed and calculated (from 0.001 to 8 kbar) compositions of plagioclase from lavas from ODP Hole 648B (labelled *a*) and DSDP Hole 396B (labelled *b*), south of the KFZ. The An contents calculated over the pressure range of 0.001 to 5 kbar for crystallization from an aggregate primary magma (labelled *c*). The range of An contents calculated for crystallization from fractional melts generated from a progressively depleted mantle at 2 kbar prior to aggregation (labelled *d*).

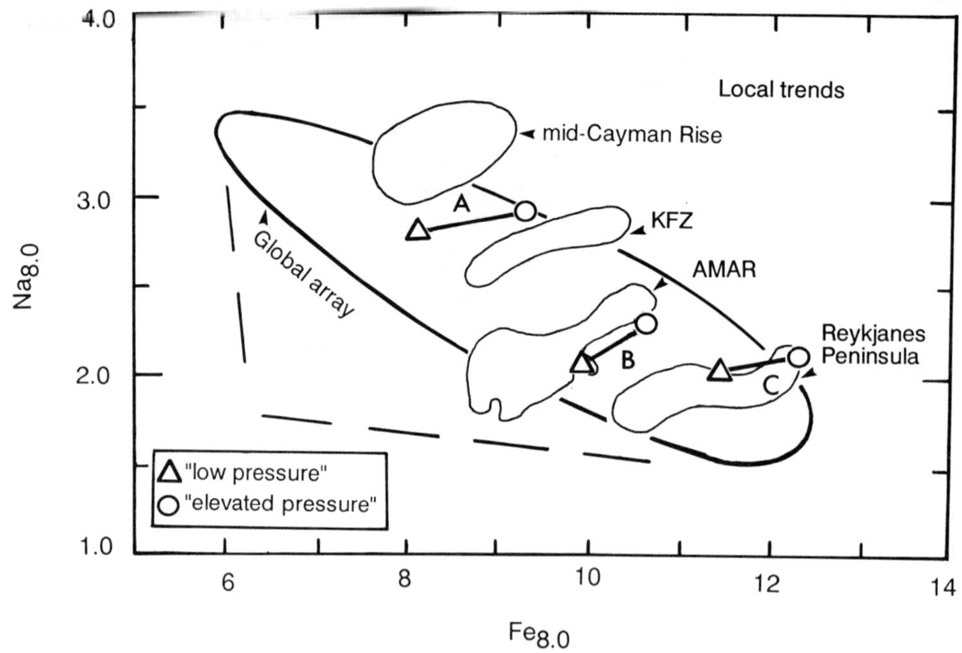

Fig. 16. A schematic representation of the $Na_{8.0}$ versus $Fe_{8.0}$ projection from Klein and Langmuir [1989]. The dark oval region outlines the bulk of the data that constitute the global array. The dashed lines near the lower left show the region covered by the rest of the global data set. The local trends are traced from Klein and Langmuir [1989, Figure 5*b*]. A, B and C refer to pairs of low pressure and elevated pressure fractionation models for selected MORB and MORB-analog compositions.

these local areas cross the global trend. Klein and Langmuir [1989] explored several mechanisms to explain the development of the local trends. They suggested that high pressure fractionation of "olivine +/- orthopyroxene +/- clinopyroxene would be expected to produce increases in iron and sodium..." but concluded that detailed knowledge of phase relations and phase compositions would be required to test models of fractionation. Klein and Langmuir [1989] preferred a model that generates the local trends by instantaneous sampling of melts at different depths in a melting column beneath the region, and they suggest that these trends provide information on melting process and melt extraction. In the following section we examine the local trends at 3 areas and show that fractional crystallization of olivine, clinopyroxene and plagioclase over a range of pressures can reproduce part of the compositional variations that characterize the local trends, if the pressure range of fractionation is from 4 to 0.001 kbar.

The choice of the $Na_{8.0}$-$Fe_{8.0}$ projection eliminates the effects of fractional crystallization from a suite of magmas that have experienced differentiation under similar, near-ocean floor conditions. However, if magmas are derived from similar parents by fractionation at different depths, the resulting compositional differences will not be removed by the Klein and Langmuir [1987] low-pressure projection. To illustrate this point we have carried out fractionation calculations in which a magma has been allowed to differentiate under two different sets of conditions. One process we designate as "low pressure", and the second we call "elevated pressure". The "low pressure" trend occurs within the oceanic crust beneath an active spreading center. The "elevated pressure" fractionation process involves an extended period of differentiation at depth. For the purpose of this discussion, we consider fractionation at pressures of 3-5 kbar, a pressure range consistent with preceding discussions.

Figure 17 illustrates the effect of "low" and "elevated pressure" differentiation on a primary magma from Kinzler and Grove [1992b]. The calculation is carried out using the method described above and in Appendix 2, and is detailed in Table 5. We allow "low pressure" and "elevated pressure" differentiation to proceed until the liquid reaches 8 wt % MgO, and then examine the results directly on Klein and Langmuir's $FeO_{8.0}$ versus $Na_{8.0}$ diagram (Figure 16, labelled A). The extent of enrichment of Na_2O and FeO is controlled by the amount of fractionation required to evolve the magma to 8 wt % MgO. The amount of fractionation required is controlled primarily by the proportion of oliv in the crystallizing assemblage. "Low pressure" oliv and oliv+plag fractionation produces a large decrease in MgO and modest increases in FeO and Na_2O (Figure 17), because the proportion of oliv in the crystallizing assemblage is high, and thus the bulk partition coefficient for MgO is large. In contrast, the crystallization of oliv+plag+augite at high pressures reduces the bulk partition coefficient for MgO. More fractionation is required to reach 8 wt % MgO in this case and the extent of enrichment of FeO and Na_2O is therefore greater. The magma that has experienced "low pressure" differentiation plots at low FeO and Na_2O, at 8 wt

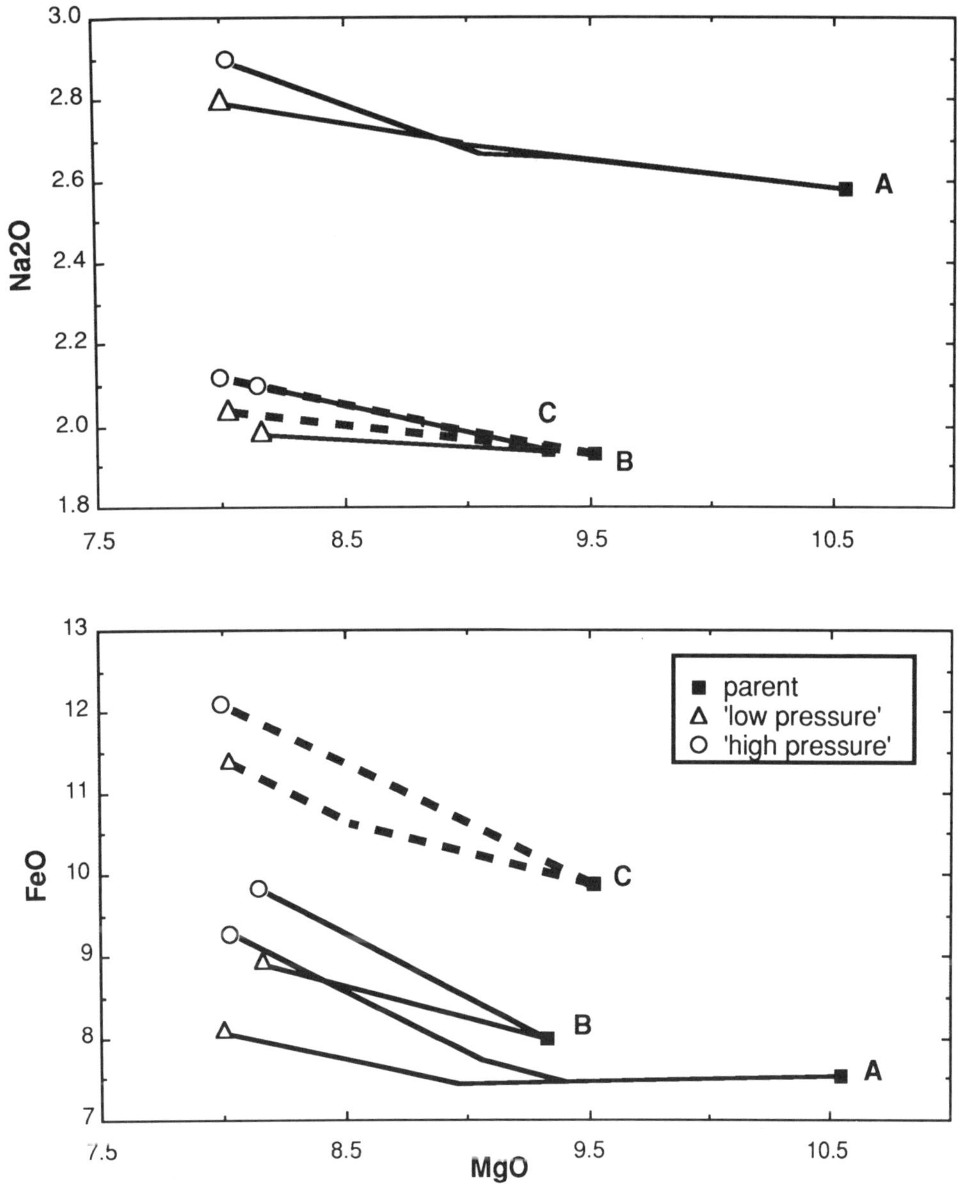

Fig. 17. Plots of MgO, Na_2O and FeO covariations for the fractionation models plotted in Figure 16 and presented in Table 5.

% MgO, and the magma that has experienced "elevated pressure" differentiation plots at higher FeO and Na_2O. The end products of these two fractionation processes are plotted on Figures 16 and 17 for the Kinzler and Grove [1992a] primary magma and define a trend in Figure 16 that crosses the global array. By varying the extent and depth of fractionation experienced by a magma as it ascends and cools in shallow upper oceanic mantle, a series of derivative liquids can be produced that will define local trends. The lava that experienced the "low pressure" fractionation path undergoes less fractionation to reach 8 wt % MgO, and it should be characterized by lower abundances of incompatible elements (La, Ce and Zr) and higher abundances of compatible elements (Ni, Cr and Sr). The lava that experienced the "elevated pressure" fractionation path from the same parent undergoes more fractionation to reach 8 wt % MgO, and should be characterized by higher abundances of incompatible elements (La, Ce and Zr) and lower abundances of compatible elements (Ni, Cr and Sr). The magma used to generate local trend A is a calculated 7.5 wt % partial melt of oceanic mantle, generated at 10 kbar, and it is a good analog of a primary magma for high Na_2O MORBs like those sampled at the mid-Cayman rise. A suite of magmas that experience fractionation at different depths or to differing extents will fill

TABLE 5. Fractional crystallization models for selected MORB and MORB-analog compositions

SiO_2	Al_2O_3	TiO_2	FeO	MgO	CaO	Na_2O	K_2O	P_2O_5	Oliv	X_{Fo}	Plag	X_{An}	Cpx	X_{En}	INC	F
Trend A Starting composition: 7.5 wt. % batch partial melt [Kinzler and Grove, 1992b]																
"elevated pressure" (5 kbar)																
50.70	16.35	0.92	7.54	10.55	11.30	2.58	0.04									
51.02	16.85	0.95	7.45	9.38	11.65	2.66	0.04		1.0	0.89					0.01	0.97
51.25	16.51	1.01	7.71	9.07	11.74	2.68	0.04		0.30	0.88	0.70	0.70			0.02	0.94
51.45	16.12	1.13	8.44	8.58	11.45	2.79	0.05		0.13	0.87	0.56	0.69	0.31	0.56	0.02	0.88
51.70	15.64	1.29	9.34	7.98	11.09	2.91	0.06		0.13	0.84	0.56	0.67	0.31	0.55	0.02	0.81
"low pressure" (0.001 kbar)																
50.70	16.35	0.92	7.54	10.55	11.30	2.58	0.04									
51.12	17.03	0.96	7.42	8.98	11.77	2.69	0.04		1.0	0.89					0.02	0.96
51.58	16.43	1.06	7.83	8.41	11.89	2.76	0.05		0.30	0.87	0.70	0.74			0.02	0.87
51.97	15.91	1.15	8.16	7.94	12.01	2.80	0.05		0.30	0.86	0.70	0.72			0.02	0.80
Trend B FAMOUS - Narrowgate. Starting composition: Sample 818-5 [Stakes et al., 1984]																
"low pressure" (0.001 kbar) back calculation to parent that would be saturated at 4 kbar with oliv+plag+cpx																
50.98	15.01	1.04	8.95	8.16	13.25	1.99	0.22	0.13								
51.02	15.21	1.02	8.88	8.29	13.24	1.99	0.22	0.13	0.30	0.85	0.70	0.77			0.02	-0.98
50.64	15.81	0.95	8.52	8.71	13.08	1.97	0.20	0.12	0.30	0.86	0.70	0.78			0.02	-0.90
50.11	16.64	0.84	7.99	9.32	12.88	1.94	0.18	0.11	0.30	0.88	0.70	0.79			0.02	-0.80
"elevated pressure" 4 kbar oliv + plag + cpx fractionation																
50.11	16.64	0.84	7.99	9.32	12.88	1.94	0.18	0.11								
50.14	16.59	0.85	8.10	9.25	12.84	1.95	0.18	0.11	0.12	0.88	0.53	0.78	0.35	0.57	0.02	-0.98
50.48	15.80	1.07	9.83	8.14	12.22	2.10	0.25	0.15	0.12	0.84	0.53	0.75	0.35	0.54	0.02	-0.80
Trend C Reykjanes Peninsula. Starting composition: Sample RE24 [Jakobbson et al., 1978]																
"elevated pressure" back calculation (3 kbar) along the liq+oliv+plag+cpx boundary to parent, related to primary magma derived from spinel-lherzolite source at mean pressure of 17 kbar																
48.70	14.57	1.80	12.32	7.76	11.76	2.13	0.18	0.17								
49.01	14.74	1.75	12.11	8.00	11.93	2.12	0.17	0.16	0.14	0.80	0.48	0.75	0.38	0.51	0.02	-0.96
49.05	15.07	1.51	10.89	8.83	12.36	2.02	0.15	0.14	0.14	0.83	0.48	0.77	0.38	0.53	0.02	-0.80
49.07	15.33	1.33	9.88	9.52	12.70	1.93	0.12	0.12	0.14	0.85	0.48	0.79	0.38	0.54	0.02	-0.68
"low pressure" (0.001 kbar)																
49.07	15.33	1.33	9.88	9.52	12.70	1.93	0.12	0.12								
49.14	15.20	1.36	9.98	9.39	12.75	1.94	0.13	0.12	0.35	0.86	0.65	0.79			0.02	0.98
49.60	14.40	1.53	10.60	8.55	13.06	1.97	0.14	0.14	0.35	0.84	0.65	0.78			0.02	0.87
49.63	14.08	1.9	11.41	8.02	12.80	2.04	0.16	0.16	0.13	0.82	0.49	0.76	0.38	0.51	0.02	0.77

See Table 4 caption for explanation of abbreviations. Positive values of F indicate a forward fractionation model (crystalline phases subtracted from the starting composition); negative values of F indicate a reverse fractionation model (crystalline phases added to starting composition).

in a local trend and span the composition space between our "low" and "elevated pressure" end members. The total width of a local array depends on magma bulk composition and reflects the depth interval over which differentiation has occurred.

The slope of this fractionation-produced local trend is similar to that observed in the local trends, and the span generally covers about half of the observed trends (Figure 16). If we increase the depth range over which fractionation occurs the span of the resulting derivative magmas is greater. However, the compositional variations exhibited by the sampled MORB do not support fractionation at pressures > ~5 kbar. Furthermore, the effect of pressure on plagioclase composition lowers the Na_2O abundance in the high-pressure derivative magmas and flattens the slope of the resulting trend. Thus, fractionation can account for only part of the observed trends.

Kinzler and Grove [1992b] show that magma compositions resulting from partial aggregation of fractional melt increments would form trends that parallel the global trend in FeO versus Na_2O. Geophysical constraints indicate a primary control of mantle temperature on the depth range of melting beneath a given spreading center, and that the mantle temperature is unlikely to vary within the time interval of the generation of the local trends. These constraints rule out processes that would create the local trend by varying the depth range over which the mantle melts. The greater span of the observed trends can be explained, if the aggregate magmas come from mantle sources that have been variably depleted prior to transport into the convecting melt region. These variably depleted sources undergo melting under similar *P-T* conditions. The major effect of melting these variable sources is to produce a greater range of FeO abundance in the local trends. The derivative magmas with highest FeO result from "elevated pressure" fractionation of the aggregate primary magma from the most depleted mantle. The derivative magmas with lowest FeO result from "low pressure" fractionation of the aggregate primary magma from the least depleted mantle.

Local trends at other spreading centers were modeled by polybaric fractional crystallization, using specific lavas that define part of the local trends. For example, trend B (Figures 16 and 17 and Table 5) is produced using sample 818-5 from Stakes et al. [1984]. Sample 818-5 is a natural MORB from the AMAR area that contains about 8 wt % MgO and that plots near the "low pressure" end of the AMAR local trend in Figure 16. When this magma composition and our predictive expressions (Appendix 2) are used to estimate the pressure of multiple phase saturation, composition 818-5 is found to plot on the 0.001 kbar OPAM boundary. We can subtract this low pressure fractionation effect by back-calculating using the method described in Appendix 2. We return the composition to one that would be saturated at a pressure of 4 kbar with oliv+plag+augite. This is accomplished by adding oliv and plag incrementally to return composition 818-5 to a magnesian parent magma that lies on the 4 kbar OPAM boundary (Table 5). The parental liquid is then subjected to 4 kbar fractionation of oliv+plag+augite until a residual liquid with 8 wt % MgO is produced. This "elevated pressure" fractionation path is reproduced in Figure 17 (trend B) and presented in Table 5. When plotted in Figure 16 the end point of the "elevated pressure" trend and the 818-5 starting composition define part of the AMAR local trend.

As a final example we developed a model of the local trend for the lavas from the Rekyjanes Plateau [Jakobsson et al., 1978]. We chose a lava that lay at the high Na_2O, high FeO "elevated pressure" end of the local trend (Sample RE24). When sample RE24 is plotted on a mineral component projection scheme, it is found to be multiply saturated at 3 kbar with oliv+plag+augite. To produce a magma that experienced "low pressure" fractionation, we returned RE24 to a more mafic parental composition. We calculated the mafic composition by adding the phases oliv+plag+augite incrementally to RE24. The calculated parent was then fractionated at 0.001 kbar (Table 5, Figures 16 and 17, trend C). As in the case of the AMAR local trend, the slope of the local trend parallels that of the model, and the model spans part of the trend defined by the lavas.

In general, the predicted trends also match the compositional spread observed for other variables (SiO_2 and CaO/Al_2O_3) in the local trends (see Kinzler and Grove [1992b]). It is important to emphasize that the scenarios we have drawn are oversimplifications of more complex natural processes. The conclusion is that if magmas in a single region experience differing depths and extents of fractionation, then the systematics of the local trends in $FeO_{8.0}$ versus $NaO_{8.0}$ can be produced. The generation of the entire span of any observed local trend requires the introduction of variably depleted peridotite to the melting regime.

SUMMARY

In this paper we have presented evidence consistent with the operation of fractional crystallization in the oceanic crust and mantle. We also have shown that evidence preserved in plagioclase xenocrysts is consistent with the operation of fractional crystallization of fractional melts prior to aggregation in some MORB. Detailed petrologic and geochemical investigations of MORB lavas in restricted portions of the active spreading ridge always reveal evidence of complexity of process and diversity in magmatic parentage. In the Pacific [Thompson et al., 1989], at the Kane Fracture Zone [Bryan et al., 1981] and at the mid-Cayman rise [Thompson et al., 1981] geochemists and petrologists recognize the existence of distinct mantle sources in the sampled lavas and the operation of multiple processes in the modification of these melts as they encounter overlying mantle and crust. For example, in the FAMOUS-Narrowgate area natural MORBs at 8 wt % MgO that correspond to the predicted "elevated pressure" end member are not easy to find. Sample 829-7 [Stakes et al., 1984, Table 1a] comes closest in major element characteristics, but there are other lavas at the high FeO end of the local trend that can not be produced by any fractionation process from the lava parental to 818-5 and 829-7. This is a common observation in MORB suites. Often, the geochemical characteristics of magmas that are most primitive in a restricted region render these basalts unsuitable as parents to more evolved magmas that are found in the same area. These observations indicate a multiplicity of mantle source processes as well as complexity in fractionation processes.

When the observations for the operation of mixing in crustal level magma reservoirs [Dungan and Rhodes, 1978; Walker et al., 1979] are combined with the processes discussed in this paper, the scenario most consistent for MORB production and evolution is the one depicted in Figure 1*c*. The dominant processes are fractional melting, aggregation/mixing and fractional crystallization. Fractional melting ends when the convecting mantle crosses its solidus and crystallization begins when ascending basaltic magma encounters below-liquidus wall rock temperatures, but aggregation/mixing processes may occur in both the melting and the crystallization regimes.

Evidence consistent with fractionation at depths up to 18 km has implications for both the distribution of isotherms in the upper oceanic mantle and estimates of the amounts of melting that occur to produce a given amount of ocean crust. If temperatures are cool enough for magmas to crystallize in the uppermost oceanic mantle then it is unlikely that the melt producing region extends to the base of the crust [cf. McKenzie and Bickle, 1988; Klein and Langmuir, 1987]. Geochemical arguments constrained from the composition of oceanic crust indicate that the mantle is depleted from 6 to 20% [Dick et al., 1984; Klein and Langmuir, 1987; Hofmann, 1988]. Estimates of the thickness of normal ocean crust are about 6 km [Reid and Jackson, 1981]. Assuming a simple one-dimensional model, a melting column of 36 km yields between 10 and 14 % accumulated fractional melt. If all of this melt were delivered to the crust the resulting thickness would be 4 to 8 km [Kinzler and Grove, 1992b]. Therefore, the mass of melt processed by fractional crystallization in the mantle below the ocean crust, could be on average 25% in a region of 6 km crust above a mantle depleted by 14% melting.

APPENDIX 1

The method for calculating fractional crystallization paths uses the mineral component projection scheme presented in Tormey et al. [1987]. An earlier version of this projection scheme is also presented in Grove et al. [1982, 1983]. In both of these papers there are errors in the equations for the recalculation scheme. In this appendix we describe the method of recalculation and provide the correct expressions for the Tormey et al. [1987] and Grove et al. [1982] projection schemes.

Recalculation Procedure

First, recalculate the composition of the lava, liquid or mineral from weight % oxides into mole %, and recalculate Al, Fe^{3+}, K, Na and P as $AlO_{1.5}$, $Fe^{3+}O_{1.5}$, $K_{0.5}O$, $Na_{0.5}O$ and $PO_{2.5}$. Next, combine molar MgO and FeO into the component FMO and combine $K_{0.5}O$ and $Na_{0.5}O$ into the component Alk. Transform to mineral components using the following expressions:

$$\mathrm{Sum} = SiO_2 - CaO - 2*Alk + Cr_2O_3 + TiO_2 + 2*PO_{2.5} + FeO_{1.5}$$

$$\mathrm{Qtz} = \{SiO_2 - 0.5*FMO - 1.5*CaO - 0.25*AlO_{1.5} - 2.75*Alk + 0.5*Cr_2O_3 + 0.5*TiO_2 + 2.5*PO_{2.5}\} / Sum$$

$$\mathrm{Plag} = \{AlO_{1.5} + NaO_{0.5} - KO_{0.5}\}/\{2.0*Sum\}$$

$$\mathrm{Oliv} = \{FMO + 0.5*(AlO_{1.5} - Alk) - CaO - TiO_2 - Cr_2O_3 + 1.667*PO_{2.5}\} / \{2.0*Sum\}$$

$$\mathrm{Cpx} = \{CaO - 0.5*AlO_{1.5} + 0.5*Alk - 1.667*PO_{2.5}\} / Sum$$

$$\mathrm{IHC} = \{Cr_2O_3 + TiO_2 + FeO_{1.5}\}/Sum$$

$$\mathrm{Or} = KO_{0.5} / Sum$$

$$\mathrm{Ap} = PO_{2.5} / \{3*Sum\}$$

These molar components are then transformed to oxygen units by multiplying the molar value by the number of oxygens in the mineral component and renormalizing to unity. Relative to the 2 oxygens in quartz the oxygen-based values become:

$$Qtz_{oxy} = Qtz_{mol}$$
$$Plag_{oxy} = 4*Plag_{mol}$$
$$Oliv_{oxy} = 2*Oliv_{mol}$$
$$Cpx_{oxy} = 3*Cpx_{mol}$$
$$IHC_{oxy} = 1.5*IHC_{mol}$$
$$Or_{oxy} = 4*Or_{mol}$$
$$Ap_{oxy} = 6*Ap_{mol}.$$

These are the coordinates used for the predictive expressions and for the projections presented in this paper (Figures 2 and 3 are exceptions, and are reproduced in original form using the authors's component choices). The error in Tormey et al. [1987] is in the Cpx equation where the term ($KO_{0.5}$ - $NaO_{0.5}$) should be ($KO_{0.5}$ +$NaO_{0.5}$) or Alk in the equations above. An error also appeared in an earlier version of the projection scheme reported by Grove et al. [1982, Figure 3 caption]. In this case, the error is in the Qtz expression, and the correct value should be:

$$\mathrm{Qtz} = \{SiO_2 - 0.5*(FeO + MgO) - 1.5*CaO - 0.25*AlO_{1.5} - 2.75*(KO_{0.5} + NaO_{0.5}) + 0.5*Cr_2O_3 + TiO_2\} / Sum.$$

To correct the expression for Qtz provided in Grove et al. [1982], Cr_2O_3 should be divided by 2 and TiO_2 has no multiplier.

APPENDIX 2

Calculation of Fractional Crystallization Paths

Three pieces of information are required to calculate the fractional crystallization path of a basalt: 1) a prediction of the phase appearance sequence as the magma cools, 2) the compositions of the crystallizing phases, and 3) the proportions of the crystallizing phases. Algebraic relations obtained from experimentally determined phase boundaries for the Oliv-Cpx-Qtz and Oliv-Cpx-Plag pseudo-ternary projections of the Oliv-Plag-Cpx-Qtz pseudo-quaternary, and constraints from mineral/melt exchange reactions derived from experimentally produced mineral-melt pairs, both described in this appendix, are used to calculate a MORB's crystallization path. The fractional crystallization modelled by this approach approximates perfect fractional crystallization.

We have written a FORTRAN program (MORBFRAC) that is available upon request, and that carries out this calculation, but the technique could be easily coded into a spreadsheet.

(1) Prediction of the phase appearance sequence as the magma cools. Step (1) requires prediction of the pertinent phase boundaries for a given lava composition at a fixed pressure. For MORB, crystallizing under anhydrous conditions at oxygen fugacities near the Quartz-Fayalite-Magnetite buffer the pertinent phase boundaries are the OPM, OPAM, and OPALM boundaries. Kinzler and Grove [1992a] have developed expressions that describe the change in composition of liquids saturated with the phases: oliv, plag, augite and low-Ca pyroxene (liquids on the OPALM boundary). They found that the composition of liquids saturated with these 4 solid phases could be expressed using the following linear relations involving the mineral components Oliv, Cpx, Plag and Qtz, and P (in kbar), T (in ^{o}C), *Mg #* (molar), *A #* (wt) and TiO_2 (wt %):

														R^2
T	=	1242	+	9	(P-0.001)	+	-120	(1-*Mg #*)	+	-88.8	*A #*	+	-6.6 TiO_2	0.88
Ol	=	.120	+	.008	(P-0.001)	+	.229	(1-*Mg #*)	+	-.335	*A #*	+	-.009 TiO_2	0.97
Cpx	=	.261	+	-.008	(P-0.001)	+	-.069	(1-*Mg #*)	+	-.104	*A #*	+	.012 TiO_2	0.95
Pl	=	.419	+	.012	(P-0.001)	+	-.135	(1-*Mg #*)	+	.691	*A #*	+	.000 TiO_2	0.95
Qz	=	.200	+	-.012	(P-0.001)	+	.000	(1-*Mg #*)	+	-.252	*A #*	+	.000 TiO_2	0.97

In the CMAS end member case, the analogous 5-phase equilibrium (fo+di+an+enstatite+melt) is univariant, and both the temperature and composition of the melt coexisting with fo+di+an+enstatite are described by pressure alone. When the pressure dependent terms go to zero at 0.001 kbar and the independent variables (1-*Mg #*), *A #*, and TiO_2 content go to zero, a set of constant terms is left in the expressions for the OPALM boundary that reflect the temperature and composition of a melt in the CMAS system coexisting with fo+di+an+enstatite at 0.001 kbar. The constant terms in the above equations were thus fixed to the values of T and melt components Ol, Cpx, Pl and Qz determined by Presnall et al. [1979] for the CMAS system at 0.001 kbar. The derivation of the above equations is further discussed in Kinzler and Grove [1992a].

Our purpose is to use these expressions to locate the position of the OPALM boundary, which we will use in turn to locate the OPAM and OPM boundaries. To locate the position of the saturation boundaries we calculate the *Mg #*, *A #* and TiO_2 content from the bulk composition of a lava and with this information use the above equations to calculate the projected position of the OPALM boundary at the desired pressure. The position of the OPAM and OPM boundaries in the Oliv-Cpx-Qtz and Oliv-Cpx-Plag pseudoternaries are then inferred or calculated. We find that it is practical and sufficient in the Oliv-Cpx-Qtz pseudoternary to simply sketch the OPAM boundary as extending from the OPALM boundary down to the Oliv-Cpx sideline parallel to the boundaries determined for the Serocki lava ALV-1690-20 (Figure 6). In the Oliv-Cpx-Plag pseudoternary, the OPAM boundary "line" projects as a "point" that lies directly on the projected position of the OPALM boundary "point". The OPM boundary in this projection is a line that extends down toward the Oliv-Plag sideline of the pseudo-ternary. Thus, the phase boundaries in the Oliv-Cpx-Plag pseudoternary provide enough information to constrain the relative roles of oliv, plag and augite fractionation from a given lava composition. To infer whether a low-Ca pyroxene is also crystallizing requires the Oliv-Cpx-Qtz pseudoternary projection, which distinguishes between the OPAM and OPALM boundaries.

We have calculated the orientation of the OPM boundary in the Oliv-Cpx-Plag pseudoternary from the 0.001-kbar experiments on AII96-18-1 and AII78-3-102 [Tormey et al., 1987] saturated with oliv and plag using the following procedure.

We use the method of Chayes [1969] for calculating the orientations of boundary curves in ternary systems. The Oliv - Cpx - Plag pseudoternary is recalculated into an orthogonal coordinate system with the following equations:

$$Y_i = \text{Oliv/[Oliv + Cpx]}$$
$$Y_j = \text{Oliv/[Oliv + Plag]}.$$

The OPM boundary "line" in the orthogonal coordinate system has the slope $4.447*Y_j$. Given the position of the OPALM boundary in the orthogonal coordinates (Y_i,Y_j), the slope - intercept formula is used to determine the equation of the boundary:

$$b = Y_i - 4.447*Y_j.$$

Computationally, one predicts the position of the OPALM boundary in the Oliv-Cpx-Plag pseudoternary at a chosen P from the *Mg #*, *A #* and TiO2 content for a given lava. In doing this one also has predicted the position of the OPAM boundary. Then one predicts the orientation of the OPM boundary "line". By comparing the plotted composition of a lava assumed to represent a liquid with the estimated OPALM, OPAM and OPM phase boundaries, the saturating phases and crystallization path may be inferred.

This method for estimating the positions of the OPALM, OPAM and OPM boundaries provides a useful tool for estimating crystallization paths followed by MORBs during fractional crystallization. The projection casts liquids into mineral components, and recalculates these components into a composition space where the phase boundaries can be shown and the fractionation path can be visualized in 2-dimensional diagrams. Oxygen units are used in the projection scheme [Grove et al., 1983] to preserve volume relationships among solid and liquid phases and to allow the crystallization assemblage to be inferred accurately. Transformation to mass units is achieved by normalizing through by the density of each phase. With mass proportions and compositions of phases crystallizing along these boundaries, the change in liquid composition can be calculated. Furthermore for each liquid, information on its coexisting phases are provided through information from the exchange relationships for Fe-Mg in olivine and augite with coexisting liquid, and for Ca-Na between plagioclase and liquid. A method for estimating augite compositional variation is presented below. The predictive expressions account for the effect of changes in bulk composition on the position of saturation boundaries and compensate for the variations in the compositions of coexisting solid phases.

(2) Estimating the compositions of the crystallizing phases. For a MORB magma under the oxygen fugacity conditions present in the oceanic crust and mantle, the phases that are likely to crystallize are oliv, plag augite and low-Ca pyroxene. Although it is likely that low-Ca pyroxene crystallization plays a role in the evolution of MORB magmas (as it is observed in oceanic gabbros), none of the calculations presented in this paper require the crystallization of a low-Ca pyroxene. This is partly due to the rather primitive nature of most of the lavas considered, either with respect to the *Mg #*, the Qtz component, or both. The composition of oliv that crystallizes from a melt is calculated using the *Mg #* of the liquid. Olivine is removed as stoichiometric $(Mg,Fe)_2SiO_4$ and the Mg/Fe is calculated using the Fe-Mg exchange distribution coefficient ($K_D^{Fe\text{-}Mg} = X_{Fe}^{xtl} * X_{Mg}^{liq} / [X_{Mg}^{xtl} * X_{Fe}^{xtl}]$). The value determined in our experiments on MORB is 0.29.

The plag composition to be removed is calculated as a function of pressure and melt compositional variables. In the case of plagioclase - liquid equilibria the exchange reaction is:

$$NaAlSi_3O_8^{xtal} + CaAl_2O_4^{liq} = CaAl_2Si_2O_8^{xtal} + SiO_2^{liq} + NaAlO_2^{xtal},$$

and the Ca-Na exchange distribution coefficient ($K_D^{Ca\text{-}Na}$) is:

$$K_D^{Ca\text{-}Na} = [X_{An}^{xtal}]\,[X_{SiO2}^{liq}]\,[X_{NaAlO2}^{liq}] / \{[X_{Ab}^{xtal}]\,[X_{CaAl2O4}^{liq}]\}.$$

We assembled an experimental data set of 171 plag - liquid assemblages produced over the pressure range of 0 .001 to 27 kbar. These plagioclase coexist with silicate liquids that range from quartz-normative to highly silica undersaturated. Data sources are: this study; Bartels et al. [1991]; Kinzler and Grove [1992a]; Grove et al. [1990]; Grove and Bryan [1983]; Fram and Longhi [1992]; Sack et al. [1987]; Tormey et al. [1987]. Mole fraction albite, anorthite and orthoclase are the components chosen for plagioclase and for liquid we used the recalculation scheme of Bottinga and Weill [1972] to recast liquid composition into the mole fraction components: SiO_2, $KAlO_2$, $NaAlO_2$, $CaAl_2O_4$, CaO, MgO, FeO, TiO_2, $PO_{2.5}$, etc. A model that accounts for effects of pressure and deviations from ideal behavior in the silicate melt provides a close approximation to the experimental data. The deviations from ideality of the liquid are approximated using empirical terms that resemble binary symmetric Margules parameters. The model is:

$$\text{Log}\, K_D^{Ca\text{-}Na} = 11.1068 - 0.0338{*}P\,(\text{in kbars}) - 4.4719{*}(1-X_{NaAlO2}^{liq})^2 - 6.9707{*}(1-X_{KAlO2}^{liq})^2.$$

The R^2 for this model is 0.88 for the 171 observations. This expression can be solved explicitly for plagioclase An content ($X_{CaAl2Si2O8}^{xtal}$):

$$A = X_{CaAl2O4}^{liq}/[X_{NaAlO2}^{liq} * X_{SiO2}^{liq}] * \text{Exp}[11.1068 - 0.0338{*}P\,(\text{in kbars}) - 4.4719{*}(1-X_{NaAlO2}^{liq})^2 - 6.9707{*}(1-X_{KAlO2}^{liq})^2],$$

$$X_{CaAl2Si2O8}^{xtal} = A/[1+A].$$

Figure 18 shows a comparison of the An content estimated by the model versus the An content measured in the MORB-experimental data used to calibrate the predictive expression. The model recovers plagioclase An content to within +/- 5 mole % An for all but 6 of the MORB-experimental data used in its calibration.

Augite shows compositional variation in response to changing pressure and temperature. In the case of MORB magmas that do not cover a significantly wide range in composition, the effect of increased pressure and temperature are coupled. As the pressure is increased, the temperature of augite saturation also increases. We use the following approximation to calculate the composition of the augite crystallizing from a MORB as a function of pressure. As pressure increases the major change that occurs is that the amount of Ca decreases and Al increases. The $K_D^{Fe\text{-}Mg}$ for augite remains constant at 0.27. The averages of augite analyses from our 0.001-kbar, 2-kbar and 8-kbar experiments are given in Table 6, recalculated in terms of molar components, and show the dominant effect of pressure on augite composition. For the purposes of a fractionation calculation we use these molar averages to predict the change in augite composition as a function of pressure. The Fe-Mg content of the augite is calculated using the liquid *Mg #* and the Fe-Mg K_D, and the augite is recalculated into weight percent.

(3) Estimating the proportions of the crystallizing phases. First the phase boundaries are estimated in the Oliv-Cpx-Plag and Oliv-Cpx-Qtz pseudoternaries, using the technique described above. Phase proportions are then estimated using an iterative approach. A first guess is made for the appropriate boundary (i.e., for the OPM boundary, 0.3 oliv+0.7 plag, in wt units, is usually a good starting point), and then tested to determine if the separation of the phases in small increments of the proportions guessed causes melt compositions at successive increments to track along the boundary in question. It is necessary to recalculate the boundary position with each increment of mass removed as the *Mg #*, *A #* and TiO_2 content of the melt change, and thus the boundary position may change. Practically speaking, as

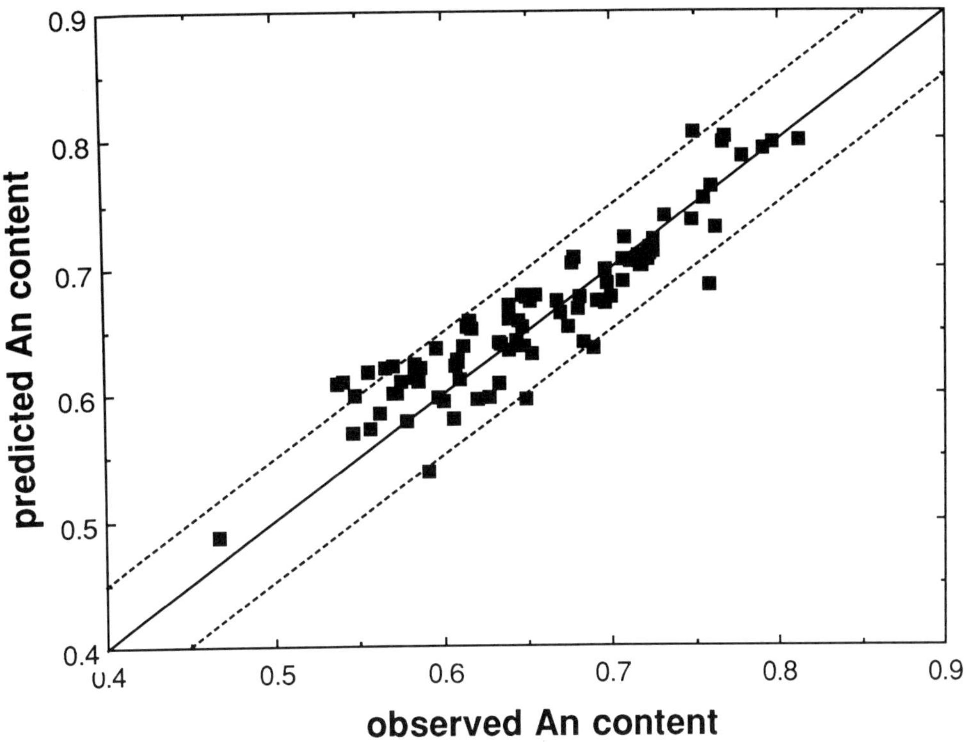

Fig. 18 Comparison of observed and predicted An contents for experimentally produced plagioclase, using the expression presented in Appendix 2. Shown are the MORB data used to calibrate the model. Pressure varies from 0.001 - 13 kbar.

the melt composition evolves from a chosen parent during fractional crystallization, the effects of the increasing *A #* and decreasing *Mg #* on the OPALM, OPAM and OPM boundaries tend to cancel each other in the Oliv-Cpx-Plag pseudoternary. Thus the boundaries may not change significantly over the compositional range spanning the MORBs evolution. In the case of the OPM boundary, the evolving melt compositions should track along the estimated OPM boundary 'line' towards the OPAM 'point'. If the trend defined by the evolving liquids does not match the estimated boundary, the proportions of oliv and plag are adjusted (generally the range is 0.25-0.35 for oliv and 0.75-0.65 for plag), and the calculation is repeated. In the case of the OPAM boundary, the evolving melt compositions should project near coincidentally in the Oliv-Cpx-Plag projection, on top of the estimated OPAM boundary (which coincides with the estimated OPALM boundary). Thus proportions of oliv, plag and augite are varied accordingly until this behavior is achieved (generally the range is 0.10-0.14 for oliv, 0.48-0.57 for plag, and 0.27-0.37 for augite). The proportions of oliv+plag+augite can also be checked in the Oliv-Cpx-Qtz pseudoternary by comparing the trend defined by the evolving liquids against the slopes sketched parallel to the experimentally determined boundaries as shown in Figure 6.

In summary:

1.) Calculate the positions of the OPALM, OPAM, and OPM boundaries using the bulk composition of the MORB in question at the pressure of interest.

2.) Plot up the predicted boundaries and compare the projected composition of the MORB to the position of the boundaries.

If the MORB lies in the Oliv phase volume on the Oliv-Plag-Cpx pseudoternary, oliv of the equilibrium composition is the only phase removed.

If the MORB lies on the OPM boundary, oliv+plag of equilibrium compositions are removed (in appropriate proportions, iterated to as described above).

If the MORB lies on the OPAM boundary, the compositions of all three phases are calculated using the appropriate K_D's, and all three phases are subtracted (in appropriate proportions, iterated to as described above).

If the MORB lies in the Plag phase volume on the Oliv-Plag-Cpx pseudoternary, plag of the equilibrium composition is removed.

In our calculation fractionation is approximated by removing equilibrium compositions of phases incrementally in small amounts. Commonly, 1 to 2 wt % total solid(s) is (are) removed. By this technique the fraction of liquid remaining (F) changes by 0.99^n to 0.98^n where n is the number of fractionation steps. The melt composition is renormalized to 100%.

TABLE 6. Variation of Augite Composition and Phase Proportions Crystallizing Along the Oliv - Plag - Augite - Liquid Boundary

P (kbar)	SiO_2	Al_2O_3	TiO_2	FMO	CaO	Na_2O
Augite Compositions[a] (mole % with FeO and MgO combined)						
0.001	48.88	1.41	0.49	29.18	19.82	0.23
2	48.89	1.45	0.61	30.64	18.09	0.30
8	48.65	2.65	0.49	30.84	16.77	0.42

P (kbar)	Oliv	Plag	Aug
Phase Proportions[b] (weight %)			
0.001	8(2)	55(2)	37(2)
2	11(3)	59(2)	30(4)
8	16(2)	57(3)	27(3)

[a] Estimated from compositions in experiments reported in Table 3.
[b] Deduced from materials balance calculations on multiply saturated MORB experiments for oliv - plag - augite crystallization. Parenthesized units show 2 sigma standard errors on phase proportion estimates.

Return to 1). The new fractionated melt composition is again used to estimate the positions of the saturation boundaries in the Oliv - Plag - Cpx pseudoternary. The predicted position of the saturation boundaries is compared with the new melt composition. If the liquid is found to lie on a saturation boundary, the crystallization assemblage is changed to the appropriate assemblage (oliv+plag, oliv+plag+augite, etc.). The compositions of the phases are calculated using K_D's and these compositions are again subtracted from the liquid in the appropriate proportions.

Acknowledgments. This work is an outgrowth of discussions and interactions at the 1st Summer RIDGE Institute. We thank Dave Christie, Henry Dick, Emily Klein, Charlie Langmuir, Peter Michael and Tom Sisson for discussion and interest. Reviews by Dave Christie and John Sinton were especially helpful. We thank Donna Blackman and Jason Phipps-Morgan for editorial assistance. This research was supported by NSF-OCE 8816903 and NSF-OCE 9115901.

REFERENCES

Ahern, J. L. and D. L. Turcotte, Magma migration beneath an ocean ridge, *Earth Planet. Sci. Lett., 45,* 115-122, 1979.

Akella, J., and G. C. Kennedy, Melting of Gold, Silver and Copper-Proposal for a new high-pressure calibration scale, *J. Geophys. Res., 76,* 4969-4977, 1971.

Albee, A. L. and L. Ray, Correction factors for electron microprobe analysis of silicates, oxides, carbonates, phosphates and sulfates, *Anal. Chem., 42,* 1408-1414, 1970.

Bartels, K. S., R. J. Kinzler, and T. L. Grove, High pressure phase relations of a near primary high alumina basalt from Medicine Lake Highland, N. California, *Contrib. Mineral. Petrol., 108,* 253-270, 1991.

Basaltic Volcanism Study Project, *Basaltic Volcanism on the Terrestrial Planets.* Pergamon Press, Inc., New York. 1286 pp, 1981.

Bence, A. E., and A. L. Albee, Empirical correction factors for the electron microanalysis of silicates and oxides, *J. Geol., 76,* 382-403, 1968.

Biggar, G. M., and D. J. Humphries, The plagioclase, forsterite diopside, liquid equilibrium in the system CaO-Na2O-MgO-Al2O3-SiO2, *Mineral. Mag., 44,* 309-314, 1981.

Bodinier, J. L., Geochemistry and petrogenesis of the Lanzo peridotite body, western Alps, *Tectonophysics, 149,* 67-88, 1988.

Boyd, F. R., and J. L. England, Apparatus for phase equilibrium studies at pressures up to 50 kbars and temperatures up to 1750ºC, *J. Geophys. Res., 65,* 741-748, 1960.

Bryan, W. B., L. W. Finger, and F. Chayes, Estimating proportions in petrographic mixing equations by least squares approximation, *Science, 163,* 926-927, 1969.

Bryan, W. B., G. Thompson, and J. N. Ludden, Compositional variation in normal MORB from 22° - 25° N; mid-Atlantic ridge and Kane Fracture Zone, *J. Geophys. Res., 86,* 11815-11836, 1981.

Chayes, F., Curve fitting in the ternary diagram, *Carnegie Inst. Wash. Yearbook, 67,* 236-239, 1969.

Detrick, R. S., P. Buhl, E. Vera, J. Mutter, J. Orcutt, J. Madsen, and T. Brocker, Multi-channel seismic imaging of a crustal magma chamber along the East Pacific Rise, *Nature, 236,* 35-41, 1987.

Detrick, R. S., J. D. Mutter, P. Bulh, and I. I. Kim, No evidence from multichannel reflection data for a crustal magma chamber in the MARK area on the mid-Atlantic Ridge, *Nature, 347,* 61-64, 1990.

Dick, H. J. B., R. L. Fisher, and W. B. Bryan, Mineralogical variability of the uppermost mantle along mid-ocean ridges, *Earth Planet. Sci. Lett., 69,* 88-106, 1984.

Dick, H. J. B., Abyssal peridotites, very slow spreading ridges and ocean ridge magmatism, *Magmatism in the Ocean Basins,* edited by A. D. Sounders, & M. J. Norry, *Geol. Soc. Spec. Publ. No. 42,* 71-105, 1989.

Dungan M. A. and J. M. Rhodes, Residual glasses and melt inclusions in basalts from DSDP Legs 45 and 46: evidence for magma mixing, *Contrib. Mineral. Petrol., 67,* 417-431, 1978.

Elthon, D., The petrogenesis of primary mid-ocean ridge basalts, *Reviews in Aquatic Sciences, 2, 1,* 27-53, 1990.

Elthon, D., Pressure of origin of primary mid-ocean ridge basalts, *Magmatism in the Ocean Basins,* edited by A. D. Saunders & M. J. Norry, *Geol. Soc. Spec. Publ. No. 42,* 125-136, 1989.

Elthon, D., Petrology of gabbroic rocks from the mid-Cayman Rise spreading center, *J. Geophys. Res., 92,* 658-682, 1987.

Fisk, M. R., J. -G. Schilling, and H. Sigurdsson, An experimental investigation of Iceland and Reykjanes Ridge Tholeiites: I. Phase Relations, *Contrib. Mineral. Petrol., 74,* 361-374, 1980.

Flower, M. F. J., W. Ohnmacht, H. -U. Schmincke, I. L. Gibson, P. T. Robinson, and R. Parker, Petrology and geochemistry of basalts from Hole 396B, Leg 46, *Initial Reports of the Deep Sea Drilling Project, 46,* Washington (U.S. Government Printing Office), p. 179-214, 1978.

Fram, M., and J. Longhi, Phase Equilibria of Dikes Associated with Proterozoic Anorthosite Complexes, in press for *Amer. Mineral.,* 1992.

Francis, D., The pyroxene paradox in MORB glasses - a signature of picritic parental magmas?, *Nature, 319,* 586-588, 1986.

Fujii, T., and H. Bougault, Melting relations of a magnesian abyssal tholeiite and the origin of MORBs, *Earth Planet. Sci. Lett., 62*, 283-295, 1983.

Fujii, T., and C. M. Scarfe, Composition of liquids coexisting with spinel lherzolite at 10 kbar and the genesis of MORBs, *Contrib. Mineral. Petrol., 90*, 18-28, 1985.

Ghiorso, M. S., Modeling magmatic systems: Thermodynamic relations, in: *Thermodynamic modeling of geological materials: Minerals, fluids and melts*, eds. P. H. Ribbe, *Reviews in Mineralogy, vol. 17*, 443-463, 1987.

Grove T. L., D. C. Gerlach, and T. W. Sando, Origin of calc-alkaline series lavas at Medicine Lake volcano by fractionation, assimilation and mixing, *Contrib. Mineral. Petrol., 80*, 160-182, 1982.

Grove, T. L., D. C. Gerlach, T. W. Sando, and M. B. Baker, Origin of calc-alkaline series lavas at Medicine Lake volcano by fractionation, assimilation and mixing: Corrections and clarifications, *Contrib. Mineral. Petrol., 82*, 407-408, 1983.

Grove, T. L., and W. B. Bryan, Fractionation of pyroxene-phyric MORB at low pressure: an experimental study, *Contrib. Mineral. Petrol., 84*, 293-309, 1983.

Grove, T. L., and T. C. Juster, Experimental investigations of low-Ca pyroxene stability and olivine - pyroxene - liquid equilibria at 1-atm in natural basaltic and andesitic liquids, *Contrib. Mineral. Petrol., 103*, 287-305, 1989.

Grove, T. L., R. J. Kinzler, and W. B. Bryan, Natural and experimental phase relations of lavas from Serocki Volcano, *Ocean Drill. Proc., Scientific Results, 106/109*, 9-17, 1990.

Hofmann, A. W., Chemical differentiation of the Earth: the relationship between mantle, continental crust, and oceanic crust, *Earth Planet. Sci. Lett., 90*, 297-314, 1988.

Humphris, S. E., W. B. Bryan, G. Thompson, and L. K. Autio, Morphology, geochemistry and evolution of Serocki volcano, *Ocean Drill. Proc., Scientific Results, 106/109*, 67-84, 1990.

Jakobsson, S. E., J. Jonsson, and F. Shido, Petrology of the western Reykjanes Peninsula, Iceland, *J. Petrol. 19*, 669-705, 1978.

Johnson, K. T. M., H. J. B. Dick, and N. Shimizu, Melting in the oceanic mantle: an ion microprobe study of diopsides in abyssal peridotites, *J. Geophys. Res., 95*, 2661-2678, 1990.

Keleman, P. B., K. T. M. Johnson, R. J. Kinzler, and A. J. Irving, High-field-strength element depletions in arc basalts due to mantle-magma interaction, *Nature, 345*, 521-524, 1990.

Kinzler, R. J. and T. L. Grove, Primary magmas of mid-ocean ridge basalts, 1: Experiments and methods, *J. Geophys. Res.*, in press, 1992.

Kinzler, R. J. and T. L. Grove, Primary magmas of mid-ocean ridge basalts, 2: Applications, *J. Geophys. Res.*, in press, 1992.

Klein, E. M., and C. H. Langmuir, Global correlations of ocean ridge basalt chemistry with axial depth and crustal thickness, *J. Geophys. Res., 92*, 8089-8115, 1987.

Klein, E. M., and C. H. Langmuir, Local versus global variations in ocean ridge basalt composition: a reply, *J. Geophys. Res., 94*, 4241-4252, 1989.

Kushiro, I., On the nature of silicate melt and its significance in magma genesis: regularity in the shift of liquidus boundaries involving olivine, pyroxene and silica minerals, *Amer. J. Sci., 275*, 411-431, 1975.

Langmuir, C. H., Geochemical consequences of in situ crystallization, *Nature, 340*, 199-205, 1989.

Langmuir, C. H. and J. F. Bender, The geochemistry of oceanic basalts in the vicinity of transform faults: observations and implications, *Earth Planet. Sci. Lett., 69*, 107-127, 1984.

Lindsley, D. H., Melting relations of plagioclase at high pressures, *New York State Museum and Science Service Memoir 18*, 39-46, 1970.

Longhi, J., Comparative liquidus trends of hypersthene-normative basalts at low pressure, *Amer. Mineral., 76*, 785-800, 1991.

Longhi, J. and V. Pan, A reconnaissance study of phase boundaries in low-alkali basaltic liquids, *J. Petrol., 29*, 115-148, 1988.

Machado, N., J. N. Ludden, C. Brooks, and G. Thompson, Fine scale isotopic heterogeneity in the sub-Atlantic mantle, *Nature, 295*, 226-229, 1983.

Mahood, G. A. and D. R. Baker, Experimental constraints on depths of fractionation of mildly alkalic basalts and associated felsic rocks: Pantelleria, Straits of Sicily, *Contrib. Mineral. Petrol., 93*, 251-264, 1986.

McKenzie, D., The generation and compaction of partially molten rock, *J. Petrol., 25*, 713-765, 1984.

McKenzie, D., The extraction of magma from the crust and mantle, *Earth Planet. Sci. Lett., 74*, 81-91, 1985.

McKenzie, D., and M. J. Bickle, The volume and composition of melt generated by extension of the lithosphere, *J. Petrol., 29*, 625-629, 1988.

Meyer, P. S. and T. Shibata, Complex zoning in plagioclase feldspars from ODP Site 648, *Ocean Drill. Proc., Scientific Results, 106/109*, 123-142, 1990.

Morse, S.A., *Basalts and Phase Diagrams*, Springer, New York, 493 p., 1980.

Nicolas, A., A melt extraction model based on structural studies in mantle peridotites, *J. Petrol., 27*, 999-1022, 1986.

Nicolas, A. and A. Prinzhofer, Cumulative or residual origin for the transition zone in ophiolites: Structural evidence, *J. Petrol., 24*, 188-206, 1983.

Nielsen, R. L., A model for the simulation of combined major and trace element liquid lines of descent, *Geochim. Cosmochim. Acta, 52*, 27-39, 1988.

O'Donnell, T. H. and D. C. Presnall, Chemical variations of the glass and mineral phases in basalts dredged from 25o-30oN along the mid-Atlantic ridge, *Amer. J. Sci., 280*, 845-868, 1980.

O'Hara, M..J., Are any ocean floor basalts primary magma?, *Nature, 220*, 683-686, 1968.

O'Hara, M. J., Data reduction and projection schemes for complex compositions, *Progress in Exptl. Petrol. 3rd report. NERC London, Publication Series D, No. 6*, 103-126, 1976.

O'Hara, M. J., Importance of the "shape" of the melting regime during partial melting of the mantle, *Nature, 314*, 58-62, 1985.

O'Hara, M. J., and R. E. Mathews, Geochemical evolution in an advancing, periodically replenished, periodically tapped, continuously fractionated magma chamber, *J. Geol. Soc. London, 138*, 237-277, 1981.

Presnall, D. C., S. A. Dixon, T. H. O'Donnell, N. L. Brenner, R. L. Schrock, and D. W. Dycus, Liquidus phase relations on the join Diopside-Forsterite-Anorthite from 1 atm to 20 kbar: Their bearing on the generation and crystallization of basaltic magma, *Contrib. Mineral. Petrol., 66*, 203-220, 1978.

Presnall, D. C., T. H. Dixon, T. H. O'Donnell, and S. A. Dixon, Generation of mid-ocean ridge tholeiites, *J. Petrol., 20*, 3-35, 1979.

Presnall, D. C., and J. D. Hoover, High pressure phase equilibrium constraints on the origin of mid-ocean ridge basalts, *Magmatic processes: physicochemical principles*, edited by B. O. Mysen, pp. 75-89, Geochemical Society, University Park, PA, 1987.

Quick, J., Petrology and petrogenesis of the Trinity peridotite, an upper mantle diapir in the eastern Klamath mountains, northern California, *J. Geophys. Res., 86*, 11837-11863, 1981.

Reid, I., and H. R. Jackson, Oceanic spreading rate and crustal thickness, *Marine Geophys. Res., 5*, 165-172, 1981.

Sinton, J. M., D. S. Wilson, D. M. Christie, R. N. Hey, and J. R. Delaney, Petrologic consequences of rift propagation on oceanic spreading ridges, *Earth Planet. Sci. Lett., 62*, 193-207, 1983.

Shi, P., and G. Libourel, The effects of FeO on the system CMAS at low pressure and implications for basalt crystallization processes, *Contrib. Mineral. Petrol., 108*, 129-145, 1991.

Sobolev, A. V. and N. Shimizu, The direct evidence for existence of ultra- depleted melts in the oceanic mantle: SIMS data on the composition of melt inclusion in olivine from MORB (abstract), *EOS, Trans. Amer. Geophys. U., 72*, 544, 1991.

Stakes, D. S., J. W. Shervais, and C. A. Hopson, The volcanic-tectonic cycle of the FAMOUS and AMAR valleys, mid-Atlantic ridge

(36°17'N): Evidence from basalt glass and phenocryst compositional variations for a steady state magma chamber beneath the valley midsections, AMAR3, *J. Geophys. Res., 89,* 6995-7028, 1984.

Stolper, E., A phase diagram for mid-ocean ridge basalts: Preliminary results and implications for petrogenesis, *Contrib. Mineral. Petrol., 74,* 13-27, 1980.

Thompson, R. N., Primary basalts and magma genesis I. Skye, Northwest Scotland, *Contrib. Mineral. Petrol., 45,* 317-341, 1974.

Thompson, R. N., Primary Basalts and magma genesis II. Snake River Plain, Idaho, U.S.A, *Contrib. Mineral. Petrol., 52,* 213-232, 1975.

Thompson, G., W. B. Bryan, and W. G. Melson, Geological and geophysical investigation of the mid-Cayman rise spreading center: geochemical variation and petrogenesis of basalt glasses, *J. Geol., 88,* 41-55, 1980.

Thompson, G., W. B. Bryan, and S. E. Humphris, Axial volcanism on the East Pacific Rise, 10 - 12°N, *Magmatism in the Ocean Basins,* edited by A. D. Sounders, & M. J. Norry, *Geol. Soc. Spec. Publ. No. 42,* 181-200, 1989.

Tormey, D. R., T. L. Grove, and W. B. Bryan, Experimental petrology of normal MORB near the Kane Fracture Zone: 22½-25½N, mid-Atlantic ridge, *Contrib. Mineral. Petrol., 96,* 121-139, 1987.

Walker, D., and S. E. DeLong, Soret Separation of Mid-Ocean Ridge Basalt Magma, *Contrib. Mineral. Petrol., 79,* 231-240, 1982.

Walker, D., T. Shibata, and S. E. DeLong, Abyssal tholeiites from the Oceanographer Fracture Zone, II. Phase equilibria and Mixing, *Contrib. Mineral. Petrol., 70,* 111-125, 1979

Weaver, J. and C. H. Langmuir, Calculation of phase equilibrium in mineral-melt systems, *Computers and Geosciences, 16,* 1-19, 1990.

Williams, D. W., Improved cold seal pressure vessels to operate to 1100 oC at 3 kilobars, *Am. Mineral., 53,* 1765-1769.

Yoder, H. S., Jr. and C. E. Tilley, Origin of basalt magmas: experimental study of natural and synthetic rocks, *J. Petrol., 3,* 342-532, 1962.

T.L. Grove and R.J. Kinzler[1], Department of Earth, Atmospheric and Planetary Sciences, Massachusetts Institute of Technology, Cambridge, MA 02139.

W.B. Bryan, Department of Geology and Geophysics, Woods Hole Oceanographic Institution, Woods Hole, MA 02543.

1 Currently at Lamont Doherty Geological Observatory and Department of Geological Sciences, Columbia University, Palisades, New York 10964

The Effects of Plate Thickening on Three-Dimensional, Passive Flow of the Mantle Beneath Mid-Ocean Ridges

DONNA K. BLACKMAN[1]

School of Oceanography, University of Washington, Seattle WA 98195

DONALD W. FORSYTH

Department of Geological Sciences, Brown University, Providence RI 02912

A new three-dimensional model of passive mantle flow beneath a ridge-transform-ridge plate boundary shows that the thickening of the plates with age has a significant effect on the details of the predicted flow pattern. The Boundary Element Method is used to compute the flow induced in a constant viscosity asthenosphere by motion of the overlying rigid plates. The spreading plates are specified to thicken in proportion to the square-root of their age and are assumed to behave rigidly. This creates a diverging wedge beneath the ridge axis which, in combination with the motion of the thick, older plate edge at a ridge-transform intersection, introduces complexities in the flow not predicted by previous passive flow models. Subaxial upwelling rates exceed the spreading rate by up to 20% and along-axis flow is required to compensate for a material deficit in the vicinity of the ridge-transform intersection where the endwall impedes upwelling and drags the asthenosphere in the direction of the older plate's spreading. The mantle temperature field associated with the predicted flow contains high gradients across the transform; high axial temperatures grade into lower values in a broad V shape that points into the ridge-transform intersection. This model can explain more of the observed mantle Bouguer anomaly patterns over ridge-transform systems than previous models but, under the assumption that density anomalies are due only to thermal expansion, a good portion of the anomaly remains. Although more realistic assumptions about melting and the reduction of density in the residual mantle could improve the fit, the surface gravity signal is fairly insensitive to the details of the thermal structure introduced by plate thickening; the distribution of melt production and the pattern of viscous stresses within the mantle are more strongly affected by plate thickening.

1. Introduction

The bathymetric and gravimetric signature of a mid-ocean ridge reflects, in part, the thermal structure of the underlying mantle. The temperature distribution results from flow in the asthenosphere and conductive cooling of the lithospheric plates. Therefore, gravity and bathymetry data can provide constraints on the pattern of flow and density variability in the mantle. Crustal structure may also play an important role in shaping the observed fields and potential data alone cannot provide unique determination of the source region for the anomalies. Nevertheless, fairly complete bathymetric and gravimetric coverage exists along several slow-spreading ridge-transform systems and an approach in which the predictions of end-member models are compared to these data leads to useful insight on subaxial processes. For models of mantle structure, we focus on the role of passive flow that is induced by spreading of the lithospheric plates to which the asthenosphere is assumed to be perfectly coupled.

Mid-ocean ridges are segmented; zones of deformation and/or strike-slip faulting offset the ridge axes which are dominated by extensional faulting and volcanism. Ridge segmentation may be the cause or the effect of along-strike variations in mantle upwelling beneath spreading centers. In this paper we investigate the extent to which offsets of the ridge can *cause* significant three-dimensionality in the passively induced mantle flow and corresponding thermal structure.

[1] Now at Institute of Geophysics and Planetary Physics, Scripps Institution of Oceanography, La Jolla, California.

Mantle Flow and Melt Generation at Mid-Ocean Ridges
Geophysical Monograph 71
Copyright 1992 American Geophysical Union

The segmented nature of mid-ocean spreading centers has been clearly illustrated in magnetic anomaly patterns [Schouten and Klitgord, 1982] and ridge flank morphology [Macdonald et al., 1988]. Almost all ridge segments vary in axial depth from a shallow, roughly central section gradually deepening towards the termination at an offset where steeper topographic gradients occur [Sleep and Biehler, 1970; Fox and Gallo, 1984; Blackman and Forsyth, 1989]. Fox and Gallo [1984] showed that the magnitude of the nodal deep at a ridge-transform intersection (RTI) increases with the age offset of the transform. They inferred that thermal contraction due to cooling near the older plate edge is primarily responsible for the greater intersection depths. Forsyth and Wilson [1984] determined that the direct thermal effect of the cold plate edge would probably only affect the ridge-transform intersection region and thus would not explain the more gradual deepening along most of the segment. Blackman and Forsyth [1989] showed that segment length and spreading rate also play significant roles in the observed 500-2500 m of deepening along the axis. The dependence of down-strike (middle of the segment towards the RTI) axial relief on segment length and spreading rate suggests a role for along-axis mantle flow.

Subaxial upwelling may be reduced near the end of a ridge segment due to viscous resistance to flow along the endwall comprised of the juxtaposed plate edge [Sleep and Biehler, 1970]. If plate accretion is relatively constant along a ridge segment, down-strike transport of material is required to balance the deficit caused by a reduction in upwelling at the endwall. Simple conduit models have shown that a few hundred meters of along-strike relief could reflect viscous stresses generated along the endwall and by the induced along-axis flow [Parmentier and Forsyth, 1985]. Forsyth [1989] presented a conduit model that includes subaxial drag due to horizontal motion of the endwall in addition to the divergent motion of the sidewalls. Along-axis velocities predicted by this model were a significant fraction of the upwelling velocity, thus potentially an important source of along-axis variability in thermal structure.

Although conduit models have been successful in explaining the gross characteristics of mid-ocean ridge structure [Lachenbruch, 1973, 1976] there are serious doubts about the validity of the assumption that the rigid boundary of a relatively narrow channel can be maintained. Phipps Morgan et al. [1987] show that, if it exists at all, this rigid boundary is

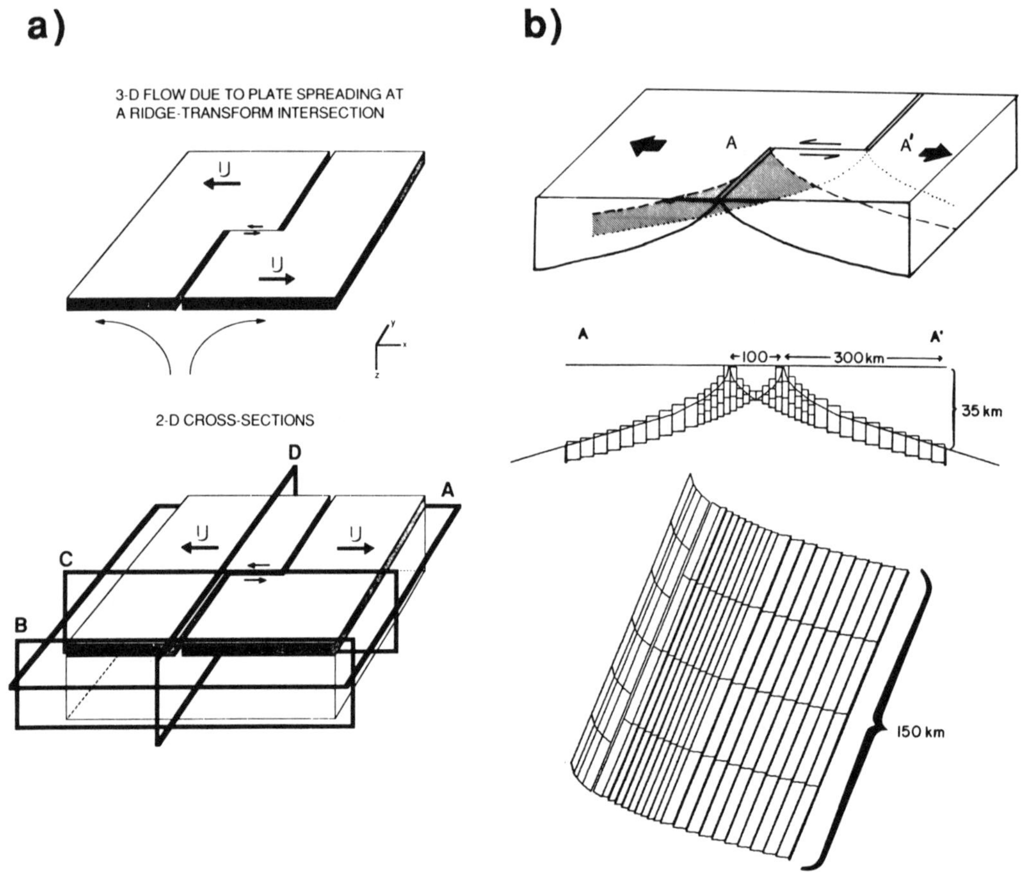

Fig. 1. Sketches of passive flow model geometries for 100 km transform offset. (a) flat-plate model of *Phipps Morgan and Forsyth* [1988]. Location of cross sections shown in Figures 2 and 3 are indicated in lower diagram. (b) thickening-plate model where shading indicates surfaces for which boundary element discretization is shown in lower diagrams.

probably not thermally controlled. They predict that isotherms deepen more smoothly, similar to the square-root of age predictions of cooling half-space models. Conduit models will over-estimate the horizontal velocities at depth in comparison to models in which the rigid plates thicken in proportion to the square-root of their age, so we use a new 3-D Boundary Element Model (BEM) to estimate the flow and thermal structure for the latter case.

In recent years the 'flat-plate' model of Phipps Morgan and Forsyth [1988] has been used by several groups to calculate the three-dimensional flow, temperature field and mantle Bouguer anomalies over a series of spreading ridges offset by transform faults. Observed mantle Bouguer anomaly patterns along ridge-transform systems are characterized by axial lows grading out to higher values off-axis [Prince and Forsyth, 1988; Kuo and Forsyth, 1989; Lin et al., 1990; Morris et al., 1991; Blackman and Forsyth, 1991]. This general pattern is in agreement with the predicted 3-D thermal structure of the mantle resulting from upwelling driven by spreading of flat plates [Phipps Morgan and Forsyth, 1988]. However, the shape of the predicted axial anomaly in the vicinity of ridge-transform intersections is more rounded than the observed anomaly, suggesting that the thickening of the plate with age alters the flow and thermal regimes significantly in this region.

Investigation of the detailed flow near ridge-transform intersections requires a more complete representation of the rigid plates whose thickness varies rapidly in this region (Figure 1). For example, a ridge spreading at a half-rate of 10 mm/yr that is offset 100 km along a transform fault will be juxtaposed to a cold plate edge 20-40 km thick at the RTI, depending on whether a mechanical (600-800°C isotherm) or thermal (1150-1350°C) definition of the plate is used. The thickness of this rigid, moving boundary is a significant fraction of the other length scales in the problem and the geometry of its position will introduce both resistance to axial upwelling and skew due to the horizontal velocities induced at depth. The magnitude of these modifications in mantle structure can be estimated using a 3-D BEM approach.

2. Models of Passive Mantle Flow at Mid Ocean Ridges

2a. Previous flow models

The gross form of mantle upwelling beneath a spreading center is predicted by a two-dimensional corner-flow model [Batchelor, 1967; Reid and Jackson, 1981]. The flow is driven by separation of two flat surfaces that move at a specified rate and are coupled to the underlying fluid. The viscous fluid rises beneath the spreading axis with maximum vertical velocities reaching about six-tenths of the spreading rate. Horizontal velocities decrease monotonically with depth.

When the finite thickness of the rigid plates is considered, or when the plates do not reach the full spreading rate until outside the axial region, the subaxial flow is affected. Phipps Morgan et al. [1987] present several 2-D models in which it is clear, although not directly discussed, that vertical velocities beneath the ridge axis can surpass the spreading rate. This is due to the narrowing-upward shape of the plate base which focuses the flow near the spreading axis. Plate rigidity is incorporated by assigning all velocities in the plate to be the spreading rate. Therefore, horizontal flow rates are somewhat higher at depth than in the corner flow models with thin, flat plates due to the fact that rates do not begin to decrease with depth until beyond the base of the plate. Lin and Parmentier [1989] corroborate these findings with iterative 2-D modeling that produces self consistent subaxial flow and thermal fields.

The three-dimensional form of passive mantle flow beneath a ridge-transform system is approximated by assuming, analogous to 2-D corner flow, that ridge axes occupy an infinitely thin line from which constant thickness plates diverge. Strike-slip motion occurs along a perpendicular line that connects the ends of the offset ridge segments. The predicted flow pattern for this model [Phipps Morgan and Forsyth, 1988] differs from the 2-D model in the vicinity of RTIs (Figure 2): vertical velocities decrease from the spreading velocity to zero as the fracture zone is crossed from the ridge to the older plate; material flows into the ridge from across the transform at rates up to 25% of the spreading velocity. Both these effects are reasonable but the prediction that they occur all the way up to the surface is not. As the plate cools with distance from the axis it behaves (ideally) as a rigid layer moving only horizontally with velocity equal to the spreading rate. Therefore across the transform from a ridge, vertical and along-axis mantle velocities should be zero within the plate. In the following section we present a model that estimates the effect that the finite plate thickness has on the pattern of mantle flow beneath a series of ridges offset by transform faults.

2b. 3-D Boundary Element flow model

The boundary element method can be used to calculate the flow in an infinite, iso-viscous medium bounded by a surface along which specified velocity or stress conditions exist [Crouch and Starfield, 1983]. The boundary is discretized into elements at whose midpoints the desired boundary conditions are exactly met. Green's functions are used to determine the flow induced in the medium by the motion of each individual element. The resultant velocity field is calculated by summing the effects of all the boundary elements. Since the Green's functions are analytic, the points at which the flow is calculated are not constrained to follow any regular pattern or spacing. The main factor in determining the accuracy of a computed flow field is the discretization interval of the boundary: the size of the boundary elements in relation to the desired resolution of the field and their ability to reasonably represent the shape of the boundary.

The interface between lithosphere and asthenosphere is discretized into boundary elements for a model of passive mantle flow beneath two spreading ridges offset by a transform fault. Away from the ridge axes, the plate base is specified to deepen in proportion to the square-root of its age [Turcotte and Schubert, 1982]:

$$x_3 = 2\sqrt{\frac{\kappa x_2}{V}}\,\mathrm{erf}^{-1}\left(\frac{T}{T_m}\right)$$

Where x_3 is depth of the plate base defined by an isotherm (T),

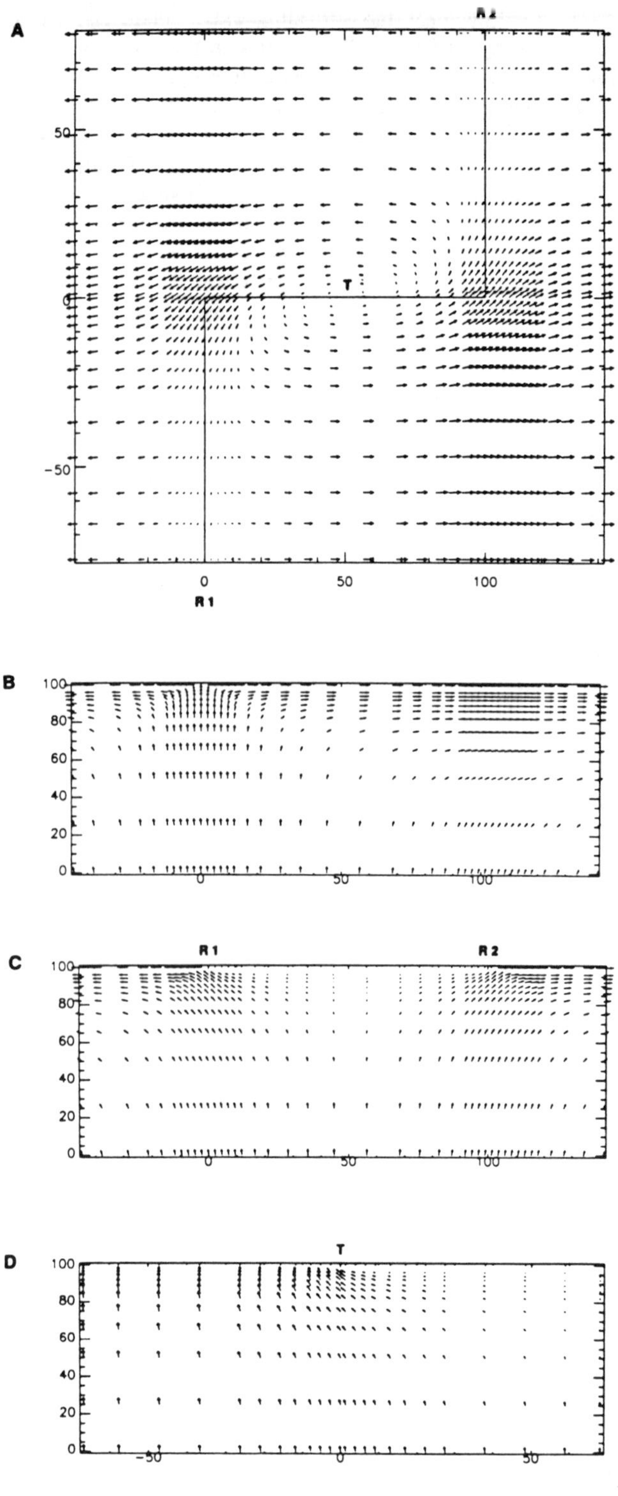

Fig. 2. Velocity vector diagrams for flat-plate model. Locations shown in Figure 1a. Distances along axes are in km. (a) map view for plane at 14 km depth. Surface position of ridges and transform are shown by solid line. (b) cross-axis plane near center of ridge segment. (c) cross-axis plane at transform. (d) along-axis plane beneath ridge 1 and older plate beyond transform.

T_m is the temperature of the asthenosphere, κ is thermal diffusivity, V is spreading rate, and x_2 is distance from axis. At the end of the ridge segment, the juxtaposed plate edge across the transform is assumed to be vertical. Similar to the flat-plate case in Figure 1*a*, the two ridges are offset 100 km along the intervening transform fault (Figure 1*b*). For the models in this paper, spreading rate is taken as 10 mm/yr, thermal diffusivity is 10^{-6} m^2/s and T_m is about 1350°C. The base of the plates corresponds to the depth of the 800°C isotherm. This temperature represents the base of the transition from elastic to ductile mantle behavior as indicated by maximum oceanic earthquake focal depths [Bergman and Solomon, 1984], plate flexure [McNutt, 1982], and rock deformation mechanisms [Parmentier and Phipps Morgan, 1990]. In the plane of the fracture zone, boundary elements are only required where the plate edge would be in contact with asthenosphere. The plate base has a stair-stepped shape where rectangular boundary elements in the fracture zone plane meet the plate base. This is necessary in order to form a complete boundary against the asthenosphere in the region of interest. Far off-axis, the boundary elements extend to a distance of 300 km from the closest ridge axis. Thus, for the region in which the flow field is calculated, the influence of the edge is negligible [Blackman, 1991]. The locations at which the velocities are computed follow the pattern used in the flat-plate model. Points are closely spaced in the vicinity of high velocity gradient, near the ridge axes and transform fault, and larger spacing is used away from these regions. The velocities are determined in a region that extends 75 km to either side of the fracture zone plane, 250 km from the ridge axes and to a depth of 100 km. Since the Boundary Element Method is formulated for an infinite medium, there is no prescribed condition at depth in the asthenosphere.

The flow pattern predicted by the BEM model is illustrated by velocity vectors for several cross-sections in Figure 3. Far from the transform, upwelling is symmetric beneath the ridge axes and horizontal flow follows the plate spreading direction. Near the transform plane, upwelling is reduced at shallow levels and horizontal velocities are skewed due to the existence and motion of the endwall. The old plate edge that forms the endwall is about 20 km thick right at the RTI for this model but upwelling mantle encounters a variable sized edge to either side of the axis since the juxtaposed plate continues to cool and thicken with age. Material is pulled away from the inside corner at the RTI and the resulting pressure drop induces along-axis flow to that region.

There are significant differences between the predicted flow pattern for the flat-plate model and the thickening-plate model. The vertical velocity component for the two models is contoured at depths of 8, 14, 25 and 35 kms in Figure 4*a*. The thickening-plate model predicts narrower regions of faster upwelling at each of these depths. A similar narrowing and increase in magnitude is obtained for the cross-axis velocity component (Figure 4*b*) due to the presence of a moving boundary that increases in depth away from the spreading axis. The thickness of the (old) plate edge at the end of a ridge

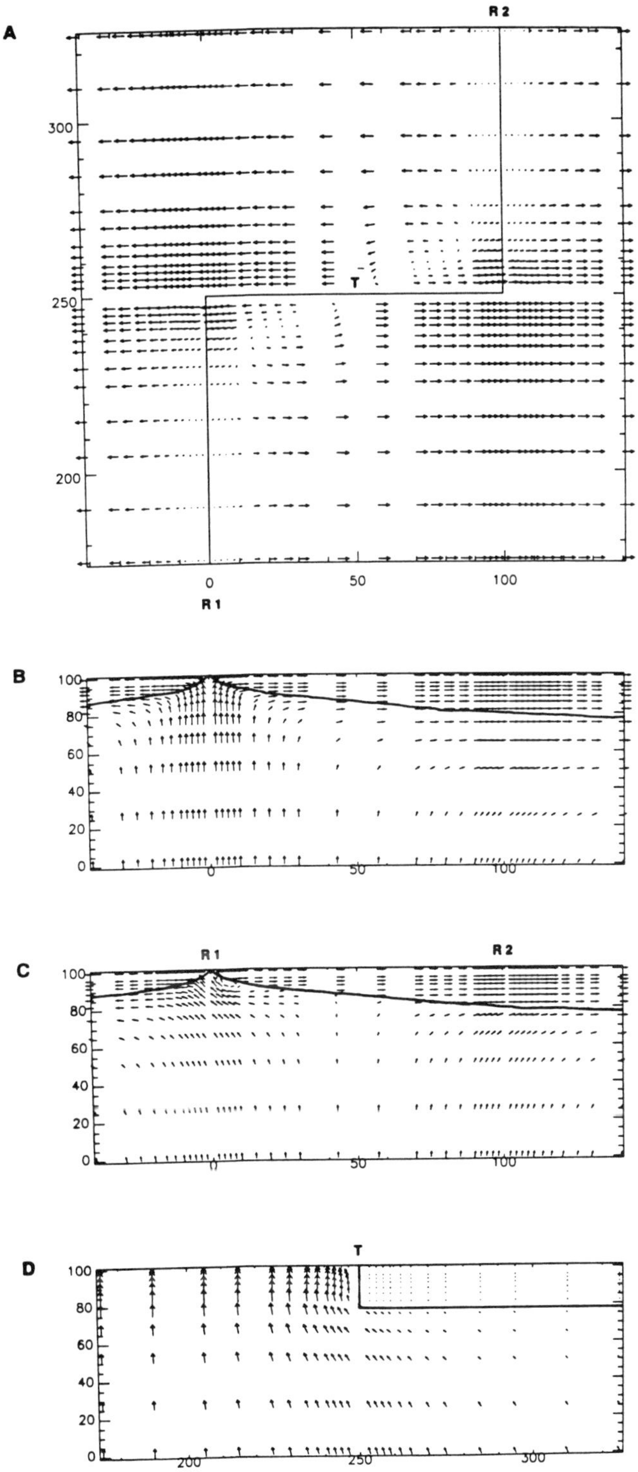

Fig. 3. Velocity vector diagrams for thickening-plate model. Locations shown in Figure 1a. Regions occupied by rigid plate are outlined by heavy line. (a) map view for plane at 14 km depth. Surface position of ridges and transform are shown by solid line. (b) cross-axis plane near center of ridge segment. (c) cross-axis plane just south of transform. (d) along-axis plane beneath ridge 1 and older plate beyond transform.

prevents flow from across the fracture zone at shallow depths. This reduction in shallow, axis-parallel velocities as well as an overall lack in symmetry of this flow component about the RTI, in comparison to the flat-plate model, is shown in Figure 4*c*. At shallow depths, the highest vertical velocities occur in a small region 10-20 km from the transform on the 'inside' of the ridge axis. Evidently, this increased upflow is required to balance the material deficit caused by reduced upwelling at the segment ends and by horizontal drag on the asthenosphere due to motion of the endwall away from the inside corner.

3. MANTLE THERMAL STRUCTURE

The temperature field associated with passive mantle flow can be obtained under the assumption that the seafloor is 0°C and mantle at 100 km depth is 1350°C. Hot mantle is advected upward beneath the spreading axis and the plates are cooled by conduction from above. The velocities calculated with the BEM are used in a finite-difference program [Phipps Morgan and Forsyth, 1988] to compute the steady state temperatures associated with the flow.

Comparison of the temperature field predicted by the flat-plate and thickening-plate models illustrates the effect of including the rigid plate in the flow, and subsequent thermal, calculations. The predicted thermal structure of the mantle for the thickening-plate model contains higher gradients than the flat-plate model in the ridge-transform intersection region and more asymmetry about the RTI. Figure 5 displays contour plots of mantle isotherms for the flat-and thickening-plate models. Both models predict high temperatures beneath the ridge axes and cooling that increases with distance off-axis. Whereas the flat-plate temperature contours are symmetric about the ridge axes and the fracture zone plane, for the thickening-plate model a significant difference in the thermal structure beneath the active transform and the inactive fracture zone trace is predicted. The active transform region is 100-200° cooler than the inactive limb at 14 km depth. In map view, the thickening-plate temperature contours form a broad V that points into the RTI in comparison with the more rounded flat-plate contours. The relevance of this pattern to observed gravity anomaly distributions is discussed in the following sections.

Beneath the axis, temperatures for the thickening-plate model are elevated more than for the flat-plate model but just off-axis the contours deepen more rapidly so that at distance the two models have similar cross-axis thermal structure. Figures 5*d* and 5e display mantle isotherms in vertical sections across- and along-axis, respectively. Thermal gradients across the fracture zone are significantly greater in the thickening-plate model where the plate rigidity inhibits flow within the older plate and upwelling rates beneath the axis are higher.

4. OBSERVED BATHYMETRY AND MANTLE BOUGUER ANOMALIES

Data along several sections of the Mid-Atlantic Ridge provide gravimetric and bathymetric constraints on the structure of the crust and upper mantle over slow-spreading

a)

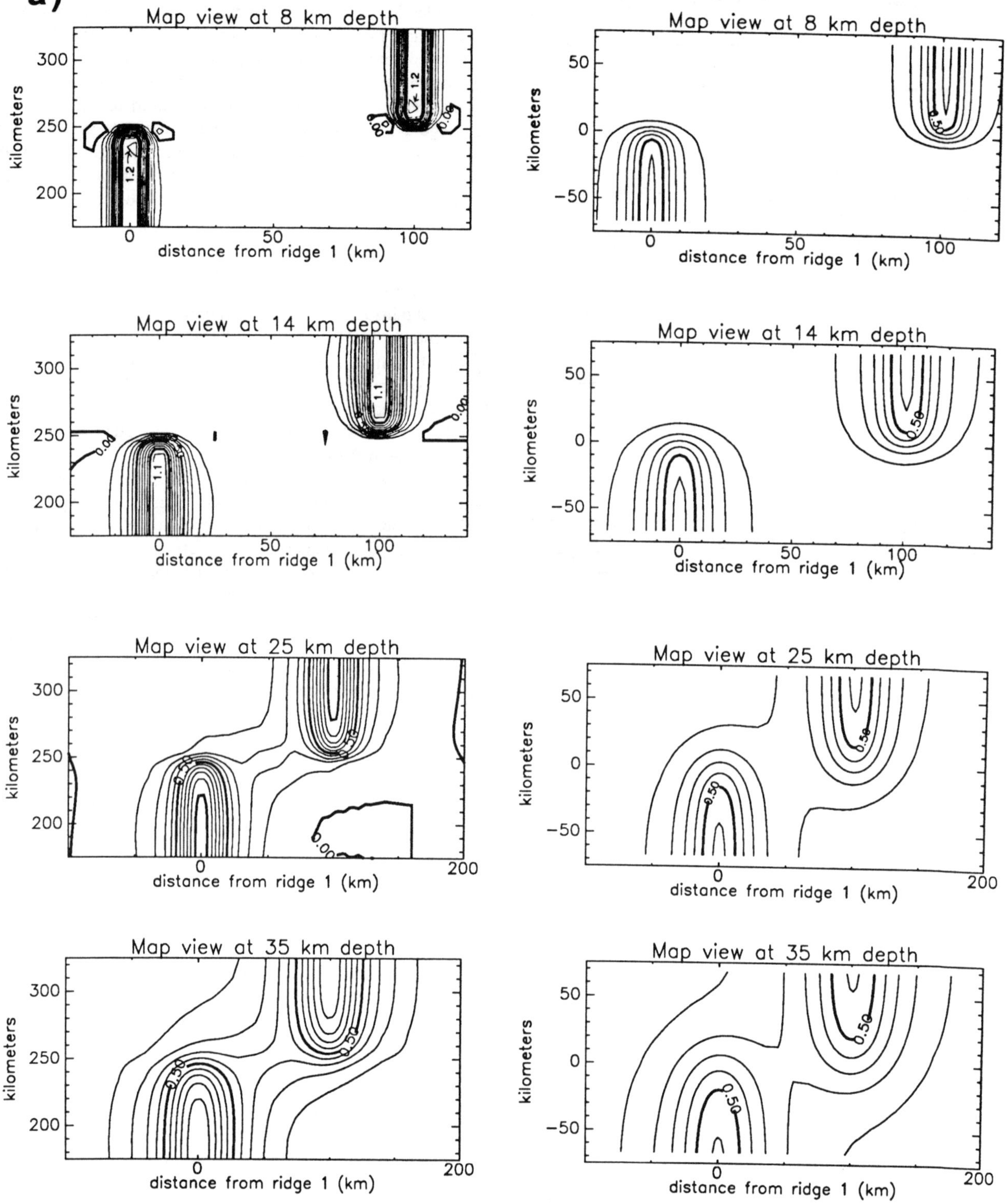

Fig. 4. Contours of velocity magnitude for thickening-plate (left) and flat-plate (right) models. Spreading rate is 10 mm/yr. Note that horizontal scales are larger for plots at shallower depths. (a) vertical velocities contoured at intervals of 1 mm/yr. (b) cross-axis velocities contoured at 1 mm/yr intervals. (c) along-axis velocities contoured at 0.3 mm/yr intervals. Direction of positive and negative are indicated by arrows. Shaded area indicates region occupied by rigid plate at 14 km depth.

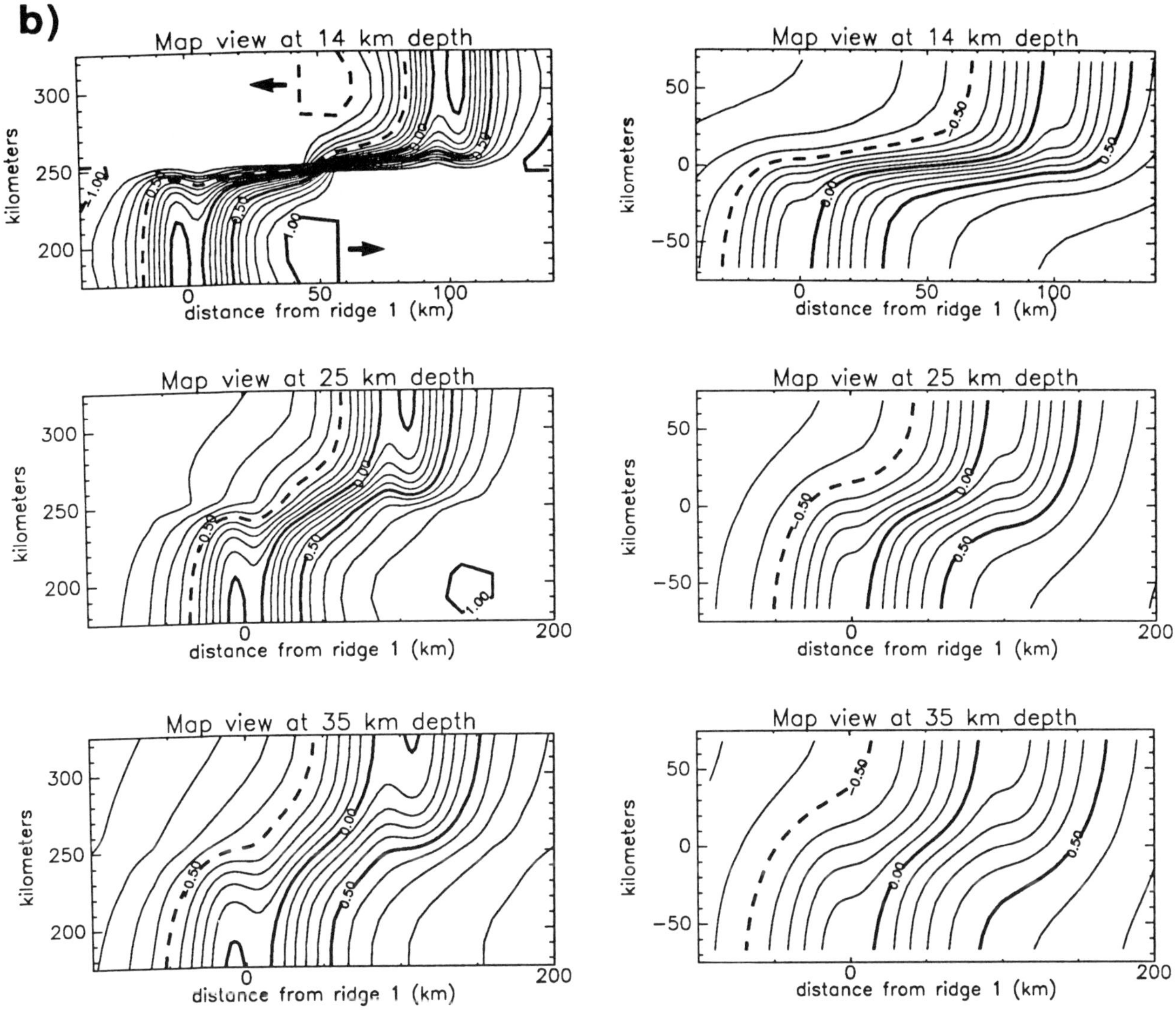

Fig. 4. Continued

ridge-transform systems [Prince and Forsyth, 1988; Kuo and Forsyth, 1989; Lin et al., 1990; Forsyth et al., 1990; Morris et al., 1991; Blackman and Forsyth, 1991]. A combination of high quality gravity measurements, coincident SeaBeam bathymetry data, closely spaced tracklines and coverage extending from the ridge axis out to crustal ages of a few million years provides the necessary resolution to map the density structure on the scale of kilometers.

Seafloor topography along a slow-spreading ridge system typically consists of a broad high that is cut at the axis by a rift valley that is 15-30 km wide and 1-2 km deep (Figure 6 shows an example from the Mid-Atlantic ridge 25°-27°30'S). Fracture zones (FZ) mark the path of ridge offsets and are characterized by depths that exceed the average seafloor by 2-3 km. The deepest FZ troughs tend to occur along the active transform. Axial depths increase from a shallow, roughly central section to the deeper ridge segment ends; the largest along-axis topographic gradients typically occur at the intersection of a ridge and the zone of deformation that characterizes an offset (either transform fault or non-transform discontinuity). At the inside corner of a RTI, the ridge flank is often shallow and is sometimes paired with a nodal basin in the transform [Fox and Gallo, 1984; Severinghaus and Macdonald, 1988].

The distribution of mantle Bouguer anomalies (MBA) over a ridge system whose seafloor topography is known can be derived from shipboard gravity measurements. The series of steps taken to obtain the MBA from the free air anomalies is straightforward. The largest contributions to the free air anomalies come from topography on the crust/water and crust/mantle interfaces. The seafloor topography is well

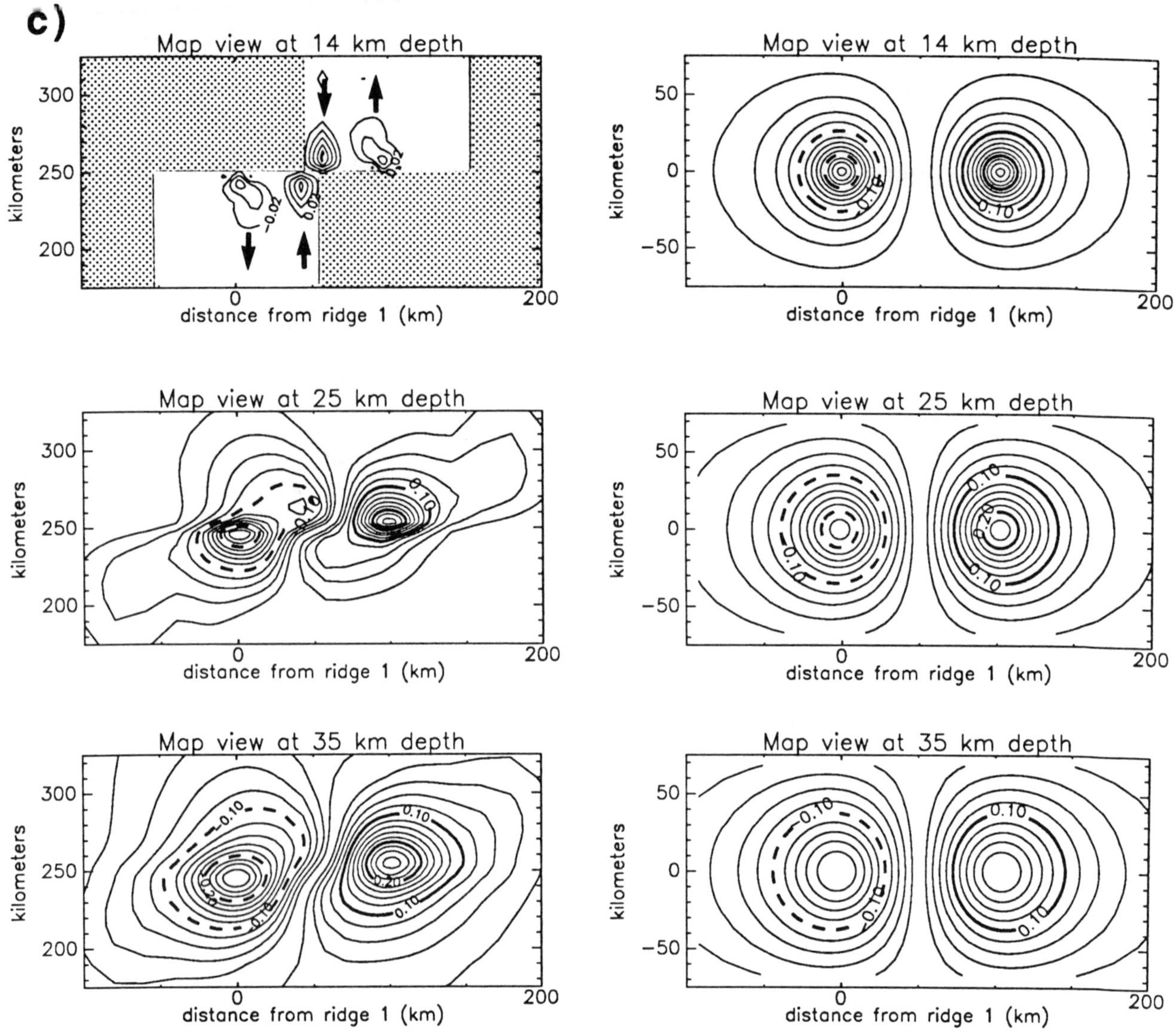

Fig. 4. Continued

represented by SeaBeam data that is interpolated to an evenly space grid. Topography on the crust/mantle interface is approximated by a surface that mimics the seafloor but resides 6 km deeper. It is assumed that the crustal structure is reasonably represented by layers of constant density. The gravitational attraction of the topography on each density interface is computed [Parker, 1972] and their sum constitutes a model of the crustal contribution to the observed free air anomalies. The predicted crustal signal is interpolated onto the original tracklines and subtracted from the along-track free air anomalies to obtain along-track MBA. These are then gridded to obtain a map of the mantle Bouguer anomalies.

The assumption of a constant thickness, constant density crust is valid to first order as indicated by numerous seismic studies throughout many ocean basins [eg. Spudich and Orcutt, 1980; White, 1984]. Removal of the effects of this simple crustal model does not mean that this model is correct in detail. It simply provides a means of reducing the range of the data so that the smaller contributions of crust or mantle density variations and deviations in the depth of the crust/mantle interface from average are visible.

Two examples of MBA distribution over a series of offset ridges are shown in Figure 7. Figure 7*a* covers the same area as Figure 6 on the southern Mid-Atlantic ridge where half-spreading rates average 18 mm/yr [Carbotte et al., 1991]. Figure 7*b* shows the Mid-Atlantic ridge at 27°-30°N where half rates are 12-14 mm/yr [Lin et al., 1990]. Gravity lows occur along all the ridge axes and a steady increase in MBA magnitude off-axis is clear in the southern Mid-Atlantic ridge area where coverage is more extensive. Positive anomalies

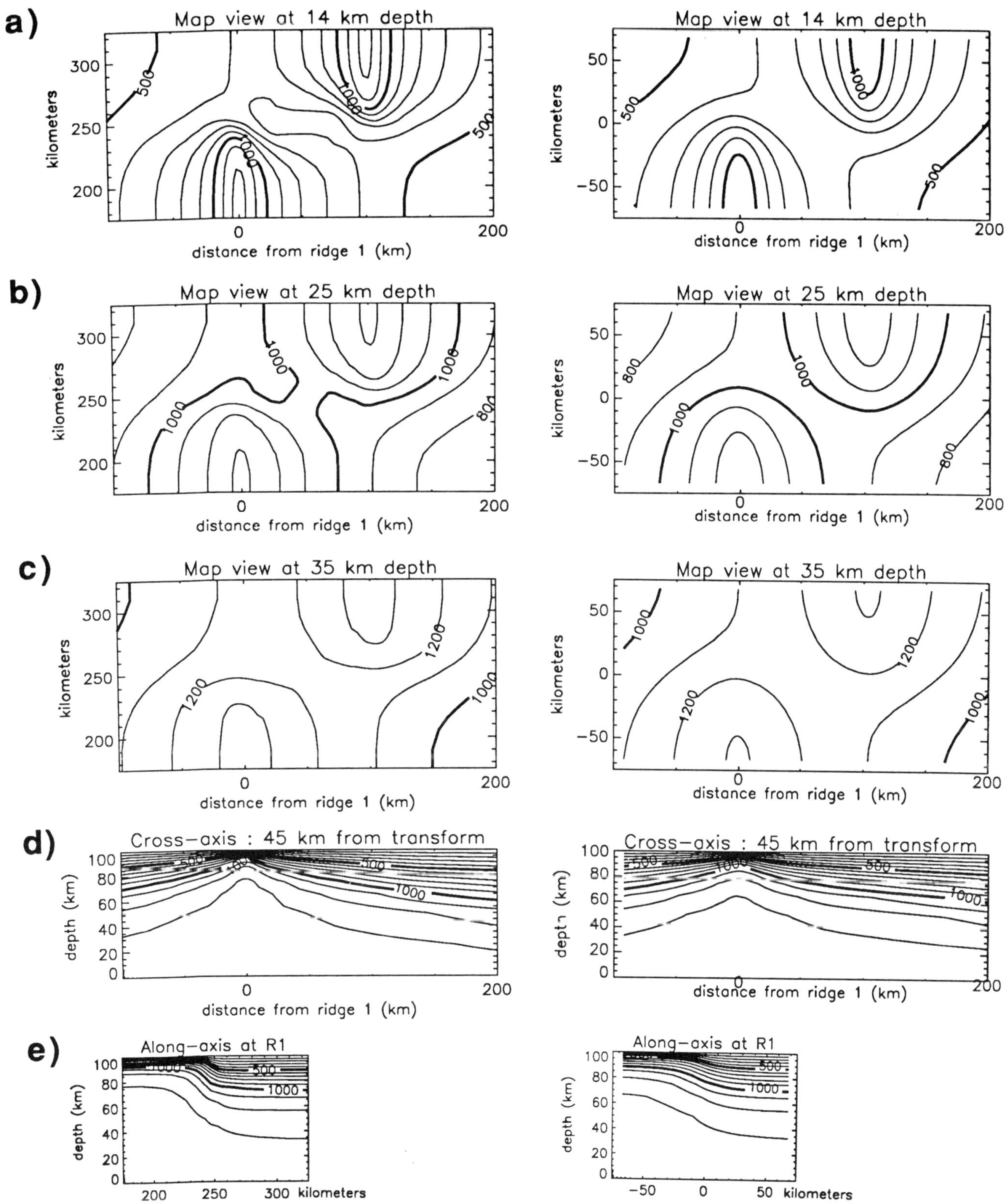

Fig. 5. Isotherm plots for thickening-plate (left) and flat-plate (right) models. Contour interval is 100°C. Surface position of ridges and transform are sketched in upper right plot. (a-c) map views at 14, 25 and 35 km depths, respectively. (d) cross-axis section 45 km from transform. (e) along-axis section beneath ridge 1 and older plate beyond transform.

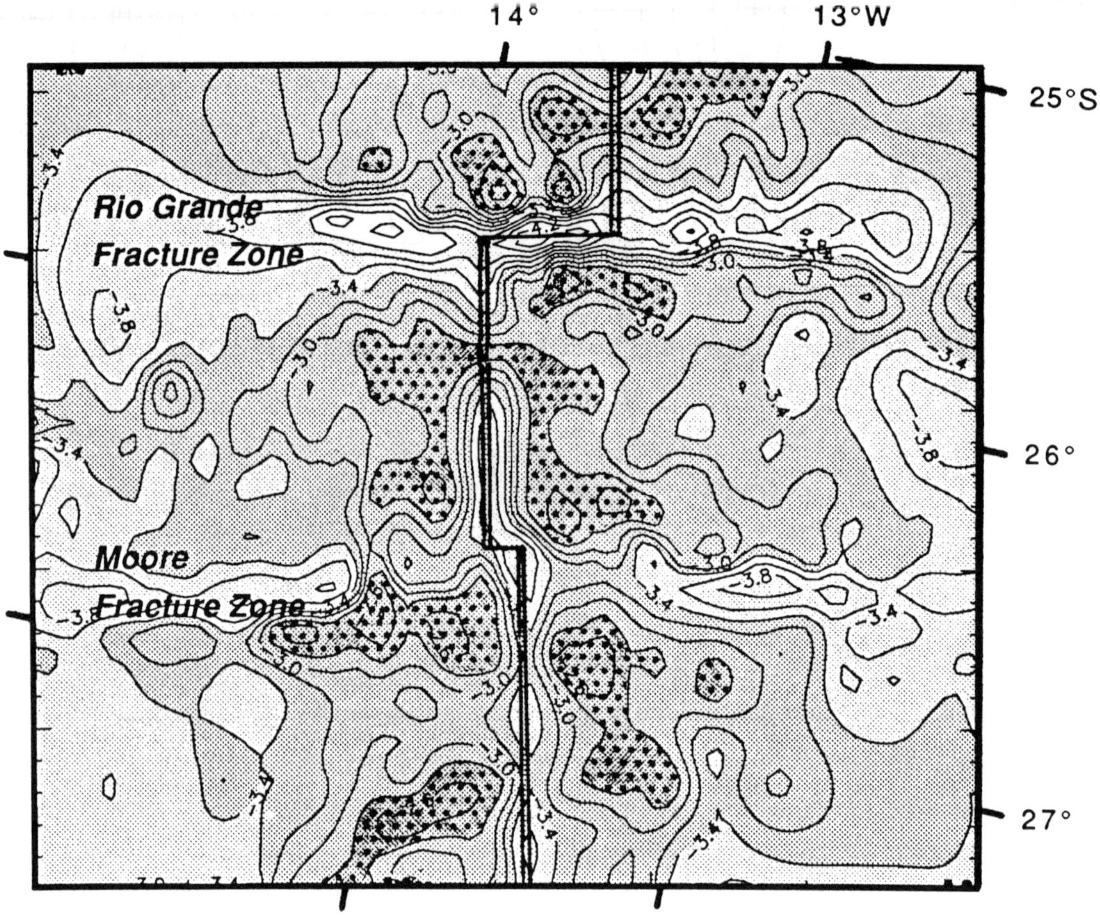

Fig. 6. Example of seafloor topography along ridge-transform system 25°-27°30'S on the mid-Atlantic ridge. Depths are in km; contour interval is 200 m. SeaBeam data was interpolated to 1 km grid and machine contoured. Fracture zones and median valleys are characterized by greater depths than average.

occur along both the Rio Grande and Atlantis FZs. At the ends of all ridge segments the gradient from low axial values to higher values in the offset zone is broadly V-shaped and points into the intersection. This type of pattern is seen in other areas as well: the Mid-Atlantic ridge 30°-36°S [Kuo and Forsyth, 1988; Forsyth et al., 1990]; southern Juan de Fuca and Gorda ridges [Holmes and Zervas, 1989]. The difference between anomaly magnitudes at the center vs the ends of segments varies but there is a tendency for greater differences (25-30 mgals) to occur where ridges intersect a large-offset transform (for example, just south of the Rio Grande and Atlantis FZs) as compared to ridges that intersect smaller offsets. Along-axis anomaly ranges of 10-15 mgals are observed at 26°30'S on the Mid-Atlantic ridge (Moore offset in Figure 7*a*) as well as several non-transform offsets between 27°N and 30°N (Figure 7*b*: southern segment 2, northern segment 3 and segment 5 intersections).

5. PREDICTED BATHYMETRY AND MANTLE BOUGUER ANOMALIES

The density distribution in the asthenosphere that results from passive upwelling of hot mantle beneath ridge axes can be determined from the predicted temperature field. Given the density distribution, the thermally produced topography and the mantle Bouguer anomalies can be computed. The volume coefficient of thermal expansion is assumed to be 3.4 x 10^{-5} /°C, temperatures at the finite-difference grid points are averaged (vertically) to obtain temperatures within each layer, local isostatic compensation is assumed in computing the thermal topography due to each layer and the gravitational attraction of each layer is calculated. The layer contributions to topography and gravity are summed to obtain the final prediction for the flat- and thickening-plate models shown in Figure 8.

The general form of the predicted MBA (Figure 8*a*) is similar for the flat- and thickening-plate models with low axial values grading to higher flank values but the latter shows more skew towards the inside corner at the RTIs. Also, the gradient into the RTI is notably higher for the thickening-plate model having an anomaly range 5-10 mgals greater than the flat-plate model over a somewhat smaller distance. Although the predicted topography for the thickening-plate model closely follows the V-shaped isotherm pattern (Figure 5), the MBA is

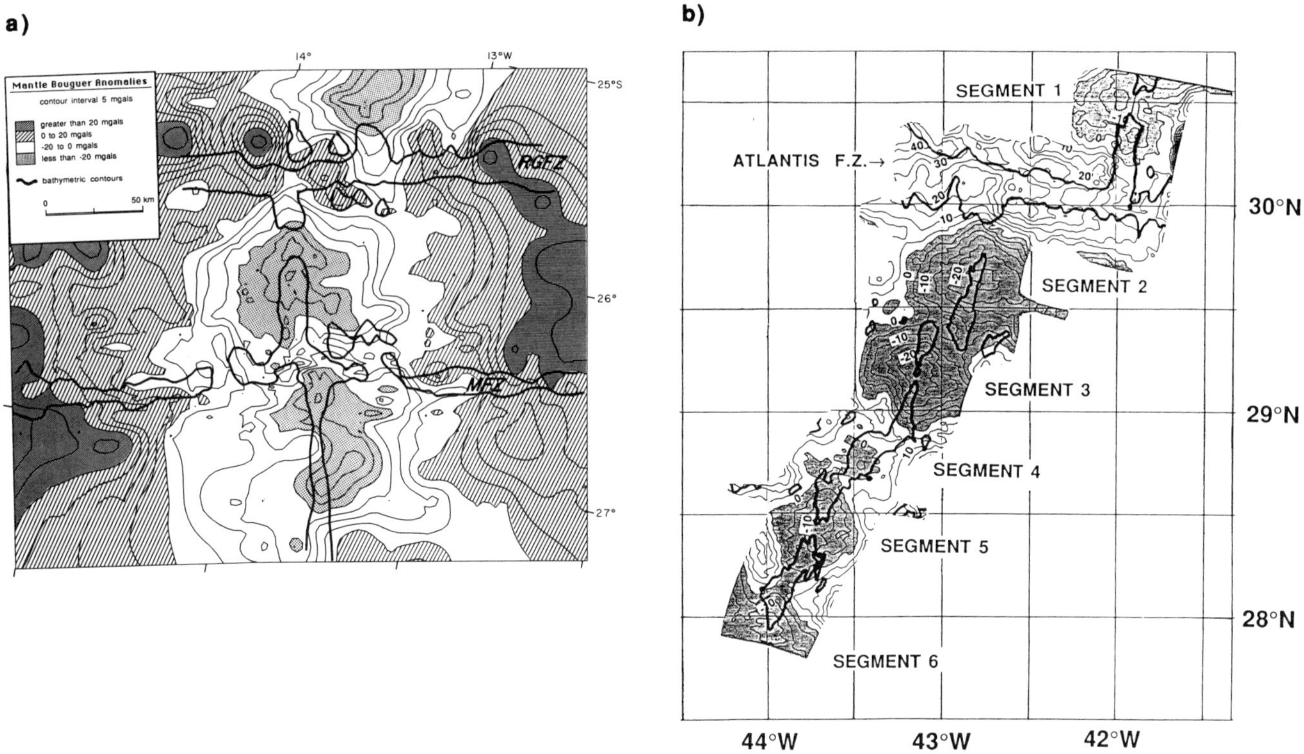

Fig. 7. Examples of mantle Bouguer anomaly pattern over ridge-transform systems. Contour interval is 5 mgal. Axes are characterized by gravity lows that grade into highs off-axis. Along-axis gradients are highest at ends of ridge segments. (a) mid-Atlantic ridge 25°-27°30'S from *Blackman and Forsyth* [1991]; (b) mid-Atlantic ridge 28°-30°30'N from *Lin et al.* [1990].

more rounded due to the broadening of the surface signal of density anomalies at depth.

The thickening-plate model predicts 800 m of along-axis topographic deepening (Figures 8*b*) with more gradual gradients away from the RTI. In comparison, the flat-plate model predicts thermal topography that deepens 600 m at a fairly consistent rate from the center of the ridge towards the RTI. For the thickening-plate model, the transform zone is predicted to be 100-200 m deeper near the RTI than the inactive FZ at an equal distance from the ridge.

6. Discussion

We have shown that the pattern of passive mantle flow beneath a ridge-transform system can be notably affected by the shape of the rigid, moving boundary that drives the flow. Several simplifying assumptions made in order to carry out the calculations at a desirable resolution make direct application of these results to the mid-ocean ridge environment a matter which requires some caution. Nevertheless, this modeling does provide insight into the extent to which the passive response of the mantle to plate driving forces plays a role in shaping the observed seafloor topography and gravity anomalies. In this section we review some of our major assumptions, qualitatively assess how more realistic properties would change our results and show how inclusion of plate rigidity in the flow and thermal calculations results in predictions of mantle Bouguer anomalies that explain more of the observed field than previous models.

6a. Shape of the plate at the axis

The depth of the boundary elements that represent the base of the lithosphere in the thickening-plate ridge-transform-ridge model corresponds to the 800°C isotherm depth predicted by a 2-D half-space cooling model. This results in a plate that increases in thickness at the same rate all along the ridge starting from zero at the axis. Contrary to these assumptions, at the ridge axis the plate will comprise a brittle crustal layer about 6 km thick and, due to variability in the subaxial thermal structure, it is unlikely that the plate cools and thickens at the same rate all along a ridge segment. Evidence for a non-zero axial plate thickness comes from seismic refraction experiments along the slow-spreading Mid-Atlantic ridge which indicate crustal thicknesses of 5-6 km [Purdy and Ewing, 1986] and microearthquake focal depths of 4-8 km near 23°N [Toomey et al., 1985]. In addition, our temperature calculations show that plate thickness should vary somewhat along strike. Below we discuss the effect that the assumed shape of the plate near the ridge axis has on the calculated flow and thermal structure.

The cross-axis shape of the plate does affect the magnitude of the predicted velocities within 10-15 km of the ridge axis but the general flow pattern is not sensitive to the specific shape assumed. Three models of a single ridge segment

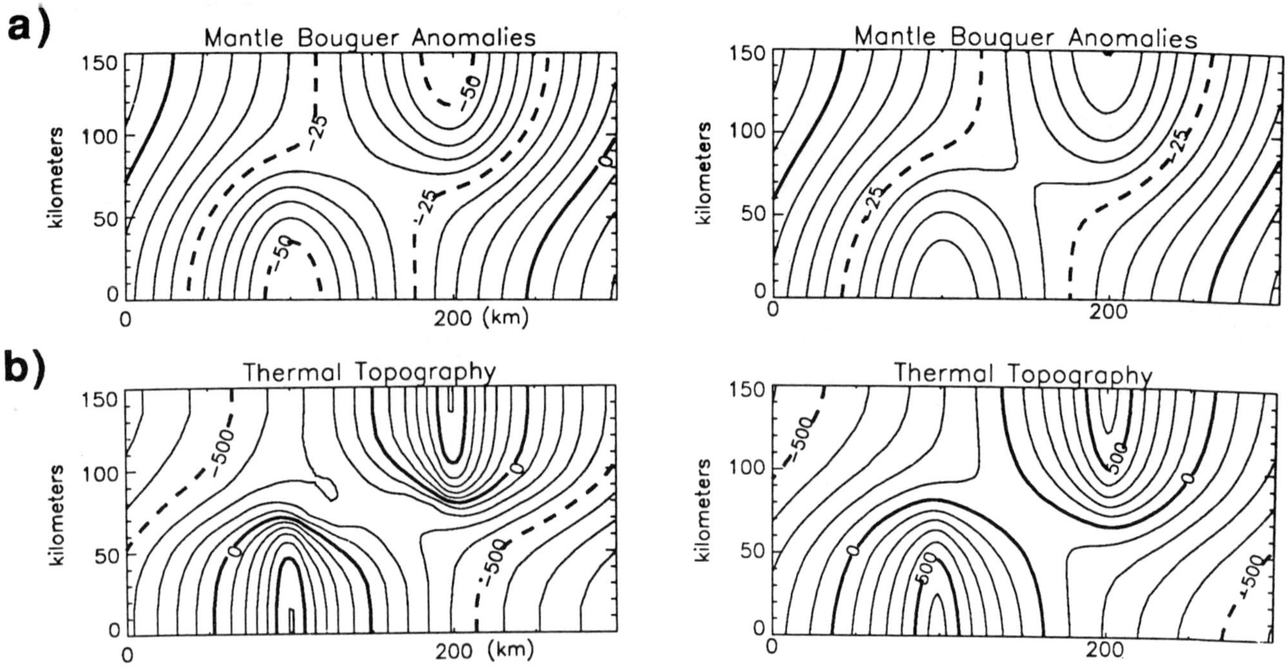

Fig. 8. Predicted thermally produced gravity and topography for thickening-plate (left) and flat-plate (right) models. Heavy dashed lines indicate region where anomalies are negative (low densities or deep seafloor). (a) mantle Bouguer anomalies contoured at 5 mgal interval; (b) topography contoured at 100 m interval. Gradients for both fields are higher near RTI for thickening-plate model.

bounded by large offset transform faults show the effect of different plate shapes at the axis on the calculated velocities. The general form of the single-segment thickening-plate model is shown in Figure 9*a*; the assumption of large transform offset allows the endwall to be represented by vertical boundary elements that extend sufficiently far below the region of interest that their termination does not affect the flow at shallower depths. The first model has the same axial plate geometry as that used in the ridge-transform-ridge models of the previous sections (Figure 9*b*). The plate base in second model also follows the 800°C isotherm but at the axis a plate thickness of 6 km is prescribed (Figure 9*c*). In the third model, a 6 km deep axial plate base is joined to the 1200°C-isotherm base off-axis (Figure 9*d*). In the first model, vertical velocities right below the axis exceed the spreading rate by up to 50% due to the narrowing-upward wedge into which asthenosphere must flow to supply material for constant plate production. In the second model, the smoothly sloping axial plate does not focus the upwelling and vertical velocities are less than or equal to the spreading rate. In the third model the steepness of the plate base forms a broad diverging wedge that results in vertical velocities up to 30% greater than the spreading velocity below the axis. The cross-axis sections in Figures 9*b*-9*d* illustrate the flow near the center of the ridge segment. The magnitudes of the along-axis velocities for the three models do not differ greatly.

The important effect of the differences in vertical upwelling rates caused by the assumed shape of the plate in the axial region is the resultant temperature distribution. The higher upwelling rates associated with the narrow axial wedges of the first and third model should introduce hotter subaxial temperatures thus creating a narrow mantle Bouguer low of greater magnitude than the second model. But, the second and third model do not explicitly account for the creation of oceanic crust, it is just defined to exist at a thickness of 6 km in the axial region. Physically, these two models presume that melt within the upwelling asthenosphere is rapidly drawn towards the surface to provide the crustal material. Based on the observed narrow zone of active volcanism at spreading centers [Macdonald, 1982], melt within the crust must also be focused about the central few kilometers of the rift. So, to determine the thermal signature for the second and third models it would be necessary to assess the contribution of shallow heat sources corresponding to the latent heat of the melt [Sleep, 1975]. This type of calculation, and the assumptions involved in specifying the distribution of the sources, are beyond the scope of this paper. Qualitatively, a narrow axial region of shallow heat sources will produce a surface temperature signal similar to that determined for the first model where faster mantle upwelling introduces a narrow region of higher temperatures.

The assumption that the plates thicken at the same rate all along a ridge segment is probably more important than the cross-axis plate geometry in controlling the pattern of flow and thermal structure in the mantle. Figure 10 shows a map view of isotherms predicted for the thickening-plate ridge-

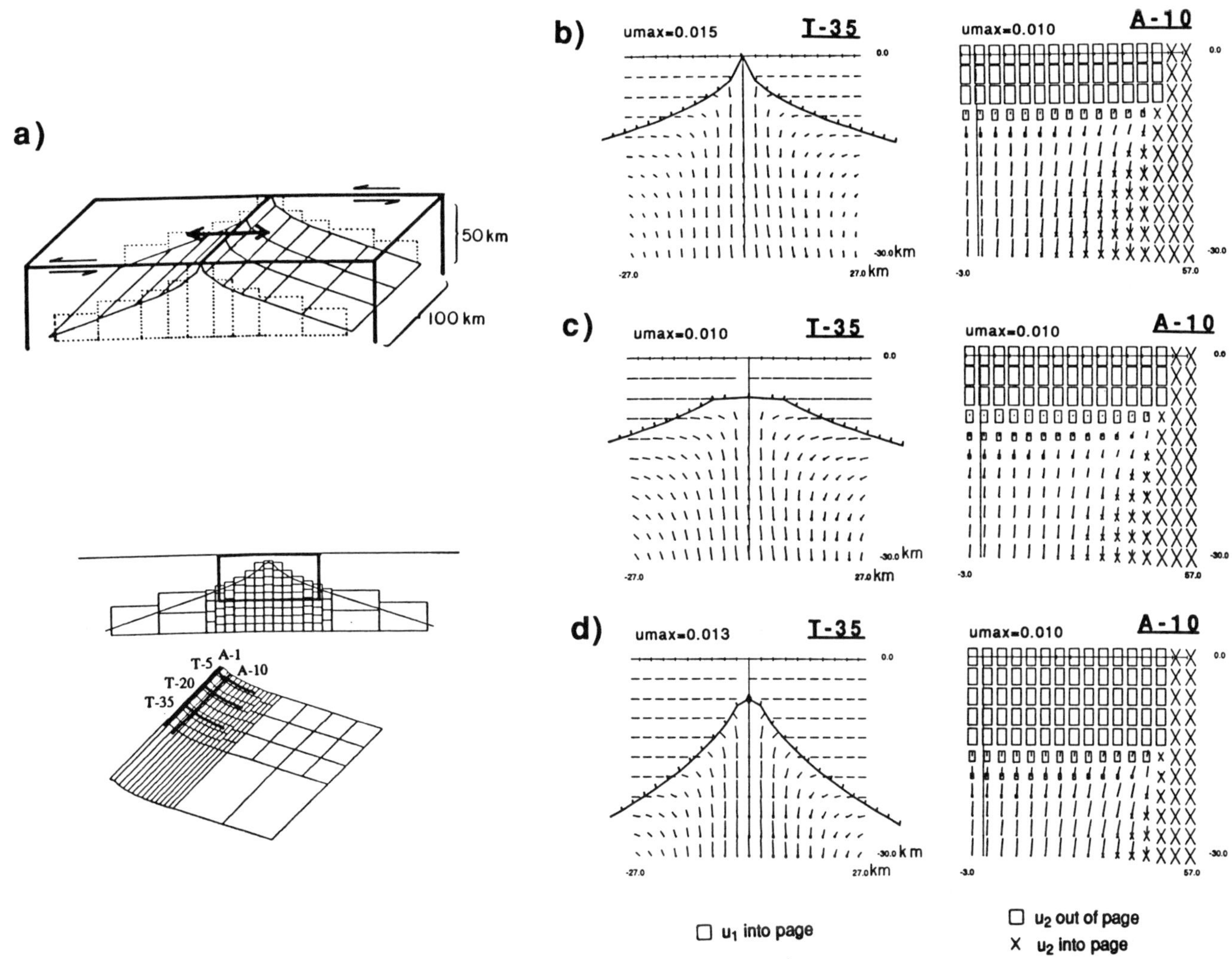

Fig. 9. Effect of assumed axial plate shape on calculated flow for model of single ridge segment bounded by large-offset transforms. (a) geometry and scale of model (upper diagram) and boundary element discretization (lower diagram). In lower diagram, heavy rectangle drawn over transform-plane boundary elements and labeled heavy lines on plate-base boundary elements indicate location of cross sections shown in b-d. (b-d) Velocity vectors show three flow components. Direction of component perpendicular to plane of plot is indicated below for cross-axis (left) and along-axis (right) sections. Cross-axis plots are taken 35 km from transform and heavy ticked line shows location of plate base. Along-axis sections are taken 10 km from the axis towards the 'inside' of the RTI. The magnitude of the largest vector for each plot is labeled in upper left (umax in m/yr). Along-axis sections are scaled the same; cross-axis vector scales vary. In-plate velocity vectors are all 0.01 m/yr. (b) zero axial thickness, 800° isotherm plate model; (c) 6 km axial thickness, 800° isotherm plate model; (d) 6 km axial thickness, 1200° isotherm plate model.

transform-ridge model at a depth of 14 km. The assumed location of the rigid plates is shaded. Away from the fracture zone, the a priori position of the 800°C isotherm agrees pretty well with that computed. Near the RTI, however, the predicted isotherm deviates considerably from the assumed position. At the end of a ridge segment the plates should cool more rapidly than assumed due to the presence of the cold juxtaposed plate edge. The axial plate would be thicker in correspondence to the deeper isotherms near the RTI. This would result in a deeper, less steeply sloping axial wedge that would augment the reduction in upwelling rates near the end of the ridge (narrower diverging wedges require higher rates). On the other hand, the delivery of melt that creates crust in the RTI region could maintain a more typical shallow thermal structure.

6b. Mantle rheology

The boundary element method restricts us to modeling a constant viscosity fluid and thus the effect of temperature on the mantle behavior cannot be incorporated in our calculations. Although the exact relationship between temperature and viscosity continues to be the subject of study and debate, it is clear that mantle viscosity should decrease with increasing temperatures. The degree to which the presence of melt in the mantle can further alter the effective viscosity is unclear but

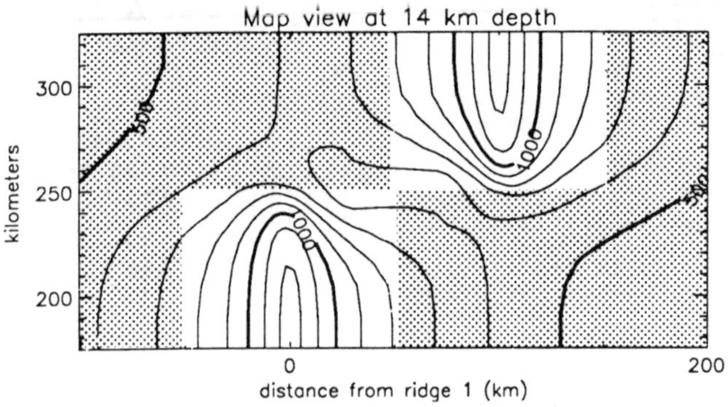

Fig. 10. Contour plot of predicted temperatures for thickening-plate model at depth of 14 km. Region occupied by rigid plate is shaded for comparison with computed position of 800°C isotherm.

could range from an increase due to stiffening of the residual [Karato, 1986], through a factor of 2-5 decrease [Cooper et al., 1989], to a decrease of a few orders of magnitude [Buck and Su, 1989].

The rheology that we assume for the BEM is one in which the mantle behaves perfectly rigidly at temperatures below 800°C and is uniformly ductile at higher temperatures. This results in the lithosphere/asthenosphere boundary being an infinitely thin, no-slip boundary rather than a more realistic boundary layer across which the mantle becomes more and more viscous as the plate cools and becomes rigid. The degree of mismatch between our simplified model and the real world depends on the details of mantle rheology that are not presently well constrained. Thus, we are content to take an end-member approach with the knowledge that our calculated velocities may be somewhat higher than actual velocities if the lithosphere and asthenosphere are not perfectly coupled due to the presence of such a boundary layer.

We predict mantle temperatures that decrease rapidly within about 20 km of a ridge-transform intersection above about 30 km depth. This should be accompanied by an increase in mantle viscosity in the vicinity of the transform zone. Higher viscosities would augment the reduction in upwelling velocity near the end of a ridge which would presumably induce stronger upwelling elsewhere and along-axis flow towards the RTI to maintain constant plate production. For temperature and pressure dependent mantle viscosity, Shen et al. [1991] determine velocity and temperature fields induced by passive flow that are generally similar to the fields predicted for the BEM herein. They show a somewhat broader transition in thermal structure across the RTI.

6c. *Fit of model results to data*

The predicted mantle Bouguer anomalies for the thickening-plate model can account for somewhat more of the observed field along slow-spreading ridges near large-offset transforms than do the MBA predicted by the flat-plate model. The increase in gradient from an axial gravity low to higher values at the RTI corresponds with the *direction* of change in the predicted field that is needed to match the observations. The range of the predicted anomaly under the assumption that mantle density variations are solely controlled by thermal expansion is not as large as the observed anomaly range at large-offset transforms (nor would it be at smaller offsets). If the effect of melting and the reduction in density of the residual mantle were included, our predicted MBA would more closely match that observed. Parmentier and Phipps Morgan [1990] show that 25% melting of the mantle can result in a residual whose density reduction is similar to that associated with 200°C temperature anomaly. The axial upwelling rates and thermal structure predicted for the thickening-plate model would result locally in somewhat higher degrees of melting than predicted for the flat-plate model [Phipps Morgan and Forsyth, 1988] so we might expect melting to reach a maximum of about 25%. The surrounding pattern of lower degrees of melting near the RTI and off-axis would follow the shape of the isotherms. Because the temperature and melting anomalies basically follow the same distribution (above the depth at which melting begins), the total gravitational effect of including the latter should be to amplify the magnitude of the MBA predicted for only the thermal expansion. Thus, our predicted anomaly range should be perhaps 10-15 mgals greater than the flat-plate model rather than the 5-10 mgal range determined for the thermal expansion effect alone.

For smaller ridge offsets or faster spreading rates the influence of the plate thickening on the flow and thermal structure of the mantle is still important in the upper several kilometers of the mantle but at greater depths the plate effect is small. For a 42 km offset of a ridge spreading at 18 mm/yr half-rate (eg. the Rio Grande FZ in the South Atlantic, Figure 6 and 7*a*), the flat-plate model predicts a 5-10 mgal difference in MBA between the center of a 100 km long ridge segment and the RTI [Blackman and Forsyth, 1991]. The thickening-plate model (which only assumes two long ridges offset along a 42 km transform as opposed to the specific ridge-transform geometry of the region) predicts an along-axis MBA difference of 10-15 mgals. Neither model can produce the observed 20-25 mgal range that is observed (Figure 7*a*) although, as mentioned above, adding the mantle density reduction due to the removal of melt to the thermal expansion could improve the fit some.

The predicted thermally supported axial topography (Figure 8*b*) is in the range of observed along-axis relief at slow spreading ridges (500-2500 m) but the response of the overlying plate to this load has not been considered. The effect of the strong plate would be to damp out some of the anomaly predicted for local compensation and to distribute the remaining signal more broadly. The BEM provides analytic expressions for the stress contribution of each boundary element so we have calculated the total vertical normal stress at the mid-points of the elements that make up the base of the plate. The load due to seafloor topography that can locally balance these stresses is computed under the assumption of a constant density difference between seawater (1030 kg/m^3) and mantle (3300 kg/m^3). The distribution and magnitude of the computed stresses is sensitive to the assumed shape of the axial plate (see section 6a) but for all models tested we predict along-axis deepening with the greatest depths on the inside of the rift valley at the RTI where nodal basins are commonly observed. The scale of this stress-supported axial deepening is directly proportional to the assumed mantle viscosity; it is of the order of several tens to a few hundred meters for a viscosity of 10^{19} Pa s [Blackman, 1991]. Thus, the combination of thermal and stress supported topography predicted by the thickening-plate model could explain much of the observed along-axis relief. The response of the overlying plate to the downward pull of viscous stresses on the edges of the plate 'inside' the RTI could produce a flexural high several kilometers back corresponding to observed inside corner highs [Figure 6; Severinghaus and Macdonald, 1988].

We have intentionally ignored the contribution of crustal thickness variations to the observed gravity and topography in order to focus on the maximum possible contribution from the passively induced mantle structure. We find that much of the observed signals at large-offset transforms can be explained by variations in mantle properties. Obviously we cannot rule out the possibility that the signals might be due in part to crustal thickness variations. In fact the temperature distribution that we predict would result in lower melt production near the RTI which could correspond to lower crustal production there if along-axis migration does not redistribute the melt significantly. For smaller or non-transform offsets and faster spreading rates our predicted thermal structure will be more similar to that of the Phipps Morgan and Forsyth model [1988] so the need to explain some of the MBA with crustal thickness variability or buoyantly driven mantle flow [Kuo and Forsyth, 1988; Lin et al., 1990] will remain.

7. Conclusions

The details of passively induced mantle flow beneath a ridge-transform-ridge plate boundary are affected by the thickening of the plates with age. The diverging axial wedge formed by the rapidly thickening plates near the axis induces upwelling rates that locally exceed the spreading rate. A significant three-dimensionality in the flow structure is caused by reduced upwelling and cross-axis flow at the ridge-transform intersection due to the finite thickness and motion of the old plate edge; along-axis flow at rates up to 25% of the spreading rate are required to compensate for the resulting material deficit near the endwall. The high velocity gradients associated with interaction of the plate and the asthenosphere will affect the pattern and amplitude of stresses in the mantle.

The thermal structure predicted for the new model contains higher gradients at a ridge-transform intersection than a model which neglects plate thickening and temperatures beneath the active transform are 100°C cooler to depths of about 25 km. In terms of the alteration in mantle thermal structure due to plate-thickening, the maximum contribution to the predicted along-axis gravity and topography at slow-spreading, large-offset ridges is 5-10 mgals and 200-300 m, respectively.

Acknowledgements. We thank Dave Pollard and Allan Rubin for providing the starting versions of the Boundary Element code that was modified for this project. Thanks, also, to Jian Lin and Jason Morgan for reviewing the manuscript.

References.

Batchelor, G.K., *An Introduction to Fluid Dynamics*, 615 pp., Cambridge University Press, Cambridge, 1967.

Bergman, E.A. and S.C. Solomon, Source mechanisms of earthquakes near mid-ocean ridges from body wave inversion: implications for early evolution of oceanic lithosphere, *J. Geophys. Res.*, *89*, 11,415-11,441, 1984.

Blackman, D.K. and D.W. Forsyth, Axial topographic relief associated with ridge-transform intersections, *Earth Planet. Sci. Lett.*, *95*, 115-129, 1989.

Blackman, D.K. and D.W. Forsyth, Isostatic compensation of tectonic features of the Mid-Atlantic ridge: 25°-27°30'S, *J. Geophys. Res.*, *96*, 11,741-11758, 1991.

Blackman, D.K., Mid-ocean ridge processes: Geophysical constraints on the variability within ridge segments, Ph.D. thesis, Brown University, Providence, Rhode Island, 1991.

Buck, W.R. and W. Su, Focused mantle upwelling below mid-ocean ridges due to feedback between viscosity and melting, *Geophys. Res. Lett.*, *16*, 641-644, 1989.

Carbotte, S.M., S.M. Welch, K.C. Macdonald, Spreading rates, rift propagation, and fracture zone offset histories during the past 5 Myr on the Mid-Atlantic Ridge: 25-27°30'S and 31-34°30'S, *Mar. Geophys. Res.*, *13*, 51-80, 1991.

Cooper, R.F., D.L. Kohlstedt, K. Chyung, Solution-precipitation enhanced creep in solid-liquid aggregates which display a non-zero dihedral angle, *Acta Mettall.*, *37*, 1759-1771, 1989.

Crouch, S.L. and A.M. Starfield, *Boundary Element Methods in Solid Mechanics*, 322 pp., George Allen and Unwin Ltd., London, 1983.

Forsyth, D.W., 3-D viscous flow in a conduit: possible explanation for inside corner highs and asymmetric nodal deeps at ridge-transform intersection, *EOS Trans. AGU*, *70*, 456, 1989.

Forsyth, D.W. and B. Wilson, Three-dimensional temperature structure of a ridge-transform-ridge system, *Earth Planet. Sci. Lett.*, *70*, 355-362, 1984.

Forsyth, D.W., D.K. Blackman, G.A. Neumann, Segmentation of the southern Mid-Atlantic Ridge: gravity signature of an unusually robust spreading segment, *EOS Trans. AGU*, *71*, 1628, 1990.

Fox, P.J. and D.G. Gallo, The tectonics of ridge-transform intersections, *Tectonophysics*, *104*, 204-242, 1984.

Holmes, M. and C. Zervas, Residual gravity anomalies on the Juan de Fuca and Gorda ridges, *Eos Trans. AGU*, *70*, 1326, 1989.

Karato, S., Does partial melting reduce the creep strength of the upper mantle?, *Nature*, *319*, 309-310, 1986.

Kuo, B.-Y. and D.W. Forsyth, Gravity anomalies of the ridge-transform system in the South Atlantic between 31 and 34.5°S: upwelling centers and variations in crustal thickness, *Mar. Geophys. Res.*, *10*, 205-232, 1988.

Lachenbruch, A.H., A simple mechanical model for oceanic spreading centers, *J. Geophys. Res.*, *78*, 3395-3417, 1973.

Lachenbruch, A.H., Dynamics of a passive spreading center, *J. Geophys. Res.*, *81*, 1883-1902, 1976.

Lin, J. and E.M. Parmentier, Mechanisms of lithospheric extension at mid-ocean ridges, *Geophys. J.*, *96*, 1-22, 1989.

Lin, J., G.M. Purdy, H. Schouten, J.-C. Sempere, C. Zervas, 1990, Evidence for focused magmatic accretion along the Mid-Atlantic Ridge, *Nature*, *344*, 627-632, 1990.

Macdonald, K.C., Mid-ocean ridges: fine scale tectonic, volcanic and hydrothermal processes within the plate boundary zone, *Ann. Rev. Earth Planet. Sci.*, *10*, 155-190, 1982.

Macdonald, K.C., P.J. Fox, L.J. Perram, M.F. Eisen, R.M. Haymon, S.P. Miller, S.M. Carbotte, M-H. Cormier, A.N. Shor, A new view of the mid-ocean ridge from the behaviour of ridge-axis discontinuities, *Nature*, *335*, 217-225, 1988.

McNutt, M.K. and H.W. Menard, Constraints on yield strength in the

oceanic lithosphere derived from observations of flexure, *Geophys. J. Roy. Astron. Soc.*, *71*, 363-394, 1982.
Morris, E. and R.S. Detrick, Three-dimensional analysis of gravity anomalies in the MARK area (Mid-Atlantic Ridge, 23°N), *J. Geophys. Res.*, *96*, 4355-4366, 1991.
Parker, R., The rapid calculation of potential anomalies, *Geophys. J.*, *31*, 447-455, 1972.
Parmentier, E.M. and D.W. Forsyth, Three-dimensional flow beneath a slow spreading ridge axis: a dynamic contribution to the deepening of the median valley toward fracture zones, *J. Geophys. Res.*, *90*, 678-684, 1985.
Parmentier, E.M. and J. Phipps Morgan, Spreading rate dependence of three-dimensional structure in oceanic spreading centres, *Nature*, *348*, 325-328, 1990.
Phipps Morgan, J., E.M. Parmentier, and J. Lin, Mechanisms for the origin of mid-ocean ridge topography: implications for the thermal and mechanical structure of accreting plate boundaries, *J. Geophys. Res.*, *92*, 12,823-12,836, 1987.
Phipps Morgan, J. and D.W. Forsyth, Three-dimensional flow and temperature perturbations due to a transform offset: effects on oceanic crustal and upper mantle structure, *J. Geophys. Res.*, *93*, 2955-2966, 1988.
Purdy, G.M. and J. Ewing, Seismic structure of oceanic crust, in P.R. Vogt and B.E. Tucholke, eds., *The Geology of North America*, vol. M, *The Western North Atlantic Region*, Geol. Soc. America, 313-340, 1986.
Prince, R.A. and D.W. Forsyth, Horizontal extent of anomalously thin crust near the Vema fracture zone from the three-dimensional analysis of gravity anomalies, *J. Geophys. Res.*, *93*, 8051-8063, 1988.
Reid I. and H.R. Jackson, Oceanic spreading rate and crustal thickness, *Marine Geophys. Res.*, *5*, 165-172, 1981.
Schouten, H. and K.D. Klitgord, The memory of the accreting plate boundary and the continuity of fracture zones, *Earth Planet. Sci. Lett.*, *59*, 255-266, 1982.
Severinghaus, J.P. and K.C. Macdonald, High inside corners at ridge-transform intersections, *Mar. Geophys. Res.*, *9*, 353-367, 1988.
Shen, Y., E.M. Parmentier, D.W. Forsyth, The effect of temperature- and pressure-dependent viscosity on a 3-D passive mantle flow beneath a ridge-transform system, *Eos Trans. AGU*, *72*, 260, 1991.
Sleep, N.H., Formation of oceanic crust: some thermal constraints, *J. Geophys. Res.*, *80*, 4037-4042, 1975.
Sleep, N.H. and S. Biehler, Topography and tectonics at the intersection of fracture zones with central rifts, *J. Geophys. Res.*, *75*, 2748-2752, 1970.
Spudich, P. and J. Orcutt, A new look at the seismic velocity structure of the oceanic crust, *Reviews Geophys. Space Phys.*, *18*, 627-645, 1980.
Toomey, D.R., S.C. Solomon, G.M. Purdy, and M.H. Murray, Microearthquakes beneath the median valley of the Mid-Atlantic ridge near 23°N: hypocenters and focal mechanisms, *J. Geophys. Res.*, *90*, 5443-5458, 1985.
Turcotte, D.L. and G. Schubert, *Geodynamics Applications of Continuum Physics to Geological Problems*, 450 pp., John Wiley and Sons, New York, 1982.
White, R.S., in *Ophiolites and Oceanic Lithosphere*, I.G. Gass, S.J. Lippard and A.W. Shelton (eds), Geol. Soc. London, 34-44, 1984.

D. Blackman, Scripps Institution of Oceanography, La Jolla, CA 92093-0225.

D. Forsyth, Department of Geological Sciences, Brown University, Providence, RI 02912.

Small-Scale Convection and Mantle Melting Beneath Mid-Ocean Ridges

DAVID R. SCOTT

Department of Earth Sciences, University of Oxford

Active flow in the mantle beneath mid-ocean ridges driven by thermal, melt and compositional buoyancy is investigated using numerical methods. The calculations employ a two-dimensional Cartesian geometry with flow confined to a vertical plane parallel to the plate motion. The rheological structure incorporates temperature- and pressure-dependent viscosity and a rigid lithospheric plate. In the region beneath a ridge axis the spreading of the plates sustains a permanent lateral temperature gradient. If the viscosity in this region is sufficiently low, steady thermal convection can occur. Calculations using a range of plate velocities and viscosities illustrate the form of this convection, and show a transition between weak and strong convective regimes. In the weak regime the thermal structure sustained by passive spreading is undisturbed; in the strong regime a convective thermal structure develops. These calculations complement previous work on small-scale thermal convection with flow confined to a vertical plane perpendicular to the plate motion. During mantle melting, melt buoyancy arises when a finite melt porosity is present, and compositional buoyancy arises because the density of the solid residue decreases as melting proceeds. The form of mantle flow driven by these sources of buoyancy is exemplified. Some consideration is given to the simulation of the delivery of melt to the surface by eruption. Further calculations combine thermal, melt and compositional buoyancy, and explore a range of plate velocities and viscosities. The implications of these calculations for the interpretation of various observables is discussed. The narrow zone of crustal formation at ridges may reflect a narrow zone of melting in the mantle. The thickness of the oceanic crust is constant down to a plate velocity (half-rate) of 1 cm yr^{-1}, but is reduced at lower plate velocities. Rapid transport of melt in fractures is required by the major element compositions of MORBs; the effects of this process are simulated and are consistent with a narrow melting region if the fractures reduce the mantle viscosity as well as limiting the melt porosity. Incompatible trace elements may be concentrated into MORBs by incipient melting in the depth range of small-scale convection. The density structure of the main melting region is not observable in the gravity field, because the melt and compositional density structures are complementary. Speculative conclusions about the spatial segmentation and temporal episodicity of magmatic activity are also offered.

1. INTRODUCTION

A significant body of work has developed on the modeling of mantle flow and melting beneath mid-ocean ridges with fluid-mechanical calculations. This introductory review emphasises the distinction between *passive flow* models in which mantle flow is a passive response to the separation of the plates, and *active flow* models in which mantle flow is also driven by buoyancy forces.

Mantle Flow and Melt Generation at Mid-Ocean Ridges
Geophysical Monograph 71
Published in 1992 by the American Geophysical Union.

Passive models. These are typically based on the classical solution for corner flow, i.e., flow in an isoviscous half-space driven by velocity boundary conditions at the free surface [Batchelor, 1967]. Reid and Jackson [1981] calculated the temperature distribution associated with corner flow for a range of plate velocities, and inferred the distribution of melting and the thickness of crust produced. Phipps Morgan and Forsyth [1988] developed a model along similar lines but generalised to three dimensions, allowing them to estimate the distribution of melting in the region of a transform offset. Spiegelman and McKenzie [1987] showed how the non-hydrostatic pressure gradient sustained by corner flow could cause lateral movement of melt towards the ridge axis. Ribe [1988] introduced a finite 'plate boundary zone' to remove the singularity from the corner flow solution and studied the effects on melt extraction and axial topography. Phipps Morgan [1987] proposed that the finite strain due to corner flow could cause anisotropic permeability, in turn causing lateral movement of melt. In models such as that presented by Bottinga and Allegre [1978] the asthenospheric upwelling is passive but confined to a vertical conduit between the plates. Given the likely rheological structure of the asthenosphere, this pattern of flow is unlikely to arise without the action of buoyancy forces.

Active models. In these models, buoyancy forces perturb the corner flow driven by the plates; they can be categorised according to the types of buoyancy acting. The effect of thermal buoyancy alone has been studied in a body of work on small-scale convection in the uppermost mantle [e.g., Parsons and McKenzie, 1978; Fleitout and Yuen, 1984; Buck and Parmentier, 1986]. These studies are not directed at an understanding of the melting processes at mid-ocean ridges, but they provide important insight into the fluid-mechanical environment within which melting occurs. Rabinowicz et al. [1984] proposed that melt buoyancy creates a 'rolling mill' effect beneath mid-ocean ridges, which is observed in mantle flow structures in ophiolites. Whitehead et al. [1984] and Crane [1985] proposed that melt buoyancy leads to along-axis segmentation through the development of Rayleigh-Taylor instabilities. Scott and Stevenson [1989] presented calculations in which the action of melt buoyancy, compositional buoyancy and percolative flow of melt produces a narrow zone of upwelling and melting beneath a mid-ocean ridge. Buck and Su [1989] showed that a large decrease in mantle viscosity due to the presence of melt would emphasise this narrowing effect. Mutter et al. [1988] included compositional and thermal buoyancy in a model of "convective melting" during the initiation of rifting Sotin and Parmentier [1989] assumed that melt buoyancy is negligible and studied the interaction between compositional and thermal buoyancy forces. In the first three-dimensional calculations of active flow beneath ridges, Parmentier and Phipps Morgan [1990] showed that compositional buoyancy alone could lead to along axis segmentation of the mantle upwelling.

The above studies are distinct from work on the shallow structure of mid-ocean ridges, i.e., the structure of the crust and shallow mantle, above the source region of melts. For example, Sleep [1975] and Kuznir and Bott [1976] calculated the thermal structure during the formation of oceanic crust, and Lin and Parmentier [1989] studied the mechanical effects of a lithosphere of finite thickness at the ridge axis.

This study. The various studies cited indicate that active flow, driven by one or more of the candidate sources of buoyancy, may help to reconcile the diversity of constraints on melting and mantle flow beneath mid-ocean ridges. A primary goal of this study is to compare and contrast the character of active flow driven by different sources of buoyancy. In order to effect an objective comparison, the basic modeling procedure is identical in all the calculations presented.

The work presented here develops an earlier study of melting and mantle flow beneath mid-ocean ridges [Scott and Stevenson, 1989; henceforth SS89]. In that study, we used simple numerical simulations to demonstrate how the buoyancy forces generated during the melting process could alter the pattern of mantle flow and hence feed back to control the location of melting. The narrowness of the zone of crustal formation at ridges was attributed to the confinement of melting to a narrow upwelling beneath the ridge, with a velocity greater than that of the plate. Here, a considerably more realistic modeling procedure is used to provide a detailed assessment of this and other effects. The major new physical effects that are included are thermal buoyancy, variable viscosity, and a realistic, thermodynamic melting behavior.

A significant limitation of the calculations in SS89 was the possibility that the outer and lower boundary conditions influenced the pattern of flow within the modeled region. In that study, these boundaries were 240 km out and down from the ridge axis and the velocity field prescribed was that of passive, isoviscous corner flow. These boundaries were certainly a comfortable distance from the region of melting (within 50 km of the ridge), but the choice of passive isoviscous corner flow as a boundary condition may be inappropriate. This concern is addressed here by a detailed study of small-scale thermal convection without melting within a few hundred kilometers of the ridge. Despite its name, the length scale of this convection is large compared to the size of the melting region. In subsequent calculations including melting, the melting region is isolated from the boundaries by the small-scale convective flow.

Section 2 describes a study of small-scale convection designed to gain an improved understanding of the environment of mantle melting beneath mid-ocean ridges. Section 3 describes a series of calculations that do not include thermal buoyancy but which clarify the differing roles of the other two possible sources of buoyancy, namely the presence of melt and the changing composition of the solid. Section 4 summarises a large number of calculations that combine thermal convection and the processes associated with melting. The results from these calculations are discussed with respect to a range of observables in Section 5. The mathematical model and some analysis of small-scale convection are described in two appendices.

2. SMALL-SCALE CONVECTION

The motivation for calculations in this section is simple; if the buoyancy due to melt porosities of a few percent can have a large effect on the mantle flow beneath a ridge, then the effect of small amounts of thermal buoyancy must also be considered. To expand on this reasoning, consider the viscosity and thermal structure expected in the uppermost mantle beneath a ridge with reference to Figure 2.1. The viscosity of the mantle is probably nowhere lower than directly beneath a ridge. Material can ascend without conductive cooling to within perhaps ten kilometers of the surface;

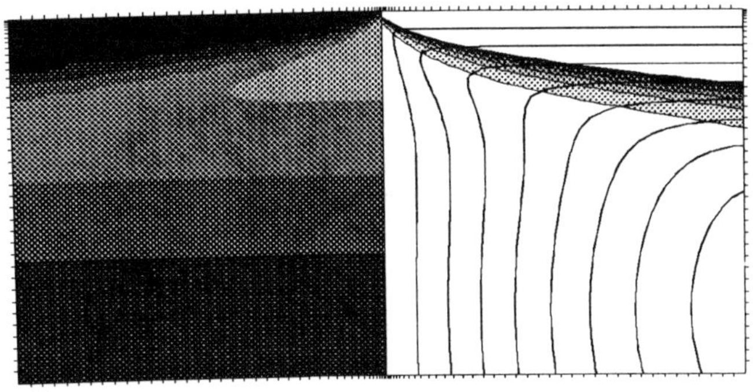

Fig. 2.1. Flow pattern, temperature distribution and viscosity for passive flow beneath a ridge. Each of this pair of panels represents a vertical section through a 400 km x 400 km region beneath a ridge, reflected about a vertical plane of symmetry. The ridge axis lies at the top center, and the oceanic plates move out to the left and right. The tick marks indicate the finite element grid, which has high resolution close to the ridge axis. The shading on the left side shows the viscosity distribution, with lighter shading indicating lower viscosity. Each shade spans an order of magnitude in viscosity. Moving up from the base of the box, the boundaries between shades are at 10^{20}, 10^{19}, and 10^{18} Pa s, so the lightest shade shows indicates viscosities below 10^{18} Pa s. The black region at the top is a rigid plate. The lines on the right side contour the stream function of the mantle flow field, which is driven by the moving plate with zero-stress boundary conditions at the outer and lower boundaries. The plate velocity is 1 cm yr^{-1}. Note that a modest amount of focussing of the flow occurs beneath the ridge, due to the viscosity structure. The shading on the right side shows contours of potential temperature (i.e. temperature with the adiabatic gradient removed), with a constant contour interval of 26 K. Only regions with potential temperature between 1560 K and 1430 K are shaded; this range is the hottest 10 % of the total temperature variation between the surface and the interior is shown. The cold plate is shown unshaded, as is the hot interior where the potential temperature is 1573 K. The buoyancy forces associated with this thermal structure are not active in this calculation.

keeping material hot while reducing the pressure leads to a lowering of the viscosity. The spreading of the plates also sustains a permanent lateral temperature gradient in this region; material ascending beneath the ridge axis will always be hotter than material at the base of the cooling plates to either side. The coincidence of low viscosity and an unstable temperature distribution suggests that convective flow will occur. The calculations presented in this section are designed to characterise and quantify this convective flow.

The equations used are the Stokes equation for the incompressible flow $\mathbf{V}$ of a fluid driven by thermal buoyancy and flows prescribed at the boundaries, and an advection-diffusion equation for the rate of change of potential temperature Θ:

$$-\frac{\partial}{\partial x_j}\left[\eta\left(\frac{\partial V_i}{\partial x_j}+\frac{\partial V_j}{\partial x_i}\right)\right]+\frac{\partial p^*}{\partial x_i}=g\rho_m\alpha(\Theta-\Theta_m)\delta_{iz} \quad (2.1)$$

$$\dot{\Theta}=-\mathbf{V}.\nabla\Theta+\kappa\nabla^2\Theta \quad (2.2)$$

where η is the mantle viscosity, p^* is a reduced pressure variable, g is the acceleration due to gravity, ρ_m is the mantle density, α is the thermal expansivity, Θ_m is the potential temperature of the mantle and κ is the thermal diffusivity. The potential temperature Θ is related to the true temperature T by a simple first-order form:

$$\Theta = T + \left.\frac{\partial T}{\partial z}\right|_S z \quad (2.3)$$

where the adiabatic temperature gradient is given by

$$\left.\frac{\partial T}{\partial z}\right|_S = \frac{g\alpha\Theta_m}{C_P} = -0.48 \text{ K km}^{-1} \quad (2.4)$$

C_P is the isobaric heat capacity. The derivation of these equations is discussed further in Appendix A. The model geometry, boundary conditions and rheological structure that were chosen are discussed below.

A finite element technique is used to solve the equations. The Stokes equation (2.1) is solved using the penalty function formulation [Hughes et al., 1979] and an augmented Lagrangian algorithm. The advection-diffusion equation (2.2) is solved using streamline-upwind shape functions [Brooks, 1981]. The finite element grid is designed to give high resolution close to the ridge axis; the element dimensions in this region are typically about 1.5 km. The element dimensions increase gradually with vertical and lateral distance from the axis (see Figure 2.1). All the elements are quadrilaterals, with a maximum aspect ratio of approximately ten.

2.1. *Model Geometry and Boundary Conditions*

The model presented here does not, by design, incorporate the entire convecting mantle; instead the model focusses on the region within a few hundred kilometers of a mid-ocean ridge. The plate motion is prescribed as a boundary condition; no attempt is made to account for the forces that move the plate. Subduction, the complement of seafloor spreading, is not included in the model.

A typical box, as shown in Figure 2.1, is 400 km deep and extends 400 km laterally from the ridge axis. The results of calculations using a larger 800 km box and a smaller 200 km box will also be presented here. By limiting the size of the box the details of flow in the region of the plate boundary can be resolved well enough to allow melting and melt segregation to be included in later calculations. The main drawback is that flow must be permitted through the lower and outer boundaries of the model; these boundary conditions must be specified in such a way as to mimic the effect of the full convecting system on the smaller region under study.

The new plate that is continuously created within the model is disposed of by a prescribed rate of flow through the outer boundary. This material must be replaced by flow into the box; this flow is constrained to be primarily into the bottom of the box. The exact flow field specified at the outer and lower boundaries is simply the pattern of flow for passive, plate-driven flow with the same rheological structure used in the calculation of active flow, and with no normal or shear stress at the lower and outer boundaries. An example of this passive flow field is shown in Figure 2.1. It basically resembles corner flow, with some focussing of flow in the low viscosity region below the ridge. New material brought into the base of the box has a constant temperature, so that isotherms at the base of the box are horizontal.

These boundary conditions are intended to reflect the view [Houseman, 1983] that ridges are passive features in the context of large-scale mantle convection. If ridges were considered to be active features in this context, i.e., always underlain by a thermally-driven mantle upwelling, then different boundary conditions on both flow and temperature would be used.

2.2. *Rheological Structure*

An example of the type of rheological structure used in the calculations is shown in Figure 2.1. It features a temperature- and pressure-dependent viscosity in the asthenosphere (resulting in a low viscosity zone) and a rigid lithosperic plate. The rheology is also time-dependent, responding to changes in temperature. Before detailing the way that the rheology is determined, the motivation for the basic features of this structure is discussed.

In Section 4 the processes of thermal convection described in this section are combined with the processes associated with melting described in Section 3. An increase in viscosity with depth in the asthenosphere is required so that these processes can be combined realistically. In general, the vigor of active flow is determined by the average value of viscosity over the depth range that the flow occupies. The depth extent of convection is greater than that of the melting region, and the increase in viscosity with depth means that the flow samples a higher average viscosity. The rigid upper layer is required to stabilise the lithospheric plate. If the plate were viscous and had a viscosity in the range used in the asthenosphere, the large negative buoyancy of the plate would dominate the flow.

The viscosity η at any point is calculated from the depth and temperature, using the following relationship.

$$\eta = \eta_0 \exp\left(\frac{E + P\Omega}{RT}\right) \tag{2.5}$$

where η_0 is a reference viscosity, E and Ω are the activation energy and volume, $P = \rho_m g z$ is the pressure and T is the temperature. A useful measure of the combined effect of the parameters in the exponent is the rate at which the viscosity increases with depth when the temperature is on the mantle adiabat. The values $E = 420$ kJ mol^{-1} and $\Omega = 17 \times 10^{-6}$ m^3 mol^{-1} are used, and Equation (2.4) gives a value of $\partial T/\partial z|_S = -0.48$ K km^{-1}. These give an increase in viscosity by approximately a factor of twenty for an increase in depth of 100 km. For comparison, Buck and Parmentier [1986] presented models with $E = 420$ kJ mol^{-1}, $\Omega = 10 \times 10^{-6}$ m^3 mol^{-1} and $\partial T/\partial z|_S = -0.3$ K km^{-1}, giving a similar rate of increase in viscosity with depth.

The same viscosity law is used in all the calculations presented here, except that the constant η_0 is varied to change the average viscosity. Three values are used: $\eta_0 = 3170$, 659, or 143 Pa s. These values give $\eta = 10^{18}$ Pa s at a depths of 50, 100 and 150 km respectively, when the temperature is on the mantle adiabat. The viscosity approaches 10^{21} Pa s at a depth of 350 km, while the viscosity in the region within 50 km of the ridge axis where melting occurs is between 10^{17} and 10^{18} Pa s. Above a depth of 200 km there is a zone where the viscosity is less than 10^{19} Pa s.

A limitation of the numerical method used is that it can only accommodate a variation in viscosity of four orders of magnitude. Hence, viscosities below 3×10^{16} Pa s (this occurs only in extreme cases), or higher than 3×10^{20} Pa s, are truncated at these values. A plate is defined, consisting of elements with the maximum viscosity that are continuously connected with the upper boundary through other elements having the maximum viscosity. This plate is then treated as rigid, i.e., all velocities are set equal to the plate velocity. An exception is made for elements that lie closer to the vertical axial plane than to the surface; these always remain viscous and are not incorporated into the plate. This is necessary to prevent large velocity gradients in the region between the rigid plates due to kinematic constraints on the flow.

Within the plate, vertical heat transport only occurs by conduction. Note that the plate grades into the low viscosity zone through layers of high viscosity. While the plate cannot become unstable as a whole, the viscous layers at its base are cooled and can form the instabilities characteristic of small-scale convection.

2.3. *Results and Discussion*

The calculations performed using the model described above are very simple. The temperature distribution in the initial state consists of a cold plate thickening as the square-root of age. With flow driven by both the movement

Fig. 2.2. Flow patterns and temperature distributions for small-scale convection. Each panel represents a vertical section through a 400 km × 400 km region beneath a ridge. From the top down, the plate velocity (half-spreading rate) is 5, 2, 1, and 0.5 cm yr^{-1}; the plate velocity is the same for all three panels in each row. Within each row the viscosity decreases from left to right, by approximately a factor of five between adjacent panels; the viscosity function is the same for all the panels in each column. The center column has the same viscosity function as is shown in Figure 2.1. The bold lines contour the stream function of the mantle flow field; the contour interval is proportional to the plate velocity, so the spacing between contours indicates the velocity relative to that of plate. The potential temperature is shown as in Figure 2.1. The panels to the top and left resemble Figure 2.1, showing little of the effects of thermal convection. Convective flow is promoted by either decreasing the viscosity (moving right) or increasing the lateral temperature gradient by decreasing the plate velocity (moving down).

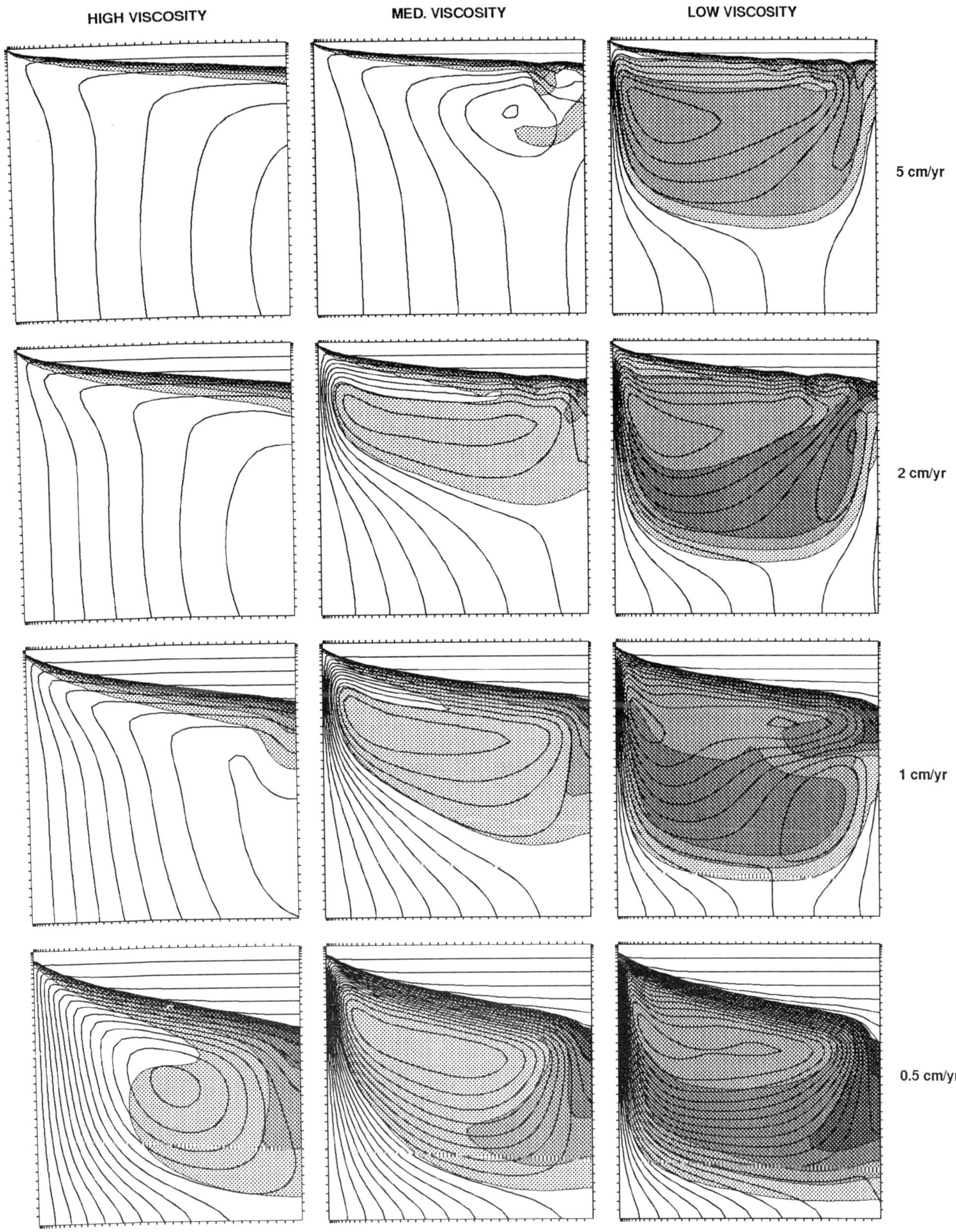
HIGH VISCOSITY
MED. VISCOSITY
LOW VISCOSITY
5 cm/yr
2 cm/yr
1 cm/yr
0.5 cm/yr

of this plate and thermal buoyancy, the system evolves to approach a steady state. This typically involves the development of convective flow in the low viscosity zone beneath the plate, accompanied by slight cooling of the convecting region. Other initial temperature distributions have been used, and were found to reach the same final states.

These calculations have been performed using three box sizes (200 km, 400 km and 800 km), the three values of the viscosity parameter η_0, and four values of plate velocity (half-rates of 0.5, 1, 2, and 5 cm yr^{-1}). Other calculations, not presented here, have established that spatial and temporal resolution used is adequate by varying the time step and the numbers and sizes of the elements. Figure 2.2 shows the results for a box size of 400 km. At the higher values of plate velocity and higher viscosities (i.e. to the top left of the figure) the flow is essentially passive and the isotherms have the familiar appearance of a thickening plate. At lower values of plate velocity and viscosity, the convective flow becomes quite vigorous. It should be noted, however, that only the very bottom of the plate becomes unstable; the thickening of the main body of the plate with age is not substantially affected.

Convective velocities beneath the ridge axis. The results from these calculations have been summarised using a convective velocity variable V_C. This is designed to provide an objective measure of the vigor of the convective flow in a box that extends 100 km down from the ridge axis and 100 km to either side. To calculate V_C, the r.m.s. velocity in this box is evaluated, excluding points that are incorporated in the rigid plate. The passive flow field for the same rheological structure is then calculated, excluding thermal buoyancy, and a second r.m.s. velocity is evaluated in the same 100 km box. V_C is the difference between these two r.m.s. velocities: note that it is small if convective flow is weak relative to plate-driven flow. Note also that the same 100 km box is used to calculate V_C for the 200 km, 400 km and 800 km total box sizes, so that these calculations can be compared objectively.

V_C is plotted as a function of the plate velocity V_P in Figure 2.3, for the three values of viscosity and three box sizes. Note that, when convection is weak, V_C has an inverse dependence on V_P; when convection is strong, V_C has an direct dependence on V_P. In the calculations with intermediate viscosity and a 400 km box, the transition between these two regimes is seen at a plate velocity of about 2 cm yr^{-1}.

An approximate analysis of the vigor of this type of small-scale convection is presented in Appendix B. Both the weak and strong regimes are considered; the difference between the two lies in the extent to which the convective flow disturbs the passive distribution of isotherms. For the weak regime, the isotherms are assumed to be undisturbed and the dependence $V_C \sim V_P^{-1/2}$ is predicted. For the strong regime a convective temperature structure is used in which the amount of cooling across the convective cell provides the correct amount of negative buoyancy to drive the cell; this leads to the dependence $V_C \sim V_P^{1/2}$. This rather simple

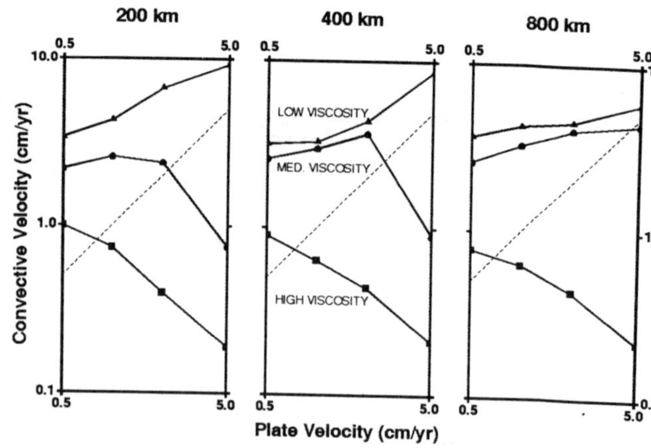

Fig. 2.3. Convective velocity as a function of plate velocity, viscosity and box size. The variable V_C on the vertical axis is a measure of the convective velocity within 100 km of the ridge axis (see text). The plate velocity V_P is shown on the horizontal axis. Both axes are logarithmic and the units are cm yr^{-1}. The box size for each set of calculations is shown above each plot. Within each plot, the triangles, circles and squares indicate calculations with low, intermediate and high viscosity respectively. Note that the center plot summarises the calculations shown in Figure 2.2. The dashed lines showing $V_C = V_P$ divide the plots into regimes of weak convection where V_C decreases with increasing V_P, and regimes of strong convection where V_C increases with increasing V_P. A transition betwen these regimes is seen in the 400 km plot for the intermediate value of viscosity.

analysis is able to reproduce the general structure of strong and weak regimes, with a transition that depends on the plate velocity and the viscosity.

Variation with box size. The basic trends shown in Figure 2.3 do not vary a great deal between the three sizes of box. The only significant effect of the size of the box is in moving the point of transition between the strong and weak convective regimes; in the cases with intermediate viscosity, the strong regime is favored when the box size is larger. This may be due to a limitation of the calculations using the 800 km box, wherein the lower half of the box has a constant viscosity (3×10^{20} Pa s). Although the viscosity law (2.5) indicates a continuing increase in viscosity with depth, the numerical method cannot accommodate viscosities more than four orders of magnitude greater than those directly beneath the ridge. This limit on viscosity stratification encourages the convection cell to expand to occupy most of the box in the calculations using an 800 km box.

It is proposed that the vigor of convective flow within 100 km of the ridge axis, where V_C is calculated, is controlled locally regardless of size of the entire box. This is reasonable since the largest lateral temperature gradients and lowest viscosities are found directly beneath the ridge axis. The practical significance of this conclusion is that in later calculations including melting the 200 km box size is used, to economise on computing time.

Ridge rolls and plate rolls. It has been established by both experimental and theoretical work [Richter, 1973; Richter and Parsons, 1975; Curlet, 1976] that small-scale convection can develop beneath a moving plate, and is characterised by rolls aligned parallel to the direction of plate motion. Rolls aligned perpendicular to the plate motion have a higher critical Rayleigh number and are less stable. Gravity lineations over younger, fast-moving plates [Haxby and Weissel, 1986], which may represent the surface expression of small-scale convection, are aligned parallel to the absolute plate motion. For these reasons, previous small-scale convection calculations in two dimensions have considered motions in a plane perpendicular to the plate motion [Buck and Parmentier, 1986].

The calculations presented here, showing motions in a plane parallel to the plate motion, are applicable only in the region of the ridge axis. In a fully three-dimensional calculation, the pair of 'ridge rolls' beneath the axis may join up with the perpendicular 'plate rolls' in a geometry that is not at present determined. Alternatively, the ridge rolls may be obliterated by plate rolls running right across the ridge. The ridge rolls should survive because they are sustained by a permanent lateral temperature gradient close to the ridge axis. The plate rolls develop from a condition of unstable equilibrium with a finite growth time. Indeed, forcing this growth time to be short enough is one constraint on the viscosity in the uppermost mantle; a low viscosity is required for the gravity lineations to develop over 5 Ma-old plates [Haxby and Weissel, 1986; Buck and Parmentier, 1986]. Ridge rolls might either increase or decrease this growth time. The growth time might increase because the the low viscosity zone is cooled by the ridge rolls; it might decrease because the ridge rolls provide a finite amplitude disturbance in an unstable thermal structure. If the latter case applied, estimates of the viscosity of the uppermost mantle could be increased.

3. Introduction to the Effects of Melting on Mantle Flow

The process of melting has two effects on the density of the whole rock relative to its initial, unmelted density. If some or all of the melt produced remains with the source rock, the density difference between solid and liquid creates a buoyancy force that is typically of order $500\phi g$ kg m^{-3} where ϕ is the melt porosity. The density of the solid residue also changes with progressive melting; in the garnet- and spinel-lherzolite stability fields, the density of the solid decreases as the denser minerals are progressively removed and the Fe/Mg ratio of the residual minerals is reduced. The compositional buoyancy force created by this effect is of order $500gF$ kg m^{-3} where F is the extent of melting. If no melt transport were to occur (i.e., batch melting), ϕ and F would be equal and the melt and compositional buoyancy would be very similar in magnitude. It would therefore appear that, since melt transport will substantially decrease ϕ without affecting F, the compositional buoyancy of the solid should be the more important of the two. This conclusion has been used to justify calculations in which the melt buoyancy has been excluded [Sotin and Parmentier, 1989; Parmentier and Phipps Morgan, 1990].

This is misleading because the nature of the two sources of buoyancy is quite different. Recall that it is lateral gradients in density that drive flow in a gravitational field. The process of melting by adiabatic ascent gives rise to relatively small horizontal gradients in the extent of melting, and a stable stratification of compositional density. The production of melt buoyancy by adabiatic ascent, on the other hand, creates a positive feedback loop; the buoyancy drives more rapid upwelling, which in turn generates more buoyancy. These statements will be clearly illustrated by calculations presented below. It is shown that active flow of any vigor associated with melting must be driven by melt buoyancy, modified but by no means dominated by compositional buoyancy.

The Stokes equation is again used to determine the mantle flow; in this section Equation (2.1) is modified by omitting thermal buoyancy but including melt and compositional buoyancy. The solid composition is parameterised by the concentration X of a perfectly compatible tracer; this is discussed in detail in Section 3.2.

$$-\frac{\partial}{\partial x_j}\left[\eta\left(\frac{\partial V_i}{\partial x_j}+\frac{\partial V_j}{\partial x_i}\right)\right]+\frac{\partial p^*}{\partial x_i}= g\left[(\rho_m-\rho_l)\phi+\frac{(\rho_m-\rho_r)}{\Delta X}\left(1-\frac{1}{X}\right)\right]\delta_{iz} \tag{3.1}$$

where ρ_l is the density of the liquid and ρ_r is the density of the solid residue after 30% melting ($\Delta X = 0.3$). In this case the 'mantle flow' $\mathbf{V}$ is defined as a porosity-weighted average of the solid and liquid velocity fields that is divergence-free.

The volume flux $\mathbf{u}$ of liquid relative to solid is determined by Darcy's law, in a simplified algebraic form.

$$u_i=\frac{k_\phi}{\mu}g(\rho_m-\rho_l)(1-\phi)\delta_{iz} \tag{3.2}$$

where k_ϕ is the permeability and μ is the melt viscosity. The choice of permeability function is discussed below. The effects of viscous compaction and pressure gradients due to mantle flow are omitted from Equation (3.2). These omissions are considered appropriate because the size of the region of melting is significantly larger than the characteristic length scales associated with these processes; this is discussed in Appendix A. The approximations made produce a system of momentum equations that can be solved using a numerical method that is very similar to that used for thermal convection in Section 2, because the Stokes equation (3.1) and Darcy's law (3.2) are not coupled through the flow fields $\mathbf{V}$ and $\mathbf{u}$.

Variations of temperature are again caused by advection and diffusion; variations of composition are caused by advection, and variations of porosity are caused by advection and compaction.

$$\dot{\Theta}=-\mathbf{V}.\nabla\Theta+\kappa\nabla^2\Theta-\frac{\Theta_m\Delta S}{C_P}\dot{\phi}_{melt} \tag{3.3}$$

$$\dot{\phi} = -(\mathbf{V} - \mathbf{u}).\nabla\phi - (1-\phi)\nabla.\mathbf{u} + \dot{\phi}_{melt} \qquad (3.4)$$

$$\dot{X} = -(\mathbf{V} - \mathbf{u}).\nabla X + \frac{X}{(1-\phi)}\dot{\phi}_{melt} \qquad (3.5)$$

where ΔS is the entropy of melting. The melting term $\dot{\phi}_{melt}$ couples the temperature, porosity and composition equations. This term is determined by a thermodynamic melting relationship that is discussed below.

3.1. *Permeability*

The permeability of a partially molten ultramafic rock has not been measured directly, but the equilibrium microscopic texture is well characterised. This texture is parameterised by the dihedral angle at solid-solid-liquid triple junctions. When the dihedral angle is less than 60°, the two-phase material resembles a sponge, with the melt phase forming a fully interconnected network. The generic ultramafic assemblage of basaltic liquid and olivine satisfies this criterion [Vaughan et al., 1982] providing justification at the microscopic level for the irrefutable geological observation that basaltic melts can escape from their source rocks.

The permeability associated with the sponge texture has been estimated by Cheadle [1989]. Noting that most of the liquid phase lies in tubes lying along triple junctions between grains, a suitable functional form for the dependence of permeability on melt porosity is:

$$k_\phi = \frac{d^2\phi^2}{b} \qquad (3.6)$$

where d is the grain diameter, ϕ is the porosity and b is a constant. Cheadle [1989] suggests $b = 3000$, so for a grain diameter of 1 mm, the permeability is approximately $3 \times 10^{-10}\phi^2 m^2$. This is the form used in all the calculations presented here.

Several effects might modify this value, such as a spectrum of grain sizes, a polymineralic assemblage with varying dihedral angle [Nakano and Fujii, 1989] and non-hydrostatic stresses. A serious concern is the possible involvement of fractures in melt transport. In this case the microscopic sponge texture no longer provides a basis for estimating the large-scale permeability of the system, as most flow over long distances will occur through fractures. However, fractures must be supplied with liquid in order to open or stay open [Sleep, 1988] so the microscopic permeability may still control the overall rate of melt extraction from the source region. An attempt is made to assess the effect of melt transport in fractures on the pattern of mantle flow in Section 5.

3.2. *Petrological Model*

The extent of melting or depletion of the mantle material plays a central role in the petrological model. The best measure of depleted solid compositions is the concentration of a perfectly compatible tracer, i.e., a trace element that does not enter the liquid phase. This quantity (X) is always positive because it is a concentration. Its value is assumed to be unity for fertile unmelted mantle; values greater than unity indicate material from which melt has been extracted, and values less than unity indicate material which has had excess melt added by *in situ* freezing. During melting, changes in X and ϕ are coupled because the quantity $X(1-\phi)$ remains constant. X is advected by the solid velocity field in the usual way and is unchanged by compaction.

In SS89 we used an extent-of-melting variable F that changes by exactly the same amount as the porosity during melting (i.e. $\delta F_{melt} = \delta\phi_{melt}$). The quantity F is not a concentration, and its behavior during advection and compaction is not as simple as that of the quantity X. Note that during batch melting, X and F are related by $X(1-F) = 1$; hence changes in X and F are equal in the limit of small extents of melting.

The rate of melting (or freezing) is determined using a simple model based on the data presented by McKenzie and Bickle [1988]. The key feature of the model is an equation for the dependence of the potential temperature of solidus of mantle material Θ_s on the composition X and the pressure, here replaced by the vertical position z (increasing upwards).

$$\Theta_s = T_{s0} + \left(\left.\frac{\partial T_s}{\partial z}\right|_X - \left.\frac{\partial T}{\partial z}\right|_S \right) z + \left.\frac{\partial T_s}{\partial X}\right|_z (X-1) \qquad (3.7)$$

where $T_{s0} = 1373$ K, $\partial T_s/\partial z|_X = -4.5$ K km^{-1} and $\partial T_s/\partial X|_z = 440$ K. The values of these constants are chosen to be consistent with the data compiled by McKenzie and Bickle [1988] in the region where $X < 0.2$.

During melting or freezing, ϕ_{melt} in Equations (3.3), (3.4) and (3.5) is chosen such that material in each element lies on the solidus if melt is present, and otherwise remains below the solidus. In a given time step, the other terms in these equations will change the temperature, porosity and composition in an element. After these changes, melting occurs if the temperature lies above the solidus; freezing occurs if the temperature lies below the solidus and melt is present.

In an attempt to be consistent with SS89, the solid density is taken to be proportional to the old extent-of-melting variable F. Hence, the variation of density with X in Equation (3.1) involves the form $(1 - 1/X)$. This approximation does not do justice to the experimental constraints on density as a function of composition [Niu and Batiza, 1991], but is considered adequate for the purposes of this modeling.

3.3. *Melting and Mantle Flow*

All the panels in Figure 3.1 show the pattern of flow and the distribution of temperature, porosity and composition in a region within 100 km of the ridge axis. The intermediate value of the parameter η_0 is used in determining the rheological structure in all the calculations shown. Passive flow velocity boundary conditions are applied at the perimeter of a 200 km box, as in the convection calculations presented in Section 2. The panels differ because different physical processes have been sequentially introduced, each modifying the patterns of the quantities shown.

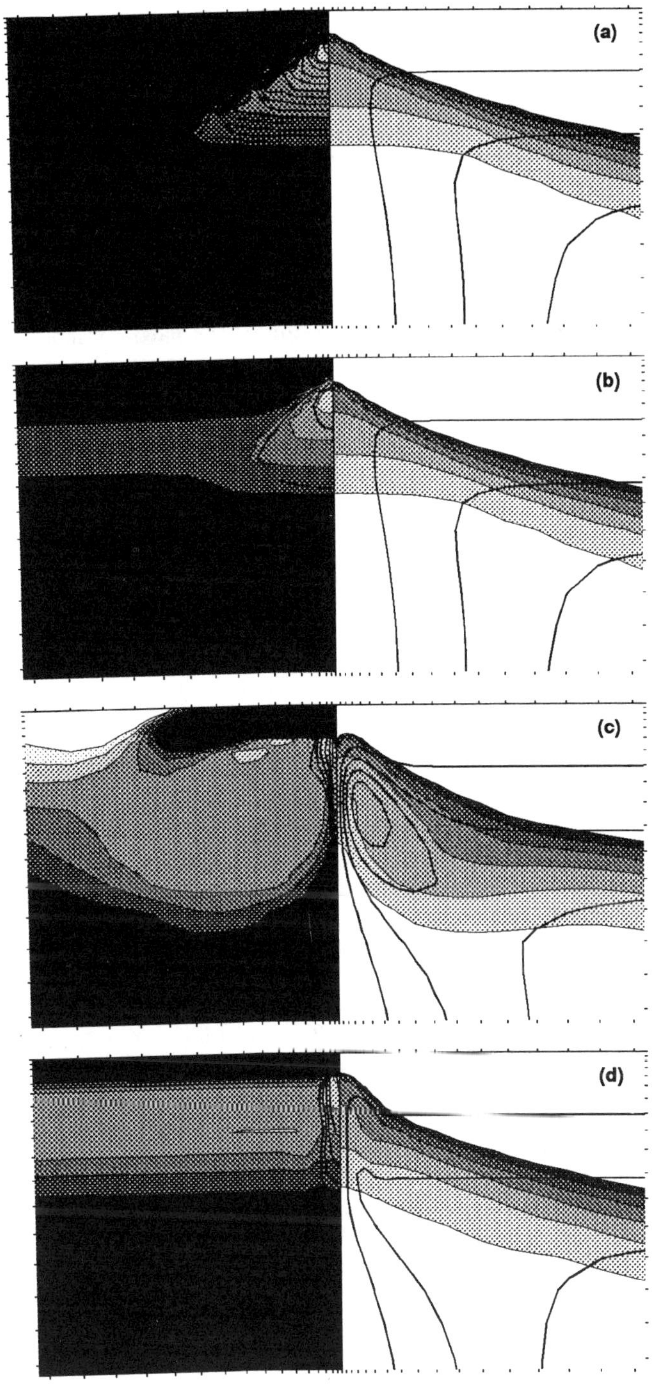

Fig. 3.1. The effects of melt and compositional buoyancy beneath a ridge. Each pair of panels shows a region 100 km deep and 100 km to either side of a ridge; in fact these plots show detail from calculations that use a 200 km x 200 km box. The quantities shown are all symmetric about the central axis. The right hand panels show the mantle flow and potential temperature as in Figure 2.1. The plate velocity is 1 cm yr^{-1} in all cases. The left hand panels show contours of the melt porosity in the region of melting directly beneath the ridge axis; the contour interval is 1 %. The shading in the left hand panels contours the composition X of the solid; the contour interval is 5 % and the lightest shades indicate the largest depletion. Each pair of panels shows a calculation which includes different physical processes: (*a*) shows passive mantle flow, melting and freezing; (*b*) adds the process of melt percolation; (*c*) adds mantle flow driven by melt buoyancy; (*d*) adds the effect of compositional buoyancy.

In Figure 3.1*a* the flow is passive, driven only by the plate. As in Figure 2.1, the pattern of the streamlines differs slightly from the familiar form of isoviscous corner flow because of the rheological structure, which causes a slight focussing of the flow beneath the ridge axis on the scale of the viscosity stratification. Melting and freezing are both occurring, but without melt transport relative to the matrix (i.e., $\mathbf{u} = 0$). As a result the patterns of porosity and composition are the same. Melting commences when material ascends without conductive cooling to a depth of 50 km. The partially molten zone extends up to a level where conductive cooling removes the latent heat carried by the melt and the material once again becomes completely solid. The composition of the solid returns to its initial value after passage through the partially molten zone; there is no compositional differentiation because there is no melt transport. (The melting process here is precisely that of batch melting). The process of *in situ* freezing is clearly not realistic, because compositional differentiation of the crust from the mantle does not occur. The temperature field shows the buffering effect of the latent heat within the partially molten region, giving horizontal isotherms in this region. The rheological structure is not shown but is for the most part similar to that shown in Figure 2.1. However, in the region of melting the temperature gradient with depth (the 'wet adiabat') is such that the viscosity does not decrease as material ascends, but remains approximately constant.

In Figure 3.1*b* a percolative flow mechanism for melt transport is introduced, using Darcy's law (3.2). Within the partially molten region, the liquid now ascends more rapidly than the solid. The porosity in the partially molten zone drops and melt arrives at the zone of freezing more rapidly than does the solid. As a result, a 'crust' forms; a layer that has experienced more enrichment by the freezing in of melt than earlier depletion by the removal of melt. A complementary depleted layer is left between the crust and the unmelted mantle. This structure was identified by Oxburgh and Parmentier [1977]; the term 'residual melting column' has been coined by Plank and Langmuir [1992] for the depleted layer. The crust is rather thick and dilute in Figure 3.1*b*, basically because the (unrealistic) process of *in situ* freezing in the mantle masks the upper part of the residual melting column. This calculation does, however, illustrate the minimal requirements for differentiation of the oceanic crust from the mantle. Note that, in a sense, gravity has been used selectively in this calculation, because the density difference between liquid and solid is used to drive percolative flow in Darcy's law (3.2) but not flow of the whole rock in the Stokes equation (3.1).

In Figure 3.1*c* the selective use of gravity is partly relaxed; the buoyancy of the liquid is used to drive mantle flow as well as percolation. This simulates a situation where the density of the solid does not change during melting. The partially molten region is buoyant and the pattern of passive flow in previous panels is modified by the addition of active upwelling beneath the ridge axis. Melt is produced at a higher rate that in the previous frame, and also depleted residual solid is produced at a higher rate. The extra residual solid cannot be carried away by the plate, and instead accumulates in the shallow mantle close to the ridge axis. Interestingly, this accumulation provides an element of negative feedback to the process of active upwelling. Closed streamlines entrain depleted material into the sides of the upwelling; this material will not melt again because its solidus temperature was increased by melt extraction during its original pass through the upwelling. This calculation is similar to those presented by Rabinowicz et al. [1984].

Note that the distribution of melting shown in Figure 3.1*c* has not reached a steady state. The newly-formed crust, shown as a black tongue extending about 60 km from the ridge axis, has not reached all the way to the outer boundary of the region shown. Beyond the end of this tongue, highly depleted material is shown in the upper layers of the plate. The composition in this region reflects the initial conditions of the calculation, in which the composition was laterally uniform and became monotonically more depleted with distance above the solidus. The associated density variations are locked into the rigid plate and have no effect on the mantle flow.

In the Earth, the depleted residue of melting is significantly less dense than the unmelted mantle and hence can modify the pattern of flow. The calculation shown in Figure 3.1*d* includes these compositional buoyancy forces, using all the terms in the Stokes equation (3.1). The residual solid cannot now be forced down into the mantle, and it can only be carried away laterally by the plate at the rate dictated by the plate velocity. Any excess residual solid generated in the upwelling is therefore forced in towards the sides of the upwelling. This constricts the upwelling, both mechanically and because the residual solid cannot be re-melted. The constriction of the upwelling reduces the rate at which residual solid is created, and a steady state is reached in which the rate of creation of residual solid balances the rate at which it can be carried away by the plate. A process of negative feedback creates a rapid but strongly focussed upwelling that is in steady state. An analysis of this feedback is provided in SS89.

3.4. *Simulation of Eruption*

In the calculations shown in Figure 3.1, the ultimate fate of the melt produced is to be frozen back into the mantle at the top of the partially molten column. By this process, the crustal layer shown in Figure 3.1*d* is formed. Also, latent heat is released during the process of freezing. This procedure has the desirable qualities of conserving mass and energy. It also has disadvantages, which can be traced to the absence of a process of 'eruption', i.e., the rapid removal of melt from the source region up to the oceanic crust. The latent heat is released somewhat below the free surface, with the effect of blanketing the partially molten column. An eruption process should involve the rapid transport of mass and heat from the source region to the surface, where the latent heat is rapidly dissipated by hydrothermal convection.

A simple way to simulate the effects of eruption on the source region of melt is simply to remove the melt and the latent heat when it reaches the top of the partially molten column. The procedure adopted here is to use the same criterion for freezing used above, based on a simple phase diagram for the system. The porosity is reduced as if *in situ* freezing had occurred but the the temperature is not increased and the solid composition is not modified. This procedure conserves mass correctly, because compaction is still treated exactly. A controlled amount of heat is lost from the system, and the final composition of the solid reflects the conditions of maximum depletion attained by each element of material.

Three examples which use eruption instead of *in situ* freezing are shown in Figure 3.2. Figure 3.2*a* is analogous to Figure 3.1*b* in that the mantle flow field is passive but differentiation by melt transport does occur. The main result of the eruption process is seen in the distribution of composition; a depleted layer now occupies the entire upper part of the lithosphere away from the ridge axis, and no crust is shown. The residual melting column is clearly exposed and not obscured by melt frozen back in to the upper part of the column. The melt lost by eruption can, in the mind's eye, be added back to the system by superimposing a crustal layer of frozen melt on the upper few kilometers of the residual melting column. This calculation is similar to those presented by Reid and Jackson [1981] and Phipps Morgan and Forsyth [1988] (the latter are in three dimensions).

Figure 3.2*b* is analogous to Figure 3.1*d*, with both melt and compositional buoyancy included. A comparison of Figures 3.2*b* (with eruption of melt) and 3.1*d* (with *in situ* freezing of melt) shows that the choice between these two processes has relatively little effect on the pattern of mantle flow and melting generated by melt and compositional buoyancy. This calculation is similar to those presented in SS89 and by Buck and Su [1989].

Figure 3.2*c* provides further insight into the system by removing the effect of the density of the liquid. This may be a reasonable assumption if the rate of removal of liquid is extremely rapid and the porosity never exceeds a fraction of a percent. The distribution of depleted solid shown in the Figure 3.2*a* is not laterally uniform and will tend to drive active mantle flow. In Figure 3.2*c* this flow is permitted; the contours of composition become very close to horizontal and a modest focussing of the mantle flow results. This calculation is similar to those presented by Sotin and Parmentier [1989] and Parmentier and Phipps Morgan [1990].

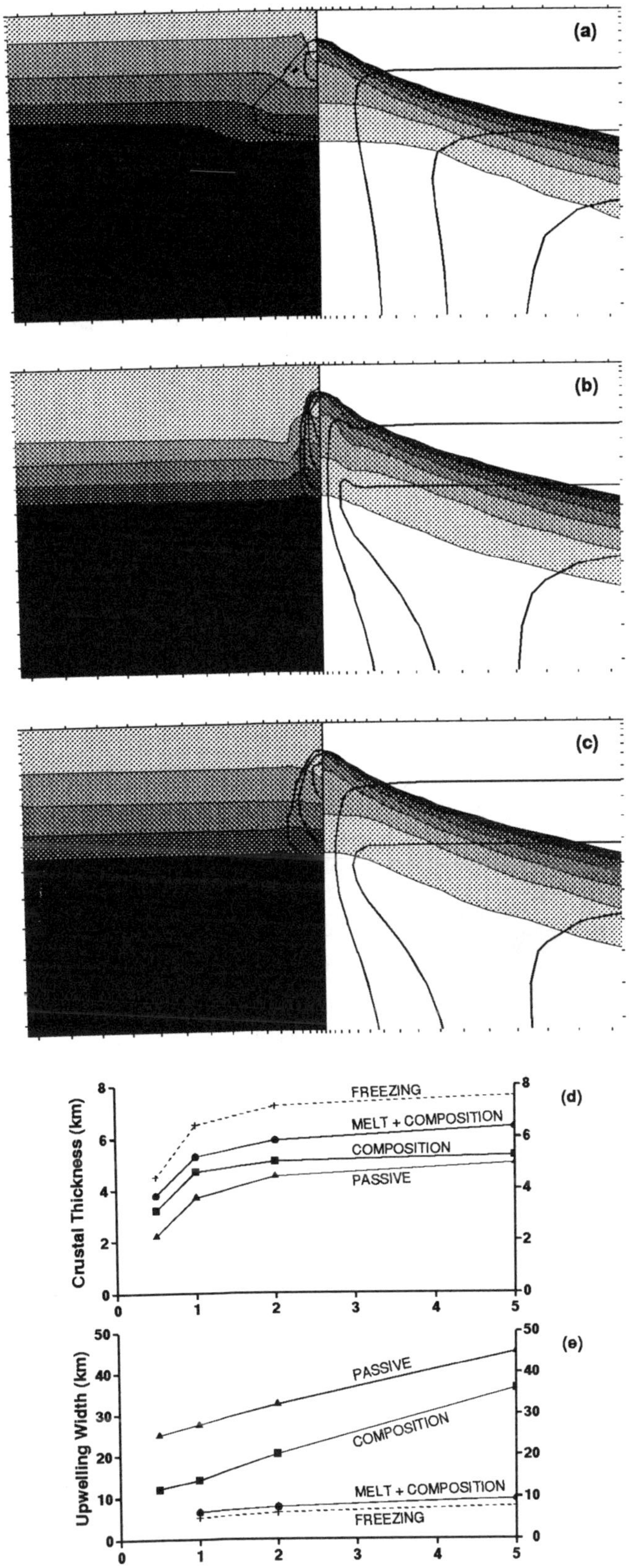

Crustal thickness and upwelling width. The graphs in Figures 3.2*d* and 3.2*e* summarise the results of calculations using a range of plate velocities. Figure 3.2*d* shows the crustal thickness z_c, assuming 100% efficiency of melt extraction; z_c is calculated as follows:

$$z_c = \frac{\dot{\Gamma}_{total}}{V_P} \tag{3.8}$$

where $\dot{\Gamma}_{total}$ is the total rate of melting per unit length along the strike of the ridge. Figure 3.2*e* shows the upwelling width x_u, calculated as in SS89:

$$x_u = \frac{z_s V_P}{V_u} \tag{3.9}$$

where $z_s = 50$ km is the depth at which melting commences, V_P is the plate velocity and V_u is the upwelling velocity on axis at a depth of 25 km. This provides a convenient, objective measure of the half-width of the region of melting, i.e., the width of the regions with porosity contours in the figures. Figures 3.2*d* and 3.2*e* show these two quantities for the examples shown in Figures 3.2*a-c*, and also Figure 3.1*d*.

The shapes of the curves showing crustal thickness have greater significance than the absolute values indicated. At a given plate velocity the amount of crust generated can be varied by changing the mantle temperature, which determines the depth at which melting commences. A mantle potential temperature of 1573 K has been prescribed, because this produces about the right amount of crust. In all cases the crustal thickness increases as the plate velocity increases from 0.5 cm yr^{-1} to 1 cm yr^{-1}, but does not increase very much above 1 cm yr^{-1}. Note also that the cases producing a greater thickness of crust also produce a more highly depleted residual melting column.

The curves showing upwelling width show more variety. In the case of passive flow, the upwelling width increases from about 25 km to about 45 km as the plate velocity increases. The reason for this variation is as follows. One would expect $V_u \propto V_P$ for passive flow, in which case Equation 3.6 would give a constant value for x_u. However, passive focussing of flow beneath the axis is more pronounced in the case of low plate velocities because the plate thickens rapidly

Fig. 3.2. Examples using the eruption mechanism. In (*a*), (*b*) and (*c*) mantle flow, potential temperature, melt porosity and compostion are presented in exactly the same way as in Figure 3.1. The physical processes included are as follows: (*a*) shows passive mantle flow, melting, melt percolation and eruption; (*b*) adds mantle flow driven by melt and compostional buoyancy; (*c*) has mantle flow driven only by compositional buoyancy. In (*c*), the only significance of the contours of porosity is in indicating the region of melting. Parts (*d*) and (*e*) show the variation of crustal thickness z_c and upwelling width x_u (both in km) with plate velocity (in cm yr^{-1}) for these examples. The upwelling width is calculated from the upwelling velocity (see text). The triangles, circles and squares correspond to Parts (*a*), (*b*) and (*c*) respectively (i.e. passive flow, melt and compositional buoyancy, only compositional buoyancy). The dashed lines corresponds to Figure 3.1*d* where melt freezes instead of erupting.

to either side of the axis. This effect is a direct result of the rheological structure and would not be observed in a passive, isoviscous calculation.

When both melt and compositional buoyancy operate the upwelling width remains below 10 km at all plate velocities. When only compositional buoyancy operates, the upwelling width is always less than in the passive case but still climbs to several tens of kilometers at the higher plate velocities.

Eruption vs. freezing. In the calculations presented in the next section the melt is disposed of by eruption instead of *in situ* freezing, a choice that should be justified. The important difference between the two concerns the fate of the latent heat absorbed during the process of melting. In the case of freezing it is released at the top of the melting column and hence affects the thermal structure in this region. In the case of eruption it is not released and does not affect the thermal structure. Hence, conductive cooling penetrates more effectively towards the melting column in the case of eruption, and the total rate of melt production is somewhat reduced. This effect is illustrated in Figure 3.2*e*, the graph of crustal thickness. The curves labeled 'freezing' and 'melt+composition' summarise calculations that are identical except that the latter uses eruption; the curves differ by about 1 km of crust. Note also that the upwelling width (Figure 3.2*d*) is not sensitive to the choice of eruption or freezing, indicating that the pattern of mantle flow is unaffected by the choice.

The eruption mechanism is arguably more consistent with the assumption that melt is extracted with high efficiency to form the crust. The latent heat of such melt should not be released within the source region; it must be released, but at very shallow levels where the effective thermal conductivity is considerably increased by hydrothermal circulation. Hence, the eruption mechanism provides a more credible model for estimating the total rate of melting and the thickness of crust produced.

4. MELTING AND SMALL-SCALE CONVECTION

The calculations presented in this section combine the process of small-scale thermal convection detailed in Section 2 with the processes associated with melting detailed in Section 3. The conclusions of the preceding two sections are used to indicate which parameters and processes have the strongest effect on the patterns of flow and melting, and which are less critical and so can be held constant or ignored. The plate velocity and mantle viscosity are treated as primary variables. The size of the box is held constant; the distance at which the passive flow boundary conditions are applied is 200 km. Also, all the thermodynamic parameters, including the mantle temperature (and hence the depth to the solidus) are held constant. The effects of varying certain of these parameters has been tested, but the results presented here are considered to be more significant. The equations used are the same as those used in Section 3, except that all the sources of buoyancy are included in the Stokes equation (see Equation (A14)).

4.1. *The Effect of Small-Scale Convection on Melting*

Figure 4.1 shows the steady-state distribution of melting and mantle flow for three values of plate velocity and with the intermediate value of the viscosity parameter η_0. Figure 4.1*b*, with a plate velocity of 1 cm yr^{-1}, is equivalent to Figure 3.2*b* with the addition of thermal buoyancy. In comparing these two calculations, it is surprising how little difference the action of small-scale convection makes to the region of melting. The flow velocity in the uppermost mantle just beneath the region of melting is several times faster when thermal buoyancy is included; this change in the lower

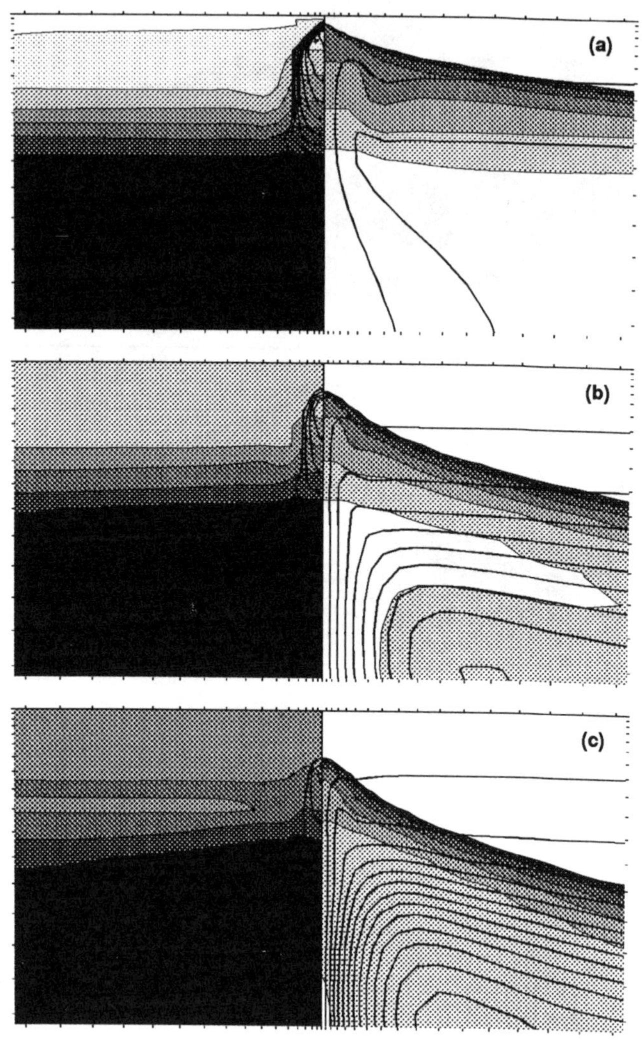

Fig. 4.1. Examples combining melting and small-scale convection, showing variations with plate velocity. The regions shown are 100 km × 100 km. In (*a*), (*b*) and (*c*) mantle flow, potential temperature, melt porosity and composition are presented in exactly the same way as in Figure 3.1. The plate velocity is 5, 1 and 0.5 cm yr^{-1} for (*a*), (*b*) and (*c*) respectively; the contour interval for the streamlines is proportional to the plate velocity, as in Figure 2.2. All the distributions shown are in steady state.

boundary conditions on the melting region is shown to be unimportant. The local rheological structure and the distribution of buoyancy forces due to melting dominate mantle flow in the region of melting.

The crustal thickness and upwelling width produced with and without thermal buoyancy can be compared in Figures 4.2*d* and *e* and Figures 3.2*d* and *e*. (The curves to compare are those labelled 'med. viscosity' in Figure 4.2 and 'melt+composition' in Figure 3.2). The upwelling width is almost the same with or without thermal buoyancy, at all plate velocities. Slightly more crust is produced when thermal buoyancy is included, but a decrease in crustal thickness is still seen below a plate velocity of 1 cm yr^{-1}.

It should also be noted that the modification to the thermal structure in the melting region has very little effect on the process of small-scale convection. Hence the two processes are effectively decoupled; one result of this decoupling is that the analyses used to investigate the melting processes (see SS89) and convective processes (see Appendix B) remain useful when the two occur simultaneously.

4.2. *Variation with Plate Velocity*

Figures 4.1*a* and 4.1*c* are equivalent to Figure 4.1*b*, but with plate velocities of 5 cm yr^{-1} and 0.5 cm yr^{-1} respectively. A variation in both the vertical and lateral extent of the melting region with plate velocity is shown in Figures 4.1*a*, *b* and *c*. While the depth at which melting begins remains fixed (this is determined by the mantle temperature), the distance of the top of the melting region from the surface is greater at the lower plate velocities. This distance is determined by competition between conductive cooling and upward advection of heat directly beneath the ridge axis. At the lower plate velocities, conduction is more effective and limits the vertical extent of the melting region.

An increase in the width of the melting region is observed with increasing plate velocity. This is symptomatic of a general expectation that the relative importance of active flow should decrease as the velocity of the passive, plate-

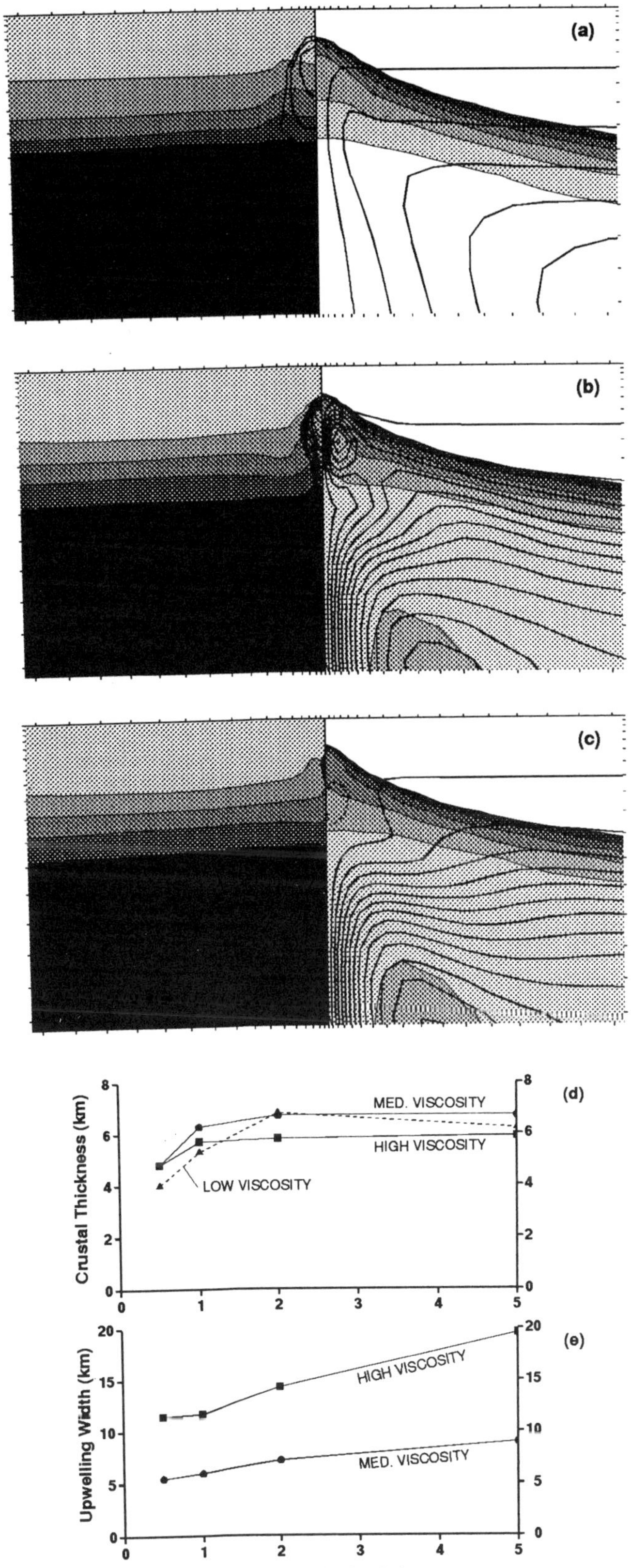

Fig. 4.2. Examples combining melting and small-scale convection, showing variations with viscosity. The regions shown are 100 km x 100 km. In (*a*), (*b*) and (*c*) mantle flow, potential temperature, melt porosity and composition are presented in exactly the same way as in Figure 3.1. The plate velocity is 1 cm yr^{-1} in (*a*), (*b*) and (*c*). The calculation shown in (*a*) uses a higher viscosity than those shown in Figure 4.1, and the distribution shown is a steady state. The calculation shown in (*b*) and (*c*) uses a lower viscosity than those shown in Figure 4.1. Under these conditions a steady state is not achieved; the flow is episodic. The two panels show the extremes of the cycle; the peak of melting and a point when almost no melt is present. The distribution of composition in the the top left corner of (*b*) and (*c*) reflects the initial condition of this calculation. Parts (*d*) and (*e*) show the variation of crustal thickness and upwelling width for the examples shown here and in Figure 4.1. For the cases with low viscosity, that show episodic behavior, the crustal thickness (shown dashed) is obtained by averaging over several cycles. No value for upwelling width is given for these cases.

driven flow is increased. The same effect was noted in SS89. Here, however, the upwelling width shows significantly less variation. This is probably due to the rheological structure used here; flow tends to be confined to the regions of lower viscosity close to the ridge axis.

4.3. *Variation with Viscosity*

Calculations with melting and small-scale convection were performed with the same three values of the viscosity parameter η_0 that were used in Section 2. Figure 4.1 shows calculations with the intermediate value, and Figure 4.2 shows calculations with the high and low values. The results from all these calculations are summarised in Figures 4.2*d* and *e*, which show the crustal thickness and upwelling width.

Figure 4.2*a* shows a calculation with the higher value of η_0 and a plate velocity of 1 cm yr^{-1}. Comparing this with Figure 4.1*b* shows that raising the viscosity by a factor of five has produced a broader upwelling, because the efficacy of the buoyancy forces has been reduced. The curves in Figure 4.2*e* show that the upwelling width has been doubled at all plate velocities. Figure 4.2*d* shows that the amount of crust produced is slightly reduced.

Lowering the viscosity is more interesting; we have reproduced the episodic behavior first reported in SS89. The following description is intended to clarify the calculation shown in Figures 4.2*b* and *c*. In calculations where a steady state is achieved, the buoyancy of the melt in the upwelling is balanced by a lateral variation in composition that places less dense material to either side of the upwelling. Episodic behavior occurs when this balance between melt and compositional buoyancy is not achieved. Imagine, as an initial condition, a state where no melt is present and the distribution of composition is laterally uniform. As there are no lateral variations in melt or composition, there is no active flow driven by these sources of buoyancy. Figure 4.2*c* shows approximately this state. When melting commences, the melt buoyancy immediately increases the rate of upwelling. The lateral variation in composition required to balance the melt buoyancy cannot be established immediately, because of the time taken for material to travel first up the upwelling and then out and down to either side. Figure 4.2*b* is in a state where the upwelling velocity and the rate of melting is very high, but only a small lateral variation in composition has developed. This state cannot, however, be sustained. Residual material is being produced more quickly than it can be disposed of by the movement of the plate; the closed streamlines seen in Figure 4.2*b* entrain refactory material into the upwelling and melting stops. The compositional density distribution produced during the melting episode now drives *downward* flow beneath the ridge axis as is readjusts to a laterally uniform state. A single streamline indicating downward flow is seen beneath the ridge axis in Figure 4.2*c*.

This behavior is seen at all values of plate velocity when the low value of viscosity is specified. It is also seen in calculations both with and without thermal buoyancy, indicating again that the melting and convective processes are effectively decoupled. The presence of convective flow was expected to favor steady over episodic behavior in the melting region, because the convective flow should help to dispose of the residual material produced. A slight tendency is seen in this direction, but it is a minor effect. Because of this decoupling, the analysis of episodic melting presented in SS89 is probably still valid.

Figure 4.2*d* shows that the amount of crust produced by the episodic melting process is, on average, very similar to that produced by the steady melting process when the viscosity is higher. In all cases, the time between episodes of melting is in the range of 0.1-0.2 z_s/V_P. During this time, the plate moves between one-tenth and one-fifth of the depth to the solidus, i.e., 5-10 km.

5. DISCUSSION

The calculations presented in the preceding three sections provide a fluid-mechanical framework within which a range of observations pertaining to mid-ocean ridges may be examined. Three forms of mantle flow beneath ridges have been presented, each with separate and distinct length and velocity scales:

1. Large-scale passive flow, driven by the separation of the plates. This can be regarded as the return flow associated with subduction of oceanic plates, so its length and velocity scales are those of the plates. The form of flow on this scale is prescribed in the calculations presented here through the boundary conditions.
2. Small-scale thermal convection at the base of the young lithosphere, in the form of a pair of rolls parallel to the ridge. The length scale of this flow is determined primarily by the viscosity stratification, and may be around 200 km. The velocity scale may be larger than the plate velocity, particularly when the plate velocity is small so that a large lateral temperature gradient is sustained close to the ridge.
3. Melt-driven circulation in the region of melting beneath the ridge axis, associated with melt and compositional buoyancy. The length scale of this flow is simply the depth to the onset of melting, around 50 km. The velocity scale may be several times larger than the plate velocity, depending on the permeability and mantle viscosity in the region of melting.

While acknowledging the limitations of the modeling presented (particularly its confinement to two dimensions), it is hard to conceive of any additional sources of buoyancy that could influence the structure of mantle flow beneath ridges. It is possible that the rheological structure used is somewhat different from reality, but again it is unlikely that reasonable changes in the rheology would introduce forms of mantle flow that differ qualitatively from those listed.

Suprisingly, the calculations show very little coupling between small-scale thermal convection and melt-driven circulation. These two smaller scales of flow are coupled to the large-scale flow in the sense that neither would occur if the plates were not continuously diverging. Because the large-scale flow is prescribed in the calculations presented, it is not possible to assess how much the smaller scales of flow influence the large-scale flow. Such influence is, however, probably negligible. The underlying paradigm that ridges are passive features of large-scale mantle flow [Houseman, 1983] remains unaltered.

Certain fluid-mechanical assumptions have been made in this work, which can now be re-appraised with hindsight. The possibility that boundary conditions influence the behavior in the melting region can probably now be discounted, because the action of small-scale convection does not affect the melting region. Lateral melt transport by non-hydrostatic pressure gradients [Spiegelman and McKenzie, 1987] may occur locally within the melting region, but will not affect the overall distribution of buoyancy that controls the mantle flow. The omission of viscous resistance to compaction is justified unless solitary waves play a major role in the melting region. If they do, the melt porosity will develop rapid spatial variations; the overall buoyancy of the region will not, however, be affected unless the waves significantly increase the average rate of melt transport. The weakest assumption is probably that of the form and magnitude of the permeability; there is no direct evidence that the microscopic texture in the partially molten region resembles an idealised sponge. This problem is partially addressed below when melt transport in fractures is discussed.

An assessment is now attempted of the extent to which the proposed patterns of mantle flow and melting beneath ridges can be constrained by observations. In this discussion some attempt is made to reconcile various proposed models of processes and structures in the mantle beneath mid-ocean ridges, within the fluid-mechanical framework developed here.

5.1. *Segmentation and Episodicity*

The multiple scales of flow identified here provide a wealth of possibilities for complex behavior in time and space. Most of the calculations presented show fairly complex spatial patterns that are in steady state. These spatial patterns may become unstable, either in time or in the third spatial dimension, along the ridge axis.

Segmentation. The large-scale passive flow will be three-dimensional when the plate boundary is divided into ridge segments, perpendicular to the direction of relative plate motion, and transform faults. Analysis of this flow [Phipps Morgan and Forsyth, 1988] shows how mantle upwelling is suppressed beneath the transform faults, down to a depth comparable to the length of the transform offset. Since both the small-scale convection and melt-driven circulation occur in response to passive upwelling, both will be suppressed in the vicinity of large transform faults. Of more interest is the question of whether either kind of active flow develops three-dimensional structure spontaneously.

The small-scale convective rolls may be stable in three dimensions, always directly reflecting the thickening of the plates. Alternatively, they may develop three-dimensional structure as perpendicular rolls develop beneath the plates away from the ridge (see Section 2.3). If so, the wavelength of the along-axis structure would probably be comparable to the depth scale of the rolls, i.e., around 200 km. There is no strong observational evidence for along axis structure on this scale at present.

Morphological segmentation of ridges of shorter wavelength is, however, ubiquitous. Ridge segments bound by features such as zero-offset transforms and overlapping spreading centers can be identified even along plate boundaries without large transforms. The same segmentation is observed in features of magmatic origin; neovolcanic features are located in the center of segments, and the crust is thicker in the center of some segments than at their ends [White et al., 1984]. The typical wavelength of segmentation is in the range of 30-80 km [Schouten et al., 1985]. The similarity between this wavelength and the depth at which major melting starts (about 50 km) suggests strongly that the segmentation is controlled by processes in the melting region. The wavelength of segmentation increases as approximately the one-third power of plate velocity, providing an additional constraint on models of the origin of segmentation.

A fluid-mechanical analysis of the development of segmentation in the melting region has been proposed by [Schouten et al., 1985] (see also Whitehead et al. [1984] and Crane [1985]). In this analysis, melt buoyancy drives a Rayleigh-Taylor instability of a horizontal cylinder representing the region of partial melting. The partial melt then forms discrete diapiric upwellings spaced apart by a preferred wavelength. The volume of the cylinder grows with time at a rate proportional to the plate velocity. By considering the growth time of instabilities of different wavelengths it is possible to produce the observed dependence of dominant wavelength on plate velocity.

Parmentier and Phipps Morgan [1990] present numerical simulations of three-dimensional flow beneath a ridge driven by the plates and compositional buoyancy produced by melting. They demonstrate that compositional buoyancy alone can produce segmentation. Significantly, they find a transition from two-dimensional flow to segmented, three-dimensional flow as either the plate velocity or the mantle viscosity is reduced. This transition occurs because changing either of these parameters changes the relative importance of passive and active flow.

The previous work cited above and the two-dimensional calculations presented here suggest that along-axis segmentation is driven by melt buoyancy with some control and modification by compositional buoyancy. Figure 5.1 shows the proposed pattern of mantle flow and melting in three dimensions; this figure was conceived as an alternative to Figure 2 in Whitehead et al. [1984]. In Figure 5.1, melting only occurs in regions where mantle material ascends above 50 km (the solidus); these regions are beneath the segments

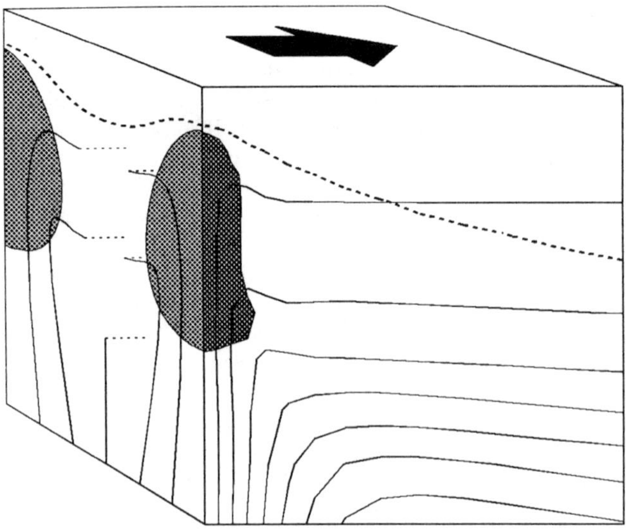

Fig. 5.1. The front face of this figure shows a 75 km × 75 km corner taken from the calculation shown in Figure 4.1*b*. The left face shows a speculative extension of the patterns of flow, temperature and melting into a vertical plane beneath (and parallel to) the ridge axis. The fine lines are streamlines of the mantle flow, with dashed extensions where flow turns out of the axial plane. The thick dashed line is a isotherm approximating the lithosphere-asthenosphere boundary. The shading shows the region of melting. Along-ridge segmentation on a lengthscale comparable to the depth extent of melting is shown; this is presumed to be controlled by the same balance between melt and compositional buoyancy that narrows the the region of melting across the ridge. The patterns shown are presumed to be in steady state.

where magmatic activity is observed. Beneath the gaps between segments, no mantle melting occurs because mantle material is prevented from crossing the solidus by residual material that is driven along the axis from the melting regions. Conceptually, this view has similarities and differences with those of both Schouten et al. [1985] and Parmentier and Phipps Morgan [1990]. To get closer to the truth, there is certainly room for further analysis and numerical modelling of mantle flow and melting in three dimensions directed towards understanding the wavelength of segmentation. What is really required, however, are better constraints on the structure of the melting region, such as might be provided by seismic or magnetotelluric methods.

Episodicity. Even if the large-scale passive flow is steady, it is possible that either the small-scale convection or the melt driven circulation may be unsteady. The method used in the calculations presented here is fully time-dependent and should permit unsteady behavior, unless the unsteady behavior depends on flow in the third dimension.

In the calculations presented in Section 2, the flow and thermal structure stabilised quite quickly and then did not show significant fluctuations. This may be misleading, because the convection is driven by quite small temperature differences (a few tens of degrees at most). Variations in the temperature of the mantle drawn in to the convective system might cause fluctuations in the vigor of the flow, on timescales of the order of millions of years. The pattern of flow focusses material in to a central upwelling between the pair of rolls, which will tend to shorten the wavelength of temperature variations and make them more effective in modifying the flow. This effect could be assessed by relaxing the isothermal lower boundary condition used in the calculations.

The period of the episodic melting process described in Section 4 is such that several kilometers of plate spreading should occur between episodes of melting. If this process operates in the Earth, a fabric should be seen in young oceanic plates running parallel to the ridge axis, with this wavelength across axis. This is a longer wavelength than has generally been reported [e.g., Pockalny et al., 1988]. On another note of caution, it is not obvious that this mechanism will operate in three dimensions, and if so for what values of plate velocity and mantle viscosity it is expected. It is certainly not clear that this particular mechanism of episodicity operates in the Earth.

Sotin and Parmentier [1989] outline a mechanism in which episodic flow results from the interaction between thermal buoyancy and the low density of the residual mantle. In their calculations, residual material becomes entrained in the small-scale convective flow. This results has not been reproduced here; all the residual material is invariably locked into the lithosphere before it cools enough to become convectively unstable. The difference between the results of these two studies reflects the difference in the rheological structures used. Sotin and Parmentier [1989] use a rigid lithosphere overlying an isoviscous asthenosphere; this is arguably less realistic than the pressure- and temperature-dependent viscosity used here.

It may be that the patterns of flow and melting are basically steady on the scale of the whole melting region. Recent work by Edwards, et al. [1991] on the distribution of fault traces on oceanic crust suggests that the pattern of segmentation on the East Pacific Rise has been stable for 2 Ma. About 100 km of crust has formed during this time, twice the depth of the melting region. There are certainly possibilities for episodic behavior in the shorter timescale processes not modeled here. A particularly likely candidate is the process of fracture transport in the melting region, which is discussed in Section 5.2 below. An instability in this process might operate on a timescale of around 0.1 Ma, intermediate between the time it takes for mantle material to pass through the melting region and the timescale of short-term crustal processes such as dike intrusion.

5.2. *The Composition of Mid-Ocean Ridge Basalts*

The calculations presented here could be extended to provide detailed predictions of the compositions of MORBs, by using the distributions of melting rate and melt porosity as inputs to a model of dynamic melting such as that presented by Williams and Gill [1989]. No attempt has been made to develop such predictions here. A multi-parameter description of the changing composition of both the melt and the solid would modify the estimates of density used here, but this modification would not change the basic fluid-mechanical effects that have been identified. There are, however, certain primary constraints provided by the composition of MORBs that must be satisfied by any model of mantle processes beneath mid-ocean ridges. The following discussion is by no means exhaustive but identifies points of particular fluid-mechanical interest.

Major element composition and fracture transport. Independent studies of the systematics of MORB compositions conclude that melts delivered to the crust sample all depths within the melting column [Klein and Langmuir, 1989; McKenzie and Bickle, 1988]. This is not consistent with the percolative flow process used in the preceding calculations. The process implicit in the choice of the magnitude and form of the permeability involves flow of melt in tubules around individual grains. In this case, the rate of diffusion of chemical species would be sufficient to keep the melt in equilibrium with the local mineral assemblage during ascent. Thus, all the melt emerging from the top of the column would reflect equilibrium at pressures of less than perhaps 5 kbar.

The unavoidable conclusion is that a fast melt transport process must operate to deliver melts from deep in the melting column to the surface without re-equilibration. This process is presumed to be buoyancy-driven transport up veins or fractures [Sleep, 1988; Lister, 1991]. The numerical simulation of such a process presents a considerable challenge, in large part because there is little or no experimental constraint on how fracturing might operate in a partially molten rock. Instead of speculating about the appropriate constitutive description, it may be more fruitful to focus on the likely effects any fracture transport process on the melting region as a whole.

One effect will certainly be a reduction in the porosity, because the large-scale permeability (i.e the permeability on a scale large compared to the fracture spacing) of the partially molten medium will increase. The fractures will presumably still rely on grain-scale percolation for the supply of melt, so it is reasonable to suggest that the porosity will remain finite. The reduction in porosity will, however, reduce the melt buoyancy driving the upwelling. The calculation shown in Figure 5.2*a* simulates this buoyancy effect by simply ignoring the effect of porosities above 2% when calculating the mantle density. In all other respects the calculation is identical to that shown in Figure 4.1*b* (e.g. the intermediate viscosity is used). As expected, the reduction of melt buoyancy leads to a broadening of the upwelling.

A second effect that is also quite likely is a reduction in the large-scale strength of the partially molten region, because of the interruption of the solid matrix by fractures. This should probably be simulated using a plastic rheology, but here a simple reduction in viscosity by a factor of ten is prescribed when melt is present. The result of a calculation with both this viscosity effect and the buoyancy effect is shown in Figure 5.2*b*. Even with the reduced buoyancy,

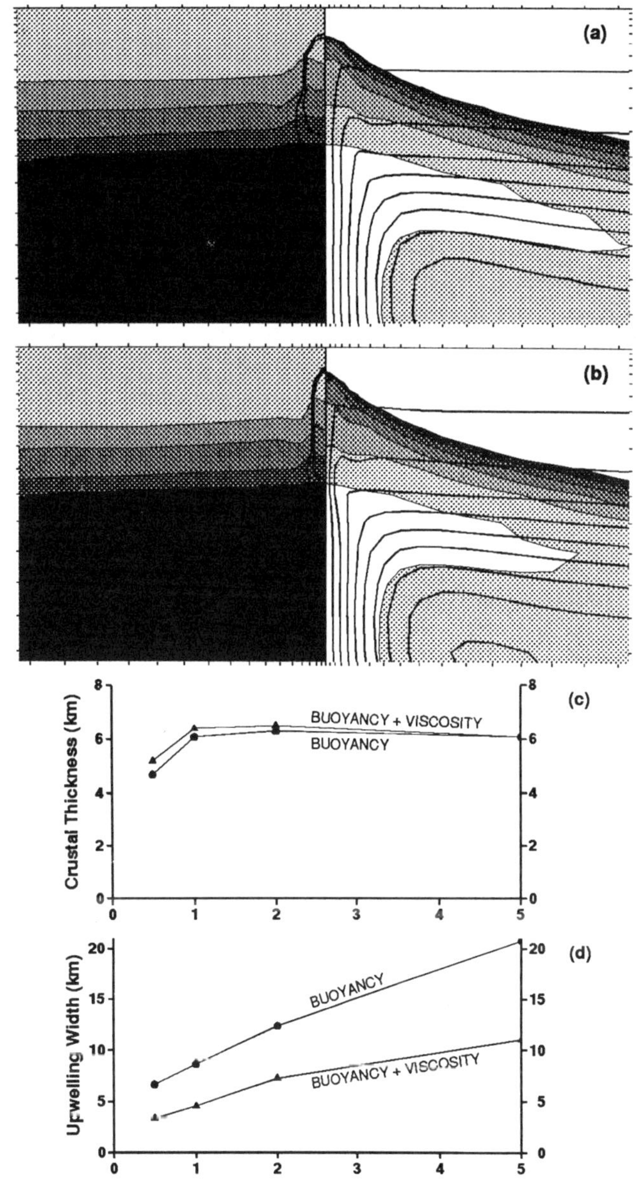

Fig. 5.2. The effects of melt transport in fractures. The regions shown are 100 km × 100 km. In (*a*) and (*b*) mantle flow, potential temperature, melt porosity and composition are presented in exactly the same way as in Figure 3.1. The plate velocity is 1 cm yr^{-1} in both cases. The calculation shown in (*a*) simulates the effect of fracture transport on buoyancy, by ignoring buoyancy forces due to porosities greater than 2%. The calculation shown in (*b*) adds the effect of fracture transport on the strength of the mantle, by reducing the visosity by an order of magnitude wherever melt is present. Parts (*c*) and (*d*) show the crustal thickness and upwelling width for similar calculations using a range of plate velocities.

a narrow upwelling is produced. Figures 5.2*c* and *d* again summarise the results of a range of calculations at different plate velocities.

The effect of a reduced viscosity in the melting region was noted ib SS89 and by Buck and Su [1989]. If this effect really occurs, then a powerful positive feedback loop should operate; a reduction in viscosity in the upwelling increases the upwelling velocity and the rate of melting, which presumably leads to a further reduction in viscosity as more fractures form to drain away the melt. The key questions (to which the answers have been guessed) are how much melt porosity is sustained in the upwelling?, and by how much the mantle viscosity is reduced?

Melt transport in fractures may have other effects. The orientation of the stress field in the solid material may influence the alignment of fractures and lead to large-scale anisotropy in the permeability. It is likely that fractures would be transient rather than steady features, giving an alternative source of episodicity in melt delivery. The establishment of constitutive forms, modeling procedures and constraints relating to fracture transport remains a challenge for future work.

Trace element composition and incipient melting. The systematics of trace elements in MORBs indicate that both the abundance and degree of fractionation of incompatible elements are unexpectedly large. A simple illustration leading to this conclusion is provided by Plank and Langmuir [1992], who plot the concentrations of Na (a moderately incompatible element) and Ce (a highly incompatible element), both corrected for the effects of low-pressure fractionation. Their data show a large spread in the concentration of Ce at a given concentration of Na, which can be modelled by mixing enriched melts that reflect very low extents of melting ($\sim 0.1\%$) with the melts expected from the main melting region. Crucially, the volume of material that melts to provide the enriched component must be several times larger than the volume that participates in the major melting.

The proposed source for the enriched component of the mixture is incipient melting of the mantle before it enters the main melting region. This incipient melting is attributed to the presence of small amounts of volatiles, and may commence at a depth of 200 km in an adiabatic upwelling. It is clearly possible to transport a greater volume of material through a source region starting at this depth than is transported through the main melting region, starting at 50 km depth. The same idea has been proposed by McKenzie [1985a] and Galer and O'Nions [1986].

One problem remains, which is reminiscent of the problem of explaining the narrow zone of crustal formation. A region of incipient melting that starts at around 200 km would be expected to have a width that is also around 200 km. The mixing model based on trace element systematics requires that the melts from this broad region be focussed in to the main melting region. The focussing mechanisms proposed to explain the narrow zone of crustal formation can be re-examined but do not seem any more probable as solutions to this problem.

The pattern of flow associated with small-scale convection may circumvent the need for such a focussing mechanism. As illustrated by Figure 5.3, the convective flow brings a large flux of material through the region from 50 km to 200 km below the ridge that does not participate in the major melting. Consider the streamlines entering the base of the box; those nearest the axis of symmetry lead into the region of major melting, but those further out only traverse the proposed region of incipient melting. This fluid-mechanical explanation is not complete because the precise trajectory of the incipient melts has not been calculated. The focussing effect of small-scale convection does, however, offer a simple and plausible framework for the interpretation of trace element systematics.

In Figure 5.3, the depth at which major melting starts is determined by the mantle temperature, but the depth at which incipient melting starts is primarily determined by the volatile content. Hence, variations in the trace element characteristics of melts can be caused by variations in temperature and volatile content, even if the trace element composition of the mantle source is constant. This idea was suggested by Schilling et al. [1980] in describing the Azores as both a 'hotspot' and a 'wetspot'. On a speculative note, recall that small-scale convection is relatively more vigorous at lower plate velocities; hence the mixture of melts might reflect a larger contribution from the incipient melt component when the plate velocity is lower.

Radioactive disequilibrium. Fresh samples of MORB display disequilibrium in the uranium radioactive decay series [Condomines et al., 1981; Newman et al., 1983]. This ob-

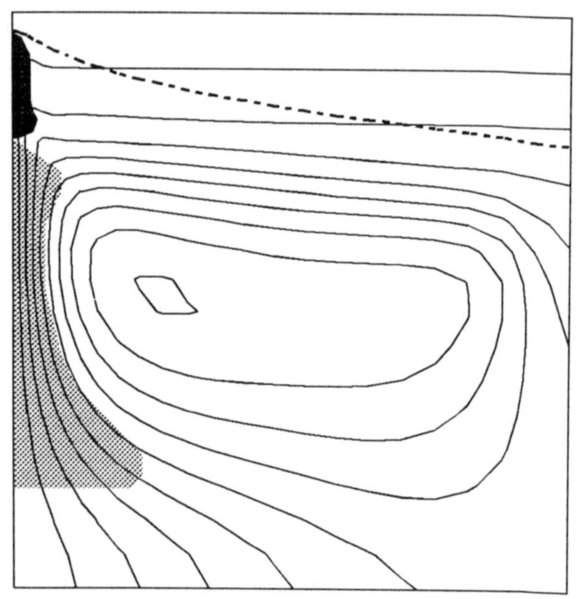

Fig. 5.3. Major and incipient melting. The region shown is 200 km x 200 km. The dark shading shows the region of major melting; the light shading shows a proposed region of incipient melting, which provides the incompatible trace elements found in MORBs. The bold dashed line is an isotherm. The mantle flow, region of major melting and temperature distribution used here are taken from the calculation shown in Figure 4.1*b*.

servation provides important constraints on both the melt porosity and the timescale of melt transport [McKenzie, 1985b; Williams and Gill, 1989]. The porosity in the melting region must be small enough to fractionate U and Th, both highly incompatible elements. The melt transport process must then be fast enough to deliver melt from the region of low porosity to the sea floor (via any crustal magma chamber) before the decay series re-equilibrates. The process of melt transport in fractures, required to produce the correct range of major element compositions, could potentially satisfy both constraints. This process will tend to limit the porosity in the melting region (see Figure 5.2), and it is also rapid; most of the delay in melt delivery would probably occur during crustal storage. This interpretation is made more complex if incompatible trace elements are drawn from a larger source region, as shown in Figure 5.3. The extent of both fractionation and radioactive disequilibrium observed in MORBs will then depend on the transport and mixing history of the incipient melts, which is rather uncertain from a fluid-mechanical standpoint.

5.3. *Gravity and Topography*

High-resolution gravity data from the Mid-Atlantic Ridge [Lin et al., 1990] and East Pacific Rise [Madsen et al., 1990] provides some of the best evidence for fundamental structural segmentation along ridge axes. Gravity lows coincide with the center of morphological and magmatic segments, and could be caused by mantle structure, crustal structure or variations in crustal thickness. Here, the extent to which melt-driven circulation and small-scale convection might influence the observed gravity field is appraised.

Figure 5.4 shows the Bouguer gravity anomaly calculated from the density distributions from a typical calculation (specifically, the calculation shown in Figure 4.1*b*). The total anomaly is dominated by the thermal contribution from the thickening of the lithosphere. A small modification is provided by density distributions due to the melt and composition. The melt contribution gives a gravity low at the axis of a few milligals, which is more than compensated for by a gravity high from the compositional contribution. Because of the fluid-mechanical balance between the melt and compositional buoyancy, the melting region produces little or no gravity anomaly.

This conclusion might not apply if the upwelling were episodic in time, as in the calculation with low viscosity shown in Figures 4.2*b* and *c*. In this case, the melt and depletion buoyancy do not reach a state of balance and a gravity low might be observable during the episodes of melting. A transient porosity of several percent would be needed to provide a clear signal, i.e., several milliGals.

Short-wavelength gravity anomalies, such as the 'bulls-eyes' with a wavelength of around 50 km observed by Lin et al. [1990], are probably not directly attributable to density structure in the main melting region (10-50 km depth). Processes in the melting region probably control the wavelength

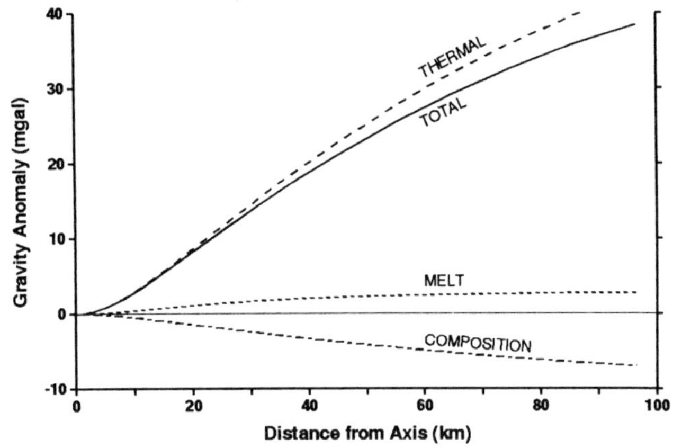

Fig. 5.4. Bouguer gravity anomalies due to thermal, melt and compositional density distribution. The distributions shown in Figure 4.1*b* were used to calculate the gravity on a profile 5 km above the free surface, by numerical integration. The calculation assumes that the density distribution at distances greater than 100 km from the ridge axis is laterally uniform. The total gravity is the sum of the components due to the melt, compositional and thermal density distributions. Each profile is plotted relative to its value directly above the ridge axis.

of these anomalies indirectly, through spatial variation in the supply of magma to the crust. The interpretation proposed by Madsen et al. [1984] is more likely. They attribute anomalies on the East Pacific Rise to shallow structure: variations in either crustal thickness or magma concentrations in the crust and upper few kilometers of the mantle. They further justify this interpretation through consideration of the likely depth of isostatic compensation of seafloor topography.

It is likely that the only way to confirm the action of small-scale convection beneath ridges directly would be through the gravity field. The signals associated with small-scale convection beneath plates are in the range of 5-20 mGal [Haxby and Weissel, 1986]. If a signal of the same magnitude is generated by rolls beneath the ridge, it may well be masked by a large long-wavelength signal due to the thickening of the lithosphere, and the short-wavelength signal due to shallow magmatic processes discussed above. Careful filtering of the gravity field would be required to resolve the details of small-scale convection in the vicinity of ridges.

A limitation of the calculations presented here is that the dynamic topography at the ridge axis cannot be determined. (This is discussed in SS89). Some comment can, however, be offered on the likely cause of the transition from smooth axial topography at fast-spreading ridges to rough topography at slow-spreading ridges. Calculations presented by Lin and Parmentier [1989] show that this transition can be caused by the development of a thick neck of strong lithosphere at the ridge axis as the plate velocity is decreased. Stretching of this neck is partly accommodated by plastic yield (i.e., faulting), giving rise to rough topography. The thermal structure above the region of melting, shown in Figure 4.1 as a function of plate velocity, shows a significantly thicker neck of cooled material at the slower plate velocities. The thermal structure in this region of the calculations is influenced by the 'eruption' process discussed in Section 3, which removes latent heat from the top of the melting region. However, calculations using *in situ* freezing also show thickening of the lithospheric neck at low plate velocities. In either case, it seems likely that the mechanism proposed by Lin and Parmentier [1989] will operate.

5.4. *Crustal Thickness and Upwelling Width*

A compilation of data by Reid and Jackson [1981] shows that the thickness of the oceanic crust created at mid-ocean ridges is around 6-7 km across a wide range of plate velocities. They also note that, at the spreading rates below 1 cm yr^{-1}, the crust is significantly thinner. This assertion is based on observations at the Arctic mid-ocean ridge [Jackson et al., 1982] and Baffin Bay [Keen and Barrett, 1972]. The model presented by Reid and Jackson [1981], using passive mantle flow, demonstrates that the constant crustal thickness at moderate and fast plate velocities reflects a constant mantle temperature. At the lowest plate velocities the penetration of conductive cooling into the melting region reduces the amount of crust formed.

The calculations presented here certainly agree qualitatively with these conclusions. The curves in Figures 4.2*d* and 5.2*c* show a constant crustal thickness above a plate velocity of 1 cm yr^{-1}, and a reduction in crustal thickness at a plate velocity of 0.5 cm yr^{-1}. These curves derive from calculations with active flow; a curve for passive flow is shown in Figure 3.2*d*, and shows somewhat more variation of crustal thickness with spreading rate.

White et al. [1984] and Fox and Gallo [1984] have noted that oceanic crust is anomalously thin near fracture zones of both large and small offset. The thinning in these regions mainly occurs in the seismic layer 3 (the gabbro layer); the thickness of the sheeted dikes that form layer 2 is more normal. These observations are completely consistent with the segmented distribution of melting illustrated in Figure 5.1. Magma is only delivered to the crust in the center of segments, and reaches the ends of segments by along-axis dike injection. This also means that the estimates of crustal thickness from two-dimensional calculations (Figures 3.2*d*, 4.2*d* and 5.2*d*) may be misleading; it is again stressed that the shape of these curves is more significant than the absolute values.

The upwelling widths calculated would also change if the form of three-dimensional flow shown in Figure 5.1 were correct. The upwelling would then be divided into discrete plumes, perhaps of oval cross-section. The across-axis width of these plumes may be larger or smaller than in the two-dimensional case; this is not clear but the question could be studied with three-dimensional calculations or perhaps analysis.

A more important question is whether the basic mechanism of balance between melt and compositional buoyancy can produce an upwelling as narrow as the seismically-observed zone of crustal formation [Detrick et al., 1987], at most a few kilometers. The calculations presented here show that this mechanism can operate at a significant level of efficiency. If the viscosity in the region of melting is substantially reduced by a population of melt-filled fractures, the mechanism may be very efficient and perform all the focussing required. At the top of the region of melting, the physical barrier presented by the lithosphere may provide a final element of focussing should the upwelling remain too wide [Spiegelman, 1991; Sparks and Parmentier, 1991]. As with the question of the origin of segmentation, this question can probably only be answered by observations of the physical structure down to 50 km beneath ridges by seismic or magnetotelluric techniques. The presence of melt, and perhaps fractures, would have to produce a reasonable contrast in elastic, anelastic or electrical properties for this to be possible.

APPENDIX A: THE MATHEMATICAL MODEL

All the calculations presented in the text are based on a single system of equations governing mantle flow, melting and melt percolation. This system is a simplified, approximate form of the equations governing percolative flow in a viscously deformable matrix [McKenzie, 1984]. In this appendix these equations are first stated in their full form; the approximations used to simplify them are then discussed. The equations used for the calculations without melting in Section 2 are produced by omitting unwanted terms and equations from the complete system. Notation and numerical values are listed in Table 1.

Following Scott [1988], the velocity variables used here are as follows:

$$\mathbf{V} = \phi \mathbf{V}_{liq} + (1-\phi)\mathbf{V}_{sol} \tag{A1}$$

$$\mathbf{u} = \phi(\mathbf{V}_{liq} - \mathbf{V}_{sol}) \tag{A2}$$

where $\mathbf{V}_{liq}$ and $\mathbf{V}_{sol}$ are the velocity fields of the liquid and solid, and ϕ is the porosity, $\mathbf{V}$ represents the mean motion of the whole rock and $\mathbf{u}$ is the volume flux of liquid relative to solid. $\mathbf{V}$ is useful when the volume change of the whole rock that accompanies melting is ignored. With this assumption, $\nabla.\mathbf{V} = 0$. Hence $\mathbf{V}$ can be treated as an incompressible flow, just like the flow field in a standard, single-phase convection calculation employing the Boussinesq approximation. By analogy to the Boussinesq approximation, the density change due to melting is used to provide a buoyancy force, but the implied volume change is ignored.

The potential temperature Θ is used as a variable. Θ is related to the true temperature T by a first-order approximation to the correct exponential form presented by McKenzie and Bickle [1988].

$$\Theta = T + \left.\frac{\partial T}{\partial z}\right|_S z \tag{A3}$$

where

$$\left.\frac{\partial T}{\partial z}\right|_S = \frac{g\alpha T}{C_P} \tag{A4}$$

This transformation absorbs the variation in temperature along the dry adiabat. Hence the potential temperature of mantle material remains constant until it loses heat by conduction or due to melting.

A.1. *Momentum Equations*

The coupled processes of percolative flow of melt and viscous circulation of the mantle are described by the following equations Scott, 1988]:

Darcy's Law:

$$\mathbf{u} = \frac{k_\phi}{\mu}\Big[g(\rho_m - \rho_l)(1-\phi)\hat{\mathbf{z}} + \eta\nabla\times\nabla\times\mathbf{V} + (\zeta + \tfrac{4}{3}\eta)\nabla\nabla.\mathbf{u} - \eta\nabla\times\nabla\times\mathbf{u}\Big] \tag{A5}$$

Stokes equation:

$$\eta\nabla\times\nabla\times\mathbf{V} + \nabla p^* = g(\rho_m - \rho_l)\phi\hat{\mathbf{z}} - (\zeta + \tfrac{4}{3}\eta)\nabla\nabla.\mathbf{u} + \eta\nabla\times\nabla\times\mathbf{u} \tag{A6}$$

where k_ϕ is the permeability, μ is the melt viscosity, g is the acceleration due to gravity, $(\rho_m - \rho_l)$ is the density difference between solid and liquid, $\hat{\mathbf{z}}$ is the unit vector pointing upwards, η and ζ are the shear and bulk viscosities of the solid matrix, and $p^* = P + g\rho_m z$ is a reduced pressure.

Following McKenzie [1984] and Spiegelman and McKenzie [1987] the following length scales are defined:

Compaction length:

$$\delta_c = \left[\frac{k_\phi(\zeta + \frac{4}{3}\eta)}{\mu}\right]^{1/2} \tag{A7}$$

TABLE 1. Notation and Numerical Values

Symbol	Description	Value
$\mathbf{V}$	mean velocity of solid and liquid	
$\mathbf{u}$	volume flux of liquid relative to solid	
$\mathbf{V}_{sol}$	solid velocity	
$\mathbf{V}_{liq}$	liquid velocity	
P	total pressure	
$p^* = P + g\rho_m z$	reduced pressure	
T	temperature	
$\Theta = T + \partial T/\partial z\|_S z$	potential temperature	
T_s	solidus temperature	
X	concentration of compatible tracer in solid	
F	extent of melting of solid	
ϕ	porosity	
z	vertical position, increasing upwards	
α	thermal expansivity	3×10^{-5} K^{-1}
g	gravitational acceleration	10 m s^{-2}
κ	thermal diffusivity	1×10^{-6} m^{2}s^{-1}
k_0	permeability constant ($k = k_0 f^2$)	3×10^{-10} m^2
ρ_m	density of unmelted mantle	3340 kg m^{-3}
ρ_r	density of residue after 30% melting	3295 kg m^{-3}
ρ_l	density of liquid	2800 kg m^{-3}
Θ_m	mantle potential temperature	1573 K
$\Theta_0 = T_0$	surface temperature	273 K
$dT/dz\|_S$	adiabatic temperature gradient	-0.48 K km^{-1}
T_{s0}	solidus temperature (for $X = 1$) at the surface	1373 K
$\partial T_s/\partial z\|_X$	gradient of solidus with depth	-5.0 K km^{-1}
$\partial T_s/\partial X\|_z$	gradient of solidus with composition	440 K
ΔS	entropy of melting	250 J K^{-1} kg^{-1}
C_P	heat capacity	1000 J K^{-1} kg^{-1}
η_0	reference viscosity	$31.7/6.59/1.43 \times 10^2$ Pa s
Ω	activation volume for viscosity	1.7×10^{-5} m^3 mol^{-1}
E	activation energy for viscosity	420 kJ mol^{-1}

Focussing length:

$$\delta_f = \left[\frac{\eta V}{g(\rho_m - \rho_l)}\right]^{1/2} \quad \text{(A8)}$$

where V is a constant mantle velocity scale. The compaction length is characteristic of the interaction between percolative flow and viscous compaction and related phenomena such as the wavelength of solitary waves [e.g., Scott and Stevenson, 1984; Richter and McKenzie, 1984]. Note that δ_c depends on ϕ through k_ϕ. The focussing length is the the length scale over which non-hydrostatic pressure gradients due to mantle flow can modify buoyancy-driven percolative flow [Spiegelman and McKenzie, 1987].

Using these scales, the momentum equations can be re-written as follows. The compaction and focussing lengths are first introduced into Darcy's Law (A5):

$$\mathbf{u} = \frac{k_\phi}{\mu} g(\rho_m - \rho_l)\left[(1-\phi)\hat{\mathbf{z}} + \delta_f^2 \nabla \times \nabla \times \tilde{\mathbf{V}}\right] \quad \text{(A9)}$$

$$+\delta_c^2\left[\nabla\nabla.\mathbf{u} - \frac{\eta}{(\zeta + \frac{4}{3}\eta)}\nabla \times \nabla \times \mathbf{u}\right]$$

where the non-dimensional variable $\tilde{\mathbf{V}} = \mathbf{V}/V$ is of order unity. The terms involving δ_f and δ_c are omitted, on the assumption that they are both small compared to the length scale of the spatial derivatives. Darcy's law then reduces to the simple algebraic form used in the calculations:

$$\mathbf{u} = \frac{k_\phi}{\mu} g(\rho_m - \rho_l)(1-\phi)\hat{\mathbf{z}} \quad \text{(A10)}$$

This form and Equation (A7) are used to define the non-dimensional variable $\tilde{\mathbf{u}} = \mathbf{u}/u(\phi)$, where

$$u(\phi) = \frac{g(\rho_m - \rho_l)}{(\zeta + \frac{4}{3}\eta)}(1-\phi)\delta_c^2 \quad \text{(A11)}$$

Again, $\tilde{\mathbf{u}}$ is of order unity. This variable is substituted into the Stokes equation (A6) to obtain:

$$\eta \nabla \times \nabla \times \mathbf{V} + \nabla p^* = g(\rho_m - \rho_l)\cdot \quad \text{(A12)}$$

$$\left\{\phi\hat{\mathbf{z}} - \left[\nabla\nabla. - \frac{\eta}{(\zeta + \frac{4}{3}\eta)}\nabla \times \nabla \times\right]\left[(1-\phi)\delta_c^2 \tilde{\mathbf{u}}\right]\right\}$$

All the terms on the right-hand side involving δ_c are now omitted on the assumption that the spatial derivatives involve a length scale much larger than the compaction length. This leaves the Stokes equation in a simple form, similar to that used for modeling of incompressible mantle circulation.

$$\eta \nabla \times \nabla \times \mathbf{V} + \nabla p^* = g(\rho_m - \rho_l)\phi\hat{\mathbf{z}} \quad \text{(A13)}$$

Two further modifications are made to this form. First, a spatially and temporally variable viscosity is allowed (the above development has assumed that the viscosity η is constant). Second, density variations due to thermal expansion and the composition of the solid are included with the melt density term $(\rho_m - \rho_l)\phi$ in Equation (A13). Changing to summation notation, Equation (A13) becomes:

$$-\frac{\partial}{\partial x_j}\left[\eta\left(\frac{\partial V_i}{\partial x_j} + \frac{\partial V_j}{\partial x_i}\right)\right] + \frac{\partial p^*}{\partial x_i} = \quad \text{(A14)}$$

$$g\left[\rho_m \alpha(\Theta - \Theta_m) + (\rho_m - \rho_l)\phi + \frac{(\rho_m - \rho_r)}{\Delta X}\left(1 - \frac{1}{X}\right)\right]\delta_{iz}$$

where ρ_m is the mantle density, ρ_r is density of the residue after 30% meting ($\Delta X = 0.3$), α is the thermal expansivity and Θ_m is the potential temperature of the mantle. The composition variable X is the concentration of a perfectly compatible tracer (see Section 3.2). The viscosity law is discussed in Section 2.2.

The small-scale convection calculations in Section 2 only include the thermal expansion term in the density variation in the Stokes equation (A14). Also, in these calculations Darcy's Law (A10) is not used because melting and percolation are not included. In the cases with melting in Sections 3, 4 and 5 selective use is made of the various contributions to the density variation in the Stokes equation (A14); this selection is specified in the text and figures.

A.2. *Conservation of Energy and the Melting Process*

In this section the equations governing the time-dependence of the temperature, porosity and composition variables are presented. The processes involved are advection and diffusion, compaction, and melting; these are considered in turn.

Advection and Diffusion. The mean velocity field $\mathbf{V}$ is used to advect potential temperature because it is a shared property of both the solid and liquid phases. The diffusive term for potential temperature is just like that for the true temperature because the two are related by a uniform spatial gradient (see Equation (A3)).

$$\dot{\Theta}_{adv/diff} = -\mathbf{V}.\nabla\Theta + \kappa\nabla^2\Theta \quad \text{(A15)}$$

Both porosity ϕ and composition X are properties of the solid matrix and so are advected by the solid velocity field; from Equations (A1) and (A2), the solid velocity is equal to $(\mathbf{V} - \mathbf{u})$.

$$\dot{\phi}_{adv} = -(\mathbf{V} - \mathbf{u}).\nabla\phi \quad \text{(A16)}$$

$$\dot{X}_{adv} = -(\mathbf{V} - \mathbf{u}).\nabla X \quad \text{(A17)}$$

Compaction. The temperature and composition are unaffected by compaction. The rate of change of porosity is given by a statement of the conservation of volume:

$$\dot{\phi}_{comp} = -(1-\phi)\nabla.\mathbf{u} \quad \text{(A18)}$$

Equations (A10) and (A11) show that, for the approximations used here, $\mathbf{u} = u(\phi)\hat{\mathbf{z}}$. Equation (A18) can therefore be simplified as follows:

$$\dot{\phi}_{comp} = -(1-\phi)\frac{\partial u}{\partial z} = -(1-\phi)\frac{\partial u}{\partial \phi}\frac{\partial \phi}{\partial z} \quad \text{(A19)}$$

Hence the compaction process can be treated numerically as

an additional advection of porosity in the z direction, with pseudo-velocity $(1-\phi)\partial u/\partial\phi$. This is particularly useful in determining the time step required to give stable numerical behavior.

Melting. The temperature in a partially molten region is constrained to lie on the solidus:

$$T_s = T_{s0} + \left.\frac{\partial T_s}{\partial z}\right|_X z + \left.\frac{\partial T_s}{\partial X}\right|_z (X-1) \quad \text{(A20)}$$

As potential temperature Θ is used elsewhere in the calculations, Equation (A20) is re-written using (A3) to give the potential temperature of the solidus:

$$\Theta_s = T_{s0} + \left(\left.\frac{\partial T_s}{\partial z}\right|_X - \left.\frac{\partial T}{\partial z}\right|_S\right) z + \left.\frac{\partial T_s}{\partial X}\right|_z (X-1) \quad \text{(A21)}$$

In any one time step the changes in the variables due to advection, diffusion and compaction are first determined. Melting then occurs if the potential temperature exceeds that of the solidus; freezing occurs if melt is present and the potential temperature lies below that of the solidus. In either case, the changes in temperature, depletion and porosity due to melting are related by:

$$\delta\Theta_{melt} = -\frac{T\Delta S}{C_P}\delta\phi_{melt} \quad \text{(A22)}$$

$$\delta X_{melt} = \frac{X}{(1-\phi)}\delta\phi_{melt} \quad \text{(A23)}$$

such that the final values $\Theta + \delta\Theta_{melt}$, $X + \delta X_{melt}$ and $\phi + \delta\phi_{melt}$ satisfy the solidus equation (or all the melt freezes out).

As with the momentum equations, the various contributions to the time derivatives have been omitted or included in different calculations. For example, in the small-scale convection calculations the porosity and composition variables are not used and temperature is only affected by advection and diffusion.

The wet adiabat. The use of equations describing melting can be illustrated by calculating the initial rate of melting on the wet adiabat. First, Equation (A21) is used to relate increments in the potential temperature, depletion and depth of material that remains on the solidus:

$$\delta\Theta = \left(\left.\frac{\partial T_s}{\partial z}\right|_X - \left.\frac{\partial T}{\partial z}\right|_S\right)\delta z + \left.\frac{\partial T_s}{\partial X}\right|_z \delta X \quad \text{(A24)}$$

Equation (A22) is used to substitute for $\delta\Theta$. At the onset of melting (i.e. when $\phi = 0$ and $X = 1$) equation (A23) gives the simple relation $\delta\phi = \delta X$ which can also be substituted into equation (A24), to give:

$$-\frac{T\Delta S}{C_P}\delta\phi = \left(\left.\frac{\partial T_s}{\partial z}\right|_X - \left.\frac{\partial T}{\partial z}\right|_S\right)\delta z + \left.\frac{\partial T_s}{\partial X}\right|_z \delta\phi \quad \text{(A25)}$$

Rearranging and taking the limit of small increments,

$$\frac{\partial\phi}{\partial z} = -\left(\left.\frac{\partial T_s}{\partial z}\right|_X - \left.\frac{\partial T}{\partial z}\right|_S\right) \Big/ \left(\frac{T\Delta S}{C_P} + \left.\frac{\partial T_s}{\partial X}\right|_z\right) \quad \text{(A26)}$$

Using the numerical values in Table 1, an initial rate of melting of 4.8×10^{-3} km^{-1} is obtained.

APPENDIX B: ANALYSIS OF SMALL-SCALE CONVECTION

B.1. *Weak Convection*

The first step in this analysis is to demonstrate that the thermal structure associated with passive flow may drive a significant component of active flow, for reasonable values of mantle viscosity. The main result from this step is to estimate the functional dependence of the convective velocity V_C on the plate velocity V_P.

The standard form for the thermal structure beneath an oceanic plate assumes that vertical conduction of heat is dominant:

$$\Theta(x,z) = \Theta_0 + (\Theta_m - \Theta_0)\,\mathrm{erf}\left(\frac{z}{2(\kappa t)^{1/2}}\right) \quad \text{(B1)}$$

where Θ_0 and Θ_m are the surface and mantle temperatures, z is depth (positive downwards in this case), $t = x/V_P$ where x is the lateral distance from the ridge axis, and κ is the thermal diffusivity. Consider the static pressure distribution at the lithosphere-asthenosphere boundary created by the variations in density due to thermal expansion. Integrating the density distribution up from great depth gives:

$$P(x,z) = \rho_m g\alpha \int_\infty^{z_L} \Theta(x,z')dz' \quad \text{(B2)}$$

where ρ_m is the mantle density, α is the thermal expansivity and z_L is the depth of the lithosphere-asthenosphere boundary ($z_L^2 \approx \kappa x/V_P$). Lateral gradients in this pressure field drive flow according to the horizontal component of the Stokes equation:

$$\eta\nabla^2 V_{Cx} = \frac{\partial P}{\partial x} \quad \text{(B3)}$$

where η is the mantle viscosity and V_{Cx} is the horizontal component of the convective flow. Noting that the temperature field (B1) must be integrated with respect to z in (B2) and differentiated with respect to x in (B3), the differentiation is performed first to give:

$$\begin{aligned}\eta\nabla^2 V_{Cx} &= \rho_m g\alpha(\Theta_m - \Theta_0)\cdot \quad \text{(B4)}\\ &\quad \int_\infty^{z_L} -\frac{z'}{4}\left(\frac{V_P}{\kappa}\right)^{1/2} x^{-3/2}\exp\left(\frac{-z'^2 V_P}{4\kappa x}\right)dz'\\ &= \rho_m g\alpha(\Theta_m - \Theta_0)\left(\frac{\kappa}{V_P x}\right)^{1/2}\exp\left(\frac{-z_L^2 V_P}{4\kappa x}\right)\end{aligned}$$

A lengthscale l for the convective flow is now introduced. A significant assumption in this analysis is that l is controlled by the rate of increase of viscosity with depth:

$$l \sim \eta\frac{\mathrm{d}z}{\mathrm{d}\eta} \sim 100\ \mathrm{km} \quad \text{(B5)}$$

Buck [1987] has argued that the depth of penetration of small-scale convection into the upper mantle is primarily determined by the stratification of viscosity.

The lengthscale l is used both to substitute for x on the right-hand side, and to approximate the Laplacian operator on the left-hand side. The convective flow is bounded by the base of the lithosphere, so the exponential factor is approximately constant and of order unity (i.e., $z_L^2 \approx \kappa x/V_P$).

$$\eta \frac{V_C}{l^2} \approx \rho_m g \alpha(\Theta_m - \Theta_0)\left(\frac{\kappa}{V_P l}\right)^{1/2} \tag{B6}$$

and hence

$$V_C \approx \frac{\rho_m g \alpha(\Theta_m - \Theta_0) l^{3/2}}{\eta}\left(\frac{\kappa}{V_P}\right)^{1/2} \tag{B7}$$

Equation (B7) indicates that $V_C \sim V_P^{-1/2}$ if the isotherms are not substantially distorted by the active flow (i.e. $V_C << V_P$).

B.2. *Strong convection*

The preceding analysis indicates that the thermal structure sustained by passive flow may drive active flow that is faster than the plate. In this case, the active flow will modify the thermal structure and the analysis loses its internal consistency. It is therefore necessary to study the case of strong active flow separately. The goal of this analysis is again to understand the functional dependence of convective velocity on plate velocity.

The approach used is a simple parameterisation of convection, based on the steady patterns of flow and thermal structure displayed in Figure 2.2. Recall that in these calculations the passive flow velocity field is imposed as a boundary condition at the bottom and outer sides of the box. Thus the number and position of streamlines entering and leaving the box are the same as in the passive case, reflecting a velocity comparable to V_P.

When active flow is strong, the streamlines entering the bottom of the box are focussed in to a flow boundary layer; this layer surrounds the inner side and top of a rotating core of material with new, closed streamlines. V_C is the incremental velocity of material in the flow boundary layer. After emerging laterally from above the core, the material from the flow boundary layer slows down to the passive flow velocity imposed at the outer boundary.

In steady state the core of the convective circulation remains at a constant temperature and is passively rotated by the flow boundary layer. The material in the boundary layer is cooled as it passes laterally below the relatively slow plate. The theme of this analysis is that that temperature drop experienced by the boundary layer in its single pass beneath the plate must be sufficient to drive the convective flow. The unknown variables considered are the convective velocity V_C, the thickness of the flow boundary layer δ and the temperature drop driving convective flow $\Delta\Theta$. The lengthscale l of the convective flow is again assumed to be controlled by the stratification of viscosity, as in Equation (B5).

Conservation of mass. Material from a swath of width l is focussed into the layer of width δ, and accelerates from a velocity of order V_P to a velocity of order $V_C + V_P$. Since the flux of material must be constant,

$$V_P l \sim (V_P + V_C)\delta \tag{B8}$$

Cooling of the flow boundary layer. The temperature in the flow boundary layer drops from the mantle temperature Θ_m in the adiabatic upwelling beneath the ridge, to $\Theta_m - \Delta\Theta$ at a lateral distance l from the ridge, where the material diverges down away from the plate. This cooling is effected by heat loss through into the conductive lid; there is very little temperature gradient between the flow boundary layer and the core. The cooling is estimated as follows:

$$\frac{\partial\Theta}{\partial t} \sim \frac{\kappa}{\delta}\frac{(\Theta_m - \Theta_0)}{z_L} \tag{B9}$$

where Θ_0 is the surface temperature and z_L is the thickness of the conductive lid. Both z_L and δ are strictly functions of x, but in this analysis it is probably adequate to treat them as constant. From the passive thermal structure at $x = l$, z_L is estimated to be:

$$z_L \sim (\kappa l/V_P)^{1/2} \tag{B10}$$

This is in part justified empirically; it is found that the $t^{1/2}$ thickening of the lid is comparatively undisturbed by the active flow. This estimate of z_L is an upper bound for the active case, hence providing a lower bound on the rate of cooling.

The time that the material in the flow boundary layer spends being cooled below the lid is simply $l/(V_C + V_P)$. Hence the total temperature drop can be estimated:

$$\Delta\Theta \sim \frac{l}{(V_C + V_P)}\frac{\kappa}{\delta}(\Theta_m - \Theta_0)\left(\frac{V_P}{\kappa l}\right)^{1/2} \tag{B11}$$

The conservation of mass constraint (B8) is used to obtain:

$$\Delta\Theta \sim (\Theta_m - \Theta_0)\left(\frac{\kappa}{V_P l}\right)^{1/2} \tag{B12}$$

This prediction that $\Delta\Theta \sim V_P^{-1/2}$ is consistent with the results for strong active flow shown in Fig 2.2, where the temperature variations in the convecting region are larger for smaller plate velocities.

Stokes flow. The system of equations is now closed by estimating the velocity fo flow driven by the temperature variation in Equation (B12). This can be done in a number of ways; perhaps the simplest is simply to assert that, in steady state, the core will be cooled to $\Theta_m - \Delta\Theta$ and hence the upwelling beneath the ridge is driven by the temperature difference $\Delta\Theta$. For a channel of width δ, the flow velocity is approximated by

$$\eta \frac{V_C}{\delta^2} \sim \rho_m g \alpha \Delta\Theta \tag{B13}$$

Substituting for $\Delta\Theta$ from (B12) and δ from (B8), and rearranging gives

$$V_C(V_C + V_P)^2 \sim \frac{\rho_m g \alpha(\Theta_m - \Theta_0)\kappa^{1/2} l^{3/2}}{\eta} V_P^{3/2} \tag{B14}$$

Hence, in the limit of strong active flow, where $V_C >> V_P$, the dependence $V_C \sim V_P^{1/2}$ is expected. In this case V_C

does indeed increase with V_P, but will reach the transitional condition $V_C = V_P$ at large V_P.

It should be noted that in the limit where $V_C << V_P$, Equation (B14) gives $V_C \sim V_P^{-1/2}$, in agreement Equation (B6) from the analysis of weak convection. The constants in Equations (B6) and (B14) are also the same. A velocity scale $\tilde{V}$ can be obtained from these constants:

$$\tilde{V} = \left(\frac{\rho_m g \alpha(\Theta_m - \Theta_0)}{\eta}\right)^{2/3} \kappa^{1/3} l \qquad \text{(B15)}$$

This provides an estimate of the transition point between the strong and weak regimes, and has a value of the order of centimeters per year.

Acknowledgements. The author was supported by an NERC Research Fellowship. This work has benefited considerably from discussions with D. J. Stevenson, B. A. Parsons, A. B. Watts and the participants at the 1990 RIDGE Workshop in Boulder, Colorado. Reviews by W. R. Buck and J. H. Davies helped to improve the manuscript.

REFERENCES

Batchelor, G. K., An introduction to fluid dynamics, Cambridge Univ. Press, Cambridge, 1967.

Bottinga, Y., and C. J. Allegre, Partial melting under spreading ridges, *Phil. Trans. R. Soc. Lond. A*, *288*, 501–525, 1978.

Brooks, A. N., A Petrov-Galerkin finite element formulation for convection dominated flows, PhD thesis, California Institute of Technology, 120pp, 1981.

Buck, W. R., Analysis of the cooling of a variable-viscosity fluid with applications to the Earth, *Geophys. J. R. astr. Soc.*, *89*, 549–577, 1987.

Buck, W. R., and E. M. Parmentier, Convection beneath young oceanic lithosphere: Implications for thermal structure and gravity, *J. Geophys. Res.*, *91*, 1961–1974, 1986.

Buck, W. R., and W. Su, Focused mantle upwelling below mid-ocean ridges due to feedback between viscosity and melting, *Geophys. Res. Lett.*, *16*, 641–644, 1989.

Cheadle, M. J., Properties of texturally equilibrated two-phase aggregates, PhD thesis, University of Cambridge, 156pp, 1989.

Condomines, M., P. Morand and C. J. Allegre, ^{230}Th-^{238}U radioactive disequilibria in tholeiites from the FAMOUS zone (Mid-Atlantic Ridge, 36°50′N): Th and Sr geochemistry, *Earth Planet. Sci. Lett.*, *55*, 247–256, 1981.

Crane, K. The spacing of rift axis highs: dependence upon diapiric processes in the underlying asthenosphere, *Earth Planet. Sci. Lett.*, *72*, 405–414, 1985.

Curlet, N. W. E., Experimental and numerical modelling of natural convection in an enclosure, PhD thesis, Massachusetts Institute of Technology, 1976.

Detrick, R. S., P. Buhl, E. Vera, J. Mutter, J. Orcutt, J. Madsen and T. Brocher, Multi-channel seismic imaging of a crustal magma chamber along the East Pacific Rise, *Nature*, *326*, 35–41, 1987.

Edwards, M. H., D. J. Fornari, A. Malinverno and W. B. F. Ryan, The regional tectonic fabric of the East Pacific Rise from 12°50′N to 15°10′N, *J. Geophys. Res.*, *95*, 7995–8017, 1991.

Fleitout, L., and D. Yuen, Steady-state, secondary convection beneath lithospheric plates with temperature- and pressure-dependent viscosity, *J. Geophys. Res.*, *89*, 9227, 1984.

Fox, P. J., and D. G. Gallo, A tectonic model for ridge-transform-ridge plate boundaries: Implications for the structure of oceanic lithosphere, *Tectonophysics*, *104*, 205–242, 1984.

Galer, S. J. G., and R. K. O'Nions, Magmagenesis and the mapping of chemical and isotopic variations in the mantle, *Chem. Geol.*, *56*, 45–61, 1986.

Haxby, W. F., and J.K. Weissel, Evidence for small-scale mantle convection from Seasat altimeter data, *J. Geophys. Res.*, *91*, 3507–3520, 1986.

Hughes, T. J. R., W. K. Liu and A. Brooks, Finite element analysis of incompressible viscous flows by the penalty function formulation, *J. Comput. Phys.*, *30*, 1–60, 1979.

Houseman, G. The deep structure of ocean ridges in a convecting mantle, *Earth Planet. Sci. Lett.*, *70*, 141–146, 1983.

Jackson, H. R., I. Reid and R. K. H. Falconer, Crustal structure near the Arctic mid-ocean ridge, *J. Geophys. Res.*, *87*, 1773–1783, 1982.

Keen, C. E., and D. L. Barrett, Seismic refraction studies in Baffin Bay: An example of a developing ocean basin, *Geophys. J. R. astr. Soc.*, *30*, 253–271, 1972.

Klein, E. M., and C. H. Langmuir, Global correlations of ocean ridge basalt chemistry with axial depth and crustal thickness, *J. Geophys. Res.*, *92*, 8089–8115, 1989.

Kuznir, N. J., and M. H. P. Bott, A thermal study of the formation of oceanic crust, *Geophys. J. R. astr. Soc.*, *47*, 83–95, 1976.

Lin, J., G. M. Purdy, H. Schouten, J.-C. Sempere and C. Zervas, Evidence from gravity data for focussed magmatic accretion along the Mid-Atlantic Ridge, *Nature*, *344*, 627–632, 1990.

Lin, J. and E. M. Parmentier, Mechanisms of lithospheric extension at mid-ocean ridges, *Geophys. J.*, *96*, 1–22, 1989.

Lister, J. R., and R. C. Kerr, Fluid-mechanical models of crack propagation and their application to magma transport in dykes, *J. Geophys. Res.*, *96*, 10049–10077, 1991.

Madsen, J. A., R. S. Detrick, J. C. Mutter, P. Buhl and J. A. Orcutt, A two- and three-dimensional analysis of gravity anomalies associated with the East Pacific Rise at 9° N and 13° N, *J. Geophys. Res.*, *95*, 4967–4987, 1990.

Madsen, J. A., D. W. Forsyth and R. S. Detrick, A new isostatic model for the East Pacific Rise, *J. Geophys. Res.*, *89*, 9997–10015, 1984.

McKenzie, D. P., The extraction of magma from the crust and mantle, *Earth Planet. Sci. Lett.*, *74*, 81–91, 1985a.

McKenzie, D. P., ^{230}Th-^{238}U disequilibrium and the melting process beneath ridge axes, *Earth Planet. Sci. Lett.*, *72*, 149–157, 1985b.

McKenzie, D. P., The generation and compaction of partial melts, *J. Petrol.*, *25*, 713–765, 1984.

McKenzie, D. P. , and M. J. Bickle, The volume and composition of melt generated by extension of the lithosphere, *J. Petrol.*, *29*, 625–679, 1988.

Mutter, J. C., W. R. Buck and C. M. Zehnder, Convective partial melting: 1. A model for the formation of thick basaltic sequences during the initiation of spreading, *J. Geophys. Res.*, *93*, 1031–1048, 1988.

Nakano, T., and N. Fujii, The multiphase grain control percolation: its implication for a partially molten rock, *J. Geophys. Res.*, *94*, 15653–15661, 1989.

Newman, S., R. C. Finkel and J. D. MacDougall, ^{230}Th-^{238}U disequilibrium systematics in oceanic tholeiites from 21°N on the East Pacific Rise, *Earth Planet. Sci. Lett.*, *65*, 17–33, 1983.

Niu, Y., and R. Batiza, In situ densities of MORB melts and residual mantle — implications for buoyancy forces beneath mid-ocean ridges, *J. Geol.*, *99*, 767–775, 1991.

Oxburgh, E. R., and E. M. Parmentier, Compositional and density stratification in oceanic lithosphere — Causes and consequences, *J. Geol. Soc. London*, *90*, 678–684, 1977.

Parmentier, E. M., and J. Phipps Morgan, Spreading rate dependence of 3-dimensional structure in oceanic spreading centers , *Nature*, *348*, 325–328, 1990.

Parsons, B., and D. McKenzie, Mantle convection and the thermal structure of plates, *J. Geophys. Res.*, *83*, 4485–4496, 1978.

Phipps Morgan, J., Melt migration beneath mid-ocean spreading centers, *Geophys. Res. Lett.*, *14*, 1238–1241, 1987.

Phipps Morgan, J., and D. W. Forsyth, Three-dimensional flow and temperature perturbations due to a transform offset: Effects on oceanic crustal and upper mantle structure, *J. Geophys. Res.*, *93*, 2955–2966, 1988.

Plank, T., and C. H. Langmuir, Influence of the melting regime on mid-ocean ridge basalt geochemistry, *J. Geophys. Res.*, in press, 1992.

Pockalny, R. A., R. S. Detrick and P. J. Fox, Morphology and and tectonics of the Kane transform from Sea Beam bathymetry data, *J. Geophys. Res.*, *93*, 3179–3193, 1988.

Rabinowicz, M., A. Nicolas and J. L. Vigneresse, A rolling mill effect in the asthenosphere beneath oceanic spreading centers, *Earth Planet. Sci. Lett.*, *67*, 97–108, 1984.

Reid, I., and H. R. Jackson, Oceanic spreading rate and crustal thickness, *Mar. Geophys. Res.*, *5*, 165–172, 1981.

Ribe, N., On the dynamics of mid-ocean ridges, *J. Geophys. Res.*, *93*, 429–436, 1988.

Richter, F. M., Convection and the large-scale circulation of the mantle, *J. Geophys. Res.*, *78*, 8735–8745, 1973.

Richter, F. M., and B. Parsons, On the interaction of two scales of convection in the mantle, *J. Geophys. Res.*, *80*, 2529–2541, 1975.

Richter, F. M., and D. P. McKenzie, Dynamical models for melt segregation from a deformable matrix, *J. Geol.*, *92*, 729–740, 1984.

Schilling, J.-G., M. B. Bergeron and R. Evans, Halogens in the mantle beneath the North Atlantic, *Phil. Trans. R. Soc. Lond.*, A*297*, 147–178, 1980.

Scott, D. R., The competition between percolation and circulation in a deformable porous medium, *J. Geophys. Res.*, *93*, 6451–6462, 1988.

Scott, D. R., and D. J. Stevenson, A self-consistent model of melting, magma migration and buoyancy-driven circulation beneath mid-ocean ridges, *J. Geophys. Res.*, *94*, 2973–2988, 1989, (referred to as SS89).

Scott, D. R., and D. J. Stevenson, Magma solitons, *Geophys. Res. Lett.*, *11*, 1161–1164, 1984.

Schouten, H., K. D. Klitgord and J. A. Whitehead, Segmentation of mid-ocean ridges, *Nature*, *317*, 225–229, 1985.

Sleep, N. H., Formation of oceanic crust: Some thermal constraints, *J. Geophys. Res.*, *80*, 4037–4042, 1975.

Sleep, N. H., Tapping of melt by dikes and veins, *J. Geophys. Res.*, *93*, 10255–10272, 1988.

Sotin, C., and E. M. Parmentier, Dynamical consequences of compositional and thermal density stratification beneath spreading centers, *Geophys. Res. Lett.*, *16*, 835–838, 1989.

Sparks, D. W., and E. M. Parmentier, Ridgeward migration of melt in a sloping, high-permeability, decompacting layer, *Trans. Am. Geophys. Un. EOS*, *72*, 265, 1991.

Spiegelman, M. W., 2-D or not 2-D: Understanding melt migration near a sloping, freezing boundary, *Trans. Am. Geophys. Un. EOS*, *72*, 265, 1991.

Spiegelman, M. W., and D. P. McKenzie, Simple 2-D models for melt extraction at mid-ocean ridges and island arcs, *Earth Planet. Sci. Lett.*, *83*, 137–152, 1987.

Vaughan, P. J., D. J. Kohlstedt and H. S. Waff, Distribution of the glass phase in hot-pressed, olivine-basalt aggregates: An electron microscope study, *Contrib. Mineral. Petrol.*, *81*, 253–261, 1982.

White, R. S., R. S. Detrick, M. C. Sinha and M. H. Cormier, Anomalous seismic crustal structure of oceanic fracture zones, *Geophys. J. R. astr. Soc.*, *79*, 779–798, 1984.

Whitehead, J. A., Jr., H. J. B. Dick and H. Schouten, A mechanism for magmatic accretion under spreading centres, *Nature*, *312*, 146–148, 1984.

Williams, R. W., And J. B. Gill, Effects of partial melting on uranium series decay, *Geochim. Cosmochim. Acta*, *53*, 1607–1619, 1983.

D. R. Scott, Department of Earth Sciences, University of Oxford, Parks Road, Oxford OX1 3PR, U. K.

Magnetotellurics and Mid–Ocean Ridge Melt Transport: A 2–D Perspective

ROB L. EVANS [1]

Bullard Laboratories, Cambridge University, England

MARK E. EVERETT [2]

Department of Physics, University of Toronto, Canada

Construction of a unified, physical model of melt transport in the upper mantle beneath the mid–ocean ridge is hampered by the virtual absence of geophysical constraints on upper mantle structure and texture beneath active spreading centres. By constructing mid–ocean ridge electrical conductivity models consistent with various proposed tectonophysical scenarios, we examine the suitability of the magnetotelluric (MT) exploration method to supply some of the missing information and to help ameliorate uncertainties associated with the migration and interconnectivity of melt beneath the ridge.

Improved two dimensional electrical models of the ocean, crust, and mantle melting zone beneath the mid-ocean ridge are developed. Taking into account various ideas suggested by RIDGE Summer Institute workers, the melt zone in the upper mantle is modeled as both a broad, prismatic region and a narrow, sheetlike region of high conductivity. Various degrees of melt interconnectivity within these regions are considered. Theoretical MT response functions are then computed for each model by a finite element approximation to the transverse electric (TE) and transverse magnetic (TM) modes of the governing Maxwell's equations, and converted into *apparent* resistivity $\rho_a(y, \ln f)$ pseudosections. If an interconnected melt network forms within the melt region, the different models generate characteristically different TE mode pseudosections in the frequency band $0.0005 < f < 0.008$ Hz, corresponding to natural field variations with periods between 2 min and 0.5 hr, and out to lateral distances y=20–60 km on either side of the ridge axis. The TM mode pseudosections do not show this effect.

Though our modeling technique is coarse and incomplete, the result for the TE mode is encouraging in that it indicates that MT, given a 2–D melt structure, is able to distinguish some end–member models for ridge melting. However, since a certain amount of non—2–D melt structure beneath mid–ocean ridges is inevitable, it is unclear whether actual MT responses would separate into TE and TM modes. An experiment based on the ideal TE mode result alone is risky. The real situation is complicated and requires considerable modeling care, including some 3–D simulations.

Where applicable, our results are consistent with those of Duba and Lilley [1972], who computed the MT apparent resistivity of a single mid–ocean ridge electrical conductivity model at the periods T=5 min, 1 hr and 1 day.

INTRODUCTION

Driven by the separation of lithospheric plates, formation of new ocean crust results from pressure release melting of upwelling mantle beneath the mid–ocean ridge. The nature of melt segregation from the residual mantle and subsequent focused migration to the crustal accretion zone at the ridge axis are relatively open geological questions. Their solution provided the major impetus for discussion at the RIDGE Summer Institute. Not entirely satisfactory, often conflicting, melt transport mechanisms have been explored recently via numerical models of porous melt flow through a permeable matrix [e.g. Spiegelman and McKenzie, 1987; Scott and Stevenson, 1989; Sotin and Parmentier, 1989], studies of a mantle flow diapir fossilized in the Oman ophiolite [Rabinowicz et al., 1987], laboratory analyses of the

1. Now at Dept of Physics, University of Toronto, Canada.
2. Now at Scripps Institution of Oceanography, La Jolla, USA.

Mantle Flow and Melt Generation at Mid-Ocean Ridges
Geophysical Monograph 71
Published in 1992 by the American Geophysical Union.

infiltration of silicate liquid into solid olivine at mantle temperatures and pressures [Riley et al., 1990], and other investigations. Reconciliation of the theoretical results, ophiolitic evidence and laboratory observations is hampered by the fact that the geometry, melt fraction and texture of the upper mantle melt zone beneath present day spreading centres is poorly constrained by the sparse geophysical data available [Shankland et al., 1981]. Participants of the RIDGE Summer Institute agreed that debates concerning the deep (50-200 km) structure of the ridge will almost certainly benefit from the input of additional seismic, gravity and magnetotelluric data from well designed experiments.

In this paper, we examine the suitability of the magnetotelluric method to address the mid–ocean ridge melt transport problem mentioned above. We do this by constructing ridge electrical models consistent with various proposed tectonophysical scenarios, and then calculating the theoretical magnetotelluric response at the seafloor for each of the models. By comparing our results to actual magnetotelluric observations collected near a ridge, in principle one can then determine which of the proposed scenarios *most nearly* corresponds to the true rheological behaviour of the mid–ocean ridge.

Ocean bottom measurements of the complex ratio $Z(\omega)=E(\omega)/H(\omega)$ of orthogonal, horizontal components of the *natural* electric and magnetic field, as a function of frequency $\omega=2\pi f$, constitutes a seafloor magnetotelluric (MT) data set. The MT response function $Z(\omega)$ is diagnostic of the earth's electrical conductivity. Apparent resistivity $\rho_a(\omega)$ is often presented and is defined as the resistivity of a homogeneous earth which would give the MT response function $Z(\omega)$ that is experimentally observed [Caignard, 1953]. Electrical resistivity is simply the reciprocal of conductivity. Complementary to the apparent resistivity is the phase of the response $Z(\omega)$. Inversion and other methods of interpreting MT data in terms of the earth's electrical conductivity structure are often based on analyzing the frequency content of apparent resistivity and phase. In this paper we present apparent resistivity results, although phase is important due to its insensitivity to static shift, an effect that sometimes plagues the correct interpretation of apparent resistivity.

The electrical conductivity of partially molten rock depends on certain physical properties of the sub–surface, in particular the amount, temperature and connectedness of partial melt and fluids [Shankland and Waff, 1977]. MT data, collected in the vicinity of the mid–ocean ridge, may therefore be analysed in order to characterize the melt zone in the upper mantle.

To be specific, electrical conductivity increases with rising temperature and degree of partial melting. Anomalously high conductivities that have been remotely sensed over many tectonically active zones have therefore been attributed to high temperatures and/or large melt fractions in the upper mantle [Shankland et al., 1981]. However, for a given temperature and melt fraction, order of magnitude variations in electrical conductivity can also be produced by relatively minor changes in the melt interconnection geometry, or texture [Schmeling, 1986]. For example, it is difficult to distinguish, on the basis of electrical measurements alone, between partially molten material with a low melt fraction but a highly interconnected melt network and material with a high melt fraction but poor melt interconnectivity. The ambiguity complicates interpretation of field data, but if electromagnetic (EM) measurements are examined in conjunction with modest *a priori* assumptions about mantle composition, partial melt textures may be constrained reasonably well. For example, Shankland and Waff, [1977], have demonstrated that petrological data on the partial melting of peridotite restricts the range of temperatures and melt fractions consistent with an observed electrical conductivity anomaly.

EM techniques, classified according to the source of energy employed, may either be artificial (controlled-source EM) or natural (e.g. MT). A thorough review of electrical exploration methods for the seafloor may be found in Chave et al., [1991]. Controlled-source electromagnetic exploration systems designed for use on the seafloor [Cox et al., 1986; Edwards and Chave, 1986] have been used to map the electrical resistivity [Cheesman et al., 1990] and constrain the texture [Cheesman et al., 1991] of unconsolidated marine sediments to 50 m depths. More powerful systems have provided electrical resistivity profiles of 25 Ma oceanic crust and uppermost mantle [Cox et al., 1986] and the top 1 km beneath an active segment of the fast-spreading East Pacific Rise [Evans et al., 1991]. The latter represents the first controlled-source sounding of the mid–ocean ridge. Although controlled-source data responds, in principle [Everett and Edwards, 1989, 1991], to the presence of the highly conductive mid-crustal axial magma chamber, sensitivity to it depends heavily on the experimental configuration and data quality. In general, artificial sources can not be made powerful enough, nor receivers sensitive enough, to resolve electrical structure below about 10–20 km. Synchronous stacking of signals can partly overcome this limitation, but this requires stable positioning of source and receivers over a long period of time. In the harsh ridge environment, remote sensing of the deep mantle melting region by controlled–source methods is therefore a formidable undertaking.

It is well–known that the usable magnetotelluric signal at the seafloor is band–limited due to attenuation at high frequencies greater than $\sim 10^{-2}$ Hz by the overlying, conductive seawater and contamination at low frequencies $<1.25\times10^{-5}$ Hz by oceanographic and tidal noise. For a given data set, the high frequency limit is determined by the sensitivity of the magnetometers used. There is usually sufficient energy density in the intermediate bandwidth 1.25×10^{-5}–10^{-2} Hz to allow resolution of the electrical resistivity structure between 50–200 km beneath the seafloor [e.g. Oldenburg, 1981]. Since this range of depths is appropriate for studying the upper mantle melting region beneath the mid–ocean ridge, MT is the preferred electromagnetic technique. MT has the additional advantage over controlled-source methods in that it does not require the construction and deployment of a transmitter. Self-contained receivers are simply placed on the seafloor, and they respond to the variations of the natural electromagnetic field. Shiptime is shared with other experiments, so that deployment and recovery of the receivers is carried out with minimum expense. Apparent earth resistivities can be extracted from the time series to <2–5% precision [A.D. Chave, personal communication] with currently available seafloor instrumentation and robust data processing, provided that data are collected for a sufficiently long time, usually at least six months. The estimate on precision is independent of the presumed melt structure. Recent, successful seafloor magnetotelluric experiments have examined the pattern of geomagnetic induction in the Tasman Sea region [Ferguson et al., 1990] and provided images of large-scale tectonic features associated with the Juan de Fuca subduction system [Wannamaker et al., 1989].

GEOLOGICAL MOTIVATION

Upper mantle melting beneath mid-ocean ridges is widely believed to be caused by adiabatic decompression of the mantle. The extent of melting is a function of depth and local conditions [Klein and Langmuir, 1987]. The site of crustal accretion is the very narrow (2–3 km wide) neo-volcanic zone found centered on all ridge axes regardless of spreading rate [Macdonald, 1982]. Emplacement of melt and subsequent crustal accretion occurs only within a region as narrow as the neo-volcanic zone itself. These observations place strong constraints on models of melt generation, segregation and migration beneath the mid-ocean ridge [Phipps Morgan, 1991].

Geochemical and petrological evidence requires most melting to occur along a polybaric melting column spanning the depth range 20–70 km beneath the ridge, with some additional melting occurring below ~70 km [Elthon, 1990]. In addition, the melt fraction within the melt region is likely to be small (less than 10 per cent everywhere) since melt is thought by petrologists to segregate rapidly from the residual matrix after formation [e.g. McKenzie, 1985 *a*, *b*]. Rapid melt extraction implies interconnection of melt pathways within the solid matrix, since interconnection provides readily accessible escape routes for the less dense, buoyant melt. Evidence that the partial melt in the mantle, even at low melt fractions, is well interconnected has come from laboratory simulations [Toramaru and Fujii, 1986; Riley et al., 1990] and studies on ophiolite analogues of present day spreading centers [Nicolas, 1989]. In situ inspection of mantle rheological structure to confirm these results is desirable, but clearly impossible. Remote sensing using geophysical techniques that are sensitive to the texture of partial melt, i.e. EM techniques such as magnetotellurics, is therefore appropriate.

The following two tectonophysical scenarios for melt transport have been considered to explain the formation of the large volume of new crust (~20 km^3 yr^{-1}), within the very narrow neo-volcanic zones, with the small melt fractions available from the mantle [Phipps Morgan, 1991]. Oceanic crust is generated either by (*a*) melting within a broad (more than 50 km wide) region of the upper mantle followed by a focused migration of melt upward and towards the ridge axis, or (*b*) the rapid vertical ascent from directly beneath the ridge axis of melt-rich mantle. Each of these scenarios is problematic. Model (*a*) incorporates rapid melt segregation, satisfying the petrological constraints, but no entirely satisfactory method of focusing the melt ridgeward has been found, despite the variety of fluid dynamical models that have been put forward [e.g. Speigelman and McKenzie, 1987; Scott and Stevenson, 1989; Sotin and Parmentier, 1989]. The methods require unrealistic mantle rheologies to achieve melt focusing. The alternate scenario (*b*) is favoured by Buck and Su, [1989] and Rabinowicz et al., [1987]. Buck and Su, in particular, solve the melt focusing problem by considering mantle viscosity to be a certain function of temperature and melt fraction, but in their models the required melt fraction within the narrow zone of ascent exceeds 10%. Other models in which the melt simply rises along with the residual solid in a narrow, low permeable zone of rapid upwelling very close to the ridge axis do not fulfill the petrological dictum that melt segregates rapidly from its host.

MODELING

We have performed calculations to test the suitability of MT apparent resistivity to discriminate between the proposed melt transport scenarios in the upper mantle beneath the mid-ocean ridge. To ease the computational burden, we assume that the resistivity of the earth is two dimensional, i.e. $\rho=\rho(y,z)$. The direction x in which resistivity is invariant is referred to as along-strike, and is of course parallel to the ridge axis. The direction z is vertical and $z=0$ corresponds to the seafloor. We compute the MT apparent resistivity using the 2-D finite element code of Wannamaker et al., [1985], for the case of the transverse electric (TE) and transverse magnetic (TM) modes of plane-wave excitation. The TE mode, also called the E-polarization mode in the literature, is characterized by the electric field vector within the earth directed along-strike ($\mathbf{E}=E_x(y,z)\hat{x}$) and the magnetic field vector lying in the orthogonal, vertical y/z-plane ($\mathbf{H}=H_y(y,z)\hat{y}+H_z(y,z)\hat{z}$). The TM (H-polarization) mode is characterized by the magnetic field vector directed along-strike and the electric field vector lying in the y/z-plane. The two modes are independent if the co-ordinate axes are correctly aligned with respect to the symmetry axis of the structure. The TE mode apparent resistivity $\rho_{xy}(\omega)$ is calculated by means of the formula [e.g. Wannamaker et al., 1989]

$$\rho_{xy}(\omega) = \frac{1}{\mu_0\omega}|Z_0|^2 \quad (1)$$

where μ_0 is the magnetic permeability of free space and $Z_0=E_x/H_y|_{z=0}$ is the complex ratio of horizontal electric and magnetic fields as measured at the seafloor. Analogous expressions apply for the TM mode apparent resistivity $\rho_{yx}(\omega)$.

Our mid-ocean ridge electrical model is shown in Figure 1. Average electrical resistivities for the various layers within the model were selected on the basis of available laboratory data on the resistivities of typical crustal [Kariya and Shankland, 1983] and upper mantle [Constable and Duba, 1990] rocks as a function of temperature and, for the layers beneath the upper mantle melt zone, resistivities are taken from an inversion of seafloor MT data [Oldenburg et al., 1984]. The value of 10^5 Ωm assigned to the off-axis zone at the bottom of the crust is based on laboratory measurements on dunite by Constable and Duba, [1990], and agrees with the results of the controlled source EM experiment by Cox et al., [1986]. Ridge isotherms, which are needed since electrical conductivity is a function of temperature, are based on the calculations by Phipps Morgan et al., [1987]. The upper

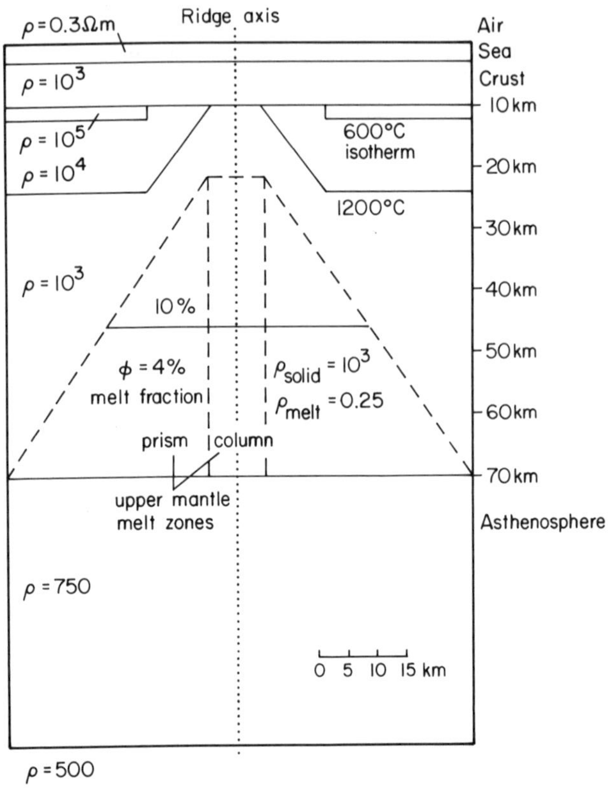

Fig. 1. An electrical model of the mid-ocean ridge. Resistivities for the various model sub-units are selected on the basis of *i*) available laboratory resistivity data as a function of temperature, *ii*) the inversion of seafloor MT data, and *iii*) theoretical calculations of the effective electric resistivity of partial melt. The MT apparent resistivity, as a function of the frequency ω of the natural electromagnetic field, is computed by a finite element method. Separate calculations are performed for both the broad, prismatic and the narrow, columnar ("sheet") upper mantle melt zones indicated by the dashed lines. No vertical exaggeration.

mantle melt zone is modeled as both a broad prism and a narrow sheet of high conductivity. These shapes, indicated by the dashed lines in Figure 1, correspond to those consistent with the narrowness of the crustal accretion zone. The prism, for example, reflects the slow, ridgeward migration of melt from a broad melting region whereas the sheet reflects a two dimensional analogue of narrow, fast mantle upwelling representative of melt–rich diapiric ascent. The sheet diapirism was chosen so that its response could be modeled using 2–D algorithms. We are not able to address the possibility that melt ascends in isolated diapirs sparsely distributed along the spreading axis. Such 3–D melt structure as isolated diapirs would almost certainly have a vanishingly small EM signature in the MT bandwidth. Our sheet model, therefore, is to be regarded as an "end–member" model in which isolated diapirs are sufficiently numerous as to merge together and form an almost completely connected sheet of melt along–strike. Finally, in our model, the melt zone has been horizontally stratified into regions of 10% and 4% partial melt, as indicated in Figure 1, in order to account for the vertical variations in the amount of melting required by petrologists.

Since the electrical properties of the melt zone depend greatly on the partial melt texture, and to facilitate a magnetotelluric field test for melt interconnectivity in the melt zone beneath the mid-ocean ridge, we have computed MT responses assuming varying degrees of melt interconnection. In fact, a pair of melt connectivity "end-member" cases are considered by placing the following bounds [Hashin and Shtrikman, 1962; Waff, 1974] on the effective electrical conductivity σ of the two-phase solid/melt material:

$$\sigma_{-} = \sigma_{\text{solid}} + \phi\left(\frac{1}{\sigma_{\text{melt}} - \sigma_{\text{solid}}} + \frac{1-\phi}{3\sigma_{\text{solid}}}\right)^{-1} \quad (2)$$

$$\sigma_{+} = \sigma_{\text{melt}} + (1-\phi)\left(\frac{1}{\sigma_{\text{solid}} - \sigma_{\text{melt}}} + \frac{\phi}{3\sigma_{\text{melt}}}\right)^{-1}. \quad (3)$$

In the above equations, σ_{solid} is the electrical conductivity of the solid phase, σ_{melt} is that of the melt phase and ϕ is the melt fraction. Large variations with texture, as exemplified for conductivity by equations (2) and (3), in the transport properties of partial melts occur because these properties depend greatly on the dihedral angle at grain solid–solid and solid–liquid surfaces [Bulau et al., 1979]. For example, if the dihedral angle is greater than 60° then melt ponds at grain triple junctions, no interconnection occurs, and the lower Hashin-Shtrikman bound σ_{-} applies. If the dihedral angle is less than 60° then most conditions are met for interconnection. In fact, the upper Hashin-Shtrikman bound applies when complete wetting along grain boundaries occurs [Waff, 1974], which is an extreme case not likely to occur in the mantle. Cheadle, [1989] has examined the relationship between pore texture and dihedral angle and suggests that a description of electrical transport based on a cubic array of cylindrical tubes may be more appropriate for a simple olivine-basalt system, which has dihedral angles of ~45° [e.g. Waff and Bulau, 1979]. In this case, wetting is along grain edges only, and Schmeling, [1986], gives an expression for the bulk conductivity, σ_{tot}, as

$$\sigma_{tot} = \frac{1}{3}\phi\sigma_{melt} + (1-\phi)\sigma_{solid}. \quad (4)$$

We consider all three cases, represented by equations (2–4), in our analysis. Finally, the electrical resistivity of the melt phase ρ_{melt} is taken to be that of basaltic melt (0.25 Ωm), as measured in the laboratory at mantle pressures by Tyburczy and Waff, [1983].

TE and TM modes.

We display results in the form of $\rho_a(y, \ln f)$ pseudosections. Theoretical interpretation of TE and TM mode pseudosections are discussed in Wannamaker et al., [1984], and Wannamaker et al., [1989], and the following remarks are based mostly on ideas propounded in those papers. Spatial resolution of conductivity structures by the magnetotelluric method is obtained since the depth of penetration of electromagnetic fields into the electrically conducting earth depends on frequency. Roughly speaking, large, deep conductivity anomalies are imaged in the pseudosections at low frequencies. Shallow small-scale structures are imaged at the higher frequencies. However, there is a complication to this simple picture, which is discussed below.

The transverse electric (TE) and transverse magnetic (TM) modes are defined only for two dimensional conductivity structures. The invariant direction, in which electrical conductivity is presumed to be constant, is referred to as along-strike. By their definition, TE mode electric currents, to satisfy boundary conditions on the electric field at interfaces, do not cross discontinuities in electrical conductivity, but instead flow parallel to them, in our case along-strike. TM mode electric currents, on the other hand, do not flow along-strike, but instead are free to cross the lateral and vertical boundaries in conductivity. Electric charge accumulates wherever TM currents cross conductivity boundaries. The boundary charges contribute to the total electric field measured at the seafloor, even as the frequency approaches zero. This "static shift" effect, which does not show up in the ideal 2-D TE mode response, makes the TM response notoriously difficult to interpret. Even small-scale extraneous features ill affect the TM mode apparent resistivity at low frequencies [Wannamaker et al., 1984], which is the region of the pseudosection where large, deep anomalies are normally imaged. Static shift can be corrected for by measuring the electric field at densely spaced points along the seafloor over the experiment site, but this procedure is expensive. Boundary electric charges do not appreciably affect the magnetic field, or the phase of $Z_0(\omega)$, at any frequency. Modeling of these quantities is important but beyond the scope of this paper.

The remarks of the previous paragraph only apply if the electrical structure of the earth is truly two dimensional. In that case *only*, the TE mode response is easier to interpret than its TM counterpart. Of course the true earth has three dimensional structure, on all spatial scales. Therefore, modeling actual data using TE mode routines, even at the mid-ocean ridge where two dimensionality is often taken for granted, leads to erroneous conclusions. These are due to the neglect of boundary charge effects arising from the finite strike-length of axial structure or from small-scale, extraneous variations in electrical conductivity along-strike. The TM mode response, by contrast, is not as greatly affected by the presence of 3-D structure since it is the boundary charges near to the measurement site which most strongly affect the response function [Wannamaker et al., 1984; Wannamaker et al., 1989]. In reality, therefore, when unknown along-strike variations in conductivity are present the TM mode is generally easier to interpret.

In summary, both TE and TM responses are susceptible to effects that hamper correct geo-electrical interpretation of MT data. The TM mode response exhibits static shift effects which are expensive to remove and obscure structure ordinarily imaged at low frequencies. The TE mode reponse is severely degraded, by the same charge effects, if even minor along-strike conductivity variations are present. Despite this, MT data have traditionally been interpreted using TE- and TM-based routines, due to the lack of a fast, reliable, fully three dimensional modeling program.

There is some evidence that seafloor MT measurements are less affected by small extraneous structure than equivalent measurements made on land would be. The relative robustness of the seafloor data is likely due to the lack of high frequencies in the seafloor magnetotelluric spectrum. Apart from frequency-independent static shift effects, high frequency components of the MT repsonse are more strongly disturbed by anomalous shallow structure than are low frequency components, in accordance with the skin effect. We therefore feel that TE and TM routines may be used with a reasonable level of confidence when modeling responses and analyzing mid-ocean ridge data. In defense of this conjecture, Wannamaker et al., [1989], after studying the EMSLAB Juan de Fuca subduction system data, concluded that both magnetotelluric modes were little disturbed by small near-surface inhomogeneities and, indeed, they reflected primarily the large scale tectonic features of interest.

RESULTS

Figures 2 and 3 show the TE mode $\rho_{xy}(y, \ln f)$ pseudosections corresponding to the mid-ocean ridge electrical model containing the prismatic melt zone, for both completely connected (Figure 2) and tube-connected (Figure 3) partial melt textures. The completely connected texture corresponds to the upper Hashin-Shtrikman bound on the electrical conductivity, $\sigma = \sigma_+$ in equation (3), while the tube-connected texture corresponds to $\sigma = \sigma_{tot}$, equation (4). An image of the prism melt zone will appear in the pseudosection, over a certain interval of frequencies, if the melt zone responds with sufficient strength to magnetotelluric excitation. If completely connected melt is modeled, Figure 2, an image of the prism does indeed appear in the pseudo-

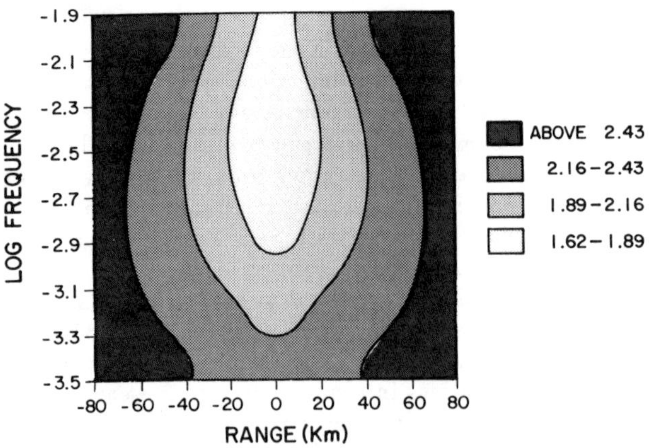

Fig. 2. TE mode apparent resistivity $\rho_{xy}(y, \ln f)$ pseudosection corresponding to the prismatic zone of partial melt assuming completely connected partial melt texture. The vertical axis is $\log_{10} f$, where $f=\omega/2\pi$ [Hz] and the horizontal axis is the distance y [m] from the ridge axis. The pseudosections display ρ_{xy} up to ± 80 km off-axis; where ρ_{xy} is given by equation (1) of the text. The large central anomalous region is an "image" of the prismatic melt zone.

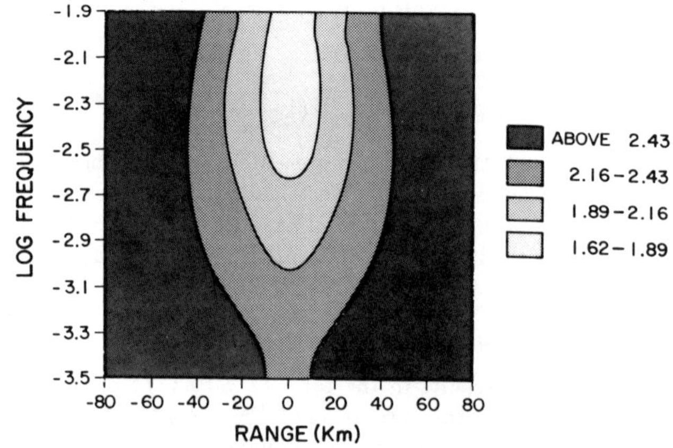

Fig. 4. TE mode pseudosection for the "sheet" melt zone assuming a completely connected partial melt texture. The image of the sheet is smaller than the corresponding prism image in Figure 2 due to the small instantaneous amounts of melt that exist beneath the ridge in the diapiric melt transport scenario.

section. The image extends over a wide band of frequencies, but due to the skin effect discussed earlier, the image is strongest in the log frequency interval $-3.3 \leq \log f \leq -2.1$, with $f=\omega/2\pi$. This interval corresponds to periods of 2–30 min. The influence of the prism on apparent resistivity extends to distances greater than $y=\pm 60$ km off-axis. If tube connectivity is modeled, the apparent resistivity anomaly extends only to $y=\pm 40$ km, and the anomalous apparent resistivity at the ridge axis itself is not as strong. The pseudosection corresponding to unconnected melt ($\sigma=\sigma_-$, the lower Hashin–Shtrikman bound), is not shown since in that case the prism is not imaged, i.e. the pseudosection is not significantly different from the response of the background, layered earth.

Figures 4 and 5 display apparent resistivity $\rho_{xy}(y, \ln f)$ TE mode pseudosections corresponding to the diapiric "sheet" melt zone model, once again with completely connected (Figure 4) and tube-connected (Figure 5) partial melt textures. Images of the "sheet" appear, as expected. The images are not as large as the prism images, due to the smaller total conductance in the sheet model, reflecting the smaller total amount of melt present at any given instant. The critical frequency band for imaging the melt zone is similar to that of the prism model. The region of anomalous apparent resistivity for the sheet model does not extend to the great lateral distances as those of the prism model, although the strength of the prism and sheet anomalies are comparable at the axis $y=0$. The largest differences

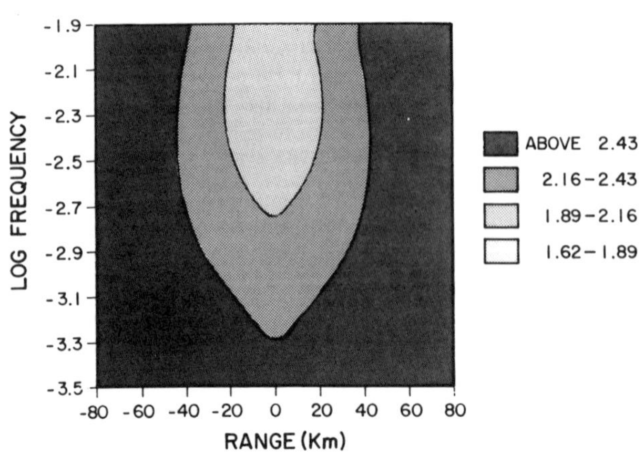

Fig. 3. TE mode pseudosection for the prism melt zone assuming a tube-connected partial melt texture. The image is smaller than the one in the previous figure, since the electrical conductivity assigned to the melt zone has decreased.

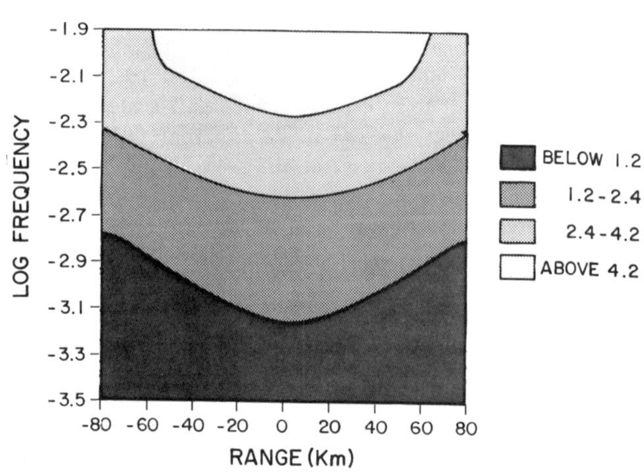

Fig. 5. TE mode pseudosection for the "sheet" melt zone assuming a tube-connected partial melt texture.

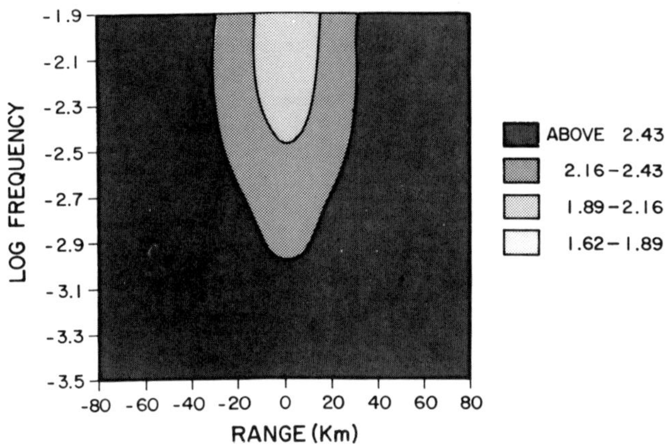

Fig. 6. Pseudosection displaying the percentage difference between TM mode apparent resistivities calculated for the prism melt zone model with two different assumed partial melt textures: unconnected and completely connected. This figure helps to illustrate the insensitivity of the TM mode to the nature of the melt transport beneath the ridge.

in apparent resistivity ρ_{xy}, evident by comparing the pseudosections, occur at distances of more than 10 km off–axis, i.e. directly above the broad flanks of the prism. Although seafloor instrumentation should be densest on–axis during an experiment, to image the critically important, narrow zone of crustal formation, sufficient instrumental coverage should extend to at least 80 km off–axis in order to delineate the horizontal extent of the mantle partial melt zone.

Analysis of TM mode shows an immeasurably small difference between the responses corresponding to unconnected and completely connected melt in the upper mantle melt zone. Figure 6 shows the percentage difference between these two extremal partial melt textures for the prismatic melt zone geometry. The maximum deviation is approximately 5% occuring at a frequencies $f>0.001$ Hz. Whether this small anomaly could be detected depends heavily on the data quality. The relative insensitivity of the TM mode indicates that the power of the MT method to resolve the melt zone is based largely in the TE mode. Duba and Lilley, [1972], also could not discern a TM mode characteristic melt zone signal from the response of their single ridge conductivity model. We can further state that the TM mode phase (not shown), a quantity not greatly affected by static shift at low frequencies, is also insensitive to large variations in partial melt texture within the upper mantle melt zone.

CONCLUSIONS

If there were no distinctive MT signatures at the coarse level of approximation implicit in our modeling, then we should recommend that ridge MT experiments to discriminate between the proposed tectonophysical models of melt transport *not* be performed. But since we do see distinctive signatures in the TE mode, and have characterized them, we have at least established motivation for comprehensive model studies. The studies should include analyses of the apparent resistivity, phase and magnetic field responses in the presence of large–scale 3-D and small–scale extraneous structure including bathymetry.

Provided the melt is well connected, the magnetotelluric TE mode apparent resistivity in the 2–30 min period range enables discrimination between the two candidate melt transport scenarios. The TM mode apparent resistivity, in contrast, does not provide responses much different to those expected from normal, plane layered oceanic lithosphere. Magnetotelluric data should be collected for at least 6 months so that recently developed robust processing techniques [Chave and Thomson, 1989] may be employed. Ideal receiver placement to image the melt zone might follow an axially symmetric square root of age rule up to $y=\pm 80$ km off-axis. During the EMRIDGE electromagnetic exploration of the Juan de Fuca ridge, Heinson et al., [in press], observed that the lowest period to obtain coherent signals with respect to a remote reference using magnetometers with a least count of 0.1 nT is 800 s, or approximately 13 min. Modest improvements in magnetometer sensitivity are therefore needed to reduce degradation in spatial resolution at periods less than 13 min. Signal/noise ratios were relatively good in the higher 13 min–2 hr 45 min range, after removal of tidal noise.

The upper mantle melt zone has been modeled as having infinite strike length. A finite strike length causes current channeling that brings in disturbing fields from considerable distances along–strike where measurements are not likely to have been made. This could be the most serious limitation of the 2-D modeling. Three dimensional modeling is required to quantify the effects on the MT response of all departures from 2–D melt structure, including finite strike length.

A disruption in the along–axis continuity of the melt, e.g. as would be the case if isolated melt diapirs are sparsely distributed along the ridge axis, reduces the magnitude of the apparent resistivity anomaly with respect to the continuous melt case. This means that the melt zone will be more difficult to detect than our pseudosections predict. To ensure a maximum anomaly, an MT transect should intersect the ridge at the center of a long continuous segment, good examples being the segment of the East Pacific Rise (EPR) north of the Clipperton transform zone between 10°–11°45′N and the EPR segment south of the Garrett Fracture Zone.

We have been circumspect about how apparent resistivities derived from actual MT field data should be geologically interpreted, we have merely pointed out that, in principle, they are characteristic of the nature of the melt transport beneath the ridge. It is well known that geophysical data sets of all types are grossly inadequate to uniquely determine the underlying structure of the Earth, and magnetotellurics is certainly no exception. Incorporation of reliable geological constraints on mid–ocean ridge properties can lessen the non–uniqueness problem considerably. Alternatively, regularised inversion algorithms [Constable et al., 1987], which have recently been extended from 1–D to 2–

D [deGroot-Hedlin and Constable, 1990; Smith and Booker, 1991], might be employed. Regularized inversion seeks models that possess minimum structure, the philosophy being that any structure appearing in the final model is that required by the data, independent of geological bias.

Acknowledgments. The authors acknowledge valuable discussions with Alan Chave, Jason Phipps Morgan, David Scott and Roger Buck during the RIDGE Summer Institute, with Nigel Edwards in Toronto and with Sarah Watson and Mike Cheadle in Cambridge. We also thank Phil Wannamaker for providing his 2-D MT forward modeling code.

REFERENCES

Buck, W.R. and W. Su, Focused mantle upwelling below mid–ocean ridges due to feedback between viscosity and melting, *Geophys. Res. Lett.*, 16, 641–644, 1989.

Bulau, J.R., H.S. Waff and J.A. Tyburczy, Mechanical and thermodynamic constraints on fluid distribution in partial melts, *J. Geophys. Res.*, 84, 6102–6108, 1979.

Caignard, L., Basic theory of the magneto–telluric method of geophysical prospecting, *Geophysics*, 18, 605–635, 1953.

Chave, A.D., S.C. Constable and R.N. Edwards, Electrical exploration methods for the seafloor, in (ed. Nabighian, M.N.) *Electromagnetic Methods in Applied Geophysics Vol.II*, Society for Exploration Geophysicists, Tulsa, OK, 1991.

Chave, A.D. and D.J. Thomson, Some comments on magnetotelluric response function estimation, *J. Geophys. Res.*, 94, 14215-14226, 1989.

Cheadle, M.J., Properties of texturally equilibrated two-phase aggregates. *PhD. Thesis, Univ. Cambridge.*, 1989.

Cheesman, S.J., R.N. Edwards and L.K. Law, A test of a short–baseline, sea–floor transient electromagnetic system, *Geophys. J. Int.*, 103, 431–437, 1990.

Cheesman, S.J., L.K. Law, and R.N. Edwards, Porosity determinations of sediments in Knight Inlet using a transient electromagnetic system, *Geo-Marine Lett.*, 11, 84–89, 1991.

Constable, S.C. and A. Duba, Electrical conductivity of olivine, a dunite, and the mantle, *J. Geophys. Res.*, 95, 6967–6978, 1990.

Cox, C.S., S.C. Constable, A.D. Chave and S.C. Webb, Controlled–source electromagnetic sounding of the oceanic lithosphere, *Nature*, 320, 52–54, 1986.

deGroot-Hedlin, C. and S.C. Constable, Occam's inversion to generate smooth 2-D models from magnetotelluric data, *Geophysics*, 55, 1613-1624, 1990.

Duba, A. and F.E.M. Lilley, Effect of an ocean ridge model on geomagnetic variations, *J. Geophys. Res.*, 77, 7100–7105, 1972.

Edwards, R.N. and A.D. Chave, A transient electric dipole–dipole method for mapping the conductivity of the seafloor, *Geophysics*, 51, 984–987, 1986.

Elthon, D., The petrogenesis of primary mid–ocean ridge basalts, *Rev. Aquat. Sci.*, 2, 27–53, 1990.

Evans, R.L., et al. Upper-crustal resistivity structure of the East Pacific Rise near 13° N. *Geophys. Res. Letts.*, 18, 1917-1920, 1991.

Everett, M.E. and R.N. Edwards, Electromagnetic expression of axial magma chambers, *Geophys. Res. Lett.*, 16, 1003–1006, 1989.

Everett, M.E., and Edwards, R.N., Theoretical controlled-source electromagnetic responses of fast spreading mid-ocean ridge models. *Geophys. J. Int.*, 105, 313-323, 1991.

Ferguson, I.J., F.E.M. Lilley and J.H. Filloux, Geomagnetic induction in the Tasman Sea and electrical conductivity structure beneath the Tasman seafloor, *Geophys. J. Int.*, 102, 299–312, 1990.

Hashin, Z. and S. Shtrikman, A variational approach to the theory of effective magnetic permeability of multiphase materials, *J. Appl. Phys.*, 33, 3125–3131, 1962.

Heinson, G.S., A. White, L.K. Law, Y. Hamano, H. Utada, T. Yukutake, J. Segawa and H. Toh, EMRIDGE: the electromagnetic expolration of the Juan de Fuca ridge, *Mar. Geophys. Res.*, in press.

Kariya, K.A. and T.J. Shankland, Electrical conductivity of dry lower crustal rocks, *Geophysics*, 48, 52–61, 1983.

Klein, E.M. and C.H. Langmuir, Global correlations of ocean ridge basalt chemistry with axial depth and crustal thickness, *J. Geophys. Res.*, 92, 8089–8115, 1987.

Macdonald, K.C., Mid–ocean ridges: fine scale tectonic, volcanic and hydrothermal processes within the plate boundary zone, *Ann. Rev. Earth Planet. Sci.*, 10, 155–190, 1982.

McKenzie, D., The extraction of magma from the crust and mantle, *Earth Planet. Sci. Lett.*, 74, 81-91, 1985*a*.

McKenzie, D., ^{230}Th–^{238}U disequilibrium and the melting processes beneath ridge axes, *Earth. Planet. Sci. Lett.* , 72, 149-157, 1985*b*.

Nicolas, A., *Structures of ophiolites and dynamics of oceanic lithosphere*, Kluwer Academic Publishers, 1989.

Oldenburg, D., Conductivity structure of oceanic upper mantle beneath the Pacific plate. *Geophys. J. Roy. Astr. Soc.*, 65, 359–394, 1981.

Oldenburg, D.W., K.P. Whittall and R.L. Parker, Inversion of ocean bottom magnetotelluric data revisited, *J. Geophys. Res.*, 89, 1829–1833, 1984.

Phipps Morgan, J., Mid–ocean ridge dynamics: observations and theory, *Rev. Geophys. Supplement April 1991*, 807–822, 1991.

Phipps Morgan, J., E.M. Parmentier and J. Lin, Mechanisms for the origin of mid–ocean ridge axial topography: implications for the thermal and mechanical structure of accreting plate boundaries, *J. Geophys. Res.*, 92, 12823–12836, 1987.

Rabinowicz, M., G. Ceuleneer and A. Nicolas, Melt segregation and flow in mantle diapirs below spreading centers: evidence from the Oman ophiolite, *J. Geophys. Res.*, 92, 3475–3486, 1987.

Riley, G.N. Jr., D.L. Kohlstedt and F.M. Richter, Melt migration in a silicate liquid– olivine system: an experimental test of compaction theory, *Geophys. Res. Lett.*, 17, 2101–2104, 1990.

Schmeling, H., Numerical models on the influence of partial melt on elastic, anelastic and electrical properties of rocks part II: electrical conductivity. *Phys. Earth Planet. Ints.*, 43, 123-136, 1986.

Scott, D.R. and D.J. Stevenson, A self–consistent model of melting, magma migration and buoyancy–driven circulation beneath mid–ocean ridges, *J. Geophys. Res.*, 94, 2973–2988, 1989.

Shankland, T.J. and H.S. Waff, Partial melting and electrical conductivity anomalies in the upper mantle, *J. Geophys. Res.*, 82, 5409–5417, 1977.

Shankland, T.J., R.J. O'Connell and H.S. Waff, Geopyhsical constraints on partial melt in the upper mantle, *Rev. Geophys. Space Phys.*, 19, 394–406, 1981.

Smith, J.T. and J.R. Booker, Rapid inversion of 2 and 3-D magnetotelluric data, *J. Geophys. Res.*, 96, 3905-3922, 1991.

Sotin, C. and E.M. Parmentier, Dynamical consequences of compositional and thermal density stratification beneath spreading centers, *Geophys. Res. Lett.*, 16, 835–838, 1989.

Spiegelman, M. and D. McKenzie, Simple 2–D models for melt

extraction at mid–ocean ridges and island arcs, *Earth Planet. Sci. Lett.*, 83, 137–152, 1987.

Toramaru, A. and N. Fujii, Connectivity of melt phase in a partially molten peridotite, *J. Geophys. Res.* , 91, 9239-9252, 1986.

Tyburczy, J.A. and H.S. Waff, Electrical conductivity of molten basalt and andesite to 25 kilobars pressure: geophysical significance and implications for charge transport and melt structure, *J. Geophys. Res.*, 88, 2413–2430, 1983.

Waff, H.S., Theoretical considerations of electrical conductivity in a partially molten mantle and implications for geothermometry, *J. Geophys. Res.*, 79, 4003–4010, 1974.

Waff, H.S. and J.R. Bulau, Equilibrium fluid distribution in an ultramafic partial melt under hydrostatic stress conditions, *J. Geophys. Res.*, 84, 6109–6114, 1979.

Wannamaker, P.E., G.W. Hohmann, S.H. Ward, Magnetotelluric responses of 3-D bodies in layered earths, *Geophysics*, 49, 1517-1533, 1984.

Wannamaker, P.E., J.A. Stodt and L. Rijo, PW2D. Finite element program for solution of magnetotelluric responses of two–dimensional earth resistivity structure. User documentation. *Earth Sciences Laboratory, University of Utah Research Institute*, Salt Lake City, UT, 1985.

Wannamaker, P.E., J.R. Booker, J.H. Filloux, A.G. Jones, G.R. Jiracek, A.D. Chave, P. Tarits, H.S. Waff, G.D. Egbert, C.T. Young, J.A. Stodt, M. Martinez, L.K. Law, T. Yukutake, J.S. Segawa, A. White and A.W. Green, Jr., Magnetotelluric observations across the Juan de Fuca subduction system in the EMSLAB project, *J. Geophys. Res.*, 94, 14111–14125, 1989.

R.L. Evans, Dept of Physics, University of Toronto, Canada.

M.E. Everett, Scripps Institution of Oceanography, La Jolla, USA.